**Table II** (continued)

| z | 0 | 1 | 2 | 3 | 4 | 5 | 6 | 7 | 8 | 9 |
|---|---|---|---|---|---|---|---|---|---|---|
| .0 | .5000 | .5040 | .5080 | .5120 | .5160 | .5199 | .5239 | .5279 | .5319 | .5359 |
| .1 | .5398 | .5438 | .5478 | .5517 | .5557 | .5596 | .5636 | .5675 | .5714 | .5753 |
| .2 | .5793 | .5832 | .5871 | .5910 | .5948 | .5987 | .6026 | .6064 | .6103 | .6141 |
| .3 | .6179 | .6217 | .6255 | .6293 | .6331 | .6368 | .6406 | .6443 | .6480 | .6517 |
| .4 | .6554 | .6591 | .6628 | .6664 | .6700 | .6736 | .6772 | .6808 | .6844 | .6879 |
| .5 | .6915 | .6950 | .6985 | .7019 | .7054 | .7088 | .7123 | .7157 | .7190 | .7224 |
| .6 | .7257 | .7291 | .7324 | .7357 | .7389 | .7422 | .7454 | .7486 | .7517 | .7549 |
| .7 | .7580 | .7611 | .7642 | .7673 | .7703 | .7734 | .7764 | .7794 | .7823 | .7852 |
| .8 | .7881 | .7910 | .7939 | .7967 | .7995 | .8023 | .8051 | .8078 | .8106 | .8133 |
| .9 | .8159 | .8186 | .8212 | .8238 | .8264 | .8289 | .8315 | .8340 | .8365 | .8389 |
| 1.0 | .8413 | .8438 | .8461 | .8485 | .8508 | .8531 | .8554 | .8577 | .8599 | .8621 |
| 1.1 | .8643 | .8665 | .8686 | .8708 | .8729 | .8749 | .8770 | .8790 | .8810 | .8830 |
| 1.2 | .8849 | .8869 | .8888 | .8907 | .8925 | .8944 | .8962 | .8980 | .8997 | .9015 |
| 1.3 | .9032 | .9049 | .9066 | .9082 | .9099 | .9115 | .9131 | .9147 | .9162 | .9177 |
| 1.4 | .9192 | .9207 | .9222 | .9236 | .9251 | .9265 | .9278 | .9292 | .9306 | .9319 |
| 1.5 | .9332 | .9345 | .9357 | .9370 | .9382 | .9384 | .9406 | .9418 | .9430 | .9441 |
| 1.6 | .9452 | .9463 | .9474 | .9484 | .9495 | .9505 | .9515 | .9525 | .9535 | .9545 |
| 1.7 | .9554 | .9564 | .9573 | .9582 | .9591 | .9599 | .9608 | .9616 | .9625 | .9633 |
| 1.8 | .9641 | .9648 | .9656 | .9664 | .9671 | .9678 | .9686 | .9693 | .9700 | .9706 |
| 1.9 | .9713 | .9719 | .9726 | .9732 | .9738 | .9744 | .9750 | .9756 | .9762 | .9767 |
| 2.0 | .9772 | .9778 | .9783 | .9788 | .9793 | .9798 | .9803 | .9808 | .9812 | .9817 |
| 2.1 | .9821 | .9826 | .9830 | .9834 | .9838 | .9842 | .9846 | .9850 | .9854 | .9857 |
| 2.2 | .9861 | .9864 | .9868 | .9871 | .9874 | .9878 | .9881 | .9884 | .9887 | .9890 |
| 2.3 | .9893 | .9896 | .9898 | .9901 | .9904 | .9906 | .9909 | .9911 | .9913 | .9916 |
| 2.4 | .9918 | .9920 | .9922 | .9925 | .9927 | .9929 | .9931 | .9932 | .9934 | .9936 |
| 2.5 | .9938 | .9940 | .9941 | .9943 | .9945 | .9946 | .9948 | .9949 | .9951 | .9952 |
| 2.6 | .9953 | .9955 | .9956 | .9957 | .9959 | .9960 | .9961 | .9962 | .9963 | .9964 |
| 2.7 | .9965 | .9966 | .9967 | .9968 | .9969 | .9970 | .9971 | .9972 | .9973 | .9974 |
| 2.8 | .9974 | .9975 | .9976 | .9977 | .9977 | .9978 | .9979 | .9979 | .9980 | .9981 |
| 2.9 | .9981 | .9982 | .9982 | .9983 | .9984 | .9984 | .9985 | .9985 | .9986 | .9986 |
| 3. | .9987 | .9990 | .9993 | .9995 | .9997 | .9998 | .9998 | .9999 | .9999 | 1.0000 |

*Notes*: 1. Enter table at $Z$, read out $P(Z \leq z)$, the shaded area.
2. For a general normal $X$, enter table at $z = (x - \mu)/\sigma$ to read $P(X \leq x)$.
3. Entries opposite 3 are for 3.0, 3.1, 3.2 . . . , 3.9.
4. For $z \geq 4$, $P(Z > z) = P(Z < -z) \doteq \dfrac{1}{\sigma\sqrt{2\pi}} e^{-z^2/2}$.

# Statistics:
# Theory and Methods

## Second Edition

**Donald A. Berry**
DUKE UNIVERSITY

**Bernard W. Lindgren**
UNIVERSITY OF MINNESOTA

**Duxbury Press**
**An Imprint of Wadsworth Publishing Company**

I(T)P®  **An International Thomson Publishing Company**

Belmont • Albany • Bonn • Boston • Cincinnati • Detroit • London • Madrid • Melbourne
Mexico City • New York • Paris • San Francisco • Singapore • Tokyo • Toronto • Washington

Editorial Assistant: *Cynthia Masow*
Marketing Manager: *Joanne Terhaar*
Advertising Project Manager: *Joseph Jodar*
Production Editor: *Julie Davis*
Print Buyer: *Karen Hunt*
Permissions Editor: *Peggy Meehan*
Copy Editor: *Lee Motteler*
Text Design: *Cloyce Wall*
Cover Design: *Ross Carron*
Compositor: *Erick and Mary Ann Reinstedt*
Printer: *Quebecor Printing Fairfield, Inc.*

For more information, contact Duxbury Press at Wadsworth Publishing Company.

Wadsworth Publishing Company
10 Davis Drive
Belmont, California 94002, USA

International Thomson Publishing Europe
Berkshire House 168-173
High Holborn
London WC1V 7AA, England

Thomas Nelson Australia
102 Dodds Street
South Melbourne 3205
Victoria, Australia

Nelson Canada
1120 Birchmount Road
Scarborough, Ontario
Canada M1K 5G4

International Thomson Editores
Campos Eliseos 385, Piso 7
Col. Polanco
11560 México D. F. México

International Thomson Publishing GmbH
Königswinterer Strasse 418
53227 Bonn, Germany

International Thomson Publishing Asia
221 Henderson Road
#05-10 Henderson Building
Singapore 0315

International Thomson Publishing Japan
Hirakawacho Kyowa Building, 3F
2-2-1 Hirakawacho
Chiyoda-ku, Tokyo 102, Japan

**Library of Congress Cataloging-in-Publication Data**
Berry, Donald A.
    Statistics : theory and methods / Donald A. Berry, Bernard W. Lindgren. — 2nd ed.
        p.   cm.
    Includes index.
    ISBN 0-534-50479-5
    1. Statistics.  I. Lindgren, B. W. (Bernard William), 1924–
    II. Title.
QA276.12.B48  1996
519.5—dc20                                                95-30979

# Contents

# Preface

## Approach and Expectations

This text is designed for a two-quarter or two-semester course in the theory of statistics and statistical methodology. The effective application of statistical methods requires an understanding of the theory behind them. So a first course in statistics for mathematically prepared students should have a solid component of theory. On the other hand, statistical theory has no reason for existence without applications. Indeed, most students find the theory dull without the associated methods and their applications to real-life problems.

The student need not have had a previous course in statistics but should have a background in differential and integral calculus. However, because these mathematical tools are often rusty or untried, we proceed rather carefully, especially in the early chapters. Occasionally we may exceed the students' knowledge of calculus, but we do so as gently as possible, motivating the mathematics by an immediate application to statistics.

Although statistical theory involves mathematics, this is a statistics text, not a mathematics text. So we do not use the "definition-theorem-proof" approach. Indeed, a rigorous development of probability and the proofs of theorems on sufficiency and large-sample properties would require considerably more mathematics than the calculus we assume as prerequisite. We do use mathematical arguments, but in some places we say, "it can be shown . . . ," and provide a reference.

## Coverage

For the most part, the topics covered are standard: distribution theory, principles of inference, and the basic methodology of one- and two-sample inference, with brief introductions to the areas of linear models, categorical data analysis, and Bayesian inference. We emphasize the importance of nonparametric methods by integrating them into the sections on inference, rather than setting them apart as optional.

## Real Data

Naturally, since our emphasis is on theory, many of the problems and examples deal with the theory. However, to illustrate methods of inference, we give many examples and problems with real-life situations and data. More than 125 of the problems and examples employ real data. A number of these are in the area of the health sciences, because health issues are of broad general interest—they are reported regularly in the news—and because the experimental designs that are involved are often of the simpler types that we deal with in this text.

## Distinguishing Features

A number of features distinguish this text from others at the same level:

• We emphasize the important distinction between the use of data as statistical *evidence* in scientific experimentation (Chapter 10) and the use of data in guiding the *decisions* that must be made by industry and regulatory agencies (Chapter 11). Whereas in the latter, it can make sense to "test" at a given level, in the former it is universal practice to report experimental results using "$P$-values." So we do not ask the student (as do most texts) to "test the null hypothesis at $\alpha = .01$" when in reality, in scientific work, researchers do not "reject" hypotheses but only report $P$-values!

• In view of the above, we include *tables* of $P$-values for $t$, $F$, $\chi^2$, Wilcoxon, and Kolmogorov-Smirnov statistics, along with the more traditional tables of percentiles.

• We place considerable emphasis on the related notions of *likelihood* and *sufficiency,* not just in connection with estimation (as in standard texts), but as basic to all modes of statistical inference: Likelihood is at the heart of estimation, testing, prediction, and inference generally. For example, it plays the central role in the Bayesian approach.

• Because Bayesian inference is being used more and more in statistical practice, we have incorporated this approach into several chapters: In Chapter 2 we introduce the basic idea in connection with Bayes' theorem, in an example dealing with a case of disputed paternity; in Chapter 8, we show how data are used to update a subjective distribution for a parameter and lead to distributions for predicting future observations; in Chapter 9, we show how the distribution of a population parameter can be used in estimation; in Chapter 10, we obtain what many users of statistics really want to know—the probability of a null hypothesis; and in Chapter 11, we use the Bayesian approach to deal with the problem of decision making in the face of uncertainty.

• In developing probability theory we introduce the notion of a joint distribution early, because even conditioning one event on another is a bivariate matter. We take independence of *variables* (rather than events) as basic, because in statistical applications of probability, the focus is almost always on the independence of variables. (Extending to independence of events is then easy using indicator variables.) Moreover, introducing the notion of independent variables means that students will know precisely what is meant by the phrase "independent trials" when we introduce the binomial distribution.

• In presenting the various important special distribution families, we emphasize and exploit the *structure* of binomial, hypergeometric, chi-square, and Erlang distributions as *sums*. This not only points up relationships among various distributions but also makes it so much easier to derive means and variances.

• We introduce the useful concept of exchangeability of random variables and exploit it in our discussions of simple random sampling.

## This Revision

In this second edition, besides much rewriting to achieve greater clarity, there are a number of changes:

• A major change in the order of topics is the incorporation of Bayesian methods into Chapters 8–11, as described above, rather than leaving them to the final chapter where they were too easily neglected. We have added a section in Chapter 1 that introduces subjective probability and how this can be elicited, preparing the way for treating parameters as random variables; and we have added a section on (Bayesian) predictive distributions, since predictions for future cases are sometimes more directly useful than parameter estimates.

• The problem sets have been reworked in several respects:

  • We have added a number of problems, including several more with real data. (As before, asterisks indicate problems with answers at the back of the book.)
  • We have relocated problem sets, placing problems that are associated with a particular section at the end of that section.
  • We have added at the end of each chapter a set of chapter review problems (with answers at the back).
  • The "Solved Problems" of the first edition are published separately in a student's study manual.

• We have given increased attention to the general availability of computers, using statistical software both as a tool in handling the more computationally intensive problems and for illustrating theoretical points and data analytic methods.

• We have added a number of computer-generated graphs, including some Q-Q plots as additional tools in exploratory data analysis.

Finally, we have submitted this revision to an exhaustive error-checking process to ensure as error-free a book as possible.

We gratefully acknowledge the help of students in our classes in spotting errors and checking answers, and of graduate students Wei Shen and Chin-Pei Tsai in detailed checking of text and problems. We also thank the following reviewers: Carmen Acuña, Bucknell University; Martin Fox, Michigan State University; C. J. Park, San Diego State University; and Mary R. Parker, University of Texas, Austin.

# Prologue

An article in the *British Medical Journal* reported on a survey of 13 consecutive issues of that esteemed periodical.[1] The survey found that of the 77 papers in those issues, 62 used statistical analyses. Of these, 32 had statistical errors of one kind or another, and 18 had fairly serious faults. The authors of the article venture the opinion that less respected journals probably show an even greater frequency of errors.

We want to make two points from this survey. First, *statistics* is more than a name for facts and figures about sports or the economy. It is a discipline that is an integral part of scientific research. Second, its methods are much misunderstood and much misused. Indeed, some of the criticisms leveled by Gore et al. can be challenged—not all statisticians agree on what is correct. One aim of this text is to give you a critical attitude toward applications of statistical techniques, along with an appreciation of their utility.

To set the stage for the step-by-step development, we offer here some accounts of situations and problems whose treatments involve probability and/or statistical ideas. These are given to whet your appetite. We describe some of the challenges but postpone any solutions until we've developed appropriate tools.

You will see from these examples that statistical ideas and methods are not restricted to medical or even to scientific applications. Business decisions, industrial production plans, governmental policies, and research in education and agriculture all rely heavily on statistical analyses.

Example **a**

## A Smoking Survey

"Cigarette smoking is at its lowest point in 20 years among high school students in Santa Maria," according to a poll taken in driver's education classes at Santa Maria

---

[1] S. M. Gore, I. G. Jones, and E. C. Rytter, "Misuse of statistical methods: Critical assessment of articles in BMJ from January to March 1976," *Br. Med. J. 1* (1977), 85–87.

High.[2] The teacher has been conducting polls on drug and alcohol use for the past 20 years, and his surveys are viewed as a barometer of trends among students in Santa Maria.

Of 173 students polled in 1986, only 16% responded that they smoked cigarettes. In a similar survey taken in 1985, 28% smoked. Other results showed that 56% drank alcohol, compared to 57% in 1985; 10% used cocaine when available, compared to 9% the previous year; 4% used LSD; and 6% abused pills.

These matters are of general concern. But are these sample percentages typical of all high school students? Or of high school students in California, or even in Santa Maria? The teacher thought the poll was "very, very accurate," because the voting was done behind closed doors. Practically all students take driver's education, so he believed that the poll included a good cross section of teenagers.          ■

## Example b

### Atomic Tests and Leukemia

In the summer of 1957, in a desert area of Nevada, the United States carried out an above-ground atomic explosion code-named "Smoky." About 3000 military men and civilians were "invited" to watch. Some 20 years later, a compensation claim by one of the participants, a leukemia victim, prompted the U.S. Centers for Disease Control (CDC) to conduct a search for Smoky participants.

The overall rate of leukemia for men of ages similar to those of the Smoky participants is 1 in 1500. The CDC tracked down 450 of the participants and found that there were eight cases of leukemia among them. Is this convincing evidence that participation in Smoky was a cause of the disease? If so, the claim should be allowed. On the other hand, it is *possible* that 450 men selected "at random" from the population at large would include eight who had leukemia. We need to know the extent of this possibility before we can decide whether or not Smoky was the culprit.          ■

## Example c

### Corneal Thickness and Glaucoma

A study was conducted to determine whether glaucoma can be linked with other eye defects—in particular, with increased corneal thickness.[3] Clearly, one would need to know about the corneal thickness of normal eyes as well as that of glaucomatous eyes. One way to get such information is to measure the corneal thickness of some glaucomatous eyes and some normal eyes. Another possibility, the one used in this study, is to use subjects with glaucoma in one eye but not in the other. The study used eight subjects, and the corneal thicknesses are shown in the following table.

---

[2]Reported in the Santa Barbara *News-Press*, Jan. 13, 1987.

[3]Reported by N. Ehlers, "On Corneal Thickness and Intraocular Pressure, II," *Acta Ophthalmologica 48* (1970), 1107–12.

| Patient | Glaucomatous Eye | Contralateral Eye |
|---------|-----------------|-------------------|
| 1 | 488 | 484 |
| 2 | 478 | 478 |
| 3 | 480 | 492 |
| 4 | 426 | 444 |
| 5 | 440 | 436 |
| 6 | 410 | 398 |
| 7 | 458 | 464 |
| 8 | 460 | 476 |

An inspection of these numbers does not indicate a clear relationship. In some cases the eye with glaucoma has a thicker cornea than the other, and in other cases the reverse is true. In the course of this text we'll develop some precise tools for analyzing such data. ∎

**Example d**

## Breaking Strength of Welds

The data in the following table are from the report of an inertia welding experiment.[4] To make a weld, the operator stops a rotating part by forcing it into contact with a stationary part. The resulting friction generates heat that produces a hot-pressure weld. Seven welds were made using four different velocities of rotation (the controlled variable, $x$). Each weld was then tested for breaking strength (the response, $Y$), with results as follows:

| Velocity $(x)$ $(10^2 ft/min)$ | Breaking Strength $(Y)$ $(ksi)$ |
|-------------------------------|-------------------------------|
| 2.00 | 89 |
| 2.5 | 97, 91 |
| 2.75 | 98 |
| 3.00 | 100, 104, 97 |

The data seem to indicate a relationship between velocity and breaking strength of the resulting weld, but different breaking strengths can occur for the same velocity of rotation, so the relationship is not perfect. In particular, there may be factors influencing the results that can't be controlled or identified. Their contribution to the results is thought of as "random error." It is necessary, somehow, to get at the relationship in the presence of such errors. ∎

[4]G. E. P. Box, W. G. Hunter, and J. S. Hunter, *Statistics for Experimenters* (New York: Wiley, 1978), 473.

These examples involve some aspects of statistical problems generally. One is that responses and measurements are **variable.** One source of variability is the existence of individual differences, reflecting "randomness" in a population of subjects or experimental units. Measurements involve variability when the measurement process itself cannot be performed in exactly the same way one time as another. The various possible measured values constitute a kind of **population** from which one draws in making a measurement.

The target population in an investigation embodies some underlying truth or status or experimental mechanism that results in variability in the data. The reason for collecting data is to learn something about this target population. In order to draw conclusions about the population, we set up mathematical models that explicitly allow for randomness. The fundamental problem of statistics is to determine which of the various possible models for a population best represent it, in light of the sample data. Probability plays an essential role in representing randomness, so we begin our study of statistics with probability, the subject of Chapter 1.

# Probability

Example **1a**

## A Case of Discrimination?

Dr. Benjamin Spock, a well-known pediatrician and author, was tried for conspiracy in 1968 for his encouragement and aid to draft resisters during the Vietnam conflict. He was convicted by an all-male jury. His lawyers felt that since he was revered by millions of mothers, Spock might have had a better chance for acquittal had the jury included women.

One argument presented in the appeal of the verdict was that women were grossly underrepresented in the list of potential jurors from which the jury was drawn. In one 30-month period, the lists of the judge in the Spock trial included just 14.4% women—86 out of 597. In that same period, the jury lists for the other six judges in the same court included 29% women. Neither percentage is close to the proportion of women in the area (over 50%), but at the very least we could expect the methods of selection of Spock's judge to be no less fair than those of the other judges.

Suppose the names on Spock's judge's list were taken at random from the same population as those on the other judges' lists (with 29% women). One might ask, what are the chances that a list of 597, randomly selected from that population, would include no more than 14.4% women? ∎

To address the question posed in this example, we'd need to calculate "chances," or **probabilities.** Typically, the phenomena we'll be describing involve uncertainties; the factors that influence the results cannot be controlled—or even identified, in some cases. We can discuss chances meaningfully provided we make assumptions about the experiment. These define a probability **model,** which includes these main ingredients: a sample space, events, and probabilities of events.

## 1.1    Sample Spaces and Events

A phenomenon whose outcome is uncertain is an **experiment of chance.** A model for it should take into account all possible outcomes.

> In a model for an experiment of chance, the **sample space** $\Omega$ is the set of all outcomes considered to be possible.

**Example 1.1a** | **Sample Space for Selecting Jurors**

To obtain potential jurors, names are often selected "at random" from lists of registered voters (as in Example 1a). In any one selection, any name on the list might be drawn. The sample space $\Omega$ is the set of all registered voters.   ■

In this chapter, we consider only **discrete** sample spaces. A discrete sample space is one whose outcomes are finite in number or, if they are infinite in number, are at most countable—can be listed in sequence (as can the integers, for instance). When defining a set by listing its outcomes, we use braces around the list. Thus, for example, $\{a, b\}$ is the set whose only two elements are the outcomes $a$ and $b$.

**Example 1.1b** | **Sample Spaces for Coin Tossing**

When tossed, a coin falls with either "heads" ($H$) or "tails" ($T$) showing. The sample space is $\{H, T\}$.

Now consider the toss of *two* coins. Sometimes two people each toss a coin to see who pays for coffee. Either their coins match, or they don't match. The sample space consisting of these two outcomes may be adequate for modeling the coffee decision: $\Omega = \{\text{same}, \text{different}\}$.

In other situations, one might be interested in how many heads show. If so, $\Omega = \{0, 1, 2\}$ is an appropriate sample space. However, if one is interested in which coin falls which way, an even more detailed sample space is needed: $\Omega = \{HH, HT, TH, TT\}$.

Thus, different purposes may call for different sample spaces, even when the physical act is the same. Of the sample spaces we've given for the toss of the two coins, the last is the most detailed and could also serve as the sample space for the other situations described.   ■

**Example 1.1c** | **An Infinite Sample Space**

Consider a stream of potential customers entering a gift shop. Starting at an arbitrary point in time, we count how many people come in before one who actually makes a purchase. This number is a nonnegative integer. However, we couldn't be certain that a purchase will be made within any specified finite number of potential customers. So we may take $\Omega$ to be (countably) infinite: $\Omega = \{0, 1, 2, 3, ...\}$.   ■

Usually, there are subsets of the sample space that are of special interest. Any subset of the sample space is called an **event**. An event can be defined by giving either (1) a list of its outcomes, or (2) a descriptive phrase or condition that characterizes its outcomes. The symbol for an event will usually be a capital letter, possibly with a subscript.

An individual outcome $\omega$, considered as a set $\{\omega\}$ whose only member is $\omega$, is an event. Another special kind of event is the whole sample space itself.

---

An **event** is a set of outcomes—a **subset** of the sample space $\Omega$. An event $E$ is said to occur if any one of its outcomes occurs.

---

**Example 1.1d** | **Events Defined by Bets in Roulette**

When a ball is rolled on a spinning roulette wheel, it eventually comes to rest in one of 38 compartments numbered 0, 00, 1, 2, ..., 36. The 0 and 00 are colored green, and half of the numbers from 1 to 36 are red and the rest black. The set of 38 compartments in which the ball may land is an obvious choice for the sample space: $\{0, 00, 1, 2, ..., 36\}$.

One can bet on any individual number or on various sets of more than one number. These sets are events. For instance, "odd" is an event; a bet on "odd" wins if any odd number turns up. Other bets are defined by the casino: "black" (which wins if any black number turns up), "first dozen" (which wins if the outcome is any integer in the event $\{1, 2, ..., 12\}$), and so on. ∎

---

## 1.2     Event Algebra

The algebra of events deals with relationships and combinations of events.

**Example 1.2a** | **Dealing a Card**

The natural sample space for the selection of a card from a standard deck of playing cards[1] is the set of the 52 cards in the deck. Some events:

| | |
|---|---|
| $R$: The card is red. | $F$: The card is a face card. |
| $H$: The card is a heart. | 3: The card is a 3. |

There are some obvious relationships among these events. For instance, every outcome of $H$ is also in $R$, but no outcome is in both $F$ and 3.

Combinations of conditions $R$, $H$, $F$, and 3 may be of interest. Consider these:

The card is a red face card (that is, both red and a face card).
The card is not a 3.
The card is a heart or a face card.

---

[1] A standard deck consists of 52 cards, 13 in each of four suits: spades, hearts, diamonds, and clubs. The 26 hearts and diamonds are red; the rest, black. The 13 denominations in each suit are 2, 3, ..., 9, 10, $J$ (jack), $Q$ (queen), $K$ (king), ace. The jack, queen, and king are "face cards."

Each statement is a condition that defines a set of outcomes—an event in the sample space. They can be thought of as combinations or operations involving $R$, $H$, $F$, and 3.                                                                                        ■

**Inclusion, equality,** and **complementation** are basic notions in the algebra of events, defined as follows:

---

   **(i)** $E$ is **included** in $F$: $E \subset F$, if and only if every outcome in $E$ is also in $F$.

   **(ii)** $E$ and $F$ are **equal:** $E = F$, if and only if $E \subset F$ *and* $F \subset E$.

   **(iii)** The **complement** of $E$ is the set $E^c$ of all outcomes *not* in $E$.

---

In working with the algebra of sets, representing events in a **Venn diagram** can be helpful in understanding relationships among events and combinations of events. In a Venn diagram, we represent the sample space as the set of points in a large rectangle in the plane. An event is then a subset of this rectangle. Figure 1-1 shows a sample space and the following events:

   $E$: set of points within the circle,
   $F$: set of points within the triangle, and
   $G$: set of points within the square.

In Figure 1-1, $F$ is a subset of $G$: $F \subset G$. However, $E$ and $F$ are not ordered by inclusion: $F$ is not a subset of $E$, and $E$ is not a subset of $F$. The complement of $E$ is the large rectangle (the whole sample space) with the circle representing $E$ deleted.

In terms of conditions that define events, the relation $E \subset F$ means that condition $E$ *implies* condition $F$. The complement of $E$ is defined by the *denial* of the condition that defines $E$.

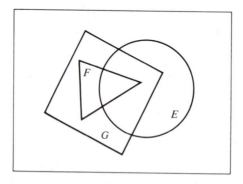

**Figure 1-1** A Venn diagram

The sample space $\Omega$ is itself an event. Its "complement"—the denial of $\Omega$—is a condition satisfied by *no* outcomes. Any condition satisfied by no outcomes in the sample space will also be called an event, denoted by $\emptyset$ and referred to as the **empty set.** Clearly, $\Omega^c = \emptyset$ and $\emptyset^c = \Omega$.

Events that have outcomes in common *intersect,* and the outcomes they have in common make up their **intersection** (as in the intersection of two streets). When an outcome is described as lying in *either* of two events (or in both), it is said to be in their **union.**

---

The **intersection** of events $E$ and $F$, denoted by $EF$, is the set of all points in *both* $E$ and $F$. The **union** of events $E$ and $F$, denoted by $E \cup F$, is the set of outcomes contained either in $E$ *or* in $F$ (or in both).

---

Two events that do not intersect are said to be **disjoint** or **mutually exclusive.** In the Venn diagram of Figure 1-2, events $E$ (circle) and $G$ (rectangle) are disjoint: $EG = \emptyset$. Events $E$ and $F$ (triangle) intersect, and their intersection $EF$ is the shaded region.

Figure 1-3 gives a Venn diagram illustrating set union. The shaded region defines the union of the events $E$, $F$, and $G$.

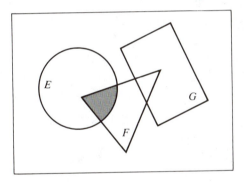

**Figure 1-2** Set intersection

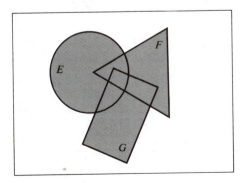

**Figure 1-3** Set union

Example **1.2b** | **Playing Cards**

Consider the sample space of Example 1.2a—the cards in a deck of playing cards. With obvious notation for king, queen, jack, for the four suits (spades, hearts, diamonds, clubs), and the two colors (black and red), we have:

$H \subset R$         (All hearts are red.)
$F3 = \emptyset$       (No card is both a 3 and a face card.)
$B = R^c$       (Cards that are not red are black.)
$SB = S$      (Cards that are spades and black are spades.)
$S \cup A$      (All the spades together with the other three aces.)
$R = D \cup H$    (Red cards are diamonds or hearts.)
$F = K \cup Q \cup J$   (Face cards are kings, queens, or jacks.)
$A = BA \cup RA$   (Aces are red aces or black aces.)
$\Omega = B \cup R = S \cup H \cup D \cup C = F \cup F^c$.  ∎

In translating verbal statements into mathematical ones, watch for the key words *and, or, implies,* and *not:*[2]

---

**Word/Symbol Equivalents:**

Condition $E$ *and* condition $F$: $\leftrightarrow$ $EF$.
Condition $E$ *or* condition $F$: $\leftrightarrow$ $E \cup F$.
Condition $E$ *implies* condition $F$: $\leftrightarrow$ $E \subset F$.
*Not* condition $E$ $\leftrightarrow$ $E^c$.

---

Like the operations of multiplication and addition of numbers, the operations of set intersection and union are **commutative:**

$$EF = FE, \text{ and } E \cup F = F \cup E. \tag{1}$$

They are also **associative:**

$$(EF)G = E(FG) = EFG, \tag{2}$$

and

$$(E \cup F) \cup G = E \cup (F \cup G) = E \cup F \cup G. \tag{3}$$

Thus, we have unambiguous definitions for intersections and unions of three events (without parentheses)—and, similarly, for intersections and unions of any finite number of events. These are special cases of general definitions, applicable even to uncountably infinite collections of events:

---

[2]English usage can be misleading. For example, "the set of hearts and diamonds" means the set of red cards—cards that are either hearts *or* diamonds. The word "and" is used in the sense that if we put together the hearts and diamonds we get all the red cards, or the set of cards each of which is a heart *or* a diamond.

The union of the events in any collection $\{E_\alpha\}$ is the set of all outcomes that are in at least one of the events, denoted by $\bigcup E_\alpha$. The intersection of the events $E_\alpha$ is the set of all outcomes that are in every event of the collection, denoted by $\bigcap E_\alpha$.

Example **1.2c**

## Uncountable Operations

Suppose $\Omega$ is the sample space of real numbers $x$. Let $\alpha$ be any positive real number. Let $E_\alpha$ denote the interval $-\alpha < x < \alpha$. There is only one number that is in every $E_\alpha$: $\bigcap E_\alpha = \{0\}$. This is because for any arbitrarily small number $\epsilon > 0$, there is an $\alpha$ such that $E_\alpha$ excludes $\epsilon$ (for instance, $\alpha = \epsilon/2$).

Let $F_\alpha$ denote the interval $0 \le x < \alpha$, where $0 < \alpha < 1$. The union of the $F_\alpha$ is the interval $0 \le x < 1$. ∎

When both intersections and unions are used in a single expression, we need a convention for the order in which operations are to be carried out. Union and intersection have many of the properties of addition and multiplication (respectively) of ordinary numbers, and we'll use the convention of ordinary arithmetic, parentheses playing the usual role. Thus, for example, $(A \cup B)(C \cup D)$ means the intersection of the events $A \cup B$ and $C \cup D$. However, with the removal of the parentheses, $A \cup BC \cup D$ means the union of the three events $A$, $BC$, and $D$.

Ordinary numbers obey the **distributive law:** $a(b + c) = ab + ac$. There is a corresponding distributive law for events—set intersection is distributive over unions:

$$A(B \cup C) = AB \cup AC. \tag{4}$$

This is evident in a Venn diagram *if* the diagram is sufficiently general, as in Figure 1-4, where the event representing the two sides of (4) is shaded.

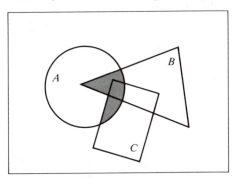

**Figure 1-4** Illustrating the distributive law

A dual distributive law also holds: Interchanging operations of union and intersection in (4), we have

$$A \cup BC = (A \cup B)(A \cup C). \tag{5}$$

[See Problem 1-6(f).] This says that the operation "union" is distributive over intersection. [See also Problem 1-6(i) for a more general distributive law.]

"Proofs by picture" can be convincing in simple cases such as (4) and (5). However, in more complicated situations, the picture one draws may have some special characteristic that makes a generally false statement true for that picture. The sure way to establish a set equality is to show that each side is a subset of the other.

| Example **1.2d** | **Verifying a Relation** |
| --- | --- |

The distributive law (4) can be argued as follows. If $\omega \in A(B \cup C)$, then $\omega \in A$ and either $\omega \in B$ or $\omega \in C$; thus, either $\omega \in AB$ or $\omega \in AC$—which is what is meant by $\omega \in AB \cup AC$. So the l.h.s. of (4) is in the r.h.s. of (4).

Now suppose $\omega \in AB \cup AC$. Then either $\omega \in AB$—in which case it is in both $A$ and $B \cup C$; or $\omega \in AC$—in which case it is in both $A$ and $B \cup C$. So either way, $\omega \in A(B \cup C)$. Thus, each point of the r.h.s. of (4) is also in the l.h.s., and equality is established.     ∎

An event $E$ and its complement $E^c$ are disjoint and (together) fill out the sample space: $E \cup E^c = \Omega$. They are said to *partition* $\Omega$, and in like manner the events $FE$ and $FE^c$ partition $F$. We next extend this notion of partition to the partitioning of a set into more than two pieces:

> A **partition** of an event $F$ is a collection of events $E_1, ..., E_k$ that are mutually disjoint and fill out $F$: $F = E_i \cup \cdots \cup E_k$.

In particular, a partition of a sample space $\Omega$ is a collection of events $E_i$ that are mutually disjoint and fill out $\Omega$: $\Omega = E_1 \cup \cdots \cup E_k$. Given such a partition of $\Omega$, the mutually disjoint events $FE_i$ constitute a partition of $F$: $F = FE_i \cup \cdots \cup FE_k$. (See Figure 1-5.)

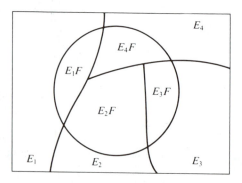

**Figure 1-5** Partitioning $F$

## Problems[3]

**∗ 1-1.**  Give an appropriate sample space for each of the following experiments:

**(a)**  Tossing a die until a 6 appears, noting how many tosses are needed.

**(b)**  In a classroom, recording a student's sex and college class.

**(c)**  Asking the political party preference of a voter selected from a population of voters.

**(d)**  Weighing a student on a scale in the health service.

**(e)**  Noting the time to failure of a lightbulb.

**∗ 1-2.**  Suppose we toss a nickel and a penny together. Let the sample space be $\{Hh, Ht, Th, Tt\}$, where $H$ and $T$ refer to the nickel and $h$ and $t$, to the penny.

**(a)**  List the outcomes in each of these events:

  $E$: Not both coins show heads.

  $F$: At least one coin shows heads.

  $G$: The penny shows heads.

**(b)**  List the outcomes in each of the following events: $EG^c$, $EFG$, $E^c \cup F^c$, $(EF)^c$, $E \cup FG$.

**(c)**  How many distinct events can be defined on this sample space? (Include $\emptyset$ and $\Omega$ in your count.)

**1-3.**  In the usual sample space for the toss of an ordinary six-sided die, consider these events:

  $E$: The outcome is not 6.

  $F$: The outcome is an even number.

  $G$: The outcome is less than 3.

List the outcomes in each of the following: $EG$, $F \cup G$, $EG^c$, $E^c \cup FG$.

**1-4.**  In the Venn diagram of Figure 1-6, let $E$ denote the set of points within the circle, $F$ the set within the triangle, and $G$ the set within the rectangle. Translate each of the following events into symbols, reproduce the diagram, and shade the set of points representing the given event:

**(a)**  $E$ and $F$ and $G$.          **(d)**  Not $E$ or not $F$.

**(b)**  $F$ but not $G$.          **(e)**  $E$ or $F$, and $E$ or $G$.

**(c)**  $E$ or both $F$ and $G$.          **(f)**  $E$ and $F$, or $E$ and $G$.

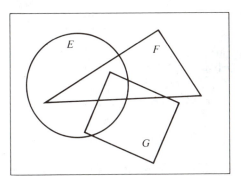

**Figure 1-6** Venn diagram for Problem 1-4

---

[3]Problems marked with an asterisk (∗) have answers at the back of the book.

∗ **1-5.**   Give each of the following as one of the events $E$, $\Omega$, or $\emptyset$:

   **(a)** $E \cup \emptyset$     **(c)** $E \cup \Omega$     **(e)** $EE$     **(g)** $E^c \cup E$

   **(b)** $E\emptyset$     **(d)** $E\Omega$     **(f)** $E \cup E$     **(h)** $E^c E$

**1-6.**   Show the following, either using a sufficiently general Venn diagram, an argument like that used in Example 1.2d, or a previously proved property:

   **(a)** $EF \subset E \cup F$.

   **(b)** $E \subset F$ implies $EF = E$.

   **(c)** $E \subset F$ and $F \subset G$ imply $E \subset G$.

   **(d)** $(EF)^c = E^c \cup F^c$.

   **(e)** $(E \cup F)^c = E^c F^c$.

   **(f)** $E \cup FG = (E \cup F)(E \cup G)$.

   **(g)** $EF \cup EF^c = E$.

   **(h)** $E \cup F = EF \cup EF^c \cup E^c F$.

   **(i)** $F(\bigcup E_\alpha) = \bigcup F E_\alpha$.

**1-7.**   Extend (d) and (e) of Problem 1-6 to the case of arbitrary collections $\{E_\alpha\}$ and so obtain *DeMorgan's laws,* of which (d) and (e) are special cases.

---

## 1.3          Probability for Experiments with Symmetries

We now complete the specification of a probability model by assigning probabilities to events. Probabilities give numerical expression to the intuitive notion of the relative likelihoods of events. Thus, if $A$ is more likely to occur than $B$, the "probability" of $A$ should be greater than the probability of $B$.

Assigning probabilities can be difficult. Indeed, any particular assignment is subject to revision in the light of actual experience with the experiment; just how to do this is a statistical problem.

There is an important class of experiments for which a particular way of assigning probabilities is quite compelling and may be generally agreed on as suitable. These are experiments with finite sample spaces, in which there is no reason to think that one outcome is more likely than another. We think of the outcomes as **equally likely,** and for this reason we *assign* the same probability to each one. This could be any number, but common practice takes it to be $1/N$, where $N$ is the number of outcomes in the sample space.

---

If the outcomes $\omega_i$ in a sample space $\Omega = \{\omega_1, \omega_2, ..., \omega_N\}$ are "equally likely," we say that each has probability $1/N$ and write

$$P(\omega_i) = \frac{1}{N}, \; i = 1, 2, ..., N. \tag{1}$$

---

There are a number of familiar experiments in which the possible outcomes are commonly assumed to be equally likely: the roll of a die, the toss of a coin, the spin of a wheel of fortune, the selection of a ticket in a lottery, etc. Because of the *symmetry,* we usually see no reason to favor one outcome as being more likely than another.

The tickets in a lottery are sold to finance the prize and are referred to as *chances*. A single ticket has "one chance in $N$" of winning. One who holds $k$ of the $N$ tickets has "$k$ chances in $N$" of winning. Two people who hold equal numbers of tickets have the same chance of winning.

We'll often use the random selection of a lottery ticket as a model for an experiment with equally likely outcomes. The sample space is the set of lottery tickets, and each ticket has probability $1/N$ of being drawn. A selection in which outcomes are equally likely is termed **random.** Thus, the phrases *random selection* and *select at random* mean that we assume each ticket has probability $1/N$—"one chance in $N$"—of being drawn.

> An individual is drawn **at random** from a population if all individuals in the population are equally likely to be the one selected.

Someone with $k$ tickets has "$k$ chances in $N$" of having a winning ticket. Those $k$ tickets constitute an event, and the probability of the event is defined to be $k/N$, the sum of the $k$ probabilities $1/N$ of the individual tickets.

> Suppose an event $E$ consists of a subset of $k$ of the outcomes in a sample space with $N$ equally likely outcomes. The **probability** of $E$ is
>
> $$P(E) = \frac{k}{N}. \qquad (2)$$
>
> The **odds on $E$** are $k$ to $N - k$. The **odds against $E$** are $N - k$ to $k$.

| Example **1.3a** | **Selecting a Card** |
|---|---|

**Selecting a Card**

In Example 1.2b we defined events for the selection of a card from a deck of playing cards. When the selection is deemed to be random, each card has probability $1/52$. From (2) the probability of the event $S$ (the card is a spade) is the number of spades divided by 52: $P(S) = 13/52$. The probability that the card is a face card is $P(F) = 12/52$, and so on. ∎

Definition (2) is appropriate only when the outcomes are equally likely, but we may adopt this model for making calculations even when we think it is only approximately correct. For example, after several shuffles of a deck of cards, the assumption of equally likely outcomes in the next deal may not be quite correct. However, the easy calculations based on the assumption of equal likelihood may be adequate. These are all we can do, not knowing the precisely correct model.

As defined by (2), the probability of an event has some important properties. First, we observe that $0 \le k \le N$, so $0 \le k/N \le 1$. Second, the sample space $\Omega$ has probability $N/N = 1$; its complement is the empty set $\emptyset$, with probability $0/N = 0$. Third, if $E$ contains $k$ and $F$ contains $m$ outcomes *and* $EF = \emptyset$, then $E \cup F$ contains $k + m$ outcomes, so

$$P(E \cup F) = \frac{k+m}{N} = \frac{k}{N} + \frac{m}{N} = P(E) + P(F).$$

These properties will be adopted as axioms in the more general setting to be described in §1.7.

---

**Properties of $P(\,\cdot\,)$:**

   **(i)** $0 \le P(E) \le 1$, for any event $E$.

  **(ii)** $P(\Omega) = 1$.

 **(iii)** $P(E \cup F) = P(E) + P(F)$ whenever $EF = \emptyset$.

---

Properties (i)–(iii) imply many other useful properties. For instance, applying (iii) with $F = E^c$, we obtain

$$1 = P(\Omega) = P(E \cup E^c) = P(E) + P(E^c),$$

or

$$P(E^c) = 1 - P(E). \tag{3}$$

We'll find this useful for finding the probability of $E$ when it is easier to calculate the probability of $E^c$.

Additivity for any finite collection of pairwise disjoint events follows by induction from (iii): Suppose it holds for $n$ events. Then since

$$E_1 \cup \cdots \cup E_n \cup E_{n+1} = (E_1 \cup \cdots \cup E_n) \cup E_{n+1},$$

we can apply additivity to the two events on the right to obtain

$$P(E_1 \cup \cdots \cup E_n \cup E_{n+1}) = \sum_{i=1}^{n} P(E_i) + P(E_{n+1}) = \sum_{i=1}^{n+1} P(E_i).$$

The induction argument then yields finite additivity for general $n$:

$$P(E_1 \cup \cdots \cup E_n) = \sum_{1}^{n} P(E_i). \tag{4}$$

For any $E$ and $F$, we know from Problem 1-6g that $F = FE \cup FE^c$. Then, since $FE$ and $FE^c$ are disjoint, it follows from (iii) that

$$P(F) = P(FE) + P(FE^c). \tag{5}$$

Extending this, it is clear that if the events $E_1$, $E_2$, ..., $E_n$ constitute a partition of $F$, then

$$P(F) = P(FE_1) + \cdots + P(FE_k). \tag{6}$$

This is called the **law of total probability**, illustrated in the next example.

**Example 1.3b** | **Survey Sampling**
In a political poll, an individual is selected at random from a population of voters. Some events of possible interest include these: $M$ (male), $F$ (female), $R$ (Republican), $D$ (Democrat), and $I$ (independent).
 Since $M \cup F = \Omega$, it follows [from (4) of §1.2] that the Republicans consist of Republican males and Republican females; and from (6), it follows further that $P(R) = P(RM) + P(RF)$. Similarly, since $R \cup D \cup I = \Omega$, we have $M = RM \cup DM \cup IM$ (upon "multiplying" through by $M$). So then,

$$P(M) = P(RM) + P(DM) + P(IM). \qquad \blacksquare$$

Using (2) to find the probability of an event $E$ requires counting the outcomes in $\Omega$ and the outcomes in $E$. The counting process can often be simplified by decomposing the experiment into a sequence of two or more simpler experiments. This brings us to the topic of the next section.

## Problems

**∗ 1-8.**  A card is drawn at random from a deck of playing cards. (See Example 1.2a for a definition of such a deck.) Find the probability that the card selected is

(a)  not a diamond.  (d)  red.

(b)  an honor card.  (e)  not a black honor card.

(c)  a red honor card.  (f)  a spade but not an honor card.

(Aces, kings, queens, jacks, and 10's are "honor cards.")

**1-9.**  The 38 equally likely positions on a roulette wheel are numbered 0, 00, 1, 2, ..., 36. (See Example 1.1d.) The 0 and 00 are green. Of the rest, half are black and half red. Find the probability that the outcome of a single spin is

(a)  red.  (c)  an odd number.

(b)  not black.  (d)  in the "third dozen" (25, ..., 36).

**∗ 1-10.**  Consider selecting *two* different letters from {A, B, C, D, E}. Assume that the ten possible pairs {AB, AC, AD, AE, BC, BD, BE, CD, CE, DE} are equally likely.

(a)  What is the probability that a particular letter (say, A) is one of those selected?

(b)  What would be the answer to the question in (a) had there been three letters selected from the five?

**1-11.**  In a population of 100 undergrads, 20 are seniors, 25 are juniors, and 30 are freshmen. Given that six of the seniors, five of the sophomores, and three of the freshmen smoke, and that 7/9 of the upperclassmen (juniors and seniors) don't smoke, find the probability that a student selected at random is

(a)  a smoker.  (b)  a sophomore.

(c)   a junior nonsmoker.

(d)   not a senior and doesn't smoke.

(e)   a freshman or a nonsmoker.

(f)   a senior or a smoking freshman.

∗ **1-12.**   You are one of ten finalists currently tied for first place in a contest. Three are to be selected at random from the ten to receive the top prizes. What are your chances of receiving a top prize?

**1-13.**   In an episode of the TV show "All in the Family," Mike claimed that he could identify different brands of cola by taste alone. He was challenged and presented with three glasses, one filled with Coke, one with Pepsi, and one with RC Cola. Suppose Mike really was *not* able to discriminate among the brands by taste.

(a)   Find the probability that none of his identifications would be correct.

(b)   Find the probability of no matches, had there been four brands of cola instead of just the three.

## 1.4     Composition of Experiments

Calculating probabilities for experiments with equally likely outcomes involves counting. Enumerating all possibilities can be quite tedious, and it is easy to miss some of them. In many important cases we can break the experiment into subexperiments that have relatively few outcomes. The following example shows how such a decomposition can be exploited to count outcomes.

**Example 1.4a** | **Throwing Two Dice**

Suppose we throw two dice, one green and one red. In many games (Monopoly and "craps," for example), the payoff or the next move is determined by the total number of points showing on the two dice. The sample space $\Omega = \{2, 3, ..., 12\}$ is adequate, as a list of the possible outcomes. But anyone experienced in throwing dice will tell you that these outcomes are *not* equally likely. A 7, for instance, is much easier to get than a 12. However, there is a more detailed sample space in which intuition suggests that its outcomes *are* equally likely.

The experiment of throwing the two dice can be thought of as consisting of two simpler experiments—tossing the green die and tossing the red die. For the composite experiment, we can take the outcomes to be the ordered pairs $(i, j)$, where $i$ is the number of points on the green die, and $j$ is the number of points on the red one. For each of the six possibilities for $i$, there are six possibilities for $j$, so the total number of pairs is $6 \times 6$ or 36. Using the simplified notation "$i\,j$" for the pair $(i, j)$, we can list the 36 possible pairs in the following array:

| | | | | | |
|---|---|---|---|---|---|
| 1 1 | 1 2 | 1 3 | 1 4 | 1 5 | 1 6 |
| 2 1 | 2 2 | 2 3 | 2 4 | 2 5 | 2 6 |
| 3 1 | 3 2 | 3 3 | 3 4 | 3 5 | 3 6 |
| 4 1 | 4 2 | 4 3 | 4 4 | 4 5 | 4 6 |
| 5 1 | 5 2 | 5 3 | 5 4 | 5 5 | 5 6 |
| 6 1 | 6 2 | 6 3 | 6 4 | 6 5 | 6 6 |

With pairs so listed, it is clear that $6 \times 6$ is the correct count.

Suppose the 36 pairs are equally likely. Then, since six of them have the total of 7 points, we have $P(7) = 6/36$. Similarly, there is just one pair with sum 12, so $P(12) = 1/36$.   ∎

The method of multiplication for counting outcomes (which yielded $6 \times 6$ in the above example) clearly extends to other composite experiments.

---

**Multiplication Principle:**

If an experiment $\mathcal{E}$ is a composite of experiments $\mathcal{E}_1$, $\mathcal{E}_2$, ..., $\mathcal{E}_k$, where the sample space of $\mathcal{E}_i$ contains $n_i$ outcomes, then the number of outcome sequences in $\mathcal{E}$ is the product $n_1 \times n_2 \times \cdots \times n_k$.

---

**Example 1.4b**   |   **Three Coin Tosses**

In a sequence of three tosses of a coin, there are two outcomes at each toss: heads (H), and tails (T). The number of outcome sequences is $2 \cdot 2 \cdot 2 = 8$:

HHH, HHT, HTH, THH, TTH, THT, HTT, TTT.

The number 8 is evident in a *tree diagram*, in which the choices at each stage are represented as branches from a node. Figure 1-7 shows the tree diagram for the three tosses of a coin.

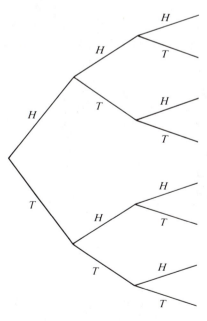

**Figure 1-7** Tree diagram for three coins

The same count (8) is correct if the three coins are tossed simultaneously, provided the coins are distinguished (for example, a nickel, a penny, and a dime), so that a "sequence" of outcomes is defined. ∎

In Examples 1.4a and 1.4b, the sample space for $\mathcal{E}_i$ does not depend on the outcomes of the other subexperiments. However, the multiplication principle applies in general if only, for each $i$, the *number* of outcomes in $\mathcal{E}_i$ (but not necessarily the outcomes themselves) is the same for each outcome of $\mathcal{E}_{i-1}$. In the next example, the outcomes constituting the sample space for $\mathcal{E}_2$ depend on the outcome of $\mathcal{E}_1$, but the *number* of outcomes in $\mathcal{E}_2$ does not.

**Example 1.4c**

## Counting Sequences of Selections

Consider a bowl with four chips, numbered 1, 2, 3, 4. Suppose we select three chips from the bowl, in sequence. The first chip we select is any one of the four chips; and no matter which it is, three chips are left for the second selection. After the first two selections, two chips remain—no matter which chips were drawn previously. The tree diagram in Figure 1-8 shows that there are $4 \times 3 \times 2 = 24$ sequences in the sample space of the composite experiment. ∎

---

In selecting $n$ objects one at a time from a group of $N$ objects, the number of possible sequences is

$$N(N-1)\cdots(N-n+1) = (N)_n, \tag{1}$$

called the *number of **permutations** of N things taken n at a time*.
[Some calculators use the notation $_NP_n$ for this number.]

---

In the special case $n = N$, *all* members of the population are included in the sample sequence. Complete sampling is seldom done, but applying (1) in this case gives a count of the number of possible arrangements of a set of distinct objects:

$$(N)_N = N(N-1)(N-2)\cdots 3 \cdot 2 \cdot 1 = N!. \tag{2}$$

The $N!$ arrangements are distinguishable provided that the objects being arranged are themselves distinguishable. When some of the objects are identical, the number of arrangements that can be distinguished is less than $N!$, as in the next example.

**Example 1.4d**

## Counting Patterns

A child has six blocks—three red (R), two blue (B), and one white (W). In how many distinct ways can these be arranged in a row?

The total number of permutations of six blocks is 6! or 720. But consider, say, this arrangement: R W R B B R. The two B's and the three R's could be rearranged while your back is turned in any of 2! × 3! or 12 ways, and you would not know the difference; the pattern would be unchanged. Similarly, there are 12 such rearrangements possible for *each* distinct pattern. So the number of distinct patterns is 720/12 or 60.                                                      ∎

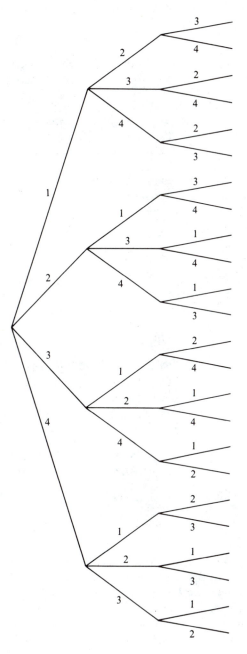

**Figure 1-8** Tree diagram for Example 1.4c

In the relatively simple setting of the example, the 60 patterns could actually be listed (and thus, counted), with a little patience. However, in making such lists it is hard to be sure of including every possibility, unless you know the total count. The method we used in Example 1.4d to count the patterns illustrates the general scheme:

---

The number of distinct sequences of $N$ objects, $m_1$ alike of type 1, $m_2$ alike of type 2, ..., and $m_k$ alike of type $k$ (where $\sum m_i = N$) is

$$\frac{N!}{m_1! m_2! \cdots m_k!} = \binom{N}{m_1, m_2, ..., m_k} \tag{3}$$

---

When there are just two types of objects or individuals in the population, we have the important special case of (3) in which $k = 2$. The number of distinct sequences of $m$ objects of one type and $N - m$ of the other is

$$\binom{N}{m, N-m} = \frac{N!}{m!(N-m)!} = \frac{N(N-1)\cdots(N-m+1)}{m(m-1)\cdots 3 \cdot 2 \cdot 1}.$$

Since the entry $N - m$ is really superfluous, a more convenient (and more common) notation drops either it or the $m$:

$$\binom{N}{m, N-m} = \binom{N}{m} = \binom{N}{N-m} = \frac{N!}{m!(N-m)!}. \tag{4}$$

**Example 1.4e** | **Arranging Objects of Two Types**

Three blocks have the letter A, and two have the letter B. In how many distinct ways can these five blocks be arranged?

The numbers are small enough that we can actually make a complete list of all possible patterns:

AAABB, AABAB, AABBA, ABAAB, ABABA,
ABBAA, BAAAB, BAABA, BABAA, BBAAA.

But by using (4), we can obtain the count 10 without having to make a list:

$$\binom{5}{3} = \binom{5}{2} = \frac{5 \cdot 4}{2 \cdot 1} = 10.$$

Viewing this problem in another way is instructive. An arrangement of three A's and two B's can be accomplished by choosing or selecting two of the five sequence positions for the B's, and then depositing the A's in the other slots. Or, we could select three of the five positions for the A's, and deposit the B's in the remaining two slots. Thus, we can accomplish the desired arrangement by a process of selecting an *un*ordered triple of sequence positions from five. The ten such selections, written so as to correspond to the order in the above list of letter sequences, are

$$123, 124, 125, 134, 135, 145, 234, 235, 245, 345.$$

(Notice that, although we have had to write down the numbers in a selection in a particular order, that order is of no consequence—we could have written 312 just as well as 123, for instance.) ∎

A selection made *without* regard to order is a **combination**. Following the reasoning of the preceding example, we have a simple way of counting combinations. The number of combinations of $n$ things chosen from $N$ is the same as the number of permutations of $N$ objects, $n$ of which are of one type and the rest of another, namely, $\binom{N}{n}$.

Another way of arriving at this number of combinations is to consider forming a sequence in the following two steps. First, make an unordered selection of objects to be used in the sequence; the name we give for the number of ways of doing this is $\binom{N}{n}$. Second, arrange the objects that have been selected; we can do this in $n!$ ways. So, using the multiplication principle, we find

$$\binom{N}{n} \cdot n! = (N)_n = N(N-1) \cdots (N-n+1).$$

Dividing by $n!$ we obtain (4) as the number of combinations.

---

The number of distinct **combinations** of $n$ objects selected from $N$ is

$$\binom{N}{n} = \frac{N!}{n!(N-n)!} = \frac{N(N-1) \cdots (N-n+1)}{n(n-1) \cdots 3 \cdot 2 \cdot 1}. \qquad (5)$$

[Another notation for $\binom{N}{n}$ is $_NC_n$.]

---

**Example 1.4f**

**Counting Poker Hands**

Dealing five cards from a well-shuffled deck of 52 playing cards can be thought of as selecting five cards at random. There are $\binom{52}{5}$ or 2,598,960 distinct hands of five cards. How many of these hands contain two aces, two 10's, and a jack?

We find the answer in a sequence of steps: Pick two aces from the four aces in the deck, then pick two 10's from the four 10's, and finally pick one of the four jacks. (We could have picked the 10's first, then the jack, and then the aces, with the same result.) The number of ways of constructing this hand is then the product of the numbers of ways of carrying out each step:

$$\binom{4}{2}\binom{4}{2}\binom{4}{1} = 6 \cdot 6 \cdot 4 = 144.$$

A poker hand like the one above is said to have "two pair"—two cards in each of two denominations together with a single card in yet a third denomination. To count *all* of the hands that would be said to have "two pair," we count how many ways the three denominations can be picked and then multiply by the 144, which applies to any particular choice of denominations: The two denominations from which the pairs are selected can be chosen from the 13 denominations in $\binom{13}{2}$ or 78 ways, and the denomination for the fifth card, in 11 ways. So there are $78 \cdot 11 = 858$ choices for the denominations. Thus,

$$P(\text{two pair}) = \frac{78 \cdot 6 \cdot 6 \cdot 11 \cdot 4}{2,598,960} = \frac{858 \cdot 144}{2,598,960} = .0475. \qquad \blacksquare$$

The number $\binom{N}{n}$, which counts the combinations of $n$ things from $N$, has another interpretation. Observe first that it is exactly the same as $\binom{N}{N-n}$, since in selecting $n$ things to use, we automatically select $N - n$ things to leave behind. That is, $\binom{N}{n}$ is the number of ways of dividing the $N$ things into two distinct groups, with $n$ in one group and $N - n$ in the other. Similarly,

$$\frac{N!}{n_1! n_2! \cdots n_k!} = \binom{N}{n_1,\ n_2,\ ...,\ n_k} \qquad (6)$$

is the number of ways of dividing or partitioning $N$ things into $k$ distinct groups, with $n_1$ in one, $n_2$ in another, and so on.

**Example 1.4g**

## Partitioning into Three Groups

Suppose we want to divide a group of 12 bridge players into 3 "tables," with 4 at each table. The number of ways of dividing the 12 into 3 *distinguishable* groups is

$$\binom{12}{4, 4, 4} = \frac{12!}{4!4!4!} = 34,650.$$

The same result can be obtained in another way: Identify the tables as Table 1, Table 2, and Table 3. First choose 4 players for Table 1 $[\binom{12}{4} = 495$ ways], then choose 4 from the remaining 8 for Table 2 (70 ways), and finally choose 4 from the remaining 4 for Table 3 (1 way). The number of ways of seating the players is then $495 \cdot 70 \cdot 1$, which is again 34,650 (of course!). (See Problem 1-33.)

Because the groups counted in this way are identified by table number, the count is appropriate only when the bridge tables are distinguishable. But if we are concerned only with which players are together at the same table, the count we've made is too high. We'd need to divide it by the number of possible arrangements of the three tables, obtaining $34,650/3! = 5775$ as the number of distinct groupings without regard to ordering of the tables. $\qquad \blacksquare$

## Problems

**∗ 1-14.** Three roads lead from town A to town B, and four lead from B to C. How many distinct routes can be followed from A to C via B?

**1-15.** In how many ways can eight six-sided dice fall? (Assume they are of different colors.)

**∗ 1-16.** A sample of four individuals is to be selected one at a time from a population of 15 individuals. The names are recorded in the order in which they are drawn.
   **(a)** How many distinct sample sequences are possible?
   **(b)** How many distinct samples (without regard to order) are possible?

**1-17.** In how many ways can five math books and four statistics books be placed on a shelf, so that the math books are together and the statistics books are together?

**∗ 1-18.** A student takes a multiple-choice exam consisting of 20 questions, with four choices per question.
   **(a)** How many answer sheets are possible?
   **(b)** How many are possible if no two successive responses are the same?

**1-19.** A club with ten members will pick a president, secretary, and treasurer.
   **(a)** In how many ways can this be done?
   **(b)** One of the ten club members wonders what his chances are of being chosen for an office. What are they, if the selections are made at random?

**∗ 1-20.** In a medical trial, each of ten subjects is assigned to one of two treatments, A and B, by the toss of a coin.
   **(a)** In how many ways can the assignments be made?
   **(b)** In how many of these assignments will exactly half of the subjects be given treatment A and the other half, B?
   **(c)** Assuming that the assignments in (a) are equally likely, find the probability that all ten subjects get the same treatment.

**1-21.** Four indistinguishable balls are to be put into three containers placed in a row. How many distinct configurations are possible? (One configuration, for instance, would be to have three balls in container #1, none in #2, and one in #3.)

**∗ 1-22.** Suppose we have 12 books and a shelf that holds only 8. In how many ways can we select 8 of the 12 books *and* arrange them on the shelf?

**1-23.** How many distinct permutations are there of the letters in
   **(a)** the word *minimum*?    **(b)** the word *lollipop*?

**∗ 1-24.** Find the number of ways three men and four women can be arranged in a row of seven chairs
   **(a)** without restriction.
   **(b)** if the men sit together.
   **(c)** if the men sit on one end and the women on the other.
   **(d)** if men and women alternate.
   **(e)** if two particular people cannot abide each other and must be separated.

**∗ 1-25.** If $n$ countries in a bloc are to exchange ambassadors, how many must be appointed?

**∗ 1-26.** In two-person cribbage, each player is dealt a hand of six cards.
   **(a)** How many distinct hands are possible for one player?

**(b)**   How many distinct deals—six cards to each player—are possible? (It would make a difference who gets which hand.)

**1-27.**   How many committees consisting of two men and two women can be chosen from a group of six men and five women?

∗ **1-28.**   A delegation of three from the ten members of a city council is to be chosen to attend a convention.

    **(a)**   In how many ways can a delegation be chosen?

    **(b)**   In how many ways, if two particular members will not attend together?

    **(c)**   In how many ways, if two particular members will either both go or neither go?

∗ **1-29.**   In how many ways can one choose a committee from a group of ten people if the size of the committee is not specified?

**1-30.**   Suppose you are to answer 9 out of 12 questions on an exam.

    **(a)**   In how many ways can you choose 9 questions to answer?

    **(b)**   In how many ways, if you must answer the first two?

    **(c)**   In how many ways, if you must answer either #1 or #2 but not both?

    **(d)**   In how many ways, if you must answer at least three of the first four?

∗ **1-31.**   In how many ways can one divide a group of six tennis players into three pairs to enter a doubles tournament?

**1-32.**   In how many ways can a group of eight individuals be divided into

    **(a)**   two groups, one of three and one of five?

    **(b)**   four teams of two each, each team with a different task?

    **(c)**   four teams of two each, all with the same task?

**1-33.**   Show the following two ways: (i) writing out in terms of factorials, and (ii) interpreting factors as counts of combinations or partitions.

    **(a)**   $\binom{n}{a,b,c} = \binom{n}{a}\binom{n-a}{b} = \binom{n}{c}\binom{n-c}{b}$, where $a+b+c=n$.

    **(b)**   $\binom{n}{a,b,c,d} = \binom{n}{a}\binom{n-a}{b}\binom{n-a-b}{c}$, where $a+b+c+d=n$.

---

## 1.5          Sampling at Random

An important type of composite experiment is that of sampling, one at a time, from a finite population. This can be done with or without replacement. In sampling *with* replacement, one replaces each individual drawn and thoroughly mixes the population before the next selection. In sampling *without* replacement, those drawn are *not* replaced before continuing.

    Suppose we sample a population of size $N$ at random, without replacement, until we obtain a sample sequence of specified size $n$. The multiplication principle of §1.4 says there are $(N)_n$ possible sample sequences. Intuition suggests that if the sampling is indeed random at each step, these $(N)_n$ sample sequences are equally likely. In the next example we assume that this is the case and examine some consequences of the assumption.

| Example **1.5a** | **Random Selection** |
|---|---|

Consider again a bowl of chips, this time six chips numbered from 1 to 6, and a sequence of three random selections of a chip, without replacement. There are $(6)_3$ or 120 possible sample sequences. Assuming these to be equally likely, we can find the probability of any event by *counting*. For instance, the number of sequences with chip $i$ in the first position is the number of ways of filling the second spot (5) times the number of ways of then filling the third spot (4), or 20. Thus

$$P(\text{first chip is } i) = \frac{20}{120} = \frac{1}{6}, \text{ for } i = 1, ..., 6.$$

That is, assuming the 120 sequences to be equally likely implies that the first chip is equally likely to be 1, 2, ...., or 6. This agrees with the original assumption that the selection of the first chip is done at random.

Another consequence of the assumption of equally likely sample sequences is not so obvious: The *second* chip drawn is also equally likely to be any of the six! The reasoning is exactly as it was for the first chip. This may seem wrong at first glance. To see intuitively why it is correct, realize that in finding the probability that the second chip is 4, you don't know which chip is removed at the first selection, and there is no reason to prefer one of the six chips over the other five for the second spot in the sequence.

If you are still not convinced, imagine that you select the first chip with your left hand but do not look at it, and then select a second chip with your right hand from those that remain. Now, is one hand more likely to contain chip 2 (say) than the other? Look at it this way: If you open your right hand first, the chip in the right hand becomes the first chip. ∎

For purposes of some probability calculations, we present the following rule, which will be established more rigorously in the next chapter (§2.3). We followed this rule in the above example to infer the equal likelihood of the sequences of three random selections (without replacement):

---

If experiment $\mathcal{E}$ consists of $\mathcal{E}_1$ followed by $\mathcal{E}_2$, ..., followed by $\mathcal{E}_k$, where in *each* $\mathcal{E}_i$ the $n_i$ possible outcomes are equally likely, then the $n_1 n_2 \cdots n_k$ possible sample sequences of $\mathcal{E}$ are equally likely.

---

In terms of tree diagrams, successive nodes (branch points) correspond to the successive subexperiments. The numbers of branches at the nodes corresponding to a particular subexperiment are assumed to be the same. And the rule asserts that the $n_1 n_2 \cdots n_k$ paths through the tree are equally likely if the branches at each node are equally likely—which is intuitively appealing.

Applying this to permutation problems, we note that a permutation of $N$ things $n$ at a time is simply a sample sequence, the result of sampling from the $N$ things one at a time. And if the individual selections are at random, the various permutations are equally likely. And this, as we shall see in the next example, implies that the various *combinations* of $n$ things are equally likely.

**Example 1.5b** | **Sampling Chips Without Replacement**

We return to the situation of Example 1.5a—selections of three from six numbered chips. When the chips are picked at random, the 120 possible sequences are equally likely. Ignoring order, the three chips selected constitute a *combination,* and there are $\binom{6}{3} = 20$ possible combinations. Each of these is made up of six sequences. And since the sequences are equally likely, the probability of any one combination is 6/120 or 1/20.

Now suppose the chips are four red and two white. We can find the probability that a selection of three chips drawn at random includes one white chip and two red ones. Since the event does not involve order, we can use the sample space of 20 equally likely combinations. Of these, the number of combinations in which two chips are red and one white is 12:

$$P(2 \text{ red}, 1 \text{ white}) = \frac{\binom{4}{2}\binom{2}{1}}{\binom{6}{3}} = \frac{12}{20} = \frac{3}{5}. \qquad \blacksquare$$

---

In a selection of $n$ objects at random from $N$, one at a time without replacement, the $\binom{N}{n}$ possible combinations are equally likely.

---

Suppose now we sample the population of size $N$ at random *with* replacement until we obtain a sample sequence of specified size $n$. The multiplication principle of §1.4 says there are $N^n$ possible sample sequences. And the above reasoning would say that these $N^n$ sequences are equally likely.

An important special case is that in which the population is the set of ten *digits*: 0, 1, 2, 3, 4, 5, 6, 7, 8, 9. We shall refer to a digit selected at random as a **random digit.** A sequence of digits selected at random with replacement defines a **random integer.** For instance, if we happen to choose the four digits 0, 1, 8, 1 (sampling with replacement permits repetitions), we can take the sequence as defining the integer 181. This is one of the $10^4$ or 10,000 equally likely possible sequences 0000 to 9999.

Random integers are used in obtaining random samples from finite populations. For this purpose, the population must be listed and (thus) numbered. A random integer defines a random selection of an individual from the population—the one assigned that number on the list. The "random integers" can be generated on a computer or read from a table of random digits, such as Table XV in Appendix I.

(We put the term random integers in quotes because a computer can only do what it is programmed to do, and a program is not exactly "random." But random number generators seem to produce streams of digits that behave, for practical purposes, as though they were generated by a process whose model is that of sampling at random with replacement from the ten digits. An alternative method would be to select chips at random from a bowl of ten chips numbered 0 to 9, but even doing this physical experiment may not be exactly represented by the ideal model we are claiming to sample.)

In taking an actual sample survey, there is not much point in interviewing the same person twice. But this can happen in sampling with replacement, so survey sampling is typically done without replacement. This can be achieved using sampling with replacement by simply ignoring any integer that has already been drawn. However, this method can be quite inefficient if the sample size is a substantial fraction of the population size. (There are more efficient ways of using random digits that we'll not go into.)

## Problems

*1-34.  A chain of pizza parlors ("Godfather's") once offered chance cards to its customers. Each card had eight numbers arranged in a circle, hidden by large dots that could be scratched away. Exactly three of the numbers were 7's, and a customer who uncovered the 7's by scratching away the dots from just three numbers would "win." (The card would be void if more than three numbers were revealed.) Find the probability that a person uncovers the three 7's by scratching out just three of the eight dots.

*1-35.  Suppose license plates are issued at random, each having three letters followed by three digits. The initial distribution of plates includes all those on which the first letter is either an A or a B.

    **(a)**  What is the probability that you'll get AAA-111? BBQ-279?

    **(b)**  What is the probability that you get a plate on which the three letters are the same and the three digits are the same?

*1-36.  I have ten socks in a drawer. Two pair are green, two pair blue, and one pair red. One blue sock has a hole in it.

    **(a)**  Dressing in the dark, I pick two socks at random. What is the probability that they are of the same color?

    **(b)**  If I pick three at random, what is the probability that at least two of the three are of the same color?

    **(c)**  If I pick four at random, what is the probability that at least two of the four are of the same color?

    **(d)**  What is the probability a random selection of four includes the sock with the hole?

1-37.  Four cards are to be dealt from a standard deck of 52 playing cards, at random and with no replacement. Find the probability that

    **(a)**  they are all from the same suit.

    **(b)**  they are half red and half black.

    **(c)**  there is one of each suit.

*1-38.  A committee of three is chosen at random from a group of eight students—two freshmen, two sophomores, two juniors, and two seniors. Find the probability

    **(a)**  that no freshmen are chosen.

    **(b)**  that the committee includes one senior, one junior, and one other.

**1-39.** Four of eight candidates for a council position are members of minority groups. If the eight names are placed on the ballot in random order, what is the probability that the names of the minority candidates head the list?

**∗ 1-40.** In Toronto, Canada, a family named Kelly bought nine tickets for a dollar each in the Interprovincial Lottery drawing. They won $13,890,588.80. In this lottery, six numbers were drawn at random from the integers 1 through 49 without replacement. To win, one had to pick all six numbers, but not necessarily in order. Find the probability that a person with nine different entries would win this lottery.

**1-41.** Explain how you could use a table of random digits (such as Table XV)

(a)  to call a Bingo game. (Numbers are selected at random and without replacement from the numbers 1 to 75.)

(b)  to obtain a sequence of random selections from a list with 3,750 names.

(c)  as a substitute for the pair of dice, in playing any game in which the players throw dice.

**∗ 1-42.** Each of four balls is put, one at a time at random, into one of ten containers. Find the probability that no container gets more than one ball.

**1-43.** Find the probability that no two among ten persons at a party have the same birthday, assuming that all $365^{10}$ possible sequences of birthdays are equally likely. [*Hint:* Find the count for the numerator as you did in the preceding problem, but with ten balls and 365 containers.]

**∗ 1-44.** In a later chapter (§10.5) we'll need to count arrangements of such things as three $X$'s and five $Y$'s. (The $X$'s refer to measured responses of treated individuals and the $Y$'s to measured responses of individuals in a "control" sample.) Suppose that all patterns of the three $X$'s and five $Y$'s are equally likely. Find

(a)  the number of possible distinct patterns.

(b)  the probability that the $X$'s will be together at one end or the other.

(c)  the probability that there will be just one $Y$ to the left of the middle.

(d)  the probability of finding exactly two $Y$'s to the left of the middle.

(e)  the probability of exactly three $Y$'s to the left of the middle.

**∗ 1-45.** Find the probability that in a five-card poker hand (random selection of five from a standard deck of cards),

(a)  three of the cards are of one denomination and the others are of two different denominations ("three of a kind").

(b)  three are of one denomination and the other two are from a second denomination ("full house").

**1-46.** Eight glasses are filled, four with New Coke and four with Classic Coke. A taster is to select four, the ones he thinks are the Classic Cokes. Suppose the taster really cannot tell them apart and is only guessing. Find the probability that among the four selected there are at least three Classic Cokes.

## 1.6   Binomial and Multinomial Coefficients

The numbers $\binom{n}{k}$ are called **binomial coefficients,** because they are coefficients in binomial expansions. We recall from algebra:

**Binomial Theorem:**

$$(x + y)^n = \sum_{k=0}^{n} \binom{n}{k} x^k y^{n-k}. \tag{1}$$

This theorem can be proved by mathematical induction on $n$.

That the coefficients are numbers of combinations is not just a coincidence: Consider the case $n = 5$. Every term in the expansion of $(x + y)^5$ is a product of five factors, each being either the $x$ or the $y$ from the corresponding factor $x + y$. There are 32 such products in the expansion because there are two choices (use either the $x$ or the $y$) for each of the five factors $(x + y)$. For instance, one product is $x \cdot x \cdot y \cdot x \cdot y = x^3 y^2$. However, there are several other products equal to $x^3 y^2$, and these "like" terms are combined in obtaining expressions such as (1). There are as many of them as there are ways of choosing three of the five factors $x + y$ from which to take the $x$ (the rest of the factors supplying the $y$). Thus, there are $\binom{5}{3}$ terms of the form $x^3 y^2$. The same reasoning applies in general to yield (1).

In similar fashion, we can obtain expansions of powers of a *multinomial*, in terms of *multinomial coefficients*. [See (6) in §1.4, page 24.] Reasoning as in the binomial case, we obtain the **multinomial theorem:**

$$(x_1 + x_2 + \cdots + x_k)^n = \sum \binom{n}{m_1, \ldots, m_k} x_1^{m_1} x_2^{m_2} \cdots x_k^{m_k}, \tag{2}$$

where the sum is taken over all possible combinations of the $k$ nonnegative integers $m_i$ that add up to $n$.

Substituting 1 for the $x$'s and $y$'s in (1) and (2) produces the following formulas:

$$\sum_{0}^{n} \binom{n}{k} = 2^n, \text{ and } \sum_{\Sigma m_i = n} \binom{n}{m_1, \ldots, m_k} = k^n. \tag{3}$$

These tell us that the total number of partitions of $n$ objects into $k$ distinct groups, of any size, is $k^n$. This result could also be reasoned as in the next example, which illustrates (3) when $k = 2$.

**Example 1.6a** | **Choosing Committees**

How many ways are there of forming a committee using individuals from a group of five when the size of the committee is not specified? Each committee represents a partition of the group into those on the committee and those not on it. The total number possible is the number with no members—the best kind—plus the number with one member, plus (and so on):

$$\binom{5}{0} + \binom{5}{1} + \binom{5}{2} + \binom{5}{3} + \binom{5}{4} + \binom{5}{5} = 32 = 2^5.$$

We can also arrive at this count by deciding, for each group member in turn, whether or not to assign that member to the committee: $2^5 = 32$. (The tree diagram has five branching stages, with two choices at each branch point.) ∎

---

## 1.7    Discrete Probability Distributions

We have defined probability for events in a sample space with equally likely outcomes. However, in many situations, the outcomes in the most natural or convenient sample space are not equally likely. A common situation is that in which there are just two possible outcomes—for instance, success or failure of some mission, or the occurrence or nonoccurrence of any event.

**Example 1.7a**    **Outcomes with Unequal Probabilities**

Suppose a thumbtack is dropped onto a flat surface. It lands either with point up (U), or with point resting on the surface (D). Without the symmetry of a coin, there is nothing to suggest that U and D are equally likely—or what the odds might be, if in fact they are not equal. Still, we'd like to assign probabilities to the outcomes U and D. ∎

In practice an experiment rarely has the types of symmetry that would suggest a model *a priori*, and in such cases we can only resort to performing the experiment in question and try to infer from the results what the odds might be—a process called *statistical inference*. In doing so, of course, we have in mind that there *is* an appropriate assignment of probabilities, albeit unknown to us.

Even though we assign unequal probabilities to outcomes, we can construct a lottery-type model for an experiment in which probabilities *are* equal and in which the original outcomes are events. For instance, if outcomes $a_1$, $a_2$, $a_3$, $a_4$ have probabilities .4, .3, .1, .2, respectively, we can use a sample space with ten equally likely outcomes $\omega_1, \ldots, \omega_{10}$. We then define the event $a_1$ in this new sample space to be $\{\omega_1, \omega_2, \omega_3, \omega_4\}$, $a_2$ to be $\{\omega_5, \omega_6, \omega_7\}$, and so on. The probability of one of these events is the sum of the probabilities assigned to its individual outcomes.

A **discrete probability model** consists of (a) a sample space $\Omega$ with at most a countable number of outcomes $\omega$, and (b) a nonnegative number $P(\omega)$ assigned to each $\omega$ as its probability, in such a way that $\sum P(\omega) = 1$. The probability of an event $E$ is defined as the sum of the probabilities of the outcomes in $E$:

$$P(E) = \sum_{\omega \in E} P(\omega). \tag{1}$$

In §1.3 we derived a number of properties of probabilities for experiments with equally likely outcomes. Properties (i)–(iii) of that section are quite basic and are satisfied when we define event probabilities using (1) above. We repeat them here:

---

**Fundamental Properties of Probability:**

For any events $E$ and $F$ in $\Omega$,

   **(i)**   $P(E) \geq 0$.
   **(ii)**   $P(\Omega) = 1$.
   **(iii)**   $P(E \cup F) = P(E) + P(F)$, if $EF = \emptyset$.

---

These properties are taken as axioms in the mathematical development of probability for arbitrary sample spaces.

In §1.3 we saw that Properties (i)–(iii) implied several other useful properties [(3)–(6) in §1.3]. Exactly the same reasoning applies to establish them for any probability space defined according to the axioms (i)–(iii). In particular, they hold for the discrete probability models defined using (1), and we repeat them here along with some further properties:

---

**Further Properties of Probability:**

   **(iv)**   $P(E_1 \cup \cdots \cup E_n) = \sum_1^n P(E_i)$ if $E_i E_j = \emptyset$ when $i \neq j$.
   **(v)**   $P(E^c) = 1 - P(E)$.
   **(vi)**   $P(\emptyset) = 0$.
   **(vii)**   If $E \subset F$, then $P(E) \leq P(F)$.
   **(viii)**   $P(E) \leq 1$.
   **(ix)**   $P(E \cup F) = P(E) + P(F) - P(EF)$.

---

Properties (iv) and (v) were demonstrated as (4) and (3) of §1.3, and the reasoning there applies here as well. Property (vi) follows from (v) and (ii) with $E = \Omega$. To show Property (vii) we observe first that $F = E \cup E^c F$, which holds when $E \subset F$; application of (iii) then yields (vii), because $P(E^c F)$ is nonnegative. Property (viii) is a special case of (vii) with $F = \Omega$. Proof of (ix) is left as an exercise, Problem 1-54. (See also Problem 1-R12.)

A property deserving special mention is the *law of total probability*, which is (6) of §1.3. Being based on (i)–(iii), it holds in this more general setting:

---

**Law of Total Probability:**

For any partition $\{E_1, ..., E_k\}$ of the sample space and any event $F$,

$$P(F) = P(FE_1) + \cdots + P(FE_k). \tag{2}$$

---

Models with equally likely outcomes require finite sample spaces, but discrete sample spaces can be (countably) infinite, as in the next example. If an event contains infinitely many outcomes, we still define its probability as the sum of the probabilities of its individual outcomes. But that sum is an infinite sum. Such sums are defined (as we learn in calculus) as limits of finite sums as more and more terms are added in.

**Example 1.7b**

## Waiting for Heads

A fair coin is tossed until heads turns up. This can happen on the first toss, but it may not happen until some subsequent toss. The sample space is

$$\Omega = \{H, TH, TTH, TTTH, TTTTH, \ldots\}.$$

The obvious probability for $H$ is 1/2, and thinking back on Example 1.4b, we might take the probability of the sequence $TTH$ to be 1/8. Similar reflection suggests that a possible probability model for $\Omega$ assigns the probabilities 1/2, 1/4, 1/8, 1/16, ..., for the outcomes as listed in $\Omega$. These are terms in a geometric series with sum

$$\frac{1}{2} + \frac{1}{4} + \frac{1}{8} + \frac{1}{16} + \frac{1}{32} + \cdots = 1.$$

The probability that it takes at least three tosses to get heads is

$$\frac{1}{8} + \frac{1}{16} + \frac{1}{32} + \cdots = \frac{1}{4}.$$

We could also calculate this as 1 minus the probability of two or fewer tosses: $1 - 3/4$, or as the probability that the first two tosses are tails. ∎

Property (iv) is referred to as the property of *finite additivity*. When probabilities are assigned according to (1) [page 32], they have the property of *countable additivity*:

$$P\left\{\bigcup_{n=1}^{\infty} E_n\right\} = \sum_{n=1}^{\infty} P(E_n), \text{ if } E_i E_j = \emptyset \text{ when } i \neq j, \tag{3}$$

because of the way we evaluate infinite sums. And this in turn would mean that the law of total probability (2) can be extended to countable partitions.

## 1.8 Subjective Probability

Strictly speaking, all probabilities are *subjective* or *personal*. One dictionary's definition of the term "subjective" reads in part: "...belonging to the thinking subject rather than to the object of thought (opposed to *objective*); personal."[4]

Probabilities seem to be properties of the object or the experiment when people tend to agree on what they are or should be, ideally. Such agreements occur usually when an experiment can be repeated over and over again. For example, experience with coin tossing reveals that "heads" turn up very close to half the time in very long sequences of tosses; and this is viewed as confirming the feeling that heads and tails are equally likely. Indeed, such long-run tendencies are sometimes taken as *defining* probabilities. Even so, the probability that a thumbtack falls with point up (Example 1.7a) is taken to be the long-run limit of the proportion of times it falls with point up in a sequence of trials, with the assumption that it exists as an objective property of the tack, albeit unknown.

Despite this common consensus about the objectivity of probabilities in some situations, people often use probabilities as subjective assessments of their personal feelings about the relative likelihoods of possible outcomes, especially of such things as horse races, elections, and so on—situations in which the experiment cannot be repeated, or studied "objectively," as in perceiving certain symmetries.

**Example 1.8a**

### Unequal Odds in Horse Races

A horse race can be thought of as an experiment of chance, in the sense that the outcome is uncertain. Someone who knows nothing about horse racing may regard the horses as being equally likely to win, but a well-educated fan knows that some horses are faster than others.

Consider a race involving just five horses: A, B, C, D, and E. Perhaps you feel that A is as likely to win as B, twice times as likely as C and as D, and five times as likely as E. For you, an appropriate assignment of probabilities would be numbers in proportion to 10:10:5:5:2, respectively. Thus, you would have in mind a model such as a lottery in which there are ten tickets marked A, ten marked B, five marked C, five marked D, and two marked E. The win probabilities for the five horses are then 10/32, 10/32, 5/32, 5/32, 2/32:

| Horse | Probability | Odds Against |
|-------|-------------|--------------|
| A | 10/32 | 10-22 |
| B | 10/32 | 10-22 |
| C | 5/32 | 5-27 |
| D | 5/32 | 5-27 |
| E | 2/32 | 2-30 |

[4]*American College Dictionary*, New York: Random House.

Your probability of the event {A, B, C}—that the winner is one of the first three horses listed—is the sum of the individual probabilities for those horses or 25/32.

But the betting board at the racetrack is apt to show odds that are different from yours and includes an amount that would be paid to you if you were to place a $2 bet on the horse that wins. It may look something like this:

| Horse | Odds Against | Payoff |
|-------|--------------|--------|
| A | 33-50 | $3.32 |
| B | 58-25 | 6.64 |
| C | 73-10 | 16.60 |
| D | 73-10 | 16.60 |
| E | 78-5 | 33.20 |

The odds in this table correspond to the payoff. For instance, a bettor puts up $2.00 on horse A against the track's $1.32, and the money odds ratio is $1.32:$2 or 33:50.

The listed odds reflect a consensus—a kind of "average" of the bettors' personal probabilities—determined by how money has been bet by the public. The "probabilities" (50/83, 25/83, etc.) implied by these odds do not sum to 1 because the bet is not fair—the track subtracts a portion of the money bet before paying off the winner. Here is how the track came up with the payoffs in the above table: Of the total amount of money bet—the "track handle"—half was on A, one-fourth on B, one-tenth each on C and D, and 1-twentieth on E, and these fractions have defined the payoffs. For instance, if the track handle were $1000, $500 was bet on A, $250 on B, $100 each on C and D, and $50 on E. The track's "take" is generally around 17%; deducting $170 leaves $830 to be paid out. If A wins, each of the 250 people who bet $2 on A gets back $830/250 = $3.32. If B wins, the return for each $2 bet is $830/125 or $6.64, and so on. ∎

People have uncertainties, not only about the outcome of an "experiment of chance," but about the way things are in any situation. They may be uncertain, as we'd be, as to the value of $p$, the probability that a thumbtack will fall with point up. They may be uncertain as to whether or not a proposed probability model is correct.

| Example **1.8b** | ## Is There a Treatment Effect? |
|---|---|

Suppose $H$ is the hypothesis that a particular drug has no effect in treating AIDS. Knowing the chemical structure of the drug and being familiar with similar chemicals and their biological effects, one scientist might say that $P(H) = .3$. Another may say $P(H) = .02$, and still a third, $P(H) = 0$. Who is right? Some statisticians say none of them is right; others say they are all right! The former feel that $P(H)$ is something that cannot be known by anyone; the latter consider probabilities as subjective, depending on the experience and information possessed by the person making the assessment. The latter view is called Bayesian, because (as we'll see in Chapter 8), Bayes' theorem is the tool used to update such probabilities in the light of new data. The first view disallows using Bayes' theorem because probabilities of hypotheses are not even defined. ∎

Does everyone *have* a personal probability for an event or situation about which he or she is uncertain? And if so, how can this probability be assessed? The existence can be shown if one assumes certain quite reasonable axioms for the relative likelihood of events. We'll not go into this but simply assume that there are subjective probabilities—probabilities that behave like objective probabilities, satisfying the axioms of §1.7.[5] The next example suggests how a subjective probability can be elicited in a particular situation.

| Example **1.8c** | ## Probability of a Hypothesis |

Consider a situation in which you are to give your probability for some hypothesis $H$, such as that in Example 1.8b. We ask you what $P(H)$ is, and you say you don't know. Not accepting this answer, we proceed as follows to elicit your $P(H)$. All we require is that you value money and that you agree that there is a fair coin—one that results in heads with probability 1/2.

We offer you this proposition: You may choose between receiving $1 if $H$ is true, or receiving $1 if the coin comes up heads. If you prefer the former, then $P(H) > 1/2$. If you prefer the latter, then $P(H) < 1/2$. And if you are indifferent, then $P(H) = 1/2$.

If, say, your $P(H)$ is greater than 1/2, we offer a second proposition (not forgetting the first, which we'll honor in any case): You may choose between receiving another $1 if $H$ is true and receiving $1 if a second (independent) coin toss also comes up heads. If you prefer the former, then $P(H) > 3/4$; if the latter, then $P(H) < 3/4$; and if you are indifferent, then $P(H) = 3/4$.

We proceed along these lines, offering you similar propositions, always splitting the interval of uncertainty in two. In this way, we can elicit your $P(H)$ to any degree of accuracy. For instance, it will cost us a maximum of $100 to identify an interval of width 1/1024 containing your $P(H)$. ∎

The procedure just outlined takes for granted an issue of some practical importance. Suppose $H$ is a hypothesis that no one will ever know to be true or false. For example, $H$ may be the hypothesis that of all brontosauruses that ever lived, more than half were female. In such a case, you'd have to be willing to take part in the elicitation just described, imagining that there will be payoffs even though you know there will be none. If you are not willing to do this, we cannot discover your $P(H)$. Another issue of some importance is that one dollar may not be enough for you to take the experiment seriously. In this case we'd have to increase the payoff.

[5]M. DeGroot, *Optimal Statistical Decisions*, New York: McGraw-Hill (1970), 77 ff.

## Problems

**∗ 1-47.** Write out the expansion of each of the following:
   **(a)** $(x + y)^5$.      **(b)** $(x + y + z)^3$.

**1-48.** Show this identity:

$$\binom{n}{k} + \binom{n}{k-1} = \binom{n+1}{k}.$$

**∗ 1-49.** Use the identity of the preceding problem to obtain the values of $\binom{6}{k}$ for $k = 1, 2, ..., 6$ from the values of $\binom{5}{k}$ given in Example 1.6a.

**∗ 1-50.** Find the coefficient of
   **(a)** $x^3 y^5$ in the expansion of $(x + y)^8$.
   **(b)** $x^2 y^4 z$ in the expansion of $(x + y + z)^7$.
   **(c)** $x^2 y^2 z^2 w^2$ in the expansion of $(x + y + z + w)^8$.

**∗ 1-51.** In the game of Monopoly, throwing a "double" (the two dice showing the same numbers of points) has special significance. (For instance, a double gets you out of "Jail" free.)
   **(a)** Find the probability of a double.
   **(b)** Following the reasoning in Example 1.7c, give a model for the experiment in which the pair of dice is cast repeatedly until a double occurs.

**∗ 1-52.** Given $P(A) = .59$, $P(B) = .30$, $P(AB) = .21$, find
   **(a)** $P(AB^c)$      **(b)** $P(A \cup B^c)$      **(c)** $P(A \cup B)$      **(d)** $P(A^c B^c)$.

**1-53.** Using properties (i)–(iii), show that $P(EF) \leq P(E \cup F) \leq P(E) + P(F)$.

**1-54.** Show Property (ix) of §1.7: $P(E \cup F) = P(E) + P(F) - P(EF)$.

**1-55.** Show by induction:

$$P(E_1 \cup \cdots \cup E_n) \leq \sum_{1}^{n} P(E_i).$$

**1-56.** Use the result of Problem 1-55 and DeMorgan's laws (Problem 1-7) to show the *Bonferroni inequality*:

$$P(E_1 E_2 \cdots E_k) \geq 1 - \sum_{i=1}^{k} P(E_i^c).$$

**1-57.** Let $\mu$ denote the mean weight of adult male emperor penguins. Use the technique of Example 1.8c to elicit a friend's prior distribution for $\mu$, to the extent of determining the median and quartiles. (Of course, there is no single "correct" answer.)

## Review Problems

**1-R1.** Simplify: **(a)** $(E \cup F)^c EF$.      **(b)** $(E \cup F)(E \cup F^c)$.

**1-R2.** Some people define $E - F$ as $EF^c$. Show the following:
   **(a)** $A(B - C) = AB - AC$.
   **(b)** If $B \subset A$, then $P(A - B) = P(A) - P(B)$.

**1-R3**. I do an experiment of chance six times and, after putting the results in numerical order, obtain: 1, 3, 7, 8, 25, 32. How many distinct sample sequences could have resulted in this particular set of ordered values?

**1-R4**. Suppose there are 12 lab animals that can be used for testing the effectiveness of three treatments. A standard procedure is to divide the 12 animals randomly into three groups of 4, and to apply treatment A to one group, B to another, and C to the third. Find the number of distinct ways of assigning animals to treatment groups.

**1-R5**. Given $P(A) = .5$, $P(B) = .3$, and $P(AB) = .21$, find $P(A \cup B^c)$.

**1-R6**. Three couples are to be seated by the random assignment of the six individuals to seats at a round table. Find the probability that Mr. and Mrs. Jones (one of the couples) are not seated next to each other. [*Hint:* Mr. A must sit someplace; seat him (anywhere) and then assign the others.]

**1-R7**. Find the term involving $p^4$ in the expansion of $(p + q)^{12}$.

**1-R8**. Individuals are selected at random, one at a time without replacement, from a group of 15 individuals, of whom 12 are white and 3 black.

(a) How many distinct ordered sequences of length 4 are possible?

(b) How many distinct combinations of 5 are possible?

(c) Find the probability that in the first 4 selected, all are white.

(d) Find the probability that the fourth one selected is black.

(e) Find the probability that a particular individual is included among the first 4 selected.

**1-R9**. Three chips are selected from eight poker chips, of which two are white, five are blue, and one is red. Assuming the selections are random, one at a time and without replacement, find the probability that

(a) the red one is included among the three picked.

(b) there are no blue chips among the three.

(c) there is at least one blue chip among the three.

(d) the third one picked is white.

(e) the chips selected are red, white, and blue in that order.

(f) there is one of each color among the three.

**1-R10**. Given $AB = \emptyset$, $P(A) = .3$, $P(B) = .5$, find $P(A^c B^c)$.

**1-R11**. Several states participate in the "powerball" lottery, in which five numbers are drawn at random (no replacement) from the numbers 1 to 45, and then a "powerball" number is drawn (independent of the first drawing) from 1 to 45. You win the lottery if the five numbers on your ticket match the five numbers drawn, without regard to order, and if the powerball number on your ticket is the same as the powerball number drawn.

(a) Find the probability that you win.

(b) You win smaller amounts for various partial matches. Find the probability that you match three of the first five numbers and also match the powerball.

(c) Find the probability that your powerball number does not match the winning number and your five other numbers are all different from the winning five.

**1-R12**. Apply Property (ix) of §1.7 three times to show

$$P(E \cup F \cup G) = P(E) + P(F) + P(G) - P(EF) - P(FG) - P(EG) + P(EFG).$$

**1-R13**. Let $\Omega$ denote the sample space of the integers from 0 to 5. Show that the following assignments of probability to integers $k$ define probability spaces:

**(a)**   $P(k) = \binom{5}{k}/32.$     **(b)** $P(k) = \binom{5}{k}(\frac{1}{3})^k(\frac{2}{3})^{5-k}.$ [*Hint:* Expand $(\frac{1}{3} + \frac{2}{3})^5.$]

## Chapter Perspective

Random sampling is basic to statistical methodology. In this chapter we have introduced, for the discrete case, the probability theory needed for the development and analysis of that methodology. In sampling *at random*, we choose each sample member in such a way that all available individuals are equally likely to be selected. The notion of *equal likelihood* of outcomes is fundamental in sampling.

When sample space outcomes are equally likely, we can calculate probabilities of events by *counting*. An essential tool in the counting process is the multiplication principle. Using this we developed shortcut methods for counting arrangements in a sequence, and counting divisions of a collection of objects into various numbers of subgroups.

You may be frustrated when you find such "combinatorial" calculations difficult to follow or to imitate. Don't give up. To be sure, in what follows, you will need to have had this experience in simple counting problems in order to deal effectively with some of the more interesting and important probability distributions and methods of inference. But there is much more to statistical theory than calculating combinations and permutations.

The basic properties of probability given in §1.7 are quite general, and apply whether one thinks of probability as a frequentist or a subjectivist. Although we have introduced these properties in connection with experiments with discrete sample spaces, they are used as axioms for probabilities defined on general sample spaces.

# Discrete Random Variables

**Example 2a**

### Pets and Recovery

A research study reported at a 1978 meeting of the American Heart Association focused on the possibility that having a pet might contribute to a patient's recovery. Patients were classified according to (1) whether or not they live at least one year after surgery, and (2) whether or not they have a pet.

With regard to whether a randomly selected patient does or does not have a pet, the observation varies from patient to patient. We say it is a *random variable.* Similarly, the observation of whether the patient survives or not is a random variable. The question is: Are these variables related? In particular, do patients who have pets live longer than those who don't? ■

We are often interested in characteristics that vary from one individual to another—sometimes focusing on one characteristic, but often focusing on relationships among several characteristics (as in the example).

A characteristic of the outcome of an experiment of chance is called a **random variable.** Random variables are **numerical** or **categorical.** The possible values of a numerical random variable are numbers; the possible values of a categorical random variable are categories in some scheme of classification. Values of a numerical variable are always ordered; values of a categorical random variable may or may not be ordered.

Numerical characteristics such as distance, time, and mass are usually thought of as numbers on a continuous scale—the possible values cannot be listed. We say that these random variables are **continuous.** When the possible values of a random variable are countable, we say that the variable is **discrete.**

**Example 2b**

A student chosen at random from the undergraduates at a university can be classified in various ways:

| Random variable | Values |
|---|---|
| Sex | Male, Female |
| Class | Fresh., Soph., Jr., Sr., Adult Special |
| Height (inches) | Numbers on the interval from 30 to 90 |
| College | Arts, Education, Technology, etc. |
| Shoe size | $3, 3\frac{1}{2}, 4, 4\frac{1}{2}, ..., 18$ |
| No. of siblings | 0, 1, 2, 3, ... |

Some of these variables are numerical (shoe size, number of siblings); some are categorical with unordered categories (sex, college); and class is categorical but with ordered categories. If measured with infinite precision, a person's height is a number on a continuous scale, but heights measured and recorded to the nearest inch are discrete. Similarly, foot length and width might be considered as measured on a continuous scale, but shoe sizes are discrete. ∎

In this and the next two chapters, we study methods for describing discrete random variables and their interrelationships. Continuous variables will be treated in Chapter 5.

## 2.1    Probability Functions

The value of a random variable depends on the outcome of the experiment being observed. In Example 2b, it depends on which student is selected. Thus, a random variable is a *function* of the outcome. We sometimes use function notation to denote a random variable, as in $X(\omega)$, but will often suppress the dependence on the outcome $\omega$ and call it simply $X$. Ordinarily, we represent random variables using uppercase letters appearing late in the alphabet $(X, Y, U, ...)$, perhaps with subscripts; their possible values are then usually denoted by corresponding lowercase letters $(x, y, u, ...)$.

Each possible value $x$ of a random variable $X$ defines an *event* in $\Omega$: the set of sample space outcomes $\omega$ assigned the value $x$ by the function $X(\omega)$. We then take the probability of that event in $\Omega$ as the probability assigned to the value $x$:

$$P[X(\omega) = x] = P[\text{set of } \omega\text{'s with } X(\omega) = x] = \sum_{X(\omega)=x} P(\omega).$$

And with probabilities so assigned to the possible values of $X$, we have made the value-space of $X$ a new probability space.

Example **2.1a**

## TV Viewers

Selecting a viewer at random from a population of TV viewers is an experiment of chance. Advertisers like to know which of the major commercial TV networks the viewer is watching at a given moment. This is a random variable. Its values have probabilities equal to the corresponding population proportions, since with random sampling, all viewers are equally likely to be selected. These probabilities are generally not known, so we give them literal names:

| Network | Probability |
|---------|-------------|
| NBC     | $p$         |
| CBS     | $q$         |
| ABC     | $r$         |
| Other   | $s$         |

All viewers are accounted for in the listed values, so the probabilities (population proportions) must sum to 1: $p + q + r + s = 1$. ■

The probability of observing the value $x$ clearly depends on $x$. It is a *function* of $x$, called the **probability function (p.f.)** of the random variable $X$ and denoted by $f(x)$. When necessary, to avoid confusion, we use a subscript to distinguish the p.f. of $X$ from other p.f.'s.

---

**Probability function** (p.f.) of the discrete random variable $X(\omega)$:

$$f(x) = P[X(\omega) = x] = \sum_{X(\omega)=x} P(\omega). \tag{1}$$

---

Example **2.1b**

## Checking Lot Quality

It is common practice in industry to decide whether a shipment of articles is acceptable on the basis of a sample taken at random from the shipment. Suppose that a shipment of 20 computer circuit boards includes 2 that are defective. (In practice, we would not know this.) When 3 circuit boards are chosen at random without replacement, the $\binom{20}{3} = 1140$ possible samples of 3 boards are equally likely. Each of these possible samples is an $\omega$, and the number of defectives in $\omega$ is a random variable $X(\omega)$, with possible values 0, 1, and 2. Since we assumed that the samples $\omega$

are equally likely, we can find the values of the p.f. by counting. For example, the probability of 0 defectives in the sample is

$$f(0) = \frac{\text{Number of samples with 0 defectives}}{\text{Number of possible samples}} = \frac{\binom{18}{3}}{\binom{20}{3}} = \frac{136}{190}.$$

In general,

$$f(x) = \frac{\binom{2}{x}\binom{18}{3-x}}{\binom{20}{3}}, \ x = 0, 1, 2.$$

When the list of possible values is short, as in this case, it is easy to give the p.f. in table form:

| $x$ | 0 | 1 | 2 |
|-----|-----|-----|-----|
| $f(x)$ | 136/190 | 51/190 | 3/190 |

Figure 2-1 gives a graphical representation of this p.f. ∎

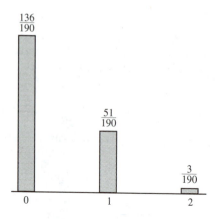

**Figure 2-1** Probability function for Example 2.1b

**Example 2.1c**

## Waiting for a Success

Example 1.7c dealt with an experiment in which we toss a coin until heads turns up. The outcomes $\omega$ are the sequences $H, TH, TTH, TTTH, \ldots$. Define $X(\omega)$ to be the number of tails before the first heads. The possible values of $X$ are 0, 1, 2, 3, .... With probabilities assigned to the sequences $\omega$ as in Example 1.7c, the p.f. for $X$ is

$$f(x) = \frac{1}{2^{x+1}} \ \text{for } x = 0, 1, 2, \ldots.$$

This function is partially shown in Figure 2-2. ∎

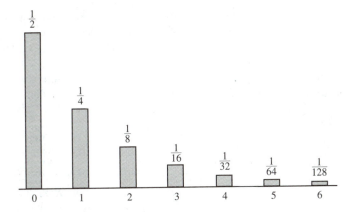

**Figure 2-2** Waiting time p.f.—Example 2.1c

The outcome $\omega$ itself is a special random variable: $X(\omega) = \omega$. The probability table for this random variable is simply the probability table for the sample space $\Omega$. The next example is a case in point.

**Example 2.1d** | **Tossing a Die**

The usual sample space for the toss of a die is the set of its six faces, identified by number: $\Omega = \{1, 2, 3, 4, 5, 6\}$. With the assumption that the faces are equally likely, $P(\omega) = 1/6$ for each face $\omega$. The probability function of the random variable $X(\omega) = \omega$ is then

$$f(x) = 1/6, \quad x = 1, 2, 3, 4, 5, 6.$$

Other random variables can be defined on $\Omega$. For instance, the *indicator function* of the event Odd $= \{1, 3, 5\}$ is $Y(\omega) = 1$ for $\omega \in$ Odd, and $Y(\omega) = 0$ otherwise. The p.f. of $Y$ is defined by

$$f_Y(1) = P(\text{Odd}) = \frac{1}{2}, \; f_Y(0) = P(\text{Even}) = \frac{1}{2}. \qquad \blacksquare$$

In the above examples we first defined probabilities in a sample space. From these we calculated the probability function of the random variable of interest. The resulting p.f. is nonnegative, because its values are probabilities. And its values sum to 1, since the events $\{X = x\}$ partition $\Omega$:

---

**Properties of a probability function $f(x)$:**

   **(i)** $f(x) \geq 0$ for all $x$.

   **(ii)** $\sum f(x) = 1$.

---

In constructing a model for a random variable $X$, it is often convenient and natural to specify its distribution by assigning probabilities directly in its value-space. Thus, the value-space is the $\Omega$, $X(\omega) = \omega$, and the p.f. we specify must satisfy (i) and (ii).

A probability function—whether given by table, graph, or formula—shows how probability is distributed among the possible values of a random variable. We refer to how it is distributed as its **probability distribution.**

## 2.2    Joint Distributions

Discovering and understanding relationships among variables is useful in applications. Changing one of two related variables may *cause* a change in the other. In such a case, we can control an effect to some extent by controlling the cause. However, the relation may not be one of cause and effect. In either case, a relationship between variables can be useful for *predicting* one variable given the values of the others.

To study a relationship between random variables $X(\omega)$ and $Y(\omega)$, we need to consider them together, as a **random vector,** $(X, Y)$. The "values" of this random vector are pairs $(x, y)$. The probability distribution of $(X, Y)$ is called **bivariate** and is defined by a **joint probability function:**

$$f(x, y) = P[X(\omega) = x, Y(\omega) = y]. \tag{1}$$

This is calculated as a sum of probabilities $P(\omega)$ for all $\omega$ such that $X(\omega) = x$ and $Y(\omega) = y$. As in the univariate case, a bivariate p.f. must satisfy certain conditions: Since it is a probability, it must be nonnegative and no larger than 1. Also, if we sum the probabilities (1) over all possible pairs $(x, y)$, the result must be 1. Each probability in $\Omega$ is assigned to one of these pairs, so all of the probability in $\Omega$ is accounted for in the sum of the $f(x, y)$.

---

**Joint probability function** for $(X, Y)$:

$$f(x, y) = P(X = x, Y = y),$$

satisfies
  (i) $f(x, y) \geq 0$ for all pairs $(x, y)$.
  (ii) $\sum f(x, y) = 1$.

---

Properties (i) and (ii) parallel corresponding properties of a univariate p.f. As in the univariate case, they follow from similar properties of probability as defined in a sample space $\Omega$. And, as in the univariate case, one can define a (bivariate) distribution by giving a function $f(x, y)$ satisfying properties (i) and (ii), taking $\omega$ as $(x, y)$ and $f(x, y)$ as $P(\omega)$.

The joint p.f. implies distributions for $X$ and $Y$ individually, which follow from the law of total probability. That is,

$$P(X = x) = P(X = x, Y = y_1) + P(X = x, Y = y_2) + \cdots.$$

**Example 2.2a** | **Class vs. Sex**

The 100 students in a class were cross-classified according to sex and class:

|  |  | $Y = class$ | | | |
|---|---|---|---|---|---|
|  |  | Fr. | So. | Jr. | Sr. |
| $X = sex$ | Male | 4 | 27 | 15 | 5 |
|  | Female | 6 | 21 | 16 | 6 |

The entry in each cell of the table is the number of students in that category of the classification.

Suppose we consider this class to be a population and select a student at random from the class. The class members constitute the $\omega$'s, and for each $\omega$, $P(\omega) = 1/100$, because of the assumption of random selection. The probability $f(x, y)$ of a given pair $(x, y)$ is then the relative frequency of that pair in the population. For instance,

$$f(\text{Male, So.}) = P(\text{male sophomore}) = \frac{\#(\text{male sophomores})}{\#(\text{students})} = .27.$$

The complete table of probabilities $f(x, y)$ is as follows:

|  |  | $Y = class$ | | | | |
|---|---|---|---|---|---|---|
|  |  | Fr. | So. | Jr. | Sr. | |
| $X = sex$ | Male | .04 | .27 | .15 | .05 | .51 |
|  | Female | .06 | .21 | .16 | .06 | .49 |
|  |  | .10 | .48 | .31 | .11 | 1 |

The subtotals of the probabilities in rows and in columns are shown in the right and lower margins. These are probabilities for the individual variables $X$ and $Y$ and define the univariate distributions of these variables. For instance, according to the law of total probability [(2) of §1.7], we have

$$f_X(\text{M}) = f(\text{M, Fr.}) + f(\text{M, So.}) + f(\text{M, Jr.}) + f(\text{M, Sr.}) = .51,$$

and

$$f_Y(\text{So.}) = f(\text{M, So.}) + f(\text{F, So.}) = .48.$$

The univariate distributions of $X$ alone and $Y$ alone, which appear in the table's margins, are called **marginal distributions.** ∎

**Marginal probability functions:**

$$f_X(x) = P(X = x) = \sum_y f(x, y), \tag{1}$$

$$f_Y(y) = P(Y = y) = \sum_x f(x, y). \tag{2}$$

**Example 2.2b**

## Selecting Two Chips at Random

From a bowl containing four chips marked 1, 2, 3, 4, one is selected at random, and then another is selected at random from the remaining three. Let $X$ and $Y$ denote the numbers on the first and second chips, respectively.

The outcomes of this experiment are pairs $(x, y)$, where $x$ and $y$ are each one of the numbers 1, 2, 3, 4. Suppose we assume that the 12 pairs $(x, y)$ with $x \neq y$ are equally likely. (We shall see that this is consistent with the way the selections were made.) The joint p.f. is

$$f(x, y) = \frac{1}{12} \text{ for } x, y = 1, 2, 3, 4 \text{ and } x \neq y,$$

and 0 otherwise. We can exhibit these probabilities in a table, including the row and column totals in the margins:

|      |        | $Y:$ 1 | 2    | 3    | 4    | $f_X(x)$ |
|------|--------|------|------|------|------|----------|
|      | 1      | 0    | 1/12 | 1/12 | 1/12 | 1/4      |
| $X:$ | 2      | 1/12 | 0    | 1/12 | 1/12 | 1/4      |
|      | 3      | 1/12 | 1/12 | 0    | 1/12 | 1/4      |
|      | 4      | 1/12 | 1/12 | 1/12 | 0    | 1/4      |
|      | $f_Y(y)$ | 1/4  | 1/4  | 1/4  | 1/4  | 1        |

The marginal entries give the univariate p.f.'s for $X$ alone and $Y$ alone. Observe that the marginal distributions are the same! ∎

We define joint probability functions for more than two random variables in the same way as for two. In the case of three variables $(X, Y, Z)$, the joint p.f. is

$$f(x, y, z) = P(X = x, Y = y, Z = z). \tag{3}$$

A three-way table giving these joint probabilities would require three dimensions, but we can exhibit them on two-dimensional paper by "layering," a technique to be illustrated in the next example. Summing entries in a three-way table along one direction, say $X$, produces the marginal joint distribution of $Y$ and $Z$. The univariate distribution for any single variable can be obtained by summing over the other two.

**Example 2.2c** | **Taste Testing**

Problem 1-13 dealt with Mike's attempt to identify three brands of cola in a taste test. We assumed there that Mike really was *not* able to discriminate among the three brands of cola and assigned probability 1/6 to each possible arrangement. (Think of the glasses of cola lined up in a sequence and his assignment of brand to glass as a permutation of the correct brand sequence.)

Let $C$ denote the event that Mike correctly identifies the Coke. The indicator function [introduced in Example 2.1d] of this event is

$$X_C = \begin{cases} 1 & \text{if he correctly identifies Coke,} \\ 0 & \text{otherwise.} \end{cases}$$

Define $X_P$ and $X_R$ similarly for Pepsi and Royal Crown, respectively. The six possible assignments (the equally likely $\omega$'s) are shown in the following table. It also shows the values of the indicator variables $X$, their joint probability function, and the values of $Y = X_C + X_P + X_R$, the total number of correct identifications. The joint p.f. is pictured in Figure 2-3.

| Actual brand: | C | P | R | $X_C$ | $X_P$ | $X_R$ | Y |
|---|---|---|---|---|---|---|---|
| Mike's | C | P | R | 1 | 1 | 1 | 3 |
| identification: | C | R | P | 1 | 0 | 0 | 1 |
| | P | R | C | 0 | 0 | 0 | 0 |
| | P | C | R | 0 | 0 | 1 | 1 |
| | R | P | C | 0 | 1 | 0 | 1 |
| | R | C | P | 0 | 0 | 0 | 0 |

More generally, a trivariate distribution is easier to exhibit using a three-way table given in layers. Here, choosing $X_R$ as the layering variable, we have the following two-way tables for $(X_C, X_P)$:

$X_R = 1$:                                    $X_R = 0$:

|  |  | $X_P$ | |
|---|---|---|---|
|  |  | 0 | 1 |
| $X_C$ | 0 | 1/6 | 0 |
|  | 1 | 0 | 1/6 |

|  |  | $X_P$ | |
|---|---|---|---|
|  |  | 0 | 1 |
| $X_C$ | 0 | 1/3 | 1/6 |
|  | 1 | 1/6 | 0 |

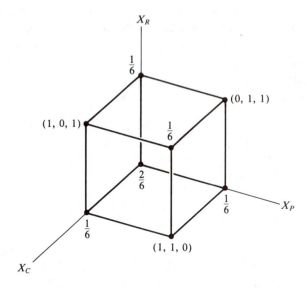

**Figure 2-3** Joint p.f. for Example 2.2c

Figure 2-3 also exhibits the joint distribution of the three $X$'s. Each of the possible triples is marked with the corresponding probability. The layers given above are easy to see in this figure. (We could have used either of the other variables as the layering variable. Because of the symmetry, the two-way layers in each case look the same as those given above.)

The distribution of the sum $Y$ is apparent in the original table, in which each row has probability $1/6$: $f_Y(3) = 1/6$, $f_Y(1) = 1/2$, $f_Y(0) = 1/3$.   ∎

## Problems

**∗ 2-1.**  Example 1.4a gave the 36 outcomes for the toss of two dice. Let $a$ and $b$ denote the numbers of points on the first die and second die, respectively. Assuming the 36 outcomes to be equally likely, construct a probability table for the random variable $X(a, b) = a + b$.

**∗ 2-2.**  Referring to Problem 2-1, construct the probability table for the random variable $Y$ $(a, b) = a - b$.

**2-3.**  A carton of 12 eggs includes 2 that are rotten. Let $X$ denote the number of rotten eggs among 3 chosen at random from the carton. Give the probability distribution of $X$. (That is, list the possible values of $X$ and give the probability of each.)

**∗ 2-4.**  Two chips are drawn without replacement from a bowl containing five chips numbered from 1 to 5. List the ten possible selections of two chips and construct probability tables for the following random variables:

  **(a)**  $S = $ sum of the numbers on the chips selected.

**(b)**  $D$ = magnitude of the difference (that is, the absolute difference).

**(c)**  $U$ = larger of the two numbers.

**\* 2-5.**  A report based on the 1970 census gives the cross-classification of state and local government employees (in 1000's) shown in the accompanying table. Find the probability that one individual selected at random from all government employees is

**(a)**  in education.

**(b)**  a state employee.

**(c)**  in education or police work.

**(d)**  a state employee not in education.

|           | State | Local | Total |
|-----------|-------|-------|-------|
| Education | 939   | 3723  | 4662  |
| Highways  | 295   | 298   | 593   |
| Health    | 443   | 465   | 908   |
| Police    | 47    | 396   | 443   |
| Other     | 604   | 1823  | 2427  |
| Total     | 2328  | 6705  | 9033  |

**2-6.**  From a bowl containing five chips numbered 1 to 5, two are selected at random, one at a time *with* replacement. Let $X$ denote the number on the first chip selected, and $Y$ the number on the second chip. Construct a table of joint probabilities and use it to find

**(a)**  $P(Y = 2)$.        **(c)**  $P(X = 1$ and $Y = 2$ or $4)$.

**(b)**  $P(X + Y = 5)$.        **(d)**  $P(|X - Y| = 2)$.

**\* 2-7.**  Repeat Problem 2-6, but with the change that the second chip is drawn *without* replacement of the first chip.

**2-8.**  There may be a relationship between the level of education of a person and his or her opinion about the death penalty proposed for skyjackers. Suppose the joint probabilities for these two variables are as follows:

|            |             | Opinion: | |
|------------|-------------|----------|--------|
|            |             | Favor    | Oppose |
| Education: | Grade school | .15     | .05    |
|            | High school  | .20     | .10    |
|            | College      | .25     | .25    |

**(a)**  Give the probability table for the variable "Education."

**(b)**  Give the probability table for the variable "Opinion."

**∗ 2-9.** Let $(X, Y)$ have the joint distribution defined by the following table of probabilities:

|  | | $X:$ | |
|---|---|---|---|
| | 1 | 2 | 3 |
| $Y:$　1 | .2 | 0 | .4 |
| 　　　2 | 0 | .3 | .1 |

(a) Find the marginal p.f.'s.
(b) Find $P(X + Y > 3)$.
(c) Find the p.f. of $Z = X + Y$.

**2-10.** Consider again the variables $S$ and $D$ in Problem 2-4.
(a) Give the table of *joint* probabilities.
(b) Find the p.f. of the variable $X = S - D$.

---

## 2.3　　Conditional Probability

Suppose we have settled on a particular probability model for some experiment of chance. If, subsequently, we learn something more about the experiment, we may need to revise the model accordingly.

**Example 2.3a** │ **Partial Information About a Card**

When a card is selected at random from a standard deck of playing cards, the probability of a spade is 1/4. Suppose, however, you learn that the card selected is red. Are the odds that the card is a spade still 1:3? Obviously not. The model for the experiment needs to be updated. All black cards are ruled out, and the appropriate sample space is now just the set of red cards—the subset of the original sample space defined by the event "red." Events in the new sample space must be given new probabilities. ■

When it is learned that the event $F$ has occurred, just how does this information change the probability of another event $E$? Consider the lottery model for the experiment—a random selection of a ticket from a lottery. To say that $F$ has occurred is to say that only the tickets with the property $F$ are now possible; the others are removed from the lottery. In calculating new probabilities, what has changed is the available pool of tickets. And if the tickets in $F$ were equally likely *before* we removed the others, it is reasonable to view them as still equally likely *after* the removal—that is, after learning that $F$ has occurred. But the denominator of the fractions defining probabilities changes from the number of tickets in $\Omega$ to the number in $F$.

When some tickets are removed, the assumption that the remaining tickets are equally likely is our rationale for defining conditional probability.

Conditional probability of an outcome in $\Omega$ given that an event $F$ with positive probability has occurred:

$$P(\omega \mid F) = \begin{cases} 0 & \text{if } \omega \text{ is not in } F \\ \frac{P(\omega)}{P(F)} & \text{if } \omega \text{ is in } F, \end{cases} \tag{1}$$

In §1.7 we defined probability for any event $E$ as the sum of the probabilities of its outcomes. And so we now define conditional probabilities for events from the probabilities given by (1):

$$P(E \mid F) = \sum_{\omega \text{ in } E} P(\omega \mid F). \tag{2}$$

Substituting from (1) into (2), we obtain the following definition of conditional probability:

**Conditional probability** of an event $E$, given $F$:

$$P(E \mid F) = \frac{P(EF)}{P(F)}, \tag{3}$$

provided $P(F) \neq 0$.

**Example 2.3b**

**Waiting for a Success**

Example 2.1c dealt with the experiment in which we toss a coin until heads turns up. With $Y$ defined as the number of trials required to get the first heads, the possible values are 1, 2, 3, ..., and the p.f. is

$$f(y) = \frac{1}{2^y} \quad \text{for } y = 1, 2, 3, \dots.$$

(The $Y$ of this example is $X + 1$, where $X$, as in Example 2.1c, is the number of tails preceding the first heads.) Suppose we'd like the probability that it takes at most five trials given that it takes at least three. Applying the defining formula (3), we get

$$P(Y \leq 5 \mid Y \geq 3) = \frac{P(3 \leq Y \leq 5)}{P(Y \geq 3)} = \frac{\frac{1}{8} + \frac{1}{16} + \frac{1}{32}}{1 - (\frac{1}{2} + \frac{1}{4})} = \frac{7}{8}. \quad \blacksquare$$

Conditional probabilities are probabilities and must satisfy properties of probabilities, in particular, (i)–(iii) of §1.7. They are clearly nonnegative, and the probability of the whole (reduced) sample space is 1:

$$P(F \mid F) = \frac{P(FF)}{P(F)} = 1.$$

Also, if $EF = \emptyset$, we have additivity:

$$P(E \cup F \mid G) = \frac{P([E \cup F]G)}{P(G)} = \frac{P(EG \cup FG)}{P(G)} = \frac{P(EG)}{P(G)} + \frac{P(FG)}{P(G)}$$

$$= P(E \mid G) + P(F \mid G),$$

where we have used the fact that $(EG)(FG) = \emptyset$.

Multiplying through the defining formula (3) by the denominator, we obtain a formula for the probability of the intersection of two events.

---

**Multiplication Rule:**

$$P(EF) = P(E)P(F \mid E). \qquad (4)$$

---

This rule may actually seem more intuitive than the definition of conditional probability to which it is equivalent. It is useful when we are constructing a bivariate model, in a situation where it is clear what the conditional probabilities must be. The following example shows how we can use it in this way.

**Example 2.3c**  |  **Selecting Cards in Succession**

Suppose we select two cards at random from a standard deck of playing cards, one at a time and without replacement. What is the probability that the first card is the king of hearts and the second, the 10 of spades?

The outcomes of this composite experiment are sequences of two cards. Since the first card is selected at random, the probability of drawing the king of hearts on this first draw is 1/52. The second card is then to be selected at random, from the 51 cards that remain. This means that the *conditional* probability of drawing the 10 of spades, given that the king of hearts was drawn first, is 1/51. The multiplication rule then gives

$$P(\text{heart king, then spade 10}) = P(\text{1st card is heart king})$$

$$\times P(\text{2nd card is spade 10} \mid \text{1st is heart king})$$

$$= \frac{1}{52} \cdot \frac{1}{51} = \frac{1}{2652}.$$

The probability of the same two cards but in reverse order is clearly (again by the multiplication rule) the same number. Thus, the probability of the pair, (heart king, spade ten) without regard to order is the sum, $\frac{2}{2652}$ or $\frac{1}{1326}$. Indeed, by the same token, the probability of *any* single particular pair of cards is $\frac{1}{1326}$.

Now, the number of *combinations* of two cards from 52 is $\binom{52}{2} = 1326$. And what we have just succeeded in showing is that, when we sample one at a time at random, the possible combinations are equally likely. ∎

The multiplication rule (4) extends easily to more than two events. For any events $E$, $F$, and $G$, we can apply (4) to the *two* events $EF$ and $G$:

$$P(EFG) = P[(EF)G] = P(EF)P(G \mid EF).$$

Applying the rule once again to substitute for $P(EF)$, we obtain

$$P(EFG) = P(E)P(F \mid E)P(G \mid EF).$$

Repeating the process yields a multiplication rule for the probability of the intersection of any finite collection of events $\{E_i\}$:

---

**Extended multiplication rule:**

$$P(E_1 \cdots E_k) = P(E_1)P(E_2 \mid E_1) \cdots P(E_k \mid E_1 E_2 \cdots E_{k-1}). \quad (5)$$

---

**Example 2.3d** | **Star Wars**
In a speech reprinted in the *Congressional Record*, we find this paragraph:

> Most of the discussions of the "Star Wars" defense assumes a many-layered defense with three or four distinct layers. The idea behind having several layers is that the total defense can be made nearly perfect in this way, even if the individual layers are less than perfect. For example, if each layer has, say, an 80 percent effectiveness—which means that one in five missiles or warheads will get through—a combination of three such layers will have an overall effectiveness better than 99 percent, which means that no more than one warhead in 100 will reach its target.

Suppose a missile meets the layers in sequence. The writer assumes that it has probability .20 of penetrating the first layer unscathed; that the conditional probability of its penetrating the second layer (given that it got through the first) is .20; and that the conditional probability of its penetrating the third layer (given that it got through the first two) is .20. With $E_i$ denoting the probability that the missile gets through layer $i$, we have

$$P(E_1 E_2 E_3) = P(E_3 \mid E_1 E_2)P(E_2 \mid E_1)P(E_1) = .2 \times .2 \times .2 = .008.$$

So the probability that the missile is foiled in its mission is $1 - .008 = .992$. ∎

Using (5) we can reason in general, as in Example 2.3c, to show:

> When any finite number of items is selected one at a time at random and without replacement, the possible combinations are equally likely.

In Chapter 1 we took this fact as intuitively evident and used it to find (by counting combinations) probabilities of events that do not involve order.

The values of a probability function $f_X(x)$ are probabilities, and these may change if one is given partial information about $X$—as in the following example.

**Example 2.3e** | **A Conditional P.F.**

Consider a selection of three balls from two white and four red balls. The selection (as we have just seen) can be specified as being made by drawing one at a time at random, without replacement, or by assuming that all combinations are equally likely. Let $X$ denote the number of white balls selected. Its p.f. is easily found (by methods of Chapter 1) to be $f(0) = .2$, $f(1) = .6$, and $f(2) = .2$.

Suppose now it becomes known that at least one white ball has been drawn. The new, conditional probabilities are given by the conditional p.f.:

$$P(X = 0 \mid X > 0) = 0, \ P(X = 1 \mid X > 0) = \frac{.6}{.8}, \ P(X = 2 \mid X > 0) = \frac{.2}{.8}.$$

A notation for this conditional p.f. is awkward; perhaps $f_{X|X>0}(x)$ is the best we can come up with.     ∎

Conditional p.f.'s most commonly arise when the condition gives the value of one or more of the components in a random vector. In particular, we are often interested in the conditional distribution of $X$ when the value of $Y$ is known. The **conditional p.f.** of $X$ given $Y = y$ is

$$f(x \mid y) = P(X = x \mid Y = y) = \frac{f(x, y)}{f_Y(y)}, \tag{6}$$

where $f(x, y)$ is the joint p.f. of the pair $(X, Y)$.

The notation $f(x \mid y)$ is in common use. But one has to be aware of the nontraditional usage: The function $f(\cdot \mid 2)$, as the conditional p.f. of $Y$ given $X = 2$, is ordinarily not the same as $f(\cdot \mid 2)$ when it is the conditional p.f. of $X$ given $Y = 2$. The notation $f_{X|Y=y}(x)$ is cumbersome but clearer.

**Example 2.3f** | **The Conditional P.F. of $Y$ Given $X$**

In Example 2.2a we gave the bivariate distribution of $(X, Y)$, where $X$ denoted the sex, and $Y$ the class of a student selected at random from a population of students. The complete table of probabilities $f(x, y)$ is repeated here:

|         |        | $Y = class$ |     |     |     |     |
|---------|--------|------|------|-----|-----|-----|
|         |        | Fr.  | So.  | Jr. | Sr. |     |
| $X = sex$ | Male   | .04  | .27  | .15 | .05 | .51 |
|         | Female | .06  | .21  | .16 | .06 | .49 |
|         |        | .10  | .48  | .31 | .11 | 1   |

The conditional distribution of $Y$ given $X = M$ is obtained by dividing each probability in the $M$ row by the row total:

| $y$           | Fr.  | So.   | Jr.   | Sr.  |
|---------------|------|-------|-------|------|
| $f_{Y|X=M}(y)$ | 4/51 | 27/51 | 15/51 | 5/51 |

Observe that these conditional probabilities are proportional to the original probabilities in the $M$ row. In like fashion, we obtain the conditional p.f. for $X$ given $Y = Fr$ by dividing the probabilities in the first column by the column total:

| $x$            | $M$  | $F$   |
|----------------|------|-------|
| $f_{X|Y=Fr}(x)$ | 4/10 | 6/10  |

Thus, the entries in the "Fr" column give the conditional odds: 4:6.       ■

The example makes an important point, one that deserves to be underscored: The conditional probabilities $f(x \mid y_0)$ are proportional to the probabilities $f(x, y_0)$, because of their definition:

$$f(x \mid y_0) = \frac{1}{f_Y(y_0)} \cdot f(x, y_0),$$

where the factor $1/f_Y(y_0)$ is simply a *constant*—because $y_0$ is a given constant. Indeed, given any set of joint probabilities as the conditioning event, the *odds* within that set are the same; the probabilities are only then altered by dividing by the probability in that set, because the total probability (given the condition) must be 1.

## Problems

\* **2-11.**  A card is selected at random from a standard deck of playing cards. Define the events $A$ (ace), $S$ (spade), and $B$ (black). Find the following:
  (a)  $P(A \mid S)$.    (c)  $P(S \mid B)$.    (e)  $P(A \mid B)$.
  (b)  $P(S \mid A)$.    (d)  $P(B \mid S)$.    (f)  $P(B \mid A)$.

**2-12.** An ordinary die (six faces, numbered 1 to 6) is tossed. Assuming the faces to be equally likely, find

**(a)** the probability of a total of 4, given that the total is an even number.

**(b)** the probability of a total of 6, given that the total is divisible by 3.

**∗ 2-13.** In the game of bridge, each of four players is dealt 13 of the 52 cards.

**(a)** A player gets a hand with just one ace. What is the probability that his partner (the player opposite) has no aces?

**(b)** Another player sees that she and her partner have the ace and queen of spades but not the king or jack. What is the probability that both the king and jack are held by the player on her left?

**2-14.** The probability that a white male born in 1973 will live to age 40 is .97, and the probability that he will live to age 65 is .66. Find the probability that if he reaches age 40, he will live to be 65.

**∗ 2-15.** Three cards are selected at random from a standard deck of playing cards. Find the probability that

**(a)** all are spades, given that all are black.

**(b)** none is an ace, given that none is a face card.

**(c)** none is a spade, given that the three cards are of different suits.

(A face card is a king, queen, or jack.)

**2-16.** Two chips are selected without replacement from a bowl with five chips numbered 1 to 5, as in Problem 2-6. Find the probability that

**(a)** the first chip is 2, given that the second is 3.

**(b)** the 2 and 3 are chosen (in any order), given that the sum is 5.

**(c)** the first chip is 2, given that the sum is 5.

**∗ 2-17.** Suppose the two chips in the preceding problem are selected *with* replacement of the first chip before the second selection. Find the probability that the 4 is drawn both times,

**(a)** given that the first one drawn is the 4.

**(b)** given that at least one of the two is a 4.

**∗ 2-18.** To see a certain executive, I have to get by two secretaries. I figure that my chances are one in five of getting past the outer secretary, and if I do, that the chances are one in five that I get past the private secretary. What is the probability (according to my reckoning) that I get to see the executive?

**2-19.** A card is drawn at random from a deck of playing cards, and another is drawn at random from those that remain. Find the probability that

**(a)** both are hearts.

**(b)** the first is not a heart but the second is a heart.

**(c)** the second is a heart.

**∗ 2-20.** I am to draw either $A$ or $B$ as my opponent in a certain game. My chances of beating $A$ are four in ten, and of beating $B$, eight in ten. If $A$ and $B$ are equally likely to be my opponent, what is the probability that I win?

**2-21.** Show that $P(EF \mid G) = P(E \mid FG)P(F \mid G)$, for any events $E$, $F$, and $G$ for which the probabilities involved are defined.

**2-22.** Three articles are drawn at random, without replacement, from ten articles, six of which are good and the rest defective. Find the probability

(a)   that the first is good and the second and third are defective.

(b)   that the sample includes exactly one good and two defective articles.  Do this in two ways:  (1) By finding [as in (a)] the probabilities of the sequences $DGD$ and $DDG$, and (2) by assuming that the 120 selections of three from ten are equally likely and counting combinations.

* **2-23.**   Consider this two-stage experiment:  First roll a regular four-sided die with outcomes labeled 1, 2, 3, and 4.  Denote the outcome by $X$.  Then, given that $X = k$, select $k$ chips at random without replacement from a bowl containing two red and two black chips.  Let $Y$ denote the number of red chips in the latter selection.

(a)   Construct a tree diagram and label the branches with the appropriate probability or conditional probability.

(b)   Construct a table of joint probabilities for $(X, Y)$.  (These are the products of the probabilities along the 12 paths.)

(c)   Obtain the marginal distribution of $Y$ in (b).

(d)   Find the probability that $X = 4$, given that $Y = 2$.

**2-24.**   Suppose the distribution of rat hairs in a jar of peanut butter is:

| Number ($\omega$) | 0 | 1 | 2 | 3 | 4 |
|---|---|---|---|---|---|
| Probability $[P(\omega)]$ | .8 | .1 | .05 | .03 | .02 |

You open a jar and happen to come across a rat hair.

(a)   For your jar, given that there is at least one rat hair in the jar, find the (conditional) distribution of the number in the jar.

(b)   What is the probability that your jar contains more than the one you have found?  (That is, given that there is at least one, what is the probability that the jar contains more than one?)

---

## 2.4   Bayes' Theorem

The multiplication rule (4) of §2.3 can be written out in two ways, with the *roles* of $E$ and $F$ simply interchanged:

$$P(EF) = P(E)P(F \mid E) = P(F)P(E \mid F).$$

Dividing by (say) $P(F)$, we obtain the essence of Bayes' theorem:

$$P(E \mid F) = \frac{P(F \mid E)P(E)}{P(F)}. \tag{1}$$

This is sometimes referred to as the *law of inverse probability,* because it expresses $P(E \mid F)$ in terms of $P(F \mid E)$, with the roles of $E$ and $F$ reversed.  If you want to find a conditional probability $P(E \mid F)$, the key to deciding whether or not Bayes' theorem can help is whether or not you know $P(F \mid E)$.

In most cases we can calculate the denominator $P(F)$ in (1) by means of the law of total probability [(2) of §1.7]:

$$P(F) = P(FE) + P(FE^c). \tag{2}$$

Applying the multiplication rule [(4) in §2.3] to each term on the right we get

---

**Bayes' theorem:**

$$P(E \mid F) = \frac{P(F \mid E)P(E)}{P(F \mid E)P(E) + P(F \mid E^c)P(E^c)}. \tag{3}$$

---

**Example 2.4a** | **Smoking and Gender**

Suppose 40% of the students in a certain college are female. Suppose further that 15% of the females and 10% of the males are smokers. What fraction of the smokers are female?

With the given information expressed in terms of probabilities, we are assuming $P(F) = .4$, $P(M) = .6$, $P(S \mid F) = .15$, $P(S \mid M) = .10$. The probability we are asked to find is $P(F \mid S)$. In the present context, Bayes' theorem gives us

$$P(F \mid S) = \frac{P(S \mid F)P(F)}{P(S \mid F)P(F) + P(S \mid M)P(M)}.$$

Upon substituting the given probabilities, we obtain

$$P(F \mid S) = \frac{.15 \times .4}{.15 \times .4 + .10 \times .6} = .5.$$

The calculation in the denominator gives $P(S) = .12$. Of this 12%, half of the smokers are female smokers (6% of the population) and half are male smokers (6% of the population).

Figure 2-4 shows a tree diagram. The main branching is into female and male, in the ratio 4 to 6. The second branching shows the proportion of smokers and non-smokers ($N$) in each case, and these branches are marked with the appropriate *conditional* probability, given that branch. The products of the probabilities along a path from left to right are the probabilities of the four outcomes $FS$, $FN$, $MS$, $MN$. The population is divided into these four subsets, in the proportions 6:34:6:54. In the subset of smokers, the ratio of females to males is 6:6.

Another way of exhibiting the calculations is in terms of the joint distribution of the variables *smoking* (with values $S$ and $N$) and *gender* (with values $M$ and $F$). The probability table is as follows:

|   | S | N |   |
|---|---|---|---|
| M | .06 | .54 | .60 |
| F | .06 | .34 | .40 |
|   | .12 | .88 | 1 |

The entries in this table are simply the path probabilities as labeled on the tree diagram. Conditional probabilities are proportional to the rows or to the columns, according to the conditioning variable. ■

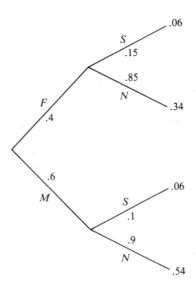

**Figure 2-4** Tree diagram for Example 2.4a

**Example 2.4b**

## A Paternity Case

Mr. G has been accused in a paternity suit of being the father of Ms. H's child. The state introduces expert testimony concerning the genetic makeup of Mr. G, Ms. H, and the child. The evidence is complicated, involving many genetic factors; we limit our discussion to blood group.

Suppose we know Mr. G is type AB, and Ms. H is type A. The datum $D$ is that the child is type B. (We now know that the mother is heterozygous AO, since if she were homozygous AA, the child could not be type B.) Let $G$ denote the event that Mr. G is the child's father. A court is interested in determining $P(G \mid D)$, the probability that Mr. G is the father in view of the datum $D$.

   Let $p = P(G)$, the "prior" probability of $G$ (that is, prior to obtaining the datum $D$). According to Bayes' theorem, we have

$$P(G \mid D) = \frac{P(D \mid G) \cdot p}{P(D \mid G) \cdot p + P(D \mid G^c)(1 - p)}. \qquad (4)$$

This is called the *posterior probability* of $G$, given $D$.

   A reasonable interpretation of $p$ is that it is a juror's personal probability that Mr. G is the father, based on evidence other than the genetic data. Unfortunately, in paternity cases in many states in the U.S. and in other countries, court-appointed pathologists always take $p$ to be $\frac{1}{2}$. They take the fact that a man is accused as reason to say that he is as likely to be the father as not! (This assumption has silly implications. For instance, if three men are suspected of being the father, the corresponding $p$'s would sum to more than 1. Moreover, it can turn out that their *posterior* probabilities (4) add up to more than 1 when there are only two potential fathers.)

   The probabilities $P(D \mid G)$ and $P(D \mid G^c)$ follow from Mendel's laws of inheritance. (You need not be concerned about the details of the genetics.) The probability that a type AB father and a type AO mother produce the child in question is $\frac{1}{4}$, and this is $P(D \mid G)$. The probability $P(D \mid G^c)$ is $\frac{1}{2}$ times the proportion of B alleles in the appropriate gene pool, typically about .08. With these probabilities substituted in (4) we have

$$P(G \mid D) = \frac{.25p}{.25p + .04(1 - p)} = \frac{p}{.16 + .84p}.$$

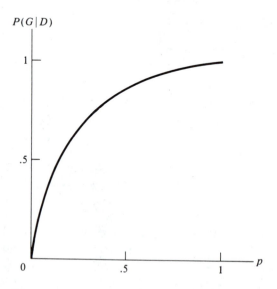

**Figure 2-5** Prior to posterior—Example 2.4b

The graph of this function of $p$ is shown in Figure 2-5. With the assumption that $p = .5$, the posterior probability that Mr. G is the father is $P(G \mid D) \doteq .86$. However, rather than say "the probability of paternity is 86%," the expert should show Figure 2-5 to the jury and explain how it converts a prior probability $p$ (based on other evidence) to a posterior probability. ∎

Suppose we know $P(F \mid E_i)$ for each of at most countably many events $E_i$ that partition $\Omega$: $\{E_1, E_2, ...\}$. (*Partition* was defined in §1.2.) We can express the denominator $P(F)$ in Bayes' formula (1) using the correspondingly more general form of the law of total probability [from (1) of §1.7]:

$$P(F) = P(F \mid E_1)P(E_1) + P(F \mid E_2)P(E_2) + \cdots . \tag{5}$$

Given the occurrence of $F$, the odds on $E_1$, $E_2$, ... are proportional to the terms in this decomposition of $P(F)$. This fact is expressed in the following version of Bayes' theorem.

---

**Bayes' theorem:**

If $\{E_1, E_2, ...\}$ is a partition of $\Omega$, then for $i = 1, 2, ...$ ,

$$P(E_i \mid F) = \frac{P(F \mid E_i)P(E_i)}{P(F \mid E_1)P(E_1) + P(F \mid E_2)P(E_2) + \cdots} . \tag{6}$$

In terms of odds,

$$\frac{P(E_i \mid F)}{P(E_j \mid F)} = \frac{P(F \mid E_i)}{P(F \mid E_j)} \cdot \frac{P(E_i)}{P(E_j)} .$$

---

**Example 2.4c** | **Delinquency and Birth Order**

Are delinquent children likely to be youngest in their families? A research study investigated the possible relationship between delinquency and birth order among high school girls. Delinquency ($D$) was somewhat arbitrarily defined according to answers on a questionnaire. In the group reported in the study, 40% were eldest ($E$), 30% middle ($M$), 20% youngest ($Y$), and 10% only children ($O$). The proportions classified as delinquent were 5% for an eldest child, 10% for a middle child, 15% for a youngest child, and 20% for an only child. (Actual percentages have been rounded for simplicity.) Thus, we have these probabilities: $P(E) = .4$, $P(M) = .3$, $P(Y) = .2$, $P(O) = .1$, and

$$P(D \mid E) = .05, \quad P(D \mid M) = .10, \quad P(D \mid Y) = .15, \quad P(D \mid O) = .20.$$

Given these probabilities, what proportion of the delinquent girls are youngest in their families? That is, what is $P(Y \mid D)$?

Using the law of total probability (5), we find the proportion of delinquents in the population studied to be

$$P(D) = P(D \mid E)P(E) + P(D \mid M)P(M) + P(D \mid Y)P(Y) + P(D \mid O)P(O)$$

$$= .05 \times .40 + .10 \times .30 + .15 \times .20 + .20 \times .10$$

$$= .02 + .03 + .03 + .02 = .10.$$

From (6), then, we have

$$P(Y \mid D) = \frac{.03}{.02 + .03 + .03 + .02} = .30.$$

Similar calculations yield $P(E \mid D) = .2$, $P(M \mid D) = .3$, $P(O \mid D) = .2$.  Figure 2-6 shows the calculations in terms of a tree diagram.  ∎

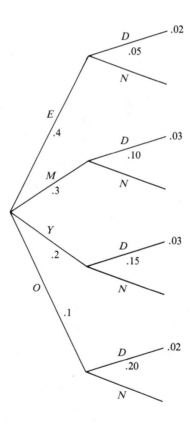

**Figure 2-6** Tree diagram for Example 2.4c

## Problems

**\* 2-25.** Suppliers A and B provide, respectively, 10% and 90% of a certain item that is used in large quantities. Suppose that 2% of those provided by A are defective and 5% of those provided by B are defective. A randomly selected item is found to be defective; find the probability that the item was supplied by A.

**2-26.** In a certain university, 10% of the undergraduates are foreign students, and 20% of the graduate students are foreign students. There are four times as many undergraduates as graduate students. Find the probability that a randomly chosen foreign student is an undergraduate.

**2-27.** A letter to the editor of a medical journal includes this argument:

Let us say that the prevalence of stenosis was fivefold that of achalasia, i.e., .030 vs. .006. Given the frequency of an absent gastric air bubble found by Orlando: 50% in achalasia, 17% in stenosis, one could conclude that in a randomly selected group of 1000 patients, stenosis with an absent air bubble would be more likely than achalasia.

Use Bayes' theorem to reproduce the writer's calculation of odds of about 5 to 3 in favor of stenosis, given an absent air bubble.

**\* 2-28.** Evidence in a paternity suit indicates that four particular men are equally likely to have been the father of the child. Blood typing in the ABO system reveals that the phenotypes of mother and child are both type O. The men are found to have blood types as shown in the following table. The table also gives the probability for each man of producing a type O child with the type O mother. (These probabilities are determined from genetic theory together with, in the case of the first two men, observed proportions of blood types in the population.) Find the probability that man 1 is the father, given all the evidence.

| Man i | Phenotype | P(O child \| man i and O-mother) |
|-------|-----------|------------------------------------|
| 1 | A | .431 |
| 2 | B | .472 |
| 3 | O | 1 |
| 4 | AB | 0 |

**2-29.** Suppose the numbers of freshmen, sophomores, juniors, and seniors in a certain college are in the proportion 6:5:4:3, respectively. Females make up 15% of the freshmen, 20% of the sophomores, 30% of the juniors, and 35% of the seniors. What fraction of the females are freshmen?

**\* 2-30.** A study on learning disability used a Bayesian analysis to estimate the probability that a child of a parent with a reading disability would also have a reading disability. Notation: $FD$ = father disabled, $CD$ = child disabled, and $CN$ = child normal. Based on sample information, it was estimated that $P(FD \mid CD) = .29$, and $P(FD \mid CN) = .04$. The population incidence of child reading disability $P(CD)$ is subject to debate, so the study report gave results for various values of $P(CD)$. Calculate $P(CD \mid FD)$

    **(a)** assuming $P(CD) = .05$.

    **(b)** assuming $P(CD) = .10$.

## 2.5          Independent Random Variables

Two random variables are **independent** when the marginal p.f. of one variable is the same as the conditional p.f. given any value of the other variable.

**Example 2.5a**

### Smoking vs. Running

Each student in one of our statistics classes was classified on the variables $X$ = smoking status, and $Y$ = running status. The categories of $Y$ were defined as $R$, for someone who runs at least five miles a week, and $R^c$, for all others. The categories of $X$ are $S$ (now smokes), $Q$ (used to but quit), and $N$ (never smoked). Here are the results:

|              |        | $X$: smoking |    |    |    |
|--------------|--------|---|----|----|----|
|              |        | S | Q  | N  |    |
| $Y$: running | R      | 2 | 4  | 16 | 22 |
|              | $R^c$  | 6 | 12 | 48 | 66 |
|              |        | 8 | 16 | 64 | 88 |

In a random selection from this class, the joint probabilities for $(X, Y)$ are obtained by dividing each table entry by 88.

Suppose we learn that a student is a runner. Given $Y = R$, the conditional probabilities for $X$ are proportional to the first-row entries, that is, in the proportion 1:2:8. Similarly, if $Y = R^c$ (the student is not a runner), the conditional probabilities for $X$ are proportional to the second row, again 1:2:8. Indeed, the marginal probabilities for $X$ are in the same proportion. Thus,

$$f(S \mid R) = f(S \mid R^c) = f_X(S) = \frac{1}{11},$$

$$f(Q \mid R) = f(Q \mid R^c) = f_X(Q) = \frac{2}{11},$$

$$f(N \mid R) = f(N \mid R^c) = f_X(N) = \frac{8}{11}.$$

So the p.f. for the variable $X$ (smoking) does not change with the information about $Y$ (running).

It is also true that the conditional probabilities for $Y$ given a value of $X$ are all equal to the corresponding unconditional probabilities for $Y$:

$$f(R \mid S) = f(R \mid Q) = f(R \mid N) = f_Y(R) = \frac{1}{4}.$$

$$f(R^c \mid S) = f(R^c \mid Q) = f(R^c \mid N) = f_Y(R^c) = \frac{3}{4}.$$

The coincidence of conditional with unconditional probabilities—rows or columns—is equivalent to the fact that each joint probability is the product of the corresponding marginal probabilities. For example,

$$f(S, R) = \frac{1}{44} = \frac{1}{11} \cdot \frac{11}{44} = f_X(S) \cdot f_Y(R).$$

You should verify that this multiplicative property holds for each entry in the probability table. ∎

When, as in the example, the distribution for $X$ does *not* change upon our learning that $Y = y$:

$$f(x \mid y) = f_X(x), \quad \text{all } (x, y),$$

the multiplication rule [(4) of §2.3] yields

$$f(x, y) = f(x \mid y) f_Y(y) = f_X(x) f_Y(y).$$

Moreover, if this factorization holds for all $x$ and $y$, the conditional p.f.'s are equal to the corresponding unconditional p.f.'s.

---

Random variables $X$ and $Y$ are **independent** if and only if for *every* pair $(x, y)$,

$$f(x, y) = f_X(x) f_Y(y). \tag{1}$$

---

Given the joint probabilities in Example 2.5a, verifying (1) shows that $X$ and $Y$ are independent. However, we usually will be using (1) to *construct* a model—defining joint probabilities from marginal probabilities as products, to incorporate independence into the model, for an experiment where we know that the variables must be independent.

Researchers may be willing to assume tentatively that two variables are independent, and this allows them to calculate joint probabilities from marginal probabilities. They seldom know for sure that variables are independent, but assumptions must be made in any scientific endeavor, or science would not advance. Scientists can make calculations under the assumption of independence and, at the same time, worry about the validity of this assumption. They may be able to check their assumptions using experimental results. (We'll show how to check independence in Chapter 13.)

| Example **2.5b** | **ABO Blood Typing** |

In a certain population, 42% have type O blood, 44% have type A, 10% have type B, and 4% have type AB. Suppose 10% have negative Rh-factor. What proportion of the population is O-negative?

Conceivably, the proportion of O-negatives is anything from 0 to 10%. However, if blood type and Rh-factor are independent, application of (1) would give the proportion of O-negatives as $.42 \times .10 = .042$, or just over 4%. Moreover, under this assumption, all joint proportions are products:

|  |  | Blood Type | | | | |
|---|---|---|---|---|---|---|
|  |  | O | A | B | AB | |
| Rh Factor | + | .378 | .396 | .090 | .036 | .900 |
|  | − | .042 | .044 | .010 | .004 | .100 |
|  |  | .420 | .440 | .100 | .040 | 1 |

The assumption of independence can be checked experimentally by comparing these joint probabilities with corresponding sample proportions.  ∎

Two random variables that are not independent are **dependent.** In the dependent case, not all of the joint probabilities factor as in (1). So all that it takes to show that two variables are dependent is one instance in which the joint probability is not the product of the corresponding marginal probabilities. In particular, an entry of 0 in a table of joint probabilities is an immediate tip-off: A joint probability $f(x, y)$ cannot be zero and equal to the product $f_X(x)f_Y(y)$ unless at least one of the marginal probabilities is zero.

**Example 2.5c**

The model for the random selection of two chips from three chips marked $1, 2, 3$ without replacement is shown in the following table of joint probabilities for $X$, the number on the first chip, and $Y$, the number on the second chip:

|  |  | Y: | | | |
|---|---|---|---|---|---|
|  |  | 1 | 2 | 3 | $f_X(x)$ |
|  | 1 | 0 | 1/6 | 1/6 | 1/3 |
| X: | 2 | 1/6 | 0 | 1/6 | 1/3 |
|  | 3 | 1/6 | 1/6 | 0 | 1/3 |
| | $f_Y(y)$ | 1/3 | 1/3 | 1/3 | 1 |

There are three 0's among the $f(x, y)$'s—any one of which would suffice to imply that $X$ and $Y$ are not independent, since the marginal entries are all different from 0. For instance, $f(1, 1) = 0 \neq f_X(1)f_Y(1)$. The dependence also shows up in the fact that the rows are not proportional and the columns are not proportional. The first-column entries, for instance, are in the proportion 0:1:1, implying conditional probabilities 0, 1/2, 1/2, given $Y = 1$. But the second- and third-column entries are proportional to 1:0:1 and 1:1:0, respectively, implying different conditional distributions given $Y = 2$ and given $Y = 3$. The unconditional probabilities for $Y$ are in the proportion 1:1:1.  ∎

We now extend the notion of independence to more than two random variables. Variables $X$, $Y$, and $Z$ are said to be independent if and only if their joint probability function is the product of the three marginal p.f.'s:

$$f(x, y, z) = f_X(x)f_Y(y)f_Z(z), \qquad (2)$$

for all possible triples $(x, y, z)$.  In particular, this condition implies that independence holds for each pair.  Thus, $X$ and $Y$ are then independent:

$$f_{X,Y}(x, y) = \sum_z f(x, y, z) = f_X(x)f_Y(y)\sum_z f_Z(z) = f_X(x)f_Y(y),$$

and so on.  However, it can happen that $X$ and $Y$ are independent, $X$ and $Z$ are independent, and $Y$ and $Z$ are independent even when (2) fails to hold (see Problem 2-51).  We then extend (1) to define independence of any finite number of random variables:

---

Random variables $X_1, X_2, ..., X_k$ are **independent** when for all points $(x_1, ..., x_k)$,

$$f(x_1, x_2, ..., x_k) = f_1(x_1)f_2(x_2) \cdot \cdot \cdot f_k(x_k), \qquad (3)$$

where $f_i$ is the p.f. of $X_i$.

---

Just as in the case of $k = 3$, the factorization (3) for all points in $k$-space implies that the variables are *pairwise independent.*  Moreover, the conditional distributions of any subset of the $k$ variables, given the values of some of the others, are the same as the corresponding unconditional distributions.

As in the case of two variables, we'll use the factorization (3) mainly to construct models for several variables that we assume are independent.

**Example 2.5d**  |  **Rolling Eight Dice**

In the carnival game "Razzle Dazzle," players roll eight dice simultaneously.  A player who throws a total of 8 wins immediately; how likely is this?

The outcome for each die is a random variable.  If the six possible faces are equally likely, the probability of a 1 is 1/6, for each die.  The only way of getting a total of 8 is for each die to show 1.  If we assume independence of the outcomes of the eight dice, we multiply the 1/6's to get the probability of 8:

$$P(\text{total of 8}) = f(1, 1, ..., 1) = f_1(1) \cdot \cdot \cdot f_8(1) = (1/6)^8 = \frac{1}{1679616}.$$

In this game, the operator claims that since there are 41 possible sums (8, 9, ..., 48), the probability of each one is 1/41. ∎

In statistical theory, the notion of independent variables is extremely important and fundamental. The notion of independence can be extended to *events*: For any event $E$, define [as in Examples 2.2c and 2.1d] the corresponding **indicator variable**:

$$X_E = \begin{cases} 1, & \text{if } E \text{ occurs,} \\ 0, & \text{if } E \text{ otherwise.} \end{cases} \qquad (4)$$

Thus $f_{X_E}(1) = P(E)$, and $f_{X_E}(0) = P(E^c)$.

---

Events $E_1$, $E_2$, ..., $E_k$ are **independent** if and only if their indicator variables $X_{E_1}$, $X_{E_2}$, ..., $X_{E_k}$ are independent.

---

In the case of *two* events $E$ and $F$, the margins of the joint probability table for their indicator variables are as follows:

|         |   | $X_E$ 1 | $X_E$ 0 |         |
|---------|---|---------|---------|---------|
| $X_F$   | 1 |         |         | $P(F)$  |
|         | 0 |         |         | $P(F^c)$ |
|         |   | $P(E)$  | $P(E^c)$ | 1      |

When the indicator variables are independent, the joint probabilities in this table are filled in as products. Thus, $f(1, 1) = P(E)P(F)$. Moreover, if this relation holds, then *all* of the entries factor. For instance,

$$P(EF^c) = P(E) - P(EF) = P(E) - P(E)P(F) = P(E) \cdot P(F^c).$$

In summary, $E$ and $F$ are independent if and only if $P(EF) = P(E)P(F)$. Moreover, $E$ and $F$ are independent if and only if $E$ and $F^c$ are independent, if and only if $E^c$ and $F$, and if and only if $E^c$ and $F^c$ are independent.

The condition $P(EF) = P(E)P(F)$ is satisfied if either $E$ or $F$ has probability zero, since $EF$ is a subset of both $E$ and $F$. Our definition implies that an event with probability zero is independent of any other event.

When $E$ and $F$ are independent, the condition $P(EF) = P(E)P(F)$ implies that if $P(F) \neq 0$, then $P(E \mid F) = P(E)$, and conversely. And, of course, the same is true with roles of $E$ and $F$ reversed. Thus, when events are independent, knowing that one has occurred (or not) does not affect the odds on the other.

In the case of more than two events, (3) implies that the probability of the intersection of independent events factors into the product of their individual probabilities. Thus, when $E$, $F$, and $G$ are independent,

$$P(EFG) = P(E)P(F)P(G).$$

However, the converse is *not* true—this factorization can hold when the three events are not independent. (See Problem 2-47.) To verify independence in a collection of events using a factorization criterion, you have to see that factorization holds for every subcollection of those events.

## 2.6    Exchangeable Random Variables

Consider an experiment with outcome $X$, a random variable with p.f. $f(x)$. A sequence of $n$ trials of the experiment results in the observation vector $(X_1, ..., X_n)$, where the p.f. of each $X_i$ is $f(x)$. When the variables $X_i$ are independent, we call the sequence of observations a **random sample.** The joint p.f. of the variables in a random sample is a product of the marginal p.f.'s:

$$f(x_1, x_2, ..., x_n) = f(x_1)f(x_2) \cdots f(x_n) = \prod_{i=1}^{n} f(x_i). \tag{1}$$

The product on the right is a *symmetric* function of the arguments $x_1, ..., x_n$: its value does not change under any permutation of those arguments. We can conclude then that the joint distribution of any subset of the $X_i$'s is the same as that of any other subset of the same number of $X_i$'s. Thus, not only do the individual $X$'s have the same distribution, every pair has the same joint distribution as every other pair, and so on.

In arriving at the last conclusion, the only property of the joint p.f. we used is that it is a symmetric function of the arguments. So the conclusion that the marginal p.f.'s of a particular dimension are all the same is valid whenever the joint probability function has this symmetry. Variables whose joint distribution has this symmetry property are said to be *exchangeable.*

Random variables $X_1$, $X_2$, ..., $X_n$ are **exchangeable** when their joint p.f. is a symmetric function of its arguments. When this is the case, the $k$-dimensional marginal distributions for any one $k$ are identical, $k = 1, ..., n - 1$.

Random sampling from a finite population yields observations that are exchangeable, whether with *or* without replacement, even though in the latter case the observations are not independent. We'll now show this in some special cases.

**Example 2.6a** | **Sampling Without Replacement**

We return to the setting of Example 2.5c—random selections of two chips from three marked from 1 to 3. The joint p.f., from that example, is repeated below. Observe that because the table is symmetric about the main diagonal, it doesn't matter whether the $x$-values are in the left margin and the $y$-values in the top margin, or vice versa. That is, the variables $X$ and $Y$ are exchangeable. Note also that the conditional probabilities for $X$ given $Y$ are the same as the conditional probabilities for $Y$ given $X$. For instance, the probability that the first chip is 1, given that the second is 3, is the same as the probability that the second chip is 1, given that the first is 3.

|   | 1 | 2 | 3 | $f_X(x)$ |
|---|---|---|---|---|
| 1 | 0 | 1/6 | 1/6 | 1/3 |
| 2 | 1/6 | 0 | 1/6 | 1/3 |
| 3 | 1/6 | 1/6 | 0 | 1/3 |
| $f_Y(y)$ | 1/3 | 1/3 | 1/3 | 1 |

■

**Example 2.6b** | **Exchangeability in Poker Hands**

A poker hand is a sample of five cards dealt without replacement. Assuming adequate shuffling (and no cheating), we may consider the successive cards as selected at random from those that remain. Let $X_i$ refer to the denomination (2, 3, ..., Q, K, A) of the $i$th card dealt. Using the multiplication rule [(5) of §2.3], we find the probability of the sample sequence (7, 2, J, 4, 7) to be

$$f(7, 2, J, 4, 7) = \frac{4}{52} \cdot \frac{4}{51} \cdot \frac{4}{50} \cdot \frac{4}{49} \cdot \frac{3}{48},$$

and for any permutation of this sequence, the same; e.g.,

$$f(J, 4, 7, 7, 2) = \frac{4}{52} \cdot \frac{4}{51} \cdot \frac{4}{50} \cdot \frac{3}{49} \cdot \frac{4}{48}.$$

Such symmetry occurs for any sequence—the factors in the numerators are simply a rearrangement of the factors for any other permutation.

Symmetry implies that the distribution for the first card drawn is the same as the distribution for each of the other cards. It also implies that the joint distribution of the first two cards drawn is the same as the joint distribution of the third and fifth, or of any other pair.

■

To see why, in sampling from a finite population without replacement, the sample observations are exchangeable, it may help to ignore the time or sequencing in the sampling and to think of a simultaneous selection. Then, upon assigning subscripts to observations arbitrarily, it should be clear that the joint distribution of $(X_1, X_2)$, say, is the same as that of $(X_4, X_3)$.

## Problems

**∗ 2-31.** You are to toss two dice. Assuming independent tosses and "fair" dice (equally likely faces), find the probability that
**(a)** both dice show a 6. **(b)** the two dice match. **(c)** the total is 4.

**∗ 2-32.** A baseball player has a batting average of .330. Assume this to be the probability that he gets a hit each time he bats, and that the times at bat are independent. (These assumptions may not be exactly appropriate, but they're not bad.) What is the probability that he gets no hits in four times at bat?

**∗ 2-33.** At one point in the 1978 National League baseball season, Pete Rose had hit in 37 consecutive games. He needed 20 more to break Joe DiMaggio's major-league record of 56 straight games. The Las Vegas odds were 99 to 1 against his doing it. See if you can approximate these odds by calculating along the lines of the preceding problem. Assume independence, an average of .330, and four "at-bats" per game.

**2-34.** Each card in a standard deck is assigned a number of "points" by the Goren system of bridge bidding: ace gets 4 points, king 3, queen 2, and jack 1. Any other card gets 0 points. Let $X$ denote the number of points for a card drawn at random from the deck, and let $Y$ be the indicator variable of the event "heart." (That is, $Y = 1$ if the card is a heart, otherwise 0.) Show that $X$ and $Y$ are independent.

**∗ 2-35.** Let $Z$ denote the total number of points (see Problem 2-34), and $W$ the number of hearts in a bridge hand—a random deal of 13 cards from a standard deck of playing cards. Are $Z$ and $W$ independent? [*Hint:* Look at the conditional distribution of $Z$ given $W = 13$.]

**2-36.** A set of Christmas tree lights has eight bulbs in "series" (all the lights go out when any one bulb fails). If the probability that any given bulb does not burn out during the first 100 hours is .9, what is the probability that the string stays lit for 100 hours? (Assume that the bulbs' lifetimes are independent.)

**2-37.** A news item in 1984 told of a female birth, the first daughter born in her family in at least 130 years. A Department of Health statistician said that the probability of such a string of male births is almost zero, and calculated the probability of ten pregnancies resulting in ten boys to be about .00098. How did he obtain this result, and what assumptions are required to get it? (*Note:* Despite the very low probability, the existence of a family with ten male births in a row is not surprising in view of the enormous number of families in the United States.)

**∗ 2-38.** To play "Russian Roulette," the player places a bullet in one of the six chambers of a revolver, spins the cylinder, points the revolver at his or her head, and pulls the trigger. The probability of surviving one play is 5/6. (We recommend that any experimentation be done with a simulation! Select one chip at random from a bowl containing six chips, one marked "bullet." Spinning the chamber corresponds to replacing the chip and mixing.) Suppose you plan to play two games in succession. Find your chances of survival if
**(a)** you spin the cylinder between plays.
**(b)** you do not spin the cylinder between plays.

**2-39.** Suppose you toss four ordinary dice independently. Find the probability that there is at least one pair. (For instance, the sequences 2 4 2 1 and 1 3 1 1 each have at least one pair.)

**∗ 2-40.** Find the probability that in a sequence of independent tosses of a die, a "6" turns up before the fourth toss.

**2-41.** Consider random variables $(X, Y)$ the joint p.f. shown on the following page:

(a)  Are $X$ and $Y$ independent?     (c)  Find $P(X = a \mid Y = 2)$.
(b)  Find $P(Y = 2 \mid X = a)$.        (d)  Find $P(X = a \mid Y \geq 2)$.

|   |   | 1 | 2 | 3 | $f_X(x)$ |
|---|---|---|---|---|---|
|   | $a$ | .1 | .2 | .1 | .4 |
| $X$ | $b$ | .2 | 0 | .2 | .4 |
|   | $c$ | 0 | .2 | 0 | .2 |
|   | $f_Y(y)$ | .3 | .4 | .3 | 1 |

**2-42.**   Let $U$ and $V$ be independent, with marginal distributions the same as those of $X$ and $Y$ (respectively) in the preceding problem. Construct the table of joint probabilities for $U$ and $V$.

**2-43.**   A card is selected at random from a standard deck of playing cards. As in Problem 2-11, define the events $A$ (ace), $S$ (spade), and $B$ (black).

(a)  Determine whether any two of the events $A$, $S$, and $B$ are independent.
(b)  The events $B^c$ and $S$ are disjoint. Are they independent?
(c)  Are the events $A$, $S$, and $B$ independent?

**\* 2-44.**   Suppose $A$, $B$, and $C$, whose union is the whole sample space, are *pairwise* independent. Given $P(A) = .6$, $P(B) = .4$, $P(C) = .6$, find $P(ABC)$.

**2-45.**   Show the following:

(a)  If $f(x, y) = f_X(x)f_Y(y)$, then $f(x \mid y) = f_X(x)$.
(b)  If $X$, $Y$, and $Z$ are independent, then $f(x \mid y, z) = f_X(x)$.

**2-46.**   Suppose events $B$ and $C$ are disjoint. Show that if $A$ is independent of $B$ and independent of $C$, then $A$ is independent of $B \cup C$.

**2-47.**   Suppose $\Omega$ has five outcomes $\omega_i$, with probabilities $P(\omega_1) = 1/8$, and $P(\omega_2) = P(\omega_3) = P(\omega_4) = 3/16$. Define events

$$A = \{\omega_1, \omega_2, \omega_3\}, B = \{\omega_1, \omega_2, \omega_4\} \; C = \{\omega_1, \omega_3, \omega_4\}.$$

Show that $P(ABC) = P(A)P(B)P(C)$, but that $A$, $B$, and $C$ are not pairwise independent.

**2-48.**   Show that if $A$, $B$, $C$, and $D$ are independent, then the events $AB$ and $C \cup D$ are independent.

**\* 2-49.**   Consider a sequence of random selections, one at a time without replacement, from an urn with eight black and five red balls. Find the probability

(a)  that the fifth one drawn is black.
(b)  that the fifth one drawn is black, given that the seventh one is red.
(c)  that the fifth one is black and the seventh one red.

**2-50.**   Let $M$ denote the number of $\omega$'s with $X(\omega) = x$ in a population of $N$ individuals, and consider a sequence of $n$ individuals selected at random without replacement. Let $X_i$ denote the outcome of the $i$th selection. Show that $P(X_i = x) = M/N$ for all $i$ by finding the number of possible sequences, and of those, the number in which $X_i = x$.

**2-51.** Suppose the distribution of $(X, Y, Z)$ assigns probability 1/4 to each of the four points $(1, 0, 0)$ $(0, 1, 0)$ $(0, 0, 1)$ $(1, 1, 1)$.

    **(a)** Show that $X$, $Y$, and $Z$ are pairwise independent but not independent.

    **(b)** Show that $X$ and $Y + Z$ are not independent.

## Review Problems

**2-R1.** Let $(X, Y)$ have the discrete distribution defined by the following table of joint probabilities:

|   |   | $Y$ | | |
|---|---|---|---|---|
|   |   | 1 | 2 | 3 |
|   | 1 | .1 | 0 | .1 |
| $X$ | 2 | 0 | .2 | .2 |
|   | 3 | .1 | .2 | .1 |

    **(a)** Give the marginal p.f.'s as probability tables.

    **(b)** Find $P(|X - Y| = 1)$.

    **(c)** Give the conditional p.f. of $Y$ given $X = 3$.

    **(d)** Are $X$ and $Y$ exchangeable?

    **(e)** Are $X$ and $Y$ independent?

    **(f)** Find the distribution of the random variable $X + Y$.

**2-R2.** Given $P(A) = .4$ and $P(B) = .5$, find $P(AB)$ and $P(A \cup B)$ when

    **(a)** $A$ and $B$ are disjoint.     **(b)** $A$ and $B$ are independent.

**2-R3.** Students in a class of 50 were cross-classified according to sex and according to whether they now smoke ($S$), did smoke but quit ($Q$), or never smoked ($N$):

|   |   | Smoking | | | |
|---|---|---|---|---|---|
|   |   | $S$ | $Q$ | $N$ | |
|   | M | 12 | 8 | 10 | 30 |
| Sex | F | 14 | 4 | 2 | 20 |
|   | Total | 26 | 12 | 12 | 50 |

For a random selection from this "population," find the probability

    **(a)** that a student so chosen is a male smoker.

    **(b)** (the marginal probability) that the student is female.

    **(c)** that the student chosen is a smoker, given that he is male.

    **(d)** that the student is female, given that the student smokes.

**2-R4.** In a certain assembly plant, equal numbers of autos are produced on each of the five working days, Monday through Friday. If 5% of those produced on Monday or Friday have major problems, and 1% of those produced on the other three days have major problems, find the probability that an auto with major defects was produced on a Monday or a Friday.

**2-R5.**  Given the following joint probability function of $X$ and $Y$:

|   |   | $X$ |   |
|---|---|---|---|
|   | 0 | 1 | 2 |
| 0 | 1/4 | 1/8 | 1/8 |
| $Y$  1 | 1/4 | 0 | 1/8 |
| 2 | 0 | 1/8 | 0 |

  **(a)**  are the events $E = [X = 0]$ and $F = [Y = 0]$ independent?

  **(b)**  are the variables $X$ and $Y$ independent? Exchangeable?

**2-R6.**  Let events $A$, $B$, $C$ be pairwise independent, such that $A \cup B \cup C = \Omega$, as in Problem 2-44. Suppose we assume $P(A) = .3$, $P(B) = .5$, and $P(C) = .6$. Proceed as in Problem 2-44 to find $P(ABC)$ and explain why this assignment of probabilities to $A$, $B$, and $C$ must be incorrect.

**2-R7.**  Given $P(EF) = .2$, $P(F) = .3$, and $P(E \cup F) = .6$,

  **(a)**  find $P(F \cup E^c)$.    **(b)**  find $P(F^c \mid E)$.

**2-R8.**  Assume that $X$ and $Y$ are independent, each with the p.f. $f(k) = 2^{-k}$ for $k = 1, 2, 3, ...$, and give the p.f. of their joint distribution.

**2-R9.**  Given that events $E$, $F$, and $G$ are independent, show that $E$ and $F \cup G$ are independent.

**2-R10.**  As in Problem 1R-9, consider random sampling without replacement from eight chips: 1 red, 5 blue, and 2 white. Find the probability that

  **(a)**  the first one drawn is white, given that the third one drawn is white.

  **(b)**  the second one is white, given that the first is blue and the third is red.

**2-R11.**  Individuals are selected at random, one at a time, from a group of 12 of whom 8 are white and 4 are black. Find the probability that the 2nd is black, given that the 4th and 5th are white,

  **(a)**  if the sampling is done with replacement and mixing.

  **(b)**  if the sampling is done without replacement.

**2-R12.**  Box A contains one black and three white balls; box B contains one white and three black balls. You pick a box at random and select a ball at random. It is *white*; in view of this, what is the probability that it came from box A?

**2-R13.**  A certain type of thumbtack lands with point up (U) or point down (D). Let $p = P(U)$. One such tack is dropped five times in succession. Give the probability of the sequence U, D, U, U, D in terms of $p$, assuming independence.

**2-R14.**  Show that if $X$ and $Y$ are independent, then

  **(a)**  $P(X + Y = k \mid Y = j) = P(X = k - j)$

  **(b)**  $P(X = j \mid X + Y = j + k) = \frac{P(X=j)P(Y=k)}{P(X+Y=j+k)}$.

**2-R15.** Given that the random pair $(X, Y)$ has the joint distribution shown in the table of probabilities below,

|   | 0  | 1  | 2  |
|---|----|----|----|
| 0 | .3 | .2 | .1 |
| 1 | .2 | .1 | 0  |
| 2 | .1 | 0  | 0  |

(a) obtain the marginal distributions. (Does it matter how the labels $X$ and $Y$ are assigned?)

(b) give the conditional distribution of $Y$ given $X = 0$.

(c) give the distribution of the variable $U = X + Y$.

(d) are the variables independent? exchangeable?

## Chapter Perspective

This chapter has introduced the language of random variables and probability functions, univariate and multivariate. Multivariate distributions are important for two reasons: (1) Researchers often seek to understand relationships among variables; (2) Even in the case of a univariate population, a sample is a set of *several* observations on this population—it is multivariate.

The notion of independence is important in representing the lack of any relationship among variables. When there are dependencies among variables, these can be exploited for purposes of control or prediction, as we'll see in Chapter 15.

Sampling a population is the basic means of learning about its unknown characteristics. Independence of the observations in a sample makes the sampling process easiest to analyze mathematically, because their multivariate model is a simple combination of their individual models—the joint p.f. is a product of the marginal p.f.'s. When sampling is done at random from a finite population without replacement, we have independence only when the sample is a small fraction of the population. But whether this is the case or not, the sample observations are at least exchangeable, in that their joint p.f. is a symmetric function of its arguments. This has important consequences, as we shall see in Chapter 8.

An important new idea of the chapter is that of conditioning—revising a probability model in the light of new information. We use the multiplication rule, involving a conditional probability, when we construct a model for an experiment that is carried out in stages, and where the model for a subsequent stage is most easily specified as conditional on what has occurred up to that point.

In predicting a random variable on the basis of what we have learned about related variables, it is its *conditional* distribution—given the values of those related variables—that we exploit. The area of "regression analysis" deals with such predictions (Chapter 15).

In the first two chapters, we have usually calculated probabilities for events, *given a probability model* for the mechanism that gives rise to data. But what researchers need is a way of drawing conclusions about the model, *given the data*. Bayes' theorem shows how to accomplish this desired reversal of the roles of mechanism and data. It is fundamental to the process of learning in that it provides a way of updating what we know in view of new data. We have touched on these ideas in Example 2.4b and in Problems 2-28 and 2-30, and will return to them in Chapters 8 and beyond.

Chapter 7 will take up ways of organizing and summarizing the data in a sample. But first we give some useful ways of describing unknown aspects of populations (Chapter 3) and then study some important, special classes of discrete random variables (Chapter 4). Continuous variables will be studied in Chapters 5 and 6.

# Averages

Example **3a**

## "Let's Make a Deal"

A contestant on the TV game show "Let's Make a Deal" had won $9,000 in prizes and was then offered the option of trading this for whatever lay behind one of the three doors. He was to choose one of the doors and would receive the prize behind that door. Behind one door was a $20,000 prize; behind another, a $5,000 prize; and behind the third, a $2,000 prize. Should the contestant take the option or stick with what he had already won?

From the contestant's point of view, the arrangement of prizes behind the three doors was random. Thus, he was equally likely to choose the $2,000 door, the $5,000 door, or the $20,000 door. The value of the door he would choose is a random variable. He must compare this random quantity with a nonrandom quantity ($9,000). Specifying a single number to take the place of a random quantity in such a situation lies at the heart of averaging. ∎

Numerical variables—volume, earnings, blood pressure, bowling scores, and the like—are often described using a single number called an *average*. In common usage, *average* suggests little more than ordinary or typical, so in this sense an average value would be one that is somewhere in the middle of the values that can occur—neither very high nor very low relative to the other values. Among various ways of making this notion of average more precise, the most usual is the *arithmetic mean*, which we take up first.

Averaging is also useful in measuring variability and other aspects of a distribution, in measuring the strength of relationships, and in defining *generating functions*—powerful tools for dealing with the sums of random variables that arise in calculating means.

| 3.1 | **The Mean** |
|---|---|

| Example **3.1a** | **Value of the Random Door** |
|---|---|

In the game show of Example 3a, the amount $X$ that the contestant would receive (in dollars) has this probability function:

| $x$ | $f(x)$ |
|---|---|
| 2,000 | 1/3 |
| 5,000 | 1/3 |
| 20,000 | 1/3 |

To assign a numerical value to this random option, imagine that the distribution of $X$ describes a "game" to be played repeatedly, say 3,000 times. Using the long-run interpretation of probabilities, we'd expect that about "one-third of the time" (about 1,000 times) the contestant would get \$2,000, one-third of the time \$5,000, and one-third of the time \$20,000. In 3,000 trials he would expect to get about

$$\$2,000 \times 1,000 + \$5,000 \times 1,000 + \$20,000 \times 1,000 = \$27,000,000.$$

Distributing this equally over the 3,000 trials gives the contestant about \$9,000 per trial. This amount per trial has been called, traditionally, the "mathematical expectation" for a single play of the game. The calculation can be written

$$\$2,000 \times \frac{1}{3} + \$5,000 \times \frac{1}{3} + \$20,000 \times \frac{1}{3} = \$9,000.$$

The contestant's mathematical expectation in playing the game happens to be precisely equal to the value of his nonrandom alternative—what he has already won and can choose to keep. The "expected" reward in choosing a door (\$9,000) is not one of the amounts he could actually win, but it lies between the extremes of \$2,000 and \$20,000.

Despite the apparent equivalence of the two options, some very rational people would distinctly prefer the sure \$9,000 and others, the random option. We return to this point in Example 3.2a. ∎

The older term *mathematical expectation* has evolved into *expected value,* and a common notation for the expected value of $X$ is $E(X)$. [When there is no possibility of confusion, we may omit the parentheses and write simply $EX$.] The terms *mean value, mean,* and *average value* are also commonly used to denote this characteristic of a distribution. The symbol $\mu$ is a generic symbol for this mean. When there is more than one random variable in a particular context, we may use subscripts for clarity, such as $\mu_X$ for $EX$.

> **Expected value** of a numerical random variable $X$ with p.f. $f(x)$:
>
> $$EX = \sum_i x_i \, f(x_i), \tag{1}$$
>
> provided the sum is absolutely convergent. This is also called the *mean value* or *average value* of $X$ and denoted by $\mu$ (or $\mu_X$).

**Example 3.1b** | **Expected Number of Correct Guesses**

In Example 2.2c we found the p.f. for $Y$, the number of correct identifications when Michael assigned cola brands to glasses of cola, in an episode of "All in the Family." The following table repeats that p.f. and includes also a column of products $y \, f(y)$, whose sum is the mean (1):

| $y$ | $f(y)$ | $y\,f(y)$ |
|-----|--------|-----------|
| 0 | 2/6 | 0 |
| 1 | 3/6 | 3/6 |
| 3 | 1/6 | 3/6 |
| Sums: | 1 | $1 = \mu$ |

So, "on average," Michael would get one correct. ∎

**Example 3.1c** | **Blood Types**

In Example 2.5b we gave a distribution for the random variable "blood type":

| $x$ | $f(x)$ |
|-----|--------|
| O | .42 |
| A | .44 |
| B | .10 |
| AB | .04 |

Blood type is not numerical, and "expected blood type" does not make sense. Only numerical random variables have mean values. ∎

The only ingredients of the formula for expected value are the various possible values and their probabilities $f(x)$; it is not necessary to refer to an underlying sample space $\Omega$. Nevertheless, the expected value *can* be found using probabilities in $\Omega$, as a weighted sum of values of the random variable with weights given by the probabilities, as follows:

---

Expected value calculated from probabilities in $\Omega$:

$$E[X(\omega)] = \sum_\omega X(\omega)P(\omega). \tag{2}$$

This is equivalent to $EX$ given by (1).

---

The equivalence of (1) and (2) is easy to see when $\Omega$ is discrete:

$$EX = \sum_i x_i f(x_i) = \sum_i x_i \left\{ \sum_{X(\omega)=x_i} P(\omega) \right\} = \sum_i \sum_{X(\omega)=x_i} X(\omega)P(\omega).$$

The double sum on the right is over indices that account for every $\omega$ in $\Omega$.

## Example 3.1d    Average Range

Consider selections $\omega$ of two chips at random from four chips numbered from 1 to 4. (This is the prototype for sampling at random from finite populations.) The *range* is the difference between the numbers on the chips we select:

$$R(\omega) = \text{larger number in } \omega - \text{smaller number in } \omega.$$

This difference does not depend on the order in which the chips were drawn, so we take the set of six possible combinations as the sample space. These are shown in the following table, which includes a column of products:

| $\omega$ | $P(\omega)$ | $R(\omega)$ | $R(\omega)P(\omega)$ |
|------|------|------|------|
| 1, 2 | 1/6 | 1 | 1/6 |
| 1, 3 | 1/6 | 2 | 2/6 |
| 1, 4 | 1/6 | 3 | 3/6 |
| 2, 3 | 1/6 | 1 | 1/6 |
| 2, 4 | 1/6 | 2 | 2/6 |
| 3, 4 | 1/6 | 1 | 1/6 |
| Sums: | 1 | | 10/6 |

The sum in the last column is the sum (2). To find the mean using (1) we first need the p.f. of $R$. This is implied in the above table; it is as follows:

| $r$ | $f(r)$ | $r\,f(r)$ |
|-----|--------|-----------|
| 1   | 3/6    | 3/6       |
| 2   | 2/6    | 4/6       |
| 3   | 1/6    | 3/6       |
| Sums: | 1    | 10/6      |

It is easy to see why the result ($ER = 10/6$) is the same using both methods. Each product $r\,f(r)$ in the second table is accounted for as the sum of one or more products in the first table. For instance,

$$\frac{4}{6} = 2 \times \frac{2}{6} = 2 \times \left\{ \frac{1}{6} + \frac{1}{6} \right\} = 2 \times P[R(\omega) = 2].$$   ∎

The calculation of $EX$ using the probabilities in $\Omega$ is useful in establishing some important properties of expected values:

---

**Properties of $E(\,\cdot\,)$:**

   (i)  $E(c) = c$, for any constant $c$.
  (ii)  $E(cX) = cEX$, for any constant $c$.
 (iii)  $E(X + Y) = EX + EY$.
 (iv)  $E(aX + bY + c) = aEX + bEY + c$.

---

[Clearly, (i)–(iii) are special cases of (iv).] Formula (iv) is seen as follows:

$$E(aX + bY + c) = \sum [aX(\omega) + bY(\omega) + c]P(\omega)$$
$$= a\sum X(\omega)P(\omega) + b\sum Y(\omega)P(\omega) + c\sum P(\omega)$$
$$= aEX + bEY + c \cdot 1.$$

We'll illustrate (iii) in Example 3.1e below, but first we note a useful analogy.

Calculating a mean value involves the same arithmetic as calculating a center of gravity (c.g.) in physics. If masses $m_1, \ldots, m_k$ are placed on a line at $x_1, \ldots, x_k$, respectively, they will balance at

$$\text{c.g.} = \frac{x_1 m_1 + \cdots + x_k m_k}{m_1 + \cdots + m_k} = \sum_{i=1}^{k} x_i \left\{ \frac{m_i}{\sum m_j} \right\}.$$

The *relative* masses $m_i / \sum m_j$ play the same role in this calculation that probabilities play in (1). Like probabilities, they are nonnegative and sum to 1.

The parallel between relative masses and probabilities is important in giving an intuitive feel for an expected value as a *balance point*. It also provides a way of roughly checking the calculation of a mean value by a visual examination of a graphical representation of a probability function. The balance property of the mean $\mu$ can be expressed in this way:

$$E(X - \mu) = \sum(x_i - \mu)f(x_i) = EX - E\mu = \mu - \mu = 0.$$

The balance property shows, incidentally, that if the distribution of $X$ is *symmetric* about the value $x = a$, then that value $a$ is the mean, $EX$.

---

If $f(x)$ is **symmetric** about $x = a$,—that is, if for all $x$,

$$f(a - x) = f(a + x),$$

then (provided $EX$ exists), $EX = a$.

---

**Example 3.1e** | **Average Total of Two Dice**

In various parlor and gambling games, players toss two dice at each turn. Let $X$ denote the number of points showing on one die, and $Y$ the number on the other die. The probability functions of $X$ and $Y$ are the same: $f(k) = 1/6$, for $k = 1, ..., 6$. This p.f. is symmetric about 3.5, which is therefore the value of $EX$ and of $EY$. The total number of points showing is the sum $X + Y$, whose mean value [from Property (iii) above] is

$$E(X + Y) = EX + EY = 3.5 + 3.5 = 7.$$

Another way to get this result is to look at the distribution of the sum, which was obtained in Problem 2-1:

| $w$ | 2 | 3 | 4 | 5 | 6 | 7 | 8 | 9 | 10 | 11 | 12 |
|---|---|---|---|---|---|---|---|---|---|---|---|
| $f(w)$ | $\frac{1}{36}$ | $\frac{2}{36}$ | $\frac{3}{36}$ | $\frac{4}{36}$ | $\frac{5}{36}$ | $\frac{6}{36}$ | $\frac{5}{36}$ | $\frac{4}{36}$ | $\frac{3}{36}$ | $\frac{2}{36}$ | $\frac{1}{36}$ |

It is apparent that the distribution is symmetrical about the value 7, which is therefore the mean value. This agrees with the value found using additivity of averages.    ∎

Conditional distributions, being distributions, may have mean values. We calculate these as we do any mean value—as the sum of the possible values weighted with corresponding probabilities, but these probabilities are then conditional. Thus, the **conditional mean** of $Y$ given $X = x$ is

$$\mu_{Y|x} = E(Y \mid x) = \sum_y y f(y \mid x) = \sum_y y \frac{f(x, y)}{f_X(x)}.$$

Being the mean of a probability distribution, this enjoys all the properties of a mean. For instance, $E[Y - \mu_{Y|x} \mid x] = 0$.

**Example 3.1f** | **Conditional Distributions**

Suppose the random pair $(X, Y)$ has the following discrete distribution:

|   | | $Y$ | | |
|---|---|---|---|---|
|   |   | 0 | 1 | 2 |
| $X$ | 1 | .3 | .2 | .1 | .6 |
|   | 2 | .1 | .1 | .2 | .4 |
|   |   | .4 | .3 | .3 | 1 |

The conditional odds for $Y$ given $X = 1$ are 3:2:1. For $Y$ given $X = 2$ they are 1:1:2. Thus, the conditional p.f.'s are as follows:

| $y$ | $f(y \mid X = 1)$ |
|---|---|
| 0 | 3/6 |
| 1 | 2/6 |
| 2 | 1/6 |

| $y$ | $f(y \mid X = 2)$ |
|---|---|
| 0 | 1/4 |
| 1 | 1/4 |
| 2 | 2/4 |

The conditional means are calculated from these as from any p.f. table. Thus, $E(Y \mid X = 1) = 0 \times 3/6 + 1 \times 2/6 + 2 \times 1/6 = 4/6$. ∎

---

**3.2**      **Expected Value of a Function of Random Variables**

After observing a random variable $X$, we are often interested in some *function* of a random variable: $g(X)$. The simplest nontrivial functions are *linear*—the type of transformation we usually encounter in changing from one scale of measurement to another, as for example, inches to centimeters or degrees Fahrenheit to degrees Celsius. Nonlinear transformations, such as the square root and logarithm, are useful in data analysis. In particular, they can reduce any adverse effect that large observations have on an analysis. More immediately, we'll encounter the nonlinear functions $x^2$ (in defining the notion of variance) and $t^x$ (in defining a generating function in §3.6).

Given the random variable $X(\omega)$ and the function $g$, the composite function $g[X(\omega)]$ is a random variable; call it $Y(\omega)$. In abbreviated notation we suppress the $\omega$ and write simply $Y = g(X)$.

Finding the expected value of $Y = g(X)$ is a straightforward application of (2) in the preceding section:

$$E[Y(\omega)] = \sum_{\omega} Y(\omega)P(\omega) = \sum_{\omega} g[X(\omega)]\, P(\omega).$$

Alternatively, we can think of the set of possible $X$-values as a sample space, with probabilities defined by the p.f. of $X$, and treat $g(X)$ as a function defined on that sample space. With this interpretation, that same relation (2) of the preceding section yields the following:

---

Mean value of a function $g$ of a random variable $X$:

$$E[g(X)] = \sum_{x} g(x) f_X(x). \qquad (1)$$

---

With $g(x) = x$, this reduces to the definition (1) of §3.1.

Since $Y$ is a random variable in its own right, we might think to find its mean value using (1) of §3.1: $EY = \sum y\, f_Y(y)$. It usually happens that this is more work than (1) above because it requires finding the p.f. $f_Y$. The two ways give the same result, as we explained in showing the equivalence of (1) and (2) in §3.1.

## Example 3.2a  |  Utility of the Random Door

The game show contestant of Example 3a had the option of choosing between the $9,000 he had already won and a random prospect of $X$ dollars with equally probable values 2,000, 5,000, and 20,000. We found that $EX = 9,000$, but raised the question of whether he would really be indifferent between the two choices—despite the equality of expectations. He could toss a coin, but it is unlikely that he'd be willing to let a coin toss make the decision for him. A rational choice would have to depend on the contestant's financial state.

The usefulness or value of money is a function of amount, but not necessarily proportional to amount. The **utility function** is a subjective, numerical assessment of the usefulness of money, for the purpose of comparing alternatives. In the theory of utility, choices among random prospects are made according to the principle of maximizing *expected* utility.

In the game show context, some might choose the certain $9,000, not wanting to risk going down to the lower amounts, even though there is a chance of ending up with $20,000. Suppose, however, the contestant must pay off a $20,000 debt the next day and has no other source of funds. The $9,000 he has with certainty would not be much more useful to him than $2,000 or $5,000. His utility function might be approximated by this:

$$u(M) = \begin{cases} 0, & M < 20,000, \\ 1, & M \geq 20,000. \end{cases}$$

This treats all dollar amounts of more than $20,000 as equally valuable or "useful," and all dollar amounts of less than $20,000 as valueless. The expected utility, from (1), is

$$E[u(X)] = \sum u(x) f(x) = u(2,000) \times \frac{1}{3} + u(5,000) \times \frac{1}{3} + u(20,000) \times \frac{1}{3} = \frac{1}{3}.$$

The utility of $9,000 is 0, so (with the assumed utility function) the contestant should take the option of choosing a door. In this simple situation, the conclusion is almost obvious, because the random prospect offers some chance of reaching the important plateau of $20,000. Problem 3-10 considers other utility functions, for which the optimal choice may be different. ∎

For another application of formula (1) for $E[g(X)]$, we observe that the conditional p.f. $f(y \mid x)$ depends on $x$. It is a *function* of $x$—a function that can play the role of $g(x)$ in (1). Thus,

$$E[f(y \mid X)] = \sum_x f(y \mid x) f_X(x) = \sum_x \frac{f(x, y)}{f_X(x)} f_X(x) = \sum_x f(x, y) = f_Y(y).$$

In words: The expected value, with respect to the distribution of $X$, of the conditional p.f. of $Y$ given $X$ is the *un*conditional p.f. of $Y$.

Similarly, the conditional mean $E(Y \mid x)$ is a function of $x$. Applying (1) to this function "$g(x)$," we calculate the mean value with respect to the distribution of $X$:

$$E[E(Y \mid X)] = \sum_x \left\{ \sum_y y \frac{f(x, y)}{f_X(x)} \right\} f_X(x) = \sum_y y \left\{ \sum_x f(x, y) \right\} = \sum_y y f_Y(y).$$

The last expression is just $EY$, so we see that averaging the conditional mean of $Y$ given $X$ yields the *un*conditional mean of $Y$.

---

For discrete random variables $X$ and $Y$,

$$E[f(y \mid X)] = f_Y(y) \text{ for all } y, \tag{2}$$

and

$$E[E(Y \mid X)] = EY. \tag{3}$$

---

We'll refer to (3) as the **iterated expectation** formula.

**Example 3.2b** | Consider again the bivariate distribution of Example 3.1f. In that example we obtained conditional p.f.'s for $Y$ given $X$, which we repeat here in the following table, along with the marginal *(unconditional)* p.f. for $Y$:

| $y$ | 0 | 1 | 2 | $f_X(x)$ |
|---|---|---|---|---|
| $f(y \mid 1)$ | 1/2 | 1/3 | 1/6 | .6 |
| $f(y \mid 2)$ | 1/4 | 1/4 | 1/2 | .4 |
| $f_Y(y)$ | .4 | .3 | .3 | 1 |

Weighting the conditional probabilities of $Y = 1$, for instance, with the probabilities $f_X(x)$, we find the unconditional probability of $Y = 1$ by summing:

$$E[f(1 \mid X)] = \sum_x f(1 \mid x) f_X(x) = \frac{1}{3} \times .6 + \frac{1}{4} \times .4 = .3 = f_Y(1).$$

(You should check the other two values of $Y$ similarly.)

In Example 3.1f we found $E(Y \mid 1) = 2/3$, and a similar calculation produces $E(Y \mid 2) = 5/4$. Weighting these with probabilities $f_X(x)$, we find

$$E[E(Y \mid X)] = \sum_x E(Y \mid x) f_X(x) = \frac{2}{3} \times .6 + \frac{5}{4} \times .4 = .9 = E(Y). \quad \blacksquare$$

In dealing with several variables we'll often need to consider a function of several random variables, such as $W = g(X, Y, Z)$. From (2) of §3.1, with $\omega = (x_1, ..., x_n)$ and $P(\omega) = f(x_1, ..., x_n)$, we obtain:

---

Expected value of a function of several variables:

$$E[g(X_1, \ ..., \ X_n)] = \sum_{(x_1,...,x_n)} g(x_1, ..., x_n) \, f(x_1, ..., x_n). \qquad (4)$$

---

Once again, there is the question of whether using this formula to find the expected value gives the same result as we'd get by first finding the p.f. of $Y = g(X_1, \ ..., \ X_n)$ and applying the usual $\sum_y f_Y(y)$. And, indeed, it does.

**Example 3.2c** | **Tetrahedral Dice**

Regular tetrahedral are marked so as to identify the four faces by the numbers 1, 2, 3, 4. When two such dice are rolled, the *a priori* model assigns equal probabilities to the 16 pairs $(i, j)$. Let $X$ and $Y$ denote the numbers showing on the two dice, and define the variable $W = \max(X, Y)$, that is, the *larger* of the two numbers. The values of $W$ for the various pairs are shown in the following array:

|   | 1 | 2 | 3 | 4 |
|---|---|---|---|---|
| 1 | 1 | 2 | 3 | 4 |
| 2 | 2 | 2 | 3 | 4 |
| 3 | 3 | 3 | 3 | 4 |
| 4 | 4 | 4 | 4 | 4 |

(Because of symmetry, labeling $X$ and $Y$ could be done either way.) Using (4), we calculate $EW$ by multiplying each table entry (value of $W$) by 1/16 and summing. If we accumulate terms in that sum that involve like values of $W$, we obtain what is essentially $\sum w \, f_W(w)$:

$$EW = 1 \times \frac{1}{16} + 2 \times \frac{3}{16} + 3 \times \frac{5}{16} + 4 \times \frac{7}{16} = \frac{50}{16}.$$

(There are seven 4's, five 3's, three 2's, and one 1 in the table.) Figure 3-1 shows the p.f. of $W$ as a bar graph; the value $50/16 = 3.125$ appears reasonable as the c.g.   ∎

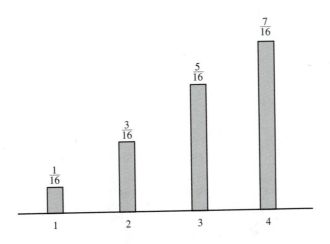

**Figure 3-1** Maximum—two tetrahedral dice

An important application of (4) is to find the expected value of a product of functions of random variables when these variables are *independent*. In the case of two random variables, let $g$ and $h$ be functions of $X$ and $Y$, respectively. Then, from (4), and if $X$

and $Y$ are independent, there follows

$$E[g(X)h(Y)] = \sum_{(x,y)} g(x)h(y)f(x, y) = \sum_x \sum_y g(x)f_X(x)h(y)f_Y(y)$$

$$= \sum_x g(x)f_X(x)\sum_y h(y)f_Y(y) = E[g(X)]E[h(Y)].$$

---

When $X$ and $Y$ are *independent*,

$$E[g(X)h(Y)] = E[g(X)] \cdot E[h(Y)]. \tag{5}$$

In particular,

$$E(XY) = EX \cdot EY. \tag{6}$$

---

The factorization of the joint p.f. of independent variables applies also for any finite number of random variables [see (1) in §2.5], so the extensions of (5) and (6) to that case are clear.

## Problems

* **3-1.** Given this p.f. for $X$:

| $x$ | 0 | 1 | 2 | 3 | 4 |
|---|---|---|---|---|---|
| $f(x)$ | .1 | .3 | .3 | .1 | .2 |

(a) Find $EX$.   (b) Find $E(X - \mu_X)$.   (c) Find $E(X^2)$.

* **3-2.** Find the expected value of each of the variables $S$, $D$, and $U$ defined in Problem 2-4 (page 50).

* **3-3.** As in Problem 2-3, let $X$ denote the number of rotten eggs in a random selection of three from a dozen, of which two are rotten. The p.f. (from Problem 2-3) is $f(0) = 12/22$, $f(1) = 9/22$, and $f(2) = 1/22$.
(a) Find $EX$, the mean number of rotten eggs among those selected.
(b) Find $E(X^2)$.

**3-4.** Let $X$ denote the number of bidding points assigned to a card selected at random from a deck of playing cards. (See Problem 2-34.) Find $EX$, given the p.f. $f(x) = 1/13$ for $x = 1, 2, 3, 4$, and $f(0) = 9/13$.

* **3-5.** When three coins are tossed, let $\omega$ denote the sequence of outcomes.
(a) List the eight $\omega$'s in the sample space $\Omega$, and give the values of the random variables $X(\omega)$, the number of heads, and $Y(\omega)$, the indicator function of the event $[X = 2]$.
(b) Find the mean values $EX$ and $EY$, assuming independent tosses and "fair" coins.

**3-6.** Ten tickets are sold in a lottery for $1 each. Three tickets are to be drawn without replacement. The first gets a $5 prize, and the second and third each get a $2 prize.

   (a)   Find the expected worth of the single ticket that you have bought.

   (b)   Find the total expected worth of four tickets.

   (c)   Is this lottery fair? [Does the ticket price equal the worth of the ticket?]

* **3-7.** Three digits are picked at random without replacement from the list 1, 2, 3, ..., 8. Let $Y$ denote the largest of the three digits selected.

   (a)   Find the probability function, $f(k) = P(Y = k)$.

   (b)   Find $P(Y \geq 5)$.

   (c)   Find $EY$.

**3-8.** The Octopus Car Wash once ran a promotion, a "happy hour" from 4 to 6 P.M., during which each customer was asked to roll a die. If the die showed 1, 2, or 3, the customer got $1 off the regular price of $7.50; if the die showed 4 or 5, the customer got $2 off; and if it showed 6, the wash was half price. Find the average cost of a wash during that period.

* **3-9.** Four men whose coats appear identical hang them up in a coat room and later pick them up at random. Let $X$ denote the number of matches (i.e., the number of men who get their own coats). Define the indicator variable $Y_i$, which is 1 if Mr. $i$ gets his own coat and 0 if he does not.

   (a)   Find $P(Y_1 = 1)$.

   (b)   Find $EY_1$.

   (c)   Note that the $Y_i$'s are exchangeable, and use this fact to find $EX$.

   (d)   Show that if there are $n$ men and $n$ coats, the expected number of men who get their own coats is the same for all $n$.

**3-10.** In Examples 3.1a and 3.2a, a contestant was offered a random prospect with probabilities 1/3 for each of the possible rewards, $2,000, $5,000, and $20,000, in exchange for a current fortune of $9,000.

   (a)   Suppose the contestant owes a murderous loan shark $9,000, due the next day. Then $2,000 and $5,000 are virtually useless to him, and the $20,000 is really no better than $9,000. Devise a suitable utility function and find the mean utility. Should he do the exchange?

   (b)   Find the expected utility, and whether the contestant should or should not trade, for each of the following utility functions:

   **(i)**   $u(M) = M$.      **(ii)**   $u(M) = M^2$.      **(iii)**   $u(M) = \sqrt{M}$.

**3-11.** Problem 1-36 dealt with selections of three socks from ten. The joint p.f. of $R$, the number of red socks, and $G$, the number of green socks among the three, is given by the following table of probabilities:

|       |       | $G$   |       |       |       |        |
|-------|-------|-------|-------|-------|-------|--------|
|       |       | 0     | 1     | 2     | 3     | $f_R$  |
| $R$   | 0     | 1/30  | 6/30  | 6/30  | 1/30  | 14/30  |
|       | 1     | 3/30  | 8/30  | 3/30  | 0     | 14/30  |
|       | 2     | 1/30  | 1/30  | 0     | 0     | 2/30   |
|       | $f_G$ | 5/30  | 15/30 | 9/30  | 1/30  | 1      |

(a)   Verify, for this distribution, that $E[f(g \mid R)] = f_G(g)$, as in (2) of §3.2.

(b)   Verify, for this distribution, that $E[E(G \mid R)] = E(G)$, as in (3) of §3.2.

## 3.3          Variability

The average value is only one characteristic of a random variable. But the "average" temperature in Minneapolis, for example, is far from the whole story. There can be large deviations from the average—it has been as hot as 106°F and as cold as $-34°F$. The amount of variability or dispersion of values about the middle value is important. We describe such variability in terms of *deviations* from the middle.

The distance of $X$ from its mean is the *absolute* deviation $|X - \mu|$. The average of this random distance is called the **mean deviation** (m.a.d.) of the distribution of $X$:

$$\text{m.a.d.} = E|X - \mu| = \sum |x - \mu| f(x). \tag{1}$$

When probability is dispersed over a wide range, there will be large deviations from the mean, and the m.a.d. will tend to be large. When probability is concentrated in a narrow range, the m.a.d. will tend to be small.

A more traditional way of characterizing variability is to average the *squared* deviations. The average squared deviation is called the **variance** of the distribution, abbreviated var $X$ and often denoted by $\sigma^2$, possibly with a subscript to identify the random variable:

$$\sigma^2 = \text{var } X = E[(X - \mu)^2].$$

An alternative formula is often convenient. To derive it, expand the square:

$$(X - \mu)^2 = X^2 - 2\mu X + \mu^2,$$

and average term by term (exploiting additivity of expectations):

$$E[(X - \mu)^2] = E(X^2) - 2\mu E X + \mu^2 = E(X^2) - \mu^2.$$

Thus, the variance is the average square minus the square of the average.

---

*Variance* of a random variable:

$$\text{var } X = E[(X - \mu)^2] = \sigma^2 \tag{2}$$

or, alternatively,

$$\text{var } X = E(X^2) - (EX)^2. \tag{3}$$

---

**Example 3.3a**

## Variance for Toss of a Die

Let $X$ denote the number of points that show at the toss of a fair die. The mean is $\mu = 3.5$; the absolute deviations from this mean (for $x = 1, ..., 6$) are

$$2.5, \ 1.5, \ .5, \ .5, \ 1.5, \ 2.5$$

Since the values of $X$ are equally likely, these six deviations are equally likely, and their average is just their sum divided by 6:

$$\text{m.a.d.} = \frac{9}{6} = 1.5.$$

The squares of the deviations are

$$6.25, \ 2.25, \ .25, \ .25, \ 2.25, \ 6.25,$$

and the average of these squared deviations is their sum divided by 6:

$$\text{var } X = \frac{17.5}{6} = \frac{35}{12}.$$

Alternatively, we can find this using (3):

$$\text{var } X = \frac{1}{6}(1^2 + 2^2 + 3^2 + 4^2 + 5^2 + 6^2) - \left(\frac{7}{2}\right)^2 = \frac{35}{12}. \qquad \blacksquare$$

It is clear from (2) that a variance is never negative. If you come up with a *negative* variance when using (3), you have made a mistake in your calculations.

Although a variance measures variability, it is awkward to interpret. The unit of measurement of var $X$ is the square of the unit of $X$. A more easily interpreted measure is the (positive) square root of the variance, called the **standard deviation** (s.d.): $\sigma = \sqrt{\sigma^2}$. (We anticipated this definition when we denoted the variance by $\sigma^2$.) The units of $\sigma$ are the same as the units of $X$ itself.

We'll want to know what happens to a standard deviation if we change units by means of a *linear* transformation: $U = a + bX$. The deviations on which variances are based do not depend on the location of the reference origin, so they do not change with the translation of the origin by $a$. Thus,

$$U - EU = a + bX - (a + bEX) = b(X - EX).$$

The translation constant $a$ drops out. But the deviation *is* multiplied by the scale factor, $b$. So the average absolute deviations—both the mean deviation and the standard deviation (a *root-mean-square* "average")—are multiplied by $|b|$, the absolute value of the scale factor.

---

*Standard deviation* of a random variable:

$$\sigma_X = \sqrt{\operatorname{var} X} = \sqrt{E(X^2) - \mu^2}. \tag{4}$$

If $Y = a + bX$,

$$\sigma_Y = |b|\sigma_X. \tag{5}$$

---

You may find it helpful to think of $\sigma$, as well as the mean deviation, as "typical" or "ordinary" deviations from the mean.

**Example 3.3b** | **Standard Deviation for the Die Toss**

In Example 3.3a, we found the variance of the number of points thrown with the toss of a die to be 35/12. The standard deviation is then

$$\sigma = \sqrt{\operatorname{var} X} = \sqrt{\frac{35}{12}} \doteq 1.71.$$

We see that the s.d. $\sigma$ is not as small as the smallest of the deviations (.5) nor as large as the largest (2.5). This is true generally: $\sigma$ is a typical deviation. In Example 3.3a we found the mean deviation to be 1.5, not far from but somewhat smaller than $\sigma$. (See Figure 3-2.) ∎

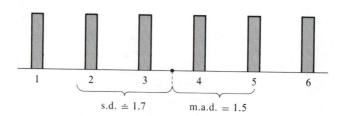

**Figure 3-2** Standard deviation and m.a.d., Example 3.3b

The idea of the mean (absolute) deviation can help in understanding the standard deviation. As in the preceding example, the mean deviation is usually somewhat smaller than the standard deviation. It is never larger. (See Problem 3-21.) It is sometimes easy to guess the value of the mean deviation; the s.d. is then just a bit larger.

In the early development of statistical methodology, there was considerable controversy as to whether the standard deviation or the mean deviation should be used to measure dispersion or variability. The standard deviation prevailed, and in current statistical practice it still dominates. However, the mean deviation has the important advantage that it is less influenced by the large deviations that often occur in actual populations.)

The mean and variance of a probability distribution are parallels of quantities in physics. As we mentioned previously, the mean is the center of gravity. The standard deviation also has a parallel, the radius of gyration about the c.g. In physics, the characteristics of distributions are *moments,* and we carry over the terminology of moments to probability distributions. The **rth moment** of a distribution about the value $c$ is defined as

$$E[(X - c)^r] = \sum (x - c)^r f(x). \tag{6}$$

The expected value is the first moment about 0. The first moment about the mean is 0:

$$E(X - \mu) = EX - \mu = 0. \tag{7}$$

Indeed, this characterizes the mean: It is that point about which the first moment is zero. In the realm of mass distributions, this is the balance property of the c.g.

The variance is the second moment about the mean, a *central* moment. This second moment is the smallest possible second moment, as we now show, using a standard formula from physics known as the *parallel axis theorem.* To obtain this, we start with a deviation about $c$ and add and subtract the mean:

$$X - c = (X - \mu) + (\mu - c).$$

Squaring both sides, and expanding the square of the binomial on the right, we obtain

$$(X - c)^2 = (X - \mu)^2 + (\mu - c)^2 + 2(\mu - c)(X - \mu).$$

Averaging yields the second moment about $c$:

$$E[(X - c)^2] = E[(X - \mu)^2] + (\mu - c)^2 + 0.$$

[The 0 is the result of averaging $X - \mu$, since $2(\mu - c)$ is a constant.]

---

**Parallel axis theorem:**

$$E[(X - c)^2] = E[(X - \mu)^2] + (\mu - c)^2. \tag{8}$$

---

The parallel axis theorem is an extremely useful tool, one that we'll be using over and over again in later chapters. At this point we can use it to show that the smallest second moment is the variance, the second moment about the mean value: The second term on the right of (8) is nonnegative, and it is 0 (its smallest value) when $c = \mu$. (The first term doesn't involve $c$.)

## Problems

**\* 3-12.**   Given $E(X^2) = 65$ and $EX = 7$, find the standard deviation of $X$.

**3-13.**   Given $EX = 3$ and $E[X(X - 1)] = 6$, find var $X$.

**\* 3-14.**   Find the standard deviation of the number of heads in three independent tosses of a fair coin.

**\* 3-15.**   In Problem 3-4, you found the mean number of points assigned to a card for bidding in bridge to be 10/13. Find the standard deviation.

**3-16.**   A census questionnaire once asked for the number of flush toilets per household. Assume these results, defining a population distribution:

| Number      | 0    | 1    | 2    | 3    | 4    |
|-------------|------|------|------|------|------|
| Proportion  | .05  | .55  | .30  | .08  | .02  |

Find the mean and standard deviation of the number per household.

**\* 3-17.**   Find the standard deviation of the random variable $X$ in Problem 3-3.

**3-18.**   I have four similar keys, just one of which opens my office door. I try them one at a time, selecting keys at random and without replacement. Let $X$ denote the number of keys I put in the lock before the door will open (including the key that works).
   **(a)**   Give the probability table for $X$.
   **(b)**   Find the mean and standard deviation of $X$.
   **(c)**   Find the probability function for $X$ if the setting is modified by making the selection random but *with* replacement.

**3-19.**   Scores on a certain hole on a golf course have been observed over a period of years, for professional golfers and for ordinary club members. Relative frequencies (which we interpret as probabilities) are as follows:

| Score       | Pro  | Member |
|-------------|------|--------|
| 2 (eagle)   | .02  |        |
| 3 (birdie)  | .16  | .03    |
| 4 (par)     | .68  | .22    |
| 5 (bogie)   | .13  | .27    |
| 6           | .01  | .27    |
| 7           |      | .13    |
| 8           |      | .05    |
| 9           |      | .02    |
| 10          |      | .01    |

For each type of golfer, find the mean and the standard deviation of the score.

* **3-20.**  A bowl contains three white and six black chips. Let $Y$ be the number of white chips in a random selection of four (without replacement). Find the mean and variance of $Y$, given the p.f. of $Y$:

$$f(k) = \frac{\binom{3}{k}\binom{6}{4-k}}{\binom{9}{4}}, \quad k = 0, 1, 2, 3.$$

**3-21.**  Show that the mean deviation (1) is never larger than the standard deviation (4). [*Hint:* Use (3) of §3.3, applied to the variable $|X - \mu|$, together with the fact that a variance is nonnegative.]

* **3-22.**  Find the mean deviation of $Y$, the number of heads in three independent throws of a coin, given the p.f. of $Y$: $f(0) = f(3) = 1/8$, $f(1) = f(2) = 3/8$.

---

## 3.4    Covariance and Correlation

To study relationships between two variables, we need to know their joint distribution. A useful characteristic of that distribution is the average product of their deviations from their respective means, called their *covariance:*

$$\text{cov}(X, Y) = E[(X - \mu_X)(Y - \mu_Y)] = \sum_{(x,y)} (x - \mu_X)(y - \mu_Y)f(x, y).$$

An alternative notation for this covariance is $\sigma_{X,Y}$.

The covariance reduces to a variance when $X = Y$:

$$\text{cov}(X, X) = E[(X - \mu_X)(X - \mu_X)] = \text{var } X.$$

And just as the variance lent itself to a sometimes simpler calculation, there is a simpler calculating formula for the covariance. We obtain it by multiplying out the product of the deviations from the mean, and averaging term by term:

---

**Covariance of $X$ and $Y$:**

$$\text{cov}(X, Y) = E[(X - \mu_X)(Y - \mu_Y)]. \tag{1}$$

Alternatively,

$$\text{cov}(X, Y) = E(XY) - \mu_X\mu_Y. \tag{2}$$

---

**Example 3.4a**     We illustrate with some simple numbers. Let $(X, Y)$ have the discrete distribution defined by the following array of joint probabilities:

|       | $Y$ |     |     |     |     |
|-------|-----|-----|-----|-----|-----|
|       | 1   | 2   | 3   | 4   |     |
| 0     | .1  | 0   | .1  | 0   | .2  |
| $X$  1 | .3  | .2  | .1  | 0   | .6  |
| 2     | 0   | .1  | 0   | .1  | .2  |
|       | .4  | .3  | .2  | .1  | 1   |

The means, calculated from the marginal probabilities, are $\mu_X = 1$, $\mu_Y = 2$. In finding the average product we observe that since 0 times anything is 0, we can ignore the first row and any products of $xy$-pairs that have 0 probability:

$$E(XY) = 1 \times .3 + 2 \times .2 + 3 \times .1 + 4 \times .1 + 8 \times .1 = 2.2.$$

Then, using (2), we find the covariance as 2.2 minus the product of the means:

$$\text{cov}(X, Y) = 2.2 - 1 \times 2 = .2.$$

To use the original definition, we rescale the $X$-values and $Y$-values by subtracting the means, 1 and 2, respectively. With these deviations as new variables, the probability table is as follows:

|        | $Y - 2$ |     |     |     |     |
|--------|------|-----|-----|-----|-----|
|        | $-1$ | 0   | 1   | 2   |     |
| $-1$   | .1   | 0   | .1  | 0   | .2  |
| $X - 1$  0 | .3   | .2  | .1  | 0   | .6  |
| 1      | 0    | .1  | 0   | .1  | .2  |
|        | .4   | .3  | .2  | .1  | 1   |

Again, in finding the average product of the deviations (1), we can ignore any row and column labeled 0, as well as terms in the sum with probability 0:

$$\text{cov}(X, Y) = (-1) \times (-1) \times .1 + (-1) \times 1 \times .1 + 1 \times 2 \times .1 = .2.$$

Of course, this is the same as the result of the earlier calculation. ∎

A covariance is a measure of concordance or association of the two variables. If $X$ tends to be large when $Y$ is large and small when $Y$ is small, then the deviations in (1) will tend to be of the same sign, and their positive products will make the covariance positive. Similarly, if $X$ tends to be small when $Y$ is large and large when $Y$ is small, then the deviations in (1) will tend to be of opposite sign, and their negative products will make the covariance negative.

Yet another possibility is that large $X$'s occur with both small and large $Y$'s, so that the corresponding deviations, both negative and positive, will tend to cancel, and the covariance may be close to 0.

As a measure of association, the covariance suffers from dependence on units of measurement. A change of units is typically a *linear* transformation; and, although the deviations from the mean do not change with a translation of the origin, they are affected by a scale factor. Thus, if $U = a + bX$ and $V = c + dY$,

$$U - EU = b(X - EX) \quad \text{and} \quad V - EV = d(Y - EY).$$

The translation constants $a$ and $c$ drop out, and the deviations are multiplied by the scale factors, $b$ and $d$. So the covariance of the new pair $(U, V)$ is

$$\text{cov}(U, V) = E[(U - EU)(V - EV)] = E[b(X - \mu_X)\, d(Y - \mu_Y)]$$
$$= bd\, E[(X - \mu_X)(Y - \mu_Y)] = bd\, \text{cov}(X, Y).$$

$$\boxed{\text{cov}(a + bX,\ c + dY) = bd\, \text{cov}(X, Y).}$$

Two variables with *zero* covariance are said to be *uncorrelated*. As we'll soon see, the relationship implied by a nonzero covariance is a very special type of relationship, and a covariance of 0 does not necessarily mean that one variable doesn't depend on the other at all. However, if two variables are *independent*, then they are uncorrelated. This is a result of (6) in §3.2, where we saw that *when X and Y are independent*, the average product is equal to the product of the averages, which means that

$$\text{cov}(X, Y) = E(XY) - (EX)(EY) = 0.$$

When $\text{cov}(X, Y) = 0$, $X$ and $Y$ are said to be **uncorrelated.**

Although independent variables are always uncorrelated, uncorrelated variables can be dependent.

**Example 3.4b**   To illustrate the fact that uncorrelated variables can be dependent, we consider the distribution of $(X, Y)$ defined by this table of probabilities:

|   |   | $Y$ |   |   |
|---|---|---|---|---|
|   |   | 0 | 2 | 4 |   |
|   | 0 | .2 | 0 | .2 | .4 |
| $X$ | 1 | 0 | .2 | 0 | .2 |
|   | 2 | .2 | 0 | .2 | .4 |
|   |   | .4 | .2 | .4 | 1 |

Symmetry makes it plain that $\mu_X = 1$ and $\mu_Y = 2$. The average product is

$$E(XY) = 1 \times 2 \times .2 + 2 \times 4 \times .2 = 2 = \mu_X \mu_Y,$$

so the covariance is 0. This value is evident when you look at the deviations from the means, and the equal weights for those pairs that have positive probabilities. For each product of deviations that is negative there is a corresponding product that is positive, and the result is complete cancellation.

But observe that the variables are *not independent!* If you know that $Y = 2$, for instance, then $X$ can only be 1. (Also, as we pointed out in §2.5, a joint probability of 0 for a pair in which both the $x$ and $y$ have positive probabilities is a tip-off that the variables are not independent.)                                                                    ■

A unitless measure of association is obtained upon dividing each deviation by its standard deviation—taking the expected product of $Z$-scores:

$$\rho = E\left\{ \frac{X - EX}{\sigma_X} \cdot \frac{Y - EY}{\sigma_Y} \right\} = \frac{\sigma_{X,Y}}{\sigma_X \sigma_Y}.$$

This covariance of the $Z$-scores is called the *coefficient of linear correlation*. Since a linear transformation introduces the scale factor above *and* below the line, that factor cancels out if it's positive, leaving the correlation *unchanged*. (If it's negative, the sign changes.)

---

**Correlation coefficient:**

$$\rho_{X,Y} = \frac{\sigma_{X,Y}}{\sigma_X \sigma_Y}. \tag{3}$$

If $U = a + bX$ and $V = c + dY$,

$$\rho_{U,V} = \begin{cases} \rho_{X,Y} & \text{if } bd > 0, \\ -\rho_{X,Y} & \text{if } bd < 0. \end{cases}$$

---

**Example 3.4c** | Let $(X, Y)$ have the joint distribution defined by these probabilities:

|   |   | $Y$ |   |   |   |
|---|---|---|---|---|---|
|   | 1 | 2 | 3 | 4 |   |
| 0 | 0 | 0 | 0 | .2 | .2 |
| 3 | 0 | .1 | .2 | 0 | .3 |
| 6 | 0 | .2 | .1 | 0 | .3 |
| 9 | .2 | 0 | 0 | 0 | .2 |
|   | .2 | .3 | .3 | .2 | 1 |

($X$ labels the rows.)

In this table the probability is mostly on the line $x = 12 - 3y$. From symmetry, the means are seen to be $\mu_X = 4.5$ and $\mu_Y = 2.5$. The average product is

$$6 \times .1 + 9 \times .2 + 12 \times .2 + 18 \times .1 + 9 \times .2 = 8.4,$$

so the covariance is $8.4 - (4.5 \times 2.5) = -2.85$. The variance of $Y$ is $7.3 - 6.25$ or $\sigma_Y^2 = 1.05$, and $\sigma_X^2 = 3^2 \sigma_Y^2 = 9 \times 1.05$. The correlation coefficient is

$$\rho = \frac{-2.85}{\sqrt{9.45}\sqrt{1.05}} \doteq -.905.$$

The correlation is negative because of the inverse nature of the relationship—$X$ tends to be large when $Y$ is small, and vice versa.    ∎

We've used the common terminology, "correlation coefficient," but it should always be kept in mind that the "relation" to which it refers is of a special kind—it is *linear*. The fact that the correlation in the above example is close to 1 in magnitude reflects the near collinearity of the distribution, as we'll explain next.

A correlation coefficient cannot exceed 1 magnitude. This fact is a form of the *Schwarz inequality*, derived as follows: Let $U$ and $V$ denote the centered variables, $U = X - EX$ and $V = Y - EY$. For any real number $z$,

$$0 \le E[(V - zU)^2] = E(V^2) - 2zE(UV) + z^2 E(U^2) = az^2 + bz + c. \quad (4)$$

The zeros of this nonnegative, quadratic function are either imaginary or, in the extreme case, identical. And this implies that the discriminant $b^2 - 4ac$ must be negative or 0, as you can see by looking at the familiar formula for the zeros of a quadratic function. So, (4) implies

$$b^2 - 4ac = 4[E(UV)]^2 - 4E(U^2)E(V^2) \le 0. \quad (5)$$

But $E(UV)$ is the covariance of $X$ and $Y$, and $E(U^2)$ and $E(V^2)$ are their variances, so

$$\sigma_{X,Y}^2 \leq \sigma_X^2 \sigma_Y^2, \text{ or } \rho^2 \leq 1. \tag{6}$$

This is (one form of) the Schwarz inequality.

When $\rho^2 = 1$, the inequality (5) is an equality, so the discriminant $b^2 - 4ac$ is 0. This means that there is a (real number) $z_0$ which is a (double) zero of the quadratic function in (4). Thus,

$$E[(V - z_0 U)^2] = \sum_{i,j}(v_j - z_0 u_i)^2 f(u_i, v_j) = 0.$$

This can be only if $V \equiv z_0 U$. In terms of $X$ and $Y$, it says that $Y$ is a linear function of $X$ with probability 1, or that all of the probability lies on the line

$$y - \mu_Y = z_0(x - \mu_X).$$

The slope $z_0$ is positive if the covariance is positive and negative if the covariance is negative, because

$$\sigma_{X,Y} = \text{cov}(U, V) = \text{cov}(U, z_0 U) = z_0 \text{ var } X.$$

Correlation coefficients will be defined for continuous distributions in §5.7 and for sample distributions in §7.6.

## 3.5   Sums of Random Variables

In many applications, statisticians add or average the observations in a sample, so we need to study the *sum* of several random variables. We know from §3.1 that the expected value of a sum of two random variables is the sum of their expected values:

$$E(X + Y) = EX + EY. \tag{1}$$

From this it follows by induction that additivity of expectations holds for any finite set of random variables:

$$E(X_1 + \cdots + X_n) = EX_1 + \cdots + EX_n. \tag{2}$$

[The step in the induction proof that takes one from $n$ to $n+1$ is carried out by writing the sum of $n+1$ $X$'s as $(X_1 + \cdots X_n) + X_{n+1}$, and applying additivity to these two terms.]

Going next to the standard deviation—is it additive? Not in general. Neither is the variance. However, in some important special cases, the variance *is* additive, as we show next. The variance of a sum is the average squared deviation about its mean:

$$\text{var}(X + Y) = E[(X + Y - \mu_{X+Y})^2].$$

Rearranging terms, we have

$$\operatorname{var}(X + Y) = E[\{(X - \mu_X) + (Y - \mu_Y)\}^2].$$

We then expand the square of the binomial in braces:

$$\{(X - \mu_X) + (Y - \mu_Y)\}^2 = (X - \mu_X)^2 + (Y - \mu_Y)^2 + 2(X - \mu_X)(Y - \mu_Y)$$

and average term by term (using additivity of expectations):

$$\operatorname{var}(X + Y) = E[(X - \mu_X)^2] + E[(Y - \mu_Y)]^2 + 2E[(X - \mu_X)(Y - \mu_Y)],$$

or

$$\operatorname{var}(X + Y) = \operatorname{var} X + \operatorname{var} Y + 2\operatorname{cov}(X, Y). \qquad (3)$$

**Example 3.5a**

## Maximum and Minimum Observations

To illustrate with easy numbers, suppose we draw a sample of two at random, without replacement, from a population of four chips numbered from 1 to 4. There are six equally likely samples:

$$(1, 2)\ (1, 3)\ (1, 4)\ (2, 3)\ (2, 4)\ (3, 4).$$

Let $U$ and $V$ denote, respectively, the larger and the smaller of the two numbers in the sample. The joint distribution is given in the following table:

|   |   | $V$ 1 | 2 | 3 | $f_U(u)$ |
|---|---|---|---|---|---|
| $U$ | 2 | 1/6 | 0 | 0 | 1/6 |
|   | 3 | 1/6 | 1/6 | 0 | 2/6 |
|   | 4 | 1/6 | 1/6 | 1/6 | 3/6 |
|   | $f_V(v)$ | 3/6 | 2/6 | 1/6 | 1 |

The means are $\mu_U = 10/3$ and $\mu_V = 5/3$, and the variances are clearly equal, with common value 5/9. (You should check these.) There are six products with nonzero probabilities, equally likely:

$$E(UV) = (2 + 3 + 4 + 6 + 8 + 12) \cdot \frac{1}{6} = \frac{35}{6}.$$

To find the covariance, we subtract the product of the means:

$$\operatorname{cov}(U, V) = \frac{35}{6} - \frac{10}{3} \cdot \frac{5}{3} = \frac{5}{18}.$$

So, the variance of the sum is

$$\text{var}\,(U + V) = \frac{5}{9} + \frac{5}{9} + 2 \cdot \frac{5}{18} = \frac{5}{3}.$$

To verify (3) in this instance, we'll calculate var$(U + V)$ from the distribution of $U + V = Z$. From the table of joint probabilities we see that $f_Z(z) = 1/6$ when $z = 3$, 4, 6, or 7, and $f_Z(z) = 2/6$ when $z = 5$. The mean is 5, and the variance is

$$\text{var}\,Z = \frac{1}{6} \cdot (4 + 1 + 1 + 4) = \frac{1}{6} \cdot 10 = \frac{5}{3}.$$

Not surprisingly, this is the same as the value calculated using (3).    ∎

Induction based on (3) yields a formula for the variance of the sum of any finite number of random variables in terms of their individual variances and covariances:

---

Variance of a sum of random variables:

$$\text{var}\,\left( \sum_{i=1}^{n} X_i \right) = \sum_{i=1}^{n} \text{var}\,X_i + \sum_{i \neq j} \text{cov}(X_i,\ X_j), \qquad (4)$$

the latter sum extending over the $n(n-1)$ pairs $(i,\ j)$ in which $i \neq j$.

---

It is evident that the variance is additive only when the summands are pairwise uncorrelated. This is the case when the summands are *independent*.

**Example 3.5b**

## Two Dice

The total number of points thrown with two dice is a random variable equal to $X + Y$, where $X$ is the number of points on one die, and $Y$ is the number on the other die. In Example 3.3a we found that the variance of the number of points on a single die is 35/12. Assuming independence, we have

$$\text{var}\,(X + Y) = \text{var}\,X + \text{var}\,Y = \frac{35}{12} + \frac{35}{12} = \frac{35}{6}.$$

The standard deviation of the sum is the square root of this, or about 2.4.    ∎

Observe that the standard deviation of the sum is not equal to the sum of the standard deviations. Rather, they combine, when the variables are uncorrelated, in the way that you find the hypotenuse of a right triangle, given the lengths of the other two sides.

$$\sigma_{X+Y} = \sqrt{\sigma_X^2 + \sigma_Y^2}.$$

> If $X_1$, $X_2$, ..., $X_n$ are **pairwise uncorrelated** (and in particular, if they are *independent*), then
>
> $$\text{var}\,(X_1 + \cdots + X_n) = \text{var}\, X_1 + \cdots + \text{var}\, X_n. \qquad (5)$$

The formulas for mean and variance of a sum take particularly simple forms when the variables are observations obtained by random sampling: Whether sampling is done with or without replacement, the observations are identically distributed, so $EX_i = \mu$ for all $i$, where $\mu$ is the population mean:

$$E\!\left(\sum X_i\right) = \mu + \mu + \cdots + \mu = n\mu.$$

The variances of the individual terms in the sum are all $\sigma^2$, the population variance. And if the observations are *independent*, then

$$\text{var}\left(\sum X_i\right) = \sigma^2 + \sigma^2 + \cdots + \sigma^2 = n\sigma^2.$$

But if the sampling is without replacement from a finite population, the $X$'s are not independent. However, they are exchangeable. And therefore the covariances of two observations are all the same—equal to the covariance of the first two. So from (4),

$$\text{var}\left(\sum X_i\right) = \sigma^2 + \sigma^2 + \cdots + \sigma^2 + n(n-1)\sigma_{12}, \qquad (6)$$

where $\sigma_{ij} = \text{cov}(X_i, X_j)$. The factor $n(n-1)$ multiplying $\sigma_{12}$ is just the number of covariances $\sigma_{ij}$ with $i \neq j$.

To finish finding the value of the variance of the sum, we need to know the common covariance, $\sigma_{12}$. We can do this with a trick: If we take the entire population as the sample, the sample sum is always the same—it is the population sum. So the sample sum does not vary, and its variance is 0:

$$0 = \text{var}\left(\sum_{1}^{N} X_i\right) = N\sigma^2 + N(N-1)\sigma_{12}.$$

Solving for $\sigma_{12}$ we get

$$\sigma_{12} = \frac{-\sigma^2}{N-1}.$$

Substituting this in (6) gives the desired variance, which is shown in (8) on the following page.

Let $(X_1, ..., X_n)$ be a sample drawn at random from a population of size $N$. Then, whether drawn with or without replacement,

$$E\left\{\sum_1^n X_i\right\} = n\mu. \tag{7}$$

If the sample is drawn without replacement,

$$\text{var}\left\{\sum_1^n X_i\right\} = n\sigma^2 \left(\frac{N-n}{N-1}\right). \tag{8}$$

If it is drawn with replacement (or if $N = \infty$),

$$\text{var}\left\{\sum_1^n X_i\right\} = n\sigma^2. \tag{9}$$

The factor $(N-n)/(N-1)$ in (8) is termed the **finite population correction factor**. Observe that as $N$ becomes infinite, with $n$ fixed, this factor tends to 1, and (8) tends to (9), the variance of the sum when the observations are independent.

## Problems

**\* 3-23.** Given the discrete bivariate distribution in the accompanying table, find
(a) $\text{cov}(X, Y)$.          (c) $\text{var}(X+Y)$.
(b) $P(X = 1 \mid X+Y = 3)$.    (d) $\text{var}(Y \mid X = 1)$.

|   |   | $Y$ | | |
|---|---|---|---|---|
| | | 0 | 1 | 2 |
| $X$ | 1 | .2 | .1 | .3 |
| | 2 | 0 | .2 | .2 |

**\* 3-24.** Given that $\text{var}\,X = \text{var}\,Y = \text{cov}(X, Y) = 1$, find
(a) $\text{var}(3-X)$.       (d) $\text{cov}(X, X)$.
(b) $\text{var}(2X+4)$.     (e) $\text{cov}(X, X+Y)$.
(c) $\text{var}(X-Y)$.      (f) $\text{var}(4X+Y-7)$.

**3-25.** Find the mean and standard deviation of the total number of points in a throw of eight dice. Assume independence and use the fact that the number of points is a sum of independent variables.

**3-26.** Given that $X$ and $Y$ are independent, with $\sigma_X = \sigma_Y = 1$, find
(a) $\text{var}(2X+Y)$.        (c) $\rho_{X,Y}$.
(b) $\text{cov}(2X+Y, X-Y)$.   (d) $\rho_{U,V}$, where $U = 2X+Y$, $V = X-Y$.

**3-27.** Show the following:

(a) $\text{cov}(aX + bY, cX + dY) = ac\sigma_X^2 + bd\sigma_Y^2 + (bc + ad)\sigma_{X,Y}$.

(b) $\text{cov}(\sum_i a_i X_i, \sum_j b_j Y_j) = \sum_i \sum_j a_i b_j \text{cov}(X_i, Y_j)$.

**\* 3-28.** Find the table of joint probabilities for $Y_1$ and $Y_2$ in Problem 3-9 (for the random distribution of coats). To check that you are on the right track, verify that the marginal probabilities agree with those found in Problem 3-9.

(a) Find the covariance of $Y_1$ and $Y_2$.

(b) Find the variance of $X$, the total number of matches. [*Hint:* Observe that $X = Y_1 + Y_2 + Y_3 + Y_4$, and use the exchangeability of the $Y$'s.]

**\* 3-29.** Find the correlation coefficient for the variables in Problem 3-23.

**3-30.** A city has streets laid out in a grid, with north-south streets running perpendicular to east-west streets. You start at an arbitrary intersection and walk in a sequence of one-block moves. At each move, you choose a direction at random from N, E, W, S. Let $X_i$ denote the number of blocks you move north at step $i$ (with values $-1, 0, 1$) and $Y_i$, the number of blocks you move east at step $i$ (with values $-1, 0, 1$). The p.f. of $(X_i, Y_i)$ is as follows:

$$f(1, 0) = f(-1, 0) = f(0, 1) = f(0, -1) = \frac{1}{4}.$$

(a) Find the variance of $X_i$.

(b) Are $X_i$ and $Y_i$ independent?

(c) Let $U_n$ be the number of blocks you have moved north of the starting point after $n$ moves, and $V_n$, the number of blocks east:

$$U_n = \sum_{i=1}^{n} X_i \quad \text{and} \quad V_n = \sum_{i=1}^{n} Y_i.$$

Find $EU_n$, $EV_n$, $\text{var}\, U_n$, and $\text{var}\, V_n$, assuming that the successive moves [and hence the successive pairs $(X_i, Y_i)$] are independent.

(d) Find the average squared distance from the starting point after $n$ steps.

---

## 3.6     Probability Generating Functions

Generating functions are mathematical tools with many important applications in statistical theory. The first one we take up generates probabilities.

Many of the random variables in the first four chapters take values in the set of nonnegative integers: 0, 1, 2, .... For such a random variable $X$, we temporarily adopt the subscript notation usually used for sequences:

$$f_X(k) = P(X = k) = p_k, \quad k = 0, 1, 2, \dots.$$

Consider the power series formed with the elements of the sequence $\{p_k\}$ as coefficients of powers of $t$:

$$\eta(t) = p_0 t^0 + p_1 t^1 + p_2 t^2 + \cdots = \sum_{0}^{\infty} p_k t^k. \tag{1}$$

This can be interpreted as the expected value of $t^X$ [see (1) in §3.2, page 86]:

$$\eta(t) = E(t^X). \tag{2}$$

It is called the **probability generating function** (p.g.f.) of $X$ or of the sequence of probabilities $\{p_k\}$.

| Example **3.6a** | **A Fair Die** |
|---|---|

The model for the throw of a fair die assigns equal probabilities to the six sides: $p_k = 1/6$ for $k = 1, 2, 3, 4, 5, 6$. The p.g.f. is a finite power series:

$$\eta(t) = \frac{1}{6}t + \frac{1}{6}t^2 + \frac{1}{6}t^3 + \frac{1}{6}t^4 + \frac{1}{6}t^5 + \frac{1}{6}t^6.$$

This can be put in more compact form using the formula for the sum of a finite geometric series:

$$\eta(t) = \frac{t - t^7}{6(1 - t)}.$$

(You can check this by multiplying $t + t^2 + \cdots + t^6$ by $1 - t$.)  ∎

---

Probability generating function (p.g.f.) of a random variable $X$ with values 0, 1, 2, 3, . . .:

$$\eta_X(t) = E(t^X) = \sum_{k=0}^{\infty} p_k t^k, \tag{3}$$

where $p_k = P(X = k)$.

---

Probability generating functions are useful in calculations involving sums of independent random variables because of a familiar law of exponents. Thus, suppose $X$ and $Y$ are independent. The p.g.f. of $X + Y$ is

$$\eta_{X+Y}(t) = E(t^{X+Y}) = E(t^X t^Y) = E(t^X)E(t^Y) = \eta_X(t)\eta_Y(t).$$

This extends easily to the sum of any finite number of independent variables:

---

The p.g.f. of the sum of independent random variables is the product of their p.g.f.'s:

$$\eta_{\Sigma X_i}(t) = \prod_i \eta_{X_i}(t). \tag{4}$$

Example **3.6b** | **Tetrahedral Dice**

In Example 3.2c we defined a tetrahedral die with sides numbered 1, 2, 3, 4. Assuming these as equally likely outcomes of a single toss, the p.g.f. of $X$, the number showing, is

$$\eta_X(t) = \frac{1}{4}t + \frac{1}{4}t^2 + \frac{1}{4}t^3 + \frac{1}{4}t^4. \qquad (5)$$

Now suppose we throw two such dice. The total number of points thrown is the sum of the two independent variables $X_1$ and $X_2$, each with the p.g.f. (5). The p.g.f. of the sum $Y = X_1 + X_2$ is

$$\eta_Y(t) = \eta_{X_1}(t)\eta_{X_2}(t) = [\eta_X(t)]^2 = \frac{1}{4^2}(t + t^2 + t^3 + t^4)^2$$

$$= \frac{1}{16}(t^2 + 2t^3 + 3t^4 + 4t^5 + 3t^6 + 2t^7 + t^8).$$

We recognize this as the p.g.f. of a discrete random variable with integer values $k$, whose probabilities are the coefficients of $t^k$:

| $k$ | 2 | 3 | 4 | 5 | 6 | 7 | 8 |
|---|---|---|---|---|---|---|---|
| $f_Y(k)$ | 1/16 | 2/16 | 3/16 | 4/16 | 3/16 | 2/16 | 1/16 |

Figure 3-3 shows this triangular distribution. A similar distribution results upon adding any two identically distributed, uniform random variables. (Cf. Example 3.1e.) ∎

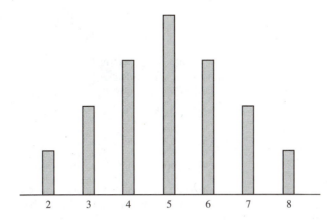

**Figure 3-3** Total points—tetrahedral dice

Example **3.6c** | **Razzle Dazzle**

Example 2.5d described a carnival game in which the player throws eight dice at each move and advances according to the total number of points thrown. A chart shows how far to advance for each possible sum (a number from 8 to 48). The game operator announces that these 41 possibilities are equally likely! We'll calculate the correct probabilities, assuming independence of the dice, using the p.g.f. From Example 3.6a, it is

$$\eta(t) = \frac{t - t^7}{6(1 - t)}.$$

The p.g.f. for the total number of points on the eight dice—the sum of the numbers of points on the eight individual dice— is the 8th power of $\eta(t)$:

$$\eta_{\Sigma X}(t) = \left\{ \frac{t}{6} \cdot \frac{1 - t^6}{1 - t} \right\}^8 = \left( \frac{t}{6} \right)^8 \sum_0^8 \binom{8}{i} (-t^6)^i \cdot \sum_0^\infty \binom{-8}{j} (-t)^j. \qquad (6)$$

Here, the first sum on the right is the binomial expansion of the 8th power of $(1 - t^6)$. The second sum on the right is the series expansion of $(1 - t)^{-8}$. The coefficients in the latter expansion are obtained formally in the same way as are ordinary binomial coefficients—a product of $j$ factors starting with $-8$ and decreasing in steps of 1, divided by $j!$:

$$\binom{-8}{j} = \frac{(-8)(-9) \cdots (-8 - j + 1)}{j!}. \qquad (7)$$

The product of sums in (6) can be written as a double sum:

$$\eta_{\Sigma X}(t) = \left( \frac{1}{6} \right)^8 \sum_{j=0}^\infty \sum_{i=0}^8 \binom{8}{i} \binom{-8}{j} (-1)^{i+j} t^{8+6i+j}. \qquad (8)$$

Each term of the double sum includes a power of $t$, and collecting the terms involving $t^k$ we can read the probability $p_k$ as the coefficient of $t^k$.

For example, we'll calculate the probability that the sum $Y$ is 13. This is the coefficient of $t^{13}$ in (8). The only term with this power of $t$ is the one in which $i = 0$ and $j = 5$:

$$P(Y = 13) = (-1)^5 \binom{8}{0} \binom{-8}{5} \left( \frac{1}{6} \right)^8 = -\frac{(-8)(-9)(-10)(-11)(-12)}{5! \, 6^8} = .00047.$$

This is a far cry from 1/41, the figure announced by the game operator.

Some sums require more work. For instance, the probability of a total of 20 is the sum of terms with $i = 0$, $j = 12$; $i = 1$, $j = 6$; and $i = 2$, $j = 0$. (These yield the probability .02184.) The complete distribution is shown graphically in Figure 3-4.[1]　∎

[1]See D. A. Berry and R. R. Regal, "Probabilities of winning a certain carnival game," *American Statistician 32* (1978), 126–29.

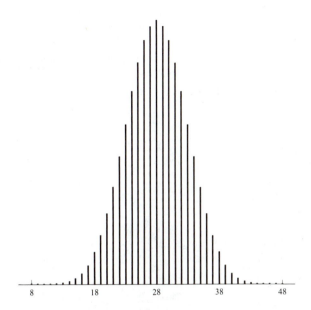

**Figure 3-4** Distribution for Razzle-Dazzle

## Problems

∗ **3-31.**   Let $X$ be the indicator variable for an event $A$ (that is, equal to 1 if $A$ occurs and to 0 otherwise).  Obtain the p.g.f. for $X$ in terms of $p = P(A)$.

∗ **3-32.**   Let $(X_1, X_2, X_3)$ be the results of three independent trials of the experiment in the preceding problem.   Their sum $Y$ is then the number of times that $A$ occurs in the three trials. Find the probability generating function of $Y$ and use it to find $P(Y = 2)$.

**3-33.**   Let $S$ denote the sum of the points thrown with two ordinary six-sided dice.  Use the result of Example 3.6a to find the p.g.f. of $S$, and use it to find the probability that $S = 5$.

∗ **3-34.**   In Example 1.7c we gave probabilities for $X$, the number of tosses of a coin required to obtain heads for the first time, as $(.5)^k$ for $k = 1, 2, 3, ....$

**(a)**   Find the p.g.f. of $X$ as a power series and (recognizing it as a geometric series) find the sum of that series in closed form.

**(b)**   The number of tosses to obtain two heads can be thought of as the sum of $X_1$, the number required to get the first, and $X_2$, the number of additional tosses required to get the second heads.  Both $X_1$ and $X_2$ have the p.g.f. found in (a), and they are independent.  Find the p.g.f. of the sum and use it to find the probability that it takes three tosses to get the two heads.

## Review Problems

**3-R1.**   Given the p.f. $f(0) = .6$, $f(1) = .1$, $f(2) = .3$, find
    **(a)**   $E(2^X)$.               **(b)**   the p.g.f., $\eta_X(t) = E(t^X)$.
    **(c)**   $E[X(X-1)]$.    **(d)**   var$(2X)$.

**3-R2.**   Let $X$ denote the number of points showing when a (fair) die is tossed. Find $E[X \mid A]$, where $A$ is the event that $X$ is an even number.

**3-R3.**   Let $(X, Y)$ have the joint p.f. given by the following table:

|        |     | $Y$  |      |
|--------|-----|------|------|
|        |     | 0    | 1    |
| $X$    | 0   | 1/4  | 1/2  |
|        | 1   | 1/12 | 1/6  |

Find:
    **(a)**   var $X$ and var $Y$.    **(c)**   cov$(X, Y)$.
    **(b)**   $E(Y \mid X = 1)$.       **(d)**   cov$(X - 2Y, 3X + Y)$.

**3-R4.**   Given var $X = 9$, var $Y = 4$, and cov$(X, Y) = -5$, find
    **(a)**   cov$(X + Y, X - Y)$.    **(c)**   cov$(X, X + 2Y)$.
    **(b)**   var$(X + 2Y)$.         **(d)**   $\rho_{X, X+2Y}$.

**3-R5.**   Given that $X$ has the p.g.f. $\eta(t) = .3 + .2t + .5t^2$,
    **(a)**   give the p.f. of $X$.    **(b)**   find var $X$.

**3-R6.**   The four sides of a tetrahedral die are numbered 1, 2, 3, 4. Let $X$ denote the number that turns up when the die is rolled. Find its variance.

**3-R7.**   The 4-sided die of the preceding problem is thrown twice; let $X$ and $Y$ denote the outcomes of the first and second roll. Find cov$(X + Y, X - Y)$.

**3-R8.**   Given that the random pair $(X, Y)$ has the joint distribution shown in the accompanying table of probabilities,
    **(a)**   Find $E(XY)$.
    **(b)**   Find the correlation coefficient.

|     | 0   | 1   | 2   |
|-----|-----|-----|-----|
| 0   | .3  | .2  | .1  |
| 1   | .2  | .1  | 0   |
| 2   | .1  | 0   | 0   |

    **(c)**   Find $E(X \mid Y = 0)$.
    **(d)**   Find $E(X + Y)$.
    **(e)**   Find var$(X + Y)$.

**3-R9.** A random variable $X$ has the p.g.f. $\eta(t) = \frac{1}{81}(16 + 32t + 24t^2 + 8t^3 + t^4)$.

   **(a)** Give the p.f. in table form.

   **(b)** Calculate $\eta'(1)$ and show that the result is the same as is obtained by calculating $EX$ from the table in (a).

**3-R10.** Let $X$ denote the smallest, and $Y$ the largest of three numbers selected at random from the integers 0 through 4, without replacement.

   **(a)** Obtain the table of joint probabilities.

   **(b)** Find $\mathrm{cov}(X, Y)$.

   **(c)** Find $\mathrm{var}(X \mid Y = 4)$.

**3-R11.** For any event $E$, let $P(E \mid X = x) = g(x)$. As an application of the law of total probability, show that $E[g(X)] = P(E)$.

**3-R12.** Given that $X_1$ and $X_2$ each has the p.g.f. $\eta(t) = .2 + .4t + .3t^2 + .1t^3$, suppose they are independent.

   **(a)** Obtain the table of joint probabilities for $(X_1, X_2)$ and from it the p.f. for the sum, $Y = X_1 + X_2$.

   **(b)** Find the p.g.f. of $Y$ from its p.f. in (a).

   **(c)** Find the p.g.f. of $Y$ by squaring $\eta(t)$, verifying your answer to (b).

## Chapter Perspective

Various kinds of averages are used to describe a distribution. The most commonly used averages are the mean, a value of the random variable that is in the middle of the distribution, and the standard deviation, a "typical" deviation from the mean. In this chapter we have defined averages for discrete random variables. In Chapter 5 we extend the definitions of this chapter to the case of continuous probability distributions. In Chapter 7 we'll apply the notion of averaging to sample distributions.

The notion of population average is important in applications. For instance, in Chapters 10, 12, and 14, we assess the effects of treatments on the basis of the average changes they produce in a population. In Chapter 15 we discuss the relationship between two variables in terms of conditional averages. Averaging also plays an important role in statistical theory. In estimation (Chapter 9), an estimate of an unknown population characteristic will be judged in terms of how well it does on average. In Chapter 12 we use averages in evaluating procedures for testing hypotheses and making decisions.

We have introduced the probability generating function as a transform of the p.f.—another application of averaging. This and related transforms are useful tools for studying the distributions that are important in statistical inference. As with any tool, you are apt to be uncomfortable with generating functions at first, but with practice you will gain in understanding and skill.

The p.g.f. has been defined here for random variables with nonnegative integer values. However, we can define the p.g.f. for *any* random variable $X$ by the same formula: $\eta_X(t) = E(t^X)$, provided that this expectation exists in the neighborhood of $t = 1$. We'll return to this idea in §5.10.

The next chapter applies the ideas and tools of the first three chapters to some particular families of discrete models that are commonly assumed in applications of statistics.

# Bernoulli and Related Variables

Categorical variables with just two categories occur frequently in statistical practice: A person selected at random smokes or does not; a product is good or defective; a mission succeeds or fails; blood pressure is lowered or not; an ICBM silo survives an enemy's preemptive nuclear strike or not; a TV viewer is or is not watching a particular program. In each of these examples, an outcome is one of two categories. We code them 0 and 1, and will refer to them generically as *failure* and *success*, respectively. In the case of the TV viewer, for instance, *success* (or 1) means that the viewer *is* watching the program in question.

A **Bernoulli population**[1] is one whose members are 0's and 1's. An observation from such a population is a *Bernoulli random variable.* In this chapter we investigate various distributions associated with Bernoulli populations. We'll be interested in the number of successes in a sample of given size from a Bernoulli population (under sampling with and without replacement) and in the sample size required to obtain a specified number of successes. Extending the process of sampling 0–1 populations in different ways will lead us to the Poisson and multinomial distributions. The generating function introduced in §3.6 will be a useful tool.

## 4.1    Sampling Bernoulli Populations

The distribution of a Bernoulli variable $X$ is completely defined by the probability of the event $X = 1$. In a finite Bernoulli population, this probability is the proportion of 1's in the population:

$$p = P(X = 1) = P(\text{success}).$$

---

[1]Named after Jakob Bernoulli (1654–1705).

The probability that $X = 0$ is then

$$q = P(X = 0) = P(\text{failure}) = 1 - p.$$

The p.f. of a Bernoulli variable $X$ can be given in a simple table:

| $x$ | $f(x)$ |
|-----|--------|
| 1   | $p$    |
| 0   | $q$    |

Because the p.f. involves the parameter $p$, we exhibit this dependence with the augmented notation $f(x \mid p)$. (We used the vertical bar previously in a notation for conditional probability. Here it indicates dependence of a distribution on the parameter that follows the bar. Later [for example, in §8.11], when we treat parameters as random variables, the two uses coincide.) It is often useful to express this p.f. $f(x \mid p)$ in terms of an equivalent algebraic formula:

---

Probability function for a Bernoulli variable with parameter $p$:

$$f(x \mid p) = p^x(1 - p)^{1-x}, \quad x = 0 \text{ or } 1. \tag{1}$$

When $X$ has this distribution, we write $X \sim \text{Ber}(p)$.

---

Moments of a Bernoulli variable are particularly easy to calculate. Indeed, we can calculate them all at once, because when $X = 1$ or $0$, $X^k = X$ for any positive integer $k$. Thus,

$$E(X^k) = EX = 1 \cdot p + 0 \cdot q = p.$$

So the mean is $EX = p$, and the mean square is $E(X^2) = p$, and therefore

$$\text{var } X = E(X^2) - (EX)^2 = p - p^2 = p(1 - p) = pq.$$

---

**Mean and variance** of a Bernoulli random variable:

$$\mu = p, \quad \sigma^2 = pq. \tag{2}$$

---

**Example 4.1a**

Suppose a population is 20% black and 80% white. For an individual selected at random, define

$$X = \begin{cases} 1 & \text{if the person selected is black} \\ 0 & \text{if that person is white.} \end{cases}$$

Then $X \sim \text{Ber}(.20)$, and from (2) we find the mean and standard deviation:

$$\mu = p = .2, \quad \sigma = \sqrt{pq} = \sqrt{.2 \times .8} = .4. \qquad \blacksquare$$

The population proportion $p$ is usually not known. To learn about it, we sample the population. The individual random variables in such a sample are *trials*. We may sample with replacement or without replacement (see §1.5).

Sampling with replacement and with thorough mixing between selections yields a sequence of *independent* observations. A sequence of independent trials of a Bernoulli experiment is a **Bernoulli process.** The joint probability function for any finite set $X_1, ..., X_n$ of such independent Bernoulli trials is the product of the p.f.'s of the individual trials [from (1) above]:

$$f(x_1, ..., x_n \mid p) = \prod_{i=1}^{n} f(x_i \mid p)$$

$$= \prod_{i=1}^{n} p^{x_i} q^{1-x_i} = p^{\Sigma x_i} q^{n - \Sigma x_i}, \; x_i = 0, 1. \qquad (3)$$

Because each $x_i$ is a 0 or a 1, the sum $\sum x_i$ is just the number of 1's in the sample sequence. For instance, if the sample sequence is (0, 1, 0, 0, 1), the sum is $0 + 1 + 0 + 0 + 1 = 2$. Similarly, the complementary count $n - \sum x_i$ is the number of 0's.

In thinking of the successive observations on a Bernoulli variable as being carried out in time sequence, we note three important characteristics of the Bernoulli process. We'll want to refer back to these when we go over to a related process that unfolds in *continuous* time.

---

**Characteristics of a Bernoulli process:**

(i) Nonoverlapping sequences of trials are independent.

(ii) The distribution for any set of consecutive trials is the same as for any other set of consecutive trials of the same length.

(iii) The process has no memory:  At any point in the sequence of trials, the distribution of future trials is independent of the results of trials up to that point.

We consider next the case of sampling *without replacement* from a finite Bernoulli population. This method of sampling is usually more practical and gives more information about $p$ than sampling with replacement. The sample is again a sequence $X_1, ..., X_n$, where each $X_i$ is 0 or 1. But now the $X$'s are not independent. The *conditional* distribution of $X_2$ given $X_1$, for instance, depends on whether $X_1$ is 0 or 1. However, the $X_i$'s are *exchangeable*, as we explained in §2.6. So all $X_i$'s have the same distribution: $X_i \sim \text{Ber}(p)$ for each $i$.

**Example 4.1b**

Consider a sample of size 5 drawn at random without replacement from a population of size $N = 20$, in which there are eight 1's and twelve 0's. The Bernoulli parameter is $p = 8/20$. To illustrate a general scheme, we'll find the probability of a particular sequence of five observations: 0, 0, 1, 0, 1.

According to the extended multiplication rule for finding the probability of a sequence of dependent events, we have

$$P(0,0,1,0,1) = \frac{12}{20} \cdot \frac{11}{19} \cdot \frac{8}{18} \cdot \frac{10}{17} \cdot \frac{7}{16} = \frac{\binom{8}{2}\binom{12}{3}}{\binom{5}{2}\binom{20}{5}}.$$

The 2 in the numerator of the last expression is the sum of the observations, and we can write the joint p.f. of the five observations, more generally, as

$$f(x_1, ..., x_5) = \frac{\binom{8}{\Sigma x_i}\binom{12}{5-\Sigma x_i}}{\binom{5}{\Sigma x_i}\binom{20}{5}}.$$

This is clearly a symmetric function of the $x_i$'s—permuting them does not change the value of the p.f. This means that the $X_i$'s are *exchangeable*. And this in turn means that the joint p.f. for $(X_3, X_5)$, for instance, is the same as the joint p.f. for $(X_1, X_2)$. So,

$$f_{X_3,X_5}(1,0) = f_{X_1,X_2}(1,0) = P(X_1 = 1) \cdot P(X_2 = 0 \mid X_1 = 1) = \frac{8}{20} \cdot \frac{12}{19}.$$

In this calculation, the 3 and 5 could be replaced by any $i$ and $j$. Similar calculations produce the p.f. of $(X_i, X_j)$. In table form it is as follows:

|        |     | $X_j$  |        |        |
|--------|-----|--------|--------|--------|
|        |     | 1      | 0      |        |
| $X_i$  | 1   | 14/95  | 24/95  | 8/20   |
|        | 0   | 24/95  | 33/95  | 12/20  |
|        |     | 8/20   | 12/20  | 1      |

Each (univariate) marginal distribution is just the Bernoulli distribution of the population: $X_i \sim \text{Ber}(8/20)$.    ∎

More generally, consider a population of size $N < \infty$, which consists of $M$ 1's and $N - M$ 0's, and a sequence of $n$ observations drawn at random from the population without replacement. Following the line of reasoning in the above example, we obtain a general formula for the joint p.f. of the observations.

---

Joint probability function of a sequence of $n$ Bernoulli trials, obtained by random sampling without replacement from a finite population of size $N$:

$$f(x_1, \ldots, x_n) = \frac{\binom{M}{\Sigma x_i}\binom{N-M}{n-\Sigma x_i}}{\binom{N}{n}\binom{n}{\Sigma x_i}}, \quad x_i = 0,\, 1. \qquad (4)$$

---

An important function of the observations in a Bernoulli process is $\sum X_i$, the number of successes in $n$ trials. The next two sections deal with the distribution of $\sum X_i$, first for sampling with replacement, and then for sampling without replacement.

---

## 4.2   The Binomial Distribution

We focus now on the number of successes in a specified number of *independent* Bernoulli trials. This number is a random variable whose distribution is called **binomial**. In the next example we illustrate the calculation of binomial probabilities.

**Example 4.2a** | **Broiled Burgers Better Bets?**

A TV commercial for Burger King (in 1982) included this statement: "Our survey shows that three-fourths of the people prefer their hamburgers broiled." The speaker (dressed as a Burger King employee) then went on to say, "I'll make you a bet. Call four of your friends; at least three of them will say they prefer them broiled." What are appropriate odds for this bet?

A sample of three of "your friends" is not apt to be a random selection from "the people," but for purposes of argument, suppose it is. Suppose further that the number of "the people" is infinite. The following sample sequences, with "broiled" coded 1, would satisfy the condition "at least three say broiled":

| Friend No. | | | | Probability |
|---|---|---|---|---|
| 1 | 2 | 3 | 4 | |
| 1 | 1 | 1 | 1 | $p^4$ |
| 1 | 1 | 1 | 0 | $p^3(1-p)$ |
| 1 | 1 | 0 | 1 | $p^3(1-p)$ |
| 1 | 0 | 1 | 1 | $p^3(1-p)$ |
| 0 | 1 | 1 | 1 | $p^3(1-p)$ |

The probability of at least three 1's is the sum of the probabilities of these five sequences:

$$P(3 \text{ or } 4 \text{ 1's}) = p^4 + 4p^3(1-p).$$

When $p = 3/4$, this probability is .738, so the odds are about 14 to 5. The Burger King worker had a good chance of winning her bet—if the assumptions are correct.

Calculations similar to the above give probabilities for the other possible numbers correct: For exactly two 1's, there are six sequences (the number of ways of selecting two sequence positions from four, for the 1's), each with probability $p^2(1-p)^2$, and so on. With the notation $q = 1 - p = P(0)$, the p.f. of the number of 1's in a sequence of four trials of Ber($p$) is this:

| $k$ | 0 | 1 | 2 | 3 | 4 |
|---|---|---|---|---|---|
| $f(k \mid p)$ | $q^4$ | $4pq^3$ | $6p^2q^2$ | $4p^3q$ | $p^4$ |

■

More generally, let $Y$ denote the number of successes in $n$ independent trials of a Ber($p$). We can write $Y = \sum X_i$, where $X_i$ is the indicator variable for success at the $i$th trial, scoring 1 for success and 0 for failure. In view of (3) of §4.1, the probability of any *particular* sequence with $Y = k$ is $p^k q^{n-k}$. And since there are $\binom{n}{k}$ such sequences, the probability of $Y = k$ is this number of sequences multiplied by the probability of a particular sequence.

---

Probability function for $Y$, the number of successes in $n$ independent trials of Ber($p$):

$$f_Y(k \mid p) = \binom{n}{k} p^k q^{n-k}, \quad k = 0, 1, ..., n. \tag{1}$$

The distribution of $Y$ is **binomial**, and we write $Y \sim \text{Bin}(n, p)$.

---

Binomial probabilities are given in Table Ia of Appendix 1, for $n = 5, 6, ..., 12$, and for selected values of $p$ from .01 to .50. Table Ib gives *cumulative* (from above) probabilities—probabilities of events of the form $Y \geq k$. We'll handle the case $n > 12$ using approximations to be developed in §4.5.

**Example 4.2b** | **Monitoring Quality**

A machine produces a stream of parts, each of which is either good or defective. Quality is monitored by taking a sample of parts from the production line. How many defective parts will there be in a sample of 12, if there is one chance in ten that a particular part is defective?

We'll assume that producing 12 parts is a sequence of 12 independent Bernoulli trials with $p = .1$, where *success* means that the part is defective. Then $Y$, the number of defective parts among the 12, is Bin(12, .1). Its p.f. as given by (1) is

$$f(k \mid .1) = \binom{12}{k}(.1)^k(.9)^{12-k}, \quad k = 0, 1, ..., 12.$$

These probabilities are to be found in Table Ia of Appendix 1. Figure 4-1 shows the p.f. as a bar graph. The distribution is skewed to the right—small values of $Y$ are the more likely ones, and probabilities tail off to the right. ■

Although Table Ia gives binomial probabilities only for $p \leq .5$, we can also use it for $p > .5$, simply by interchanging the roles of 1 and 0. That is, if there are $Y$ successes, there are $n - Y$ failures. This implies that when the number of successes is $Y \sim \text{Bin}(n, p)$, the number of failures is $n - Y \sim \text{Bin}(n, 1 - p)$. Then, for instance, if we want the probability of exactly four successes, we can look in Table Ia under the heading "$p$" $= 1 - p$, opposite $n - 4$ in the left-hand column.

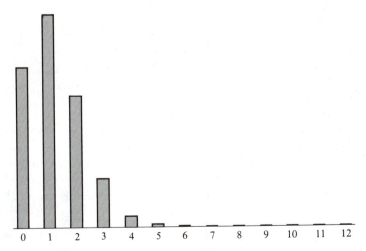

**Figure 4-1** Binomial probabilities, $n = 12$, $p = .1$

**Example 4.2c**  Suppose we want the probability of at least seven successes among nine independent trials when $p = .8$. It is

$$P(Y \geq 7 \mid p = .8) = \binom{9}{9}(.8)^7(.2)^2 + \binom{9}{8}(.8)^8(.2)^1 + \binom{9}{9}(.8)^9.$$

Now, to say that there are at least seven successes is the same as saying that there are at most two failures: $[Y \geq 7]$ is the same event as $[9 - Y \leq 2]$. But $9 - Y$, which counts the number of failures, is distributed as $\text{Bin}(9, .2)$. So with this new value $.2$ as the "$p$," we have

$$P(9 - Y \leq 2 \mid \text{"}p\text{"} = .2) = \binom{9}{2}(.2)^2(.8)^7 + \binom{9}{1}(.2)^1(.8)^8 + \binom{9}{0}(.8)^9 = .7382.$$

Observe that this sum is exactly the same sum as the first one given, since $\binom{9}{k}$ is the same as $\binom{9}{9-k}$. We can use Table Ib to find it, if we go to the block of entries for $n = 9$ and look under $p = .2$, opposite $k = 3$. There we find that $P(9 - Y \geq 3) = .2618$, so $P(9 - Y \leq 2) = 1 - .2618 = .7382$.  ∎

Binomial probabilities get their name from the binomial theorem [(1) of §1.6]. They are terms in a binomial expansion:

$$(p + q)^n = \sum_{k=0}^{n} \binom{n}{k} p^k q^{n-k} = \sum_{k=0}^{n} f(k \mid p). \tag{2}$$

Since $p + q = 1$, this $n$th power of $(p + q)$ is just 1, showing that the binomial probabilities $f(k \mid p)$ add up to 1, as of course they must.

In the particular case where $p = q = 1/2$, the product $p^k q^{n-k}$ is $(1/2)^n$ for any $k$. This tells us that the binomial probabilities are proportional to the combination counts $\binom{n}{k}$:

$$f(k \mid p = .5) = \binom{n}{k}(.5)^n.$$

And because $\binom{n}{k} = \binom{n}{n-k}$, it follows that the binomial distribution with $p = .5$ is *symmetric* about $n/2$. However, when $p \neq .5$, the distribution is skewed—to the left when $p > .5$, and to the right when $p < .5$ (as in Figure 4-1).

We'll be needing to know the first two moments of a binomial distribution. Of course, we could use $\sum y f(y)$ to calculate the mean, and $\sum y^2 f(y)$ to calculate the mean square (and from it, the variance). But it is much simpler, and more insightful, to derive formulas for these moments using the structure of the binomial variable as the sum of independent Bernoulli variables. That is, the number $Y$ of successes is the sum of the 0's and 1's in the sequence of Bernoulli trials on which it's based:

$$Y = X_1 + \cdots + X_n, \tag{3}$$

where $X_i \sim \text{Ber}(p)$ for each $i$. Since averaging is an additive operation and since $E(X_i) = p$ for each $i$ [from (2) of §4.1, page 106], it follows that

$$E(Y) = E(\sum X_i) = \sum E X_i = p + \cdots + p = np.$$

Moreover, since the $X_i$'s are *independent,* and each has variance $pq$, the variance of the sum is $n$ times $pq$ [see (9) of §3.5, page 106]:

$$\text{var } Y = \text{var}\left(\sum X_i\right) = \sum \text{var } X_i = pq + \cdots + pq = npq.$$

---

When $Y \sim \text{Bin}(p)$, the mean and variance of $Y$ are given by

$$EY = np, \quad \text{var } Y = npq. \tag{4}$$

where $q = 1 - p$.

---

The formula $np$ for the mean is intuitively appealing and therefore easy to remember. For example, in 100 tosses of a coin you "expect" to get heads half the time: $np = 100 \times .5 = 50$. In rolling two dice, you'd expect two "sevens" in twelve trials: $np = 12 \times 1/6 = 2$.

**Example 4.2d** | **Monitoring Quality (Continued)**

In Example 4.2b we considered monitoring a production line in terms of the number of defectives in a sample of 12, assuming that defectives occurred at the rate of one in ten: $p = .1$. The average number $Y$ of defectives among the 12 is

$$E(Y) = np = 12 \times .1 = 1.2.$$

The standard deviation is

$$\sigma = \sqrt{npq} = \sqrt{12 \times .1 \times .9} \doteq 1.04.$$

The graph in Figure 4-1 suggests that these are reasonable as mean and s.d. of the distribution. (Routine visual checks help to build your intuition as well as guard against mistakes in arithmetic.) ∎

The binomial distribution has an important reproductive property. Suppose $Y_1 \sim \text{Bin}(n_1, p)$, and $Y_2 \sim \text{Bin}(n_2, p)$, and suppose further that $Y_1$ and $Y_2$ are *independent.* We may think of $Y_1$ as the number of successes in $n_1$ independent trials of $\text{Ber}(p)$, and $Y_2$ as the number of successes in $n_2$ independent trials of the same $\text{Ber}(p)$. So the sum $Y_1 + Y_2$ is just the number of successes in $n_1 + n_2$ independent trials of that Bernoulli experiment—and hence, binomial.

If $Y_1 \sim \text{Bin}(n_1, p)$, $Y_2 \sim \text{Bin}(n_2, p)$, and $Y_1$ and $Y_2$ are *independent*, then $Y_1 + Y_2 \sim \text{Bin}(n_1 + n_2, p)$.

The binomial distribution is applicable as the distribution of the number of successes in a fixed number of trials when sampling is *with* replacement from a finite Bernoulli population. When sampling is done *without* replacement, however, the number of successes is no longer binomial. Its distribution is the topic of the next section.

## Problems

**∗ 4-1.** Suppose each of ten subjects is to be assigned to treatment A or to treatment B according to the toss of a coin:  heads to A, tails to B. Find
- **(a)** the probability that exactly five get each treatment.
- **(b)** the probability that at most three get treatment A.
- **(c)** the mean and s.d. of the number who get treatment A.

**∗ 4-2.** A basketball player's free throws are made 90% of the time. Assume that free throws are independent trials, each with probability $p = .90$ of being made. For a sequence of eight free throws, find
- **(a)** the probability of exactly one miss.
- **(b)** the probability of at most one miss.
- **(c)** the probability of at least one miss.
- **(d)** the expected number of misses.

**4-3.** To test a new type of golf ball, 20 golfers are paired in such a way that both golfers in a pair have the same handicap. One golfer in each of the ten pairs is assigned the new ball, and the other a standard ball, by the toss of a coin. The pairs play a round of golf (including a tiebreaker if necessary). Let $X$ denote the number of pairs in which the player using the new type of ball wins. Suppose, however, that the new ball in fact has precisely the same performance characteristics as the old, so that which ball wins is like the toss of a coin. Find the following:
- **(a)** The distribution of $X$.    **(b)** $P(X = 10)$.
- **(c)** $P(X \le 3)$.                **(d)** $P(X \ge 5)$.

**∗ 4-4.** A multiple-choice quiz consists of 20 questions, each with four choices. A student who guesses has one chance in four of being correct for each question. Let $U$ denote the student's score when each question is worth five points.
- **(a)** Find the expected number of correct answers and the expected score.
- **(b)** Find the s.d. of $U$.
- **(c)** Find the probability that the student scores exactly 50.
- **(d)** Find the probability that the student scores at most 20.

**4-5.** A random sample of 100 is drawn from the population of Washington, D.C. Assume that 55% are black. Let $X$ denote the number of blacks included in the sample and find the mean and standard deviation of $X$.

* **4-6.** Suppose that male and female births occur with the same frequency, in the population at large. Suppose further that the sexes of successive children in a family are independent. Among 160 families with four children each, how many would be expected to have 0, 1, 2, 3, and 4 boys?

**4-7.** Consider the covariance of $X_1$ and $X_2$, the first and second observations in a selection without replacement from a finite population.

   **(a)** For the situation of Example 4.1b, calculate this covariance.

   **(b)** Obtain a general formula for this covariance in terms of the population size $N$ and $M$, the number of 1's in the population.

* **4-8.** An archer hits the bull's-eye one time in ten on average. She will give even odds that she will hit the bull's-eye at least once in $n$ tries. How large must $n$ be for this bet to be favorable for her? (Giving "even odds" means that she will put up \$1 against each \$1 bet.)

**4-9.** Find the value of the following (without calculating each term):

$$\sum_{k=0}^{20} k \binom{20}{k} (.05)^k (.95)^{20-k}.$$

---

## 4.3     **Hypergeometric Distributions**

Consider sampling at random and *without* replacement from a finite Bernoulli population. Successive observations are not independent, because the pool of individuals available at any selection depends on preceding observations. Let $N$ denote the population size (before any selections are made), and $M$ the number of successes in the population. The population proportion of successes is $p = M/N$. Again, as in the preceding section, we focus on the number of successes among $n$ selections. And as before, this number is $Y = \sum X_i$, the sum of the 0's and 1's in the sample.

For a sample of given size $n$, the p.f. of the sample sequence, as given by (4) of §4.1 (page 119) is

$$f(x_1, ..., x_n) = \frac{\binom{M}{\Sigma x_i}\binom{N-M}{n-\Sigma x_i}}{\binom{N}{n}\binom{n}{\Sigma x_i}}, \quad x_i = 0, 1. \tag{1}$$

In order to find $f_Y(k)$, the p.f. of $Y$, we sum the probabilities for all of the various sequences with $k$ successes. The probability (1) is the same for all sequences $(x_1, ..., x_n)$ that have the same number of 1's—that is, the same sum, $\sum x_i$. And since each of those sequences has the probability (1) with $\sum x_i = k$, we just multiply (1) by $\binom{n}{k}$, the number of sequences of $n$ 0's and 1's in which there are exactly $k$ 1's:

$$f_Y(k \mid M) = P(Y = k) = \frac{\binom{M}{k}\binom{N-M}{n-k}}{\binom{N}{n}}. \tag{2}$$

The distribution defined by this p.f. is called **hypergeometric.**

The formula for the p.f. $f_Y$ is valid provided that $0 \le k \le M$ and $0 \le n - k \le N - M$. If these inequalities fail in any way, it means that there is something like an $\binom{i}{j}$ in which $i < j$. If we *define* such a binomial coefficient to be 0, the formula is correct for integers $k$ on the range from 0 to $n$.

| Example **4.3a** | ## Sampling Inspection |

As a way of checking on the quality of a shipment without a complete inspection of every item in the shipment, it is common practice to draw a small sample at random and check each item in the sample. Of course, there is no point to replacing an item once it is checked, so we assume the sampling is random but without replacement.

To make the results easy to display, we'll suppose $N = 8$ and $n = 2$. The formula for the p.f. of the number of defectives in the sample is

$$f_Y(k \mid M) = \frac{\binom{M}{k}\binom{8-M}{2-k}}{\binom{8}{2}}.$$

The following table shows, in a column for each value of $M$, the p.f. values multiplied by $\binom{8}{2} = 28$:

|   |   |   |   | $M$: |   |   |   |   |   |
|---|----|----|----|----|----|----|----|----|----|
| $Y$ | 0 | 1 | 2 | 3 | 4 | 5 | 6 | 7 | 8 |
| 0 | 28 | 21 | 15 | 10 | 6 | 3 | 1 | 0 | 0 |
| 1 | 0 | 7 | 12 | 15 | 16 | 15 | 12 | 7 | 0 |
| 2 | 0 | 0 | 1 | 3 | 6 | 10 | 15 | 21 | 28 |

[There is a symmetry in this table resulting from the fact that $8 - M$ is the number of good items in the shipment, and $2 - Y$ is the number of good items in the sample, so $f(k \mid M) = f(2 - k \mid 8 - M)$.] ∎

---

### Hypergeometric probability function:

When $n$ items are selected at random without replacement from a population of size $N$, $M$ of which are of type 1 and $N - M$ of type 0, the p.f. of the number of 1's in the sample is

$$f_Y(k \mid M) = \frac{\binom{M}{k}\binom{N-M}{n-k}}{\binom{N}{n}}, \tag{3}$$

for $k = 0, 1, ..., n$, where $\binom{a}{b} = 0$ when $b > a$ or $b < 0$.

Exploiting the structure of a hypergeometric variable $Y$ as the sum of Bernoulli variables $X_i$, we can get formulas for the mean and variance. These follow from (7) and (8) in §3.5. The former says that the mean of the sum of identically distributed variables is $n$ times their common mean—which in the present context is the population proportion, $p = M/N$:

$$EY = E\left(\sum_{i=1}^{n} X_i\right) = n \cdot \frac{M}{N}. \qquad (4)$$

And (8) in §3.5 says that the variance of the sum is $n$ times the variance of the common distribution of the $X$'s multiplied by a "finite population correction factor:"

$$\text{var } Y = \text{var}\left(\sum_{i=1}^{n} X_i\right) = n \cdot \frac{M}{N}\left(1 - \frac{M}{N}\right) \cdot \frac{N-n}{N-1}. \qquad (5)$$

**Example 4.3b**

## Sampling Inspection (Continued)

Returning to the shipment of size $N = 8$, and a sample of size $n = 2$, we can calculate the mean for a particular $M$ using the probability table in Example 4.3a and the defining formula for a mean value. For instance, when $M = 3$,

$$E(Y \mid M = 3) = \sum k\, f(k \mid 3) = 0 \times \frac{10}{28} + 1 \times \frac{15}{28} + 2 \times \frac{3}{28} = \frac{21}{28} = 2 \cdot \frac{3}{8} = np.$$

And for the variance, we can first find

$$E(Y^2 \mid M = 3) = \sum k^2 f(k \mid 3) = 0 \times \frac{10}{28} + 1^2 \times \frac{15}{28} + 2^2 \times \frac{3}{28} = \frac{27}{28}.$$

The variance is this average square minus the square of the average:

$$\text{var } Y = \frac{27}{28} - \left(\frac{3}{4}\right)^2 = \frac{45}{112} = 2 \cdot \frac{3}{8} \cdot \frac{5}{8} \cdot \frac{8-2}{8-1} = npq \cdot \frac{N-n}{N-1}.$$

So we see that (4) and (5) are indeed correct in this instance.

The variance we've just found is 6/7 of the binomial variance $npq$, which would be correct under sampling with replacement. Replacing items that have been selected increases the variability. ∎

---

Mean and variance of a hypergeometric variable $Y$:

$$EY = np, \quad \text{var } Y = npq \cdot \frac{N-n}{N-1}, \qquad (6)$$

where $Y$ is the number of successes in a sample of size $n$ drawn at random without replacement from a population of size $N$, and $p = M/N$ is the population proportion of successes.

We haven't given a table of hypergeometric probabilities in the Appendix because such a table takes more room than it's worth, depending as it does on two parameters ($M$ and $N$) in addition to $n$. When the numbers are small, the calculations are not hard. And when $N$ or $n$ is large, there are some approximations that help us.

Suppose $N$ is very large in comparison with the sample size $n$, so that the population correction factor in (6) is nearly equal to 1. Then, not only is the hypergeometric variance nearly equal to the binomial variance, the hypergeometric probabilities themselves are nearly equal to the corresponding binomial probabilities. This is not surprising, since when the number of items removed from the population is small compared with the size of the population, the population proportion of successes is changed only slightly.

The larger the $N$ (for a given $n$), the better the binomial approximates the hypergeometric. Whether you should use this approximation depends on the accuracy you want. A rule of thumb is that $N$ should be at least ten times as large as $n$. When $N = 10n$, the finite population correction factor is about .9, and the hypergeometric s.d. is about 95% of the binomial s.d. The following example compares binomial and hypergeometric probabilities in a particular case.

**Example 4.3c**

Consider samples of size 2 from a Bernoulli population of size $N$ in which the proportion of 1's is $p = 1/4$. The following table gives hypergeometric probabilities for selected values of $N$, along with the limiting binomial probabilities:

|  | | | | $N$ | | | |
|---|---|---|---|---|---|---|---|
| $k$ | 4 | 8 | 16 | 32 | 64 | 128 | $\infty$ |
| 0 | .500 | .536 | .550 | .556 | .560 | .5610 | .5625 |
| 1 | .500 | .429 | .400 | .387 | .381 | .3780 | .3750 |
| 2 | 0 | .036 | .050 | .056 | .060 | .0610 | .0625 |

When $N$ is 16, eight times the sample size, the approximation is not terribly bad (to two decimal places); but, of course, it gets better as $N$ increases. ∎

The mathematical basis of the binomial approximation is that the *limit* of a hypergeometric probability, as $M$ and $N$ become infinite in a fixed ratio, is a binomial probability. We can see this as follows:

$$\frac{\binom{M}{k}\binom{N-M}{n-k}}{\binom{N}{n}} = \frac{(M)_k}{k!} \cdot \frac{(N-M)_{n-k}}{(n-k)!} \cdot \frac{n!}{(N)_n} = \frac{n!}{k!(n-k)!} \cdot \frac{(M)_k}{(N)_k} \cdot \frac{(N-M)_{n-k}}{(N-k)_{n-k}}$$

$$= \binom{n}{k} \cdot \frac{M}{N} \cdot \frac{M-1}{N-1} \cdots \frac{M-k+1}{N-k+1}$$

$$\cdot \frac{N-M}{N-k} \cdot \frac{N-M-1}{N-k-1} \cdots \frac{N-M-n+k+1}{N-n+1}$$

As $N$ and $M$ become infinite, the $k$ factors starting with $M/N$ each tend to $p$. The last $n - k$ factors each tend to $1 - p$. For instance,

$$\frac{N - M - 2}{N - k - 2} = \frac{1 - p - 2/N}{1 - k/N - 2/N} \rightarrow 1 - p.$$

So, as claimed, the limiting value is $\binom{n}{k} p^k (1 - p)^{n-k}$, the binomial p.f.

In the usual way of applying a limit theorem, we'll use the limiting binomial probability as an approximation to a hypergeometric probability when $N \gg n$. But if $n$ itself is large, even that binomial probability may be awkward to calculate, as in the next example.

| | |
|---|---|
| **Example 4.3d** | **Sampling Voters** |

Suppose 40% of the 100,000 voters in a certain city are Republicans. When a poll of 500 voters is conducted by sampling at random without replacement, what is the probability that the proportion of Republicans in the sample is at least 38%?

Let $Y$ denote the number of Republicans in the sample. This number has a hypergeometric distribution, with $N = 100{,}000$, $n = 500$, $M = 40{,}000$, and $p = M/N = .4$. Since $N$ is much larger than $n$, the distribution of $Y$ is approximately binomial: $Y \approx \text{Bin}(500, .4)$. The probability called for is

$$P(Y \geq 190) = \sum_{190}^{500} \frac{\binom{40{,}000}{k}\binom{60{,}000}{500-k}}{\binom{100{,}000}{500}} \doteq \sum_{190}^{500} \binom{500}{k}(.4)^k(.6)^{500-k}.$$

Calculating the binomial approximation at the right is not a trivial task (although it is possible if you have a table of logarithms of factorials). In §4.5 we'll take up an approximation to binomial probabilities with large $n$ that will greatly simplify the task.

Incidentally, the proportion of Republicans in a sample of size 500 would have the same approximate binomial distribution as the one we've just come up with if the city involved had been one with five million voters! In sampling from large populations, it is the sample size $n$ that determines probabilities and accuracy, and *not* the ratio of the sample size to population size. ∎

---

## 4.4        Inverse Sampling

| | |
|---|---|
| **Example 4.4a** | **Monopoly** |

In the classic game of *Monopoly*, players throw a pair of dice at each turn. When a player is in "jail," it takes a "double" (a matched pair) to get out of jail without paying a fine. How many turns will it take to get out of jail if the player is allowed to keep throwing the dice indefinitely instead of paying the fine? (This requires a relaxation of rules.) The number of turns required—call it $Z$—is a random variable. We'll now derive its probability distribution.

The probability of a double on any one throw of the dice is $p = 1/6$. [See Problem 2-31(b).] A double does *not* occur on any one throw with probability $q = 5/6$. With the assumption of independent throws, the probability of $Z = 2$, or failing on the first throw and then succeeding on the second, is $qp = (5/6)(1/6)$.

For the p.f. of $Z$ for a general $k$, the reasoning is similar: The probability of $k - 1$ failures followed by success on the $k$th throw is

$$f_Z(k) = P(Z = k) = P(\text{first double on } k\text{th toss}) = \left(\frac{5}{6}\right)^{k-1}\left(\frac{1}{6}\right), \quad k = 1, 2, \dots.$$

Since each $f_Z(k)$ is obtained from the preceding by multiplying by $1/6$, these probabilities are the successive terms in a geometric series. ∎

In conducting a sequence of independent Bernoulli trials, the number of trials it takes to get a success is thought of as a "waiting time." The waiting time distribution is called **geometric,** because (as in the example) the values of the p.f. are terms in a geometric series:

$$f_Z(k \mid p) = P(\text{first success on } k\text{th toss}) = q^{k-1}p, \quad k = 1, 2, \dots.$$

These probabilities sum to 1, as we see from the standard formula for the sum of an infinite geometric series:

$$\sum_{k=1}^{\infty} f(k \mid p) = p + pq + pq^2 + pq^3 + \cdots$$

$$= p(1 + q + q^2 + q^3 + \cdots) = \frac{p}{1 - q} = 1.$$

The mean of a geometric distribution can almost be guessed. For instance, when there is one chance in six of throwing a double in the Monopoly example above, it would seem only fair that on average it would take six trials to get one. And this is correct:

$$EZ = \sum_{k=1}^{\infty} k\, f(k \mid p) = 1 \cdot p + 2 \cdot pq + 3 \cdot pq^2 + 4 \cdot pq^3 + \cdots$$

$$= p + pq + pq^2 + pq^3 + \cdots$$
$$+ pq + pq^2 + pq^3 + \cdots$$
$$+ pq^2 + pq^3 + \cdots$$
$$+ \cdots$$

From each row of this infinite array we take out a common factor, as a multiplier of the quantity

$$1 + q + q^2 + \cdots = \frac{1}{1 - q} = \frac{1}{p}.$$

Summing the rows yields the mean:

$$EZ = p\left(\frac{1}{p}\right) + pq\left(\frac{1}{p}\right) + pq^2\left(\frac{1}{p}\right) + \cdots = 1 + q + q^2 + \cdots = \frac{1}{p}.$$

Finding the variance is more work, and we'll address this task in §4.9, where we obtain it with the aid of a generating function.

---

The distribution of $Z$, the number of independent Bernoulli trials required to get a success, is **geometric:** $Z \sim \text{Geo}(p)$, with p.f.

$$f(k \mid p) = q^{k-1}p, \ k = 1, 2, 3, \dots. \tag{1}$$

The mean and variance are

$$EZ = \frac{1}{p}, \ \text{var } Z = \frac{q}{p^2}. \tag{2}$$

---

Still in the context of independent Bernoulli trials, suppose we want to know how many trials are required to obtain $r$ successes. This is a random variable; call it $W$. It is the sum of $r$ independent waiting times, because after each success the future looks the same as it did initially: It is the number of trials to get the first success, plus the number of additional trials to get the second success, and so on:

$$W = Z_1 + Z_2 + \cdots + Z_r,$$

where each $Z_i$ is distributed as $\text{Geo}(p)$. The distribution of W is called **negative binomial,** a family of distributions indexed by the two parameters $r$ and $p$: $W \sim \text{Negbin}(r, p)$.

**Example 4.4b** | **Waiting for Girls**

Suppose a couple, planning a family, wants three girls. How many children must they have to fulfill this desire?

We can answer this question in probabilistic terms—provided (as usual) that we make enough assumptions! Suppose there is a constant probability $p$ that a baby will be a girl, and that sexes are independent from one child to another. The proportion of female births in the population at large is about .48. But it could be argued that this proportion is not the probability of a girl for a given couple, that the $p$ may change with time, and that the trials are not independent. We put these arguments aside and proceed as though we were dealing with independent Bernoulli trials with $p = .48$.

The variable $W$, the (minimum) number of trials required, is an integer that must be at least 3: $W = 3, 4, \dots$. Using the assumed independence, we calculate probabilities by multiplication. The easiest is

$$P(W = 3) = P(G, G, G) = p^3.$$

For $W = 4$, we reason that in the first three trials there must be exactly two girls and one boy, in any order, followed by a girl:

$$P(W = 4) = \left\{ \binom{3}{2} p^2 q \right\} \cdot p.$$

For $W = k$, there must be two girls in the first $k - 1$ trials, and then a girl:

$$P(W = k) = \left\{ \binom{k-1}{2} p^2 q^{k-3} \right\} \cdot p, \quad k = 3, 4, \dots.$$

With $p = .48$, these probabilities are as follows:

| $k$ | 3 | 4 | 5 | 6 | 7 | 8 | $\cdots$ |
|---|---|---|---|---|---|---|---|
| $f(k)$ | .111 | .173 | .179 | .156 | .121 | .088 | $\cdots$ |

About 54% of the couples who have the goal of three girls would need to have at least six children.

　　In making these calculations, we have assumed an "ideal" couple. Few would be willing to have more than 20 children to fulfill the goal of three girls, and none could have 100. To be more realistic, we might put a limit of 20 on $W$ and recalculate the probabilities with that condition.　　　　　　　　　　　　　　　　　■

　　The reasoning leading to the p.f. in the general case is just like that of the example, with $p$ in place of .48, and $r$ in place of 3. And the mean and variance of the distribution follow immediately from the formulas for mean and variance of Geo($p$), because $W$ is the sum of $r$ independent geometric variables:

$$EW = \frac{1}{p} + \frac{1}{p} + \cdots + \frac{1}{p} = \frac{r}{p}, \quad \operatorname{var} W = \frac{q}{p^2} + \frac{q}{p^2} + \cdots + \frac{q}{p^2} = \frac{rq}{p^2}.$$

---

The number $W$ of independent Ber($p$) trials required to get the $r$th success has a **negative binomial** distribution, with p.f.

$$f(k \mid p) = \binom{k-1}{r-1} p^r q^{k-r}, \quad k = r, r+1, r+2, \dots \tag{3}$$

The mean and variance are $EW = \frac{r}{p}$, $\operatorname{var} W = \frac{rq}{p^2}$. $\tag{4}$

In some real-life situations, it makes good sense to specify a number of successes and to continue sampling until that number is reached. This is called **inverse sampling.** The next example suggests a setting in which inverse sampling would be preferred to sampling to a preset number of trials.

**Example 4.4c**

## Inverse Sampling to Control Adverse Events

A serum being tested on dogs may induce a fatal reaction. To estimate the probability of a fatal reaction, one could administer the serum to 20 dogs and see how many survive. If 19 out of 20 die, much information has been obtained—but at great cost.

Instead, one could try the serum on one dog at a time and stop the experiment when, say, two dogs have died. With such a sampling scheme, the number of fatalities is under control. (In practice, it may be advisable to stop after (say) the twentieth dog if there are fewer than two fatalities thus far.) ∎

Inverse sampling can be carried out even when the sampling is done without replacement from a finite Bernoulli population. Again let $W$ denote the number of observations required to obtain $r$ successes. Because successive observations are not independent, $W$ does not have a negative binomial distribution. To derive its p.f., we reason as we did in sampling with replacement, using instead the multiplication rule for *dependent* events [(4) in §2.3].

The event $W = k$ can be expressed as $EF$, where

$E = \{r\text{th success occurs on the } k\text{th trial}\}$, and
$F = \{r - 1 \text{ successes in the first } k - 1 \text{ trials}\}$.

The probability of $F$ is hypergeometric:

$$P(F) = \frac{\binom{M}{r-1}\binom{N-M}{k-r}}{\binom{N}{k-1}}.$$

And, given $r - 1$ successes in the first $k - 1$ trials, the conditional probability that the $k$th trial results in success is

$$P(E \mid F) = \frac{M - r + 1}{N - k + 1}.$$

We then apply the multiplication rule:

$$f_W(k) = P(EF) = P(F)P(E \mid F) = \frac{\binom{M}{r-1}\binom{N-M}{k-r}}{\binom{N}{k-1}} \cdot \frac{M - r + 1}{N - k + 1}. \qquad (5)$$

This result can be simplified a bit to yield a formula that we can also obtain with different reasoning:

Imagine that all $N$ population items are arranged in a random sequence. There are $\binom{N}{M}$ ways of assigning the label *success*. The $r$th success occurs in the $k$th position when there are $r - 1$ successes in the preceding $k - 1$ positions and $M - r$ successes in the remaining $N - k$ positions:

Probability function for a hypergeometric waiting time:

$$f_W(k) = \frac{\binom{k-1}{r-1}\binom{N-k}{M-r}}{\binom{N}{M}}, \quad k = r, r+1, ..., M. \tag{6}$$

The equivalence of (5) and (6) is easy to check. Simply express each of the combination symbols in terms of factorials. [The distribution defined by (6) is sometimes called *negative hypergeometric.*]

## Problems

* **4-10.**  A coffee taster tries to identify 20 cups of coffee. The taster knows only that 15 of the cups were brewed from regular coffee and is to pick out the 5 brewed from instant coffee. Let $X$ be the number of cups among the 5 selected by the taster that actually are instant, and suppose that the taster really can't tell which is which—is only guessing. Find:

  **(a)** $EX$.    **(b)** var $X$.    **(c)** $P(X = 3)$.    **(d)** $P(X \geq 3)$.

**4-11.**  A bridge hand is a random selection of 13 cards from a standard deck of 52 playing cards.

  **(a)**  Find the probability that a bridge hand contains at most 1 ace. (The deck contains 4 aces.)

  **(b)**  Find $EX$, where $X$ is the number of red cards in a hand. (Half of the cards in the deck are red.)

  **(c)**  Find var $X$, for the $X$ in (b).

* **4-12.**  A carton contains 12 articles of which 4 are defective and 8 good. Three are selected at random (without replacement). Let $X$ denote the number of defectives in the sample of 3. Find

  **(a)** $P(X \leq 1)$.

  **(b)** $EX$.

  **(c)**  the probability that the first 2 selected are good and the 3rd is defective.

**4-13.**  Suppose that of the 10,000 individuals at a sporting event, 5,500 are male and 4,500 are female. Five ticket stubs are drawn at random (without replacement) to receive prizes.

  **(a)**  Find the exact probability that three males and two females win.

  **(b)**  Approximate the probability in (a) using the binomial formula.

* **4-14.**  In a state "Megabucks" lottery, you pay $1 and pick any six of the numbers 1 to 36 with no duplication. Order is immaterial. The six winning numbers are selected randomly. When the winning selection is announced, you win the grand prize of several million dollars if you have them all correct, $400 if you have any five correct, and $40 if you have any four correct.

  **(a)**  Find the probability of winning the grand prize.

  **(b)**  Find the probability of winning $400.

  **(c)**  Find the probability of winning $40.

  **(d)**  In one such lottery, we know someone who said he was very unlucky because he had no matches at all. How probable is this?

**(e)** Calculate the expected value of your ticket if the grand prize is ten million dollars, and no one else has your combination, so that you don't have to share the prize. (The state will typically skim off about half of the total money paid in, so a $1 ticket would be worth about 50 cents, except for the fact that money not won in previous weeks is added to the next week's prize.)

**★ 4-15.** Suppose the archer of Problem 4-8, who hits the bull's-eye once in ten tries on average, keeps shooting until she hits it.

    **(a)** Find the probability that she first hits it on the fifteenth try.

    **(b)** Find the expected number of tries to get the first hit.

    **(c)** Find the expected number of tries to get the third hit.

**4-16.** Referring to Problem 4-12, find the probability that, in selecting one at a time, the first defective item is found on the third try, if they are drawn at random

    **(a)** with replacement.

    **(b)** without replacement.

**★ 4-17.** The game of "Russian Roulette" was defined in Problem 2-38, page 73. Find the average length of the game (as a number of attempts) for one who plays it repeatedly,

    **(a)** if the cylinder (with six positions) is spun between attempts.

    **(b)** if the cylinder is not spun, but just advances to the next position.

**4-18.** To estimate the number of fish in a pond (call it $N$), 20 fish are caught, tagged, and replaced in the pond. A few days later, ten fish are caught from the pond. Assume that this catch is a random selection without replacement.

    **(a)** Find the distribution of the number of tagged fish in the second catch, assuming that there were originally 100 fish in the pond.

    **(b)** Give an expression for the expected number of tagged fish in the second catch, in terms of $N$.

    **(c)** If you found four tagged fish in your catch of ten, what would you estimate $N$ to be?

**★ 4-19.** Find the value or an approximate value of each of the following, without using a calculator:

    **(a)** $\displaystyle\sum_{k=0}^{8} k^2 \cdot \frac{\binom{10}{k}\binom{10}{8-k}}{\binom{20}{8}}$     **(b)** $\displaystyle\frac{\binom{500}{4}\binom{500}{4}}{\binom{1000}{8}}$     **(c)** $\displaystyle\sum_{12}^{\infty} n \binom{n-1}{11} (.5)^n$.

**4-20.** Find the conditional distribution of $X$, the number of successes in the first $m$ of a sequence of independent Bernoulli trials, *given* that there are $c$ successes in the first $m+n$ trials.

**★ 4-21.** At one time, peanut M&M's came in just four colors, mixed by the manufacturer in equal proportions. Find the expected number of random selections of one M&M, from an inexhaustible supply, that would be needed to get at least one of each color.

**4-22.** Verify the asserted equality of (5) and (6) in §4.4.

---

## 4.5      Approximating Binomial Probabilities

In this section we introduce two different approximations for binomial probabilities. Both approximations are for large samples. Binomial probabilities are cumbersome to deal with when $n$ is large, especially if it is a cumulative probability that we need. Many hand calculators have a factorial key, but they may overflow at 70! (the first factorial greater than $10^{100}$). There are tables of binomial probabilities with values of $n$ up to 100, but these are not readily available and are necessarily limited to selected values of $p$. Statistical software for computers will usually include the capability of providing binomial probabilities, but these are apt to be based on approximations.

Complicated calculations and formulas for finite $n$ sometimes become simpler in the limit as $n$ becomes infinite. The simpler limiting formulas provide approximations for situations in which $n$ is finite but large.

The first approximation we take up applies when $n$ is large and the parameter $p$ is neither close to 1 nor close to 0. The second is also for large $n$ but for values of $p$ that are close either to 1 or to 0.

A formula for approximating binomial probabilities was found by the mathematicians DeMoivre (in 1733, for the case $p = 1/2$) and Laplace (in 1812, for general $p$). They showed that the cumulative binomial probability $P(Y \leq k)$ can be quite well approximated using a smooth function that is now readily available in tables and computer software:

---

If $Y \sim \text{Bin}(n,\ p)$, where $npq$ is moderately large (say, $npq > 5$), then

$$P(Y \leq k) \doteq \Phi\left(\frac{k + \frac{1}{2} - np}{\sqrt{npq}}\right), \tag{1}$$

where values of $\Phi(\,\cdot\,)$ are given in Table IIa of Appendix I.

---

The function $\Phi(z)$ is actually the area to the left of $z$ under the "normal curve"—the graph of an exponential function with a quadratic exponent:

$$\frac{1}{\sqrt{2\pi}}\ \exp(-z^2/2).$$

We'll return to this function in Chapter 6, using it there to define an important model for continuous variables. For the present, all we need is the table of values of $\Phi(\,\cdot\,)$, concentrating on the mechanics of the approximation. In Chapter 8 we'll discuss the approximation in greater detail, as an application of a remarkable theorem, the *central limit theorem*. Observe that the $np$ and $npq$ in (1) are the mean and variance of the binomial variable $Y$.

| Example 4.5a | **Sampling a Large Electorate** |
|---|---|

An opinion poll uses a random selection of 400 individuals from the adult population of a very large city. Assume that the city is half male and half female. What is the probability that among those selected there are no more than 190 women?

Let $Y$ denote the number of women among the 400 selected. For a very large city, sampling 400 without replacement is practically equivalent to sampling with replacement. So we assume that $Y \approx \text{Bin}(400, .5)$. Substituting $np = 200$ and

$\sqrt{npq} = 10$ in (1) yields

$$P(Y \leq 190) \doteq \Phi\left(\frac{190.5 - 200}{10}\right) = \Phi(-.95).$$

To find this in Table II, look opposite $-.9$ in the leftmost column and under the .5 in the top row—the top row gives the second decimal place. There we find $\Phi(-.95) = .1711.$   ∎

The condition $npq > 5$, is a "rule of thumb," saying that if $p$ or $q$ is close to 0, $n$ has to be very large to compensate for the small size of $pq$. The reason for this is that if $p$ is close to 0 or 1, the binomial distribution is quite skewed—unless $n$ is very large, whereas the approximating normal curve is symmetric about 0. In the preceding example, $npq = 100$, so we can expect the approximation (1) to be quite good.

The approximating curve is centered at 0, and the difference $Y - np$ has mean 0, so one may wonder about the $\frac{1}{2}$ in (1). It usually improves the approximation and is called a *continuity correction*. This will be explained in §8.8, after we've studied continuous distributions and normal curves.

The rule of thumb is arbitrary, of course, but has been found adequate for assuring a degree of accuracy in a calculated probability that is usually sufficient. But the approximation can be surprisingly good even when the rule of thumb is not followed, provided $p$ is neither close to 0 nor close to 1.

**Example 4.5b** | **A Check on Accuracy**

To check the accuracy of the approximation (1), let us consider random samples of size $n = 8$, and the two cases $p = .5$ and $p = .1$.

When $p = .5$, the distribution of $Y$ (the number of successes) is symmetric about its mean. The mean is $np = 4$, and the s.d. is $\sqrt{npq} = \sqrt{2}$. Thus,

$$P(Y \leq k) = \sum_{0}^{k} \binom{8}{i} .5^i .8^{8-i} \doteq \Phi\left(\frac{k + .5 - 4}{\sqrt{2}}\right).$$

The exact and approximate probabilities are as follows:

|  | *Binomial probability* | *Normal approximation* |
|---|---|---|
| $Y \leq 0$ | .0039 | .0067 |
| $Y \leq 1$ | .0351 | .0386 |
| $Y \leq 2$ | .1445 | .1444 |
| $Y \leq 3$ | .3632 | .3632 |
| $Y \leq 4$ | .6466 | .6368 |

(We've omitted values for $k > 4$ because the distribution is symmetric.)

When $p = .1$, the mean number of successes is $np = .8$, and the standard deviation is $\sqrt{8 \times .1 \times .9} = \sqrt{.72}$:

$$P(Y \le k) = \sum_{0}^{k} \binom{8}{i} .1^i .9^{8-i} \doteq \Phi\left(\frac{k + .5 - .8}{\sqrt{.72}}\right).$$

The exact and approximate probabilities are

|  | *Binomial probability* | *Normal approximation* |
|---|---|---|
| $Y \le 0$ | .4305 | .3619 |
| $Y \le 1$ | .8131 | .7953 |
| $Y \le 2$ | .9619 | .9774 |
| $Y \le 3$ | .9950 | .9992 |
| $Y \le 4$ | .9996 | 1.0000 |

(Beyond $Y = 4$, both probabilities are 1 to four decimal places.) Even though the rule of thumb is violated, the approximations for $p = .5$ are quite good. But when $p = .1$, the approximations are clearly not as accurate. ∎

When $p$ is near 0 or near 1 and $npq < 5$, the normal approximation is usually not very good. However, there is another approximation to binomial probabilities that works quite well when $n$ is large and $p$ is small. The approximation was proposed by Poisson in about 1837.

---

*Poisson approximation* to binomial probabilities:

If $Y \sim \text{Bin}(n, p)$, where $n$ is large and $p$ is small,

$$P(Y = k) \doteq \frac{(np)^k}{k!} e^{-np}. \tag{2}$$

---

The approximating quantities in (2) are given in Table IV in Appendix 1 for selected values of $m = np$. The table entries are *cumulative*, used in approximating cumulative binomial probabilities:

$$P(Y \le c) = \sum_{k=0}^{c} \binom{n}{k} p^k q^{n-k} \doteq \sum_{k=0}^{c} \frac{(np)^k}{k!} e^{-np}. \tag{3}$$

If on occasion we should need the probability of a single value, we can get this by taking the difference of successive entries:

$$P(Y = c) = P(Y \le c) - P(Y \le c - 1).$$

(The probability of a single value is easily computed from (2) in most cases, using a hand calculator.)

Table IV stops at $m = 15$, because if $np > 15$ and $p$ is small, $npq$ is apt to be large enough for the normal approximation to work well. This is not to say that the Poisson approximation would not work, but in such cases the cumulative Poisson probabilities are themselves well approximated by areas under a normal curve.

**Example 4.5c** | **An Uncommon Blood Type**

People with type AB blood constitute about 4% of the U.S. population. What is the probability of finding more than ten such individuals in a random selection of 100 people from the U.S.?

Because the population is very large compared with the sample size, we can use binomial probabilities in place of the more exact hypergeometric probabilities. Let $Y$ denote the number of individuals who have type AB blood in the sample of 100. The desired probability is

$$P(Y > 10) = 1 - P(Y \le 10) \doteq 1 - \sum_0^{10} \binom{100}{k} (.04)^k (.96)^{100-k}$$

Using (3) and entering Table IV with $m = np = 100 \times .04 = 4$, we find

$$P(Y > 10) \doteq 1 - \sum_0^{10} \frac{4^k}{k!} e^{-4} = 1 - .997 = .003.$$

(The exact probability is .00284.)                                         ■

To justify (2), we'll show that the limit of the binomial probability on the left is the expression on the right. So let $np = m$ be fixed, which means that as $n$ becomes infinite, $p$ tends to 0. Then,

$$\binom{n}{k} p^k (1 - p)^{n-k} = \frac{n(n-1) \cdots (n-k+1)}{k!} \left(\frac{m}{n}\right)^k (1 - p)^{n-k}$$

$$= \frac{n}{n} \cdot \frac{n-1}{n} \cdots \frac{n-k+1}{n} \cdot \frac{m^k}{k!} (1 - p)^{-k} (1 - p)^n.$$

As $n$ becomes infinite, each of the first $k$ factors tends to 1. The factor $m^k/k!$ doesn't change, not depending on $n$ or $p$. The factor $(1 - p)^{-k}$ tends to 1, and the last factor is

$$(1 - p)^n = \{(1 - p)^{-1/p}\}^{-m} = \{(1 + x)^{1/x}\}^{-m},$$

where $x = -p$. As $n$ becomes infinite, $x$ tends to 0, and the limiting value of $(1 + x)^{1/x}$ defines the constant $e = 2.71828....$ So the last factor approaches $e^{-m}$, and we see that (2) is justified.

Both the Poisson and normal approximations lead to probability models that are important in their own right. The normal curve defines an important continuous model that will be taken up in detail in Chapter 6. The Poisson formula defines a discrete model that is the topic of the next section.

## Problems

**\* 4-23.**  Find, approximately, the probability of more than 15 heads in 25 tosses of a fair coin.

**4-24.**  Check the accuracy of the normal approximation to the binomial p.f. in the case $n = 5$ and $p = .5$, by comparing exact and approximate values of $P(X \leq k)$ when $k = 0, 1, ..., 5$.

**\* 4-25.**  One-third of the voters in a large population favor a certain proposition. Approximate the probability that fewer than 18 favor it in a random selection of 72 voters.

**4-26.**  In a certain population, 40% have type O blood. Use a normal approximation to find the probability that there are more than 25 people with type O blood in a random selection of size 50 from the population.

**\* 4-27.**  Use the normal table to approximate $\binom{100}{48}(.5)^{100}$.

**4-28.**  A certain type of inoculation has been found to be about 99% effective. Find the probability that it is effective for at least 399 among a group of 400 individuals who are inoculated.

**\* 4-29.**  Of the resistors made by a certain company, .5% are defective. Find the probability that in a shipment of 300,

  **(a)**  none is defective.
  **(b)**  exactly one is defective.
  **(c)**  at most three are defective.

**4-30.**  The company in Problem 4-29 must decide how many resistors to package in a box. Its policy is to refund the purchase price of an entire box if any one resistor in it is defective. Find the maximum number they can put in each box if refunds are to be given for no more than 10% of the boxes sold.

**\* 4-31.**  A company will accept a shipment of 1,000 items if no more than one defective item is found in a random selection of 20.

  **(a)**  Give an exact expression for the probability that the shipment is accepted when it actually contains 50 defectives.
  **(b)**  Give a binomial approximation for the probability in (a).
  **(c)**  Give a Poisson approximation for the probability in (a).
  **(d)**  For comparison with (c), find a normal approximation for the probability in (a). (Which works better?)

**4-32.**  Use the Poisson table to approximate the value of $\binom{100}{8}(.94)^{92}(.06)^8$.

## 4.6                    Poisson Distributions

The Poisson formula [(2) of §4.5] serves to define a family of distributions that is useful for approximations to a binomial, but is also important in its own right as a model we'll use in connection with certain continuous-time processes. We first verify that the quantities given by the Poisson formula

$$f(k \mid m) = \frac{m^k}{k!}\, e^{-m}, \quad k = 0, 1, 2, ..., \tag{1}$$

define a discrete model. They are nonnegative, and their sum involves a familiar series—the Maclaurin series for the exponential function:

$$\sum_{k=0}^{\infty} f(k \mid m) = e^{-m} + m e^{-m} + \frac{m^2}{2!} e^{-m} + \frac{m^3}{3!} e^{-m} + \cdots$$

$$= e^{-m}\left\{ 1 + m + \frac{m^2}{2!} + \frac{m^3}{3!} + \cdots \right\} = e^{-m} e^m = 1.$$

So (1) defines a p.f. When $X$ has this p.f., we say it has a **Poisson distribution** and write $X \sim \text{Poi}(m)$.

The mean and variance of the distribution can be guessed from the way we arrived at (1) in a limiting process. The binomial mean is $np$, and as $n \to \infty$ and $p \to 0$ with $np$ held fixed at the value $m$, the mean tends (in a trivial sense—it is constant) to $m$. The binomial variance is $np(1 - p)$, and in the limit this is also $m$. But we can derive these results formally. First the mean:

$$E(X) = 1 \cdot m e^{-m} + 2 \cdot \frac{m^2}{2!} e^{-m} + 3 \cdot \frac{m^3}{3!} e^{-m} + \cdots$$

$$= m e^{-m}\left\{ 1 + m + \frac{m^2}{2!} + \frac{m^3}{3!} + \cdots \right\} = m e^{-m} e^m = m.$$

For the variance we have to work a little harder. It is convenient to find first the expected value of $X^2 - X$:

$$E[X(X - 1)] = \sum_{k=0}^{\infty} k(k - 1) f(k \mid m) = \sum_{k=2}^{\infty} k(k - 1) \frac{m^k}{k!} e^{-m}.$$

Canceling the $k$ and $k - 1$ [leaving $(k - 2)!$ in the denominator] and factoring out two of the $m$'s, we find

$$E[X^2 - X] = m^2 e^{-m}\left\{ 1 + m + \frac{m^2}{2!} + \frac{m^3}{3!} + \cdots \right\} = m^2 e^{-m} e^m = m^2.$$

From this we can find the variance:

$$\text{var } X = E(X^2) - [EX]^2 = E(X^2 - X) + EX - m^2 = m^2 + m - m^2 = m.$$

*Poisson probability function:*

$$f(k \mid m) = \frac{m^k}{k!} e^{-m}, \quad k = 0, 1, 2, \dots . \tag{2}$$

When $X \sim \mathrm{Poi}(m)$,

$$EX = m = \mathrm{var}\, X. \tag{3}$$

The Poisson distribution arises in the context of a continuous-time process that generalizes the Bernoulli process of §4.1. In observing a Bernoulli process, we are concerned with the occurrence of some event or phenomenon, which we've referred to generally as *success* and coded "1." When the probability of a success is small, successes occur infrequently overall, but can come close together or far apart.

Consider, for instance, the random digits in Table XIV of Appendix 1. Let the 1's stand and replace the other digits (0, 2, 3, ..., 9) with 0's. The result is a sequence of independent Bernoulli trials with $p = .1$. One row of the table gives this sequence:

00100 00100 10000 11000 00001 10000 00000 00000 00000 00100.

The 1's come at unpredictable points in the sequence. This particular sequence of length 50 happens to have eight 1's (16%), but over longer sequences the 1's will make up close to 10% of the total—no matter where the sequence starts.

A Bernoulli process has characteristics that we want to preserve in going to a continuous-time process:

**(a)**   The probability of a given number of 1's in a sequence of specified length does not depend on where the sequence starts.
**(b)**   Events having to do with sequences that don't overlap are independent.
**(c)**   If $p$ is small, the probability of a 1 in a short sequence is approximately proportional to the length of the sequence.
**(d)**   If $p$ is small, 1's rarely occur right next to each other.

The independence we've assumed in defining a Bernoulli process implies (a) and (b). As to (d), the probability of two 1's in two trials is $p^2$, an order of magnitude smaller than the probability of one 1 in two trials if $p$ is small:

$$2p(1 - p) \doteq 2p.$$

To illustrate (c), when $p = .05$,

$$P(\text{one 1 in two trials}) = .095, \quad \text{and} \quad P(\text{one 1 in four trials}) \doteq .17.$$

The second "interval" is twice as long as the first, and the second probability is nearly twice the first.

We want to construct a process that unfolds in continuous time with characteristics analogous to (a)–(d) above. Phenomena with these characteristics do occur in nature: Some rather infrequent events occur at "random times"—with irregular spacing, occasionally close together, sometimes far apart, and with no apparent periodicities. Over any two long periods of time of equal length there are about the same number of occurrences, so the *rate* of occurrence is constant over time. Some examples: fatal automobile accidents at a busy intersection, nuclear reactor melt-downs, goals in a hockey game, emissions of $\alpha$-particles from a radioactive substance, meteors striking the earth, arrivals of customers at a service facility.

In each of these settings it may be reasonable to assume properties like (a)–(d) of a Bernoulli sequence. Specifically, using the term *event* for the phenomenon we watch for, we may assume:

**(a)**   The probability of a given number of events in any region of size $t$ does not depend on the location of that region.

**(b)**   Events in regions that do not overlap are independent.

**(c)**   In a small region, the probability of occurrence of an event is approximately proportional to the size of the region.

**(d)**   The probability of *more* than one event in a small region is negligible as compared with the probability of one.

We have given these characteristics using the term "region" instead of the word "time" because the distribution we'll end up with is also applicable in situations in which a linear or spatial dimension takes the place of time. Some examples: bacterial colonies on a Petri dish, raisins in cookie batter, flaws in a wire or on a sheet of photographic paper, flat tires in driving (as a function of miles covered rather than time), and fish in a lake.

To understand assumptions (a)–(d), it helps to see what kinds of processes they exclude. Consider fatal accidents at an intersection: If accidents are more likely to occur in daylight than at night, assumption (a) is violated, although it may apply as an approximation. Consider flaws in a wire: A flaw might be rare overall; but when one occurs, this may be because the production process has developed a malfunction. In such a case, it is more likely that the next 10 feet of wire will have a flaw than had the first flaw not occurred, so assumption (b) is violated. Consider fish in a lake: For species of fish that move in schools, both (b) and (d) do not hold.

In deriving the Poisson p.f. from assumptions (a)–(d), we'll reason from the Bernoulli process, with trials occurring in discrete time, to a continuous-time process. Assumption (d) implies that in a time interval of length $h$, there is either no event or exactly one event (as an approximation). Thus, the random variable

$$X = \text{number of events in } (t,\, t+h)$$

is approximately Ber($p$), with

$$p = P(X = 1) \doteq \lambda h. \tag{4}$$

As in the case of a Bernoulli process, there are several random variables that might be of concern in the analogous continuous-time process. In Chapter 6 we'll study the (continuous) distribution of the waiting time from any point in time until the next occurrence of an event. Here we consider the discrete random variable

$$Y_t = \text{Number of events in an interval of length } t.$$

To obtain its distribution for an arbitrary $t$, we take the interval as starting at time 0. [According to (a), the starting point is irrelevant.]

We divide the interval $(0, t)$ into $n$ subintervals at length $h = t/n$. An approximate Bernoulli variable $X_i$ is defined by (4) for the $i$th subinterval, $i = 1, 2, ..., n$:

$$X_i \approx \text{Ber}(\lambda h).$$

The variables $X_i$ are independent, according to (b). And $Y_t$ is the sum of these Bernoulli variables, so it is (approximately) $\text{Bin}(n, \lambda h)$:

$$P(Y_t = k) \doteq \binom{n}{k} (\lambda h)^k (1 - \lambda h)^{n-k}.$$

We showed in §4.5 that as $n$ becomes infinite with $np = n\lambda h = \lambda t$ fixed, this converges to the Poisson probability function (1) with $m = \lambda t$:

$$f_{Y_t}(k) = P(Y_t = k) = \frac{(\lambda t)^k}{k!} e^{-\lambda t}, \ k = 0, 1, .... \tag{5}$$

This result can also be reached by a differential approach, in which the assumptions (a)–(d) lead to a system of differential equations for (5).[2]

[The term "Poisson process," in the theory of stochastic processes, is used to mean a family of random variables $\{Y_t\}$, indexed by the real number $t$, whose increments over the interval $(s, t)$ are Poisson variables with parameter $\lambda(t - s)$, and independent for nonoverlapping intervals. Our use of the term "process" is less formal; we use it to refer either to the actual process being modeled or to our model for it. And we'll say, for instance, "Assume that customers arrive in accordance with a Poisson process," meaning that (a)–(d) are satisfied, so that (5) defines the distribution of the number of arrivals in an interval of width $t$.]

In the distribution defined by (5), the Poisson parameter is $m = \lambda t$, so from (3), the mean number of events in a region of size $t$ is $\lambda t$. In particular, the mean number of events in a region of size 1 is $\lambda$. Saying it another way:

---

In a Poisson process, the **rate parameter** $\lambda$ is the *average* number of events in a *unit* interval (or region of unit size).

---

**Example 4.6a** | **Customer Arrivals**

Suppose customers arrive at a service counter according to a Poisson process at the rate of five per hour: $\lambda = 5$. The number of arrivals in a one-hour period has a

---

[2]See B. W. Lindgren, *Statistical Theory*, 4th Ed., Chapman & Hall (1993), 161.

Poisson distribution with parameter $\lambda t = 5 \cdot 1 = 5$. The p.f. for this number $Y_1$ is

$$f(k \mid 5) = P(Y_1 = k) = \frac{5^k}{k!} e^{-5}.$$

And then, for example,

$$P(Y_1 \geq 2) = 1 - P(Y_1 \leq 1)$$
$$= 1 - f(0 \mid 5) - f(1 \mid 5) = 1 - e^{-5}(1 + 5) \doteq .96.$$

To find this probability from the Poisson table (Table IV, Appendix 1), we enter the column headed $m = 5$ and go down to the entry opposite $c = 1$. That entry is .040, the probability of $Y_1 \leq 1$.

We can also calculate the probabilities for numbers of arrivals in intervals of other lengths. For instance, let $Y_2$ denote the number in a two-hour period. The average per hour is 5, so the average in a two-hour period is twice that, or $\lambda t = 2 \times 5$. The probability of, say, at most six arrivals in a two-hour period is a Poisson probability with $m = 10$:

$$P(Y_2 \leq 6) = \sum_0^6 \frac{10^k}{k!} e^{-10} = .130,$$

according to Table IV $[m = 10, c = 6]$.

In these calculations, we chose an hour as the unit of time. We could have chosen any time period as one unit. For example, had we chosen a minute as the unit of time, $\lambda$ would be $1/12$. The answers would not change, since the number of arrivals in an hour is now $Y_{60}$, a Poisson variable with $\lambda = 1/12$ and $t = 60$; so $m = \lambda t = 5$, as before. ∎

In this example, the Poisson process is one that unfolds with time. As stated earlier, the Poisson distribution may also describe such things as the locations of defects along a wire, or the distribution of flaws in sheets of metal, or the distribution of particles in a volume of air.

**Example 4.6b**    **Flaws in Cloth**

Suppose flaws in bolts of a certain cloth are distributed according to a Poisson law, with an average of one flaw per 20 square yards. Taking a square yard as the unit of area means that $\lambda = 1/20$. The expected number of flaws in a piece of 10 square yards is $10\lambda = .5$. Then, for instance,

$$P(\textit{no} \text{ flaws in 10 sq. yds.}) = P(Y_{10} = 0) = e^{-.5} = .607,$$

from (5). To find the probability of two or more flaws in a piece containing 50 square yards, first find the expected number:  $50/20 = 2.5$. Table IV (when we interpolate for $m = 2.5$ and $c = 1$) gives the probability of $Y_{50} \leq 1$ as .287, so the desired probability is $P(Y_{50} \geq 2) = .713$. [We could also use the defining formula (5) and a hand calculator:

$$P(Y_{50} \geq 2) = 1 - P(Y_{50} = 0 \text{ or } 1) = 1 - e^{-2.5}(1 + 2.5) = .713.]$$    ■

Recalling from §4.2 that adding two independent binomial variables with the same $p$ produces a binomial variable, we'd expect that adding independent Poisson variables produces a Poisson variable—for a similar reason: The sum of the numbers of events in two adjacent intervals is the number in the combined interval, which is Poisson. A mathematical proof of this using generating functions will be given in §4.9, and the extension to any finite number of independent Poisson variables follows by induction:

---

If $X_1$, ..., $X_k$ are independent Poisson variables with corresponding parameters $m_1$, ..., $m_k$, then their sum $X_1 + \cdots + X_k$ has a Poisson distribution with parameter $m = m_1 + \cdots + m_k$.

---

**Example 4.6c** | **Combining Different Customer Types**

Three different types of customers arrive according to independent Poisson processes with averages 3 per hour, 5 per hour, and 10 per hour, respectively. What is the probability of no arrivals of any type during a five-minute period?

   The total number of customers per hour is the *sum* of the numbers of the three types, so it is a Poisson process with $\lambda = 3 + 5 + 10 = 18$. The mean number of customers in the five-minute period is $m = \lambda t = 18 \times 5/60 = 3/2$, and the number of customers (of any type) in that period has a Poisson distribution with $m = 3/2$. So the answer is $f(0 \mid 1.5) = e^{-1.5} = .223$.    ■

## Problems

**∗ 4-33.** Phone calls arrive at an exchange at a rate of eight per minute. Assuming a Poisson arrival process, find the probability of
  **(a)** at most eight calls in a one-minute period.
  **(b)** no calls in a 15-second period.
  **(c)** more than ten calls in a 30-second period.
  **(d)** at least two calls in a 10-second period.

**4-34.** A large metropolitan bus company experiences bus breakdowns at the rate of two per day. Assuming a Poisson distribution, find the probability of
  **(a)** no breakdowns in a particular day.
  **(b)** no breakdowns in two consecutive days.
  **(c)** no breakdowns in a particular day, given that there were three the previous day.
  **(d)** more than two breakdowns in a particular day.
  **(e)** more than five breakdowns in a particular five-day week.

**★ 4-35.** A radioactive substance emits particles according to a Poisson process. If the probability of no emissions in a one-second interval is .165, find

(a)  the expected number of emissions per second.

(b)  the probability of no emissions in a two-second interval.

(c)  the probability of at most two emissions in a four-second interval.

(d)  the probability of more than five emissions in five seconds.

**4-36.** Flaws appear at random on sheets of a certain type of photographic paper. The average number of flaws is 1/2 per sheet. Assuming a Poisson distribution, find the probability

(a)  that a sheet has no flaws.

(b)  of no flaws in a batch of five sheets.

(c)  that there is at least one flawless sheet in a batch of five sheets.

(d)  of at most two sheets with more than one flaw, in a box of 100 sheets.

**★ 4-37.** Find a numerical value for the following sum:

$$\sum_{k=0}^{\infty} k^2 \left( \frac{2^k}{k!} \right).$$

**4-38.** In a Poisson arrival process, let $X$ and $Y$ denote the numbers of arrivals in nonoverlapping intervals of lengths $s$ and $t$, respectively. Find the conditional distribution of $X$ given that $X + Y = c$.

**4-39.** Consider a Poisson process with rate parameter $\lambda$. Let $p(k)$ denote the p.f. of the number of events in a small increment of time $\Delta t$. Show:

(a)  $\frac{p(1)}{\Delta t} \to \lambda$ as $\Delta t \to 0$. [Assumption (c) in §4.6.]

(b)  $\frac{1-p(0)-p(1)}{\Delta t} \to 0$ as $\Delta t \to 0$. [Assumption (d) of §4.6.]

---

## 4.7  Sample Proportions and the Law of Averages

Consider any repeatable experiment. For a particular event $A$ in its sample space, let $p = P(A)$. Let $X$ denote the indicator variable [see (4) in §2.5]:

$$X = \begin{cases} 1 & \text{if } A \text{ occurs,} \\ 0 & \text{if } A \text{ fails to occur.} \end{cases}$$

Then $X \sim \text{Ber}(p)$. To learn about $p$, we perform the experiment $n$ times in independent trials. The results $X_1, ..., X_n$ constitute a random sample. The sum $Y = \sum X_i$ is the number of times $A$ occurs in the $n$ trials, a binomial variable with mean $np$ and variance $npq$ (§4.2).

The relative frequency or sample proportion of occurrences of $A$ among the $n$ trials, $Y/n$, is usually denoted by $\hat{p}$ ("p-hat"), or by $\hat{p}_n$, when we want to show its dependence on $n$. From (ii) of §3.1, we have

$$E(\hat{p}) = E\left(\frac{Y}{n}\right) = \frac{EY}{n} = \frac{np}{n} = p. \tag{1}$$

And from (5) of §3.3,

$$\sigma_{\widehat{p}} = \sigma_{Y/n} = \frac{1}{n}\sigma_Y = \frac{1}{n}\sqrt{npq} = \sqrt{\frac{pq}{n}}. \qquad (2)$$

From (2) we see that the s.d. of $\widehat{p}$ tends to 0 as $n$ becomes infinite. So the distribution of $\widehat{p}$ becomes more and more tightly concentrated around its mean $p$. In this sense, the random variable $\widehat{p}$ tends to the constant $p$. This tendency is called the *law of averages* or *law of large numbers*. There are formal mathematical theorems expressing this kind of law; we'll give a proof of one such theorem in §8.8. For the present, we give this informal statement:

---

**Law of Averages (Law of Large Numbers):**

In a sequence of independent Bernoulli trials, the proportion of successes converges to the probability of success as the number of trials becomes infinite.

---

**Example 4.7a**  To see the law of large numbers in action, toss a coin repeatedly and keep track of $\widehat{p}_n$, the sample proportion of heads after $n$ trials. It may take a lot of perseverance to convince yourself that $\widehat{p}$ really tends to 1/2. You could also program a computer to simulate tossing a coin. We have done this and have plotted $\widehat{p}_n$ as a function of $n$ in Figure 4-2.

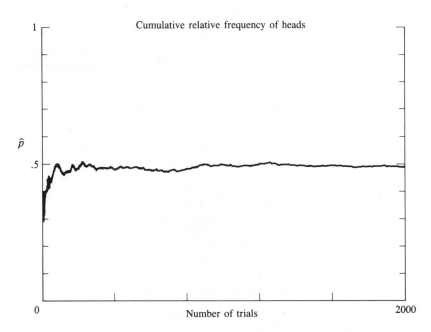

**Figure 4-2** Illustrating convergence of $p$

The decreasing variability in the sequence reflects the fact that the variance of $\widehat{p}_n$ tends to 0. However, it is never completely clear from a finite sequence that the limiting value will be exactly $p = 1/2$.                                              ∎

Some people think that the law of averages says that a success is very likely after a string of failures. This is a misconception. Things do even out "in the long run," and an imbalance would indeed be corrected *in the limit*. However, any finite string of observations (no matter how long!) has no effect on the limiting value of $\widehat{p}_n$. The law of averages assumes independent trials, so the result of any one trial is *not* affected by the result of any other trial. In particular, the mechanism producing the results does not "remember" previous results, so it cannot adapt itself to make up for any short-run imbalance in them. It does so "in the long run" because any imbalance is finite, and the long run is infinite.

---

## 4.8        Multinomial Distributions

We now extend the notion of a Bernoulli variable, which has two possible values, to categorical variables that have any finite number of values. With this generality, a numerical coding is not helpful—the categories may not even be ordered. Let the values of such a variable $X$ be denoted by $A_j$, and the corresponding probabilities by $p_j$, where $p_j \geq 0$ and $\sum p_j = 1$:

| Value of X | $A_1$  $A_2$ $\cdots$ $A_k$ | (1) |
|------------|-----------------------------|-----|
| Probability | $p_1$  $p_2$ $\cdots$ $p_k$ | |

In a sequence of $n$ independent trials of this experiment, we are interested in the frequencies of occurrence of the various categories—of $Y_j$, the number of times $A_j$ occurs in the $n$ trials, for $j = 1, 2, ..., k$. The frequencies $Y_j$ are random variables, and we now derive their joint distribution. The following example will lead to the general formula for the joint p.f.

**Example 4.8a** | **Roulette**

In roulette, a steel ball is thrown on a spinning wheel with 38 slots, presumed to be equally likely as final resting spots for the ball. Of these, 18 are red, 18 are black, and 2 are green (in most casinos). So the color resulting from a spin of the wheel is a random variable with three categories ($k = 3$) and the following probabilities:

| Color (X) | Red | Black | Green |
|-----------|-----|-------|-------|
| Probability | 18/38 | 18/38 | 2/38 |

Suppose we spin the wheel five times. If the spins are independent, what is the probability that two spins result in red, two in black, and one in green?

We proceed as we did in deriving the binomial formula, in the Bernoulli case (two categories). The probability for a particular sequence of results, say, $RRBBG$, is the product of probabilities of the individual results:

$$P(RRBBG) = \left(\frac{18}{38}\right)^2 \left(\frac{18}{38}\right)^2 \left(\frac{2}{38}\right)^1. \tag{2}$$

This is also the probability of any other sequence in which there are 2 reds, 2 blacks, and 1 green. Multiplying (2) by $\binom{5}{2,2,1}$, the number of sequences of that type, we obtain the probability that answers the question:

$$P(2 \text{ red, } 2 \text{ black, } 1 \text{ green}) = \binom{5}{2,\ 2,\ 1} \left(\frac{18}{38}\right)^2 \left(\frac{18}{38}\right)^2 \left(\frac{2}{38}\right)^1 = .0795.$$

The joint distribution of $(Y_R, Y_B, Y_G)$, where $Y_i$ is the frequency of color $i$ in $n$ independent trials, is called **trinomial.** ∎

The **multinomial formula** for the general case of $k$ categories is derived exactly as we have derived it for the special case of the preceding example.

---

**Multinomial p.f.:** If $Y_j$ is the frequency of $A_i$ among $n$ independent trials of the experiment (1), and $p_j = P(A_j)$,

$$P(Y_1 = y_1, ..., Y_k = y_k) = \binom{n}{y_1, ..., y_k} p_1^{y_1} p_2^{y_2} \cdots p_k^{y_k}. \tag{3}$$

---

Since the frequencies $Y_j$ must sum to $n$, any one of them is uniquely determined by the other $k - 1$ frequencies. Thus, a trinomial distribution is determined by the distribution of $(Y_1, Y_2)$, since $Y_3 = 1 - Y_1 - Y_2$; so it is really bivariate—just as a *bi*nomial variable is univariate.

Suppose now we focus on just one of the frequencies, say $Y_1$, and regard the outcomes that are not $A_1$ as lumped together in a second category:

$$A_1^c = A_2 \cup \cdots \cup A_k.$$

Then $Y_1 \sim \text{Bin}(n, p_1)$. By the same token, $Y_j \sim \text{Bin}(n, p_j)$. And similar reasoning shows that the marginal distributions of any proper *subset* of the category frequencies in a multinomial distribution is again multinomial (with fewer categories).

**Example 4.8b** | **A Survey**

Suppose a question is asked in a sample survey with possible answers "For," "Against," and "Neutral." The population proportions of these categories are their probabilities, $p_F$, $p_A$, and $p_N$. In a random sample of size $n$, the corresponding frequencies $(Y_F, Y_A, Y_N)$ have a trinomial distribution. If we are interested only in the number or proportion of "For" responses, we can lump the other two in a single category. Thus, $Y_F \sim \text{Bin}(n, p_F)$. For instance, if $n = 600$ and the population proportions are 60% for, 30% against, and 10% neutral, then $Y_F \sim \text{Bin}(600, .6)$. ∎

## Problems

**\* 4-40.** Among the 5,000 people in a certain town, 45% have type O blood, 40% have type A, 10% have type B, and 5% have type AB. Let $W$, $X$, $Y$, and $Z$ denote the frequencies of type O, type A, type B, and type AB, respectively, in a random selection of ten from the population.
   **(a)** Find $P(X = 4, Y = 3, Z = 0)$.
   **(b)** Find $P(X = 4, Y = 3)$.
   **(c)** Find $P(W = 4)$.
   **(d)** Find $P(Z \geq 1)$.
   **(e)** Approximate the answer to (d) using the Poisson formula.

**4-41.** In a throw of eight dice, let $X_j$ denote the number of dice that show $j$. Find the following:
   **(a)** $EX_j$.
   **(b)** $\text{var } X_j$.
   **(c)** $P(X_1 = X_2 = X_3 = X_4 = 2)$.
   **(d)** The distribution of $X_1$, given $X_2 = 3$.
   **(e)** The joint distribution of $X_1$ and $X_2$, given $X_3 = 0$.

**\* 4-42.** An ordinary six-sided die is painted red on two sides, white on two sides, and blue on two sides. Find the probability that in 24 tosses, we get
   **(a)** an equal number of reds, whites, and blues.
   **(b)** ten reds and eight whites.
   **(c)** ten reds.

**4-43.** Find the distribution of the number of brown M&M's in a small packet of 20 peanut M&M's (see Problem 4-21), given that there are six each of the green and yellow ones.

**\* 4-44.** Given that $(X_1, X_2)$ is trinomial with parameters $(n; p_1, p_2)$,
   **(a)** what is the distribution of $X_1$?
   **(b)** what is the distribution of $X_1 + X_2$?
   **(c)** find the distribution of $X_1$ given $X_2 = k$.
   **(d)** find the distribution of $X_1$ given $X_1 + X_2 = k$.

---

**4.9**  **Applying the Probability Generating Function**

In §3.6 we saw that the probability generating function of a sum of *independent* random variables is the product of their p.g.f.'s. We now use this fact to derive or rederive some properties of the various families of distributions that we've introduced.

**Example 4.9a** | **Sum of Bernoulli's**

The p.g.f. of Ber($p$) is

$$\eta(t) = E(t^X) = t^1 \cdot P(X = 1) + t^0 \cdot P(X = 0) = pt + q. \tag{1}$$

A sequence of $n$ independent trials $(X_1, ..., X_n)$ is a sequence of 1's and 0's. In §4.2 we found that the distribution of the sum $Y = \sum X_i$ is binomial. We now derive this anew using p.g.f.'s.

Since $Y$ is the sum of *independent* Bernoulli variables, its p.g.f. is the product of $n$ factors; and since the $X_i$'s are *identically distributed*, those factors are the same, each being given by (1):

$$\eta_Y(t) = \prod_1^n (pt + q) = (pt + q)^n. \tag{2}$$

When we apply the binomial theorem to expand the $n$th power of the binomial $pt + q$, (2) becomes

$$\eta_Y(t) = \sum_{k=0}^{n} \binom{n}{k} (pt)^k q^{n-k} = \sum_{k=0}^{n} \left\{ \binom{n}{k} p^k q^{n-k} \right\} t^k. \tag{3}$$

According to (1) in §3.6, the coefficient of $t^k$ in (3) is the probability that $Y = k$. This coefficient is precisely the binomial probability we found as (1) in §4.2.    ∎

**Example 4.9b** | **Sum of Poisson's**

In §4.6 we claimed that the sum of independent Poisson variables is a Poisson variable, appealing to the corresponding result for binomials. We now give a proof using p.g.f.'s.

Let $Y \sim \text{Poi}(m)$. The p.g.f. of $Y$ [see (1) of §3.6] is

$$\eta_Y(t) = E(t^Y) = \sum_{k=0}^{\infty} t^k f_Y(k \mid m) = \sum_{k=0}^{\infty} t^k \cdot \frac{m^k}{k!} e^{-m}$$

$$= e^{-m} \sum_{k=0}^{\infty} \frac{(mt)^k}{k!} = e^{-m} e^{mt} = e^{-m(1-t)}. \tag{4}$$

Now suppose $Y_i \sim \text{Poi}(m_i)$, for $i = 1, 2, ..., k$. If these variables are independent, the p.g.f. of their sum [by (4) of §3.6] is the product of their p.g.f.'s:

$$\eta_{\Sigma Y}(t) = E(t^{\Sigma Y_i}) = \prod e^{-m_i(1-t)} = e^{-\Sigma m_i(1-t)}.$$

In view of (4), this is the p.g.f. of a Poisson variable with parameter $\sum m_i$. So, $\sum Y_i \sim \text{Poi}(\sum m_i)$. In particular, if the $Y_i$ all have the *same* parameter $m$, then the sum is Poi($km$).    ∎

Although we defined the p.g.f. $\eta$ for random variables with values 0, 1, 2, ..., the formula for $\eta$ as an expected value defines it for any discrete variable:

$$\eta_X(t) = E(t^X) = \sum_x t^x f_X(x). \tag{5}$$

And we'll now show how we can use this function to find moments of a distribution. Differentiating $\eta$ with respect to $t$, we get

$$\eta'_X(t) = \frac{d}{dt}\sum_x t^x f_X(x) = \sum_x \frac{d}{dt}(t^x)f_X(x) = \sum_x x t^{x-1} f_X(x).$$

This is the formula for $E[Xt^{X-1}]$, and upon substituting 1 for $t$, we find

$$\eta'_X(1) = \sum_x x f_X(x) = EX. \tag{6}$$

So the mean (when it exists) can be calculated in this way from $\eta(t)$.

If we differentiate $\eta_X$ *again*, we obtain

$$\eta''_X(t) = \frac{d^2}{dt^2} E(t^X) = E[X(X-1)t^{X-2}].$$

And then, with the substitution $t = 1$, we find

$$\eta'_X{}'(1) = E[X(X-1)]. \tag{7}$$

This is called the second **factorial moment** of $X$. Differentiating $\eta(t)$ $k$ times and substituting $t = 1$ yields the $k$th factorial moment:

$$\eta^{(k)}(1) = E[X(X-1)(X-2)\cdots(X-k+1)] = E[X_{(k)}]. \tag{8}$$

The probability generating function $\eta$ generates probabilities as power series coefficients only in the special case in which the random variable has values that are nonnegative integers. Because (8) applies more generally, the p.g.f. is sometimes called the *factorial moment generating function* (f.m.g.f.).

**Example 4.9c**  |  **Variance Using the Factorial Moment Generating Function**

We'll find the variance of the Poisson distribution using the f.m.g.f. Let $Y$ be Poi($m$). Its f.m.g.f. (p.g.f.) is given by (4) in Example 4.9b: $\exp[-m(1-t)]$. The $k$th derivative is

$$\eta_Y^{(k)}(t) = m^k e^{-m(1-t)}.$$

Substituting $t = 1$ yields the factorial moments:

$$E(Y_{(k)}) = \eta_Y^{(k)}(1) = m^k.$$

So $EY = m$ (as we found earlier), and $E[Y(Y-1)] = m^2$. From the latter we can find the variance:

$$\text{var } Y = E(Y^2) - (EY)^2 = E[Y(Y-1)] + EY - (EY)^2 = m^2 + m - m^2 = m.$$

Higher-order central moments can be calculated from the factorial moments in similar fashion. ∎

In the following table we give the p.g.f.'s for some of the distributions discussed in this chapter. We have derived some of these in this section; the problems call for derivations of the others. (Notice that, with $1 - tq$ in a denominator, we'd need $t \neq 1/q$. But this is not an essential restriction, since to find derivatives of $\eta(t)$, we only need it to be defined in a neighborhood of $t = 1$. If $q > 0$, there's no problem.)

| Model | Notation | p.f. | f.m.g.f. |
|---|---|---|---|
| Bernoulli | $\text{Ber}(p)$ | $p^x(1-p)^{1-x}, x = 0, 1$ | $pt + q$ |
| Binomial | $\text{Bin}(n, p)$ | $\binom{n}{x} p^x q^{n-x}, x = 0, 1,..., n$ | $(pt+q)^n$ |
| Geometric | $\text{Geo}(p)$ | $pq^{x-1}, x = 1, 2, ...$ | $\frac{tp}{1-tq}$ |
| Negative binomial | $\text{Negbin}(r, p)$ | $\binom{x-1}{r-1} p^r q^{x-r}, x = r, r+1, ...$ | $\left(\frac{tp}{1-tq}\right)^r$ |
| Poisson | $\text{Poi}(m)$ | $\frac{m^x}{x!} e^{-m}, x = 0, 1, 2, ...$ | $e^{-m(1-t)}$ |

## Problems

**4-45.** Find the factorial moment generating function (in closed form) for the geometric distribution: $f(x \mid p) = p(1-p)^{x-1}$, for $x = 1, 2, ....$ [You will need the formula for the sum of a geometric series: $1 + a + a^2 + \cdots = \frac{1}{1-a}$.]

**4-46.** Suppose $Y_1 \sim \text{Bin}(n_1, p)$ and $Y_2 \sim \text{Bin}(n_2, p)$, and suppose further that these variables are independent. Use generating functions to show that their sum $Y_1 + Y_2$ is binomial.

**\* 4-47.** Let X denote the number of trials needed to obtain the $r$th success in a Bernoulli process. This is the sum of $r$ independent, geometric variables. Its distribution is negative binomial. Use the result of Problem 4-45 to find the f.m.g.f. of $X$.

**4-48.** Use the f.m.g.f. of the distribution of the number of points thrown with an ordinary fair die to find its mean and variance.

**4-49.** Calculate the mean and variance of the geometric distribution (see Problem 4-45) using the f.m.g.f.

**4-50.** Calculate the mean and variance of the negative binomial distribution (see Problem 4-47) using the f.m.g.f.

**4-51.** Given: $X$ has the f.m.g.f. $\eta(t) = 1 - p + \frac{1}{2} pt + \frac{1}{2} pt^{-1}$ for $t$ near 1.

**(a)** Interpret this as an expected value of $t^X$ and so deduce the probability function of this discrete variable.

**(b)** Calculate the mean and variance both from the f.m.g.f. and from the probability function.

## Review Problems

**4-R1.** Blacks constitute 25% of the population in a certain city. Find the probability that in a random selection of 300 individuals for a jury pool, there are at most 67 blacks.

**4-R2.** Let $X \sim \text{Ber}(.04)$. If this experiment is repeated in independent trials,

    **(a)** how many trials, on average, would be required to get six successes?

Let $X_1$ denote the number of successes in 50 trials and $X_2$, the number of successes in 50 additional trials. Let $Y = X_1 + X_2$.

    **(b)** Find the mean and variance of $Y$.

    **(c)** Give the distribution of $Y$ (name and parameter value).

    **(d)** Find $P(Y \leq 5)$.

    **(e)** Find the probability that if there are six successes in the 100 trials, these are evenly divided between the first 50 and the second 50 trials.

**4-R3.** A committee of five is chosen at random from 25 individuals—10 males and 15 females. Find the probability that

    **(a)** the number of males in the committee is equal to the expected number.

    **(b)** the committee includes at most two females.

**4-R4.** Consider four independent observations on $X$, where $X \sim \text{Geo}(p)$.

    **(a)** Find the p.g.f. of $Y = X_1 + \cdots + X_4$.

    **(b)** Find the mean and variance of $Y$ in (a), in terms of $p$.

    **(c)** Find $P(Y = 7)$.

**4-R5.** Suppose typographical errors occur according to a Poisson distribution at the rate of 1/2 per page.

    **(a)** Find the mean and variance of the number of errors in a 20-page paper.

    **(b)** Find the probability of at most three errors in six pages.

**4-R6.** Find the probability that in nine tosses of two coins, three turned up both tails, three turned up both heads (and the other three, one of each).

**4-R7.** Candidates George, Ross, and Bill are the choices of 35%, 35%, and 30%, respectively, of the voters in a state. Consider a random sample of ten.

    **(a)** Find $P(G = 4, R = 4, B = 2)$, where $X$ denotes the number in the sample who favor candidate X.

    **(b)** Find $E(G)$.

    **(c)** Give the distribution of $G + R$.

**4-R8.** There are 40 questions on a multiple-choice exam, each with choices A, B, C, and D. A student selects one answer at random for each question.

    **(a)** What is his expected score?

    **(b)** Find the probability that he gets at most eight correct.

**4-R9.** Calls arrive at a switchboard at the rate of three per minute. Assuming they follow a Poisson law, find

    **(a)** the probability of at least five calls in a two-minute period.

    **(b)** the probability of two calls in a 20-second period.

(c)   the probability, given that there were two calls in a 20-second period, that one was in the first ten seconds and the other was in the second ten seconds.

(d)   the probability of at most six calls in a four-minute period, given that there were eight calls in the preceding five-minute period.

**4-R10.**   Obtain approximate values of the following:

(a)   The probability that a sample of size 5 from a population of 10,000, of which 3,000 are Republicans and 7,000 Democrats or others, includes exactly two Republicans.

(b)   The probability that a sample of 200 individuals from a large population that is 4% black includes at least ten blacks.

**4-R11.**   The odds on winning the "Powerball" lottery conducted in many states are 1:54,979,154. For simplicity, round this off and let the probability that a single ticket wins be one chance in 55 million.

(a)   Suppose 55,000,000 tickets are sold. What is the probability that no one wins? That there is exactly one winner? At least one?

(b)   Answer the questions in (a) if there are only 20,000,000 tickets sold.

## Chapter Perspective

Bernoulli trials are simple, but they are fundamental in many applications of statistics. Different sampling schemes for Bernoulli trials lead to different distributions. When the trials are independent, the number of successes in a specified number of trials is binomial; the geometric and negative binomial distributions apply to the number of trials required to obtain a specified number of successes.

The analogous distributions for sampling without replacement (in which case the trials are dependent) are the hypergeometric and the hypergeometric waiting-time distributions. These distributions have applications in statistical inference for $p$, the proportion of successes in a finite population.

This chapter introduced the "normal curve" as a tool in approximating binomial probabilities for numbers of trials larger than provided for in the binomial tables. This is merely one application of this important continuous distribution, to be treated in detail in Chapter 6.

Although the Poisson distribution provides a convenient approximation for some binomial probabilities, it can serve as a useful model in its own right, as we saw in §4.6. In Chapter 6 we'll return to the Poisson process and take up some continuous distributions for waiting times.

The multinomial distribution is useful for describing category frequencies for discrete phenomena with a finite number of categories. We'll see in Chapter 13 that the distribution also applies as models for summaries of sample data from continuous variables, when the data are summarized in class intervals (as is commonly the case).

We have exploited the p.g.f. in finding distributions of sums of independent random variables and in finding moments of some discrete distributions. Chapter 5 will introduce the moment generating function, which is closely related to the p.g.f. These functions give us another way of characterizing a distribution, one that will be useful in getting at distributions of sample sums.

# Continuous Random Variables

In Chapter 2 we gave some examples of continuous random variables. Because they have uncountably and infinitely many possible values, it is not feasible to define a model by listing possible values and their probabilities.

In building a model for a continuous variable, we'll preserve the basic properties of a probability model [(i)–(iii) of §1.7], taking them as axioms. But we need new building blocks and new tools. With these and a new definition for mean value, we can define moments, generating functions, and independence, much as in the discrete case.

Suppose $\Omega$ is a probability space—a sample space in which we have defined events and probabilities of events $E$ satisfying these axioms:

---

**Axioms for probability:**

   **(i)** $P(\Omega) = 1$.
  **(ii)** $P(E) \geq 0$.
 **(iii)** $P(\bigcup E_n) = \sum P(E_n)$ when $E_i E_j = \emptyset$ for $i \neq j$.

---

These axioms imply various other properties, just as they did in the discrete case—namely, those given as (iv)–(ix) in §1.7 (page 36). We've given the additivity axiom (iii) in the form called *countable additivity*, where $\{E_n\}$ is a *sequence* of events—is countable. With this axiom we have the following additional property:[1]

---

If $E_1 \supset E_2 \supset E_3 \supset \cdots$, then

$$\lim_{n \to \infty} P(E_n) = P(\bigcap E_n).$$  \hfill (1)

---

[1] See B. W. Lindgren, *Statistical Theory*, 4th Ed., New York: Chapman & Hall, 29.

Example **5a**

## Waiting Times

Buses go past a campus bus stop every ten minutes throughout the day. We set out to take the bus without knowing the precise schedule. How long will we have to wait for the next bus?

The waiting time $T$ (in minutes) is a *continuous* random variable: It can't be predicted with certainty (so it's random), and it can take on any value on the continuous timescale from 0 to 10 minutes (so it's continuous). In its decimal representation, a value of $T$ is an infinite sequence of digits. Sequences starting with 0 (such as 0.2471...) represent waiting times of less than one minute; sequences starting with 1 (such as 1.902...) represent waiting times of between one and two minutes; and so on.

In representing our ignorance of the exact arrival times, we might feel that the ten one-minute intervals 0 to 1, 1 to 2, and so on, are equally likely. Further, we might feel the ten .1-minute intervals within any particular one-minute interval are equally likely, and so on. These assumptions imply that the digits in the sequence defining a point on the interval 0 to 1 are independent random digits (§1.5). On this basis, we can calculate the probability of *any* interval of $T$-values. For example, the probability that $T$ starts out with the three digits 364 (in that order) is

$$P(T = 3.64 \cdots) = P(\text{1st digit} = 3, \text{2nd digit} = 6, \text{3rd digit} = 4)$$

$$= P(\text{1st digit} = 3) \cdot P(\text{2nd digit} = 6) \cdot P(\text{3rd digit} = 4) = .1^3.$$

On the other hand, the probability of any *particular value* of $T$ is 0. For example, intuition would suggest that the probability that $T = \frac{1}{3} = .333 \cdots$ is

$$P\left(T = \frac{1}{3}\right) = \lim_{n \to \infty} P(T \text{ has 3 in its first } n \text{ places}) = \lim_{n \to \infty} .1^n = 0.$$

Now suppose we want the probability that the wait is at least two minutes but not as long as four minutes. The decimal representation of a $T$ in this interval is a 2 or a 3 followed by any sequence of digits after the decimal point. Thus, we find

$$P(2 \leq T < 4) = P(T \text{ starts out 2 or 3}) = \frac{1}{10} + \frac{1}{10} = .2 .$$

And because single values have probability 0, the intervals $2 < T < 4$, $2 < T \leq 4$, and $2 \leq T \leq 4$ all have this same probability .2. Similarly, the probability that $T$ is between 1.42 and 3.61 is the same as the probability that $1.42 < T \leq 3.61$, which in turn is the probability that the sequence of the first three digits of $T$ is one of the 219 sequences 142, 143, ..., 361:

$$P(1.42 < T \leq 3.61) = 219 \times (.1)^3 = \frac{3.61 - 1.42}{10}.$$

The same kind of calculation shows that the probability of the interval $(a, b]$ is $(b - a)/10$ [its width divided by 10], for any $a < b$ between 0 and 10.

In this way we have defined a probability model for the continuous variable $T$ as a limit of discrete models with equally likely outcomes.  ∎

**5.1**                           **The Distribution Function**

The events of interest for continuous variables are *intervals,* rather than isolated values, and we'll now take the probability of an interval as fundamental in defining a *general* probability model on the real line. The probabilities of semi-infinite intervals of the form $[X \leq x]$, where $x$ is any real number, determine the probabilities of any other intervals or any countable union of intervals. For example, finite intervals can be expressed as the "difference" [Problem 1-R2] between two semi-infinite intervals. Thus, if $b > a$, then

$$[a < X \leq b] = [X \leq b] - [X \leq a].$$

And since $[X \leq a] \subset [X \leq b]$, it follows that

$$P(a < X \leq b) = P(X \leq b) - P(X \leq a). \tag{1}$$

So if we know the probability $P(X \leq x)$ for all $x$, we can find the probability of any half-closed interval $(a, b]$.

The quantity $P(X \leq x)$, as a function of $x$, is called the *distribution function* of the random variable $X$. It is also called the *cumulative distribution function* and is commonly referred to as the "c.d.f."

---

Distribution function (c.d.f.) of the random variable $X$:

$$F(x) = P(X \leq x). \tag{2}$$

The probability of an interval $(a, b]$ is obtained as a difference:

$$P(a < X \leq b) = F(b) - F(a). \tag{3}$$

---

[Observe that (3) is simply a restatement of (1) using (2).]

**Example 5.1a** | **Waiting Times (Continued)**
In Example 5a we considered the random variable $T$, the time we have to wait until the next bus comes along, assuming that the next bus arrives at a time which is "equally likely" to be anywhere in the next ten minutes. We found the probability that the bus arrives in any subinterval $(a, b)$ of the interval from 0 to 10 to be proportional to its length:

$$P(a < T < b) = \frac{b - a}{10} \quad \text{for } 0 < a < b < 10.$$

And in particular, $P(T = a) = 0$, since the single point $a$ is an "interval" of length 0.

So $<$ is the same as $\leq$ . Thus, for any $t$ in the interval $(0,\ 10)$,

$$P(T \leq t) = P(0 \leq T \leq t) = \frac{t}{10}.$$

If $t$ is negative, the probability of $T \leq t$ is 0, and if $t > 10$, the event $T \leq t$ happens with probability 1. So the complete specification of the c.d.f. is:

$$F(t) = P(T \leq t) = \begin{cases} 0, & t < 0, \\ t/10, & 0 \leq t \leq 10, \\ 1, & t > 10. \end{cases}$$

Figure 5-1 shows the graph of this, a "ramp" function.    ■

The distribution in this last example is quite special, in that there are no preferred regions in the set of possible values. The probability of a subinterval is proportional to its length. In such cases we say that the distribution is **uniform.** When a uniform distribution is defined on an interval $(a,\ b)$, we say that $T$ is *uniformly distributed* there, and write $T \sim \mathcal{U}(a,\ b)$.

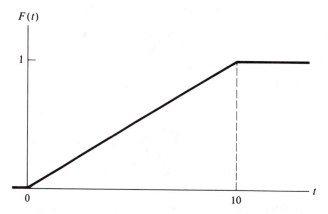

**Figure 5-1** Waiting time c.d.f.—Example 5.1a

Some properties of the c.d.f. in Figure 5-1 apply to distribution functions generally. First, $F(x)$ is a probability, so its values are on the interval from 0 to 1. Applying (1) in the introduction shows the values of $F$ "at" $\pm \infty$ to be

$$F(-\infty) = \lim_{x \to -\infty} F(x) = \lim_{n \to -\infty} P(X \leq n) = P\left\{ \bigcap_{n=1}^{\infty} (-\infty, -n] \right\} = 0,$$

and

$$F(\infty) = \lim_{x \to \infty} F(x) = \lim_{n \to \infty} P(X \le n) = P\left\{ \bigcup_{n=1}^{\infty}(-\infty, n] \right\} = 1.$$

Also, $F(x)$ is nondecreasing: Since the interval $(-\infty, x]$ widens as $x$ increases, its probability $F(x)$ cannot get smaller. [See Property (vii) in §1.7.] Moreover, a c.d.f. is *continuous* from the right, as can be shown using the countable additivity axiom [(iii) in the chapter introduction].

---

Properties of a c.d.f. $F(x)$:

   **(i)**   $F(-\infty) = 0$, $F(\infty) = 1$.

   **(ii)**  $F(a) \le F(b)$  whenever $a < b$.

   **(iii)** $\lim_{x \to a+} F(x) = F(a)$.  ($F$ is continuous from the right.)

---

If a function $F(x)$ has Properties (i)–(iii), and we define $P(X \le x)$ to be $F(x)$, this determines a distribution on the sample space of the real line that satisfies the probability axioms (i)–(iii) given at the beginning of this chapter.

   The analogy between probability and mass is especially helpful in understanding a distribution function. Imagine that probability is represented by a unit measure of dust or snow distributed along a line (give the line some width if you like). Then picture starting at the extreme left (at $-\infty$, if necessary), and sweeping up the dust or shoveling the snow as you move from left to right along the line. The amount collected up to any point is $F(x)$. Since you can only *add* dust as you sweep, the collected amount $F(x)$ cannot decrease.

   Starting from the *left*—using $P(X \le x)$ to describe a distribution—was an arbitrary choice. We could as well start from the right, using the function

$$R(x) = P(X > x) = 1 - F(x). \tag{4}$$

This complementary function is sometimes more natural to consider. As we proceed from left to right along the $x$-axis, the value of $R(x)$, the *un*collected amount of probability, can only decrease or remain constant. So $R(x)$ is a nonincreasing function. In view of (4), it is clear that knowing $R(x)$ is equivalent to knowing $F(x)$.

**Example 5.1b** | **Waiting Times in a Poisson Process**

Suppose arrivals at a service center follow a Poisson law (§4.6) with an average of six arrivals per minute:  $\lambda = 6$. The number of arrivals in an interval of $t$ minutes has a Poisson distribution, given by (1) of §4.6, with mean $m = 6t$:

$$P(k \text{ arrivals in time } t) = \frac{(6t)^k}{k!} e^{-6t}, \;\; k = 0, 1, 2, \ldots. \tag{5}$$

A variable of interest is $T$, the elapsed time from a given point $t_0$ to the next arrival. [The initial value $t_0$ is immaterial in view of the time-invariance in assumption (a) of §4.6.] The variable $T$ is continuous. The function $R(t)$ defined by (4) is easy to find by expressing the event $[T > t]$ in terms of numbers of arrivals. For any positive number $t$,

$$R(t) = P(T > t) = P(0 \text{ arrivals in time } t) = e^{-6t}.$$

The c.d.f. is therefore

$$F(t) = 1 - R(t) = 1 - e^{-6t}, \quad t > 0.$$

Clearly, $F(t)$ is 0 when $t < 0$. Figure 5-2 shows this c.d.f. The distribution is one of a family of distributions to be studied in more detail in §6.2.     ∎

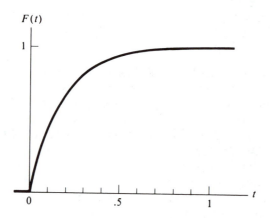

**Figure 5-2** Waiting time c.d.f.—Example 5.1b

The definition of a distribution function given by (2) is general, applying to any distribution on the real line—continuous, discrete, or neither. In (3), we were careful to express the probability $F(b) - F(a)$ as $P(a < X \leq b)$, so that it holds whether the single values $a$ and $b$ have positive probabilities or not. (In Example 5.1a, individual values had probability 0.)   So (3) is valid for the discrete distributions of the preceding chapters.

If a single value $x = k$ does have a positive probability, the c.d.f. will suddenly increase or *jump* by that amount as $x$ passes through the value $k$:

$$P(X = k) = P\{\bigcap [k - \frac{1}{n} < X \leq k]\}$$

$$= \lim_{n \to \infty} P(k - \frac{1}{n} < X \leq k) = \lim_{n \to \infty} [F(k) - F(k - \frac{1}{n})] = F(k) - F(k-),$$

which is the size of the *jump* in the function $F$ at $x = k$. [The value of $F$ *at* a jump point is the limit from the right because of the way we defined $F(x)$ as the probability to the left of or *at* the point $x$.]

**Example 5.1c** | **A Binomial Cumulative Distribution Function**

Consider the distribution of the number of successes in four independent trials of a Bernoulli experiment with $p = 1/3$: $X \sim \text{Bin}(4, 1/3)$. The probabilities of the individual possible values are given in the following table, along with the probabilities accumulated from the left up to the given point:

| $k$ | $f(k)$ | $\sum_0^k f(j)$ |
|---|---|---|
| 0 | 16/81 | 16/81 |
| 1 | 32/81 | 48/81 |
| 2 | 24/81 | 72/81 |
| 3 | 8/81 | 80/81 |
| 4 | 1/81 | 1 |

The probabilities $f(k)$ are amounts of the jumps; the values of $F(x)$ are the cumulative probabilities (last column). Figure 5-3 shows the graph of $F$.     ∎

In contrast with the c.d.f. in Figure 5-3, the c.d.f.'s in Examples 5.1a and 5.1b are continuous function—smooth except at $t = 0$. This continuity is typical of distributions we term *continuous*. Continuity of a c.d.f. means that as $x$ moves from left to right, probability is never added in discrete lumps. For technical reasons (beyond our scope), we assume further that the c.d.f. of a continuous distribution is *differentiable* nearly everywhere.

> When the c.d.f. of a random variable $X$ is continuous at each $x$, and it is differentiable except possibly at finitely many points $x$, we say that $X$ has a **continuous distribution.**

(The definition could allow for more points at which $F$ is not differentiable, but the way we've given it is adequate for our purposes.)

So, the c.d.f. of a continuous distribution has *no jumps*. The c.d.f. of a discrete

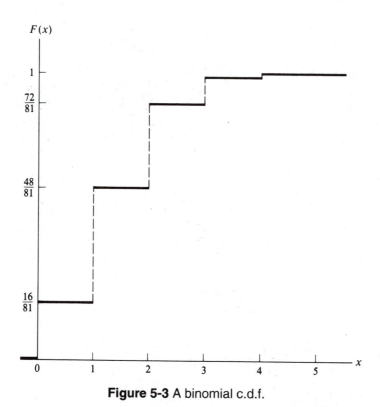

**Figure 5-3** A binomial c.d.f.

distribution, on the other hand, increases *only* in jumps. But there are occasions where a mixture of these two types of model is useful:

**Example 5.1d**

## A Distribution of Mixed Type

Let $T$ denote the time to failure of some electrical appliance. We may think of this as a continuous random variable, except that there could be a positive probability that the appliance will not work when first plugged in—that is, that $P(T = 0) > 0$. The c.d.f. will jump by the amount of that positive probability at $t = 0$, and the c.d.f. may look something like the graph in Figure 5-4. It allots a positive probability to the single value 0, but distributes the remaining probability continuously over the positive real line. ■

   In Chapter 3 we studied transformations of a discrete random variable (§3.2 and §3.3). When $X$ is continuous, we can obtain the c.d.f. of the new random variable $Y = g(X)$ from the c.d.f. of the old. If the transforming function is strictly increasing,

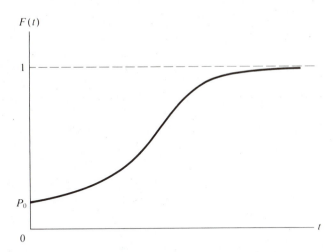

**Figure 5-4** C.d.f. for Example 5.1d

the inverse function is defined and single valued: $x = g^{-1}(y)$. The event $[g(X) \leq y]$ is then equivalent to $[X \leq g^{-1}(y)]$:

$$F_Y(y) = P(Y \leq y) = P[g(X) \leq y] = P[X \leq g^{-1}(y)].$$

The probability on the right defines the c.d.f. of $X$ at the value $g^{-1}(y)$, so

$$F_Y(y) = F_X[g^{-1}(y)]. \tag{6}$$

If $g$ is strictly decreasing, we find (similarly) that

$$F_Y(y) = P[g(X) \leq y] = P[X > g^{-1}(y)] = 1 - F_X[g^{-1}(y)]. \tag{7}$$

**Example 5.1e** | **Linear Transformation of a Uniform Variable**
Suppose $X \sim \mathcal{U}(0, 1)$.  The c.d.f. is a ramp function on the unit interval:

$$F(x) = P(X \leq x) = \begin{cases} 0, & x < 0, \\ x, & 0 \leq x \leq 1, \\ 1, & x > 1. \end{cases}$$

Now suppose we transform $X$ to $Y = 6 + 4X$.  The point $x = 0$ goes over to $y = 6$, and $x = 1$, to $Y = 10$.  We find the inverse transformation by solving for $X$:

$X = (Y - 6)/4$. So then,

$$F_Y(y) = P(Y \leq y) = \begin{cases} 0, & y < 6, \\ (y-6)/4, & 6 \leq y \leq 10, \\ 1, & y > 10. \end{cases}$$

This is a ramp function on the interval $(6, 10)$, so $Y \sim \mathcal{U}(6, 10)$. The transformation has simply stretched the unit interval to one of width 4, by multiplying by 4, and then moved or translated it to the right 6 units by adding the constant 6.

More generally, for $c < d$, the linear transformation $Y = c + (d - c)X$ takes $X \sim \mathcal{U}(0, 1)$ into $Y \sim \mathcal{U}(c, d)$, with c.d.f.

$$F_Y(y) = \begin{cases} 0, & y < c, \\ \frac{y-c}{d-c}, & c \leq y \leq d, \\ 1, & y > d. \end{cases} \tag{8}$$

This is a ramp function on the interval $(c, d)$.  ■

When the transformation defined by $y = g(x)$ does not have a *unique* inverse, the approach we used above still works but requires some care. Each inverse must be taken into account, as we illustrate in the next example.

**Example 5.1f** | **A Double Valued Inverse**

Suppose $X \sim \mathcal{U}(-1, 1)$. The c.d.f. is $F_X(x) = \frac{1}{2}(x + 1)$, $-1 < x < 1$, from (8). Now consider the nonlinear transformation $Y = X^2$. The function $g(x) = x^2$ has two inverses: $g^{-1}(y) = \pm \sqrt{y}$, as seen in Figure 5-5.

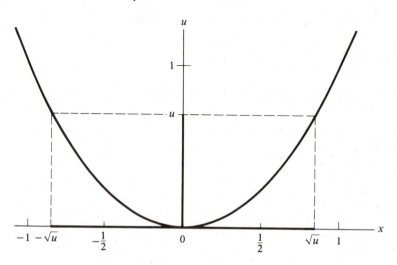

**Figure 5-5** The transformation $y = x^2$

For $0 < y < 1$,

$$F_Y(y) = P(Y \le y) = P(X^2 \le y) = P(-\sqrt{y} \le X \le \sqrt{y}),$$

which we evaluate using the c.d.f. of $X$ and (3) (page 169):

$$F_Y(y) = F_X(\sqrt{y}) - F_X(-\sqrt{y}) = \frac{1}{2}(\sqrt{y}+1) - \frac{1}{2}(-\sqrt{y}+1) = \sqrt{y}.$$

$[F_Y(y) = P(X^2 \le y)$ is 0 when $y < 0$, and is 1 when $y > 1$.] Figure 5-6 shows the new c.d.f. ∎

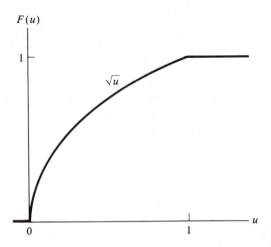

**Figure 5-6** C.d.f. of $Y$, Example 5.1f

In the next section we take up another way of describing a continuous distribution of probability—the distribution of a continuous variable $X$.

## Problems

∗ **5-1.**   Suppose $X \sim \mathcal{U}(-1, 1)$, that is, uniformly distributed on $(-1, 1)$. The c.d.f. is given in Example 5.1f:  $F_X(x) = \frac{1}{2}(x + 1)$ for $-1 < x < 1$. Find:

    **(a)**  $P(X > 3/4)$.          **(c)**  $P(X^2 \le 1/4)$.

    **(b)**  $P(X > 3/4 \mid X > 1/2)$.    **(d)**  $x_0$, such that $P(X \le x_0) = .75$.

**5-2.**   Let $F(x) = x^2$, for $0 < x < 1$. Find the following:

    **(a)**  $P(X < 1/2)$.           **(c)**  $P(\sqrt{X} < 1/2)$.

    **(b)**  $P(X < 1/4 \mid X < 1/2)$.    **(d)**  $P(\sqrt{X} < y)$, as a function of $y$.

[The probability in (d) is the c.d.f. of the variable $Y = \sqrt{X}$.]

**★ 5-3.** Given $F_V(v) = 1 - \cos v$ for $0 < v < \pi/2$,
  **(a)** deduce the value of $F_V(v)$ for $v \leq 0$ and for $v \geq \pi/2$.
  **(b)** find $P(V > \pi/4)$.

**5-4.** Verify that the following function:

$$F(x) = \frac{1 - e^{-4x}}{1 - e^{-4}}, \ 0 \leq x \leq 1,$$

is a c.d.f., when suitably defined outside the unit interval.

**★ 5-5.** A point $Q$ is chosen at random within a circle of radius 1 unit. Interpret this to mean that probability is proportional to area.
  **(a)** Find the c.d.f. of $X$, the distance from $Q$ to the center of the circle.
  **(b)** Find the c.d.f. of $Y = 12X$ (the distance in inches).

**★ 5-6.** Suppose $X \sim \mathcal{U}(0, 1)$. Find the following:
  **(a)** $P(\,|X - \frac{1}{2}| < \frac{1}{4})$.
  **(b)** $F_Y(y)$, where $Y = 2X - 1$.
  **(c)** $F_Z(z)$, where $Z = \sqrt{X}$.

**5-7.** Suppose $X$ has the c.d.f. $F(x) = x^4$ for $0 \leq x \leq 1$. Find the following:
  **(a)** $P(X > .5)$.        **(c)** The c.d.f. of $Y = X^2$.
  **(b)** $P(X^2 < .5)$.       **(d)** The c.d.f. of $Z = X^4$.

**★ 5-8.** The variable $X$ has the c.d.f. defined (in part) as follows:

$$F(x) = \begin{cases} x/3, \ 0 < x < 1, \\ (x+1)/3, \ 1 \leq x < 2. \end{cases}$$

Find the following (a sketch of $F$ will help):
  **(a)** $P(X < \frac{1}{2})$.    **(c)** $P(X \geq 1)$.        **(e)** $P(\frac{1}{2} < X < 3)$.
  **(b)** $P(X = 1)$.    **(d)** $P(X > 1 \mid X \geq 1)$.    **(f)** $P(X = \frac{3}{2})$.

**5-9.** Suppose $X$ has a strictly increasing c.d.f., $F(x)$. [Then $F$ has a unique inverse function, $F^{-1}$, so that the inequalities $F(x) \leq u$ and $x \leq F^{-1}(u)$ are equivalent.] Show that the random variable $U = F(X)$ is $\mathcal{U}(0, 1)$.

**★ 5-10.** A point is selected at random on the unit interval, dividing that interval into two pieces with total length 1. Find the probability that the ratio of the length of the shorter piece to the length of the longer piece is less than 1/4.

---

**5.2**          **Density and the Probability Element**

---

A c.d.f. describes the distribution of probability for a numerical random variable $X$ by giving the fraction of the total probability that lies at or to the left of each $x$. In the discrete case, an alternative description (the p.f.) gives the list of possible values and corresponding probabilities. When $X$ is a continuous variable, an alternative description of its distribution gives the *density* or concentration of probability at each point $x$, which we introduce next.

The probability that a continuous random variable $X$ falls in the interval from $x$ to $x + \Delta x$ is given in terms of the c.d.f. by (3) of §5.1:

$$P(x < X < x + \Delta x) = F(x + \Delta x) - F(x) = \Delta F(x). \tag{1}$$

When $F(x)$ is differentiable at $x$, and $\Delta x$ is small, we can approximate this increment $\Delta F$ by the *differential* of $F$ (see Figure 5-7.):

$$dF(x) = F'(x)\Delta x. \tag{2}$$

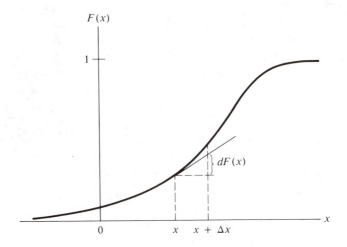

**Figure 5-7** The probability element, $dF$

The differential $dF$ is termed the **probability element** at $x$. It approximates the value of $\Delta F$ using the assumption that $F$ is nearly *linear* over the infinitesimal interval from $x$ to $x + \Delta x$, with slope $F'(x)$.

The derivative $F'$ is the rate of increase in $F$—the rate at which probability is being added at $x$ as we move from left to right along the $x$-axis. Where the c.d.f. is steep, $F'$ is large, and probability is being added at a high rate. When the c.d.f. is nearly flat, $F'$ is small, and probability is being added at a low rate.

The ratio of the amount of probability in the interval $(x, x + \Delta x)$ to the length of the interval $\Delta x$ is the **average density** of probability in that interval:

$$\text{Average density} = \frac{P(x < X < x + \Delta x)}{\Delta x} = \frac{\Delta F(x)}{\Delta x}. \tag{3}$$

Its limit, as $\Delta x \to 0$, defines the density of probability *at* the point $x$:

$$\text{Density at } x = \lim_{\Delta x \to 0} \frac{\Delta F(x)}{\Delta x} = F'(x).$$

This derivative is called the **probability density function** (p.d.f.) of $X$ (or of its distribution) and is commonly denoted by $f(x)$. Like the c.d.f., a p.d.f. may have a subscript to indicate the random variable to which it refers.

---

Probability density function of a continuous random variable $X$:

$$f(x) = \frac{dF(x)}{dx}, \tag{4}$$

where $F$ is the c.d.f. The *probability element* at $x$ is

$$f(x)\,dx \doteq P(x < X < x + dx). \tag{5}$$

---

**Example 5.2a**

## A Uniform Density

In Example 5.1a we found the c.d.f. of the waiting time $T$ at a bus stop to be $F(t) = t/10$ for $0 < t < 10$. The p.d.f. (density function) is the derivative, which is constant ($F' = 1/10$) on that interval and 0 outside it:

$$f(t) = \begin{cases} 1/10, & 0 < t < 10, \\ 0, \text{ elsewhere.} \end{cases}$$

This rectangular p.d.f. is shown in Figure 5-8. The distribution is *uniform* on the interval from 0 to 10: $T \sim \mathcal{U}(0, 10)$.  ∎

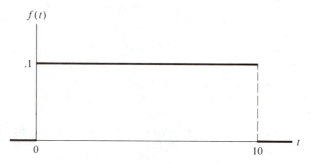

**Figure 5-8** A uniform density function

Many of our examples, like the ones above, will involve p.d.f.'s that are 0 outside some region, and in giving the definition we'll follow the convention of giving the formula(s) for $f$ only for the region where it is not 0, with the understanding that $f = 0$ outside that region.

> The **support** of the distribution of a continuous variable $X$ is the region on which $f_X(x) \neq 0$.

**Example 5.2b** | **Waiting Time in a Poisson Process**

In the Poisson process of Example 5.1b, we found the c.d.f. of the time to the next arrival (after any given point in time) to be

$$F(t) = 1 - e^{-6t}, \quad t > 0$$

(and 0 for negative $t$). The corresponding p.d.f. is the derivative:

$$f(t) = 6e^{-6t}, \quad t > 0,$$

which exists except at $t = 0$. (That $f = 0$ when $t < 0$ is actually implied in the above formula, since the integral of $f$ over $t > 0$ is 1, leaving 0 probability for $t < 0$.) This density function is shown in Figure 5-9. The probability element at $t = .10$ (for instance) is $f(.10)\, dt$. The probability of the interval $(.10, .12)$ is then approximately

$$f(.10)\, dt = 6e^{-.6} \times dt \doteq 3.293 \times .02 = .0659.$$

(See Figure 5-9.)  The actual probability of that interval is

$$P(.10 < X < .12) = (1 - e^{-.72}) - (1 - e^{-.60}) = .0621. \qquad \blacksquare$$

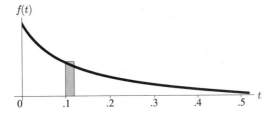

$f(t)$

$t$

**Figure 5-9** Probability element—Example 5.2b

Not all continuous, increasing functions are differentiable at enough points to be useful as a c.d.f.—defining a model for a random variable. The restriction that $F'$ exists except possibly at a finite number of points is sufficient to guarantee that a c.d.f. can be recovered by integrating its derivative:

$$F(x) = \int_{-\infty}^{x} F'(u)\, du = \int_{-\infty}^{x} f(u)\, du. \qquad (6)$$

(The integrand function in a Riemann integral can be undefined or arbitrarily defined at finitely many points without affecting the value of the integral.)

Condition (6) will always be satisfied when we construct a model by specifying a p.d.f. $f$ as a nonnegative, integrable function with the property that the area under its graph is equal to 1. We then *define* the c.d.f. as the integral (6), which is a continuous, nondecreasing function of the upper limit $x$. Moreover, when $F$ is defined in this way, $F(\infty) = 1$ and $F(-\infty) = 0$, so it is indeed a c.d.f. And if we differentiate it to find the p.d.f., we end up where we started, according to the *fundamental theorem of integral calculus:*

$$F'(x) = \frac{d}{dx} \int_{-\infty}^{x} f(u)\,du = f(x), \tag{7}$$

at all continuity points of $f(x)$.

Figure 5-10 shows a typical c.d.f. and its derivative function, $f(x)$. The area under $f$ *to the left* of $x$ is a number (of square units) equal to the height (in linear

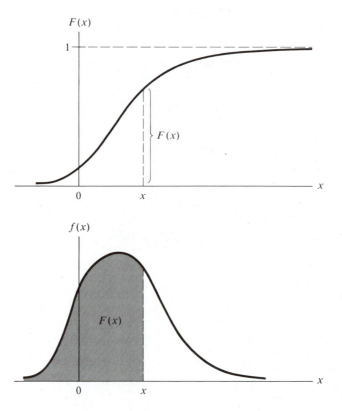

**Figure 5-10** Relating the c.d.f. and p.d.f.

units) of the c.d.f. *at* $x$. And the *height* of the p.d.f. at $x$ is equal to the *slope* of the c.d.f. at that point.

---

Any integrable, nonnegative function $f$ that satisfies the condition

$$\int_{-\infty}^{\infty} f(x)\, dx = 1$$

defines a distribution for a real-valued random variable $X$, with c.d.f.

$$F(x) = \int_{-\infty}^{x} f(u)\, du. \tag{8}$$

---

**Example 5.2c** | **A Cauchy Density**

The continuous (hence integrable) function $(1 + x^2)^{-1}$ is nonnegative, and the area under its graph is finite:

$$\int_{-\infty}^{\infty} \frac{dx}{1 + x^2} = \text{Arctan}(\infty) - \text{Arctan}(-\infty) = \pi.$$

(Here "Arctan" denotes the principal value of the inverse-tangent function: $-\pi/2 < \text{Arctan}\,\theta \leq \pi/2$.) Dividing by $\pi$ produces a function whose integral over the whole $x$-axis is 1:

$$f(x) = \frac{1/\pi}{1 + x^2}. \tag{9}$$

This is the **Cauchy** p.d.f., shown in Figure 5-11. The Cauchy c.d.f. is obtained using (8):

$$F(x) = \int_{-\infty}^{x} \frac{1/\pi}{1 + u^2}\, du = \frac{1}{\pi} \text{Arctan}\, x + \frac{1}{2}, \tag{10}$$

also shown in Figure 5-11.                                                                        ∎

In §5.1 we noted that the probability of an interval is given by the amount of increase in the c.d.f. over that interval:

$$P(a < X \leq b) = F(b) - F(a).$$

This probability can now be expressed in terms of the p.d.f., since both $F(b)$ and

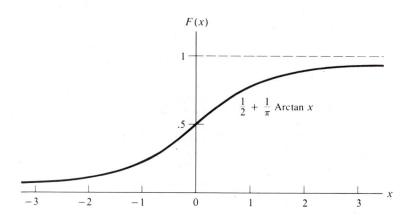

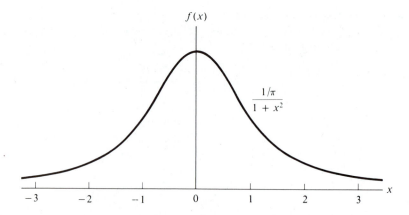

**Figure 5-11** Cauchy c.d.f. and p.d.f.

$F(a)$ are integrals of the p.d.f.:

$$P(a < X < b) = \int_{-\infty}^{b} f(x)\,dx - \int_{-\infty}^{a} f(x)\,dx = \int_{a}^{b} f(x)\,dx. \qquad (11)$$

This is the area under the graph of $y = f(x)$, between $x = a$ and $x = b$, as shown for a typical density in Figure 5-12.

We may also think of the integral (11) as approximately equal to the sum of the probability elements corresponding to subdivisions of the interval from $a$ to $b$:

$$P(a < X < b) \doteq \sum f(x_i)\,\Delta x_i.$$

The integral (11) is the limit of such approximating sums as the number of subdivisions increases and $\Delta x_i$ tends to 0.

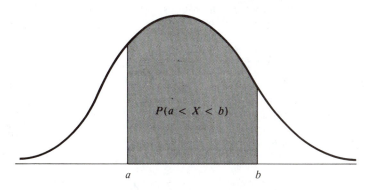

**Figure 5-12** Probability as area

---

When $X$ is continuous with p.d.f. $f(x)$,

$$P(a < X < b) = \int_a^b f(x)\,dx = F(b) - F(a). \qquad (12)$$

---

**Example 5.2d** | **Waiting Time in a Poisson Process (Continued)**

In Example 5.2b we considered the p.d.f. $6e^{-6x}$ for $x > 0$. If we start with this nonnegative function as a p.d.f. and *define* a c.d.f. as its integral:

$$F(x) = \int_0^x 6e^{-6t}dt = -e^{-6t}\Big|_0^x = -e^{-6x} + 1,$$

we obtain a function whose derivative is the p.d.f. with which we started.

The probability of any interval is also a definite integral. For example,

$$P(.1 < T < .3) = \int_{.1}^{.3} 6e^{-6t}dt = -e^{-6t}\Big|_{.1}^{.3} = -e^{-1.8} + e^{-.6} \doteq .384.$$

This can also be expressed in terms of $F$:

$$P(.1 < T < .3) = F(.3) - F(.1) = (1 - e^{-1.8}) - (1 - e^{-.6}),$$

which of course is the same result.　　　　　　　　　　　　　　■

The p.d.f. is a differential coefficient—the coefficient of $dx$ (or $\Delta x$) in a probability element, so we can sometimes derive a p.d.f. using a differential approach. We'll be doing this several times in later chapters and illustrate the method in the next example. The method is to find first the probability of an infinitesimal interval of length $dx$. When this is, at least approximately, proportional to $dx$, the coefficient of $dx$ is the desired p.d.f.

| Example **5.2e** | **Illustrating the Differential Method** |

Let $X$ denote the time to the $r$th arrival in a Poisson process with parameter $\lambda$. For any given $x > 0$, the value of $X$ will lie in the interval $(x, x + dx)$ if and only if (a) there are exactly $r - 1$ arrivals before time $x$, and (b) there is exactly one arrival between $x$ and $x + dx$. Since the events defined as (a) and (b) are independent (involving nonoverlapping intervals), the probability of their intersection is the product of their probabilities:

$$P(x < X < x + dx) = P[r - 1 \text{ arrivals in } (0, x)] \cdot P[1 \text{ arrival in } (x, x + dx)].$$

The first factor on the right is a Poisson probability with $m = \lambda x$, and the second factor is $\lambda dx$, according to one of the Poisson postulates [(c) on p. 142]:

$$P(x < X < x + dx) \doteq \left\{ \frac{(\lambda x)^{r-1}}{(r - 1)!} e^{-\lambda x} \right\} \cdot \lambda dx = \left\{ \frac{\lambda^r}{(r - 1)!} x^{r-1} e^{-\lambda x} \right\} dx.$$

The approximation becomes exact as $dx$ tends to 0, and the coefficient of $dx$ in the probability element is the p.d.f.:

$$f_X(x) = \frac{\lambda^r}{(r - 1)!} x^{r-1} e^{-\lambda x}, \ x > 0. \tag{13}$$

(Example 5.2b treated the special case $r = 1$.)    ∎

In §5.1 we showed, when $Y = g(X)$, how to obtain the c.d.f. of $Y$ from the c.d.f. of $X$. So, we could find the p.d.f. of $Y$ by first finding its c.d.f. and then differentiating it. But given the p.d.f. of $X$, the differential method of finding the p.d.f. of $Y$ is a bit more direct. We'll now use that method to obtain a general formula for the p.d.f. of the transformed variable $Y$ in terms of the p.d.f. of $X$.

Suppose first that $g(x)$ is an increasing function of $x$ with a single-valued inverse function $x = g^{-1}(y)$. Consider an increment in $y$, from $y$ to $y + \Delta y$, produced by an increment in $x$, from $x$ to $x + dx$. The probability of $\Delta y$ is

$$P(y < Y < y + \Delta y) \doteq f_Y(y) \, dy.$$

But this probability is induced from the distribution of $X$ as the probability of the corresponding increment in $x$:

$$P(y < Y < y + \Delta y) = P(x < X < x + \Delta x) \doteq f_X(x) \, \Delta x.$$

The approximation becomes exact as $\Delta x$ tends to 0, so

$$f_Y(y) \, dy = f_X(x) \, dx = \left\{ f_X(x) \frac{dx}{dy} \right\} dy. \tag{14}$$

The quantity in braces, which multiplies $dy$, is the p.d.f. of $Y$. Replacing $x$ by $g^{-1}(y)$, we can write (14) as

$$f_Y(y) = f_X[g^{-1}(y)] \frac{dg^{-1}(y)}{dy}.$$

If $g$ is *decreasing* rather than increasing, the $dy$ produced by a positive $dx$ is negative, and we need to take absolute values. Both cases are covered in the following formulas:

---

If $Y = g(x)$ is one-to-one, with inverse function $g^{-1}(y)$, the p.d.f. of $Y$ is obtained from the p.d.f. of $X$ as

$$f_Y(y) = f_X[g^{-1}(y)] \left| \frac{dg^{-1}(y)}{dy} \right|. \qquad (15)$$

In particular, if $Y = a + bX$,

$$f_Y(y) = f_X\left(\frac{y-a}{b}\right) \cdot \frac{1}{|b|}. \qquad (16)$$

---

The next example illustrates (15) and also an adaptation of the basic idea to a case in which the inverse function $g^{-1}$ is multiple-valued.

**Example 5.2f**

## A Double-valued Inverse

Let $V$ have the p.d.f. $f(v) = 2v$ for $0 < v < 1$, and consider $U = V^2$. On the interval of support, the function $v^2$ has a single-valued inverse, $v = \sqrt{u}$, with derivative $1/(2\sqrt{u})$. The range of $v^2$ is also the unit interval, and for any $u$ on that interval, we have [from (15)]

$$f_U(u) = f_V[g^{-1}(u)] \cdot \frac{dv}{du} = f(\sqrt{u}) \cdot \frac{1}{2\sqrt{u}} = 1.$$

Thus $U \sim \mathcal{U}(0, 1)$. The density of $V$ is $2v$, which increases as $x$ moves to 1, so probability is bunched up toward 1. But $V^2$ is smaller than $V$, and squaring pushes the probability toward 0 just enough to make the density uniform.

Now let's apply the same transformation (squaring) to $X \sim \mathcal{U}(-1, 1)$, to obtain $Y = X^2$, as shown in Figure 5-13 with the inverses $X = \pm\sqrt{Y}$. This transformation is *not* one-to-one, since $X$ can be negative. The probability increment at any point $y$ on $0 < y < 1$ comes from corresponding increments at both $-\sqrt{y}$ and

$+\sqrt{y}$. (See Figure 5-13.)  The p.d.f. of $X$ is $f_X(x) = 1/2$ when $0 < x < 1$, so

$$P(y < Y < y + dy) \doteq f_X(-\sqrt{y}) \cdot \frac{dy}{2\sqrt{y}} + f_X(\sqrt{y}) \cdot \frac{dy}{2\sqrt{y}} = \frac{dy}{2\sqrt{y}}.$$

The p.d.f. of $Y$ is the coefficient of $dy$, $1/(2\sqrt{y})$, and this agrees with what we obtained using c.d.f.'s in Example 5.1d.  ■

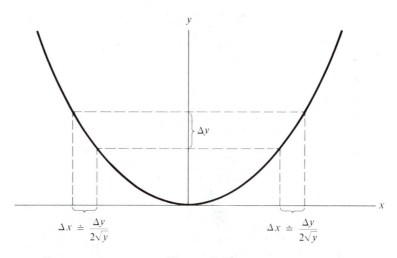

**Figure 5-13**

## Problems

**∗ 5-11.**   The random variable $X$ has the c.d.f. $F(x) = x/2$ for $0 < x < 2$.  Find the p.d.f. and sketch both the c.d.f. and the p.d.f.

**5-12.**   Let $X$ have the following c.d.f.:

$$F(x) = \begin{cases} 2x^2, & 0 < x < 1/2 \\ 1 - 2(1-x)^2, & 1/2 \le x < 1. \end{cases}$$

(Define it appropriately outside the unit interval.)  Find and sketch the p.d.f.

**∗ 5-13.**   Let $X$ have the c.d.f. shown in Figure 5-14.
  **(a)**   Find and sketch the p.d.f.
  **(b)**   Find $P(X > 2 \mid X > 1)$ by finding areas under the graph of the p.d.f.

**5-14.**   Let $X$ have the p.d.f. shown in Figure 5-15.
  **(a)**   Find $P(X < -1)$.
  **(b)**   Find $P(X < -1 \mid X < 1)$.
  **(c)**   Find $P(|X| < 1/2)$.

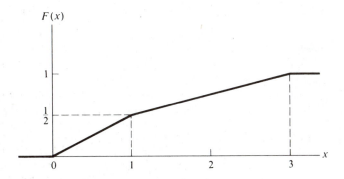

**Figure 5-14** C.d.f. for Problem 5-13

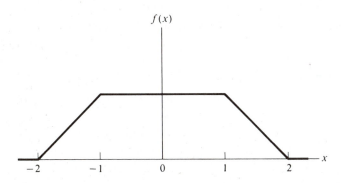

**Figure 5-15** P.d.f. for Problem 5-14

* **5-15.** Let X have the p.d.f. $f(x) = c(1 - |x|)$ for $|x| < 1$, where $c$ is a positive constant. Find the following:
  (a) The value of $c$.       (b) $P(|X| > 1/2)$.       (c) The c.d.f. of $X$.

**5-16.** Suppose $X$ has the p.d.f. $30x^2(1 - x)^2$ on $0 < x < 1$. Find
  (a) the c.d.f. of $X$.        (c) $P(X < 1/4 \mid X < 1/2)$.
  (b) $P(1/4 < X < 3/4)$.

* **5-17.** Let $X$ have the distribution given in Problem 5-14 (and Figure 5-15). Find the p.d.f. of the random variable $Y = |X|$.

**5-18.** Let $T$ denote the time to the *second* arrival in a Poisson process.
  (a) Derive the c.d.f. after verifying that the event $[T > t]$ is equivalent to the event that there is either no arrival or exactly one arrival in the interval from 0 to $t$.
  (b) Differentiate the c.d.f. found in (a) to verify that it agrees with the p.d.f. as given in Example 5.2e, when $r = 2$.

∗ **5-19.** Use the probability element to find the approximate probability of the interval $0 < X < .1$, where $X$ has the Cauchy p.d.f. in Example 5.2c.

**5-20.** Suppose $X \sim \mathcal{U}(0, 1)$. Find the distribution of $Y = \theta X$, where $\theta$ is a positive constant.

∗ **5-21.** Suppose $X \sim \mathcal{U}(-1, 1)$. Find the p.d.f. of each of the following:

(a)   $X$.    (b)   $Y = |X|$.    (c)   $Z = X^2$.

**5-22.** Let $X$ have the distribution defined in Problem 5-15.

(a)   Find the probabilities of the four subintervals determined by the points $-1/2, 0, 1/2$.

(b)   Suppose we classify the results of $n$ independent trials of $X$ according to the intervals of (a) into which they fall. Let $Y_1, ..., Y_4$ denote the numbers of $X$'s in four subintervals. Identify the joint distribution of the $Y_i$'s.

---

**5.3**         **The Median and Other Percentiles**

---

The median of a continuous distribution is a number such that half the *probability* is to its left and half to its right. We can always find such a number, since the c.d.f. is continuous and therefore takes on every value between 0 and 1. Any value $x$ at which $F_X(x) = 1/2$ is a median value.

---

A **median** of a continuous distribution with c.d.f. $F$ is any number $m$ such that

$$F(m) = \frac{1}{2}. \tag{1}$$

---

There may be more than one number $m$ that satisfies (1), so the median may not be uniquely defined. For instance, suppose a random variable has half of its probability on the interval $(0, 1)$ and the other half on the interval $(2, 3)$. Then the c.d.f. is $1/2$ for all values between 1 and 2, and any such number is a median.

Because $1/2 = 50\%$, we also refer to a median as a *50th percentile*. More generally, a $100p$th percentile of a continuous distribution is a number $x_p$ (not necessarily unique) such that the proportion of probability to its left is $p$:

$$F(x_p) = P(X \leq x_p) = p. \tag{2}$$

Figure 5-16 shows an example, both in terms of the c.d.f. and in terms of the p.d.f. The area under the p.d.f. to the left of $x_p$ is equal to $F(x_p)$, which is $p$, and the area to the right is $1 - p$.

The 25th percentile is also called the first *quartile*, and the 75th percentile, the third *quartile*. (The second quartile would be the 50th percentile, or the median.) The various percentiles, quartiles, and deciles (10th, 20th, etc., percentiles) are referred to also as *quantiles*.

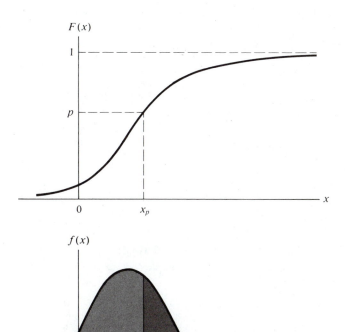

**Figure 5-16** Illustrating a percentile

Example **5.3a**

**Quantiles for a Waiting Time**

In the Poisson process of Example 5.1b, we found that the time $T$ to the next arrival after any specified point in time has the c.d.f.

$$F(t) = 1 - e^{-6t}, \quad t > 0.$$

To find the median waiting time, we set $F$ equal to .5 (Here, and in what follows, "log" means logarithm to the base $e$):

$$1 - e^{-6t} = .5, \quad \text{or } e^{-6t} = .5, \quad \text{or } 6t = -\log(.5) \doteq .693.$$

Division by 6 produces the median, $.693/6 \doteq .116$. The first quartile, or 25th percentile, is a value of $t$ such that

$$1 - e^{-6t} = .25 \quad \text{or } t = -\frac{1}{6}\log.75 \doteq .048. \qquad \blacksquare$$

The median is especially easy to find for a **symmetric** distribution. A distribution is symmetric if there is a point such that the pattern of probability on one side is a reflection of the pattern on the other side:

---

A distribution is **symmetric about $c$** when, for every $x$,

$$F(c - x) = 1 - F(c + x), \tag{3}$$

or, in terms of densities,

$$f(c - x) = f(c + x). \tag{4}$$

If it is symmetric about $\tilde{x}$, then $\tilde{x}$ is a median of the distribution.

---

**Example 5.3b** | **Quantiles of a Cauchy Distribution**

Translating the distribution of Example 5.2c by an amount $\theta$, we obtain a distribution with p.d.f.

$$f(x \mid \theta) = \frac{1/\pi}{1 + (x - \theta)^2}. \tag{5}$$

This is also termed a Cauchy density. It is symmetric about $\theta$, so $\theta$ is the median of the distribution. There is only the one median, and to use (1) to find it, we first find the c.d.f.:

$$F(x) = \int_{-\infty}^{x} \frac{du/\pi}{1 + (u - \theta)^2} = \frac{1}{\pi} \left[ \operatorname{Arctan}(x - \theta) - \operatorname{Arctan}(-\infty) \right]$$

$$= \frac{1}{\pi} \operatorname{Arctan}(x - \theta) + \frac{1}{2}.$$

Setting this equal to 1/2, we see that the first term must be 0, which means that $\theta$ is the only median.

To find, say, the first quartile of the distribution, we set the c.d.f. equal to 1/4, and find

$$\operatorname{Arctan}(Q_1 - \theta) = -\frac{\pi}{4}, \quad \text{or} \quad Q_1 - \theta = -1.$$

The first quartile is therefore $\theta - 1$. Because of the symmetry, the third quartile is $\theta + 1$. ∎

---

## 5.4     Expected Value

We define the mean value or expected value of a *continuous* random variable much as we defined it for discrete variables in §3.1, as a "sum" of possible values weighted according to their probabilities. Thus, we multiply each value $x$ by $f(x)\,dx$, the probability element at $x$, and "sum" these products—the way we sum over a continuous index, by *integrating.*

| Example **5.4a** | **An Approximate Integration** |

In practice, integrations must often be done using numerical approximations. To illustrate such an approximation, suppose $X$ has the triangular p.d.f. $f(x) = 2x$ for $0 < x < 1$. We can approximate this distribution with that of a discrete variable $X^*$, obtained by defining ten subintervals of equal width with tenths as endpoints and rounding each value of $X$ to the center of the interval in which it lies. The subintervals are listed in Table 5-1 according to those midpoint values: .05, .15, .25, etc.

Each subinterval has width .1, and the probability of a subinterval is approximately given by the probability element:

$$P(X^* = x) \doteq f(x)\,\Delta x = 2x\,\Delta x = .2\,x.$$

Table 5-1 gives these probabilities, as well as a column of values of the products $x\,f(x)\Delta x$. The mean value of $X^*$ is the sum of those products, or .6650, and this is an approximating sum for the definite integral

$$\int_0^1 x\,f(x)\,dx = \int_0^1 2x^2\,dx = \frac{2}{3} = .66\overline{6}.$$

**Table 5-1**

| $x$ | $f(x)\,\Delta x$ | $xf(x)\,\Delta x$ |
|-----|------------------|-------------------|
| .05 | .01 | .0005 |
| .15 | .03 | .0045 |
| .25 | .05 | .0125 |
| .35 | .07 | .0245 |
| .45 | .09 | .0405 |
| .55 | .11 | .0605 |
| .65 | .13 | .0845 |
| .75 | .15 | .1125 |
| .85 | .17 | .1445 |
| .95 | .19 | .1805 |
|     | 1.00 | .6650 |

We'd get better and better approximations to this integral by taking ever smaller subintervals, since the corresponding approximating sums converge to the definite integral as the number of subintervals becomes infinite (and $\Delta x \rightarrow 0$). ∎

As in the example, we define the mean value of a continuous random variable as a definite integral. In calculus, the integral over a finite range is defined as the limit of approximating sums like the one in the above example. So when $X$ has a distribution whose support is the finite interval $(a, b)$, we define the mean or expected value of $X$ as

$$EX = \int_a^b x\, f(x)\, dx. \tag{1}$$

When the support is not finite, the integral over the support region is an improper integral:

$$\int_{-\infty}^{\infty} x\, f(x)\, dx, \tag{2}$$

and this may or may not "exist."

The value of an integral that is improper because of the infinite range of integration is defined as the limit of a proper integral over a finite interval $(a, b)$ as the endpoints tend independently to $-\infty$ and $+\infty$, respectively:

$$EX = \lim_{\substack{b \to \infty \\ a \to -\infty}} \int_a^b x\, f(x)\, dx = \lim_{a \to -\infty} \int_a^0 x\, f(x)\, dx + \lim_{b \to \infty} \int_0^b x\, f(x)\, dx. \tag{3}$$

If each limit on the right is finite, the improper integral (2) is said to *converge,* and the mean is defined as (3). If the integral in (3) is of the form $-\infty + C$ or $C + \infty$ for some finite $C$, we say that the mean value is infinite. If it is of the form $-\infty + \infty$, the integral is said not to exist, and the mean value is not defined—we say that the distribution does not have a mean value.

As in the case of discrete variables, the terms *mean, average, expectation,* and *expected value* are synonymous. The notations $EX$, $\mu$, and $\mu_X$ will be used interchangeably.

---

The **mean value** of a continuous random variable with p.d.f. $f$ is

$$EX = \int_{-\infty}^{\infty} x\, f(x)\, dx, \tag{4}$$

when this integral exists.

---

If the support of the distribution is not the whole infinite line, the limits of the integral (4) can be replaced by the endpoints of the supporting interval.

Example **5.4b** | **Mean Waiting Time**

Suppose $X$ has the p.d.f. $6e^{-6x}$ for $x > 0$, as in Example 5.2b. The mean value is

$$EX = \int_0^\infty x \cdot 6e^{-6x}\,dx = \lim_{b \to \infty} \int_0^b 6xe^{-6x}\,dx = \frac{1}{6}.$$

(The integration can be done "by parts" or by looking in a table of integrals.) Figure 5-17 shows the p.d.f. and the mean value, as well as the median value ($\tilde{x} = .116$, from Example 5.2b). As is true generally for distributions that are skewed to the right, the mean is to the right of the median. ∎

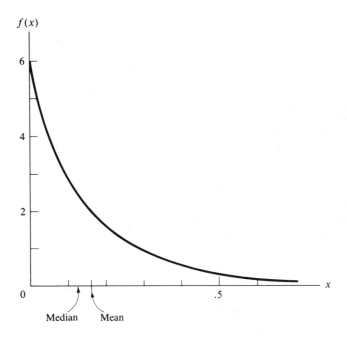

**Figure 5-17** Mean and median waiting time

When $f(x)$ is a mass density, normalized so that the total mass is 1, the formula (4) for the mean value is the same as the formula for the center of gravity. And this suggests that if the distribution is symmetric, the point of symmetry is the mean value. To see that this is indeed the case, we'll assume (without loss of generality) that the point of symmetry is 0. [If it is something other than 0, say $x = k$, then $Y = X - k$ is symmetric about 0, and $EY = 0$ would imply that $EX = k$.]

Breaking the range of integration into two pieces, $(-\infty, 0)$ and $(0, \infty)$, we can write (4) as the sum of two integrals:

$$EX = \int_{-\infty}^{0} x\, f(x)\, dx + \int_{0}^{\infty} x\, f(x)\, dx.$$

In the first integral we change the variable of integration to $-x$:

$$EX = \int_{+\infty}^{0} (-x)\, f(-x)(-dx) + \int_{0}^{\infty} x\, f(x)\, dx.$$

We then use one minus sign to reverse the direction of the first integral:

$$EX = \int_{0}^{\infty} (-x)\, f(-x)\, dx + \int_{0}^{\infty} x\, f(x)\, dx = \int_{0}^{\infty} x\, \{f(x) - f(-x)\}\, dx.$$

Symmetry about 0 means that $f(x) - f(-x) = 0$, so $EX = 0$.

---

If the distribution of $X$ is symmetric about $x^*$, then either $EX = x^*$, or the mean value does not exist.

---

**Example 5.4c** | **Cauchy Distributions Have No Mean Value**

In Example 5.3b, we considered the Cauchy p.d.f. centered at $\theta$:

$$f(x \mid \theta) = \frac{1/\pi}{1 + (x - \theta)^2}.$$

As we pointed out there, the median is $\theta$, the point of symmetry. However, the integral we'd write for the mean value does not exist:

$$\lim_{a \to -\infty} \int_{a}^{0} \frac{x/\pi}{1 + (x - \theta)^2}\, dx + \lim_{b \to \infty} \int_{0}^{b} \frac{x/\pi}{1 + (x - \theta)^2}\, dx. \tag{5}$$

The first integral diverges to $-\infty$, because the integrand behaves like $1/x$ at infinity, and the integral of $dx/x$ is divergent. Similarly, the second integral diverges to $+\infty$, so (5) is of the form $\infty - \infty$. The mean does not exist.   ∎

---

## 5.5    Expected Value of $g(X)$

A function of a continuous random variable is a random variable. We can find its mean value using a formula that parallels the formula we used in the discrete case (§3.2). The probability element plays the role of the probability of $x$, and we multiply the value of $g(x)$ by the probability element and "sum":

When $X$ has the p.d.f. $f$,

$$E[g(X)] = \int_{-\infty}^{\infty} g(x)\, f(x)\, dx. \tag{1}$$

An alternative way of calculating $E[g(X)]$ is to find the p.d.f. of the random variable $Y = g(X)$, if this is continuous, and apply the definition (4) of §5.4. [If it is discrete, we'd find the p.f. and apply (1) of §3.1.] The advantage of (1) above is that it avoids the extra step of finding the p.d.f. of $Y$ (or p.f. if $Y$ is discrete).

If $g$ is monotonic, showing the equivalence of the two methods of finding $E[g(X)]$ amounts to changing the variable in the integral (1): $y = g(x)$. In general, a rigorous proof of the equivalence requires mathematics beyond what we are assuming.

**Example 5.5a**  Suppose $X \sim \mathcal{U}(0, 1)$, and let $Y = X^3 + 1$. The range of $Y$-values is from 1 (when $X = 0$) to 2 (when $X = 1$). Using (1), we find

$$E(X^3 + 1) = \int_0^1 (x^3 + 1) \cdot 1\, dx = \frac{1}{4} + 1 = \frac{5}{4}.$$

To use the alternate method we first need to find the p.d.f. of $Y$. The transformation from $X$ to $Y$ is monotonic, so we can use (15) of §5.2:

$$f_Y(y) = f_X[(y-1)^{1/3}] \frac{d}{dy} (y-1)^{1/3} = \frac{1}{3(y-1)^{2/3}}, \quad 1 < y < 2.$$

We then find the mean value of $Y$ in the usual way:

$$EY = \int_{-\infty}^{\infty} y\, f_Y(y)\, dy = \int_1^2 \frac{y\, dy}{3(y-1)^{2/3}} = \frac{5}{4}.$$

The last integration is easy to carry out if you first make the substitution $y = u + 1$. And if, instead, you make the change of variable $(y - 1)^{1/3} = x$, the integral reduces to the above $x$-integral. ∎

**Example 5.5b**  A function of a continuous variable $X$ may define a discrete random variable $Y$. Perhaps the simplest example of this is $Y = I_{(a,b)}(X)$, the indicator function of the interval from $a$ to $b$. Using (1), we find

$$EY = \int_{-\infty}^{\infty} I_{(a,b)}(x)\, f_X(x)\, dx = \int_a^b f_X(x)\, dx = P(a < X < b).$$

For the other method we need the distribution of $Y$. This variable takes on one of only two values, 1 when $X \in (a, b)$ and 0 when $X \notin (a, b)$. So it is a Bernoulli variable with $p = P(a < X < b)$, and its mean value is $p$.     ■

Formula (1), together with the fact that the definite integral operates *linearly* on the integrand, gives us the following property, which parallels the property of linearity of expectations in the discrete case:

Given functions $g$ and $h$ of a random variable $X$ and any constants $a$, $b$, and $c$,

$$E[ag(X) + bh(X) + c] = aE[g(X)] + bE[h(X)] + c, \qquad (2)$$

provided the expected values exist.

## Problems

* **5-23.**  Suppose $Y$ has the p.d.f. $f(y) = y/2$ for $0 < y < 2$. Find
  (a)   the median.
  (b)   the mean.
  (c)   the probability that $Y$ exceeds its 10th percentile.
  (d)   the probability that $Y$ falls between its 10th and 90th percentiles.

**5-24.**  Let $X$ have the p.d.f. $3(1 - x^2)/4$ for $|x| < 1$. Without integrating, find the mean and the median of the distribution.

* **5-25.**  Given that $W$ has the triangular p.d.f. $f(w) = 1 - |w|$ for $|w| < 1$, find
  (a)   $P(|W| < 1/2)$     (b)   $EW$.     (c)   $E|W|$.     (d)   $E(W^2)$.

**5-26.**  Let $X$ have the constant density $1/2$ on $(0, 1)$ and the constant density $1/4$ on $(1, 3)$.
  (a)   Verify that the density outside those two intervals must be 0.
  (b)   Find $P(1/2 < X < 2)$.
  (c)   Find $EX$.

* **5-27.**  Suppose $Y$ has the c.d.f. $F(y) = y^3$ for $0 < y < 1$.
  (a)   Find the median value.
  (b)   Find $E(Y - Y^2)$.
  (c)   Find the probability that $Y$ falls between its 20th and 65th percentiles.

**5-28.**  Given $U \sim \mathcal{U}(0, 1)$. Find
  (a)   $E(4U - 1)$.     (b)   $E[(U - 1/2)^2]$.     (c)   $E(e^{2U})$.

**5-29.**  Consider the *logistic* distribution defined by the c.d.f.

$$F(x) = \frac{1}{1 + e^{-\pi x/\sqrt{2}}}.$$

  (a)   Show that the distribution is symmetric about $x = 0$.
  (b)   Find the median and quartiles of the distribution.

**∗ 5-30.** Given that $X$ has the p.d.f. $\sin x$ for $0 < x < \pi/2$, find
   **(a)** $E(\cos X)$.
   **(b)** $EX$. (Use a table of integrals or integrate by parts.)

**∗ 5-31.** The Maxwell distribution arises in mathematical physics. Its p.d.f. is

$$f(y) = \sqrt{\frac{2}{\pi}}\, y^2 e^{-y^2/2}, \quad y > 0.$$

Find the mean of the distribution. (Integrate by parts, perhaps first making the change of variable $z = y^2/2$.)

**5-32.** One finds this definite integral in every book of mathematical tables:

$$\int_0^\infty x^n e^{-x}\, dx = n!.$$

It follows that $f(x) = \frac{1}{24} x^4 e^{-x}$ is a p.d.f. on $(0, \infty)$. Find the following:
   **(a)** $EX$     **(b)** $E(X^2)$

---

**5.6**  ## Average Deviations

As in the case of a discrete random variable, we describe variability in terms of "average" deviation from the mean. The difference here is that in calculating expected values, we use an integral in place of a sum.

---

**Variance** of a continuous random variable $X$:

$$\sigma^2 = \text{var}\, X = E[(X - \mu)^2] = \int_{-\infty}^{\infty} (x - \mu)^2 f(x)\, dx. \qquad (1)$$

Alternatively,

$$\sigma^2 = E(X^2) - \mu^2 = \int_{-\infty}^{\infty} x^2 f(x)\, dx - \mu^2. \qquad (2)$$

---

The equivalence of (1) and (2) is a special case of the **parallel axis theorem:**

$$E[(X - c)^2] = \text{var}\, X + (\mu - c)^2. \qquad (3)$$

The proof of (3) is precisely the same as in the discrete case. And as in that case, it shows that the *smallest* second moment is the one taken about the mean. That is, the smallest second moment is the variance.

The positive square root of the variance is (as before) the **standard deviation** (s.d.), whose units are those of $X$. The more natural, though less commonly used

**mean deviation** (m.a.d.) is defined, as for discrete variables, as the expected value of the absolute deviation $|X - \mu|$:

---

Standard deviation of $X$:

$$\sigma = \sqrt{\operatorname{var} X}. \tag{4}$$

Mean deviation of $X$:

$$\text{m.a.d.} = E|X - \mu| = \int_{-\infty}^{\infty} |x - \mu|\, f(x)\, dx. \tag{5}$$

---

As in the discrete case, we have these properties:

(i)    m.a.d. $\leq \sigma$

(ii)   $E|X - c|$ is smallest when $c$ is the *median* of the distribution.[2]

Because of (ii), the mean deviation is sometimes defined as the average deviation about the median. The proof of (i) is as in the discrete case (Problem 3-21). It follows from the fact that a variance is always nonnegative; with this fact, and (2) applied to the random variable $|X - \mu|$, we have

$$\operatorname{var}|X - \mu| = E[(X - \mu)^2] - [E|X - \mu|]^2 \geq 0.$$

With the transposition of the second term to the right side, this says that the variance of $X$ is at least as large as the square of the m.a.d., and this is equivalent to (i).

Both the standard deviation and the mean deviation are to be thought of as "average" or "typical" deviations from the mean. They are both no larger than the largest deviation nor smaller than the smallest deviation.

As a rough check on your calculations, it is usually a good idea to guess the s.d. even before you do any integrating. But it is generally easier to guess the value of the m.a.d. You can do this by covering either half of the p.d.f. (say, to the left of $\mu$) and estimating the mean of the uncovered portion. The distance from this to $\mu$ is an estimate of the m.a.d. Increasing it slightly gives a ballpark estimate of the s.d. If your calculation of the s.d. results in something far from your guess, recheck the work. And if your calculation turns out to give a number larger than the largest or smaller than the smallest deviation, you know you're wrong.

**Example 5.6a** | **The Variance of $\mathcal{U}(0, 1)$**

If you cover the left half of a uniform p.d.f., the mean of the right half is at the midpoint, so the mean deviation here has to be one-quarter. To verify that the m.a.d.

---

[2]For a proof of (ii), see Cramér [5], 178–79.

is 1/4, we can find it by doing an integration, breaking the range of integration into the interval where $x - .5$ is positive and the interval where it is negative:

$$\text{m.a.d.} = \int_0^1 |x - .5| \, dx = \int_0^{.5} (.5 - x) \, dx + \int_{.5}^1 (x - .5) \, dx = .25.$$

To calculate the variance, we use (2), for which we need the average square:

$$\int_0^1 x^2 \cdot 1 \, dx = \frac{1}{3}.$$

Because the distribution is symmetric about $1/2$, the mean is $1/2$. And then,

$$\sigma^2 = E(X^2) - \mu^2 = \frac{1}{3} - \left(\frac{1}{2}\right)^2 = \frac{1}{12}.$$

The s.d. is the square root: $\sigma = 1/\sqrt{12} \doteq .289$, a bit larger than the mean deviation. Figure 5-18 shows the p.d.f., m.a.d., and standard deviation. ∎

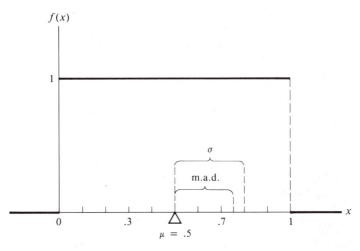

**Figure 5-18** Standard deviation and m.a.d.—Example 5.6a

Equation (2) of the preceding section, showing the linearity of the expectation operator, tells us that with a linear transformation of $X$ to $Y = a + bX$, the mean is transformed in the same way. As in the discrete case (and for the same reason) the standard deviation is only affected by the scale factor $b$, and this is also true of the m.a.d.:

If $Y = a + bX$, then

$$\mu_Y = a + b\mu_X, \tag{6}$$

and

$$\sigma_Y = |b|\sigma_X, \text{ and m.a.d.}(Y) = |b| \cdot \text{m.a.d.}(X).$$

There is a particular transformed scale that is especially useful. The origin or reference point on this new scale is at the mean of $X$, and the unit for the new scale is the standard deviation of $X$. The variable obtained by this linear transformation is called a **standard score** or **$Z$-score:**

$$Z = \frac{X - \mu_X}{\sigma_X}. \tag{7}$$

The interpretation of this $Z$-score is that $X$ lies $Z$ standard deviations to the right of its mean. (A negative $Z$ means that $X$ lies to the left of the mean.) So, as we see by solving (7) for $X$,

$$X = \mu_X + Z\,\sigma_X. \tag{8}$$

We say that $Z$ is a "standard" score because it has what we term the standard parameters, $\mu = 0$ and $\sigma = 1$:

$$EZ = \frac{E(X - \mu_X)}{\sigma_X} = 0, \text{ and } \sigma_Z = \frac{1}{\sigma_X}\sigma_X = 1.$$

**Example 5.6b**

## Standardizing GRE Scores

In any given year, hundreds of thousands of college students take the Graduate Record Exam (GRE), and the collection of their scores can be approximated by a continuous distribution. Suppose the mean score is 521 and the s.d. is 123 (as was the case for the "analytical ability" scores in one particular year). Figure 5-19 shows a density function with a scale of $Z$-scores along side the scale of GRE scores.

The $Z$-score for a student whose raw score is 562 is

$$Z = \frac{562 - 521}{123} = \frac{1}{3}.$$

This means that 562 is one-third of a standard deviation above average. The $Z$-score corresponding to a raw score of 320 is

$$Z = \frac{320 - 521}{123} = -\frac{67}{41} \doteq -1.63.$$

So 320 is 1.63 s.d.'s below the average.

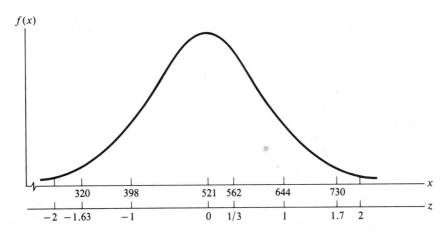

**Figure 5-19** $Z$-scale—Example 5.6b

Going the other way, we can find the GRE corresponding to a given $Z$-score using (8). For instance, when $Z = 1.7$ (see Figure 5-19), the raw score is

$$X = 521 + 1.7 \times 123 = 730.$$

■

## Problems

∗ **5-33.** Find the standard deviation of $X$, defined in Problem 5-24, with p.d.f.

$$f(x) = \frac{3}{4}(1 - x^2), \ |x| < 1.$$

**5-34.** Given that $X$ has the c.d.f. $F(x) = x^4, 0 \le x \le 1$, as in Problem 5-7, find the standard deviation of $X$.

∗ **5-35.** Find the mean deviation of the random variable in Problem 5-33.

**5-36.** Given $Y$ with p.d.f. $f(y) = 1 - |y|$ for $|y| < 1$, as in Problem 5-25, find
(a) the standard deviation of $Y$.       (b) the mean deviation of $Y$.

∗ **5-37.** Using the answer to Problem 5-27 for $E(Y - Y^2)$, find var $Y$, where $Y$ has the c.d.f. $F(y) = y^3$ on the interval $(0, 1)$.

∗ **5-38.** Suppose ACT scores of a population of students have mean 26 and standard deviation 2.4.
(a) Find the $Z$-score of a student whose ACT score is 29.
(b) Find the ACT score if the $Z$-score is $-1.6$.

∗ **5-39.** Suppose $X$ has the density $f(x) = 20x^3(1 - x)$ for $0 < x < 1$. Find
(a) the c.d.f., $F(x)$.        (c) var $X$.
(b) $EX$.                     (d) $P(X < 1/4 \mid X < 1/2)$.

**5-40.**   Find the variance of $Y = 3X - 5$, given $\sigma_X = 2$.

$\ast$ **5-41.**   Find the variance of $X^2$ when $X \sim \mathcal{U}(0, 2)$.

**5-42.**   The function $(1 + |x|^3)^{-1}$ has a finite integral over $(-\infty, \infty)$, so with an appropriate constant multiplier, it defines a p.d.f. Show that the mean and mean deviation exist but the variance does not.

| 5.7 | **Bivariate Distributions** |
|---|---|

We studied discrete *bivariate* distributions in §2.2. Here we take up continuous bivariate probability distributions.

A bivariate p.d.f. $f(x, y)$ gives the *density* of probability, or the probability *per unit area* at the point $(x, y)$. The probability that $(X, Y)$ lies in a small rectangle of dimensions $\Delta x$ and $\Delta y$ at $(x, y)$ is approximately the product of the density at that point and the area of the rectangle:

$$P(x < X < x + \Delta x, \, y < Y < y + \Delta y) \doteq f(x, y)\, \Delta x\, \Delta y. \tag{1}$$

This is the bivariate **probability element** at $(x, y)$. Its graphical representation is the *volume* of the infinitesimal cylinder whose base in the $x$-$y$ plane is a rectangle with sides $\Delta x$ and $\Delta y$, and whose top is the surface $z = f(x, y)$, as shown in Figure 5-20.

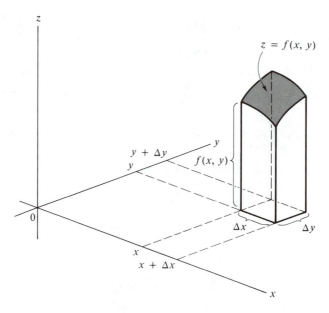

**Figure 5-20** Bivariate probability element

The probability that $(X, Y)$ lies in a region $A$ in the plane is the "sum" of the probability elements in $A$.

The notion of "summing" can be made precise in two dimensions as in one, by a limiting process in which $\Delta x$ and $\Delta y$ tend to 0. The result is a *double integral* over $A$:

$$P[(X, Y) \in A] = \lim_{\Delta x, \Delta y \to 0} \sum_{\Delta y} \sum_{\Delta x} f(x, y) \Delta x \Delta y = \iint_A f(x, y)\, dx\, dy.$$

Interpreted graphically, this integral is approximated by the sum of the volumes of all rectangular cylinders corresponding to probability elements in $A$. Its value is thus the volume under the surface $z = f(x, y)$, above the $x$-$y$ plane, and within a cylinder with base $A$. (See Figure 5-21.)

When the region $A$ is the entire $x$-$y$ plane, its probability must of course be 1. So the volume under the surface defined by a p.d.f. in two dimensions must be 1.

---

A function $f(x, y)$ is a joint p.d.f. if $f(x, y) \geq 0$ and

$$\iint_{x\text{-}y \text{ plane}} f(x, y)\, dx\, dy = 1. \qquad (2)$$

The probability in a region $A$ is the double integral of the p.d.f. over that region:

$$P(A) = \iint_A f(x, y)\, dx\, dy. \qquad (3)$$

---

Any nonnegative integrable function of two variables can serve as the basis of a joint p.d.f., provided the volume under its graph is finite: Dividing the function by that volume produces a function whose double integral is 1.

**Example 5.7a**

## A Uniform Bivariate Density

Suppose $(X, Y)$ is uniformly distributed in the unit square, that is, with p.d.f.

$$f(x, y) = 1, \quad 0 < x < 1 \text{ and } 0 < y < 1.$$

The graph of $f$ is a flat surface above the square, like a square tabletop. The volume under the surface is 1, the volume of a unit cube. The probability of a subregion $A$ of the unit square, represented as the volume of a cylinder of base $A$ and constant

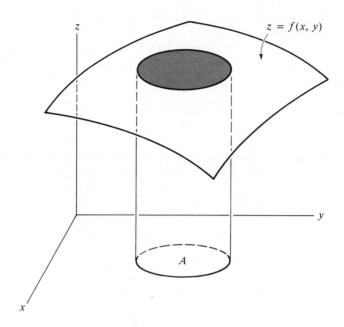

**Figure 5-21** Probability as volume

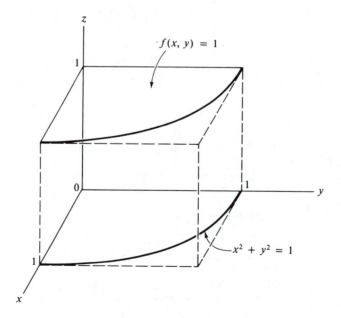

**Figure 5-22** Probability proportional to area

height 1, is simply the area of $A$. For instance, the probability that $X^2 + Y^2 < 1$ is the area of the quarter circle in Figure 5-22:

$$P(X^2 + Y^2 < 1) = \frac{\pi}{4}.$$
◼

Whenever the joint p.d.f. is *constant*, as in Example 5.7a, calculating volumes amounts to calculating areas in the $xy$-plane. Indeed, the probability of $A$ is just the area of $A$ divided by the area of the support set. But calculating a double integral whose integrand is *not* constant is seldom that easy. When the joint p.d.f. is given by a sufficiently simple formula, the value of the double integral can be found by a process of repeated single integrations, as in the following example.

**Example 5.7b**

## Volume Under an Exponential Surface

Let $X$ and $Y$ have the p.d.f. $f(x, y) = e^{-x-y}$ for $x > 0$ and $y > 0$. The graph over the first quadrant is a surface that meets the coordinate planes in the decaying exponential curves $e^{-x}$ and $e^{-y}$. (See Figure 5-23.)

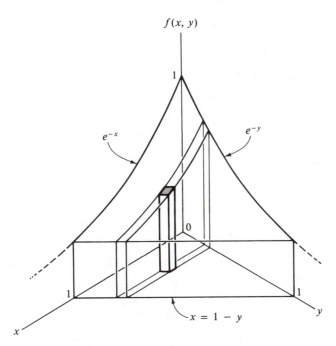

**Figure 5-23** Density surface for Example 5.7b

The probability of the event $X + Y < 1$, for example, is the volume under the p.d.f. surface, above the region where the inequality holds. We write this as a double integral over that region:

$$P(X + Y < 1) = \iint\limits_{x+y<1} e^{-x-y} \, dy \, dx.$$

We can find this volume by adding the probability elements within $A$ systematically. Adding them first in the $x$-direction at a fixed $y$, we find the volume in a slab of width $dy$. We then add these slab volumes in the $y$-direction to find the desired volume. (Again see Figure 5-23).

The first summation defines (in the limit) a single integration with respect to $x$ with $y$ held fixed. It extends from $x = 0$ to the value of $x$ at the boundary line: $x = 1 - y$. The second summation also defines (in the limit) a single integration, this time with respect to $y$, from $y = 0$ to $y = 1$:

$$P(X + Y < 1) = \int_0^1 \left\{ \int_0^{1-y} e^{-x-y} dx \right\} dy = \int_0^1 e^{-y} \left\{ \int_0^{1-y} e^{-x} dx \right\} dy$$

$$= \int_0^1 e^{-y} [1 - e^{-(1-y)}] \, dy = 1 - 2e^{-1}. \qquad \blacksquare$$

**Example 5.7c**  |  **A Case Requiring Numerical Integration**

The function $\exp[-(x^2 + y^2)/2]$ is nonnegative, so dividing by an appropriate constant $K$ (namely, the integral of this function over the plane) produces a bivariate p.d.f. Suppose we want the probability of $X + Y < 1$ when $(X, Y)$ has this density. It is given, formally, by this double integral:

$$P(X + Y < 1) = \frac{1}{K} \int_{-\infty}^{\infty} \int_{-\infty}^{1-y} e^{-(x^2+y^2)/2} dx \, dy.$$

$$= \frac{1}{K} \int_{-\infty}^{\infty} e^{-y^2/2} \left\{ \int_{-\infty}^{1-y} e^{-x^2/2} dx \right\} dy.$$

The indefinite integral of $\exp(-x^2/2)$ is not an "elementary function"—the standard formulas of integral calculus fail. In §6.1 we'll see that $K = 2\pi$, but the $x$-integral here requires numerical methods. $\qquad \blacksquare$

Suppose we are interested in just one of the two variables $X$ and $Y$ whose joint distribution is defined by a given joint p.d.f. The distribution of $X$ alone is called its *marginal distribution*, as in the discrete case. When the distribution of $(X, Y)$ is discrete, we find the marginal p.f. of $X$ at $x$ by summing the joint probabilities in the row corresponding to $X = x$—that is, by "summing out" the $y$ from the joint p.f. Similarly, in the continuous case, we find the probability element $f_X(x) \, \Delta x$ at a given value $x$ by summing the probability elements $f(x, y) \Delta x \, \Delta y$ with respect to $y$ in the strip between $x$ and $x + \Delta x$:

$$f_X(x) \Delta x = P(x < X < x + \Delta x) \doteq \sum_y f(x, y) \Delta y \, \Delta x.$$

The (approximate) marginal p.d.f. of $X$ is the coefficient of $\Delta x$:

$$f_X(x) \doteq \sum_y f(x, y) \Delta y. \tag{4}$$

The limit of the right-hand side of (4), an ordinary single integral of $f(x, y)$ in which $x$ is held fixed, is the *marginal* p.d.f. at $x$. The marginal p.d.f. of $Y$ is defined similarly.

---

**Marginal** (univariate) density functions for $X$ and $Y$:

$$f_X(x) = \int_{-\infty}^{\infty} f(x, y) \, dy, \quad f_Y(y) = \int_{-\infty}^{\infty} f(x, y) \, dx \tag{5}$$

---

Interpreted geometrically, the marginal p.d.f. of $X$ at $x = x_0$ is the *area* under the graph of $f(x_0, y)$. This graph is the intersection of the joint density surface $f$ with the plane $x = x_0$, as we illustrate in the following example.

**Example 5.7d**   | **Uniform Distribution on a Triangle**

Suppose the joint distribution of $(X, Y)$ is uniform on the triangle bounded by the coordinate axes and the line $x + y = 1$. The joint p.d.f. is

$$f(x, y) = 2, \quad x + y < 1, \; x > 0, y > 0.$$

The graph of $f$ is a plane surface parallel to the $xy$-plane, 2 units above the support region, as shown in Figure 5-24.

Using (5), we find the marginal p.d.f. of $X$ at the point $x$ by integrating out the $y$:

$$f_X(x) = \int_0^{1-x} 2 \, dy = 2(1 - x), \quad 0 < x < 1.$$

This integral is the area of the rectangular cross section (at $x$) of the solid region under $f$ and above the triangular support, shown shaded in Figure 5-24. It is just the product of the height (2) times the width $(1 - x)$.

Thus, the (univariate) distribution of $X$ is triangular on the unit interval. It is clear from symmetry that the marginal distribution of $Y$ is the same as that of $X$:
$f_Y(\cdot) = f_X(\cdot)$.   ∎

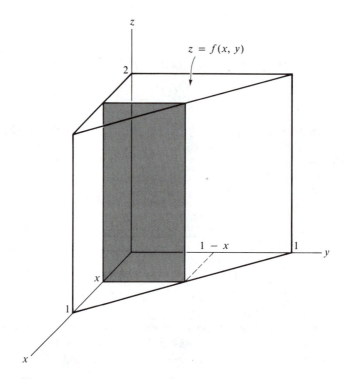

**Figure 5-24** Marginal p.d.f. as cross-section area

Perhaps we should verify that the marginal p.d.f.'s obtained using (5) are indeed density functions: The integral over $y$ of the marginal p.d.f. of $Y$ is

$$\int_{-\infty}^{\infty} f_Y(y)\, dy = \int_{-\infty}^{\infty} \left\{ \int_{-\infty}^{\infty} f(x,\, y)\, dx \right\} dy = \iint_{x\text{-}y \text{ plane}} f(x,\, y)\, dx\, dy = 1.$$

The next example illustrates this in a particular case.

**Example 5.7e**   Suppose that the random pair $(X,\, Y)$ has the joint p.d.f.

$$f(x,\, y) = 6(x - y), \quad 0 < y < x < 1.$$

The constant 6 is determined by the condition that the (double) integral of a bivariate p.d.f. is 1. Figure 5-25 shows the density surface (shaded).

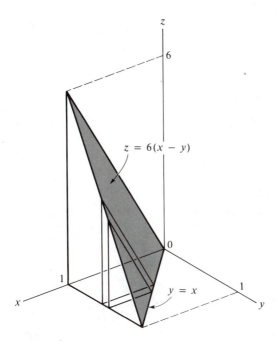

**Figure 5-25** Density surface for Example 5.7e

We find the marginal p.d.f. of $Y$ by integrating the joint p.d.f. with respect to $x$ with $y$ fixed. If $0 < y < 1$, the joint p.d.f. is 0 outside $y < x < 1$:

$$f_Y(y) = \int_y^1 6(x - y)\, dx = (3\,x^2 - 6xy)|_y^1 = 3(1 - y)^2,\ 0 < y < 1.$$

Similarly,

$$f_X(x) = \int_0^x 6(x - y)\, dy = 3x^2,\ 0 < x < 1.$$

Figure 5-25 shows the cross section at $y$, whose area is the value of $f_Y(y)$. [This area, $3(1 - y)^2$, is that of a triangle with base $1 - y$ and height $6(1 - y)$.] It also shows a slab of thickness $\Delta y$ and area $3(1 - y)^2$; the summation on $y$ adds the volumes of these slabs to yield the total volume under the joint p.d.f., which is 1. ∎

Paralleling our calculation of $E[g(X)]$, we find the expected value of a function $g(X, Y)$ as the sum of the products of the value of $g$ at $(x, y)$ and the probability element at that point. But the "sum" is an integral, which in the bivariate setting is a double integral:

---

Expected value of a function of $(X, Y)$:

$$E[g(X, Y)] = \iint\limits_{x\text{-}y\ plane} g(x, y)\, f(x, y)\, dx\, dy. \tag{6}$$

---

Applying (6) to the function $g(x, y) = x$ would give us another definition of $EX$, but of course it is equivalent to the definition based on $f_X$:

$$EX = \int_{-\infty}^{\infty} \left\{ \int_{-\infty}^{\infty} x\, f(x, y)\, dy \right\} dx$$

$$= \int_{-\infty}^{\infty} x \left\{ \int_{-\infty}^{\infty} f(x, y)\, dy \right\} dx = \int_{-\infty}^{\infty} x\, f_X(x)\, dx.$$

**Example 5.7f**    **An Expected Product**

We calculate $E(XY)$ for the distribution of the preceding example, where the joint p.d.f. was $6(x - y)$ for $0 < y < x < 1$:

$$E(XY) = \int_0^1 \int_0^x xy \cdot 6(x - y)\, dy\, dx$$

$$= 6 \int_0^1 x \left\{ \int_0^x (xy - y^2)\, dy \right\} dx = \int_0^1 x^4 dx = \frac{1}{5}.$$

Incidentally, the mean of $Y$ is 1/4 and the mean of $X$ is 3/4. So we see that the average of the product is not the product of the averages.    ■

---

**5.8**              **Several Variables**

---

The concepts and formulas for joint distributions of more than two continuous variables are quite analogous to those for two variables. For random variables $X_1$, $X_2$, ..., $X_n$, the joint p.d.f. is a function $f$ of $n$ arguments that is nonnegative and whose $n$-fold integral over the whole $n$-space is 1:

The joint p.d.f. of $(X_1, X_2, ..., X_n)$ is a function $f$ satisfying

**(i)** $f(x_1, \cdots, x_n) \geq 0,$

**(ii)** $\displaystyle\int_{-\infty}^{\infty} \cdots \int_{-\infty}^{\infty} f(x_1, ..., x_n)\, dx_1 \cdots dx_n = 1.$

We'll not be actually calculating $n$-dimensional integrals. Suffice it to say that their definition parallels that of two-dimensional integrals, but the graphical representations in more than three dimensions are a bit elusive.

Some notational conventions: It is convenient to omit limits in integrals such as that in (ii), when the integral is taken over the whole of $n$-space. Also, we'll be using a boldface $\mathbf{x}$ to represent $(x_1, ..., x_n)$, and $d\mathbf{x}$ to represent the "volume" element $dx_1 \cdots dx_n$.

The *probability element* in $n$ dimensions is the probability of an infinitesimal rectangular parallelepiped:

$$P(x_1 < X_1 < x_1 + dx_1, ..., x_n < X_n < x_n + dx_n) \doteq f(\mathbf{x})\, d\mathbf{x}. \tag{1}$$

For a given region $A$ in $n$-space we define

$$P(A) = \underset{A}{\int \cdots \int} f(\mathbf{x})\, d\mathbf{x}. \tag{2}$$

In particular, if the density is *uniform* over a region $R$ in $n$-space, then the probability in that region is proportional to its *volume*.

With more than two variables, there are marginal distributions of various dimensions. Thus, when $n = 5$, there are 1-, 2-, 3-, and 4-dimensional marginal distributions, obtained by "integrating out" the dummy variables in the joint p.d.f. that correspond to the unwanted variables. The marginal distributions of single variables are obtained by integrating out all the other variables. For instance,

$$f_{X_1}(x_1) = \int_{-\infty}^{\infty} \underset{(n-1)}{\cdots} \int_{-\infty}^{\infty} f(x_1, ..., x_n)\, dx_2 \cdots dx_n.$$

Given a function $g$, we calculate the expected value of $g(X)$ as

$$E[g(\mathbf{X})] = \int \cdots \int g(\mathbf{x}) f(\mathbf{x})\, d\mathbf{x}. \tag{3}$$

Applying (3) when $g$ is a linear function of its arguments, we see that the linearity of expectations [(iv) in §3.1, for the discrete case] will extend to the case of continuous random vectors:

---

**Linearity of Expectations:** Given random variables $(X_1, X_2, ..., X_n)$,

$$E\left\{\sum_{i=1}^{n} a_i X_i\right\} = \sum_{i=1}^{n} a_i E(X_i). \tag{4}$$

---

When the joint p.d.f. of $\mathbf{X}$ is a *symmetric* function of its arguments, the joint p.d.f. is the same for any rearrangement of the variables. This means (as in the discrete case) that the univariate marginal distributions are identical. It also implies that the joint marginal distributions for any subset of $k$ of the $X$'s are the same as those of all other subsets of $k$ $X$'s.

---

Random variables $X_1, ..., X_n$ are **exchangeable** when

$$f(x_1, ..., x_n) = f(x_{i_1}, ..., x_{i_n})$$

for every permutation $(i_1, \ldots, i_n)$ of $(1, 2, \ldots, n)$.

---

**Example 5.8**

### Uniform Distribution in a Tetrahedron

Suppose the p.d.f. of $(X, Y, Z)$ is constant within the tetrahedron defined by the plane $x + y + z = 1$ and the coordinate planes:

$$f(x, y, z) = 6, \quad x + y + z < 1, \; x > 0, \; y > 0, \; z > 0.$$

(The density has to be 6, because the volume of the tetrahedron is 1/6.) This is a symmetric function: Permuting the variables in $(x, y, z)$ does not change the value of the function $f$. So $X$, $Y$, and $Z$ are exchangeable. In particular, they are identically distributed, with common p.d.f.

$$f_X(u) = f_Y(u) = f_Z(u) = \int_0^{1-u} \int_0^{1-u-y} 6 \, dz \, dy = 3(1-u)^2, \; 0 < u < 1.$$

The bivariate marginals are also identical, with common p.d.f.

$$f(u, v) = \int_0^{1-u-v} f(u, v, w) dw.$$

And this in turn implies, for instance,

$$E(XY) = E(XZ) = E(YZ) = \int_0^1 \int_0^{1-u} \int_0^{1-u-v} 6uv \, dw \, dv \, du = \frac{1}{20}. \quad \blacksquare$$

## Problems

**∗ 5-43.** Let $(X, Y)$ have the bivariate p.d.f. $f(x, y) = 2$ in the portion of the first quadrant bounded by $x + y = 1$. Find the following:

    **(a)**  the marginal p.d.f.'s.    **(d)**  $P(X < 1/2)$.

    **(b)**  $E(XY)$.    **(e)**  $P(X < 1/2, Y < 1/2)$.

    **(c)**  $E(X + Y)$.    **(f)**  $P(X + Y < 1/2)$

**∗ 5-44.** Suppose the joint distribution of $X$ and $Y$ concentrates probability 1 uniformly on the line segment $x = y$ in the unit square, $0 < x < 1$, $0 < y < 1$. Using your intuition, determine the marginal distributions of $X$ and $Y$.

**5-45.** Given that $(X, Y)$ has the joint p.d.f. $4xy$ for $0 < x < 1$ and $0 < y < 1$,

    **(a)**  find the marginal means and variances.    **(b)**  find $E(XY)$.

**5-46.** Find the marginal densities, given the joint p.d.f. $f(x, y) = e^{-y}$ when $0 < x < y$.

**∗ 5-47.** Let $(X, Y)$ have the joint density $x + y$ on the unit square.

    **(a)**  Find the marginal p.d.f.'s.    **(c)**  Find $P(X + Y < 1)$.

    **(b)**  Find $E(X + Y)$.    **(d)**  Find $P(X + Y < z)$ for $0 < z < 1$.

**5-48.** Determine whether or not the variables are exchangeable

    **(a)**  in Problem 5-43.    **(b)**  in Problem 5-45.    **(c)**  in Problem 5-47.

**∗ 5-49.** Suppose $(X, Y)$ is uniformly distributed on the region consisting of two unit squares, one in the first quadrant, where $0 < x < 1$ and $0 < y < 1$, and the other in the third quadrant, where $-1 < x < 0$ and $-1 < y < 0$.

    **(a)**  Find $P(X + Y < 1)$.

    **(b)**  Find the marginal p.d.f.'s.

    **(c)**  Are the variables exchangeable?

**5-50.** Suppose $(X, Y, Z)$ has the joint p.d.f. $8xyz$ for $(x, y, z)$ in the unit cube where all coordinates are between 0 and 1. Find $f_X(x)$ by integrating out $y$ and $z$. Find $f_Y$ and $f_Z$ without further integrations.

---

**5.9**      ## Covariance and Correlation

Covariance and correlation are defined for a continuous bivariate pair $(X, Y)$ just as they were defined, in terms of expected values, in the discrete case. The difference is that for continuous distributions, expected values are calculated as integrals.

---

**Covariance and Correlation** of $X$ and $Y$:

$$\sigma_{X,Y} = E[(X - \mu_X)(Y - \mu_Y)], \quad \rho_{X,Y} = \frac{\sigma_{X,Y}}{\sigma_X \sigma_Y}. \tag{1}$$

Moreover, since expectations have the linearity property here as in the discrete case, the various relations developed in §3.4 for the discrete case remain valid:

---

Properties of covariance:

**(i)** $\operatorname{cov}(X,\,Y) = E(XY) - \mu_X\mu_Y.$        (2)

**(ii)** $\operatorname{cov}(X,\,Y) = \operatorname{cov}(Y,\,X).$        (3)

**(iii)** $\operatorname{cov}(aX,\,bY) = ab\operatorname{cov}(X,\,Y).$        (4)

**(iv)** $\operatorname{cov}(X+Y,\,Z) = \operatorname{cov}(X,\,Z) + \operatorname{cov}(Y,\,Z).$        (5)

---

**Example 5.9a**   Consider once again the joint p.d.f. from Example 5.7e: $f(x,\,y) = 6(x-y)$, for $0 < y < x < 1$. There we found the marginal p.d.f. of $Y$: $3(1-y)^2$, $0 < y < 1$, and the marginal p.d.f. of $X$ is

$$f_X(x) = \int_0^x 6(x-y)\,dy = 3x^2,\ \ 0 < x < 1.$$

From these we find the marginal means and variances in the usual way; they are $\mu_X = 3/4$, $\mu_Y = 1/4$, and $\sigma_X^2 = \sigma_Y^2 = 3/80$. In Example 5.7f, we found the expected product to be $1/5$. So from (2), we have

$$\sigma_{X,Y} = \frac{1}{5} - \frac{3}{4}\cdot\frac{1}{4} = \frac{1}{80}\ \text{ and }\ \rho_{X,Y} = \frac{1/80}{3/80} = \frac{1}{3}. \qquad\blacksquare$$

**Example 5.9b**   Applying (3)–(5) we can find, for instance, the covariance of the sum and difference of two variables:

$$\operatorname{cov}(X+Y,\,X-Y) = \operatorname{cov}(X,\,X) - \operatorname{cov}(X,\,Y) + \operatorname{cov}(Y,\,X) - \operatorname{cov}(Y,\,Y)$$

$$= \sigma_X^2 - \sigma_{X,Y} + \sigma_{Y,X} - \sigma_Y^2 = \sigma_X^2 - \sigma_Y^2.$$

If, as sometimes happens, the variances are *equal*, this covariance is 0—which means that then the variables $X+Y$ and $X-Y$ are uncorrelated.      $\blacksquare$

The correlation we calculated in Example 5.9a is between $-1$ and $+1$. We showed in the context of discrete variables that the correlation coefficient *always* lies in that interval: $\rho^2 \le 1$. The proof we gave in §3.4 is just as valid here. And so, as before, if $\rho^2 = 1$, there is a straight line in the plane that carries all of the probability—one variable is a linear function of the other with probability 1.

In §3.4 we also showed that the magnitude of a correlation is preserved under linear transformations of the variables. This holds here as well, and for the same reasons.

---

Properties of correlation:

**(i)** $\rho^2 \leq 1$.                                                                                (6)

**(ii)** If $U = a + bX$ and $V = c + dY$, then

$$\rho_{U,V} = \begin{cases} \rho_{X,Y} & \text{if } bd > 0, \\ -\rho_{X,Y} & \text{if } bd < 0. \end{cases}$$                                (7)

---

**Example 5.9c**    Problem 5-44 dealt with a bivariate distribution that is uniform on the line segment $x = y$ in the unit square, $0 < x < 1$, $0 < y < 1$. Since all of the probability is on a straight line, the correlation coefficient must be 1 ( + 1 because the slope of the line is positive). And of course, the linear relationship between the variables is $X = Y$. So $\text{cov}(X, Y) = \text{cov}(X, X) = \text{var}\,X = \text{var}\,Y$. And the defining formula (1) then does give us $\rho = 1$.                                                                   ∎

We'll be needing formulas, extending (4) and (5) above, for variances and covariances of sums and other linear combinations of several random variables. One such formula was given as (5) in §3.5, for discrete variables; its derivation applies in the same way to continuous variables. The idea in that derivation gives us a general formula for the covariance of one sum or linear combination of variables with another. The essence of the proof is that a covariance is an expected product, and that the product of one sum with another is found by multiplying each term in the one sum by each term in the other.

---

For any random variables $X_1$, ..., $X_m$ and $Y_1$, ..., $Y_n$ with finite second moments, and constants $a_i$ and $b_j$,

$$\text{cov}\left\{\sum_i a_i X_i, \sum_j b_j Y_j\right\} = \sum_i \sum_j a_i b_j \, \text{cov}(X_i, Y_j).$$                        (8)

In particular,

$$\text{var}\left\{\sum_i X_i\right\} = \sum_i \text{var}\,X_i + \sum_{i \neq j} \text{cov}(X_i, X_j).$$                        (9)

---

The right-hand side of (8) is the sum of the $mn$ terms obtained when we pair each term of the first sum on the left with each term of the second—that is, each $a_i X_i$ with each $b_j Y_j$—and add the covariances. The covariance of two such terms is

$$\text{cov}(a_i X_i, \, b_j Y_j) = a_i b_j \, \text{cov}(X_i, \, Y_j).$$

Formula (9) is a special case of (8) in which the $a$'s and $b$'s are all 1's, and all the $Y$'s are $X$'s, because $\text{cov}(X_i, \, X_i) = \text{var}\, X_i$. Observe that if no two of the $X$'s are correlated, the variance is additive:

---

If variables $X_1, \, ..., \, X_m$ are *pairwise uncorrelated,* then

$$\text{var}\left\{\sum_i X_i\right\} = \sum_i \text{var}\, X_i. \tag{10}$$

---

**Example 5.9d**   We illustrate the use of (8) to find the covariance of two linear combinations of $(X, Y, Z)$, namely, $2X + 3Y - Z$ and $X - 2Z$. If we were to multiply these, we'd multiply each term of the first factor by each term of the second, and this would result in six terms. We do the analogous thing with covariances, taking the covariance of each term of the first factor with each term of the second:

$$\begin{aligned}
\text{cov}\,(2X + 3Y - Z, \, X - 2Z) = \, & \text{cov}(2X, \, X) + \text{cov}(2X, \, -2Z) + \\
& \text{cov}(3Y, \, X) + \text{cov}(3Y, \, -2Z) + \\
& \text{cov}(-Z, \, X) + \text{cov}(-Z, \, -2Z).
\end{aligned}$$

Then, removing constant factors and using the fact that the covariance of a variable with itself is its variance, we simplify this to

$$\text{cov}\,(2X + 3Y - Z, \, X - 2Z) = 2\sigma_X^2 - 4\sigma_{X,Z} + 3\sigma_{Y,X} - 6\sigma_{Y,Z} - \sigma_{Z,X} + 2\sigma_Z^2.$$

We could also combine the second and fifth terms, since $\sigma_{Z,X} = \sigma_{X,Z}$. ∎

---

## 5.10        Independence

We define **independence** for two continuous random variables by analogy with the definition in the discrete case, with density functions replacing probability functions:

Continuous variables $X$ and $Y$ are **independent** whenever their joint p.d.f. is the *product* of the marginal p.d.f.'s:

$$f(x, y) = f_X(x)f_Y(y) \ \text{ for all pairs } (x, y). \tag{1}$$

The criterion (1) can be used to check for independence in a given bivariate model, but we'll use it most often to *construct* a joint distribution for variables that are assumed to be independent.

**Example 5.10a** | **Life-lengths of Husband and Wife**
A sociologist is interested in whether the lengths of life of the husband and wife in a married couple are independent. Let $f$ denote the joint p.d.f. and $f_H$ and $f_W$ denote the (marginal) p.d.f.'s of the husband's and wife's length of life. If they are independent, then [by (1)] their joint p.d.f. is the product:

$$f(x, y) = f_H(x)f_W(y) \ \text{ for all pairs } (x, y).$$

The statistical problem for the sociologist is to determine whether the model so defined is supported or contradicted by available data. (We'll consider such inferential questions in Chapters 13 and 15.) ∎

**Example 5.10b** | Suppose the random pair $(X, Y)$ has this joint p.d.f.:

$$f(x, y) = 24xy, \ \ x + y < 1, \ x > 0, \text{ and } y > 0.$$

The variables are exchangeable, because of the symmetry of $f$, so the marginal p.d.f.'s are the same:

$$f_Y(u) = f_X(u) = 12u(1 - u)^2, \ \ 0 < u < 1.$$

The product of $f_X(x)$ and $f_Y(y)$ is *not* equal to the joint p.d.f., so the variables are not independent. So, even though "$xy$" is a product of the symbols $x$ and $y$, the *functions*—including their supports—are not related by the factorization condition (1) for all $x$ and $y$. ∎

Consider a set $A$ of $x$-values and a set $B$ of $y$-values. As an event in the $xy$-plane, the set $A$ defines a strip or cylinder set—a strip parallel to the $y$-axis. Similarly, $B$ defines strip or cylinder set parallel to the $x$-axis. The intersection $C$ of these two cylinder sets is a **product set,** which we denote by $C = A \times B$.

Figure 5-26 shows how intervals $A$ and $B$ define such strips. The product set is a rectangle with sides parallel to the coordinate axes:

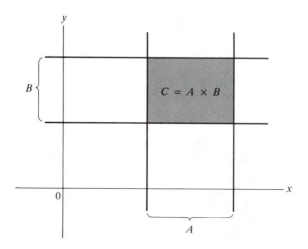

**Figure 5-26** A product set

If either interval is infinite, the product rectangle is infinite; for instance, the product of the sets defined by the conditions $X > 0$ and $Y > 0$ is the first quadrant.

**Example 5.10c** | **Uncorrelated, Yet Dependent**

Given the p.d.f. $f(x, y) = 45x^2y^2$ for $|x| + |y| < 1$, we see that the formula for $f$ is a product of a function of $x$ and a function of $y$ on the given support. But the support region is a square with corners at $(1, 0)$, $(-1, 0)$, $(0, 1)$, and $(0, -1)$. This is not a product set! It is rectangular, but the sides are not parallel to the coordinate axes. So the variables are *not* independent.

However, the variables are *uncorrelated*: The means are clearly 0, because of the symmetry about each axis. And the expected product is also 0:

$$E(XY) = 2\int_0^1 \left\{ \int_{y-1}^{1-y} xy\,(45x^2y^2)\,dx \right\} dy = 90 \int_0^1 y^3 \left\{ \int_{y-1}^{1-y} x^3\,dx \right\} dy.$$

This is 0, because the inner $x$-integral is 0, being the integral of an "odd" function over an interval that is symmetric about the origin.   ∎

Next we'll show that functions of independent variables are independent. To do this we'll use an alternative criterion for independence. Suppose $X$ and $Y$ are independent, having a joint p.d.f. satisfying (1). Consider any set $A$ of $x$-values and any set $B$ of $y$-values. The probability that $X$ is in $A$ *and* $Y$ is in $B$ is the double integral of their joint p.d.f. over the product set: $C = A \times B$. We can evaluate the double integral over $A \times B$ as a succession of single integrals, first over $A$ with respect to $x$ and then over $B$ with respect to $y$. Using (1), we have

$$P[(X, Y) \in C] = \int\int_C f(x, y)\, dx\, dy = \int\int_C f_X(x) f_Y(y)\, dx\, dy$$

$$= \int\int_{B\ A} f_X(x) f_Y(y)\, dx\, dy = \int_B f_Y(y) \left\{ \int_A f_X(x)\, dx \right\} dy$$

$$= \int_A f_X(x)\, dx \int_B f_Y(y)\, dy = P(X \in A) \cdot P(Y \in B).$$

That is, when $X$ and $Y$ are independent, the probability of a product set is the product of the probabilities of the marginal sets that define it. Conversely, taking $X = [-\infty, x]$ and $Y = [-\infty, y]$, this factorization of $A \times B$ implies that the joint c.d.f. factors into the product of the marginal c.d.f.'s, which in turn implies independence of $X$ and $Y$.

---

Random variables $X$ and $Y$ are independent if and only if

$$P[(X, Y) \in A \times B] = P(X \in A) \cdot P(Y \in B) \qquad (2)$$

for every choice of sets $A$ and $B$.

---

Suppose now that $X$ and $Y$ are independent and consider functions of these: $U = g(X)$ and $V = h(Y)$. For arbitrary sets of real numbers $C$ and $D$, denote by $A$ the set of points $x$ such that $g(x) \in C$ and by $B$, the set of points $y$ such that $h(y) \in D$. Then, applying (2) we find

$$P[(U, V) \in C \times D] = P[(X, Y) \in A \times B] = P(X \in A) \cdot P(Y \in B)$$

$$= P[g(X) \in C] P[h(Y) \in D] = P(U \in C) \cdot P(V \in D).$$

And this factorization, obtained for arbitrary $C$ and $D$, implies [by (2), a sufficient condition for independence] that $U$ and $V$ are independent.

---

When $X$ and $Y$ are independent, so are $g(X)$ and $h(Y)$.

---

The notion of independence extends readily to the case of *several* continuous random variables:

Random variables $X_1, ..., X_n$ with multivariate p.d.f. $f(x_1, ..., x_n)$ are independent if and only if, at *all* points $(x_1, ..., x_n)$,

$$f(x_1, ..., x_n) = f_{X_1}(x_1) \cdots f_{X_n}(x_n). \tag{3}$$

The reasoning which shows that functions of independent variables are independent applies equally well to the case of more than two variables.

When the random variables $X_1, ..., X_n$ are independent as defined by (3), the variables in any subset of them are independent. For instance, if we integrate out the variables $x_3$ through $x_n$, what remains as the joint p.d.f. of $X_1$ and $X_2$ is the *product* $f_{X_1}(x_1)f_{X_2}(x_2)$, which factors out of the $n-2$ integrations. And in particular, then, it follows that any two of the variables, being independent, are automatically uncorrelated.

If $X_1, ..., X_n$ are independent, then

  **(i)** $g_1(X_1), ..., g_n(X_n)$ are also independent.

  **(ii)** the variables in any subset of them are independent.

  **(iii)** $\operatorname{var}\left\{\sum_i a_i X_i\right\} = \sum_i a_i^2 \operatorname{var} X_i.$         (4)

If, further, they are *identically distributed* with common variance $\sigma^2$,

  **(iv)** $\operatorname{var}\left\{\sum_i X_i\right\} = n\sigma^2.$         (5)

**Example 5.10d** | **Elevator Passengers**

Suppose a population of potential elevator riders has mean $m = 150$ lb and s.d. $\sigma = 18$ lb. We select 20 passengers at random from this population and assume that their weights are independent. Consider now the random variable $W$, the total weight of all 20 passengers. Applying (5) we find

$$EW = 20\mu = 3{,}000 \text{ lb, and } \operatorname{var} W = 20 \times 18^2.$$

So $\sigma_W = \sqrt{20} \times 18$, or only about 4.47 times the s.d. of a single weight—*not* 20 times as large. ∎

## Problems

**∗ 5-51.**  In Problem 5-45 you found the first and second moments of the joint distribution with p.d.f. $4xy$ on the unit square. Find $\rho$.

**5-52.**  Suppose $(X, Y)$ has the joint p.d.f. $x + y$ on the unit square.
  **(a)**  Are the variables exchangeable?
  **(b)**  Find the marginal p.f.'s.
  **(c)**  Find the marginal means and variances.
  **(d)**  Are the variables independent?
  **(e)**  Find the correlation coefficient $\rho$.

**∗ 5-53.**  Problem 5-43 dealt with a uniform bivariate distribution supported on the triangle bounded by the coordinate axes and the line $x + y = 1$.
  **(a)**  For each $z$ between 0 and 1, find $P(X + Y < z)$, the c.d.f. of $X + Y$, and from it obtain the p.d.f. of $Z = X + Y$.
  **(b)**  Verify that $\operatorname{var}(X + Y) = \operatorname{var} X + \operatorname{var} Y + 2 \operatorname{cov}(X, Y)$.
  **(c)**  Find $\rho$.

**∗ 5-54.**  Given $\sigma_X = \sigma_Y = \sigma$, find the correlation of $X$ and $Y - X$ in terms of the correlation of $X$ and $Y$.

**5-55.**  Given that $X$ and $Y$ are independent, each with $\mu = 0$ and $\sigma = 1$, find the correlation coefficient of $X$ and $X + Y$.

**5-56.**  Find $\operatorname{cov}(X, Y)$ for the distribution in Problem 5-49.

**∗ 5-57.**  Suppose $(X, Y)$ to be uniformly distributed in the region $|x| + |y| < 1$.
  **(a)**  Find the marginal p.d.f.'s.
  **(b)**  Are $X$ and $Y$ independent? Exchangeable?
  **(c)**  Find $E|X|$, the mean deviation of $X$.
  **(d)**  Find the c.d.f. of $X + Y$ without a formal integration, using geometry.

**5-58.**  Give the joint p.d.f. of variables $X^*$ and $Y^*$ which are independent and have the same distributions, respectively, as $X$ and $Y$ in Problem 5-53.

**∗ 5-59.**  Suppose $X, Y$, and $Z$ are independent and identically distributed, each being uniform on $(0, 1)$. Find
  **(a)**  $E(X + Y + Z)$.        **(c)**  $\operatorname{var}(X - Y)$.
  **(b)**  $\operatorname{var}(X + Y + Z)$.      **(d)**  $\operatorname{cov}(X + Y - Z, 2X - Y + 3Z)$.

**5-60.**  Given the joint p.d.f. $f(x, y) = 2 \exp(-2x - y)$ for $x > 0, y > 0$,
  **(a)**  find the marginal p.d.f.'s.     **(c)**  find $E(X + Y)$.
  **(b)**  check for independence.       **(d)**  find $\operatorname{var}(X - Y)$.

**∗ 5-61.**  Find the joint p.d.f. of a sequence of $n$ independent, identically distributed variables with common p.d.f.
  **(a)**  $f(x) = 2x$ for $0 < x < 1$.     **(b)**  $f(x) = e^{-x}$ for $x > 0$.

## 5.11    Conditional Distributions

In the context of discrete bivariate distributions in §2.3, we studied conditional probability distributions, defining the p.f. of $Y$ given $X = x$ to be

$$f(y \mid x) = P(Y = y \mid X = x) = \frac{P(X = x \text{ and } Y = y)}{P(X = x)} = \frac{f(x, y)}{f_X(x)}, \qquad (1)$$

provided $f_X(x) \neq 0$. In the setting of continuous random variables, defining $P(Y \leq y \mid X = x)$ with this approach would lead to an indeterminate $0/0$. So, instead, we work with probability elements, or (since the "$dx$'s" would cancel anyway) with probability densities. The definition is as follows:

> The p.d.f. of the conditional distribution of $Y$ given $X = x$ is
>
> $$f(y \mid x) = \frac{f(x, y)}{f_X(x)}, \quad \text{when } f_X(x) > 0. \qquad (2)$$

[As in the discrete case, the common notation $f(y \mid x)$ is a little too casual, but the context will usually make clear what is meant; if not, we may want to write $f_{Y \mid X=x}(y)$, although this is a bit awkward.]

The function $f(y \mid x)$ defined by (2) is a p.d.f: It is nonnegative, and the area under its graph is 1:

$$\int_{-\infty}^{\infty} f(y \mid x)\, dy = \int_{-\infty}^{\infty} \frac{f(x, y)}{f_X(x)}\, dy = \frac{1}{f_X(x)} \int_{-\infty}^{\infty} f(x, y)\, dy = \frac{f_X(x)}{f_X(x)} = 1.$$

Moreover, averaging with respect to $X$ yields the marginal p.d.f. of $Y$:

$$E[f(y \mid X)] = \int_{-\infty}^{\infty} f(y \mid x) f_X(x)\, dx = \int_{-\infty}^{\infty} f(x, y)\, dx = f_Y(y). \qquad (3)$$

[This is the analog of (2) in §3.2 (page 87), which gives the average of a conditional p.f. as the corresponding marginal p.f.]

## Example 5.11a | Uniform Distribution on a Disc

Suppose $(X, Y)$ is uniformly distributed on the disc of radius 2 centered at the origin. The area of the disc is $4\pi$, so the constant density is $1/(4\pi)$ within the circle $x^2 + y^2 = 4$. Figure 5-27 shows the p.d.f. surface.

If it becomes known that $Y = 1$, the only possible points $(x, y)$ are those on the chord $I$ of the support circle at $y = 1$. So the conditional distribution of $X$ is

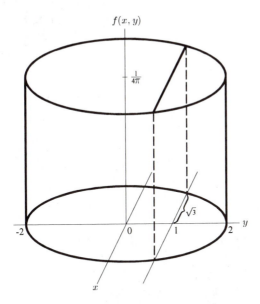

**Figure 5-27** Density surface for
Example 5.11a

restricted to that chord, whose length is $2\sqrt{3}$. The marginal p.d.f. of $Y$ at $y = 1$ is

$$f_Y(1) = \int_I \frac{1}{4\pi} \, dx = \frac{2\sqrt{3}}{4\pi},$$

which is the cross-section area $y = 1$ at (base $\times$ height). The conditional p.d.f. of $X$ is *proportional* to $f(x, 1)$:

$$f(x \mid Y = 1) = \frac{f(x, 1)}{f_Y(1)} = \frac{1/(4\pi)}{2\sqrt{3}/(4\pi)} = \frac{1}{2\sqrt{3}}, \quad -\sqrt{3} < x < \sqrt{3}.$$

That is, the conditional distribution of $X$ is *uniform* on the chord $I$ defined by the condition $y = 1$, as is evident in Figure 5-27. Dividing the joint p.d.f. by the marginal p.d.f. of $Y$ at $y = 1$ serves to make the integral of the conditional p.d.f. equal to 1. ∎

---

If the joint distribution of $(X, Y)$ is *uniform* on a region $R$ of the plane, every conditional distribution (given the value of $X$ or the value of $Y$) is uniform. The support is the intersection of $R$ with the line defined by the given value of the conditioning variable.

A conditional distribution is a probability distribution. The modifier "conditional" refers only to its origin as deriving from a joint distribution by conditioning. So a conditional distribution may have moments. In particular, the mean of a conditional is an important parameter, one that will be used in a later chapter for predicting one variable given a value of the other.

---

The **regression function of $Y$ on $X$** is the conditional mean

$$E(Y \mid X = x) = \int_{-\infty}^{\infty} y\, f(y \mid x)\, dy. \tag{4}$$

---

**Example 5.11b**   Suppose $(X, Y)$ is distributed with p.d.f. $6(x - y)$ for $0 < y < x < 1$, as in Example 5.7e. The marginal p.d.f. of $X$ was found in that example to be $3x^2$ for $0 < x < 1$. The conditional p.d.f. of $Y$, given $X = x_0$, is then

$$f(y \mid x_0) = \frac{6(x_0 - y)}{3x_0^2}, \quad 0 < y < x_0.$$

The conditional mean of $Y$, given $x = x_0$, is

$$E(Y \mid x_0) = \int_0^{x_0} y\, f(y \mid x_0)\, dy = \frac{2}{x_0^2} \int_0^{x_0} y(x_0 - y)\, dy = \frac{x_0}{3}.$$

The "regression line" $y = x/3$ is the locus of these conditional means. Figure 5-28 shows the surface for the bivariate p.d.f., the cross section that defines the conditional p.d.f. given $x_0$, and the regression function $E(Y \mid x)$.  ■

The regression function defines a new random variable, $E(Y \mid X)$. If we average this with respect to the distribution of $X$, we obtain an iterated expectation formula, analogous to (3) in §3.2:

$$E[E(Y \mid X)] = \int \left\{ \int y\, f(y \mid x)\, dy \right\} f_X(x)\, dx$$

$$= \int\int y\, f(y \mid x)\, f_X(x)\, dy\, dx = \int\int y\, f(x, y)\, dy\, dx = EY.$$

By analogy, one might expect that averaging the conditional variance of $Y$ with respect to the conditioning variable $X$ would produce the unconditional variance. This is not the case. The correct relationship follows upon applying the parallel axis theorem to the conditional distribution of $Y$:

$$\mathrm{var}(Y \mid x) = E(Y^2 \mid x) - [E(Y \mid x)]^2.$$

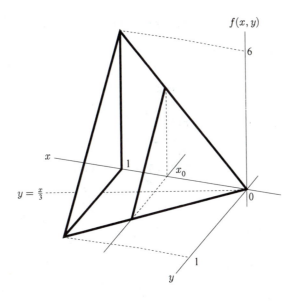

**Figure 5-28** Regression function—
Example 5.11b

Using this to define a relation between random variables, by replacing $x$ by $X$, and taking expected values, we obtain

$$E[\text{var}(Y \mid X)] = E[E(Y^2 \mid X)] - \{\text{var}[E(Y \mid X)] + [EE(Y \mid X)]^2\}$$

$$= E(Y^2) - \text{var}[E(Y \mid X)] - (EY)^2 = \text{var}\, Y - \text{var}[E(Y \mid X)].$$

Averaging conditional p.d.f.'s and moments:

  (i)  $E[f(y \mid X)] = f_Y(y)$.         (5)

  (ii)  $E[E(Y \mid X)] = EY$.         (6)

  (iii)  $E[\text{var}(Y \mid X)] = \text{var}\, Y - \text{var}[E(Y \mid X)]$.         (7)

**Example 5.11c** | Given the p.d.f. of the preceding example: $f(x, y) = 6(x - y)$ for $0 < y < x < 1$, the marginal p.d.f.'s are $f_X(x) = 3x^2$ and $f_Y(y) = 3(1 - y)^2$, each on $(0, 1)$. The means are, respectively, 3/4 and 1/4, and the variances are equal, both 3/80. In the preceding example, we found $E(Y \mid x) = x/3$. Then

$$\text{var}[E(Y \mid X)] = \text{var}\left(\frac{X}{3}\right) = \frac{\text{var } X}{9} = \frac{3/80}{9} = \frac{1}{240}.$$

To get the variance of $Y$ given $X = x$, we first find the conditional mean square:

$$E(Y^2 \mid x) = 2\int_0^x y^2 \cdot \frac{x - y}{x^2}\, dy = \frac{x^2}{6}.$$

Subtracting the square of the conditional mean yields the conditional variance:

$$\text{var}(Y \mid x) = \frac{x^2}{6} - \left(\frac{x}{3}\right)^2 = \frac{x^2}{18}.$$

With these, we can verify (6) and (7) for this particular case:

$$E[E(Y \mid X)] = E\left(\frac{X}{3}\right) = \frac{1}{3}(EX) = \frac{1}{4} = EY,$$

and

$$E[\text{var}(Y \mid X)] + \text{var}[E(Y \mid X)] = E\left\{\frac{X^2}{18}\right\} + \frac{1}{240} = \frac{3}{80} = \text{var } Y. \qquad \blacksquare$$

## Problems

**∗ 5-62.** For the joint distribution in Problem 5-47, where $f(x, y) = x + y$ for $0 < x < 1$, $0 < y < 1$,
  (a)  find the conditional p.d.f. of $X$ given $Y = y$.
  (b)  find the regression function of $X$ on $Y$.

**5-63.**  Problem 5-46 dealt with the p.d.f. $e^{-y}$ for $0 < x < y$.
  (a)  Find the conditional p.d.f. of $Y$ given $X = x$.
  (b)  From (a), obtain $E(Y \mid x)$.
  (c)  Verify that $E[E(Y \mid X)] = EY$ by calculating both sides of the equation.
  (d)  Verify that $\text{var } Y = E[\text{var}(Y \mid X)] + \text{var}[E(Y \mid X)]$.

**∗ 5-64.**  Find the conditional p.d.f. of $Y$ given $X = x$ for the distribution of Problem 5-53, with constant density on the triangle bounded by the coordinate axes and the line $x + y = 1$.

**5-65.**  Show that if $X$ and $Y$ are independent, the conditional p.d.f.'s are the same as the corresponding marginal p.d.f.'s.

**∗ 5-66.**  Suppose $f_{X\mid y}(x) = ye^{-yx}$ for $x > 0$, and $f_Y(y) = e^{-y}$ for $y > 0$.
  (a)  Find the joint p.d.f. $f(x, y)$.
  (b)  From (a), find the marginal p.d.f. of $X$.
  (c)  From (a) and (b), find the conditional p.d.f. of $Y$ given $X = x$.
  [Note: This amounts to a continuous-variable version of Bayes' theorem.]

**5-67.**   Assume the distribution of Problem 5-57 (uniform within $|x| + |y| < 1$).
   **(a)**   Find the conditional distribution of $Y$ given $X = x$.
   **(b)**   Find the regression function, $E(Y \mid x)$.

★ **5-68.**   Given the joint p.d.f. of $X$ and $Y$, $f(x, y) = 6x$ for $0 < x < y < 1$, find
   **(a)**   the marginal p.d.f. of $Y$.
   **(b)**   the conditional p.d.f. of $X$ given $Y = 1/2$.

| | |
|---|---|
| **5.12** | **Moment Generating Functions** |

In §3.6 and §4.9 we used the factorial moment generating function in dealing with sums of certain independent, discrete random variables. The definition of the f.m.g.f. as the expected value of $t^X$ can be applied in the continuous case as well. However, a closely related function, the **moment generating function** (m.g.f.), handles sums in the same way but is more convenient for most of the particular continuous distributions that we'll encounter.

---

Moment generating function of $X$:

$$\psi_X(t) = E(e^{tX}). \tag{1}$$

---

(As with p.d.f.'s, we may drop the subscript on $\psi$ when there is no risk of confusion.) The m.g.f. is related to the factorial moment generating function:

$$\psi(t) = E[(e^t)^X] = \eta(e^t). \tag{2}$$

The variable $X$ in (1) can be either discrete or continuous—or involve a mixture of these two types. When it is continuous, the expectation in (1) is an integration:

$$\psi(t) = \int_{-\infty}^{\infty} e^{tX} f(x)\, dx, \tag{3}$$

provided that the integral converges. When it does, it transforms a function $f$ into a function $\psi$.

Similar transformations have long been found useful in many areas of applied mathematics. In a typical application, the scheme is to find the transformed functions of $t$, do some mathematics on these, and then go back to the original functions—the inverse transforms. The beauty of the scheme is that the necessary mathematics is easier in the realm of the transforms than in the realm of the original functions.

Moment generating functions are mysterious for many students. Don't look for physical interpretations for $\psi$ or for the dummy variable $t$. Rather, view the m.g.f. as a powerful mathematical tool that is useful in finding distributions of sums of independent random variables.

**Example 5.12a** | **A Waiting-Time Distribution**

The waiting-time distribution of Example 5.2b has the density function $6e^{-6x}$ for $x > 0$. Using (3), we find the m.g.f. to be

$$\psi(t) = \int_0^\infty e^{tx}\, 6e^{-6x}\, dx = 6\int_0^\infty e^{-x(6-t)}\, dx = \frac{6}{6-t} = \frac{1}{1-t/6}\,.$$

The calculation is valid only for $t < 6$, where the integral converges.   ■

**Example 5.12b** | **Binomial m.g.f.**

Let $Y \sim \text{Bin}(n,\, p)$. In §4.9 we found the p.g.f. (or f.m.g.f.) to be

$$\eta(t) = (pt + q)^n,$$

where $q = 1 - p$. Applying (2), we see that the m.g.f. is

$$\psi(t) = \eta(e^t) = (pe^t + q)^n.$$

Of course a direct calculation yields the same result:

$$\psi(t) = E(e^{tY}) = \sum_{k=0}^n e^{tk} \binom{n}{k} p^k q^{n-k} = \sum_{k=0}^n \binom{n}{k} (e^t p)^k q^{n-k} = (pe^t + q)^n.   ■$$

It will be important to know what happens to the m.g.f. of $X$ when $X$ undergoes a linear transformation. Suppose $Y = a + bX$. Then, according to the laws of exponents,

$$\psi_Y = E(e^{tY}) = E[e^{t(a+bX)}] = E[e^{at}e^{(bt)X}].$$

The factor $e^{at}$, being a constant, can be moved out of the expectation.

---

The m.g.f. of $Y = a + bX$ is

$$\psi_{a+bX}(t) = e^{at}\psi_X(bt).   \tag{4}$$

---

**Example 5.12c** | **The $Z$-Score**

Suppose $X$ has the m.g.f. $\psi_X(t)$. We'll often be "standardizing"—that is, transforming to a scale on which the mean is 0 and the standard deviation 1:

$$Z = \frac{X - \mu}{\sigma}\,.$$

We use (4) to find $\psi_Z$ in terms of $\psi_X$, with $a = -\mu/\sigma$, $b = 1/\sigma$:

$$\psi_Z(t) = e^{-\mu t/\sigma}\psi_X(t/\sigma).$$   ∎

Like the p.g.f., the m.g.f. is important in statistical theory for studying sums of independent variables. In particular, it provides a convenient means for deriving the distribution of the sum of the observations in a random sample. The more obvious approach is to find the c.d.f. of the sum and then differentiate to get the p.d.f., but this has its problems. For, consider the simplest possible sum, that of two independent observations $(X_1, X_2)$ from a population with p.d.f. $f(x)$.

To find the c.d.f. of $U = X_1 + X_2$, we express it as the probability of a region in the $x_1x_2$-plane:

$$F_U(u) = P(X_1 + X_2 \le u) = \iint\limits_{x_1+x_2\le u} f(x_1)f(x_2)\,dx_1dx_2.$$

Figure 5-29 shows the region of integration:

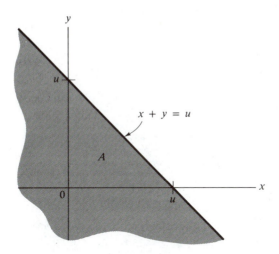

**Figure 5-29**

We evaluate the double integral as a succession of single integrals, first (say) with respect to $x_1$:

$$F_U(u) = \int_{-\infty}^{\infty}\int_{-\infty}^{u-x_2} f(x_1)f(x_2)\,dx_1dx_2.$$

To find the p.d.f., we differentiate with respect to $u$. To accomplish this, we first move the derivative inside the outer integral (which has constant limits); we then differentiate the inner integral, where $u$ appears only in the upper limit, by substituting that limit for $x_1$ in the integrand:

$$f_U(u) = \int_{-\infty}^{\infty} f(u - x_2) f(x_2)\, dx_2.$$

This is called a *convolution* of $f$ with itself.

For a sum of $n$ independent variables, we'd need an $n$-fold convolution of the density—an awesome prospect! So we won't use this approach. Rather, we exploit the properties of m.g.f.'s, and in so doing we are able to circumvent the difficulties of convolutions.

Consider first two random variables, $X$ and $Y$. The m.g.f. of their sum, by definition, is

$$\psi_{X+Y}(t) = E(e^{t(X+Y)}) = E(e^{tX}e^{tY}).$$

Now, if $X$ and $Y$ are *independent*, then $e^{tX}$ and $e^{tY}$ are independent, and the last expected value factors into the product of the expected values:

$$\psi_{X+Y}(t) = E(e^{tX})E(e^{tY}) = \psi_X(t)\psi_Y(t). \tag{5}$$

This kind of factorization of the m.g.f. holds for any finite number of independent summands, because the factorization of the joint p.d.f. into the product of the marginals holds in general:

---

If $X_1, ..., X_n$ are *independent*, the m.g.f. of their sum is the *product* of their individual m.g.f.'s:

$$\psi_{\Sigma X_i}(t) = \psi_{X_1}(t) \cdots \psi_{X_n}(t). \tag{6}$$

If, additionally, the $X_i$'s all have the same distribution with common m.g.f. $\psi(t)$, then

$$\psi_{\Sigma X_i}(t) = [\psi(t)]^n. \tag{7}$$

---

**Example 5.12d**   Consider $n$ independent observations on $X$, where $X$ has the waiting-time distribution of Example 5.12a, with p.d.f. $6e^{-6x}$ for $x > 0$. There we found that $\psi(t) = (1 - t/6)^{-1}$. Using (7), we obtain the m.g.f. of the sum of the $n$ observations:

$$\psi_{\Sigma X}(t) = \frac{1}{(1 - t/6)^n}.$$

Having found this, there remains the question—what *is* the distribution of the sum? To answer, we need to have some way of going from m.g.f.'s back to density functions.   ∎

Suppose we just happen to find a distribution whose m.g.f. is the same as the $\psi_{\Sigma X}$ we found in the example. Is it necessarily the distribution we were looking for? It is, if a moment generating function uniquely determines a distribution. And there *is* such a uniqueness theorem for m.g.f.'s (with certain regularity assumptions).[3]

When learning integral calculus, you may have noticed that integral formulas were the result of formulas for derivatives. That is, to "integrate" an expression, you looked for a function whose derivative you knew to be given by that expression. So, here, to find the "inverse" transform of an expression, we look for a function whose direct transform is that expression. We'll see several examples of such inversion in the next chapter.

Before going to the next chapter, we should explain the term "moment generating." As it suggests, a m.g.f. can be used to generate moments of a distribution (if it has moments) as follows. Suppose we differentiate a m.g.f., moving the operation of differentiation inside the expected value (an integral):

$$\frac{d}{dt}\,\psi(t) = E\left(\frac{d}{dt}\,e^{tX}\right) = E(Xe^{tX}).$$

(Interchanging derivative and integral is not always legitimate, but sufficient conditions for the interchange to be valid hold in all of the cases we'll encounter.) If we now substitute $t = 0$ in the derivative, we get the mean:

$$\psi'(0) = E(Xe^0) = EX.$$

A second differentiation yields

$$\psi''(t) = \frac{d}{dt}\,E(Xe^{tX}) = E\left\{\frac{d}{dt}\,(Xe^{tX})\right\} = E(X^2e^{tX}),$$

and with $t = 0$, we get the average square:

$$\psi''(0) = E(X^2).$$

Continuing, we have as a general formula

$$\psi^{(n)}(0) = E(X^n). \tag{8}$$

Another way of obtaining this result is by use of the Maclaurin series expansion of the exponential function:

$$e^{tX} = \sum_{n=0}^{\infty} \frac{(tX)^n}{n!}.$$

The expected value of this infinite sum is the sum of the expected values:

$$\psi(t) = E(e^{tX}) = \sum_{n=0}^{\infty} E(X^n)\frac{t^n}{n!}.$$

---

[3]See Feller [7], 430.

Thus, the moments of $X$ are coefficients of $t^n/n!$ in the power series expansion of the moment generating function. So, if we happen to know the power series for $\psi$, we can read the moments from the coefficients.

---

The moment $E(X^n)$ can be found from the m.g.f. of $X$ either as

$$E(X^n) = \psi^{(n)}(0), \tag{9}$$

or as the coefficient of $t^n/n!$ in the power series for $\psi(t)$.

---

**Example 5.12e** | **Moments of $\mathcal{U}(0, 1)$**

Suppose $X \sim \mathcal{U}(0, 1)$. The moment generating function is

$$\psi(t) = \int_0^1 e^{tx}\, dx = \frac{e^t - 1}{t} = \frac{(1 + t + t^2/2! + t^3/3! + \cdots) - 1}{t}$$

$$= 1 + \frac{t}{2!} + \frac{t^2}{3!} + \cdots = 1 + \frac{1}{2}t + \frac{1}{3}\cdot\frac{t^2}{2!} + \cdots.$$

The mean is $EX = 1/2$, and the mean square is $E(X^2) = 1/3$. The pattern is clear: $E(X^n) = 1/(n+1)$.

Using (9) to find the moments is possible, but in this case a bit more complicated. The first derivative of $\psi$ is

$$\psi'(t) = \frac{te^t - (e^t - 1)}{t^2}.$$

Substituting $t = 0$ produces an indeterminate form, 0/0, but the limit as $t$ approaches 0 can be found using L'Hospital's rule.   ∎

The m.g.f. always exists at $t = 0$ [$\psi(0) = 1$], but for the m.g.f. to exist for $t \neq 0$, it is necessary that all moments exist. In Example 5.4c we encountered the Cauchy distribution, which has no moments of any order, and the integral (3) that would define the m.g.f. does not exist.

A more generally useful tool is the **characteristic function:**[4]

$$\phi(t) = E(e^{itX}) = E(\cos tX) + iE(\sin tX),$$

where $i$ is the imaginary unit, $\sqrt{-1}$. (In mathematics, the m.g.f. is known as a [bilateral] Laplace transform. The characteristic function is a Fourier transform. Using the characteristic function involves complex analysis.) This function exists for every distribution because the sine and cosine functions are bounded. It is related to the

---

[4]See Feller [7], Chapter 15.

m.g.f., when the latter exists, by the relation

$$\phi(t) = \psi(it). \tag{10}$$

We'll be using only the m.g.f. (which is adequate for our purposes) because it is simpler and easier to deal with mathematically.

## Problems

**\* 5-69.** Given that the m.g.f. of $X$ is $\psi(t) = (1 - t^2)^{-1}$,
   **(a)** obtain a formula for $E(X^k)$.
   **(b)** use $\psi$ from (a) to find var $X$.

**5-70.** Suppose $X$ is singular at $x = b$, that is, $P(X = b) = 1$. Find the m.g.f. of $X$ and use it to find a formula for the $k$th moment.

**\* 5-71.** Apply (4) and the result in Example 5.12e to find the m.g.f. of a uniform distribution on the interval $(0, \theta)$.

**5-72.** Apply (4) to find the m.g.f. of $Y = X - \mu_X$ when $X \sim \mathcal{U}(0, 1)$, and find the central moments of $X$ as the corresponding moments of $Y$ (about 0).

**\* 5-73.** Use the m.g.f. to find the moments of $X$ with p.d.f. $e^{-x}$ for $x > 0$.

**5-74.** Given that $X$ is symmetric about 0: $f(x) = f(-x)$, show: $\psi(t) = \psi(-t)$.

**5-75.** Find the m.g.f. of a Poisson variable with mean $m$.

**5-76.** Let $\psi(t) = 1 - p + \frac{1}{2}(e^t + e^{-t})$. Find the distribution of $X$ and obtain its mean and variance. (*Hint:* $X$ is discrete.)

**5-77.** Is $\psi(t) = (1 + t^2)^{-1}$ a m.g.f.? (*Hint:* Expand $\psi$ in a geometric series.)

## Review Problems

**5-R1.** Let $X$ have the c.d.f. $F(x) = 1 - \frac{1}{2x}$ for $x \geq 1$ and 0 for $x < 1$. Find
   **(a)** $P(X = 1)$.     **(b)**  $P(X = 2)$.      **(c)**  $P(1 < X < 2)$.

**5-R2.** Given that $Y$ has the p.d.f. $f(y) = \begin{cases} 2y/3 \text{ for } 0 < y < 1 \\ 2/3 \text{ for } 1 < y < 2, \end{cases}$

   **(a)** find and sketch the c.d.f., $F(y)$.     **(b)**  find the first quartile.
   **(c)** find the median.     **(d)**  find $EY$.     **(e)**  find var $Y$.

**5-R3.** Given the c.d.f. $F_V(v) = v/3$ for $0 < v < 3$, find
   **(a)** $P(V < 1 \mid V < 2)$.     **(b)**  $f(v)$, the p.d.f.     **(c)**  the c.d.f. of $U = V/3$.

**5-R4.** Given $X \sim \mathcal{U}(0, 1)$, find the p.d.f. of $Y = \sqrt{X}$.

**5-R5.** Given the density function $f_X(x) = 3x^2$, $0 < x < 1$, find the p.d.f. of the variable $Y = -\log X$.

**5-R6.** Let $U$ have the p.d.f. $f(u) = 3(2 - u^2)/10$, $-1 < u < 1$. Find
   **(a)**  $EU$.     **(b)**  $\sigma_U$.     **(c)**  $E(U^3)$.     **(d)**  $E(3U + 5)$.

**5-R7.**   Suppose $X$ has the m.g.f. $\psi(t) = (1 - 2t)^{-1/2}$. Find the following:
  **(a)**  $E(e^{-X})$.    **(b)**   var $X$.

**5-R8.**   Suppose $(X, Y)$ is uniformly distributed on the "unit circle" (center at the origin, radius 1). Find the following:
  **(a)**  $P(X < Y)$.    **(b)**  $f_X(x)$.    **(c)**  $\sigma_{X,Y}$.

**5-R9.**   Let $U$ and $V$ be independent and identically distributed with common p.d.f. $f(x) = xe^{-x}$, $x > 0$. Find
  **(a)**  the joint p.d.f., $f_{U,V}(u, v)$.    **(b)**  $\rho_{U,V}$.    **(c)**  $\psi_{U+V}(t)$.

**5-R10.**   Given: $(X, Y)$ is uniformly distributed on the region $0 < y < x < 1$. Find the following:
  **(a)**  $f_X(x)$.    **(b)**  $E(Y \mid x)$.    **(c)**  $P(X + Y < 1)$.    **(d)**  $E(XY)$.

**5-R11.**   Assume that $(X, Y)$ is uniform on the square with corners $(0, 1)$, $(1, 0)$, $(-1, 0)$, $(0, -1)$. (Are $X$ and $Y$ independent?) Find
  **(a)**  $EX$.    **(b)**  $f_X(1/2)$.    **(c)**  $f_{Y|X=\frac{1}{2}}(y)$.    **(d)**  $\sigma_{X,Y}$.    **(e)**  var$(X + 2Y)$.

**5-R12.**   The joint p.d.f. of $(X, Y)$ is $f(x, y) = xye^{-(x^2+y^2)/2}$ for $x > 0$, $y > 0$.
  **(a)**  Show that $X$ and $Y$ are independent.
  **(b)**  Find $F_X(x)$, the c.d.f. of $X$.
  **(c)**  Find $E(Y^2)$.

**5-R13.**   If $X$ has the logistic distribution (Problem 5-29), use the probability element to approximate the probability that $|X| < .02$.

**5-R14.**   Suppose $(X, Y, Z)$ is uniformly distributed in the tetrahedron with vertices at $(0, 0, 0)$, $(1, 0, 0)$, $(0, 1, 0)$, and $(0, 0, 1)$.
  **(a)**  Find the value of the p.d.f. on the support region.
  **(b)**  Find the conditional p.d.f. of $Z$, given $X = x$ and $Y = y$.
  **(c)**  Find the marginal densities of $X$, $Y$, and $Z$.

**5-R15.**   Let $X$ have the m.g.f. $\psi_X(t) = 1 + t + t^2 + \frac{2}{3}t^3 +$ (terms you won't use).
  **(a)**  Find the variance of $X$.    **(b)**   Find var$(3X - 2)$.

**5-R16.**   Given that $X$ has the c.d.f. $F(x) = 1 - .9e^{-x}$ for $x \geq 0$ and that $P(X < 0) = 0$, find
  **(a)**  $P(X = 2)$.    **(b)**   $P(X = 0)$.    **(c)**   $P(X > 2)$.    **(d)**   the median.

**5-R17.**   Given continuous random variables $(X, Y)$, let $A$ denote an event on the $X$-space. Define $g(y) = P(A \mid Y = y)$ and show that $P(A) = E\{g(Y)\}$. [We write this as $P(A) = E\{P(A) \mid Y\}$. This same averaging of a conditional probability with respect to the distribution of the conditioning variable also produces the corresponding unconditional probability when $X$ is discrete and $Y$ is continuous; but to show this is beyond our scope.]

**5-R18.**   Let $\{A_i\}$ be a partition of $\Omega$, and let $f$ denote either a p.f. or a p.d.f. for $X$. Show the following in each case:
  **(a)**  $f(x) = \sum f_{X|A_i}(x)P(A_i)$. [Hint for the continuous case: First show the equality with $f$ replaced by the c.d.f. $F$, and differentiate.]
  **(b)**  $EX = \sum E(X \mid A_i)P(A_i)$.

## Chapter Perspective

Continuous distributions provide simple representations for variables of a type observed in a wide variety of phenomena. The simplicity is provided by methods of calculus, with integration playing the role of summation.

When a random variable has a continuous distribution, single values have probability zero. So we've described such distributions by giving either the probabilities of intervals or the density of probability at each point. Expected values are defined in terms of integration with respect to the density as a weighting function, and the concepts and formulas for discrete variables in terms of expected value are then carried over to the continuous case.

In dealing with two or more continuous variables, we need continuous multivariate distributions, defined by a multivariate p.d.f. We've seen how to obtain marginal and conditional p.d.f.'s from the joint p.d.f., and how to incorporate independence into a multivariate model. The covariance and correlation coefficient provide means for examining pairwise relationships of a particular type.

Because sums of independent variables are so commonly encountered in statistical applications, the generating function is an important theoretical tool that we often use in the development to follow.

The distribution theory we need as a theoretical basis for the various methods of inference is nearly complete. Before turning to these methods, we take up, in the next chapter, several families of continuous distributions that often serve to represent real phenomena (at least approximately). We'll also introduce some families of distributions that will be used in studying the sampling variation of a sample statistic. In working with these various special families, you will be applying and solidifying the ideas of this and earlier chapters, as well as preparing for the theory and methods of inference to follow.

## CHAPTER 6

# Families of Continuous Distributions

In developing tools for continuous distributions in Chapter 5, we introduced a number of particular continuous distributions as examples. Many of these belong to parametric *families* of distributions—families whose individual members are identified by the value of a parameter. Here we study these distribution families further and introduce some new, useful families of continuous distributions.

## 6.1 Normal Distributions

One of the most important and most used families of distributions in statistics and probability goes by the names **normal** and **Gaussian,** after the German mathematician Karl Friedrich Gauss (1777–1855), who used the distribution extensively in his theory of errors. (The term "Gaussian" is used mainly in the physical sciences.) In §4.5 we encountered a particular member of this family, the *standard normal* distribution, in approximating binomial probabilities.

The distributions in the general normal family all have the same basic shape as the standard normal distribution: Any member of the family can be transformed into any other by translating the origin and changing the scale by a linear transformation. There are at least two reasons for the importance of this class of distributions. (1) Many populations encountered in practice are well approximated by a normal distribution. (2) Many of the commonly used random variables encountered in statistical inference have distributions that are either exactly or (for large samples) approximately normal.

In §4.5 we gave the p.d.f. of the standard normal distribution, and we repeat it here, along with the c.d.f.

**Standard Normal p.d.f.:**

$$f(z) = \frac{1}{\sqrt{2\pi}}\, e^{-z^2/2}. \tag{1}$$

**Standard Normal c.d.f:**

$$\Phi(z) = \int_{-\infty}^{z} \frac{1}{\sqrt{2\pi}}\, e^{-u^2/2}\, du. \tag{2}$$

The constant $1/\sqrt{2\pi}$ makes the area under the p.d.f. 1, as we'll show at the end of this section. The graph of the standard normal p.d.f. is shown in Figure 6-1. The graph of the c.d.f. is shown in Figure 6-2.

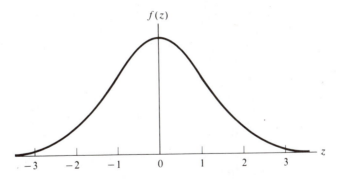

**Figure 6-1** Standard normal p.d.f.

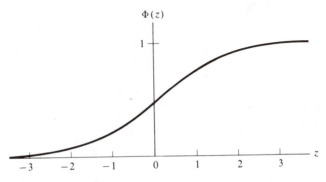

**Figure 6-2** Standard normal c.d.f.

Observe that we have not given the c.d.f. in terms of "elementary" functions—because it can't be done. A special table for this function [prepared with the aid of numerical integrations to evaluate the integral in (2)] is given as Table IIa of Appendix I. The table gives values of $\Phi(z)$ for selected $z$-values.

The standard normal distribution is clearly symmetric about $z = 0$, since $f(-z) = f(z)$. One consequence of this is that if $Z$ is standard normal, so is the variable $-Z$:

$$f_{-Z}(z) = f_Z(-z) \cdot |-1| = f_Z(z).$$

Another consequence of the symmetry is that the median of a normal distribution is 0. Moreover, the mean value is 0, because the integral defining it does exist. (The exponential in the integrand dominates any power of $z$ that may multiply it.)

All the higher-order moments also exist and are finite. We can find them using the moment generating function:

$$\psi(t) = \int_{-\infty}^{\infty} e^{tz} \frac{1}{\sqrt{2\pi}} e^{-z^2/2}\, dz = \frac{1}{\sqrt{2\pi}} \int_{-\infty}^{\infty} e^{tz - z^2/2}\, dz.$$

To evaluate the latter integral, we first complete the square in the exponent:

$$tz - \frac{1}{2} z^2 = -\frac{1}{2}(z^2 - 2tz + t^2) + \frac{t^2}{2} = -\frac{1}{2}(z - t)^2 + \frac{t^2}{2}.$$

Making this replacement in the exponent, we have

$$\psi(t) = \frac{1}{\sqrt{2\pi}} \int_{-\infty}^{\infty} e^{-(z-t)^2/2} e^{t^2/2}\, dz = e^{t^2/2} \left\{ \int_{-\infty}^{\infty} \frac{1}{\sqrt{2\pi}} e^{-(z-t)^2/2}\, dz \right\}.$$

The integrand in the integral in braces is just a standard normal curve displaced by an amount $t$; but such a translation does not change the *area*, which is still 1. So, the m.g.f. is the function in front of the braces.

---

Moment generating function of a standard normal $Z$:

$$\psi_Z(t) = e^{t^2/2}. \tag{3}$$

---

To find the moments of $Z$, we'll expand $\psi_Z$ using the power series for the exponential function:

$$e^x = 1 + x + \frac{x^2}{2!} + \frac{x^3}{3!} + \cdots.$$

Setting $x = t^2/2$, we obtain

$$\psi_Z(t) = e^{t^2/2} = 1 + t^2/2 + \frac{(t^2/2)^2}{2!} + \frac{(t^2/2)^3}{3!} + \cdots.$$

To be able to read out the coefficient of $t^k/k!$, we rewrite this to put those coefficients on display:

$$\psi_Z(t) = 1 + 0 \cdot t + 1 \cdot \frac{t^2}{2!} + 0 \cdot \frac{t^3}{3!} + 3 \cdot \frac{t^4}{4!} + 0 \cdot \frac{t^5}{5!} + 5 \cdot 3 \cdot \frac{t^6}{6!} + \cdots.$$

Thus, all moments of odd order vanish—there are no odd powers of $t$ in the expansion. The even-order moments are

$$EZ = 0, \ E(Z^2) = 1, \ E(Z^4) = 3, \ E(Z^6) = 5 \cdot 3, \ldots.$$

---

Moments of the standard normal variable $Z$:

$$EZ = 0, \ \text{and} \ \text{var} \ Z = E(Z^2) = 1. \tag{4}$$

More generally,

$$E(Z^n) = \begin{cases} (n-1)(n-3) \cdots 5 \cdot 3 \cdot 1 \ \text{for } n \text{ even,} \\ 0 \ \text{for } n \text{ odd.} \end{cases} \tag{5}$$

---

We turn now to the general normal distribution—to the family of distributions related to the standard normal distribution by linear transformations. Variables such as heights, weights, exam grades, blood pressures, IQ's, and the like have "bell-shaped" densities, resembling that of the standard normal density; however, their means and variances are generally not the "standard" parameters, $\mu = 0$ and $\sigma = 1$. A suitable linear transformation of the measurement scale—combining translation (so that the new mean is 0) and change of scale (so that the new unit is one standard deviation)—creates variables with these standard parameters, while preserving the bell shape.

A variable $X$ with mean $\mu$ and standard deviation $\sigma$ is said to have a *normal distribution* if the $Z$-score, $(X - \mu)/\sigma$, is standard normal. Another way of saying this is that a random variable $X$ is normal if for some constants $a$ and $b$, $X = a + bZ$, where $Z$ is standard normal. It follows that $a$ is the mean of $X$, and $|b|$ is the standard deviation of $X$.

A random variable $X$ has a **normal distribution** if and only if

$$X = \mu + \sigma Z, \tag{6}$$

where $Z$ is standard normal and $EX = \mu$, $\operatorname{var} X = \sigma^2$. We then write $X \sim \mathcal{N}(\mu, \sigma^2)$.

**Example 6.1a**

## GRE Scores

In Example 5.6b we considered the GRE "analytical ability" scores. Each year, students taking GRE exams number in the hundreds of thousands. It turns out that the distribution of such scores is fairly well approximated by a normal distribution. In the academic year 1977–1978, $X \approx \mathcal{N}(521, 123^2)$. ∎

Using (6) we can obtain the c.d.f., p.d.f., and m.g.f. of a normally distributed variable $X$ in terms of the corresponding expressions for the standard normal distribution. Thus,

$$F(x) = P(X \le x) = P(\mu + \sigma Z \le x) = P\left(Z \le \frac{x - \mu}{\sigma}\right) = \Phi\left(\frac{x - \mu}{\sigma}\right).$$

And, differentiating, from this we get the p.d.f. by means of the formula for the p.d.f. of a linear function of a random variable [(16) in §5.2]:

$$f(x) = \Phi'\left(\frac{x - \mu}{\sigma}\right) \cdot \frac{1}{\sigma} = \frac{1}{\sigma\sqrt{2\pi}} \exp\left\{-\frac{(x - \mu)^2}{2\sigma^2}\right\}.$$

We get the m.g.f. by writing out its definition for $X$:

$$\psi_X(t) = E(e^{t(\mu + \sigma Z)}) = e^{t\mu}\psi_Z(t\sigma) = \exp\left\{\mu t + \frac{\sigma^2 t^2}{2}\right\},$$

where the factor $e^{t\mu}$, being a constant, factors out of the expected value.

If $X \sim \mathcal{N}(\mu, \sigma^2)$, its p.d.f., c.d.f., and m.g.f. are as follows:

$$f_X(x) = \frac{1}{\sigma\sqrt{2\pi}} \exp\left\{-\frac{(x - \mu)^2}{2\sigma^2}\right\}. \tag{7}$$

$$F_X(x) = \Phi\left(\frac{x - \mu}{\sigma}\right). \tag{8}$$

$$\psi_X(t) = \exp\left\{\mu t + \frac{\sigma^2 t^2}{2}\right\}. \tag{9}$$

Example **6.1b**

### Probability of an Interval

Blood pressures vary from individual to individual, as well as within the same individual at different times. The blood pressure of an individual selected at random from some population may be considered to be a random variable. Its distribution is the distribution of blood pressures in the population. Suppose systolic blood pressures $X$ in a certain population are normally distributed with mean $\mu = 117$ and s.d. $\sigma = 12$ (in mm Hg): $X \sim \mathcal{N}(117, 12^2)$. The proportion of individuals with pressures between, say, 100 and 135 is the probability that an individual picked at random has a blood pressure in this interval. To find it, we use (8), which gives us $F_X$ in terms of the function $\Phi$:

$$P(100 < X < 135) = F_X(135) - F_X(100)$$
$$= \Phi\left(\frac{135 - 117}{12}\right) - \Phi\left(\frac{100 - 117}{12}\right)$$
$$\doteq \Phi(1.50) - \Phi(-1.42) \doteq .9332 - .0778 \doteq .855.$$

Figure 6-3 shows this as a shaded area under the graph of the p.d.f.  ∎

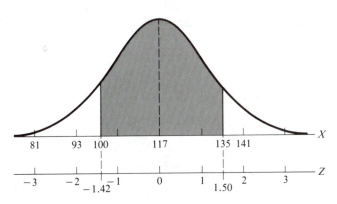

**Figure 6-3** Interval probability, Example 6.1b

Example **6.1c**

### Finding a Percentile

We mentioned in Example 6.1a that in 1977–78, GRE scores on the analytic ability portion are well approximated by $\mathcal{N}(521, 123^2)$. If we assume that distribution for $X$, what is the score of a student at the 63rd percentile?

The closest entry to .6300 in Table II is .6293, corresponding to $Z = .33$. Interpolating between that entry and the next larger one, we find the 63rd percentile of $Z$ to be $z_{.63} = .332$. The corresponding $X$-score, from (6), is

$$x = \mu + z_{.63}\,\sigma = 521 + .332 \times 123 \doteq 562.$$

Using (8), we can find the percentile rank for a given score. For instance, if a student's score is 485, the proportion of students with $X < 485$ is

$$P(X \leq 485) = \Phi\left(\frac{485 - 521}{123}\right) = \Phi(-.293) \doteq .39.$$

So, 485 is at about the 39th percentile.                                                                 ■

We'll find it useful to know that a function which is an exponential with a quadratic exponent whose leading term is negative can serve, with a suitable constant multiplier, as a normal p.d.f. The next example illustrates this and shows how to find the parameters of the distribution.

**Example 6.1d**   Consider the function $\exp(-2x^2 - 8x)$. We rewrite the exponent by the process of "completing the square," familiar from algebra. First, factor out the 2:

$$-2x^2 - 8x = -2(x^2 + 4x \quad\quad).$$

If we insert into the blank space the square of half the coefficient of $x$, the quantity inside the parentheses becomes a perfect square; but to pay for this addition of $-8$, we have to add 8:

$$-2x^2 - 8x = -2(x + 2)^2 + 8.$$

So,

$$e^{-2x^2 - 8x} = e^8 e^{-2(x+2)^2}.$$

Except for a constant factor, this is the p.d.f. of $\mathcal{N}(-2, 1/4)$, where we found the 1/4 by setting $1/(2\sigma^2) = 2$.                                     ■

---

Properties of normal random variables:

(i) If $X \sim \mathcal{N}(\mu, \sigma^2)$,

$$E[(X - \mu)^n] = \begin{cases} (n-1)(n-3)\cdots 5 \cdot 3 \cdot \sigma^n \text{ for } n \text{ even,} \\ 0 \text{ for } n \text{ odd.} \end{cases} \quad (10)$$

(ii) If $X \sim \mathcal{N}(\mu, \sigma^2)$, then $a + bX \sim \mathcal{N}(a + b\mu, b^2\sigma^2)$.

(iii) If $X \sim \mathcal{N}(\mu, \sigma^2)$, $Y \sim \mathcal{N}(\nu, \tau^2)$, and $X$ and $Y$ are independent,

$$X \pm Y \sim \mathcal{N}(\mu \pm \nu, \sigma^2 + \tau^2).$$

(iv) If $X_1, ..., X_n$ are independent, and $X_i \sim \mathcal{N}(\mu_i, \sigma_i^2)$, then

$$\sum_i a_i X_i \sim \mathcal{N}\left(\sum a_i \mu_i, \sum a_i^2 \sigma_i^2\right).$$

The central moments in (i) follow immediately from the moments of the standard normal $Z$ as given by (5), upon dividing through by $\sigma^n$.

Property (ii) says that a linear transformation of a normal variable produces a normal variable: If $X$ is a linear function of $Z$ and $Y$ is a linear function of $X$, then $Y$ is a linear function of $Z$—and therefore is normal.

Properties (iii) and (iv) are easiest to derive using the m.g.f. We'll write out the proof of (iii): The m.g.f.'s of $X$ and $Y$, from (9), are

$$\psi_X(t) = \exp\left\{\mu t + \frac{\sigma^2 t^2}{2}\right\}, \quad \psi_Y(t) = \exp\left\{\nu t + \frac{\tau^2 t^2}{2}\right\}.$$

And if $X$ and $Y$ are independent, the m.g.f. of their sum is the product of their m.g.f.'s:

$$\psi_{X+Y}(t) = \psi_X(t)\psi_Y(t) = \exp\left\{(\mu + \nu)t + \frac{\sigma^2 + \tau^2}{2}t^2\right\}.$$

This is the m.g.f. of a normal distribution whose mean is $\mu + \nu$ and variance is $\sigma^2 + \tau^2$. Because only one distribution has a given m.g.f., it follows that the sum $X + Y$ has this normal distribution. And the difference $X - Y$ can be written as a sum: $X - Y = X + (-Y)$, where $-Y$ is normal with variance $\tau^2$. [Property (iv) is proved in much the same way as Property (iii).]

Returning to the matter of the constant $1/\sqrt{2\pi}$ in (1), we give just an outline of a derivation. (A rigorous proof is a tricky because the integrals are improper.) Denote the value of the integral of $\exp(-z^2/2)$ by $K$. Then,

$$K^2 = \int e^{-x^2/2}dx \int e^{-y^2/2}dy = \int\int e^{-(x^2+y^2)/2}dx\,dy.$$

Introducing polar coordinates: $x = r\cos\theta$, $y = r\sin\theta$, we have (with $u = r^2/2$)

$$K^2 = \int_0^{2\pi}\int_0^\infty e^{-r^2/2}r\,dr\,d\theta = 2\pi\int_0^\infty e^{-u}du = 2\pi.$$

So, $K = \sqrt{2\pi}$.

## Problems

* **6-1.** Find the standard scores ($Z$-scores) for
  (a)  $X = 14.2$, when $\mu_X = 12.6$ and $\sigma_X = 0.4$.
  (b)  $Y = 135$, when $\mu_Y = 150$ and $\sigma_Y^2 = 100$.

* **6-2.** The height of female students at the University of Baroda follows an approximately normal distribution, with mean 60 inches and s.d. 2 inches. Find the probability that the height of a female student selected at random is
  (a)  less than 58 in.      (b)  between 58 and 62 in.

**6-3.** A candy bar wrapper says "Net weight 1.4 oz." The bars actually vary in weight. To be reasonably confident that most bars weigh at least 1.4 oz, a manufacturer may adjust the production so that the mean weight is 1.5 oz. Assuming weights to be approximately normally distributed with s.d. .05 oz, find the proportion of bars weighing less than the advertised 1.4 oz.

* **6-4.** ACT scores for a large group of entering freshmen in a particular year averaged 22.2 with standard deviation 4.6. Suppose ACT scores are normally distributed with these parameter values.

    **(a)** What proportion of scores are less than 18?

    **(b)** Find the first and third quartiles of the distribution.

    **(c)** What fraction of the scores would lie between the quartiles in (b)?

    **(d)** The quartiles were reported to be $Q_1 = 19$, $Q_3 = 26$. What fraction of the scores lie between these reported quartiles if the distribution is actually normal with mean 22.2 and s.d. 4.6?

**6-5.** GRE scores of two students are reported as follows:

    Student 1:  760 (96th percentile)
    Student 2:  520 (49th percentile).

Find the mean and s.d. of all GRE scores, assuming a normal distribution.

* **6-6.** The distribution of increases in human heart rate (in beats per minute) after taking a certain drug is given approximately by the p.d.f.

$$f(x) \propto \exp\left\{-\frac{1}{128}(x+6)^2\right\}.$$

    **(a)** Find the mean and variance of the distribution.

    **(b)** Find the m.g.f. of the distribution.

**6-7.** Given the p.d.f. $f(x) \propto \exp(-x^2 - 5x)$, find the mean and variance.

* **6-8.** Given $X \sim \mathcal{N}(2, 1)$, find $E(X^3)$

    **(a)** using the m.g.f.

    **(b)** using (10), the formula for central moments, together with the identity

$$X^3 = (X - 2 + 2)^3 = (X - 2)^3 + 6(X - 2)^2 + 12(X - 2) + 8.$$

**6-9.** Suppose the m.g.f. of $X$ is $\psi(t) = \exp(2t + t^2)$.

    **(a)** Find the mean and variance of $X$.     **(b)** Give the p.d.f. of $X$.

    **(c)** Find $E(e^X)$.     **(d)** Find the m.g.f. of $2X + 1$.

* **6-10.** Suppose $X \sim \mathcal{N}(4, 1)$, $Y \sim \mathcal{N}(5, 4)$, $Z \sim \mathcal{N}(2, 2)$, and that the three variables are independent. Find the distribution of each of the following:

    **(a)** $X + Y + Z$.     **(b)** $X - Y$.     **(c)** $2X - Y - Z$.

**6-11.** For the variables $X$ and $Y$ as defined in the preceding problem, find

    **(a)** $P(X + Y > 10)$.     **(b)** $P(X < Y)$.

* **6-12.** The heights of male students at the University of Baroda are approximately normal, with mean 64 in. and s.d. 2.5 in. Suppose a male student and a female student (see Problem 6-2) are selected independently and at random. Find the probability that the male student is taller than the female student.

**6-13.** Use generating functions to show that if $X$ is normal, so is $a + bX$.

## 6.2   Exponential Distributions

In Example 5.2d we derived the p.d.f. of the distribution of the time to the first "event," in a Poisson process, after an arbitrary point in time. In that example we assumed the rate parameter to be $\lambda = 6$. The derivation did not depend on the particular value assigned to $\lambda$, and leaving the rate as $\lambda$, we have a class or family of distributions called exponential, with c.d.f. and p.d.f. just as in Example 5.2d, but with 6 replaced by $\lambda$.

---

When $X$ has the distribution defined by the p.d.f.

$$f(x \mid \lambda) = \lambda e^{-\lambda x}, \quad x > 0, \tag{1}$$

or by the corresponding c.d.f.

$$F(x \mid \lambda) = 1 - e^{-\lambda x}, \quad x > 0, \tag{2}$$

the distribution is called **exponential**, and we write $X \sim \text{Exp}(\lambda)$.

---

The mean of an exponential distribution is easy to find by integration—easy, that is, if you have a table of definite integrals or don't mind integrating "by parts":

$$EX = \int_0^\infty x \left( \lambda e^{-\lambda x} \right) dx = \frac{1}{\lambda}. \tag{3}$$

The median $m$ is found by equating the c.d.f. to 1/2:

$$\frac{1}{2} = 1 - e^{-\lambda x}, \text{ or } e^{-\lambda x} = \frac{1}{2}.$$

Taking (natural) logs, we find $-\lambda x = -\log 2$. So $m = (\log 2)/\lambda$.

**Example 6.2a** | **Particle Emissions**

Alpha particles emitted by carbon-14 are recorded using a Geiger counter. Suppose emissions occur according to a Poisson process at the rate $\lambda = 16$ per second. The time $X$ to the next emitted particle, measured from any given point in time, is then exponential with mean $1/\lambda$:   $X \sim \text{Exp}(16)$. The mean time to the next particle emission is $1/\lambda = 1/16$ or .0625 sec. The median of the distribution is .693/16 or about .0433 sec. ∎

To find the variance of an exponential distribution, we could integrate (again by parts, twice), as we did to find the mean. However, we can find all the moments with a single integration, by calculating the moment generating function:

$$\psi(t) = \int_0^\infty e^{tx} (\lambda e^{-\lambda x}) dx = \lambda \int_0^\infty e^{-x(\lambda - t)} dx = \frac{1}{1 - t/\lambda}. \tag{4}$$

This calculation is valid so long as $t < \lambda$, because then the exponent in the integrand is negative, and the integral converges. Moreover, when $t/\lambda < 1$, the power series expansion of $(1 - t/\lambda)^{-1}$ converges:

$$\frac{1}{1 - t/\lambda} = 1 + \frac{t}{\lambda} + \left(\frac{t}{\lambda}\right)^2 + \left(\frac{t}{\lambda}\right)^3 + \cdots$$

$$= 1 + \left(\frac{1}{\lambda}\right) t + \left(\frac{2!}{\lambda^2}\right) \frac{t^2}{2!} + \left(\frac{3!}{\lambda^3}\right) \frac{t^3}{3!} + \cdots.$$

We can now read out the coefficient of $t^k/k!$ as the $k$th moment, $E(X^k)$:

$$EX = \frac{1}{\lambda}, \quad E(X^2) = \frac{2!}{\lambda^2}, \quad E(X^3) = \frac{3!}{\lambda^3}, \tag{5}$$

and so on. The variance is the average square minus the square of the average:

$$\text{var } X = \frac{2}{\lambda^2} - \left(\frac{1}{\lambda}\right)^2 = \frac{1}{\lambda^2}. \tag{6}$$

---

When $X \sim \text{Exp}(\lambda)$, its median, mean, s.d., and m.g.f. are

$$m = \frac{\log 2}{\lambda}, \tag{7}$$

$$\mu = \frac{1}{\lambda}, \tag{8}$$

$$\sigma = \frac{1}{\lambda}, \tag{9}$$

$$\psi(t) = \frac{1}{1 - t/\lambda}. \tag{10}$$

---

The exponential distribution is called *memoryless*, for the following reason. Suppose the waiting time $X$ is $\text{Exp}(\lambda)$, and suppose also that we have been waiting for time $c$, and the "event" (arrival, emission, or whatever) has not yet occurred:

$X > c$. How much longer do we have to wait? The probability that we have to wait an additional time $y$ (given $X > c$) is

$$P(X > c + y \mid X > c) = \frac{P(X > c + y)}{P(X > c)}$$

$$= \frac{1 - F_X(c+y)}{1 - F_X(c)} = \frac{e^{-\lambda(c+y)}}{e^{-\lambda c}} = e^{-\lambda y}.$$

This is precisely the tail-probability of $\text{Exp}(\lambda)$. And this says that the conditional distribution given $X > c$ is independent of $c$, and is the same as the distribution of the time to the first event after time 0, no matter how long we've been waiting!

The memoryless feature of the exponential distribution derives from the Poisson process, which is a continuous-time version of a Bernoulli process. The discrete-time analog of the exponential distribution is the geometric distribution, which also has the memoryless property. Indeed, the geometric distribution is the only discrete distribution with this property, and the exponential is the only continuous distribution with this property.

Consider a Poisson arrival process with rate $\lambda$, and let $T_1$ be the time to the first arrival, $T_2$ the time from the first to the second arrival, and so on, so that $T_1, T_2, ..., T_r$ are successive interarrival times. It might be anticipated, in view of the Poisson postulate, that these times are independent. We'll use the differential technique (see §5.2) to show that this is the case.

Consider the infinitesimal increment $dt_i$ in $T_i$, for $i = 1, ..., r$, and the intersection of the events $[t_i < T_i < t_i + dt_i]$. These $r$ conditions are satisfied if there is exactly one arrival in each incremental interval, $(t_i, t_i + dt_i)$ and no other arrivals between 0 and $\sum t_i$. (See Figure 6-4.)

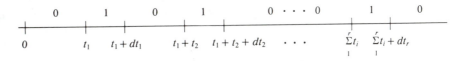

**Figure 6.4**

The probability of an arrival in a particular one of the incremental intervals is approximately $\lambda \, dt_i$, and the probability of no arrival in an interval of approximate length $t_i$ is $\exp(-\lambda t_i)$. The numbers of arrivals in these various intervals are independent because the intervals do not overlap. Thus,

$$P\left\{ \bigcap_1^r [t_i < T_i < dt_i] \right\} = \prod_1^r (\lambda \, dt_i) \cdot \prod_1^r \exp(-\lambda t_i)$$

$$= \left\{ \prod_1^r f_{T_i}(t_i) \right\} dt_1 \cdots dt_r.$$

The product in the braces—the coefficient of the differential element—is the joint p.d.f. of the $T_i$'s. And it is the product of the marginal p.d.f.'s of the $T_i$'s, so these interarrival times are independent.

> In a Poisson process with rate parameter $\lambda$, the successive interarrival times $T_1$, ..., $T_r$ are independent, with $T_i \sim \text{Exp}(\lambda)$.

## 6.3 Gamma Distributions

We now extend the notion of the waiting time in a Poisson process and study the waiting time to the $r$th event after a given point in time. This, in turn, will lead us to a new (and useful) family of continuous distributions.

Consider a Poisson arrival process with rate parameter $\lambda$, and the random variable $Y$, the time from an arbitrary point ($t = 0$) to the $r$th arrival. The time to that $r$th arrival is the sum of the $r$ successive waiting times, $T_1$ to the first arrival, $T_2$ from the first to the second arrival, etc.:

$$Y = T_1 + T_2 + \cdots + T_r. \tag{1}$$

These $T_i$'s are just the variables shown to be independent at the end of the preceding section. So, $Y$ is the sum of $r$ independent and identically distributed exponential variables with parameter $\lambda$. From the structure of $Y$ as given by (1), we easily obtain the mean and variance, as the sum of the means and the sum of the variances, respectively:

$$EY = rET = \frac{r}{\lambda}, \text{ and } \text{var}\, Y = r\, \text{var}\, T = \frac{r}{\lambda^2}.$$

We'll find the p.d.f. using the differential method, finding the probability of an infinitesimal interval from $t$ to $t + dt$. We do this by expressing the event $[t < Y < t + dt]$ in terms of numbers of arrivals in nonoverlapping intervals:

$$P(t < Y < t + dt) = P[r - 1 \text{ arrivals in } (0, t) \text{ and } 1 \text{ in } (t, t + dt)]$$
$$= P[r - 1 \text{ arrivals in } (0, t)] \cdot P[1 \text{ arrival in } (t, t + dt)]$$
$$= \frac{(\lambda t)^{r-1}}{(r - 1)!} e^{-\lambda t}(\lambda\, dt).$$

The coefficient of $dt$ is the desired p.d.f. The distribution is sometimes called the *Erlang distribution*.

We can find an expression for the c.d.f. by expressing the event $[Y > y]$ in terms of an event with a Poisson probability:

$$[Y > y] = [\text{at most } r - 1 \text{ arrivals in time } y].$$

For, if we have to wait longer than $y$ for the $r$th event, it is because there are at most $r - 1$ arrivals in the interval $(0, y)$; and, conversely, if there are no more than $r - 1$ arrivals in $(0, y)$, then the time to the $r$th exceeds $y$. Hence,

$$P(Y > y) = P(\text{at most } r - 1 \text{ arrivals in time } y) = \sum_{0}^{r-1} \frac{(\lambda y)^k}{k!} e^{-\lambda y}.$$

From this tail-probability, we find the c.d.f. as $1 - P(Y > y)$.

---

Distribution of $Y$ the waiting time to the $r$th event in a Poisson process with rate $\lambda$: The density function is

$$f_Y(y) = \frac{\lambda^r y^{r-1}}{(r-1)!} e^{-\lambda y}, \tag{2}$$

and the c.d.f.,

$$F_Y(y) = 1 - \sum_{0}^{r-1} \frac{(\lambda y)^k}{k!} e^{-\lambda y}. \tag{3}$$

The mean and variance are

$$EY = r/\lambda \quad \text{and} \quad \text{var } Y = r/\lambda^2. \tag{4}$$

---

The formula in (2) assumes that $r$ is a positive integer. We now want to replace $r$ by a positive real number $\alpha$ and consider the function $x^{\alpha-1} e^{-\lambda x}$ as a potential p.d.f. For this we need to know how to handle integrals of the form

$$\Gamma(\alpha) = \int_{0}^{\infty} x^{\alpha-1} e^{-x} dx. \tag{5}$$

This is a convergent improper integral, provided $\alpha > 0$. Its value depends on the parameter $\alpha$, so it is a function of $\alpha$. As such it is called the **gamma function.** (The $\Gamma$ is an uppercase Greek gamma.)

The gamma function appears in disguise as various other (useful) definite integrals, obtained from (5) by change of variable. And each such integral implies an integration formula:

Equivalent expressions for $\Gamma(\alpha)$:

$$\Gamma(\alpha) = \lambda^\alpha \int_0^\infty y^{\alpha-1} e^{-\lambda y} dy. \tag{6}$$

$$\Gamma(\alpha) = \frac{1}{2^{\alpha-1}} \int_0^\infty z^{2\alpha-1} e^{-z^2/2} dz. \tag{7}$$

Integral formulas:

$$\int_0^\infty y^{\alpha-1} e^{-\lambda y} dy = \frac{\Gamma(\alpha)}{\lambda^\alpha}. \tag{8}$$

$$\int_0^\infty z^{2\alpha-1} e^{-z^2/2} dz = 2^{\alpha-1}\Gamma(\alpha). \tag{9}$$

For most values of $\alpha$, it is necessary to use a numerical integration to find the value of $\Gamma(\alpha)$. However, if we ever find the value for one $\alpha$, we can easily find the value at $\alpha$ + an integer $k$ $(k > 0)$, using the recursion identity:

$$\Gamma(\alpha+1) = \alpha\,\Gamma(\alpha). \tag{10}$$

(Problem 6-23 calls for a demonstration of this identity.) For instance, it is clear that $\Gamma(1) = 1$. Then (10) says $\Gamma(2) = 1 \cdot 1$, and that $\Gamma(3) = 2\Gamma(2) = 2 \cdot 1$, and $\Gamma(4) = 3\Gamma(3) = 3 \cdot 2 \cdot 1$, and so on. In general,

$$\Gamma(n) = (n-1)!. \tag{11}$$

So, $\Gamma(\alpha)$ is a generalized factorial function.

The values of the gamma function halfway between successive integers are all implied by its value at $\alpha = 1/2$. And we get this from (7) above and (2) in §6.1:

$$\Gamma\left(\frac{1}{2}\right) = \sqrt{2} \int_0^\infty e^{-z^2/2} dz = \frac{1}{\sqrt{2}} \int_{-\infty}^\infty e^{-z^2/2} dz = \sqrt{\pi}.$$

We can then find the value of the gamma function at any odd multiple of 1/2 by means of the recursion relation (10). For instance,

$$\Gamma\left(\frac{9}{2}\right) = \frac{7}{2}\cdot\Gamma\left(\frac{7}{2}\right) = \frac{7}{2}\cdot\frac{5}{2}\cdot\Gamma\left(\frac{5}{2}\right) = \frac{7}{2}\cdot\frac{5}{2}\cdot\frac{3}{2}\cdot\Gamma\left(\frac{3}{2}\right) = \frac{7}{2}\cdot\frac{5}{2}\cdot\frac{3}{2}\cdot\frac{1}{2}\cdot\Gamma\left(\frac{1}{2}\right),$$

or

$$\Gamma\left(\frac{9}{2}\right) = \frac{105}{16}\sqrt{\pi}. \tag{12}$$

A sketch of the gamma function is shown in Figure 6-5.

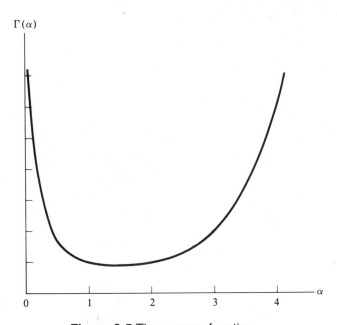

**Figure 6-5** The gamma function

The integrand of the integral (8) for $\Gamma(\alpha)$ is a nonnegative function with a finite integral. Dividing by the value of the integral produces a density function—rather, a two-parameter family of densities on the positive real line:

$$f(y \mid \alpha, \lambda) = \frac{\lambda^{\alpha}}{\Gamma(\alpha)} y^{\alpha-1} e^{-\lambda y}, \quad y > 0. \tag{13}$$

The distributions defined by these p.d.f.'s are called **gamma distributions.** The waiting-time distributions defined by the densities (2) are special cases in which the parameter $\alpha$ is a positive integer. When a variable $Y$ has the p.d.f. (13) for a particular $\alpha$ and $\lambda$, we write $Y \sim \text{Gam}(\alpha, \lambda)$.

The *moment generating function* of a gamma distribution is defined as

$$\psi(t) = E(e^{tY}) = \int_0^\infty e^{ty} \cdot \frac{\lambda^{\alpha}}{\Gamma(\alpha)} y^{\alpha-1} e^{-\lambda y} dy = \frac{\lambda^{\alpha}}{\Gamma(\alpha)} \int_0^\infty y^{\alpha-1} e^{-(\lambda-t)y} \, dy.$$

We can evaluate the integral on the right using (8), replacing the $\lambda$ by $\lambda - t$:

$$\psi(t) = \frac{\lambda^{\alpha}}{\Gamma(\alpha)} \cdot \frac{\Gamma(\alpha)}{(\lambda - t)^{\alpha}} = \frac{1}{(1 - t/\lambda)^{\alpha}} \cdot$$

It is a straightforward matter (Problem 6-22) to find the first two moments as $\psi'(0)$ and $\psi''(0)$, and from them, the variance. [When $\alpha$ is an integer, these agree with what we found earlier for that case by exploiting (1).]

---

When $Y \sim \text{Gam}(\alpha, \lambda)$, the p.d.f. is

$$f(y \mid \alpha, \lambda) = \frac{\lambda^{\alpha}}{\Gamma(\alpha)} y^{\alpha-1} e^{-\lambda y}, \quad y > 0. \tag{14}$$

The mean and variance are

$$EY = \frac{\alpha}{\lambda}, \quad \sigma_Y^2 = \frac{\alpha}{\lambda^2}, \tag{15}$$

and the m.g.f. is

$$\psi_Y(t) = \frac{1}{(1 - t/\lambda)^{\alpha}} \cdot \tag{16}$$

---

We have not given a "formula" for the c.d.f. of a gamma distribution because, except in the special case where $\alpha$ is an integer, the c.d.f. is not expressible in elementary functions. The c.d.f. for integer $\alpha$ was derived at the beginning of this section and given by (3).

**Example 6.3a**

**Relation to the Normal**

The even-order moments of a standard normal distribution can be expressed in terms of gamma functions. By definition, we have

$$E(Z^{2k}) = \frac{1}{\sqrt{2\pi}} \int_{-\infty}^{\infty} z^{2k} e^{-z^2/2} dz = \frac{2}{\sqrt{2\pi}} \int_{0}^{\infty} z^{2k} e^{-z^2/2} dz.$$

The value of the integral on the right is given by (9), with $\alpha = k + 1/2$:

$$E(Z^{2k}) = \frac{1}{\sqrt{\pi}} \cdot 2^k \Gamma\left(k + \frac{1}{2}\right).$$

With $k = 4$, for example, this together with (12) above verifies the earlier formula [(5) in §6.1] for the 8th moment of $Z$:

$$E(Z^8) = \frac{1}{\sqrt{\pi}} \cdot 2^4 \, \Gamma\left(\frac{9}{2}\right) = 7 \cdot 5 \cdot 3 \cdot 1. \qquad \blacksquare$$

## Problems

**∗ 6-14.** Fuses in an electric circuit fail when there is an overload. Assume that in a particular application, overloads follow a Poisson process with a mean time between occurrences of 6 months. Find the probability that

 **(a)** a fuse will last at least 12 months.

 **(b)** a fuse will fail before 3 months.

 **(c)** there will be more than two failures in a one-year period, assuming that fuses are replaced immediately after they fail.

**6-15.** During the busy periods of the day, customer arrivals at an airport car-rental counter are Poisson, averaging one in two minutes. Find

 **(a)** the probability that no customers arrive in the next minute.

 **(b)** the mean time for six more customers to arrive.

 **(c)** the probability that fewer than four customers arrive in the next minute.

**∗ 6-16.** A computer is subject to major breakdowns and minor breakdowns. Suppose these two types of breakdown follow independent Poisson distributions. The average time to failure is ten days for major breakdowns and two days for minor breakdowns.

 **(a)** Find the average time to the first breakdown of either type.

 **(b)** Find the probability that no breakdowns occur in the next week.

 **(c)** Find the probability that the next breakdown is minor.

**6-17.** For the fuses of Problem 6-14, find

 **(a)** the probability that the time to the third failure exceeds one year.

 **(b)** the mean time to the third failure.

 **(c)** the mean time from the third failure to the seventh failure.

 **(d)** the probability that the time to the third failure is within one standard deviation of its mean.

**∗ 6-18.** Find the joint p.d.f. of a sequence of $n$ independent variables, each of which is $\text{Exp}(\lambda)$.

**∗ 6-19.** Calculate the following:

 **(a)** $\Gamma(15/2)$.  **(b)** $\int_0^\infty x^9 e^{-2x}\,dx$.  **(c)** $\int_0^\infty x^{5/2} e^{-x/2}\,dx$.

**6-20.** Find (to the nearest hundredth) the median of $\text{Gam}(2, 1)$. *(Hint:* Use integration by parts and solve iteratively or by trial and error.)

**∗ 6-21.** Suppose $X_1 \sim \text{Gam}(\alpha_1, \lambda)$ and $X_2 \sim \text{Gam}(\alpha_2, \lambda)$, and that these are independent variables. Find the distribution of $X_1 + X_2$ and deduce a result for the sum of $n$ independent variables with gamma distributions.

**6-22.** Obtain the formulas for the mean and variance of a gamma distribution by calculating $\psi'(0)$ and $\psi''(0)$.

**6-23.**   Write the integral defining $\Gamma(\alpha+1)$ and integrate by parts, to derive the recursion relation $\Gamma(\alpha+1) = \alpha\,\Gamma(\alpha)$.

## 6.4        Chi-Square Distributions

A subfamily of the family of gamma distributions known as "chi-square" is important for its numerous applications in statistical inference.

The simplest chi-square distribution is that of the square of a standard normal variable. If $Z \sim \mathcal{N}(0, 1)$, the m.g.f. of $Z^2$ is

$$\psi_{Z^2}(t) = E(e^{tZ^2}) = \frac{1}{\sqrt{2\pi}}\int_{-\infty}^{\infty} e^{tz^2}e^{-z^2/2}dz = \frac{1}{\sqrt{2\pi}}\int_{-\infty}^{\infty} e^{-(1-2t)z^2/2}dz.$$

Substituting $(1-2t)^{1/2}z = u$ and $dz = du/\sqrt{1-2t}$, we obtain

$$\psi_{Z^2}(t) = \frac{1}{\sqrt{1-2t}}\int_{-\infty}^{\infty} \frac{1}{\sqrt{2\pi}}\,e^{-u^2/2}\,du = \frac{1}{\sqrt{1-2t}}. \tag{1}$$

Referring to (16) in the preceding section, we see that this is the m.g.f. of $\mathrm{Gam}(\frac{1}{2}, \frac{1}{2})$.

Now consider the sum of $k$ squares of independent, standard normal variables; let

$$\chi^2 = Z_1^2 + Z_2^2 + \cdots + Z_k^2. \tag{2}$$

The m.g.f. of this sum of independent variables is the product of their m.g.f.'s, which (because they are identically distributed) is just the $k$th power of the m.g.f. of a single term:

$$\psi_{\chi^2}(t) = [(1-2t)^{-1/2}]^k = (1-2t)^{-k/2}. \tag{3}$$

This is the m.g.f. of $\mathrm{Gam}(\frac{k}{2}, \frac{1}{2})$. So we can get the p.d.f., mean, and variance of $\chi^2$ as special cases of the corresponding formulas for the gamma distribution [(14) and (15) of §6.3]. However, it should be noted that the mean and variance follow directly from the *structure* of $\chi^2$ as the sum of squares of independent, standard normal variables (see Problem 6-25).

Tail-areas and percentiles of the chi-square distribution for degrees of freedom up to 30 are given in Table Va of Appendix I. For degrees of freedom above 30, one could use the asymptotic distribution of $\chi^2$, and we'll see in Chapter 8 that for large $k$, $\mathrm{chi}^2(k) \approx \mathcal{N}(k, 2k)$. However, a somewhat better approximation is obtained from the fact (which we'll take on faith) that

$$\sqrt{2\chi^2} - \sqrt{2k-1} \approx \mathcal{N}(0, 1). \tag{4}$$

The footnote to Table Va shows how to use this to obtain approximate percentiles of chi-square from standard normal percentiles.

The **chi-square distribution with $k$ degrees of freedom** is the distribution of the sum of squares of $k$ independent, standard normal variables. The chi-square p.d.f. is

$$f_{\chi^2}(x) = \frac{1}{2^{k/2}\Gamma(\frac{k}{2})}\, x^{k/2-1}e^{-x/2}, \quad x > 0. \tag{5}$$

When $Y$ has this density we write $Y \sim \text{chi}^2(k)$. The mean and variance are

$$EY = k, \quad \text{var}\, Y = 2k. \tag{6}$$

Graphs of chi-square p.d.f.'s for several values of the parameter $k$ are shown in Figure 6-6.

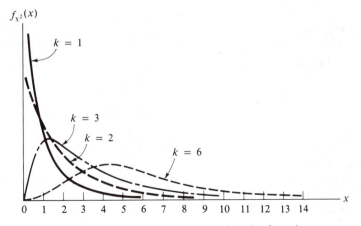

**Figure 6-6** Some chi-square density functions

The chi-square distribution has an important reproductive property, which follows from the structure of $\chi^2$. Thus, if $X \sim \text{chi}^2(k)$ and $Y \sim \text{chi}^2(m)$, and if $X$ and Y are independent, it follows [from (2)] that their sum,

$$X + Y = Z_1^2 + \cdots + Z_k^2 + Z_{k+1}^2 + \cdots + Z_{k+m}^2,$$

is the sum of $k + m$ squares of independent, standard normal variables. So $X + Y$ is $\text{chi}^2(k + m)$. [This result is, of course, a special case of the reproductive property of gamma distributions, given in Problem 6-21.]

> If $X \sim \text{chi}^2(k)$ and $Y \sim \text{chi}^2(m)$, and $X$ and $Y$ are independent, then
>
> $$X + Y \sim \text{chi}^2(k + m).$$
>
> Also, if $X \sim \text{chi}^2(k)$ and $X + Y \sim \text{chi}^2(n)$, and $X$ and $Y$ are independent, then $Y \sim \text{chi}^2(n - k)$.

We can show the second of these properties, which we'll need in later chapters, by using m.g.f.'s:

$$\psi_Y(t) = \frac{\psi_{X+Y}(t)}{\psi_X(t)} = \frac{(1 - 2t)^{-n/2}}{(1 - 2t)^{-k/2}} = (1 - 2t)^{-(n-k)/2}.$$

This is the m.g.f. of $\text{chi}^2(n - k)$, and uniqueness then implies $Y \sim \text{chi}^2(n - k)$.

## 6.5　Distributions for Reliability

**Reliability theory** deals with the operating life or time to failure of a unit of system. This time $L$ is a random variable, varying from unit to unit. The **reliability function** is the probability that the operating life exceeds $x$:

$$R(x) = P(L > x) = 1 - F_L(x). \tag{1}$$

For example, suppose $L \sim \text{Exp}(\lambda)$. The reliability function is

$$R(x) = 1 - F_L(x) = 1 - (1 - e^{-\lambda x}) = e^{-\lambda x}.$$

We saw in §6.2 that when $L$ has an exponential distribution, the *future* life after any point in time does not depend on how long the unit has been operating up to that point. In such a case, replacing old units with new ones as a means of preventive maintenance is pointless. However, the exponential distribution is *not* appropriate (at least, not exactly) for the lifetime of a system that ages or wears out.

Given the p.d.f. of the system life $L$, we can find the *mean* life in the usual way, but we can also calculate it directly from the reliability function. To see this, we integrate $R(x)$ by parts, with $u = R(x)$, $du = -f_L(x)\,dx$, $dx = dv$, and $x = v$:

$$\int_0^\infty R(x)\,dx = x\,R(x)\big|_0^\infty + \int_0^\infty x f_L(x)\,dx.$$

If $x\,R(x) \to 0$ as $x \to \infty$, then

$$\int_0^\infty R(x)\,dx = \int_0^\infty x\,f_L(x)\,dx = EL. \tag{2}$$

[If $R(x)$ is of the order of $1/x$ at infinity, the mean does not exist.]

An intuitive quantity useful in judging system reliability is the **hazard function.** This is the logarithmic derivative of $R$:

$$h(x) = \frac{d}{dx} \{ - \log R(x) \} = \frac{-R'(x)}{R(x)} = \frac{f(x)}{R(x)}. \tag{3}$$

And then, if $R(0) = 1$,

$$R(x) = \exp \left\{ - \int_0^x h(u) \, du \right\}. \tag{4}$$

To understand the hazard function, consider a large number of identical systems. The reliability function is the proportion of systems surviving to time $x$. The hazard is thus the rate of failure relative to the number surviving at $x$:

$$h(x) \, dx = P(x < L < x + dx \mid L > x) = \frac{P(x < L < x + dx)}{P(L > x)} \doteq \frac{f(x) \, dx}{R(x)}.$$

**Example 6.5a** | **Constant Hazard**
The reliability function for a system whose time to failure is exponential is $e^{-\lambda x}$ for $x > 0$. So the hazard is constant:

$$h(x) = \frac{f(x)}{R(x)} = \frac{\lambda e^{-\lambda x}}{e^{-\lambda x}} = \lambda.$$

Conversely, if the hazard is constant for all $x > 0$: $h \equiv \lambda$, then by (4),

$$R(x) = \exp \left\{ - \int_0^x \lambda \, du \right\} = e^{-\lambda x}, x > 0,$$

and the time to failure is exponential. Having constant hazard is equivalent to having no memory (see §6.2):  Lack of memory is the same as not "aging"—that is, not wearing out with time.                                                                 ∎

A hazard that increases with time may be appropriate in cases where failure is the result of wear, as in the case of the life of a manufactured article. Suppose the hazard function is a power of $x$:

$$h(x) = \frac{\alpha x^{\alpha-1}}{\beta^\alpha}, \quad x > 0,$$

where $\alpha$ and $\beta$ are positive parameters. Then, by (4),

$$R(x) = e^{-(x/\beta)^\alpha}, \text{ and } F(x) = 1 - e^{-(x/\beta)^\alpha}.$$

This defines the family of **Weibull** distributions. The Weibull p.d.f. is

$$f(x \mid \alpha, \beta) = \frac{\alpha x^{\alpha-1}}{\beta^\alpha} e^{-(x/\beta)^\alpha}.$$

(When $\alpha = 1$, the hazard is constant, and this an exponential density.) The Weibull distribution is often used in modeling system reliability.

    Another setting in which hazard increases with time is that of lifetimes in human populations.

| | |
|---|---|
| Example **6.5b** | **The Gompertz Distribution** |

Let $L$ be a person's age at death. Insurance companies want to know the chance that a 40-year old man, for example, will die in the next year, *given that he has reached the age of 40.* This is the hazard at $t = 40$. (Actuaries call hazard the *force of mortality.*) It is estimated as the ratio of the number of deaths among 40-year-old males to the number of 40-year-old males "at risk." Figure 6-7 shows the approximate hazard function for males of ages 15–70. (The graph is plotted from tables in the 1982 *Transactions of the Society of Actuaries,* which gives the figures for ages 15–100.) The bump in the hazard function in the late teens corresponds to the time that people start to drive automobiles!

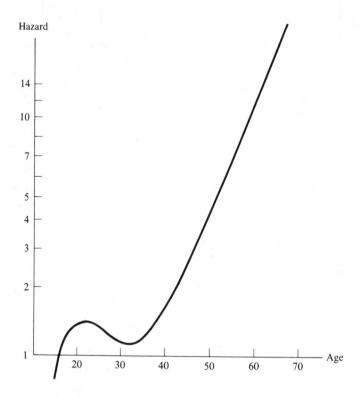

**Figure 6-7** Hazard function for males

Past age 40, the hazard curve is nearly linear, when plotted (as in Figure 6-7) on a logarithmic scale. This suggests that the hazard is exponential: $h(x) = ae^{bx}$, for some constants $a$ and $b$. According to (4), this would mean

$$R(x) = \exp\left\{-\frac{a}{b}\left(e^{bx} - 1\right)\right\},$$

and

$$f(x) = h(x)R(x) = a \exp\left\{bx - \frac{a\,e^{bx}}{b} + \frac{a}{b}\right\}, \quad x > 0.$$

This is the p.d.f. of the *Gompertz distribution*, important in actuarial science.   ∎

The reliability of a system consisting of several units is derivable from reliabilities of the individual units, when we can assume that their times to failure are independent. Particularly easy to analyze are series and parallel combinations. Components interconnected so that the system fails when any unit fails are said to be *in series*. The time to failure of a series system is the *minimum* of the times to failure of its components. If those times are independent and the component reliabilities are $R_i$, the system reliability is

$$R(x) = P(\text{system life} > x) = P(\text{all units survive to } x) = \prod R_i(x). \qquad (5)$$

Components interconnected so that the system fails only when all component units fail are said to be *in parallel*. The time to failure of a parallel system is the *maximum* of the times to failures of its components. If those times are independent, the reliability function for a parallel system is

$$R(x) = 1 - P(\text{all units fail before } x) = 1 - \prod[1 - R_i(x)]. \qquad (6)$$

Figure 6-8 gives schematic diagrams of series and parallel combinations of two components.

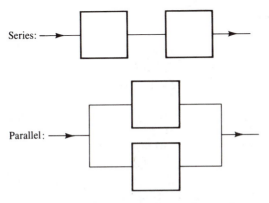

**Figure 6-8** Schematic for series and parallel combinations

The next example considers a system that is neither a series nor a parallel combination.

**Example 6.5c** | Suppose three units are connected according to the schematic diagram of Figure 6-9, and assume that times to failure of the units are independent. Units 2 and 3 are connected in series, so the reliability function of the lower branch is $R_2 R_3$. The reliability function of the parallel combination of the upper and lower branches is

$$R = 1 - (1 - R_1)(1 - R_2 R_3).$$

If the individual hazards are constant—say, 2 for unit 1, 4 for unit 2, and 5 for unit 3, then

$$R(x) = 1 - (1 - e^{-2x})(1 - e^{-4x} e^{-5x}) = e^{-2x} + e^{-9x} - e^{-11x}.$$

The mean time to failure is the integral of $R$, or $\frac{1}{2} + \frac{1}{9} - \frac{1}{11} \doteq .52.$    ∎

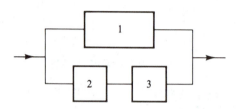

**Figure 6-9** Schematic for Example 6.5c

## Problems

**6-24.**   Given that $Z_1, Z_2 \ldots$ are independent, standard normal variables, use the chi-square table to find

* **(a)**   $P\left\{\sum_1^4 Z_i^2 > 10\right\}.$    **(b)**   $P\left\{\sum_1^{18} Z_i^2 < 26\right\}.$

**6-25.**   Use the structure of a chi-square variable, in terms of standard normal variables, to derive the formulas for its mean and variance.

* **6-26.**   Suppose $X_1, \ldots, X_n$ are independent, each distributed as Exp($\lambda$).
   **(a)**   Find the m.g.f. of their sum.
   **(b)**   Find the m.g.f. of $V = 2\lambda \sum X_i$, and identify the distribution of $V$ as a chi-square distribution.
   **(c)**   Using (b), find $P(\sum X_i < 3)$ when $n = 9$ and $\lambda = 5$.

**6-27.**   Approximate the probability in Problem 6-24(b)
   **(a)**   using the normal table, and the fact that $\chi^2(k) \approx \mathcal{N}(k, 2k)$.
   **(b)**   using the approximation (4) in §6.4.

* **6-28.** Given the following joint p.d.f. for $(X, Y, Z)$:

$$f(x, y, z) \propto \exp\{-\frac{1}{2}(x^2 + y^2 + z^2)\},$$

find the distribution of $X^2 + Y^2 + Z^2$, the squared distance from the origin to the point $(X, Y, Z)$. [*Hint:* The joint p.d.f. factors.]

**6-29.** Let $L$ denote the length of time (in minutes) an automobile battery randomly selected from a certain production line will continue to crank an engine. Assume that $L \sim N(10, 4)$.

(a) Find the probability that the battery will crank the engine longer than $10 + x$ minutes, given that it is still cranking at 10 minutes, as an expression depending on $x$ and involving the standard normal c.d.f. $\Phi$.

(b) Evaluate the probability in (a) for $x = 2$ and for $x = 6$.

(c) Explain why an exponential distribution for $L$ is not apt to be correct.

* **6-30.** A system has the following p.d.f. for its effective lifetime $T$:

$$f(t) = (t + 1)^{-2}, \quad t > 0.$$

(a) Find the reliability function, $R(t)$.

(b) Find $ET$.

(c) Calculate $P(T > t_0 + t \mid T > t_0)$ where $t > 0$ and $t_0 > 0$.

(d) Find the hazard function. Observe whether it increases or decreases with time, and see how this bears on the answers to (a) and (c).

**6-31.** Buses in a certain fleet have independent exponential times to breakdown, with mean 3 months. Suppose there are $n$ buses in the fleet. Find the mean time to the first breakdown.

* **6-32.** A system consists of two components connected in series. The operating life of each component is exponential with mean one hour, and their times to failure are independent. Find the p.d.f. and mean value of $T$, the time to breakdown of the system.

**6-33.** Repeat the preceding problem with this change: The two components are connected in parallel.

* **6-34.** Three units are connected to form a system according to the block diagram in Figure 6-10. Find the system reliability function $R$ in terms of the reliability functions $R_i$ for unit $i$.

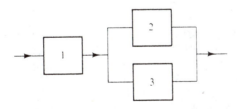

**Figure 6-10** Schematic for Problem 6-34

**6-35.** When exposed to high temperatures, a system fails at time $T$, with reliability function $R(t) = 8/(2 + t)^3$, $t > 0$.

(a) Find the c.d.f. and p.d.f. of $T$.

(b) Find the mean and variance of $T$.

## Review Problems

**6-R1.**   Evaluate:

   (a)   $\int_{-\infty}^{\infty} e^{-(x-3)^2} dx.$     (b)   $\int_{0}^{15} x^{7/2} e^{-x/2} dx.$

**6-R2.**   Given the p.d.f. $f_X(x) = K \exp[-(x-2)^2/8]$, where $K$ is a constant.
   (a)   Find $K$.
   (b)   Give the distribution of $Y = \frac{1}{2}(X - 2)$.
   (c)   Find $E(e^X) = \psi_X(1)$.

**6-R3.**   Suppose $X_1, ..., X_n$ are independent, and that $X_i \sim \mathcal{N}(5, 4)$ for each $i$.
   (a)   Give the distribution of $\sum X_i$.
   (b)   Give the joint p.d.f. of the $X$'s.
   (c)   Find $\mathrm{cov}(2X_1 - X_2, X_1 + 2X_2)$.

**6-R4.**   Consider a Poisson process as a model for arrivals at a service facility. Given that the probability of no arrivals in a ten-minute period is .135, find
   (a)   the mean time to the sixth arrival after a particular point in time $t_0$.
   (b)   the probability that the time (measured from $t_0$) to the next arrival after $t_0$ is less than 5 minutes, given that there were no arrivals in the preceding 10-minute interval.
   (c)   the probability that the time to the sixth arrival exceeds one-half hour.

**6-R5.**   Given that $X_1, X_2, X_3$ are independent standard normal variables, find
   (a)   $P(X_1 + X_2 + X_3 > 4)$.
   (b)   $P(X_1^2 + X_2^2 + X_3^2 > 4)$.
   (c)   the conditional distribution of $X_1$ given $X_2 = X_3$.

**6-R6.**   Given that $X$ has the m.g.f. $\psi(t) = \exp[2(t + t^2)]$, find
   (a)   its mean and variance.     (b)   $P(X < 0)$.

**6-R7.**   Given that $Y$ has the m.g.f. $(1 - 4t)^{-3}$, identify the distribution. [Give the answer as a named distribution with parameter value(s).]

**6-R8.**   Given $X \sim \mathcal{N}(5, 4)$,
   (a)   find $E(X^2)$.
   (b)   give the distribution of $Y = 2(X + 2)$.
   (c)   give the distribution of $Z = \left(\frac{X-5}{2}\right)^2$

**6-R9.**   Suppose that system failures follow an exponential distribution with mean time to failure 3 weeks. Assume that the system is immediately put back on line after each failure, and that successive times to failure are independent.
   (a)   Find the probability that the time to the fourth failure exceeds 12 weeks.
   (b)   Find the probability that the time to failure from $t = t_0$ is at most 6 weeks, given that it has been (at time $t_0$) 4 weeks since the most recent failure.

**6-R10.**   Consider $X_1, X_2, X_3$, independent observations on $X$, whose distribution is Exp(1).
   (a)   Give the joint p.d.f. of the three sample observations.
   (b)   Give the p.d.f. of the sample *sum*: $Y = X_1 + X_2 + X_3$.

## Chapter Perspective

In Chapters 2 through 6, we have studied population models—both discrete and continuous—in some detail. We have introduced various ways of characterizing and describing distributions and discussed several special families of distributions. These are important both for providing models for real variables and, as we'll see in the chapters to follow, for describing how quantities calculated from samples vary from sample to sample.

Several of the distribution families are related to the Bernoulli and Poisson processes, and these are interrelated:  The Poisson distribution is for the number of "events" in a fixed interval of time, for a Poisson process; the binomial distribution is for the number of events in a fixed "time" (number of trials) in a Bernoulli process. The exponential distribution is for the time until the next event in a Poisson process; the geometric distribution is for the "time" to the next event in a Bernoulli process. The Erlang distribution (gamma, with integer $\alpha$) is for the time to the $r$th event after a given point in a Poisson process; the negative binomial distribution is for the "time" (number of trials) to the $r$th event after a given point in a Bernoulli process.

The normal family of models is perhaps the most useful and most commonly used of all. Having infinite support, a "normal curve" cannot represent a real-life variable exactly, but it provides a good approximation in many cases. And as we'll see, the theory of inference for normal population models gives rise to a number of important distribution families: chi-square, $t$, gamma, beta, and $F$. (The latter three are coming up shortly.)

We turn next to the main task of the statistician, that of learning about a population model by sampling—gathering data by doing the experiment the model is intended to represent. In Chapter 7 we develop various ways of looking at, characterizing, and summarizing sample information. Some of these ways are familiar to you from what you have seen in the various print and other media; others may be new. We'll define a sample distribution as a special case of a discrete distribution and, in describing samples, apply some of what we know about discrete distributions from Chapter 3.

In Chapter 8 we'll see that with *random* sampling we are able to analyze the extent to which a sample gives information about population characteristics. And we'll investigate how best to utilize the information in a random sample for the purpose of drawing inferences about the population from which it was drawn. This will prepare us for the study of various modes of inference and their application in learning about effects, differences, and relationships, when there is an element of uncertainty in the observations we can make.

# Organizing and Describing Data

In the first six chapters we have developed the notion of probability for describing or modeling an experiment of chance. Probability models describe population distributions. They represent the experiment of selecting a population member at random. (Even though an experiment does not involve sampling from a tangible population, it is useful to think of the experiment as sampling a population—the conceptual population of possible outcomes.)

In practice, populations are seldom known or understood completely. The fundamental problem of statistical inference is to learn what we can about a population from *sample* data. Before tackling this problem of inference, we present in this chapter various ways of organizing and describing sample data using tables, graphs, and summary statistics. We first treat data from discrete populations, univariate and bivariate. We then take up methods for displaying and summarizing data from continuous populations.

## 7.1 Frequency Distributions

In Chapter 4 we considered sampling from categorical populations. We summarized sample results by giving the numbers of times the various categories occurred. These numbers are the category **frequencies;** the ratios of frequencies to sample size are **relative frequencies.** The list of categories and corresponding frequencies or relative frequencies is a **frequency distribution.**

Example **7.1a** | **Statistics Students' Statistics**

Table 7-1 gives data collected from the male students in one of our statistics classes. The variables are height (Ht), class (Cla), number of siblings (Sibs), use of marijuana (Marij), and social behavior motivation (Beh).

Some of the variables are numerical (height, number of siblings); others are categorical (Behavior, Marijuana). The number of siblings is discrete, with values 0, 1, 2, .... Height is essentially continuous, but as recorded it is discrete, being rounded to the nearest inch.

**Table 7-1**  Survey of males in a statistics class

| No. | Ht | Cla | Sibs | Marij | Beh | No. | Ht | Cla | Sibs | Marij | Beh |
|---|---|---|---|---|---|---|---|---|---|---|---|
| 1 | 76 | 3 | 5 | Y | B | 25 | 68 | 4 | 1 | Y | B |
| 2 | 70 | 4 | 7 | N | B | 26 | 69 | 4 | 3 | N | B |
| 3 | 72 | 3 | 1 | Y | A | 27 | 72 | 3 | 2 | N | B |
| 4 | 73 | 4 | 3 | N | B | 28 | 73 | 3 | 2 | Y | B |
| 5 | 74 | 3 | 2 | N | M | 29 | 66 | 4 | 3 | Y | B |
| 6 | 68 | 3 | 5 | Y | B | 30 | 62 | 4 | 2 | N | M |
| 7 | 71 | 4 | 2 | Y | B | 31 | 73 | 3 | 10 | Y | B |
| 8 | 68 | 3 | 4 | N | B | 32 | 74 | 3 | 3 | N | M |
| 9 | 69 | 2 | 7 | Y | M | 33 | 68 | 3 | 2 | N | M |
| 10 | 72 | 4 | 2 | Y | B | 34 | 77 | 3 | 6 | Y | B |
| 11 | 68 | 3 | 1 | Y | B | 35 | 70 | 3 | 8 | N | M |
| 12 | 70 | 2 | 1 | Y | M | 36 | 70 | 5 | 2 | N | B |
| 13 | 70 | 5 | 2 | Y | M | 37 | 74 | 4 | 4 | Y | B |
| 14 | 70 | 3 | 8 | Y | A | 38 | 71 | 5 | 1 | N | B |
| 15 | 70 | 3 | 2 | Y | B | 39 | 71 | 5 | 5 | N | M |
| 16 | 74 | 2 | 3 | Y | M | 40 | 70 | 5 | 4 | Y | B |
| 17 | 69 | 3 | 3 | N | M | 41 | 70 | 4 | 3 | Y | M |
| 18 | 70 | 2 | 2 | N | B | 42 | 75 | 5 | 1 | N | B |
| 19 | 73 | 4 | 2 | N | M | 43 | 68 | 3 | 3 | Y | B |
| 20 | 72 | 3 | 2 | N | M | 44 | 69 | 2 | 5 | Y | M |
| 21 | 70 | 2 | 1 | N | B | 45 | 70 | 4 | 1 | Y | M |
| 22 | 74 | 3 | 4 | N | M | 46 | 73 | 3 | 2 | N | M |
| 23 | 68 | 3 | 1 | N | M | 47 | 66 | 2 | 6 | N | M |
| 24 | 76 | 3 | 3 | Y | A | 48 | 70 | 5 | 3 | Y | B |

The classification according to use of marijuana has two categories. The questionnaire we used defined the category "yes" (Y) to mean "more than just experimented once or twice." The question on social behavior motivation was worded this way: "Would you describe your behavior with respect to socializing with the opposite sex as motivated or guided mostly by moral considerations (M), by fear of AIDS (A), or by some of both (B)?"

For each of the individual variables, we find category frequencies by counting. We found frequencies for Marijuana and Behavior by counting. (You should check our counts.) The resulting frequency distributions for Marijuana and Behavior are as follows:

| Marijuana? | Freq. | Relative Freq. |
|---|---|---|
| Yes | 25 | 25/48 = .521 |
| No | 23 | 23/48 = .479 |
| Total | 48 | 1 |

| Motivation | Freq. | Relative Freq. |
|---|---|---|
| Moral | 20 | 20/48 = .417 |
| AIDS | 3 | 3/48 = .063 |
| Both | 25 | 25/48 = .521 |
| Total | 48 | 1 |

In §2.2 we saw that studying relationships between two random variables requires their *joint* probability distribution. The sample analog is a joint *frequency* distribution. In the case of categorical variables, joint frequencies can be given in a two-way array called a **contingency table.** In a contingency table, each combination of a category of one variable with a category of the other defines a **cell.** The number we put in a particular cell is the frequency of this combination in the sample.

Row totals and the column totals are given in the margins of the table. These are frequencies of categories of the corresponding variables, considered separately, and define univariate distributions called **marginal** distributions.

**Example 7.1b**

In the data of Table 7-1 there are two categories for Marijuana and three for Behavior, making six cells in the contingency table for these two variables. When all students in the sample are tallied, the number of tally marks in each cell is its frequency. The result is as follows:

|  |  | *Marijuana* | | |
|---|---|---|---|---|
|  |  | Y | N |  |
|  | M | 7 | 13 | 20 |
| *Behavior* | A | 3 | 0 | 3 |
|  | B | 15 | 10 | 25 |
|  |  | 25 | 23 | 48 |

The marginal totals give marginal frequency distributions for Behavior (right-hand margin) and for Marijuana (bottom margin). These marginal distributions are the univariate distributions shown in Example 7.1a.

To take into account Class as a third variable, we could make a two-way table like the above for each category of Class:

*Soph (2):*

|  | Y | N |  |
|---|---|---|---|
| M | 4 | 1 | 5 |
| A | 0 | 0 | 0 |
| B | 0 | 2 | 2 |
|  | 4 | 3 | 7 |

*Jr (3):*

|  | Y | N |  |
|---|---|---|---|
| M | 0 | 9 | 9 |
| A | 3 | 0 | 3 |
| B | 8 | 2 | 10 |
|  | 11 | 11 | 22 |

*Sr (4):*

|  | Y | N |  |
|---|---|---|---|
| M | 2 | 2 | 4 |
| A | 0 | 0 | 0 |
| B | 5 | 3 | 8 |
|  | 7 | 5 | 12 |

*Adult Special (5):*

|  | Y | N |  |
|---|---|---|---|
| M | 1 | 1 | 2 |
| A | 0 | 0 | 0 |
| B | 2 | 3 | 5 |
|  | 3 | 4 | 7 |

Together, these tables constitute a *three-way* contingency table. Imagine that the four tables are stacked, and think of each table as a *layer* in the three-dimensional array of frequencies.

The three-way table could also be layered according to either of the variables Behavior and Marijuana. For example, the layer for Behavior = M is

|  | | Class | | | |
|---|---|---|---|---|---|
|  | | 2 | 3 | 4 | 5 |
| Marijuana | Y | 4 | 0 | 2 | 1 | 7 |
|  | N | 1 | 9 | 2 | 1 | 13 |
|  | | 5 | 9 | 4 | 2 | 20 |

The right-hand column gives the frequencies for $M$ that appear in the first row of the first table in this example. ∎

In some cases the categories of a discrete variable are numerical, and the natural ordering of numbers permits a graphical representation of data as a bar graph on a numerical scale. In the case of two such variables, each data pair $(X, Y)$ is plotted as a point in the $xy$-plane. If there are repeated pairs, the frequency of occurrence of that pair in a data set can be plotted as a bar graph, with bars perpendicular to the plane.

**Example 7.1c** | **Siblings**

The number of siblings an individual has is a discrete, numerical variable. The only possible values are integers. Counting the number of students in Table 7-1 with siblings numbering 0, 1, 2, ..., we obtain this frequency distribution:

| No. of Siblings | 1 | 2 | 3 | 4 | 5 | 6 | 7 | 8 | 10 |
|---|---|---|---|---|---|---|---|---|---|
| Frequency | 9 | 14 | 10 | 4 | 4 | 2 | 2 | 2 | 1 |

Figure 7-1 shows this distribution as a bar graph in which the heights of the bars are proportional to the number of siblings.

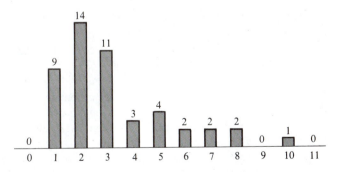

**Figure 7-1** Bar Graph for Siblings

It would be somewhat surprising to find a relationship between the variables Class and Number of Siblings, but any such relationship would only show up in a cross-classification of the 48 students on these two variables. Several of the possible pairs in Table 7-1 occur more than once, so we summarize the data in a **contingency table,** with counts for each combination of a category of Class with a category of Sibs:

|  | | Number of Siblings | | | | | | | | | |
|---|---|---|---|---|---|---|---|---|---|---|---|
|  | | 1 | 2 | 3 | 4 | 5 | 6 | 7 | 8 | 10 | |
|  | 2 | 2 | 1 | 1 | 0 | 1 | 1 | 1 | 0 | 0 | 7 |
| Class | 3 | 3 | 7 | 4 | 2 | 2 | 1 | 0 | 2 | 1 | 22 |
|  | 4 | 2 | 4 | 4 | 1 | 0 | 0 | 1 | 0 | 0 | 12 |
|  | 5 | 2 | 2 | 1 | 1 | 1 | 0 | 0 | 0 | 0 | 7 |
|  | | 9 | 14 | 10 | 4 | 4 | 2 | 2 | 2 | 1 | 48 |

Figure 7-2 shows a bivariate bar diagram for these data. Pictures like this can be awkward in that some bars will often hide others. Computer software is available that can overcome such problems by drawing the diagrams from various perspectives.

A different way of showing bivariate frequencies is a **star plot**[1] (or sunflower plot): At each point $(X, Y)$ that occurs once in the data, a point is plotted; at points that occur more than once, a point is plotted together with a short ray from the point to represent each occurrence of that pair. A frequency of 2 has two rays, in opposite directions; a frequency of 3, three rays in equally spaced directions; and so on. A star plot for Class versus Number of Siblings is shown in Figure 7-3. ∎

## 7.2  Data on Continuous Variables

Variables such as time, temperature, and velocity are thought of as continuous. However, to measure and record values of such variables, we use a finite, discrete scale. So some of the devices described in §7.1 for categorical variables—bar charts and star plots—are also useful for continuous variables.

When there is not much data, we can convey the information they contain by simply listing the observations, but graphical representations are useful no matter how small the data set. For instance, simply marking a *dot* (or an "x") on a numerical scale, for each observation, produces a plot that can be quite revealing. We refer to such a plot as a **dot diagram.** Such diagrams are awkward when the marks begin piling up on repeated values. The remainder of this section is concerned with methods specifically adapted to displaying larger data sets.

---

[1]See W. S. Cleveland, *The Elements of Graphing Data* (Pacific Grove, CA: Wadsworth Advanced Books and Software, 1985).

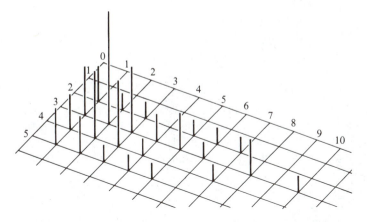

**Figure 7-2** Bivariate bar graph—Example 7.1c

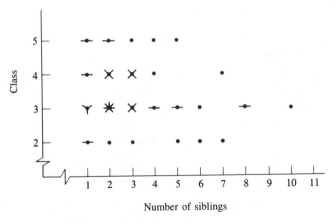

**Figure 7-3** Star plot—Example 7.1c

A first step in organizing numerical data is to put them in numerical order. Ordering a large set of data is not easy without some sort of system. One such is the **stem-leaf diagram,** which orders the data numerically and provides a visual display. We explain this easy-to-learn device with an example.

Example **7.2a** | **Spot-Weld Strength**
Strength of spot-welds made by a particular welding tool and operator vary. The following are weld strengths (in psi) in a sample of 50 welds:[2]

---

[2]From a consulting file.

| | | | | | | | | | |
|---|---|---|---|---|---|---|---|---|---|
| 400 | 395 | 398 | 421 | 445 | 389 | 372 | 400 | 398 | 401 |
| 399 | 386 | 423 | 364 | 394 | 414 | 390 | 412 | 398 | 363 |
| 388 | 431 | 392 | 438 | 411 | 399 | 399 | 408 | 390 | 420 |
| 400 | 389 | 430 | 426 | 388 | 406 | 431 | 411 | 404 | 424 |
| 450 | 416 | 397 | 404 | 388 | 405 | 392 | 405 | 379 | 419 |

To make a stem-leaf diagram, take the first two digits as "stems" and record measurements according to the last digits as "leaves" on the appropriate stems. Thus, 400 is a 0-leaf on the 40-stem, 395 is a 5-leaf on the 39-stem, and so on. The result is shown, together with a diagram in which the leaves have been ordered, in Table 7-2. ∎

**Table 7-2**

| Stem | Leaves | | Stem | Leaves |
|---|---|---|---|---|
| 36 | 43 | | 36 | 34 |
| 37 | 29 | | 37 | 29 |
| 38 | 968988 | | 38 | 688899 |
| 39 | 5889408299072 | | 39 | 0022457888999 |
| 40 | 0018064455 | | 40 | 0001445568 |
| 41 | 421169 | | 41 | 112469 |
| 42 | 13064 | | 42 | 01346 |
| 43 | 1801 | | 43 | 0118 |
| 44 | 5 | | 44 | 5 |
| 45 | 0 | | 45 | 0 |

The stems of a stem-leaf diagram constitute categories of nearby observations. Counting the leaves on each stem yields a *frequency distribution.* Stem frequencies are proportional to stem length—if one uses typewriter spacing, so that all leaves take the same amount of space. The outline formed by the leaves shows the frequency distribution as a bar diagram in which the stems define categories. Figure 7-4 outlines the stem-leaf diagram from Table 7-2.

Not all data sets lend themselves usefully to display in a stem-leaf diagram. For example, the student heights in Table 7-1 are given as two-digit numbers, and the first digit is either a 6 or 7; there would only be the two stems. At the other extreme, data such as the populations of the 50 states differ in not just the last digit or two, but in the last five to seven digits. There are ways of adapting stem-leaf diagrams to such awkward cases.[3]

---

[3]See J. Tukey, *Exploratory Data Analysis* (Reading, MA: Addison-Wesley, 1977).

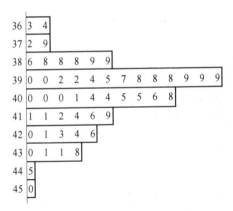

```
36 | 3  4
37 | 2  9
38 | 6  8  8  8  9  9
39 | 0  0  2  2  4  5  7  8  8  8  9  9  9
40 | 0  0  0  1  4  4  5  5  6  8
41 | 1  1  2  4  6  9
42 | 0  1  3  4  6
43 | 0  1  1  8
44 | 5
45 | 0
```

**Figure 7-4** Stem-leaf diagram as bar graph

In a stem-leaf diagram, the choice of stems is restricted by the decimal system, but there are other ways of grouping values that result in frequency distributions: Given the range of values that might be encountered, partition that range into an arbitrary number of subintervals, preferably of equal size. These are called **class intervals.** Tallying the data in these class intervals as "categories" and counting the tally marks in these categories yields a frequency distribution.

Any frequency distribution loses some of the sample detail, whereas a stem-leaf diagram preserves all the data (albeit not in the original order of observation). However, the details are often unimportant. The usual reason for sampling is to find out about a population, and a second sample from the same population can be quite different in detail. But the two sample distributions will be roughly similar in broad outline, especially when both sample sizes are large.

A frequency distribution depends on the choice of class intervals, and that choice is arbitrary. Indeed, with the same set of data, different choices of partition points are apt to produce distributions that look different. Some guidelines are useful: The intervals should be wide enough that most of them get more than a few observations, but there should be a reasonable number of class intervals—perhaps between 5 and 20.

It is common practice to mark each class interval with its midpoint—its **class mark.** When we calculate sample parameters in the next section, the class mark will be taken as representing all the values that fell in that class interval. This amounts to *rounding* sample values to class marks.

The summary of a sample from a continuous population in a frequency distribution is commonly represented graphically by a **histogram**—a bar diagram in which the categories are class intervals. The height of each bar is such that the area of the bar is proportional to the frequency of the corresponding class interval. When the support of a continuous distribution consists of all real numbers in an interval (which is the usual case in practice), there are no gaps where observations cannot fall; so adjacent class intervals are contiguous, and the corresponding histogram bars should touch.

Example **7.2b** | **Histogram for Weld Strengths**

The spot-weld strengths in Example 7.2a range from the 360's to the 450's. Suppose we divide this range into nine equal subintervals, each about 11 units wide. We arbitrarily take the first interval to be 361–371 (inclusive). Its class mark is the midpoint, 366. The class mark for the next interval (372–382) is 377; and so on. Table 7-3 shows the tallies and frequencies. Figure 7-5 gives the histogram for the frequency distribution in Table 7-3. Because the class intervals are of equal width, bar heights are proportional to bar areas, which in turn are proportional to class frequencies. The measurements were rounded to the nearest psi, but strength is inherently continuous. So the class interval 361–371, for example, includes all numbers in the interval from 360.5 to 371.5, which were rounded to integers from 361 to 371. The base of the corresponding bar is the interval from 360.5 to 371.5. The second bar covers the interval from 371.5 to 382.5, and so on. (This way, none of the rounded observations falls on a boundary point.) ∎

**Table 7-3** Distribution of spot-weld strengths (psi)

| Class mark | Class interval | Frequency |
|:---:|:---:|:---:|
| 366 | 361–371 | 2 |
| 377 | 372–382 | 2 |
| 388 | 383–393 | 10 |
| 399 | 394–404 | 15 |
| 410 | 405–415 | 8 |
| 421 | 416–426 | 7 |
| 432 | 427–437 | 3 |
| 443 | 438–448 | 2 |
| 454 | 449–459 | 1 |

A **histogram** is a bar graph for representing a frequency distribution for a set of observations on a continuous variable in which the *areas* of the bars are proportional to class frequencies.

Although we'd recommend using class intervals of equal sizes in most cases, frequency distributions are sometimes constructed with unequal class intervals. In such cases, it is necessary to adjust the heights of the bars to avoid giving the wrong

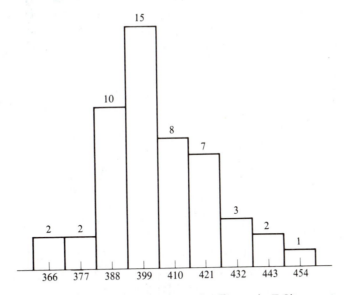

**Figure 7-5** Histogram for Example 7.2b

visual impression. The eye perceives *area,* so we want one unit of area to represent one observation—in every class interval.

Example **7.2c**  |  **Plasma Clearance**

The next table summarizes plasma clearance of a certain drug (in ml/min/kg) for 44 smokers.[4]

| Clearance | Frequency |
|-----------|-----------|
| 4.5 – 8.5 | 6 |
| 8.6 –12.6 | 18 |
| 12.7–16.7 | 14 |
| 16.8 –20.8 | 4 |
| 20.9–24.9 | 1 |
| 25.0 –29.0 | 1 |

Suppose we were to combine the last three class intervals, making one interval from 16.8 to 29.0. This is three times as wide as the others, so the height that represents

[4]From the consulting practice of one of the authors.

one observation in that interval should be only one-third the height that represents one observation in the others. This is equivalent to using the average of the three frequencies (4, 1, 1) as the new height. Figure 7-6 shows the histogram with this modification. ∎

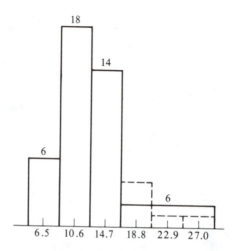

**Figure 7-6** Histogram with unequal class intervals

To organize and display continuous *bivariate* data, we interpret the two variables as $x$- and $y$-coordinates of a point in the plane. For data sets of moderate size, marking a dot or other symbol at each point provides an excellent visual representation of the data. Such a plot is called a **scatter diagram.** Such diagrams are helpful in studying relationships between the two variables. If data points occur more than once in a data set (as they might, because of round-off), one can mark each point with its frequency or any other convenient code such as that used in the star plot of Figure 7-3.

**Example 7.2d** | **LSAT Versus GPA**

Table 7-4 gives average LSAT scores $(X)$ and average GPA's $(Y)$ for entering students in each of 15 law schools in 1973.[5] (The LSAT is a national aptitude test taken by those who want to enter a law school. GPA is undergraduate grade point average.) Figure 7-7 shows these $(x, y)$-pairs in a scatter diagram.

As might be expected, high LSAT scores tend to be associated with high GPA's. This is only a *tendency,* however. For instance, the first pair in the data set doesn't fit this description very well. ∎

[5]Taken from B. Efron, "Computers and the theory of statistics," *SIAM Review* 21 (1979), 460–80.

**Table 7-4**

| $x$ | $y$ | $x$ | $y$ | $x$ | $y$ |
|---|---|---|---|---|---|
| 576 | 3.39 | 580 | 3.07 | 653 | 3.12 |
| 635 | 3.30 | 555 | 3.00 | 575 | 2.74 |
| 558 | 2.81 | 661 | 3.43 | 545 | 2.76 |
| 578 | 3.03 | 651 | 3.36 | 572 | 2.88 |
| 666 | 3.44 | 605 | 3.13 | 594 | 2.96 |

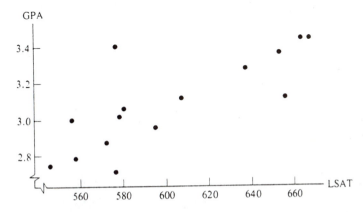

**Figure 7-7** Scatter plot for Example 7.2d

A third variable that is categorical can be included in a bivariate display by using its category labels, or by using any other system of symbols (such as triangles for one category and circles for another), in place of a dot, to mark the location of an $(x, y)$-pair. Or a scatter diagram can be made for each category.

Two-dimensional frequency distributions for large data sets employ a system of class intervals on the $x$-axis and another on the $y$-axis. Each interval defines a strip (perpendicular to the axis), and the two systems of strips intersect to form rectangular cells (like graph paper). The cells play the role played by the class intervals in a univariate distribution. To summarize a set of data, we can construct a two-way table of cell frequencies. A *histogram* for such a frequency distribution consists of a column above each cell, with volume proportional to the cell frequency. As in the case of discrete data (Example 7.1c), such pictures are awkward without computer graphics.

Frequency distributions and their graphical representations convey the sense of a data set and also give a rough picture of the population from which they are drawn. The following example shows two sample histograms that by themselves offer convincing evidence that the populations sampled are different.

Example **7.2e** | **Platelet Count**

Figure 7-8 shows two histograms, one representing blood platelet count data for 35 healthy males, and the other, data for a group of 153 male cancer patients.[6] No one looking at these graphs could fail to see a difference between the two kinds of individuals.    ■

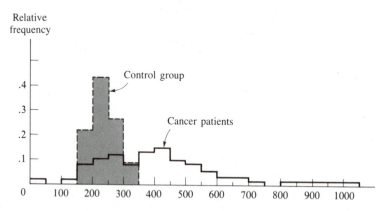

**Figure 7-8** Histograms for Example 7.2e

Conclusions drawn from histograms are not always this obvious. Histograms are visually appealing but difficult to work with analytically. It is usually easier to work with numbers, sample characteristics or **statistics** that are relevant to a particular problem.

A **statistic** is a number calculated from the observations in a sample.

## Problems

**7-1.**  For the variables in Table 7-5,

(a)  construct frequency tables for Class (Cla), Marijuana use (Marij), and Number of Siblings (Sib).

✱(b)  give the joint distribution of (Cla, Marij) as a frequency table.

✱ **7-2.**  From the table in Problem 7-1(b), find

(a)  the proportion of junior girls who have not used marijuana.

(b)  the proportion of senior girls who have not used marijuana.

(c)  the proportion of girls who have not used marijuana.

---

[6]S. Silvis et al., "Thrombocytosis in patients with lung cancer," *J. Amer. Med. Assn. 211* (1970), 1852.

**Table 7-5**   Survey of females in a statistics class

| No. | Ht | Cla | Sibs | Marij | Beh | | No. | Ht | Cla | Sibs | Marij | Beh |
|---|---|---|---|---|---|---|---|---|---|---|---|---|
| 1 | 67 | 3 | 4 | N | M | | 24 | 64 | 4 | 1 | N | M |
| 2 | 72 | 3 | 2 | N | B | | 25 | 64 | 2 | 7 | Y | M |
| 3 | 64 | 2 | 3 | N | M | | 26 | 63 | 3 | 7 | Y | T* |
| 4 | 67 | 4 | 3 | Y | B | | 27 | 67 | 1 | 3 | Y | T |
| 5 | 66 | 3 | 3 | N | B | | 28 | 72 | 1 | 5 | N | M |
| 6 | 66 | 3 | 1 | Y | B | | 29 | 59 | 3 | 5 | N | M |
| 7 | 67 | 3 | 3 | Y | B | | 30 | 69 | 4 | 3 | Y | M |
| 8 | 67 | 3 | 1 | N | M | | 31 | 63 | 4 | 5 | N | M |
| 9 | 69 | 4 | 13 | N | M | | 32 | 65 | 1 | 9 | N | M |
| 10 | 66 | 4 | 8 | N | M | | 33 | 62 | 3 | 2 | Y | M |
| 11 | 63 | 3 | 2 | N | B | | 34 | 62 | 2 | 0 | Y | M |
| 12 | 64 | 3 | 2 | N | M | | 35 | 68 | 3 | 3 | N | M |
| 13 | 66 | 4 | 1 | Y | B | | 36 | 70 | 2 | 5 | N | M |
| 14 | 64 | 4 | 3 | N | B | | 37 | 68 | 3 | 1 | Y | M |
| 15 | 65 | 3 | 2 | N | M | | 38 | 64 | 2 | 3 | Y | T |
| 16 | 66 | 3 | 7 | N | B | | 39 | 74 | 2 | 6 | N | M |
| 17 | 66 | 2 | 5 | N | B | | 40 | 65 | 2 | 6 | N | M |
| 18 | 69 | 4 | 2 | Y | B | | 41 | 68 | 2 | 3 | N | M |
| 19 | 69 | 3 | 1 | N | B | | 42 | 63 | 2 | 13 | Y | |
| 20 | 66 | 3 | 2 | N | B | | 43 | 62 | 5 | 1 | N | M |
| 21 | 64 | 2 | 1 | N | M | | 44 | 69 | 2 | 2 | N | M |
| 22 | 63 | 5 | 1 | N | B | | 45 | 64 | 3 | 1 | Y | |
| 23 | 63 | 3 | 8 | N | B | | | | | | * T = "tossup" | |

**7-3.**   Refer to Table 7-1 (page 257).

**(a)**   Construct a two-way table for Class and Marijuana for the males.

**(b)**   The table in (a) together with the table in Problem 7-1(b) constitute a three-way cross-classification of the whole class, with Sex as the layering variable. Construct another three-way table using Class as the layering variable.

* **7-4.**   Obtain the two-way classification on Class and Sex from the three-way table in Problem 7-3(b).

**7-5.**   A group of 129 grade-school children was cross-classified[7] on race (black or white), sex (M or F), and whether their placement in reading groups was below (BGL), at (GL), or above grade level (AGL). The result:

_Black:_

| | F | M |
|---|---|---|
| BGL | 6 | 4 |
| GL | 14 | 12 |
| AGL | 9 | 13 |

_White:_

| | F | M |
|---|---|---|
| BGL | 15 | 11 |
| GL | 9 | 16 |
| AGL | 6 | 14 |

[7]Data from L. Grant, "Black females' 'place' in desegregated classrooms," _Sociology of Education_ 57 (1984), 98–110.

(a)   Recast these data so that Sex is the layer variable. (That is, make two two-way tables, one each for males and females.)

(b)   What fraction of those below grade level are black?

∗ **7-6.**   Regroup the spot-weld data of Example 7.2a (page 262) using eight class intervals, as in Example 7.2b, but starting instead with the class interval 358–370. Observe that the table you get is different from the table given as Table 7-3 in Example 7.2b. (Which one is correct?)

**7-7.**   The slag (leftover solid waste) from a lead-smelting process was checked for concentration of lead, a toxic pollutant. The following concentrations $X$ were obtained:[8]

| | | | | | | | | |
|------|------|------|------|------|------|------|------|------|
| 2.83 | 209  | .53  | 1300 | 3.86 | 141  | 543  | 21.9 | 125  |
| 17.9 | 4.88 | 31.8 | 774  | 493  | 409  | 1.58 | 21.2 | 16.6 |
| 35.3 | 15.7 | 59.1 | 2.26 | 291  | 4.18 | 380  | 39.4 | 340  |

(a)   Consider various ways of representing these data:  dot diagram, stem-leaf diagram, frequency distribution, .... What problems do you see?

(b)   Try a *transformation:*  Let $Y = \log X$ (any convenient base). Transform each observation and summarize in a dot diagram or frequency table.

**7-8.**   For the heights of the females in Table 7-5,

(a)   make a stem-leaf diagram with split stems: (0 to 4, 5 to 9).

(b)   make a stem-leaf diagram with split stems: (0-1, 2-3, 4-5, 6-7, 8-9).

(c)   construct a histogram for the frequency distribution implied in the stem-leaf diagram of (b).

**7-9.**   Birthweights (in ounces) of 30 babies were recorded in a maternity ward, as follows:[9]

| | | | | | | | | | |
|-----|-----|-----|-----|-----|-----|-----|-----|-----|-----|
| 131 | 120 | 112 | 88  | 114 | 128 | 133 | 104 | 108 | 94  |
| 133 | 124 | 84  | 132 | 107 | 144 | 93  | 114 | 116 | 120 |
| 128 | 132 | 108 | 107 | 86  | 123 | 116 | 92  | 106 | 108 |

∗ (a)   Make a stem-leaf diagram for these data.

(b)   Construct a histogram for these data.

**7-10.**   Construct a frequency table and a graphical representation for these times to burnout for a type of light bulb, expressed in thousands of hours:[10]

| | | | | | | | | | |
|------|------|------|------|------|------|------|------|------|------|
| 1.07 | 0.86 | 1.16 | 1.02 | 0.92 | 0.52 | 0.93 | 1.00 | 0.90 | 1.00 |
| 1.19 | 0.82 | 0.84 | 1.04 | 1.02 | 1.13 | 1.00 | 0.61 | 0.99 | 1.07 |
| 1.24 | 1.44 | 1.12 | 1.22 | 1.13 | 1.16 | 1.17 | 1.02 | 1.24 | 1.01 |
| 1.08 | 1.21 | 0.80 | 1.03 | 1.49 | 1.11 | 0.76 | 1.25 | 1.02 | 1.11 |

---

[8]N. Woodley and J. V. Walters, "Hazardous waste characterization extraction procedures ...," *Environmental Progress* (1986), 12–17.

[9]From a student's class project.

[10]Taken from D. J. Davis, "An analysis of some failure data," *J. Amer. Stat. Assn.* 47 (1952), 142.

**7-11.**  Two judges of piano contestants turned in these ratings:

|        |   | \multicolumn{8}{c}{Performer} |
|--------|---|----|----|----|----|----|----|----|----|
|        |   | 1  | 2  | 3  | 4  | 5  | 6  | 7  | 8  |
| Judge  | 1 | 92 | 90 | 85 | 96 | 92 | 88 | 96 | 88 |
|        | 2 | 89 | 90 | 88 | 93 | 90 | 85 | 95 | 90 |

Represent the data in a scatter diagram.

**7-12.**  Given the weight and height of each of ten students (lb, in.):

$(133, 65)$   $(125, 68)$   $(155, 71)$   $(150, 70)$   $(155, 68)$

$(155, 65)$   $(125, 67)$   $(105, 62)$   $(170, 71)$   $(175, 72)$

Make a scatter diagram.

## 7.3  Order Statistics

In statistical applications, sample observations are often regarded as exchangeable, so that the order in which they were collected is not important. (Just when this is justifiable will be discussed in §8.3.)  However, many important statistics depend on the *numerical* order of the sample observations.

Given a sample $(X_1, ..., X_n)$ from a continuous population, we'll denote the smallest observation by $X_{(1)}$, the second smallest by $X_{(2)}$, and so on. The $n$th smallest is the largest. Each $X_{(i)}$ is a statistic—a number calculated from the sample. The vector of these ordered values, $(X_{(1)}, ..., X_{(n)})$ is called **the order statistic.** Any function of the components is called *an* order statistic.

The most commonly used order statistic is the number in the middle of the ordered observations—the **sample median,** $\widetilde{X}$ . If $n$ is odd, there is a middle order statistic, so that (for instance) when $n = 15$, the median is $X_{(8)}$—the eighth smallest as well as the eighth largest observation. If $n$ is even, there are two numbers in the middle; somewhat arbitrarily, we take their average (half their sum) as the sample median. For example, if $n = 24$, then $X_{(12)}$ and $X_{(13)}$ are in the middle, and the median is the average of these two observations. There are then 12 observations in either side of the median.

*Sample median:*

$$\widetilde{X} = \begin{cases} X_{([n+1]/2)} \text{ for } n \text{ odd,} \\ \frac{1}{2}\left[X_{(n/2)} + X_{(n/2+1)}\right] \text{ for } n \text{ even.} \end{cases} \tag{1}$$

Ordering the leaves on the individual stems of a stem-leaf diagram orders all the data, and this makes it easier to find the sample median.

Example **7.3a**

## Median of Spot-Weld Data

Table 7-6 repeats the stem-leaf diagram for the spot-weld data in Examples 7.2a and 7.2b, with an added column called *depth*. The depth is a cumulative count of the observations from the top and from the bottom, toward the middle. The depth for the 400-stem is special. In cumulating from the bottom of the diagram (the larger numbers), adding 10 to the count for the leaves on the 400-stem yields 27, which is more than half of $n = 50$. From the top, adding the 10 on that stem yields 33, again more than halfway. So, the 400-stem includes the median. We stop short of that stem and isolate it between horizontal lines.

According to (1), the median is the average of the 25th and 26th from either end. There are 23 leaves above the 400-stem; the 24th, 25th, and 26th smallest numbers are thus the first three leaves on the 400-stem (400, 400, 400). The median is the average of the 25th and 26th: 400.

(When using a stem-leaf diagram in which the leaves have not been ordered, enter a stem at the smallest number counting from the low end of the scale and at the largest number counting from the high end.)    ∎

**Table 7-6**

| Depth | Stem | Leaves |
|---|---|---|
| 2 | 36 | 34 |
| 4 | 37 | 29 |
| 10 | 38 | 688899 |
| 23 | 39 | 0022457888999 |
| 10 | 40 | 0001445568 |
| 17 | 41 | 112469 |
| 11 | 42 | 01346 |
| 6 | 43 | 0118 |
| 2 | 44 | 5 |
| 1 | 45 | 0 |

$n = 50$ (braces grouping stems 39, 40, 41)

The median divides a set of data in "half." That is, roughly one-half the observations are smaller than the median and one-half larger. When $n$ is large, it is sometimes useful to divide the data set further and identify *quartiles*. When $n$ is a multiple of 4, the first quartile $Q_1$ has one-quarter of the observations to its left and three-quarters to its right. The second quartile is the median, and the third quartile $Q_3$ has three-quarters of the observations to its left and one-quarter to its right. When $n$ is not a multiple of 4, such division can only be approximate. One convenient convention is to take the first and third quartiles as the medians of the left and right halves, respectively, of the data. (If $n$ is odd, include the median in both halves.) The first and third quartiles are sometimes referred to as *hinges*.

## Example 7.3b

### Quartiles

In the Example 7.3a we found the median of 50 spot-weld strengths to be $\widetilde{X} = 400$. The median of the 25 numbers smaller than 400 is the 13th smallest: $Q_1 = X_{(13)} = 392$. Similarly, $Q_3 = X_{(38)} = 416$. In this case the quartiles do not divide the sample into four groups of exactly equal numbers: 12 observations are smaller and 37 are larger than $Q_1$, and 12 are larger and 37 are smaller than $Q_3$. (There are other ways to define quartiles; none can get around the fact that 50 is not divisible by 4.)   ■

The word *range* could refer to the interval from the smallest to the largest observation in a sample, but its technical meaning in statistics is that it is the *width* of that interval. Another measure of dispersion in a sample is the **interquartile range**— the width of the interval between first and third quartiles.

Besides the median, some other measures of "middle" or location, based on the ordered observations, are the **midrange** (middle of the range) and the **midhinge** (middle of the interval between $Q_1$ and $Q_3$).

---

Some statistics based on order:

Range: $R = X_{(n)} - X_{(1)}$.         Interquartile range: $\text{IQR} = Q_3 - Q_1$.

Midrange: $\frac{1}{2}\left[X_{(n)} + X_{(1)}\right]$.         Midhinge $= \frac{1}{2}\left(Q_3 + Q_1\right)$.

---

The *five-number summary* consisting of the median, the first and third quartiles, and the maximum and minimum observations, gives a crude but informative picture of how the sample observations are distributed. This summary is often represented in a **box plot** (or box-and-whisker plot). It is constructed alongside a number scale as a box whose ends are at the first and third quartiles, with whiskers extending from those ends to the smallest and largest observations. A line through the box indicates the median. (There are variants of this scheme that are also called box plots.)

## Example 7.3c

### Box Plot for the Spot-Weld Data

The five-number summary for the data in Table 7-6 is as follows:

$$X_{(1)} = 363, \; Q_1 = 392, \; \widetilde{X} = 400, \; Q_3 = 416, \; X_{(n)} = 450.$$

From these we also can find other order statistics:

$$R = 450 - 363 = 87, \ \ IQR = 416 - 392 = 24,$$
$$\text{Midrange} = \tfrac{1}{2}\,(450 + 363) = 406.5.$$

The box plot is shown in Figure 7-9.                                       ■

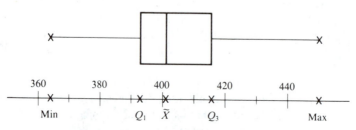

**Figure 7-9** Box plot for Example 7.3c

A box plot gives a rough, visual feel for the data. So do histograms, but the box plot, although cruder in detail, also displays the median and quartiles. This can be quite useful in data analysis in comparing distributions.

**Example 7.3d** | **Measuring Viscosity**

Two hundred measurements of viscosity were made with an Ostwald viscosimeter,[11] summarized in the computer-generated stem-leaf diagram in Figure 7-10. (The computer software split the stems, indicating the second half-stem with an asterisk.)

From the stem-leaf diagram, one can reconstruct the original data—but not in the order in which they were collected. In inspecting the raw data, it was noticed that numbers early in the list seemed to be larger than those toward the end. So the data (in the original order) were divided into four groups, and the four side-by-side box plots were drawn by computer. These are shown in Figure 7-11. (The software employs a commonly used version of a box-and-whisker plot, in which the whiskers stop at "inner fences" at 1.5 IQR's outside the quartiles. The asterisk in the second plot marks the location of an "outlier," outside the inner fences but inside "outer fences" at 3 IQR's outside the quartiles. The observation so marked is 26.8, quite different from the bulk of the 25 observations in the second subsample, and in fact, it is the smallest observation among all 200 in the whole sample. We'll say more about outliers in the next section.)

The figure seems to suggest a shifting center of the distribution, as well as a change in variability, as the sampling proceeded. That is, the population may have been changing as the observations were being made and recorded. This possibility is not apparent in the stem-leaf diagram, which looks quite like an ordinary sample from

---

[11]From a consulting file. Figures 7-10 and 7-11 were obtained using the software *Statistix*, Analytical Software (PO Box 12185, Tallahassee, FL 32317).

```
       STEM  LEAVES
   1    26.  8
   1    27*
   3    27.  66
   4    28*  4
  10    28.  777889
  16    29*  001233
  29    29.  5555678889999
  43    30*  00000012222334
  70    30.  5556666777777777788888889999
  94    31*  000000122223333333333444
 (23)   31.  55566666677777888899999
  83    32*  000000000000111122333444
  61    32.  55555677788888999
  44    33*  011333344
  35    33.  5556677888899
  22    34*  1223444
  15    34.  5556889
   8    35*  034
   5    35.  599
   2    36*  2
   1    36.
   1    37*  0
```

**Figure 7-10** Stem-leaf diagram for viscosity

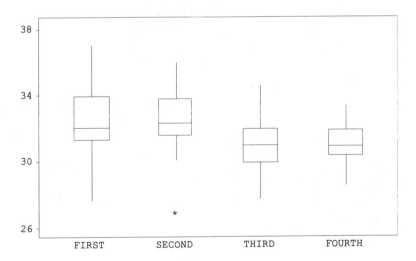

**Figure 7-11** Side-by-side box plots—Example 7.3d

a possibly normal population. Summaries that ignore the order in which the observations were obtained would obscure any trend—any change in population parameters with time. Models that allow for such changes may be needed, but these and the concomitant statistical inferences are beyond our scope. The random samples with which we are primarily concerned comprise identically distributed observations, in which case the population being sampled does *not* change as sampling proceeds.

(We'll see in the next chapter that with this assumption, the order of observation can be safely ignored.) ∎

When a sample consists of hundreds of observations, **percentiles** may be useful. We defined percentiles for continuous distributions in §5.3, and this definition can be used to define percentiles (albeit not uniquely) for discrete distributions and hence for samples, if we think of a sample as a distribution of "mass" 1/n at each sample observation. The 37th percentile of a sample, for example, is a number such that 37% of the observations are smaller and 63% larger—approximately. As in the case of populations, the median and upper and lower quartiles of a sample are the 50th, 75th, and 25th percentiles, respectively; and these are essentially what we gave at the outset as median and quartiles.

Percentiles are special cases of **quantiles.** The "$p$-quantile," or quantile of order $p$, is a number such that (approximately) the fraction $p$ of the sample observations are not larger. Like the notion of a quantile for a discrete population, the notion of a sample quantile is awkward to define because of the discreteness, unless the sample is really large.

When a c.d.f. is continuous and strictly increasing, we find the $p$-quantile of the population by locating the value $x$ (of $X$) such that $F(x) = p$. On the graph of the c.d.f., we move to the right from the $F$-axis at the height $p$ until we hit the c.d.f., and then go down to the $x$-axis to read the uniquely defined $p$-quantile. But when we try to do this in the case of a discrete distribution, either we hit a level portion—a step tread, which means that any value of $x$ with $F(x)$ at that level could be called the $p$-quantile; or we hit the riser of a stair step and go down to the value of $x$ at that jump point. In the latter case, the value we get is a sample observation. And that same observation would be counted (by this method) as the $p$-quantile for other values of $p$.

To illustrate, consider the sample consisting of these five observations: (1.49, 2.06, 2.25, 3.33, 3.60). The c.d.f. of the sample distribution is shown in Figure 7-12. To define, say, the 30th percentile or .3-quantile, we'd go from .3 on the vertical axis over to the d.f. (as shown), and find 2.06—the second smallest observation—as the .3-quantile. But we'd get this same value as any quantile with order above .2 up to and including .4.

Various ways of specifying a unique value $p$ such that $X_{(k)}$ is the quantile of order $p$ have been proposed and used, including these:

$$\frac{k}{n+1}, \quad \frac{k-.5}{n}, \quad \frac{k-\frac{3}{8}}{n+\frac{1}{4}},$$

numbers that are all between $(k-1)/n$ and $k/n$. With one of these as the value of $p$, we may define $X_{(k)}$ to be the quantile of order $p$. Thus, using the first-listed fraction for our sample of five, $X_{(k)}$ is the quantile of order $k/6$.

Sample quantiles as defined by one of the above schemes are sometimes used for comparing a sample with a population (Chapter 13), and for comparing one sample with another. This is done with the aid of a scatter plot of corresponding, selected quantiles—a quantile-quantile or "Q-Q plot." In comparing samples, the simplest

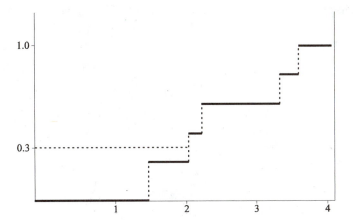

**Figure 7-12** 30th percentile of a discrete distribution

case is that in which the sample sizes are equal, because then the corresponding ordered observations ($x_{(k)}$ and $y_{(k)}$) are the corresponding quantiles—whichever definition of quantile is used. The next example illustrates such use. (When sample sizes are not equal, one can make a Q-Q plot by a process of interpolation, which we'll not go into.)

If two samples are from the same population, the corresponding quantiles would tend to be equal, and the Q-Q plot would more or less follow the line $y = x$. If one sample consists of responses of treated individuals and the other untreated individuals, and if there is no treatment effect, then the two samples are from the same population. But if a treatment effect is additive, the response of a treated individual being always a constant more than it would be without the treatment: $Y = X + a$, then the corresponding quantiles would be in the same relation, and the Q-Q plot would be a straight line with slope 1 and intercept $a$. If the effect is multiplicative: $Y = bX$, the Q-Q plot would be a straight line with slope $b$ and intercept 0.

**Example 7.3e** | **Viscosity**

The preceding example used side-by-side box plots to exhibit an apparent change in the way viscosities were being measured as the data were being collected. With the same aim in mind, we used computer software to generate a Q-Q plot of the last 50 of the 200 viscosity measurements against the first 50. The plot is shown in Figure 7-13.

If the populations sampled were the same for these two groups of data, one would expect that the corresponding quantiles would be more or less equal. In Figure 7-13, we have drawn in the line $x = y$. The fact that all points of the plot are to the right of this line suggests that the first 50 observations came from a population shifted to the right of the population sampled for the last 50. A greater variability among the first 50 than among the last 50 is evident in the fact that the data pairs are close to a line with slope less than 1/2, suggesting a scale factor as well as an additive constant, which would mean that the $x$'s are spread out more than the $y$'s. Although these same changes were suggested by the box plots, a Q-Q plot has the possible advantage that all observations are shown rather than simply summarized in quartiles and extreme values.    ■

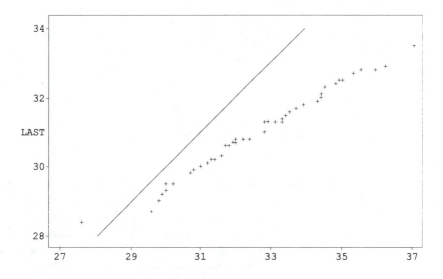

**Figure 7-13** Q-Q plot—first 50 vs. last 50

---

**7.4**        **Data Analysis**

When we get to the theory of statistical inference in subsequent chapters, we'll need familiarity with the various means of summarizing sample information being presented in this chapter. However, these techniques are also useful when one is simply trying to comprehend a mass of data, to study them from various points of view, noting what features are worthy of attention and what suggestions one might find for further investigation. So we may look for peculiarities, anomalies, unexpected occurrences, possible relationships—not for proving something or obtaining definite answers, but for pointing to next directions in an ongoing investigation. The various methods of graphical presentation that we've presented so far (including box plots and Q-Q plots), as well as sample summaries of various types, are among the tools used in "exploratory data analysis."[12]

Example **7.4a**    **Plastic Tubing**

One of our students with a part-time job collected (as a class project) data on the time for an extruding machine to process polyethylene tubing. The rating of the machine

---

[12]J. Tukey, *Exploratory Data Analysis* (Reading, MA: Addison-Wesley, 1977).

was 45 minutes for 6000 feet. He recorded these times:

45, 93, 45, 50, 51, 49, 45, 48, 40, 45, 45, 65, 55, 40.

A dot plot of these times shows the second observation (93) to be out of line with the rest—what is termed an "outlier." Its unusualness suggests that one should look into the possibility that it is faulty in some way—an error in recording, or caused by some change in experimental conditions. The student realized this, and a check back revealed that the 93 included time for changing the screen pack during the run.  ■

The term "outlier" does not have a clear definition. Roughly speaking, an outlier is an observation that is "far" from the bulk of the data. You might consider an observation to be an outlier that others would not. Some statistical software packages include a program to identify outliers. Outliers in univariate data can usually be spotted visually in graphical displays—without the need for the fancy rules that are often presented in textbooks and applied by computer software. Some software packages go so far as to delete outliers automatically before carrying out statistical procedures. (Don't use such software if it does this!)

An outlier may reflect an aspect of the population that is important for addressing the scientific question of interest. An example of such an outlier is a patient who dies after taking a headache remedy. One should try to understand as much as possible about such an outlier (certainly not delete it). It may be indicative of an important population characteristic, and we'd certainly want to know this.

An outlier may involve something that crept in during the process of collecting data, such as a misplaced decimal point in recording an observation, or a power surge that affected the process of measurement and momentarily changed the population being sampled. If such a cause can be identified, the observation may be properly deleted. But an outlier may be simply the "luck of the draw"—as when you happen to get an observation far out in the tail of a distribution.

Studying data to discover its unusual features, with whatever devices one might use, is sometimes referred to as "data dredging." This is legitimate and useful, provided you realize that even an ideally random sampling process is apt to produce results that are peculiar in some respect. If you look hard enough at any sample, you will find something that seems unusual. This is what we did with the viscosity data in Examples 7.3d and 7.3e, thereby uncovering what might have been a trend—a mean that shifted with time, something that would warrant looking into further. Here's another example:

**Example 7.4b** | **The 1969 Draft Lottery**

In 1969, a lottery was conducted to determine the order in which, according to birthdays, 18-year-olds would be drafted for military service. Each of the 366 dates in a year was marked on a slip of paper and inserted in a capsule 1.5 inches long and 1 inch in diameter. The capsules were placed in a box, mixed, and subsequently poured from the box into a 2-foot-deep bowl from which they were selected, one at a time, until a permutation of the 366 dates was obtained.

After the lottery, 13 men sued the Selective Service System, contending that the lottery was biased against birthdays near the end of the year. Pentagon officials had said that men with numbers in the highest third (mid 200's to 366) would probably escape the draft entirely. A U.S. District Court judge, although denying a temporary restraining order, said, "I find that the selection made in the December 1 drawing was not a perfectly random selection." He defined "randomness" as that quality that makes any one sequence of birthdays as likely as the selection of any other sequence. So far, so good. But he went on to conclude that "there is a substantial discrepancy between a perfect selection, on the one hand, and the selection which *resulted* from the December 1 drawing, on the other." (Italics are ours.)

If all sequences are equally likely, the selection process is "random," no matter which sequence is the result of a particular drawing. Moreover, *any* particular sequence has some kind of quirk—if you look long and hard enough, you will find one. The real question is whether steps were taken in preparing the actual lottery so that the method of selection would be as close to "random" as possible.

(For the 1970 lottery, the final order of dates was determined by a selection of one date from the 366, which was then matched with a second selection of an order number, from 1 to 366. Perhaps it was felt that this double randomness would lend an air of fairness to the proceedings and preclude more lawsuits.)     ■

Relationships between two variables are of interest, in essentially every field of application. In some cases a relationship is a simple "law" of nature, abeit often obscured by factors that we think of as introducing random errors into what we can observe experimentally. In other cases, data suggest relationships between variables that, although not "perfect" and not explained in terms of a known "law," can be useful for purposes of predicting one variable from a knowledge of another.

In searching for relationships between two variables, we like to think that real relationships are simple—linear, or quadratic, or exponential, or perhaps a power type: $y = \alpha x^\beta$. Even when just exploiting the value of one variable to predict another, simple relationships are easier to interpret and to use.

| Example **7.4c** | ## Carbon in Clay |
|---|---|

The amount of carbon in clay can be measured directly, but it can be estimated by an indirect method, combining the amounts of its constituents in a suitable way. The direct method is more costly and time-consuming, involving heating until all carbon compounds are burned and collecting and measuring resulting $CO_2$. The following measurements on 25 clays from South Devonshire, England,[13] are shown in the scatter diagram of Figure 7-14:

---

[13]Taken from C. A. Bennett and N. L. Franklin, *Statistical Analysis in Chemistry and the Chemical Industry* (New York: John Wiley & Sons, 1954), Table 6.4 on page 218.

| Direct | Indirect | Direct | Indirect | Direct | Indirect |
|--------|----------|--------|----------|--------|----------|
| 1.53 | 2.46 | .87 | 1.54 | .28 | .70 |
| .27 | − .40 | 3.07 | 4.82 | .25 | .30 |
| .25 | .64 | .29 | .78 | .12 | .12 |
| 1.50 | 2.36 | 1.31 | 2.14 | .31 | .08 |
| .14 | − .01 | 2.98 | 4.53 | 6.84 | 9.94 |
| 2.15 | 3.68 | 1.35 | 1.84 | .40 | .97 |
| 4.18 | 6.14 | .22 | .52 | .16 | .35 |
| 5.06 | 7.49 | .86 | 1.41 | .16 | − .50 |
| 11.43 | 15.80 | | | | |

(The data were reported in this way, with negative entries; of course, these are impossible and should be replaced by 0.)

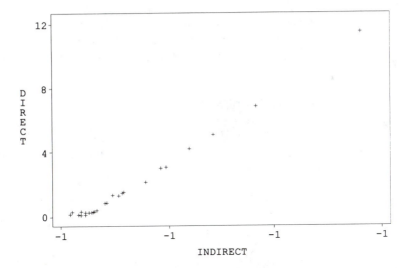

**Figure 7-14** Direct vs. indirect measurement

We see a strong linear relationship in the scatter plot, which suggests that a linear function of the indirect measurement could be used to replace the direct measurement—except when the amount of carbon is very small. To see the situation at the low end of the scale more clearly, we magnified the low end by replotting the data with the rightmost seven cases omitted.

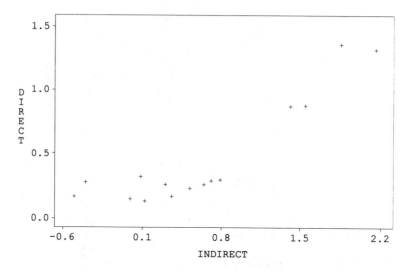

**Figure 7-15**

From the plot in Figure 7-15, it is apparent that the relationship is very fuzzy when the direct measurement is less than .5. Using a straight line (or any other curve) to estimate the direct from the indirect measurement is apt to be quite inaccurate in terms of relative error.    ■

Example 7.4d | **A Study in Ophthalmology**

In studying the reaction rate of a certain synthetase of a bovine lens, it has been found that the rate is related to the substrate concentration. In one study, two determinations of reaction rate were obtained for each of several substrate concentrations.[14] The investigators recorded data as follows:

| Reciprocal Substrate Concentration (x) | Reciprocal Reaction Rate (y) | |
|:---:|:---:|:---:|
| 24 | .429 | .444 |
| 20 | .293 | .293 |
| 16 | .251 | .268 |
| 12 | .207 | .218 |
| 8 | .239 | .218 |
| 6 | .180 | .199 |
| 2 | .156 | .167 |

[14]Private communication.

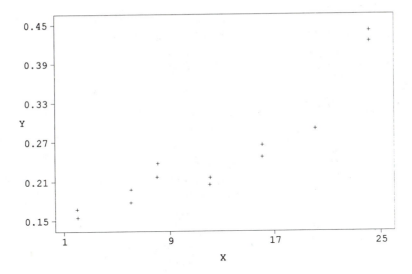

**Figure 7-16** Reciprocal rate vs. reciprocal concentration

The scatter plot in Figure 7-16 suggests a relationship that is somewhat more curved than linear. In seeking a linear relationship, investigators often transform one or the other variable (just as they did here in choosing to relate the reciprocals). In this instance, we might try a logarithm or a power of one variable; suppose we simply take the reciprocal of $y$ (that is, $1/y = y^{-1}$). This reciprocal is then the reaction rate. A plot of the reaction rate vs. $x$ is shown in Figure 7-17. The relationship is visibly closer to linear than before the transformation. ∎

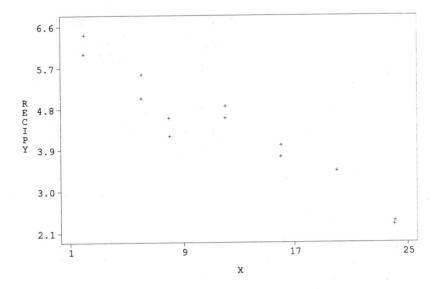

**Figure 7-17** Reaction rate vs. reciprocal concentration

We have given some of the tools and basic ideas of exploratory analysis, but there is no mathematical theory of exploration. A productive exploration of data does require familiarity with the standard methods of inference and their supporting theory, as well as familiarity with the substantive field in which the data arise.

## Problems

**★ 7-13.** Plasma clearances of a certain drug (in ml/min/kg) for 32 nonsmokers were recorded as follows:[15]

| | | | | | | | | | | |
|---|---|---|---|---|---|---|---|---|---|---|
| 3.1 | 3.7 | 4.2 | 4.4 | 4.8 | 4.8 | 5.4 | 5.7 | 5.8 | 6.0 | 6.0 |
| 6.4 | 7.0 | 7.2 | 7.2 | 7.3 | 7.4 | 7.5 | 7.8 | 8.0 | 8.2 | 8.6 |
| 9.0 | 9.4 | 10.2 | 10.3 | 11.2 | 13.2 | 13.6 | 14.5 | 15.5 | 26.0 | |

   **(a)**   Construct such graphical representations as might exhibit any anomalies and comment.
   **(b)**   Find the median and midrange.

**★ 7-14.**  Use the stem-leaf diagram to find the median of the birthweights given in Problem 7-9.

**7-15.**  Construct side-by-side box plots for the heights of males in Table 7-1 (page 257) and the heights of females in Table 7-5 (page 269).

**★ 7-16.**  Find the median weight and median height in Problem 7-12.

**★ 7-17.**  A population density per square mile in 1980 was reported by the Bureau of the Census for each of the 50 states, as follows. (The order is alphabetical by states' names.)  Find the median. Is this the median density for the United States?  (Explain.)

| | | | | | | | | | |
|---|---|---|---|---|---|---|---|---|---|
| 77 | 1 | 24 | 44 | 151 | 28 | 638 | 308 | 180 | 94 |
| 150 | 12 | 205 | 153 | 52 | 29 | 92 | 95 | 36 | 429 |
| 733 | 163 | 51 | 53 | 71 | 5 | 21 | 7 | 102 | 987 |
| 11 | 371 | 120 | 9 | 263 | 44 | 27 | 264 | 989 | 103 |
| 9 | 112 | 54 | 18 | 55 | 135 | 62 | 81 | 87 | 5 |

**★ 7-18.**  Construct a box plot for the data in
   **(a)**   Problem 7-13.    **(b)**   Problem 7-9.

**7-19.**  The median age for each state in 1980, as reported by the Census Bureau, was as follows (same order as in Problem 7-17):

| | | | | | | | | | |
|---|---|---|---|---|---|---|---|---|---|
| 29.3 | 28.6 | 28.4 | 30.1 | 31.2 | 29.0 | 27.4 | 30.1 | 28.9 | 29.8 |
| 26.1 | 32.0 | 27.6 | 29.1 | 28.9 | 29.7 | 31.9 | 30.2 | 30.1 | 29.8 |
| 29.1 | 29.7 | 29.9 | 27.2 | 29.2 | 30.3 | 29.6 | 32.1 | 28.2 | 30.4 |
| 30.6 | 34.7 | 29.2 | 30.4 | 27.7 | 30.1 | 28.3 | 31.8 | 24.2 | 29.4 |
| 29.2 | 28.7 | 30.0 | 30.3 | 30.9 | 32.2 | 29.9 | 28.2 | 29.4 | 27.1 |

---

[15]Data from a consulting project.

Use a stem-leaf diagram to find the *median* of these 50 medians. Explain why this is not necessarily the median age in the United States.

**7-20.**   Among the data collected from families participating in a California study were children's birthweights. Birthweights (in lbs.) of children of mothers in the age range 23–25 years are shown in the stem-leaf diagrams of Table 7-7.[16]

(a)   Construct a box plot for these data.

(b)   The smallest weight might seem to be an "outlier."  Should one be suspicious of this observation?

**Table 7-7**

| Stem | Leaves (tenths) |
|------|------------------|
| 2 | 3 |
| 2 | |
| 3 | |
| 3 | 9 |
| 4 | 1 |
| 4 | |
| 5 | 3344 |
| 5 | 666899 |
| 6 | 111233344 |
| 6 | 5677788889999 |
| 7 | 01122333333444 |
| 7 | 566677899 |
| 8 | 33333 |
| 8 | 67 |
| 9 | 034 |
| 9 | 678 |
| 10 | |
| 10 | 5 |

---

## 7.5          The Sample Mean

The **sample mean** is perhaps the most commonly used measure of the middle of a set of numerical data. If we think of the sample observations $(X_1, ..., X_n)$ as constituting a population of size $n$ from which one member is selected at random, each sample

---

[16]The family data are reported in Hodges, Krech, and Crutchfield, *Stat Lab* (New York: McGraw-Hill, 1975).

value has weight $1/n$. The (population) mean value of this "sample distribution" is the sample mean:

$$\overline{X} = \sum_{i=1}^{n} X_i \cdot \frac{1}{n} = \frac{1}{n}\sum_{i=1}^{n} X_i, \tag{1}$$

the arithmetic average of the $n$ $X_i$'s—their sum divided by their number.

Samples often include repeated values and may then be summarized in a frequency distribution, using an arbitrary scheme of class intervals. Formula (1) counts an $X$-value once each time it occurs in the sample, so a value that occurs $f$ times has total weight $f/n$. If there are $k$ class intervals, we denote the *distinct* values by $x_1$, ..., $x_k$, and the frequency of $x_j$ by $f_j$. The calculation of the sample mean can be carried out as a sum over the class intervals with weights $f_j/n$. These weights correspond to the probability weights in the formula for the mean of a probability distribution. (The long-range limit of the relative frequency $f_j/n$ is the probability of $x_i$.)

---

Sample mean for a data set $X_1$, ..., $X_n$:

$$\overline{X} = \frac{1}{n}\sum_{i=1}^{n} X_i. \tag{2}$$

For data given in a frequency distribution with distinct values $x_j$ and corresponding frequencies $f_j$, $j = 1$, ..., $k$:

$$\overline{X} = \frac{1}{n}\sum_{1}^{k} x_i f_i. \tag{3}$$

---

Because $\overline{X}$ can be thought of as an expected value, the various properties of expected values from §3.1 will apply. In particular, the average deviation about the mean is 0, so that

$$\sum_{1}^{n}(X_i - \overline{X}) = 0. \tag{4}$$

Again, this is the *balance* property, and a graphical plot of the data—dot diagram or histogram—should be referred to, to see that the calculated mean looks reasonable as a balance point.

**Example 7.5a** | A student counted the number of pieces of licorice candy in each of 25 boxes. His results are shown in the following frequency table.

The column of products $x_j f_j$ aids in calculating the sample sum, which is just the total number of pieces in the 25 boxes. Dividing by 25, in a sense, divides the 326 pieces equally among the 25 boxes: $\bar{X} = 326/25 = 13.04$ per box. Figure 7-18 shows a bar graph representing the frequency distribution with the mean indicated, which appears reasonable as the center of gravity.  ∎

| $x_j$ | $f_j$ | $x_j f_j$ |
|-------|-------|-----------|
| 11 | 3 | 33 |
| 12 | 3 | 36 |
| 13 | 10 | 130 |
| 14 | 8 | 112 |
| 15 | 1 | 15 |
| Sums: | 25 | 326 |

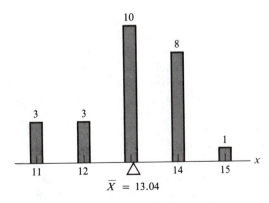

**Figure 7-18** Bar graph for Example 7.5a

When data have been *rounded* in the process of constructing a frequency distribution, the $\bar{X}$ calculated using (3) is only an approximate value of the mean. This is because the original observations are replaced by the midpoints of the class intervals in which they lie. But if the original observations are available, the value of $\bar{X}$ can be calculated from them, using (2).

**Example 7.5b** | **Spot-Weld Data**

The sum of the 50 observations in the data set of Example 7.2a is 20,202, so from (2), the mean value is $20{,}202/50 = 404.04$. If the raw data were not available, we could approximate the value of $\bar{X}$ from the frequency distribution in Table 7-3 using (3). That distribution is repeated below, along with products $x_j f_j$:

| $x_j$ | $f_j$ | $x_j f_j$ |
|-------|-------|-----------|
| 366 | 2 | 732 |
| 377 | 2 | 754 |
| 388 | 10 | 3880 |
| 399 | 15 | 5985 |
| 410 | 8 | 3280 |
| 421 | 7 | 2947 |
| 432 | 3 | 1296 |
| 443 | 2 | 886 |
| 454 | 1 | 454 |
| Sums: | 50 | 20214 |

Thus, $\overline{X} \doteq 20214/50 = 404.28$, very close to the exact value—perhaps close enough for practical purposes, inasmuch as reading and/or recording observations can easily introduce inaccuracies in the data. ∎

Both the median and mean are measures of the middle of a distribution. Which one is better? The answer depends in part on what it's going to be used for and also in part on the nature of the population. Both the mean and median are "typical" observations. The median is relatively unaffected by extreme observations. Thus, for example, whether the most expensive house in a city has a market value of $1 million or $10 million, the median market value of a house in the city is the same. Not so the mean, because the total market value in the city is $9 million larger in the second case. Of great practical importance is the possibility of gross errors in the process of recording or measuring. The mean is influenced by such errors, whereas the median is much less apt to be affected. However, when a population does not contain "extreme" values, the mean has the advantage of using all the data and using them efficiently.

A procedure that may well become more commonplace, because it results in a compromise between the mean and the median without having the disadvantages of either, is to delete a specified number of the largest and the same number of the smallest observations in a data set and then calculate the mean of what's left. (This is common practice in some athletic competitions where a score is calculated based on the ratings of several judges.) The result is a **trimmed mean.** In the special case in which the largest 25% and the smallest 25% of the observations are trimmed, the trimmed mean is called the **midmean**—the mean of the middle half of the data.

One reason for calculating a sample mean is to estimate the mean of the sampled population. How successful this estimation will be depends on the sample size and on the amount of population variability. (This will be studied in more detail in Chapter 9.) One clue as to population variability lies in sample variability. The next section takes up this sample characteristic.

## 7.6   Measures of Dispersion

In §7.3 we defined the range $(R)$ and the interquartile range (IQR) as measures of the width or spread of a sample distribution. Here we take up measures of variability based on deviations from the mean, measures that usually make better use of the data and are more widely used.

The traditional way of averaging deviations about the sample mean, as in the case of populations (§5.6), is to take the square root of their average square. We tentatively define the **variance** of a sample as we defined the mean, by relating it to a corresponding calculation for populations. As in §7.5, think of selecting at random from the $n$ observations in the sample. This assigns weight $1/n$ to each observation, or in the case of a frequency distribution, weights $f_j/n$ to each of the *distinct* values $x_j$. This assignment of weights defines a distribution, whose variance is

$$V = \frac{1}{n}\sum_{i=1}^{n}(X_i - \overline{X})^2. \tag{1}$$

In the case of data grouped in $k$ class intervals,

$$V = \frac{1}{n}\sum_{j=1}^{k}(x_j - \overline{X})^2 f_j. \tag{2}$$

As so defined, $V$ is a special case of the variance as defined in §3.3, applied here to the "population" of sample values. So the general properties of variance apply to $V$. In particular, the following formulas provide alternatives to (1) and (2), respectively:

$$V = \frac{1}{n}\sum_{i=1}^{n}X_i^2 - \overline{X}^2. \tag{3}$$

$$V = \frac{1}{n}\sum_{i=1}^{k}x_j^2 f_j - \overline{X}^2. \tag{4}$$

Formula (3) is a special case of the parallel axis theorem [(8) of §3.3], which reads as follows in the present context:

$$\frac{1}{n}\sum_{1}^{n}(X_i - c)^2 = \frac{1}{n}\sum_{1}^{n}(X_i - \overline{X})^2 + (\overline{X} - c)^2. \tag{5}$$

As a function of $c$, this is clearly smallest when $c = \overline{X}$: The average squared deviation is smallest when taken about the mean.

We gave (1) as a "tentative" definition of variance. The usual statistical convention is to replace $n$ in the denominator with $n - 1$:

**Sample variance:**

$$S^2 = \frac{1}{n-1}\sum_{1}^{n}(X_i - \overline{X})^2. \tag{6}$$

Comparing (1) and (6), we see that

$$S^2 = \frac{n}{n-1}V. \tag{7}$$

So the difference between using $S^2$ and $V$ is slight, especially for large $n$. Although $V$ seems more natural and simple, we'll conform to the standard in statistical practice (and computer software) and use $S^2$ as the sample variance.

(The "$n-1$" is called the "degrees of freedom" in $S^2$. This refers to the fact that the $n$ deviations whose squares are summed are not completely free, but satisfy the one constraint that they sum to 0. In other estimates of variance to be encountered in Chapters 14 and 15, the degrees of freedom to be used as a divisor may be something other than $n-1$. There is no theoretical basis for preferring to divide by degrees of freedom,[17] but if we use this divisor, the same statistical tables that are needed in important applications will serve a wide variety of purposes. And this is sufficient justification.)

With sample variance defined by (6), we define the *standard deviation* as its square root, given by various equivalent formulas:

**Sample standard deviation,** for a sample $(X_1, ..., X_n)$:

$$S = \sqrt{\frac{1}{n-1}\sum_{1}^{n}(X_i - \overline{X})^2} = \sqrt{\frac{\sum X^2 - (\sum X)^2/n}{n-1}}. \tag{8}$$

When the sample is summarized as a frequency distribution with $k$ distinct values or class marks $x_j$ and frequencies $f_j$,

$$S = \sqrt{\frac{1}{n-1}\sum_{1}^{k}(x_j - \overline{X})^2 f_j} = \sqrt{\frac{\sum x^2 f - (\sum xf)^2/n}{n-1}}. \tag{9}$$

The second formulas in (8) and (9) are usually easier to use in calculations. (To make them less cluttered, we suppressed the subscripts.)

---

[17]B. W. Lindgren, "To bias or not to bias," *Amer. Statistician* 37 (1983), 254.

The mean deviation given by (1) of §3.3, when applied to the "population" of sample values (each assigned probability $1/n$), defines the **sample mean deviation:**

$$\text{m.a.d.} = \frac{1}{n}\sum_1^n |X_i - \overline{X}|. \tag{10}$$

At present, this statistic is not in common use in statistical practice. The mean deviation is sometimes defined as the mean absolute deviation from the median, because this is the smallest average absolute deviation (as compared with what would be obtained using deviations about other points).

**Example 7.6a**

### 6-MP and Leukemia

The drug 6-MP (mercaptopurine) has been used in therapy to extend remission in leukemia patients. The lengths of remission (in weeks) for 21 patients receiving this treatment were observed as follows:[18]

10, 7, 32, 21, 22, 6, 16, 34, 32, 25
11, 20, 19, 6, 17, 35, 6, 13, 9, 6, 10

The mean length of remission is $357/21 = 17$ weeks. A patient about to undergo therapy would be interested in deviations from the average:

$-7, -10, 15, 4, 5, -11, -1, 17, 15, 8$
$-6, 3, 2, -11, 0, 18, -11, -4, -8, -11, -7$

The mean of these 21 deviations is 0, as it must be because of the balance point property of the mean. The mean of the *absolute* deviations is 174/21, or about 8.29. The plot in Figure 7-19 suggests that this is about right as an average distance from 17.

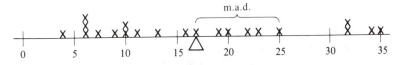

**Figure 7-19** Data plot—Example 7.6a

To find the standard deviation of a set of numbers using the first formula in (8), we square the deviations and find the sum of the squares to be 1980:

$$S = \sqrt{\frac{1980}{20}} = 9.95.$$

---

[18]E. J. Freireich et al., "The effect of 6-Mercaptopurine on the duration of steroid-induced remissions in acute leukemia...," *Blood* 21 (1963), p.699. (We have altered one observation slightly (the 21 was 23) to make the mean an integer, so as to avoid decimals.)

Alternatively, we can use the second formula, first finding the sum of the squares of the observations to be 8049:

$$S = \sqrt{\frac{1}{20}\left(8049 - \frac{357^2}{21}\right)} = 9.95.$$

This is a bit larger than the mean deviation (8.29)—as it must be.  ■

Example **7.6b**   **Spot-Weld Data**

The histogram for the frequency distribution of the spot-weld data from Examples 7.2a and 7.2b was shown in Figure 7-5 (page 265). The frequency table is repeated here with additional columns of products $xf$ and $x^2f$:

**Table 7-8**

| $x_j$ | $f_j$ | $x_j f_j$ | $x_j^2 f_j$ |
|-------|-------|-----------|-------------|
| 366   | 2     | 732       | 267912      |
| 377   | 2     | 754       | 284258      |
| 388   | 10    | 3880      | 1505440     |
| 399   | 15    | 5985      | 2388015     |
| 410   | 8     | 3280      | 1344800     |
| 421   | 7     | 2947      | 1240687     |
| 432   | 3     | 1296      | 559872      |
| 443   | 2     | 886       | 392498      |
| 454   | 1     | 454       | 206116      |
| Sums: | 50    | 20214     | 8189598     |

Before calculating $S$, try to guess its value from the histogram. It may help to think of it as just a little larger than the average distance from the sample mean.

Substituting the appropriate column sums from Table 7-7 in the second formula of (9), we have

$$S^2 = \frac{1}{49}\left(8,189,598 - \frac{20,214^2}{50}\right) = 356.8.$$

The standard deviation is the square root, $S = 18.89$. How does this compare with your guess? If it is not close, your intuition may need adjusting. Making a guess as to the value of $S$ before calculating it can serve as a crude check on your calculations. It will also help teach you the meaning of an s.d. In this instance we do have the original data (that is, before grouping into class intervals) and can use the second formula in

(8):  The sample sum is 20,202, and the sample sum of squares, 8,179,568, with the result $S \doteq 18.71$.

It should be noted, however, that because of the subtraction in the second formula of (8) or (9), rounding the sum and sum of squares can cause a problem. For instance, if we round 20,202 to 20,200 and 8,179,568 to 8,180,000 (which is not all that severe), the calculated standard deviation is 24.4, rather far from 18.71.    ∎

Calculating a standard deviation is easy for hand calculators and computers. Yet just *entering* data on a calculator or computer keyboard involves risks of error related to the complexity of the data. Sometimes, performing a simple linear transformation of the data will substantially eliminate this source of error. For this and other reasons, we present next a method involving linear transformations of data.

We saw in §3.3 that when a random variable undergoes the linear transformation $Y = a + bX$, its mean undergoes the same transformation, and its standard deviation is multiplied by the scale factor $|b|$. The same is true for the sample mean and standard deviation, as special cases:

If observations $X$ are transformed to $Y = a + bX$, then

$$\overline{Y} = a + b\overline{X}, \quad \text{and} \quad S_Y = |b|S_X. \tag{11}$$

These relations can be used to advantage in choosing a scale of values that will simplify the calculations of mean and variance, especially when data are grouped into equal-sized class intervals.

**Example 7.6c**

## Using a Transformation to Simplify Arithmetic

To show how a linear transformation of the measurement scale can simplify calculations, we recalculate the mean and s.d. of the data in the preceding example. First we choose any class mark near the center of the data, say 399, as the new reference point; call this 0 on the new scale. Then we take the spacing between class marks as a unit on the new scale, as follows:

$$Y = \frac{X - 399}{11}, \quad \text{or} \quad X = 399 + 11Y.$$

Thus, $Y$ is the number of class marks to the right of 399—positive if to the right, and negative if to the left. Table 7-9 repeats the distribution, but now with columns of the coded values ($y$'s) and the products needed for use in (9). Using the sums in the last row of the table, we find the mean and s.d. of the $Y$'s to be $\overline{Y} = .48$ and $S_Y = 1.717$. Then from (11), with $a = -399/11$ and $b = 1/11$, we find the mean and s.d.: $\overline{X} = 404.28$ and $S_X = 18.89$ (as before).    ∎

**Table 7-9**

| $x_j$ | $f_j$ | $y_j$ | $y_j f_j$ | $y_j^2 f_j$ |
|-------|-------|-------|-----------|-------------|
| 366   | 2     | $-3$  | $-6$      | 18          |
| 377   | 2     | $-2$  | $-4$      | 8           |
| 388   | 10    | $-1$  | $-10$     | 10          |
| 399   | 15    | 0     | 0         | 0           |
| 410   | 8     | 1     | 8         | 8           |
| 421   | 7     | 2     | 14        | 28          |
| 432   | 3     | 3     | 9         | 27          |
| 443   | 2     | 4     | 8         | 32          |
| 454   | 1     | 5     | 5         | 25          |
| Sums: | 50    |       | 24        | 156         |

# Problems

**7-21.** Find the mean and standard deviation of the times to burnout given in Problem 7-10 from the frequency table obtained in that problem.

* **7-22.** Find the mean of the spot-weld data of Example 7.2b,
 **(a)** from the table you obtained in Problem 7-6.
 **(b)** from Table 7-3, page 264.

**7-23.** Find the mean and standard deviation of the weights in Problem 7-12.

* **7-24.** Find the standard deviation of these numbers: 1, 1, 3, 3. [Note that all distances from the mean are 1. The mean deviation has to be 1, but using the variance with the $n - 1$ divisor gives an s.d. that is a bit larger; with the divisor $n$, the s.d. would be 1.]

* **7-25.** For the plasma clearance data of Problem 7-13,
 **(a)** find the mean and midmean.
 **(b)** find the standard deviation.
 **(c)** find the interquartile range.

**7-26.** Find the mean and s.d. of the birthweights in Problem 7-9.

* **7-27.** Find the s.d. of the spot-weld data of Example 7.2b from the table you obtained in Problem 7-6.

* **7-28.** Given that the mean of 100 observations is 42.5 and that their s.d. is 6.313, find the sum of the squares of the observations.

**7-29.** For the birthweight data in Problem 7-9,
 **(a)** calculate the *mean deviation* about the mean and compare with the s.d. found in Problem 7-26. (See also Problem 7-41.)
 **(b)** find the mean deviation about the *median.*

**7-30.** A sample of rivet head diameters is summarized in the accompanying frequency table. Use the coding $Y = (X - 13.39)/.09$ to find the mean and s.d. of the rivet head diameters.

| Class interval | Frequency |
|---|---|
| 13.17-13.25 | 1 |
| 13.26-13.34 | 7 |
| 13.35-13.43 | 10 |
| 13.44-13.52 | 5 |
| 13.53-13.61 | 6 |
| 13.62-13.70 | 1 |

* **7-31.** A sample of ten temperature readings in degrees Fahrenheit has mean 50 degrees and s.d. 9 degrees. Find the mean and s.d. in degrees Celsius, where $C = 5(F - 32)/9$.

**7-32.** The parallel axis theorem [(5) of §7.5] shows algebraically why the sample variance is the smallest second moment. Show this minimum property by using the method of calculus to minimize $\sum(X_i - c)^2$ with respect to $c$.

---

## 7.7    Correlation

In §7.2 we explained how to make a graphical display of bivariate data in a *scatter diagram*. Such pictures show, qualitatively, the extent and nature of a relationship between two variables. However, for purposes of quantitative inference and for succinct summaries, it is useful to have a numerical measure of the relationship. The most commonly used measure is the "product moment correlation coefficient," or "coefficient of linear correlation," commonly referred to simply as the **correlation coefficient.**

We defined a correlation coefficient for populations in §3.4. To apply this definition to a sample, we imagine (as usual) a distribution that assigns weight $1/n$ to each sample pair. With this assignment, $\mu_X$ and $\mu_Y$ become $\overline{X}$ and $\overline{Y}$, $\sigma_X^2$ and $\sigma_Y^2$ become $V_X$ and $V_Y$, and the covariance $\sigma_{X,Y}$ becomes

$$C_{X,Y} = \frac{1}{n}\sum(X_i - \overline{X})(Y_i - \overline{Y}) = \frac{1}{n}\sum X_i Y_i - \overline{X}\,\overline{Y}. \tag{1}$$

The sample correlation coefficient $r$ (or $r_{X,Y}$) is the covariance divided by the square root of the product of the $V$'s:

$$r_{X,Y} = \frac{C_{X,Y}}{\sqrt{V_X V_Y}} = \frac{\sum(X_i - \overline{X})(Y_i - \overline{Y})}{\sqrt{\sum(X_i - \overline{X})^2 \sum(Y_i - \overline{Y})^2}}. \tag{2}$$

(We canceled $1/n$'s in the numerator and the denominator to get the last formula. Notice that if we define a sample covariance $S_{X,Y}$ with $n - 1$ as a divisor in place of

$n$, we could write $r$ as this covariance divided by the product of the sample standard deviations. The following equivalent formula for $r$ is useful in calculating $r$ (and avoids the $n$ vs. $n - 1$ issue):

**Sample correlation coefficient,** for data pairs $(X_1, Y_1)$, ..., $(X_n, Y_n)$:

$$r = \frac{\sum XY - \frac{(\sum X)(\sum Y)}{n}}{\sqrt{\left(\sum X^2 - \frac{(\sum X)^2}{n}\right)\left(\sum Y^2 - \frac{(\sum Y)^2}{n}\right)}} . \qquad (3)$$

In the next example we calculate $r$ for a typical bivariate sample. The two examples that then follow illustrate important special cases—symmetry and collinearity.

**Example 7.7a**

## LSAT Scores

Example 7.2d gave average LSAT scores $(X)$ and average GPA's $(Y)$ for entering students in each of 15 law schools in 1973. We calculated the following sums, needed to calculate $r$ using (3):

$$\sum X_i = 9{,}004, \quad \sum Y_i = 46.42, \quad \sum X_i Y_i = 27{,}975,$$

$$\sum X_i^2 = 5{,}429{,}256, \quad \sum Y_i^2 = 144.4846.$$

Substituting these values in (3) yields $r \doteq .776$. Now look back at Figure 7-7 (page 267), showing the scatter diagram for these data. There seems to be a tendency toward linearity, but the first data point is rather far from what would otherwise seem

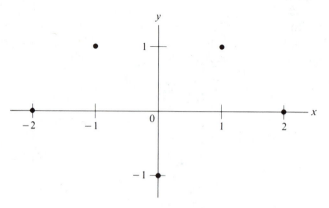

**Figure 7-20** Scatter plot—Example 7.7b

to be the "trend line." Indeed, when we recalculated $r$ without that data point, we found $r = .893$—an appreciably higher degree of linear correlation. This shows how one (possibly erroneous) "outlier" point can influence the correlation coefficient as well as the trend line. ∎

Example **7.7b**

## Zero Correlation

Consider the five points listed in the table below and shown in Figure 7-20. The average $X$ is 0 and the average product is 0, so the numerator of (3) is 0. So, $r = 0$. The $X$'s and $Y$'s are uncorrelated. This is because the data points are symmetrically located about the $y$-axis. Each positive product of deviations in the numerator is canceled by a negative product.

| | $X$ | $Y$ | $XY$ | $X^2$ | $Y^2$ |
|---|---|---|---|---|---|
| | $-2$ | 0 | 0 | 4 | 0 |
| | $-1$ | 1 | $-1$ | 1 | 1 |
| | 1 | 1 | 1 | 1 | 1 |
| | 0 | $-1$ | 0 | 0 | 1 |
| | 2 | 0 | 0 | 4 | 0 |
| Sum: | 0 | 1 | 0 | 10 | 3 |

∎

Example **7.7c**

## Perfect Correlation

Consider the five data points $(X, Y)$ in the following table:

| | $X$ | $Y$ | $X^2$ | $Y^2$ | $XY$ |
|---|---|---|---|---|---|
| | 0 | $-2$ | 0 | 4 | 0 |
| | 2 | $-1$ | 4 | 1 | $-2$ |
| | 4 | 0 | 16 | 0 | 0 |
| | 6 | 1 | 36 | 1 | 6 |
| | 8 | 2 | 64 | 4 | 16 |
| Sum: | 20 | 0 | 120 | 10 | 20 |

With these sums, we find

$$\sum X^2 - \left(\sum X\right)^2 / n = 120 - 400/5 = 40,$$

$$\sum Y^2 - \left(\sum Y\right)^2 / n = 10 - 0 = 10, \quad \sum XY - \sum X \sum Y / n = 20.$$

So, $r = 20/\sqrt{40 \times 10} = 1$. The scatter plot in Figure 7-21 shows that the five data points lie on a straight line with positive slope. In §3.4 we showed that a correlation coefficient is 1 when and only when the distribution is concentrated on a straight line with positive slope. ∎

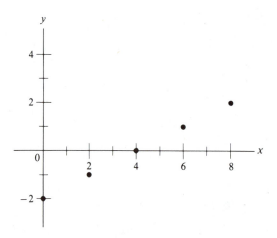

**Figure 7-21** Scatter plot—Example 7.7c

Calculating correlations by hand can be tedious when $n$ is large. Some calculators are preprogrammed to calculate $r$. Even with this aid, however, simplifying the numbers to be entered can help avoid mistakes. Both variances and covariances are based on *deviations*, and these do not change when a constant is added to or subtracted from all the $X$'s or all the $Y$'s. And changing the scale of either variable introduces the same constant multiplier (except possibly for sign) in both numerator and denominator of (2). This implies that transformations such as changing units from degrees Fahrenheit to degrees Kelvin or kilograms to pounds have no effect on $r$.

If data $(X_i, Y_i)$ are transformed to $U_i = a + bX_i$, $V_i = c + dY_i$ with $b > 0$ and $d > 0$, the new correlation coefficient is the same as the old:

$$r_{U,V} = r_{X,Y}.$$

Figures 7-22 through 7-25 give four examples, each with 500 computer-generated data points, to give a visual notion of various values of $r$.

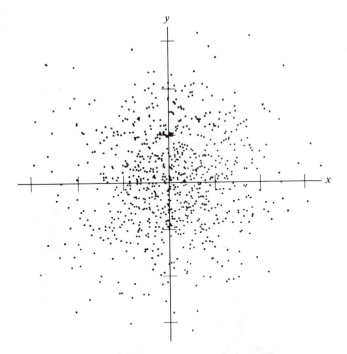

**Figure 7-22** Data with correlation .05

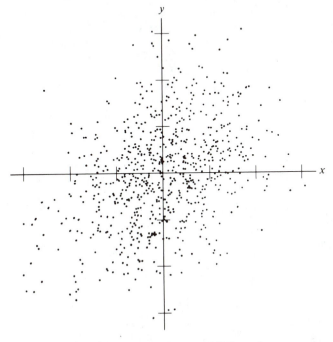

**Figure 7-23** Data with correlation .3

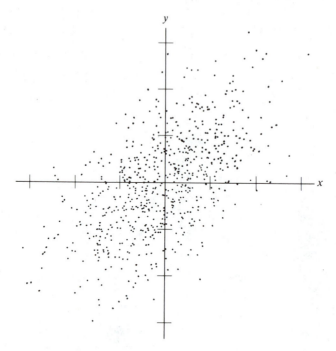

**Figure 7-24** Data with correlation .6

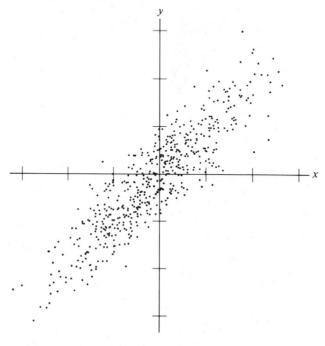

**Figure 7-25** Data with correlation .9

## Problems

**∗ 7-33.** Find the correlation coefficient for the data in Problem 7-11.

**7-34.** Find the correlation coefficient for the height and weight of the ten students in Problem 7-12.

**∗ 7-35.** Suppose that in Problem 7-12 the heights had been measured in centimeters, and the weights in kilograms. Find the correlation coefficient of the resulting data (from your answer to the preceding problem).

**7-36.** Given $\overline{X} = 10$, $\overline{Y} = 5$, $S_X = 3$, $S_Y = 2$, and $r = .5$, find the corresponding statistics for $(U, V)$, where $U = 32 + 9X/5$ and $V = 2.2Y$.

**∗ 7-37.** Plot these points in a scatter diagram: $(1, 1)$, $(1, 2)$, $(2, 1)$, $(2, 2)$, $(4, 4)$ and calculate the correlation coefficient

   **(a)** for all five points.

   **(b)** for the first four points [omitting $(4, 4)$]. In this case you may be able to find $r$ by inspection.

(The point of this exercise is to show the tremendous influence one point can have on the correlation coefficient, but the possibility of such influential points is not restricted to small data sets.)

**7-38.** Ten states were selected at random (using a random number table). For these states, the per capita personal income (in dollars) and the divorce rate (per thousand) were given as follows in the 1980 census:

| State | Income | Divorce rate |
|---|---|---|
| Tennessee | 7,720 | 6.8 |
| California | 10,938 | 5.8 |
| Massachusetts | 10,125 | 2.9 |
| Wisconsin | 9,348 | 3.7 |
| Michigan | 9,950 | 4.4 |
| Nebraska | 9,365 | 4.1 |
| Mississippi | 6,580 | 5.5 |
| Colorado | 10,025 | 6.4 |
| Pennsylvania | 9,434 | 3.0 |
| North Dakota | 8,747 | 3.3 |

Plot income-divorce rate pairs in a scatter diagram and calculate $r$. How would $r$ change if both Tennessee and Mississippi were omitted? [The correlation for all 50 states, the population from which these ten were drawn, is about .02.]

**7-39.** Examples 7.2d and 7.7a dealt with data on 15 American law schools. The data can be simplified somewhat by subtracting 500 from each LSAT score and subtracting 3 from each GPA and multiplying by 100: $U = X - 500$, $V = 100(Y - 3)$. Do this, calculate $r$ from the pairs $(U, V)$, and so verify that $r$ is unchanged.

**7-40.** Defining the sample distribution by putting mass $1/n$ at each data point $(X, Y)$, we interpret an expected value as an arithmetic mean. With this in mind, recast Schwarz's inequality (6) of §3.4 as an inequality involving sums of squares and products.

**7-41.** Show that the mean deviation is never larger than the standard deviation, when the latter is defined with divisor $n$ (rather than $n - 1$),

    **(a)**  using the inequality obtained in Problem 7-40.

    **(b)**  as the corresponding fact was shown for probability distributions in Problem 3-22.

**7-42.** Referring to the data in Example 7.4c, calculate the linear correlation coefficient of the direct and indirect measurements and also with the seven cases omitted as in Figure 7-15. Can you explain, in view of the scatter diagrams, why the latter is smaller than the former?

## Review Problems

**7-R1.** Given $n = 10$, $\sum X_i = 5$, and $S^2 = 3/2$, find $\sum X_i^2$.

**7-R2.** Given $\sum X^2 = 198$, $S^2 = 12$, $n = 10$, find two possible values of $\overline{X}$.

**7-R3.** Show that the second sample moment $V = \frac{n-1}{n} S^2$ can be calculated using all differences:

$$V = \frac{1}{2n^2} \sum_i \sum_j (X_i - X_j)^2.$$

**7-R4.** The following are numbers of cars per minute, recorded by a student on a county road in a metropolitan suburb in 20 one-minute periods between 12 noon and 12:30 P.M.:

    29, 22, 10, 25, 26, 21, 30, 24, 16, 27, 23, 31, 17, 15, 21, 21, 30, 18, 33, 19.

    **(a)**  Make a stem-leaf diagram with split stems.

    **(b)**  Find the median, range, and IQR, and make a box plot.

    **(c)**  Find the mean and standard deviation.

**7-R5.** Aerial surveys [reported at a 1963 conference on wildlife and resources] on each of 20 days in a certain area of the Alaska Peninsula yielded counts of the number $(Y)$ of black bears sighted, and the average wind velocity $(X)$:

| X | Y | X | Y | X | Y | X | Y |
|---|---|---|---|---|---|---|---|
| 2.1 | 99 | 11.8 | 76 | 10.5 | 79 | 30.6 | 47 |
| 16.7 | 60 | 23.6 | 43 | 18.6 | 57 | 13.5 | 73 |
| 21.1 | 30 | 4.0 | 89 | 20.3 | 54 | 14.0 | 72 |
| 15.9 | 63 | 21.5 | 49 | 11.9 | 69 | 6.9 | 84 |
| 4.9 | 82 | 24.4 | 36 | 6.9 | 87 | 27.2 | 23 |

    **(a)**  Make a scatter plot.

    **(b)**  Use a computer or statistical calculator to find summary statistics (means, s.d.'s, correlation).

**7-R6.** Exposure to Agent Orange resulted in high levels of a certain dioxin in blood and fat tissue. The following are reported TCDD levels of 14 Vietnam veterans (parts/trillion) who may have been exposed.[19]

Plasma: 2.5, 3.5, 6.8, 4.7, 4.6, 1.8, 2.5, 3.1, 3.1, 3.0, 6.9, 1.6, 20, 4.1
Fat:      4.9, 6.9,  10, 4.4, 4.6, 1.1, 2.3, 5.9, 7.0, 5.5, 7.0, 1.4, 11, 2.5.

**(a)** Devise a graphical display for comparing the distributions of dioxin levels in blood plasma and in fat.
**(b)** The data are actually paired: (2.5, 4.9) for subject 1, (3.5, 6.9) for 2, etc. Make a scatter plot. (And if this is done on a computer, calculate the correlation coefficient.)
**(c)** Calculate the 14 differences in level and construct a graphical display.

**7-R7.** Two methods of determining fat content of meat were tried on eight samples of meat, with these results:

| Sample: | 1 | 2 | 3 | 4 | 5 | 6 | 7 | 8 |
|---|---|---|---|---|---|---|---|---|
| A: | 23.1 | 23.2 | 26.5 | 26.6 | 27.1 | 48.3 | 40.5 | 23.0 |
| B: | 22.7 | 23.6 | 27.1 | 27.8 | 27.4 | 46.8 | 40.4 | 24.9 |

**(a)** Find the difference in sample means in two ways.
**(b)** Would you expect the results of the two methods to be correlated? Find the correlation coefficient.

**7-R8.** The correlation coefficient calculated from ten pairs $(X, Y)$ is $r = .8$. Suppose we transform the data to $U_i = X_i + 7$ and $V_i = 2Y_i - 3$. Given $S_X = 3$ and $S_Y = 5$, find
**(a)** $S_U$ and $S_V$.     **(b)** $r_{U,V}$.

**7-R9.** Given the hemoglobin readings in the accompanying stem-leaf diagram,

| Stem | Leaves (tenths) |
|---|---|
| 9 | 1 |
| 10 | 7917 |
| 11 | 0434463 |
| 12 | 793104 |
| 13 | 562555818673 |
| 14 | 8622280826 |
| 15 | 6100407 |
| 16 | 393 |

**(a)** find the median.
**(b)** find the interquartile range.
**(c)** construct a box plot.

---

[19]A. Schecter et al., "Partitioning of 2, 3, 7, 8-Chlorinated Dibenzo-p-Dioxins..." *Chemosphere* **20** No. 709 (1990), 954–55.

**7-R10.**  Among the information collected by a California HMO on 72 mothers and their new-born children were the following birthweights (lbs), arranged according to whether the mother never smoked or had smoked:[20]

Never:   3.3, 5.3, 5.6, 5.6, 5.6, 6.1, 6.1, 6.1, 6.5, 6.6, 6.6, 6.6, 6.6, 6.7,
        6.7, 6.9, 7.1, 7.1, 7.1, 7.1, 7.3, 7.4, 7.4, 7.6, 7.8, 7.8, 7.8, 7.8,
        7.8, 7.9, 8.5, 8.6, 8.6, 8.6, 8.8, 9.2, 9.2, 10.9

Smoked:  4.5, 5.4, 5.4, 5.6, 5.9, 6.0, 6.1, 6.4, 6.6, 6.6, 6.6, 6.6, 6.8, 6.8,
        6.9, 6.9, 6.9, 7.1, 7.2, 7.2, 7.3, 7.4, 7.5, 7.6, 7.6, 7.8, 8.0, 9.9

**(a)**  Construct back-to-back stem-leaf diagrams. [Omit the depth columns and extend leaves for one to the right and for the other to the left of the stem.]

**(b)**  Construct side-by-side box plots.

**7-R11.**  Show that the sum $\sum |X_i - a|$ has its minimum value at a median of the sample distribution. [*Hint:* Differentiate the sum, using

$$\frac{d|x|}{dx} = \frac{d\sqrt{x^2}}{dx} = \frac{x}{|x|} = \text{sign of } x,$$

and examine the value of the sum of the derivatives of $|X_i - a|$ as $a$ increases through the sample values.]

## Chapter Perspective

This chapter is different in many ways from those that precede it. It assumes a given data set—everything is known and nothing is random. Thus, the word *probability* appears in this chapter only to relate the calculation of sample moments to those of population moments.

Just as population parameters give us important features of a population, the descriptive methods developed in this chapter are useful for simply understanding the nature of a data set. The various displays, graphs, and sample statistics help us comprehend what the data are like. They are useful in *exploratory data analysis*—in an informal examination of data with various tools, highlighting features that may (or may not) be features of the population that are interesting and important. These can then be investigated further with additional experimentation, to see if they are merely peculiar to a particular sample or are something "real."

The usual reason for gathering sample information is to learn something about the population that was sampled. The various sample characteristics that we've introduced will be used in the following chapters as the basis for drawing conclusions about the population from which the samples are drawn.

When the result of doing an experiment or of observing a characteristic of an individual selected from a population involves uncertainty—uncertainty that we represent with a probability model—the result is a random variable. So each observation in the sample is a random variable, and there is variation from sample to sample. Any

---

[20]Reported in *Stat Lab*, by Hodges, Krech, and Critchfield (New York: McGraw-Hill, 1975).

one sample will not be exactly like any other sample, and neither will be just like the population. Yet, there is information about the population in any particular sample, so it is reasonable to expect that one can infer something about the population from the sample, and perhaps act on that basis. Such inference is of course subject to error, because the sample is an incomplete picture of the population. But there is no way around that possibility, so all we can hope to do is to try to assess the error and to minimize it by a judicious choice of procedure.

Extrapolating from sample characteristics to unknown population characteristics is the fundamental problem of statistical inference. It is this problem to which we now turn. In Chapter 8 we investigate how much and what kind of information is available, what sample statistics are most appropriate to a particular problem of inference, and how these sample characteristics vary from sample to sample when we assume random sampling.

# Samples, Statistics, and Sampling Distributions

Scientists, market analysts, and others sample populations to learn about various population characteristics. A politician wants to know the opinion of constituents; a dairy scientist is interested in the butterfat content of milk from cows that have been fed a certain diet; a physician needs to know how patients of a certain type respond to a treatment; and so on. Because it is usually impractical or even impossible to examine the whole population, they must use sample information. Based on what a sample can say about the population, they may want to estimate population parameters, test hypotheses about population characteristics, predict responses of future subjects, or make decisions. These processes (estimating, testing, predicting, and so on, based on sample information) are examples of **statistical inference.**

In most applications, the statistician is concerned with a particular population characteristic, rather than with every aspect of a population distribution. A *population* characteristic is called a **parameter;** one example is the population mean. Samples also have descriptive characteristics, and a characteristic of a sample is called a **statistic.**

How best to summarize sample information in a way that is appropriate for a given problem of inference is seldom completely clear. Intuition suggests that when a problem is stated in terms of a particular population parameter, the corresponding sample characteristic (if there is one) is an obvious choice of summary statistic. Intuition will also give us the notions of *likelihood* and *sufficiency* (§8.2 and §8.3), which will help us find appropriate statistics—for whatever specific type of inference we are to make.

The value of any sample statistic depends on the particular sample one happens to obtain. Typically, it *varies from sample to sample.* In mathematical terms, a sample statistic is a *function* of the sample observations:

---

A function $g$ of the sample observations $X_1, ..., X_n$ defines a **statistic:**

$$T = g(X_1, ..., X_n).$$

---

A statistic $T$, depending as it does on random variables $X_1, ..., X_n$, is a random variable—or a random vector, if $T$ is a vector of functions. As such, a statistic has a probability distribution, called its **sampling distribution.** Much of this chapter is devoted to obtaining and describing sampling distributions of commonly used statistics. If we are to use a statistic $T$ in a problem of inference, we need to know the nature and extent of its variation from sample to sample. An essential tool in our development of sampling distributions is the *moment generating function* (§5.12).

## 8.1   Random Sampling

A sample is considered representative if it accurately reflects the population characteristics of interest. Unfortunately, no method of sampling short of taking the whole population can guarantee a representative sample. Some methods are better than others at producing nearly representative samples, but many (including some in common use) produce samples that are usually quite unlike the population. For instance, samples are apt to be biased if they contain only individuals with similar characteristics, such as having the same religion, coming from the same hometown, or being at a cocktail party together. And samples made up of those who volunteer to be included are notoriously misleading, but are in common use.

**Example 8.1a**  | **A Voluntary Response Sample**

In early 1988, "Dear Abby" asked her readers to write to her about whether they cheated on their spouses. She received 210,336 responses—from 149,786 females and 60,550 males. "The results were astonishing," she writes. "There are far more faithfully wed couples than I had surmised....The marriage vow...is still honored by 85% of the females and 74% of the males who responded." She seems to assume that her voluntary response sample is representative of married folks generally, but it is almost surely not. ∎

Roughly speaking, *random sampling* (as introduced in Chapter 1) has a good chance of producing a sample that is close to representative—especially if the sample is large. Unfortunately, no sampling method can guarantee perfect accuracy of representation, but random sampling has the advantage that it allows us to assess the *degree* of accuracy.

Chapter 1 introduced the idea of sampling at random from a finite population. In the lottery model for such a process, each ticket represents a population member.

Consider selecting one ticket at a time, drawing blindly from the available tickets. In random sampling *with replacement,* we replace each ticket and mix the tickets thoroughly before selecting the next one. In random sampling *without replacement,* tickets selected are not replaced; so they are not available for subsequent selections.

---

**Random sampling** from a finite population:   Individuals are selected at random, one at a time from those available, either with replacement and mixing or without replacement.

---

We saw in Chapter 1 that random sampling (either with or without replacement) has these properties:

**(i)**   All possible samples are equally likely, and (as a result)
**(ii)**   All population members have the same chance of being included.

In dealing with finite populations, we use the number of possible samples as the denominator in calculating probabilities; this number may be quite different under sampling with replacement than under sampling without replacement.

The population of interest is sometimes too unwieldy for random sampling. For example, although one can imagine selecting an individual at random from the population of the United States, it is almost impossible actually to do so. Opinion polls, TV viewer surveys, and so on, employ methods that are harder to describe but easier to carry out than random sampling. For instance, poll takers may divide the population into *strata*—say by locale, age, or income level—and sample at random within each stratum. Another technique is to *cluster* into convenient groups such as households or precincts, sample the clusters, and then sample some or all individuals from each cluster selected. These variations are used partly out of practical necessity and partly to reduce sampling error.[1]

We return to (strictly) random sampling. Suppose we are interested in a particular measurement $X$ on the individuals in a population. The value of $X$ for an individual selected at random from the population is a random variable, and we refer to its distribution as the **population distribution.** A sample of size $n$ is a sequence: $\mathbf{X} = (X_1, X_2, \dots, X_n)$, where $X_i$ is the value of $X$ for the $i$th individual selected. The variables $X_i$ are *exchangeable* (§2.6 and §5.8), whether sampling is done with or without replacement. In particular, the individual variables $X_i$ have a common distribution—the population distribution.

**Example 8.1b** | **Checking Lot Quality**

Consider sampling from a lot consisting of 20 articles—a "population" of size 20. Suppose that two of the 20 are defective, and the rest good. To determine whether or

---

[1]See W. G. Cochran, *Sampling Techniques,* 2nd Ed. (New York: Wiley, 1963) for more precise descriptions of such variants of sampling and their advantages and disadvantages.

not an article is defective, it must be tested. Let

$$X = \begin{cases} 1 & \text{if the selected article is defective,} \\ 0 & \text{if it is good.} \end{cases}$$

(This is the indicator function of the event "defective.") The distribution of $X$ is the population distribution, with p.f. $f(1) = \frac{2}{20}$, $f(0) = \frac{18}{20}$.

A random sample of three articles produces a sequence of 1's and 0's: $(X_1, X_2, X_3)$. In sampling with *or* without replacement, the $X_i$ are exchangeable, and each has the distribution of $X$. With no replacement, the possible sample sequences are as follows: 110, 101, 011, 001, 010, 100, 000. (The sequence 111 is impossible in this case.) The sampling scheme implicitly defines the probability of each of these sample sequences. For instance, by the multiplication rule of §2.3,

$$P(001) = \frac{18}{20} \cdot \frac{17}{19} \cdot \frac{2}{18}.$$

Similar calculations yield the probabilities of the other possible samples, and these are shown (to three decimal places) in the table below.

In sampling *with* replacement, there are eight possible sequences, since it is then possible to observe 111. In this case, the sequence elements are not only exchangeable but also independent. The probability of each sequence is the product of three factors: a factor $\frac{18}{20}$ for each 0 and a factor $\frac{2}{20}$ for each 1. These sequence probabilities are also shown in the following table:

| Sequence | 000 | 001 | 010 | 100 | 110 | 101 | 011 | 111 |
|---|---|---|---|---|---|---|---|---|
| *With Replacement* | .729 | .081 | .081 | .081 | .009 | .009 | .009 | .001 |
| *Without Replacement* | .716 | .089 | .089 | .089 | .005 | .005 | .005 | 0 |

Sampling finite populations with replacement is not commonly done. Indeed, in some settings it is impossible—tests can be destructive, as in testing automobile tires, bullets, or flashbulbs. However, as we saw in the case of Bernoulli populations in §4.3, the simpler formulas based on sampling with replacement are usually adequate when the sample size is much smaller than the population size. In most of this chapter we assume random sampling with replacement.

We have seen that in sampling with replacement and mixing, the sample observations are independent and have a common distribution. When the "population" is only a conceptual one, it is often reasonable to assume (at least as a first approximation) that successive observations are independent and identically distributed. Examples of this sort are repeated measurements of time, mass, temperature, and the like—in general, the results of repeated trials of the same experiment. We refer to any sequence of independent variables with a common distribution as a *random sample.* (Although our definition of *random sample* is one commonly used, a sample produced by "random sampling" is a random sample in this sense only if sampling is done *with* replacement.) Whether or not there is a real, tangible population, we refer to that common distribution as the *population distribution*.

| Example **8.1c** | ## Standard Weights |
|---|---|

One of the standard weights used by the National Bureau of Standards is made of a chrome-steel alloy and has a nominal weight of 10 grams—about the weight of two nickels. Such standard weights are used in calibrating the various standard weights at the state and local levels. To assess the variability in its weighing apparatus, the Bureau weighs this 10-gram weight repeatedly (once a week). Freedman, Pisani, and Purves[2] report 100 measurements made in the years 1962–1963. The first 20 of these are as follows, given as numbers of micrograms below the nominal weight of 10 grams:

409   400   406   399   402   406   401   403   401   403

398   403   407   402   401   399   400   401   405   402

Successive measurements are assumed to be the result of sampling at random from the "population" of all possible measurements.   ■

---

A *random sample* of size $n$ is a sequence of independent observations $\mathbf{X} = (X_1, X_2, \ldots, X_n)$, each of which has the distribution of the population being sampled.

---

The mathematical description of a random sample is especially simple: Because of the independence, the joint p.d.f. (continuous case) or p.f. (discrete case) of the sample observations is the *product* of their marginals. The marginal distribution of each observation is the population distribution.

---

The joint p.d.f. (or p.f.) of the observations in a random sample is

$$f(x_1, \ldots, x_n) = f_X(x_1) \cdots f_X(x_n) = \prod_{i=1}^{n} f_X(x_i), \qquad (1)$$

where $f_X$ is the population p.d.f. (or p.f.).

---

[2]*Statistics* (New York: W. W. Norton, 1978).

| Example **8.1d** | **Normal Population with Known Variance** |
| --- | --- |

Consider a random sample $(X_1, \ldots, X_n)$ from $\mathcal{N}(\mu, 1)$. The population p.d.f. [from (7) of §6.1] is

$$f_X(x) = \frac{1}{\sqrt{2\pi}} \, e^{-(x-\mu)^2/2}.$$

The joint p.d.f. of the sample observations is the product

$$f(x_1, \ldots, x_n) = \prod_1^n \frac{1}{\sqrt{2\pi}} \, e^{-(x_i-\mu)^2/2} = \frac{1}{(2\pi)^{n/2}} \exp\left\{-\frac{1}{2}\sum_1^n (x_i - \mu)^2\right\}. \qquad \blacksquare$$

| Example **8.1e** | **Sampling Without Replacement** |
| --- | --- |

Suppose we draw a sample of size $n$ by sampling without replacement from a Bernoulli population of size $N$ that includes $M$ 1's and $N - M$ 0's. The joint p.f. is the probability of $[X_1 = x_1, \ldots, X_n = x_n]$, where each $x_i$ is 0 or 1. Since the $X$'s are exchangeable, this probability is the same for each possible sample sequence that has a specified number of 1's. The probability of a specified number of 1's is hypergeometric, so the probability of a particular sequence is a hypergeometric probability divided by the number of sequences with that same number of 1's, that is, by $\binom{n}{\Sigma x}$:

$$f(x_1, \ldots, x_n \mid M) = \frac{\binom{M}{\Sigma x}\binom{N-M}{n-\Sigma x}}{\binom{n}{\Sigma x}\binom{N}{n}}.$$

For instance, if $N = 10$, $M = 3$, and $n = 5$, the probability of the sequence $(1, 0, 1, 0, 0)$ is

$$f(1, 0, 1, 0, 0 \mid M) = \frac{3}{10} \cdot \frac{7}{9} \cdot \frac{2}{8} \cdot \frac{6}{7} \cdot \frac{5}{6} = \frac{\binom{3}{2}\binom{7}{3}}{\binom{5}{2}\binom{10}{5}}. \qquad \blacksquare$$

## 8.2 Likelihood

The term *likelihood* has a specialized meaning in statistics, something like its ordinary meaning in English (likely: "reasonably to be believed or expected").

| Example **8.2a** | **Unusual Dice** |
| --- | --- |

I have two pairs of dice. Pair #1 is ordinary, the six faces of each die being marked with dots numbering from 1 to 6, and we assume these outcomes to be equally likely. Pair #2 is specially made: One die has five dots on each face, and the other has two dots on four faces and six dots on the other two faces. Assuming the faces of pair #2 to be equally likely, I can throw only a 7 (with probability 4/6), or an 11 (with probability 2/6).

You watch as I toss a pair of dice, and I throw a 7. Which pair did I use? The probability of the 7 is 4/6 with the special pair but only 1/6 with the ordinary pair. Knowing the two possible explanations of what happened, and knowing that the probability with the special pair is four times what it is with the ordinary pair, we say that the special pair is four times as "likely" as the ordinary pair to be the one I threw.

The numbers 4/6 and 1/6 are probabilities, but they are probabilities in different models. They don't add up to 1, and indeed, their sum is irrelevant. To calculate a *probability* that I used the special dice, given what you saw, you'd need to know how I selected the pair to use. For instance, if you knew I selected a pair at random, you could use Bayes' theorem. (You did this kind of calculation in Chapter 2.) ∎

We shall deal mainly with models (population distributions) that are identified by a parameter $\theta$ (say), and write $f(x \mid \theta)$ for the population p.d.f. or p.f. A parameter $\theta$ indexes a family of models in the sense that each possible value of $\theta$ defines a different model.

As in Example 8.2a, having observed $\mathbf{x}$, we take the likelihood of a discrete population model $\theta$ to be the probability in that model of obtaining $\mathbf{x}$: $L(\theta) = f(\mathbf{x} \mid \theta)$. (See page 203 for the boldfaced notation $\mathbf{x}$.)

Likelihoods can be compared by forming ratios. Thus, we may say, in view of certain data, that $\theta_1$ is twice as "likely" as $\theta_2$. In a ratio of the likelihoods of two values of $\theta$, any common factor not involving $\theta$ cancels. For this reason, we modify the definition slightly and consider two likelihoods the same if they differ only by a factor not involving $\theta$. Thus, likelihood is defined only up to a multiplicative constant. With this understanding we'll refer to "the" likelihood function, perhaps dropping any factors not depending on $\theta$.

When the population is continuous, the probability of getting what has been observed is 0. However, in a limiting sense, the probability "at" $\mathbf{x}$ is $f(\mathbf{x} \mid \theta)d\mathbf{x}$, and we drop the $d\mathbf{x}$ to define the likelihood for the continuous case.

---

Given a sample $\mathbf{x} = (x_1, ..., x_n)$ from a population with p.d.f. or p.f. $f(x \mid \theta)$, the **likelihood function** $L$ is any function of $\theta$ proportional to the joint p.d.f. or p.f.:

$$L(\theta) \propto f(x_1, ..., x_n \mid \theta). \tag{1}$$

---

**Example 8.2b** | **Bernoulli Trials**

Consider a sequence $(X_1, X_2, ..., X_{10})$ of ten independent Bernoulli trials with parameter $p$. If we observe six successes followed by four failures, what is the likelihood function $L(p)$? It is the probability, as a function of $p$, of getting what we got: $p^6(1 - p)^4$.

More formally, the p.f. of a Bernoulli variable [(1) of §4.1] is

$$f(x \mid p) = p^x(1-p)^{1-x}, \quad x = 0, 1.$$

The joint p.f. of the sequence $\mathbf{X}$ is the product of the individual p.f.'s:

$$L(p) = f(x_1, ..., x_{10} \mid p) = \prod_{i=1}^{10} f(x_i \mid p) = p^6(1-p)^4, \ 0 \le p \le 1.$$

The graph of this likelihood function is shown in Figure 8-1. Observe that the values of $p$ near .6 have the highest likelihoods.

Suppose we were informed only that in ten trials there were six successes: $Y = 6$, where $Y = \sum X_i \sim \mathrm{Bin}(n, p)$. This includes the possibility that the first six trials resulted in success, but there are other possibilities as well. The binomial p.f. at $y = 6$ is

$$f_Y(6 \mid p) = \binom{10}{6} p^6(1-p)^4, \ 0 \le p \le 1.$$

This differs from the likelihood function we found above by the factor $\binom{10}{6}$. But it is the same according to our convention of ignoring constant factors.

$L(p)$

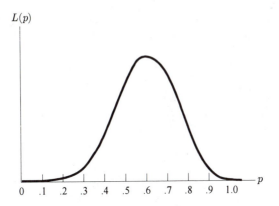

**Figure 8-1** $L(p)$, six successes in ten Bernoulli trials

Now, suppose that instead of conducting trials to the specified number 10, we had planned to carry out as many trials as needed to obtain six successes—and found that it took ten trials. The probability function for $W$, the number of trials needed, is negative binomial [(3) of §4.4]:

$$f_W(10 \mid p) = \binom{9}{5} p^6(1-p)^4, \ 0 \le p \le 1.$$

The likelihood function is again $p^6(1-p)^4$. The binomial and negative binomial experiments use different stopping rules; but with six successes and four failures, the likelihood function is the same in both cases.     ■

**Example 8.2c**

## Normal $X$ with Known Variance (Continued)

As in Example 8.1d, let $(X_1, \ldots, X_n)$ be a random sample from $\mathcal{N}(\mu, 1)$. The joint p.d.f. of the $X$'s given in that example is the likelihood function:

$$L(\mu) = f(\mathbf{x} \mid \mu) = \frac{1}{(2\pi)^{n/2}} \exp\left\{-\frac{1}{2}\sum_1^n (x_i - \mu)^2\right\}.$$

We can then drop the factor $(2\pi)^{-n/2}$ in writing the likelihood function. The dependence on $\mu$ becomes clearer when we expand the square in the exponent and complete the square in $\mu$ by adding and subtracting $n\bar{x}^2/2$:

$$\exp\left\{-\frac{1}{2}\sum_1^n (x_i - \mu)^2\right\} = \exp\left\{-\frac{1}{2}\sum_1^n x_i^2 + \mu\sum_1^n x_i - \frac{1}{2}n\mu^2\right\}$$

$$= \exp\left\{-\frac{n}{2}(\mu^2 - 2\mu\bar{x} + \bar{x}^2) - \frac{1}{2}\sum_1^n x_i^2 + \frac{n}{2}\bar{x}^2\right\}.$$

The last two terms in the exponent do not involve $\mu$, so they define factors that are constant (in $\mu$) and can be ignored in defining likelihood. The likelihood function can thus be written in the following simple form:

$$L(\mu) = \exp\left\{-\frac{n}{2}(\mu - \bar{x})^2\right\}. \tag{2}$$

We see from (2) that the shape of the likelihood function is that of a normal p.d.f. centered at the value $\mu = \bar{x}$. The graph is shown in Figure 8-2 for the case $n = 9$ and $\bar{x} = 2$.     ■

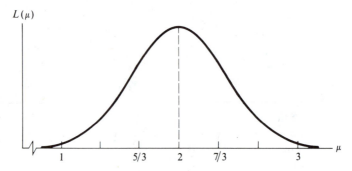

**Figure 8-2** A normal likelihood

**Example 8.2d** | **Multinomial Sampling**

The biologist Gregor Mendel (1822–1884)—considered the father of modern genetics—crossed round yellow pea plants with wrinkled green pea plants, obtaining plants bearing peas in one of four categories. The categories and the numbers of plants in each category are as follows:

| Category | Frequency | Probability |
|---|---|---|
| Round yellow | 315 | $p_1$ |
| Round green | 108 | $p_2$ |
| Wrinkled yellow | 101 | $p_3$ |
| Wrinkled green | 32 | $p_4$ |

The probability of the experimental result, given category probabilities $p_i$, is multinomial [(3) of §4.8]:

$$\binom{556}{315,\ 108,\ 101,\ 32}\ p_1^{315} p_2^{108} p_3^{101} p_4^{32}.$$

We can take this as the likelihood function—or drop the constant multiplier. More generally, for a variable with $k$ categories and sample frequencies $f_i$, the likelihood function is

$$L(\mathbf{p}) = p_1^{f_1} \cdots p_k^{f_k}. \tag{3}$$

∎

**Example 8.2e** | **Sampling $\mathcal{U}(0, \theta)$**

A biologist gathers $n$ individual organisms to learn about the maximum weight of such organisms. She assumes that they grow to their maximum linearly with time, and then die. (All assumptions in science are subject to criticism and revision; this particular one may be more open to objection than most, but she has to assume something to get started.) She assumes further that the individual weights constitute a random sample from a population whose distribution is $\mathcal{U}(0, \theta)$, where $\theta > 0$. The population p.d.f. is then

$$f_X(x \mid \theta) = \frac{1}{\theta}, \quad 0 < x < \theta, \tag{4}$$

and each sample observation $X_i$ has this population distribution.

Given a random sample of size $n$, the likelihood function is the joint p.d.f. of the observations:

$$L(\theta) = \prod_1^n f_X(x_i \mid \theta) = \begin{cases} 1/\theta^n, & 0 < x_i < \theta \text{ for all } i, \\ 0, & \text{if any } x_i > \theta. \end{cases}$$

The condition "$0 < x_i < \theta$ for all $i$" is satisfied if and only if the *largest* of the $x_i$'s is less than $\theta$, or $\theta > x_{(n)}$; thus, the likelihood function can be written as

$$L(\theta) = \begin{cases} 1/\theta^n, & \theta > x_{(n)}, \\ 0, & \text{otherwise.} \end{cases}$$

This is shown in Figure 8-3 for $n = 5$ and $x_{(n)} = .9$. (To be continued.) ■

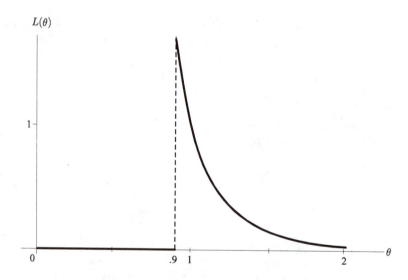

**Figure 8-3** Likelihood function for Example 8.2e

Each possible sample $(x_1, ..., x_n)$ defines the likelihood $L(\theta)$ as a function. Before we observe a particular one of the possible sample sequences, we think of the sample as a sequence of random variables: $(X_1, ..., X_n)$. The likelihood function is then a random function—a *statistic*. Each particular sample sequence results defines a corresponding likelihood function.

It is the conviction of many statisticians that everything in the sample relevant to the unknown $\theta$ is embodied in the likelihood function. They accept the following:

> **Likelihood principle:** If different experiments based on the model defined by $\theta$ result in the same likelihood function, one should draw the same inference— come to the same conclusion or take the same action.

(Example 8.2b is an example in which different experiments lead to the same likelihood.)

The notion of likelihood will play an important role throughout the following chapters. In the next section we show how an examination of the form of a likelihood function can lead to appropriate summary statistics.

## Problems

**8-1.** Write the joint density function of four observations in a random sample from a population whose distribution is

* **(a)** $\mathcal{U}(-1, 1)$ [uniform on the interval from $-1$ to $1$].
* **(b)** $\text{Exp}(2)$ [with p.d.f. $2e^{-2x}$ for $x > 0$].
  **(c)** $\mathcal{N}(10, 2)$ [normal with mean 10 and variance 2].

**8-2.** Write an expression for the joint probability function of five independent observations on a random variable whose distribution is

  **(a)** $\text{Bin}(3, \frac{1}{4})$ [binomial with $n = 3$ and $p = \frac{1}{4}$].
  **(b)** $\text{Geo}(\frac{1}{2})$ [geometric with $p = \frac{1}{2}$].

* **8-3.** Suppose, in sampling from $\text{Exp}(\lambda)$, you obtain the sample $(2, 4, 7)$. Find and sketch the likelihood function, $L(\lambda)$.

**8-4.** For each of the following populations, write the likelihood function, assuming a random sample of size $n$:

  **(a)** $\mathcal{N}(0, \theta)$.    **(b)** $\mathcal{N}(\mu, \theta)$.

* **8-5.** Suppose we interview people one at a time, recording $X_1, X_2, \ldots$, where $X_1$ is the number of interviews it takes to find a person with type B blood, $X_2$ is the number of additional interviews it takes to find a second person with type B blood, and so on. Consider each interview as a Bernoulli trial in which $p$ is the population proportion with type B blood.

  **(a)** Give the joint probability function of the observations $(X_1, \ldots, X_n)$.
  **(b)** Suppose we find that it takes 30 interviews to locate three people who are type B. Give the likelihood function $L(p)$.

**8-6.** Write the likelihood function for a random sample of size 10 from a population whose distribution is

  **(a)** Cauchy, with location parameter $\theta$: $f(x \mid \theta) = \frac{1}{\pi} [1 + (x - \theta)^2]^{-1}$.
  **(b)** Gamma, with parameters $(\alpha, \lambda)$: $f(x \mid \alpha, \lambda) = \frac{\lambda^\alpha}{\Gamma(\alpha)} x^{\alpha-1} e^{-\lambda x}$, $x > 0$.
  **(c)** Truncated exponential: $f(x \mid \theta) = \frac{e^{-x}}{1 - e^{-\theta}}$ for $0 \le x \le \theta$, and 0 elsewhere.

* **8-7.** Consider sampling one at a time at random, without replacement, from a carton of 12 items including an unknown number $M$ of defective items.

**(a)**   You decide to stop sampling after drawing four items and find two defective items among the four. Give the likelihood function $L(M)$.

**(b)**   You decide to stop after finding a second defective item and note that it has taken four selections to do this. Give the likelihood function $L(M)$.

**(c)**   Find the value of $M$ for which $L(M)$ in (a) has its maximum value.

**8-8.**   Consider two persons, each observing a Poisson process with rate parameter $\lambda$ (counts per hour). One plans to watch until ten events have occurred, recording the total elapsed time $T$ (in hours); the other plans to watch for one hour and record $X$, the number of events in that hour. Suppose the first person finds $T = 1$ and the second person observes $X = 10$. Compare the likelihood functions.

---

## 8.3        Sufficient Statistics

At the end of the preceding section it was suggested that all we need from a sample, in drawing inferences about a model, is the likelihood function—that this is sufficient, and that we don't need any more detailed information, such as the individual observations in the sample. This is at the heart of the notion of *sufficiency*.

**Example 8.3a** | **Sampling $\mathcal{N}(\mu, 1)$**

In Example 8.2c we found the likelihood function for a random sample of size $n$ from $\mathcal{N}(\mu, 1)$:

$$L(\mu) = \exp\left\{ -\frac{n}{2}(\mu - \overline{X})^2 \right\}.$$

This depends on the sample only through the value of $\overline{X}$ (and $n$). So we could throw away the raw data, keeping only the value of $\overline{X}$—and still calculate the likelihood function. In this case it is enough to know $\overline{X}$. However, the sample mean is not always sufficient for inference about a population mean, as the next example shows.   ■

**Example 8.3b** | **An Unusual Scale**

Imagine a scale that registers only either a pound more or a pound less than a person's actual weight, so that these erroneous readings are equally likely. Let $\theta$ denote an actual weight, so that the reading $X$ has the value $\theta + 1$ or the value $\theta - 1$, each with probability $\frac{1}{2}$. The mean value of $X$ is $\mu = \theta$.

Given (say) 25 observations $X_i$, consider two candidates for estimating $\theta$: the sample mean, and the sample midrange (halfway between the largest and smallest). If all the $X_i$ are equal, which happens with probability $2/2^{25}$, the two estimates would be the same, both wrong by a pound, one way or the other. If they are not all equal, then the midrange gives the correct weight $\theta$, but the sample mean is wrong. For instance, if the sample includes eleven 144.4's and fourteen 142.4's, the sample mean is 143.28—which is wrong. The midrange is 143.4—which is the true weight $\theta$.   ■

The law of large numbers suggests that the mean of a random sample would be a good estimate of the population mean. And so it is, for large samples. But in the preceding example there is a better way to estimate the mean. One reason is that in that setting, the mean does not have the property of *sufficiency,* which we are about to define.

Most statistical methods involve reducing the sample data to just a few numbers—to a statistic of low dimension. Such reductions may lose important information. Roughly speaking, a statistic is **sufficient,** despite such reduction as it may involve, provided it contains all of the useful information. However, if $T$ contains this information, so does any one-to-one function of $T$: If you know one, you can calculate the other—both are sufficient, if one is.

A statistic $T$ is a function $t(X_1, \ldots, X_n)$ of the sample observations, and each possible value $t_0$ identifies a set of possible samples—the set of all **X** such that $t(\mathbf{X}) = t_0$, called the pre-image of the value $t_0$. The union of all these disjoint pre-images is the whole sample space, so the collection of all pre-images is a *partition* of the sample space. [Recall the definition of "partition" from §1.3, page 15.] Two statistics that define the same partition are thereby in one-to-one correspondence, so they are equivalent in their usefulness for inference. Sufficiency is thus a property of partitions, but the term is applied to any statistic defining a sufficient partition.

The definition we are about to give is motivated by the notion that the likelihood function $L(\theta)$ embodies all of the information about $\theta$ contained in the sample, so that it is sufficient to know the likelihood function.

---

A statistic $T = t(\mathbf{X})$ is **sufficient** for a family of distributions $\{f(\mathbf{x} \mid \theta)\}$ if and only if the likelihood function depends on **X** through the value of $T$:

$$L(\theta) = g[t(\mathbf{X}), \theta].\tag{1}$$

---

**Example 8.3c** | **Bernoulli Trials**

For a random sample of size $n = 3$ from Ber($p$), there are eight possible sample sequences:   111, 110, 101, 011, 001, 010, 100, 000. Given a particular sample sequence, the likelihood function is the joint probability of that sequence as a function of $p$:

$$L(p) = f(x_1, x_2, x_3 \mid p) = \prod_1^3 p^{X_i}(1 - p)^{1 - X_i} = p^{\Sigma X}(1 - p)^{3 - \Sigma X}.$$

Clearly, this likelihood function is determined by the value of the sample sum, so $Y = \Sigma X_i$ is sufficient for $p$.

Let $Y$ denote the number of successes, which is the sample sum. The partition defined by $Y$ consists of these four partition sets:

$$\{111\}, \ \{110, 101, 011\}, \ \{001, 010, 100\}, \ \{000\}.$$

Any other statistic in one-to-one correspondence with $Y$ would define the same partition, and is also sufficient for $p$. In particular, the mean is sufficient: $\overline{X} = Y/3$.  ∎

### Example 8.3d | The Order Statistic

Let $\mathbf{X}$ denote a sample obtained by random sampling, with or without replacement, from a population whose distribution has p.d.f. or p.f. $f(x \mid \theta)$. The joint p.d.f. or p.f. of $\mathbf{X}$ is a symmetric function of its arguments. That is, the components $X_i$ are *exchangeable* (see §2.6). Rearranging the arguments in the joint p.d.f. does not then change its value, so in particular,

$$L(\theta) = f_{\mathbf{X}}(X_1, ..., X_n \mid \theta) = f_{\mathbf{X}}(X_{(1)}, ..., X_{(n)} \mid \theta),$$

where $X_{(i)}$ is the $i$th smallest observation. Since this likelihood can be calculated if we know the order statistic $(X_{(1)}, ..., X_{(n)})$ and $\theta$, the order statistic is sufficient for $\theta$.

The practical meaning of this is that if the sampling is indeed random, we lose nothing of importance—nothing about $\theta$—when we put the observations in numerical order.  ∎

### Example 8.3e | The Exponential Family

Suppose a random variable $X$ has p.d.f. or p.f. in the following special form:

$$f(x \mid \theta) = B(\theta)h(x)e^{Q(\theta)R(x)}. \tag{2}$$

The family of distributions so defined is called the **exponential family.** It includes as special cases the family of binomial distributions, the family of exponential distributions, the family of normal distributions with known variance—and many others, including most of the special families we studied in Chapters 4 and 6.

The joint p.d.f. (p.f.) of a random sample of size $n$ from (2) is a product:

$$f(\mathbf{x} \mid \theta) = \prod B(\theta)h(x_i)e^{Q(\theta)R(x_i)} = B^n(\theta)e^{Q(\theta)\sum R(x_i)}\prod h(x_i).$$

Since the product of the $h(x_i)$ does not involve $\theta$, the likelihood function is

$$L(\theta) = B^n(\theta)e^{Q(\theta)\sum R(X_i)} = g\left[\sum R(X_i), \theta\right].$$

So we can calculate the likelihood function if we know $\sum R(X_i)$, which means that this statistic is sufficient for $\theta$.

As a special case, consider the p.f. for Ber($p$), defined (for $x = 0$ and 1) by the formula

$$f(x \mid p) = p^x(1-p)^{1-x} = \left(\frac{p}{1-p}\right)^x(1-p) = (1-p)\exp\left\{x\log\frac{p}{1-p}\right\}.$$

This is clearly a special case of (2), in which

$$\theta = p, \; B(p) = 1-p, \; Q(p) = \log\frac{p}{1-p}, \; R(x) = x, \text{ and } h(x) = 1.$$

It follows, then, that the statistic $\sum R(X_i) = \sum X_i$, the sample sum, is sufficient for $p$, as we saw in a particular case in Example 8.3c above.                    ■

**Example 8.3f**

**Sampling $\mathcal{U}(0, 1)$ (Continued)**
In Example 8.2e we found the likelihood function based on a random sample of size $n$ from $\mathcal{U}(0, \theta)$ to be

$$L(\theta) = \begin{cases} 1/\theta^n, & \theta > X_{(n)}, \\ 0, & \text{otherwise.} \end{cases}$$

Clearly, when given the value of $\theta$ and the value of $X_{(n)}$, we can calculate $L(\theta)$. That is, $L(\theta)$ is a *function* of those two variables:  $g[X_{(n)}, \theta]$. Thus, according to the criterion (1), the largest observation $X_{(n)}$ is sufficient for $\theta$.                    ■

**Example 8.3g**

**Sampling $\mathcal{N}(\mu, \sigma^2)$**
The family of normal distributions is indexed by two parameters: $\mu$ and $\sigma^2$. Given a random sample of size $n$, the likelihood function is

$$L(\mu, \sigma^2) = \exp\left\{ -\frac{n}{2}\log\sigma^2 - \frac{1}{2\sigma^2}\sum(X_i - \mu)^2 \right\}.$$

According to the parallel axis theorem [(5) of §7.6, with $c = \mu$],

$$\sum(X_i - \mu)^2 = nV + n(\overline{X} - \mu)^2,$$

where

$$V = \frac{1}{n}\sum(X_i - \overline{X})^2.$$

So we may write the likelihood function as follows:

$$L(\mu, \sigma^2) = \exp\left\{ -\frac{n}{2}\log\sigma^2 - \frac{1}{2\sigma^2}\left[ nV + n(\overline{X} - \mu)^2 \right] \right\}.$$

If we know $\overline{X}$ and $V$ (as well as $\mu$ and $\sigma^2$), we can calculate $L(\mu, \sigma^2)$—it is a function of the pair $(\overline{X}, V)$. It follows that $(\overline{X}, V)$ is sufficient for $(\mu, \sigma^2)$.    ■

In Example 8.2b, which involved Bernoulli trials, we saw that the likelihood function was the same whether we observed the sample data or the value of the sufficient statistic $\sum X_i$. This happens when the statistic is sufficient. We'll demonstrate this in the discrete case.

Consider a random sample $\mathbf{X}$ from a discrete population with indexing parameter $\theta$, and suppose that the statistic $T = t(\mathbf{X})$ is sufficient for $\theta$. When we observe that $\mathbf{X}$ has a value $\mathbf{x}$ such that $t(\mathbf{x}) = t_0$, the likelihood function is

$$L(\theta) = g[t(\mathbf{X}), \theta] = g(t_0, \theta).$$

On the other hand, suppose we observe just that $T = t_0$, rather than **X**. Then, because $T$ is sufficient, and using (1), we have

$$P(T = t_0 \mid \theta) = \sum_{T=t_0} f(\mathbf{x} \mid \theta) = \sum_{T=t_0} g[t(\mathbf{x}),\, \theta]\, h(\mathbf{x})$$

$$= g(t_0, \theta) \sum_{T=t_0} h(\mathbf{x}).$$

This is proportional to $L(\theta)$, so it is also $L(\theta)$, as asserted.

The converse is also true:  If $T$ is *not* sufficient, the likelihood function determined by **X** is different from the likelihood function determined by $T$.

A somewhat more general definition of sufficiency is the following:  The statistic $T$ is sufficient for $\theta$ if the conditional distribution in the sample space given the value of $T$ is independent of the parameter $\theta$. This is perhaps more intuitively appealing as a definition, as we'll explain with the aid of the example that follows. Subject to certain regularity conditions, the two definitions are equivalent, but it is usually easier to show sufficiency by showing (1). Calculations of the conditional distribution in the sample space given $T$ can be quite difficult when the population is continuous. Calculations in certain discrete cases are called for in Problems 8-12 and 8-16, and one is given in the next example.

The equivalence of the two definitions is usually expressed in what is commonly referred to as a "factorization criterion" for sufficiency.  This states that (under certain fairly mild conditions) $T = t(\mathbf{X})$ is sufficient in the more basic sense just given if and only if the p.f. or p.d.f. of the sample **X** is expressible in the factored form

$$f(\mathbf{x} \mid \theta) = g[t(\mathbf{x}, \theta]h(\mathbf{x}). \tag{3}$$

Since the function $h$ does not involve $\theta$, this is equivalent to (1).

| | |
|---|---|
| **Example 8.3h** | **Bernoulli Trials (Continued)** |

As in Example 8.3c, consider three independent trials of Ber($p$). We'll calculate the probability in the sample space of the observations $(X_1, X_2, X_3)$, *given the value of* $Y$, the number of successes among the three outcomes. Thus, if we know $Y = 2$, the sample sequence is one of those in the partition set found in Example 8.3c: $\{110, 101, 011\}$, and the probability of each of these sequences is $p^2 q$. The probability that $Y = 2$ is binomial: $3p^2 q$, so the conditional probability of the sequence 110 is

$$f(1,\, 1,\, 0 \mid Y = 2) = \frac{p^2 q}{3p^2 q} = \frac{1}{3}.$$

And the result is the same for the other two sequences with two successes. For each sequence in any of the partition sets corresponding to $Y = 0$, 1, or 3, the probability (given $Y = 2$) is 0. Thus, when we observe $Y = 2$, only the points $\{110, 101, 011\}$ have positive probability, and these are equally likely—*independent of the value of $p$.*

Similar calculations show that for each given value of $Y$, the conditional distribution in the sample space does not depend on $p$, so the sample sum is sufficient according to the "general definition" given above.

To see the intuition in this definition, observe that we can think of a sample sequence as being generated in either of two equivalent ways: (1) We can carry out the Bernoulli experiment in three independent trials, resulting in $(x_1 \, x_2, \, x_3)$; or (2) we can carry out the binomial experiment Bin($p$) to get $Y = y$, and then, from the partition set defined by $y$, select at random from its points to get $(x_1 \, x_2, \, x_3)$. In the latter approach, it should be clear that there is nothing to be gained by carrying out the latter selection because it does not involve the parameter $p$. That is, it is enough to find the value of $Y$. Saying it another way, it is enough to know which partition set the sample result is in—it would tell us nothing about $p$ to find out which point of that partition set is the actual outcome. ■

## Problems

* **8-9.** Show that the exponential family of distributions [(2) of §8.3] includes the following, and give a sufficient (one-dimensional) statistic in each case, assuming a random sample of size $n$:

    **(a)** $\mathcal{N}(0, \, \theta)$.     **(b)** Exp($\lambda$).     **(c)** Poi($\lambda$).

**8-10.** Show that the exponential family [(2) of §8.3] includes the following, and give a sufficient (one-dimensional) statistic in each case:

    **(a)** The geometric family with parameter $p$.

    **(b)** The gamma family with $\alpha = 2$ and unknown $\lambda$.

    **(c)** The family of distributions with p.d.f. $f(x \mid \theta) = \theta x^{\theta - 1}$ for $0 < x < 1$.

* **8-11.** Find a two-dimensional sufficient statistic for $\theta$ based on a random sample from the bivariate population with p.d.f.

$$f(x, \, y) = e^{-\theta x - y/\theta} \text{ for } x > 0 \text{ and } y > 0.$$

[The sample is a set of $n$ pairs, $\{(X_1, \, Y_1), \, ..., \, (X_n, \, Y_n)\}$].

**8-12.** Consider a random sample $(X_1, \, ..., \, X_n)$ from a Poisson distribution with parameter $m$. Let $T = \sum X_i$ and find the conditional joint probability function in the sample space given $T = t$.

* **8-13.** Find a sufficient statistic for the parameter pair $(\theta_1, \theta_2)$ based on a random sample from $\mathcal{U}(\theta_1, \theta_2)$ [that is, uniform on $\theta_1 < x < \theta_2$].

**8-14.** Consider a random sample of size $n$ from $X \sim$ Poi($\lambda$). Let $Y = \sum X_i$, which is sufficient for $\lambda$, according to Problem 8-9(c). Find the likelihood function $L(\lambda)$ given the sample of $X$'s and compare it with the likelihood function based on a single observed value from Poi($n\lambda$), the distribution of $Y$.

* **8-15.** For random samples of size $n$, find a sufficient statistic for the parameter $M$ in the discrete distribution with p.f. $f(x \mid M) = 1/M$ for $x = 1, 2, ..., M$.

**8-16.** Given a random sample $\mathbf{X} = (X_1, \, ..., \, X_n)$, find the conditional joint p.f. of $\mathbf{X}$ given that $\sum X_i = t_0$, when

    **(a)** $X \sim$ Geo($p$): $f(x \mid p) = p(1 - p)^{x-1}$, $x = 1, 2, ....$

    **(b)** $X \sim$ Ber($p$): $f(x \mid p) = p^x (1 - p)^{1-x}$, $x = 0$ or $1$.

[Observe that in both cases the answer does not depend on $p$.]

* **8-17.** Find a sufficient statistic for the unknown number $M$ of 1's in a finite Bernoulli population, when sampling is at random but without replacement and mixing. [See Example 8.1e.]

| 8.4 | **Sampling Distributions** |
|---|---|

A statistic, being a function of random variables, is itself a random variable—its value varies from sample to sample. Its probability distribution, which describes this variation, is called its **sampling distribution.** The sampling distribution of a particular statistic depends on the statistic, on the population being sampled, and on the method of sampling.

When the population distribution is known, it may be possible to find a sampling distribution mathematically. Indeed, we have actually derived some sampling distributions in earlier chapters, without calling them by that name. For instance, suppose we sample a Bernoulli population at random. For a sample of given size, the sample *sum* is a statistic of interest (being sufficient for $p$), and its sampling distribution is binomial (if sampling is with replacement) or hypergeometric (if sampling is without replacement). (See §4.2 and §4.3.) The next example compares these two cases.

| Example **8.4a** | **Sampling Inspection** |
|---|---|

As in Example 8.1b, consider a lot of 20 articles, $M$ of which are defective. To learn about this unknown parameter, we select three at random without replacement. The number $Y$ of defective articles in the sample is sufficient for $M$ (Problem 8-17). The sampling distribution of this statistic $Y$ is hypergeometric, with p.f.

$$f(y \mid M) = P(y \text{ defectives}) = \frac{\binom{M}{y}\binom{20-M}{3-y}}{\binom{20}{3}}, \; y = 0, 1, 2, 3.$$

In practice, it is usually pointless to replace an article once it has been tested; but for comparison, suppose the sampling is done with replacement. In this case, $Y$ is again sufficient, but now its sampling distribution is binomial:

$$f(y \mid M) = \binom{3}{y}\left(\frac{M}{20}\right)^y\left(1 - \frac{M}{20}\right)^{3-y}, \; y = 0, 1, 2, 3.$$

In Example 8.1b we gave the probability table for each of these sampling distributions for the particular case $M = 2$. ∎

| Example **8.4b** | **A Service Counter** |
|---|---|

Suppose customer arrivals follow a Poisson model with unknown rate $\lambda$ per hour. To learn about $\lambda$ one may count arrivals in, say, 5 consecutive one-hour periods, recording these counts $(X_1, ..., X_5)$ as a random sample of size 5. The sample sum

$Y = \sum X_i$ is sufficient (Problem 8-9), and in Example 4.9b we showed the distribution of $Y$—its sampling distribution—to be Poi($5\lambda$):

$$f(k \mid \lambda) = \frac{(5\lambda)^k}{k!} e^{-5\lambda}, \quad k = 0, 1, \ldots.$$

[This p.f. is (1) in §4.6.]                                                     ■

As in these two examples, the sample sum is often a sufficient statistic, and we have studied the distribution of a sum of independent, identically distributed (i.i.d.) random variables in various other special cases. The next example treats quite a different type of statistic.

**Example 8.4c** | **The Sample Maximum**

In Example 8.2e we considered a setting in which the population distribution is uniform on an interval with unknown right-hand endpoint. The population c.d.f. is

$$F_X(x \mid \theta) = \frac{x}{\theta}, \quad 0 < x < \theta. \tag{1}$$

We found that the likelihood function, for a random sample of size $n$, depends on the observations through the value of the largest observation—which is therefore sufficient for $\theta$.

We'll find the sampling distribution of $Y = X_{(n)}$ by calculating its c.d.f.:

$$F_Y(y \mid \theta) = P(Y \le y \mid \theta) = P(\text{largest observation} \le y \mid \theta).$$

Now, the largest observation is less than or equal to $y$ if and only if *each* observation is less than or equal to $y$:

$$F_Y(y \mid \theta) = P(X_1 \le y, \ldots, X_n \le y \mid \theta).$$

Since the $X$'s are independent, this probability factors; and because they are identically distributed, the factors are the same:

$$F_Y(y \mid \theta) = \prod_{i=1}^{n} P(X_i \le y \mid \theta) = \prod_{i=1}^{n} F_{X_i}(y \mid \theta) = [F_X(y \mid \theta)]^n. \tag{2}$$

Substituting from (1), we obtain the c.d.f. of the sampling distribution of $Y$ as

$$F_Y(y \mid \theta) = \left(\frac{y}{\theta}\right)^n, \quad 0 < y < \theta.$$

Differentiating this c.d.f. produces the p.d.f. of $Y$:

$$f_Y(y \mid \theta) = \frac{d}{dy} F_Y(y \mid \theta) = \frac{n y^{n-1}}{\theta^n}, \quad 0 < y < \theta. \qquad ■$$

## 8.5    Simulating Sample Distributions

Although the sampling distributions of some statistics can be found analytically, by mathematical manipulations, others are not tractable. When any distribution is not known, an empirical approach is to carry out the experiment it represents many times, and to take the sample distribution of the results as an approximation to the unknown distribution. This process can sometimes be carried out to approximate the sampling distribution of a statistic whose mathematical derivation is intractable.

Consider any statistic of interest, $T = t(X_1, ..., X_n)$. The sample space of this statistic and its distribution define a new population—a population of $T$-values. This population of $T$-values derives from a "population" of samples from the target population $X$. Imagine drawing a large number of samples (thousands, say) from $X$ and calculating the value of $T$ from each sample. This "sample" of $T$-values should be nearly representative of the population of $T$-values, so the histogram of $T$-values should be close to the p.d.f. (or p.f.) of the distribution of $T$.

**Example 8.5a**

### Distribution of the Sample Range

The sample range is a statistic. We can (and will, in §8.6) derive its sampling distribution in terms of the population distribution; but to illustrate the method just described, we drew 400 samples of size $n = 5$ from $\mathcal{U}(0, 1)$. For each sample, we calculated the range $R$ and tallied these 400 $R$-values using a system of 20 class intervals. Figure 8-4 shows the corresponding histogram, along with a smooth curve drawn in by eye. This curve is a rough estimate of the density function of $R$. The exact p.d.f. (as we shall see in §8.6) is

$$f_R(r) = 20r^3(1 - r), \ \ 0 < r < 1.$$

The smooth curve in the figure closely approximates this p.d.f.  ∎

In order to approximate a sampling distribution by repeated sampling as in the example, one has to simulate the underlying population of $X$-values. For instance, suppose we'd like to generate an observation from $\mathcal{U}(0, 1)$. One way is to spin a pointer with a scale from 0 to 1. Another way is to use random digits to obtain a random integer, as described in §1.5; putting a decimal point in front results in a random number on the unit interval—or more precisely, in a discrete approximation thereto. Also, computer software can be used to generate these random numbers.

For populations other than $\mathcal{U}(0, 1)$, samples can be generated using a transformation of uniform random numbers. The software MINITAB has a routine called "NRAND" that produces, in this way, a sequence of random numbers from a normal population with mean and s.d. of your choosing. For other populations, you may have to program an appropriate transformation.

Suppose you need a sequence of observations from a continuous population with c.d.f. $F$. The random variable $U = F(X)$ is uniformly distributed on the unit interval, as shown in Problem 5-9. Conversely, if $U$ is uniform on $\mathcal{U}(0, 1)$, the transformed

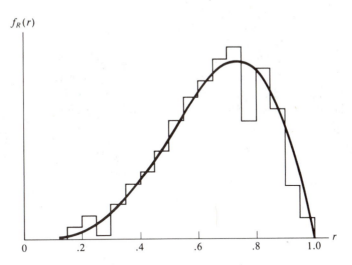

**Figure 8-4** Empirical p.d.f. for a sample range

variable $X = F^{-1}(U)$ has the c.d.f. $F(x)$:

$$P(X \leq x) = P[F^{-1}(U) \leq x] = P[U \leq F(x)] = F(x).$$

This means that if $(U_1, \ldots, U_n)$ is a random sample from $\mathcal{U}(0, 1)$, the vector of the $n$ transformed quantities $[F^{-1}(U_1), \ldots, F^{-1}(U_n)]$ is a random sample from a population with c.d.f. $F$.

**Example 8.5b** | **Simulating Exponentials**

To simulate a random sample of size $n = 10$ from $\text{Exp}(\frac{1}{2})$, we first obtained a sequence of ten (rounded) observations from $\mathcal{U}(0, 1)$ using Table XV in Appendix I:

.4224   .0663   .9728   .8801   .4410   .9582   .0402   .2316   .9240   .5073.

Given the population c.d.f. $F(x) = 1 - e^{-x/2}$, we solve $u = F(x)$ for $x$ to obtain the inverse function:  $F^{-1}(u) = -2\log(1 - u)$. Applying this transformation to each of the above $U$'s yields the desired ten observations from $\text{Exp}(\frac{1}{2})$:

1.098   .137   7.209   4.242   1.163   6.350   .082   .527   5.154   1.416.

Figure 8-5 shows the two sets of observations and their correspondence via the exponential c.d.f. ∎

Although one can always simulate a random sample from a given population as we did in the example, sometimes there are more efficient methods. For example, a common method of generating normal random numbers is to use the "Box-Muller transformation," which generates them in pairs.[3]  If $U$ and $V$ are independent, and each is $\mathcal{U}(0, 1)$, the variables

$$
\begin{cases}
X = \sigma\sqrt{-2\log U}\,\cos\left(2\pi V\right) + \mu \\
Y = \sigma\sqrt{-2\log U}\,\sin\left(2\pi V\right) + \mu
\end{cases}
\tag{1}
$$

are independent, and each is $\mathcal{N}(\mu, \sigma^2)$. So (1) can be used to transform $2n$ independent observations on $\mathcal{U}(0, 1)$ into $2n$ independent normal random numbers.

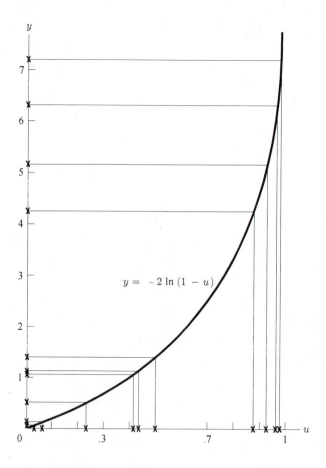

$$y = -2\ln\left(1 - u\right)$$

**Figure 8-5** Transforming uniform to exponential variates

[3]See B. W. Lindgren, *Statistical Theory*, 4th Ed. (New York: Chapman & Hall, 1993), 416.

| Example **8.5c** | **Generating Standard Normal Observations** |
|---|---|

To obtain a random sample of size 10 from $N(0, 1)$, we start with a random sample from $U(0, 1)$. For purposes of illustration, we'll use the ten uniform observations from the preceding example:

.4224   .0663   .9728   .8801   .4410

.9582   .0402   .2316   .9240   .5073.

With the first row as $U$'s and the second as $V$'s, we apply (1) to obtain

$X$:  1.2678   2.2557   .0271   .4489   $-1.2783$

$Y$:  $-.3409$   .5822   .2333   $-.2323$   .0587.

(You could and should reproduce some of these using a scientific calculator.)  These ten numbers constitute a random sample from $N(0, 1)$.    ∎

## Problems

* **8-18.**  Six families have incomes as follows (in thousands of dollars):  22, 24, 24, 28, 32, 50. Treating this as a population, suppose a sample of two is selected at random. For this sample, taken without replacement,

   **(a)**  Find the sampling distribution of $\overline{X}$.

   **(b)**  Using (a), calculate the mean and variance of $\overline{X}$.

**8-19.**  When three chips are selected at random without replacement from a bowl containing five chips numbered from 1 to 5, define the variables

   $Y =$ the largest number among the three drawn,

   $S =$ the *sum* of the numbers drawn,

   $M =$ the *smallest* of the numbers drawn.

   **(a)**  List the ten equally likely samples.

   **(b)**  Obtain the sampling distributions of $Y$ and $M$.

  * **(c)**  Find the distribution of $S$.

   **(d)**  Give the joint p.f. of $Y$ and $M$ as a two-way table of probabilities.

**8-20.**  Consider a sample of size $n = 3$ from the bowl in Problem 8-19 when sampling is done *with* replacement and mixing. Find the p.f.'s of the largest observation and of the sample sum. (Compare these with the corresponding p.f.'s in Problem 8-19.)

**8-21.**  As a check on the answers to Problem 8-20, actually perform the experiment 50 times. [You could use five coins or buttons of equal size marked with numbers 1 to 5, taking care to mix thoroughly after each selection of three.]  Record the values of $S$ and $M$ and compare relative frequencies with the probabilities found in Problem 8-20.

* **8-22.**  Determine the sampling distribution of the sample sum for random samples of size $n$ from each of the following populations, giving the answer as a named distribution family with appropriate parameter values:

   **(a)**  $N(\mu, \theta)$.    **(c)**  $Exp(\lambda)$.    **(e)**  $Geo(p)$.

   **(b)**  $Poi(m)$.     **(d)**  $Bin(k, p)$.

**\* 8-23.**  Find the mean value of the largest observation in a random sample of size $n$ from a population with c.d.f. $F(x) = x^k$, $0 < x < 1$.

**\* 8-24.**  Find the probability that the largest observation in a random sample of size $n = 5$ from $\mathcal{N}(0, 1)$ is less than 1.2.

**8-25.**  Starting at an arbitrary point, take ten consecutive numbers from Table XV as a random sample from $\mathcal{U}(0, 1)$. After suitable transformations, use them as in Example 8.5b to produce

(a)   another random sample from $\text{Exp}(\frac{1}{2})$.

(b)   a random sample from a population with $F(x) = x^2, 0 < x < 1$.

**8-26.**  Use statistical computer software to generate 50 samples of size 5, 50 samples of size 15, and 50 samples of size 50 from $\mathcal{U}(0, 1)$. As in Example 8.5b, convert these into a sample from $\text{Exp}(1)$, obtain the sample means, and examine the resulting (empirical) sampling distributions of the sample mean.

---

## 8.6          Order Statistics

---

**Example 8.6a**  |  ### $R$-Charts

A standard tool for monitoring the variability in a production process is the $R$-chart. (It is used in conjunction with an $\overline{X}$-chart, but for this example we focus on the $R$-chart.) An $R$-chart is a record of the ranges of samples taken at regular intervals from the production line. The chart has lines to tell the operator when the process variability is "in control." The placement of these lines depends on the sampling distribution of the range, $R = X_{(n)} - X_{(1)}$.

The distribution of $R$ can be found approximately using a simulation, as described in the preceding section, or (in some cases) derived mathematically from the joint distribution of $X_{(1)}$ and $X_{(n)}$.                                ■

In this section we show how to derive the distributions of order statistics and functions of order statistics, such as $R$ in the preceding example, for random sampling from a specified continuous population.

We could find the distribution of the $k$th smallest observation $X_{(k)}$ by using the approach of Example 8.4c—deriving its c.d.f. and differentiating. However, the differential method introduced in §5.2 yields the p.d.f. a little more easily. To apply it, we find the probability that $X_{(k)}$ is in the infinitesimal interval from $y$ to $y + dy$. The coefficient of $dy$ in the expression for this probability will be the desired p.d.f.

Each observation $X_i$ lies in one of the intervals $(-\infty, y)$, $(y, y + dy)$, and $(y + dy, \infty)$. The frequencies that count the numbers of $X_i$'s in these intervals have a *trinomial* distribution. To find it, we need the interval probabilities. The probability of $(-\infty, y)$, of course, is $F(y)$. The probability of the interval $(y + dy, \infty)$ is approximately the probability of $(y, \infty)$, or $1 - F(y)$. The probability of the interval $(y, y + dy)$ is approximately the probability element:

$$P(y < X < y + dy) \doteq f(y)dy,$$

where the $f$ and $F$ refer to the population variable $X$. In summary, we have

| Outcome | Approximate Probability |
|---|---|
| $X < y$ | $F(y)$ |
| $y < x < y + dy$ | $f(y)\,dy$ |
| $X > y + dy$ | $1 - F(y)$ |

The $k$th smallest $X_i$, denoted by $X_{(k)}$, is in the interval $(y, y + dy)$ if and only if there are $k - 1$ $X$'s to the left of $y$, exactly one $X$ is between $y$ and $y + dy$, and the remaining $n - k$ $X$'s are in the interval $(y + dy, \infty)$. (See Figure 8-6.)  The probability of this event is a trinomial probability (§4.8):

$$P(y < X_{(k)} < y + dy) \doteq \frac{n!}{(k-1)!1!(n-k)!}\,[F(y)]^{k-1}[f(y)dy]^1[1 - F(y)]^{n-k}.$$

The coefficient of $dy$ is the desired p.d.f.

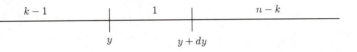

**Figure 8-6**

---

The p.d.f. of the $k$th smallest observation in a random sample of size $n$ from a continuous population with c.d.f. $F$ and p.d.f. $f$ is

$$f_{X_{(k)}}(y) = \frac{n!}{(k-1)!1!(n-k)!}\,[F(y)]^{k-1}[1 - F(y)]^{n-k}f(y). \qquad (1)$$

---

In Example 8.4c we derived the p.d.f. of the largest observation in a random sample; when $k = n$, formula (1) reduces to the p.d.f. of $X_{(n)}$:

$$f_{X_{(n)}}(y) = n[F(y)]^{n-1}f(y). \qquad (2)$$

The formula for the smallest observation is obtained when we set $k = 1$:

$$f_{X_{(1)}}(y) = n[1 - F(y)]^{n-1}f(y). \qquad (3)$$

And with (1), we can find the mean value of any $X_{(k)}$ in the usual way, as an integral with respect to its p.d.f.:

$$E[X_{(k)}] = \int_{-\infty}^{\infty} y f_{X_{(k)}}(y \mid \theta) \, dy.$$

In §7.3 we defined the sample *median* $\tilde{X}$ in terms of components of the order statistic. When $n$ is odd, the median is the middle observation—the $k$th smallest when $k = (n+1)/2$. To obtain the p.d.f. of $\tilde{X}$ in this case, we set $k = (n+1)/2$ in (1):

$$f_{\tilde{X}}(u) = \frac{n!}{[\frac{1}{2}(n-1)]! \, [\frac{1}{2}(n-1)]!} \, [F(u)]^{(n-1)/2} [1 - F(u)]^{(n-1)/2} f(u). \qquad (4)$$

(When $n$ is even, the sample median is the average of two order statistics, a complication we'll not go into.)

| Example 8.6b | ## Order Statistics for $\mathcal{U}(0, 1)$ |
|---|---|

If $X \sim \mathcal{U}(0, 1)$, the population c.d.f. is $F(x) = x$, and the p.d.f. is $f(x) = 1$, both for $0 < x < 1$. Substituting in (2) gives us the p.d.f. of the largest observation:

$$f_{X_{(n)}}(u) = n u^{n-1}, \quad 0 < u < 1.$$

The mean value of $X_{(n)}$ is

$$E[X_{(n)}] = \int_0^1 u \left[ n u^{n-1} \right] du = \frac{n}{n+1},$$

and the average square:

$$E[X_{(n)}^2] = \int_0^1 u^2 n \left[ u^{n-1} \right] du = \frac{n}{n+2}.$$

From these we find the variance:

$$\text{var } X_{(n)} = \frac{n}{n+2} - \left( \frac{n}{n+1} \right)^2 = \frac{n}{(n+1)^2(n+2)}.$$

The variance tends to 0 as $n$ becomes infinite, inversely proportional to $n^2$. (This is faster than the rate at which the variance of a sample mean tends to 0, namely, $1/n$.)

Because the population is symmetric about the middle of the support interval, it follows that the mean of the smallest observation is

$$E[X_{(1)}] = 1 - E[X_{(n)}] = \frac{1}{n+1}.$$

The variance, again because of the symmetry (but you can verify this by calculating it), is the same as the variance of the largest observation.

$$\text{var}\,[X_{(1)}] = \text{var}\,[X_{(n)}] = \frac{n}{(n+1)^2(n+2)}.$$

For $n = 15$ (for instance), the p.d.f. of the *median*, from (4), is

$$f_{\tilde{X}}(y) = \frac{15!}{7!7!}\, y^7(1-y)^7,\ \ 0 < y < 1.$$

This distribution is symmetric about $1/2$, so the mean and median are both equal to $1/2$. (The symmetry is the result of the population symmetry: Problem 8-29 asks you to show that if $f_X$ is symmetric, so is $f_{\tilde{X}}$.)  ■

We return now to the problem of finding the sampling distribution of the sample range, which is a function of the smallest and largest sample observations. For this we need the joint distribution of those two order statistics, which we derive next, using the differential method.

Given values $u$ and $v$ with $u < v$, Figure 8-7 shows how we divide the axis into five subintervals, to calculate the probability element defined by the infinitesimal intervals $(u, u + du)$ and $(v, v + dv)$ for $X_{(1)}$ and $X_{(n)}$ respectively. The intervals, with approximate probabilities calculated much as for $X_{(k)}$, are shown in the following table:

| Outcome | Approximate Probability |
|---|---|
| $X < u$ | $F(u)$ |
| $u < x < u + du$ | $f(u)\,du$ |
| $u + du < X < v$ | $F(v) - F(u)$ |
| $v < X < v + dv$ | $f(v)\,dv$ |
| $X > v + dv$ | $1 - F(v)$ |

If the smallest observation is to be in $(u, u + du)$ and the largest in $(v, v + dv)$, the frequencies of the five intervals must be as shown in Figure 8-7:

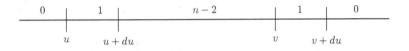

**Figure 8-7**

The joint distribution of the class frequencies is multinomial, so

$$P(u < X_{(1)} < u + du, \; v < X_{(n)} < v + dv)$$
$$= C[F(u)]^0[f(u)\, du]^1[F(v) - F(u)]^{n-2}[f(v)\, dv]^1[1 - F(v)]^0, \qquad (5)$$

where $C$ is the multinomial coefficient:

$$C = \binom{n}{0,\, 1,\, n-2,\, 1,\, 0} = \frac{n!}{(n-2)!} = n(n-1).$$

The desired joint p.d.f. is the coefficient of $du\, dv$ in (5):

---

Joint p.d.f. of $U = X_{(1)}$ and $V = X_{(n)}$ in a random sample of size $n \geq 2$ from a continuous population with c.d.f. $F$ and p.d.f. $f$:

$$f_{U,V}(u, v) = n(n-1)[F(v) - F(u)]^{n-2}f(u)f(v), \; u < v \qquad (6)$$

---

We can now use (6) to find the p.d.f. of any function of the smallest and largest observations, and in particular, of the sample range, $R$. The p.d.f. of $R$ is an integral involving $F$ and $f$, which we derive next in a particular case.

**Example 8.6c**

## Sample Range for $\mathcal{U}(0, 1)$

Suppose $X \sim \mathcal{U}(0, 1)$. The p.d.f. is $f(x) = 1$, and the c.d.f. is $F(x) = x$, both for $0 < x < 1$. In this case, (6) becomes

$$f_{U,V}(u, v) = n(n-1)(v - u)^{n-2}, \quad 0 < u < v < 1 \qquad (7)$$

The c.d.f. of the sample range $R$ is

$$F_R(r) = P(R \leq r) = P(V - U \leq r).$$

To find this, we'd need to evaluate a double integral of the joint p.d.f. (6) over the subset of the support region in which $v - u \leq r$, or where $0 < u < v < u + r$ and $v < 1$. Having thus found the c.d.f., we'd differentiate with respect to $r$ to find the p.d.f. The result of doing this mathematics is

$$f_R(r) = n(n-1)r^{n-2}(1 - r), \; 0 < r < 1.$$

When $n$ is large, the graph of this p.d.f. behaves like $1 - r$ near $r = 1$ and has a high order of tangency to the $r$-axis at $r = 0$. Most of the distribution is crowded near $r = 1$, so the range of a large sample is very likely to be near the population range. (See also Example 8.5a.) ∎

We turn, finally, to the joint distribution of all components of the order statistic, deriving its p.d.f. by use of the differential method. For given values $y_1 < \cdots < y_n$, the probability of finding exactly one sample observation in each subinterval $(y_i, y_i + dy_i)$ is multinomial:

$$n![f(y_1)\,dy_1] \cdots [f(y_n)dy_n].$$

The coefficient of the "volume" element is the desired p.d.f.:

> The joint density of the ordered observations $(X_{(1)}, ..., X_{(n)})$ in a random sample from a continuous distribution with p.d.f. $f$ is
> $$f^*(y_1, ..., y_n) = n!f(y_1) \cdots f(y_n) \ \text{ for } y_1 < \cdots < y_n. \qquad (8)$$

**Example 8.6d**  Consider a random sample of size $n \geq 2$ from $\mathcal{U}(0, 1)$. The population p.d.f. is $f(x) = 1$ for $0 < x < 1$, so the joint p.d.f. of the ordered observations in a random sample, from (8), is

$$f^*(y_1, ..., y_n) = n!, \ \ 0 < y_1 < \cdots < y_n < 1.$$

The distribution is thus uniform in a simplex in $n$-space.  ∎

## Problems

\* **8-27.**  Consider a random sample of size $n = 5$ from $X \sim \mathcal{U}(0, 1)$.

(a)  Find the sampling distribution of $\widetilde{X}$, the sample median.

(b)  Find the mean and variance of $\widetilde{X}$.

(c)  Interpret each sequence of five digits in Table XV of Appendix 1 as the first five decimal digits of an observation from $\mathcal{U}(0, 1)$. Each $5 \times 5$ block of digits in the table can be considered a random sample of size 5. Find the median of each sample in 40 successive blocks, starting at an arbitrary point in the table. Find the (sample) mean and variance of the 40 $\widetilde{X}$-values and compare with the parameters found in (b). [They won't be the same, but should be close.]

**8-28.**  (Continuing Example 8.2e.)  Find the mean and variance of the largest observation from $\mathcal{U}(0, \theta)$. [A practical implication of the rapid approach to 0 of the variance is that the biologist has some assurance that the largest observation in a sample of moderate size is a good approximation to $\theta$.]

**8-29.**  Use (a) of Problem 8-27 to find the p.d.f. of the sampling distribution of the median of a random sample of size 5 from $\mathcal{U}(a, b)$, where $a < b$.

**8-30.**  Use the differential method to find the joint p.d.f. of the second smallest and second largest observations in a random sample from a population with c.d.f. $F$ and p.d.f. $f$.

\* **8-31.**   Use (6) of §8.6 and the transformation $Y = (X - a)/(b - a)$ to obtain the joint p.d.f. of the smallest and largest observations in a random sample from a uniform distribution on $(a, b)$.

**8-32.**   Consider a continuous variable $X$ whose p.d.f. is symmetric about its median. Show that the median of a random sample of size $n$ is also symmetrically distributed about this value when $n$ is odd. [The result is also true when $n$ is even.]   Recall the definition: $f$ is symmetric about   $x = a$   when,   for all   $x$,   $f(a - x) = f(a + x)$   or   (equivalently)   $F(a - x) = 1 - F(a + x)$.

---

**8.7**                        ## Moments of Sample Means and Proportions

Many problems in the application of statistics hinge on knowing—as well as can be known—the value of a population mean $\mu$. Intuition suggests that the sample mean $\overline{X}$ should be close to $\mu$, and in many cases it is sufficient for $\mu$. The sampling distribution of $\overline{X}$ depends on the population being sampled and on the method of sampling. In §8.9 we'll show how (in principle) to derive the distribution of the sample mean from that of the population, in the case of random samples. For the large sample inferences of the next few chapters, we'll need only the expected value and variance.

When sampling is done at random, with or without replacement, the sample observations are identically distributed. In particular, all observations $X_i$ have the same mean $\mu$ and the same variance $\sigma^2$. In §3.5 and §5.10, using the linearity of expectations, we found the expected value of the sample sum to be $n\mu$. So,

$$E(\overline{X}) = E\left\{\frac{1}{n}\sum X_i\right\} = \frac{1}{n} E\left\{\sum X_i\right\} = \frac{n\mu}{n} = \mu. \tag{1}$$

The variance of a sum and (hence) the variance of the mean, depend on whether sampling is done with or without replacement. In either case,

$$\sigma_{\overline{X}}^2 = \text{var}(\overline{X}) = \text{var}\left\{\frac{1}{n}\sum X_i\right\} = \frac{1}{n^2} \text{var}\left\{\sum X_i\right\}. \tag{2}$$

We'll return to the case of sampling without replacement, but for now, suppose the sample is a *random sample* (sampling with replacement if the population is finite). The variance of the sum is the sum of the variances [(9) of §3.5 and (5) of §5.10], so (2) becomes

$$\text{var}\,\overline{X} = \frac{1}{n^2} \text{var}\left\{\sum X_i\right\} = \frac{1}{n^2}(n\sigma^2) = \frac{\sigma^2}{n}. \tag{3}$$

Because it is easier to interpret the standard deviation as a measure of variability, we usually focus on the square root of (3):

For a *random sample* from a population with mean $\mu$ and s.d. $\sigma$,

$$E(\overline{X}) = \mu, \tag{4}$$

$$\sigma_{\overline{X}} = \frac{\sigma}{\sqrt{n}}. \tag{5}$$

Formula (5) shows that the sampling distribution of $\overline{X}$ narrows as $n$ increases: *The larger the sample, the less the variability in the sample mean.*

**Example 8.7a**

## Rolling Three Dice

The total number of points showing at the roll of three dice is the sum of the numbers on the individual dice: $Y = X_1 + X_2 + X_3$. The average is $\overline{X} = Y/3$. The distribution of $X_i$ is uniform on the integers from 1 to 6, and we found in Examples 3.1e and 3.3b that $\mu = 7/2$ and $\sigma^2 = 35/12$. Thus, $E(\overline{X}) = \mu = 7/2$ and, if we assume independence of the numbers on the three dice,

$$\sigma_{\overline{X}} = \frac{\sigma}{\sqrt{3}} = \sqrt{\frac{35}{36}} \doteq .986.$$

This is only $1/\sqrt{3}$, or 58% as large as the standard deviation for a single die.

In this instance you could verify these results by finding the distribution of the sum, by counting how many of the 216 possible outcomes result in each of the possible sums. (These outcomes are equally likely.) For instance, the sum is 7 for the samples 2 2 3 (in three arrangements), 1 2 4 (in six arrangements), 1 1 5 (in three arrangements), and 1 3 3 (in three arrangements). This makes a total of $3 + 6 + 3 + 3$ or 15 out of the 216 samples:

$$P(Y = 7) = \frac{15}{216}.$$

[You could also find such probabilities using the probability generating function, as shown in §3.6. We don't ask you to do this, but you would find that $\mu = 7/2$ and $\sigma^2 = 35/36$, as we found using (4) and (5).] ∎

Consider now the important special case of sampling from a Bernoulli population. The population mean is $\mu = p$, the population proportion of successes [(2) of §4.1]. A parallel calculation shows that the mean of a sample $(X_1, ..., X_n)$ is the sample proportion of successes:

$$\overline{X} = \frac{1}{n}\sum_{i=1}^{n}X_i = \frac{1}{n}(\text{\# of 1's}) = \widehat{p}.$$

The population variance [again from (2) of §4.1] is $\sigma^2 = pq$, where $q = 1 - p$. Applying (4) and (5) we obtain the following:

---

The mean and variance of $\hat{p}$, the proportion of successes in a random sample from a Bernoulli population with probability $p$ of success, are

$$E(\hat{p}) = p, \tag{6}$$

$$\sigma_{\hat{p}} = \sqrt{\frac{pq}{n}}. \tag{7}$$

---

We have seen closely related formulas in connection with the binomial distribution: The sample proportion is just $1/n$ times $Y$, the number of successes in the sample, and we know that $Y \sim \text{Bin}(n, p)$. So (6) and (7) in fact follow at once from formulas (4) of §4.2, for the mean and variance of a binomial variable.

**Example 8.7b** | **A Sample Survey**

Suppose voter sentiment in a certain district is evenly divided on a proposal to increase the school levy. What kind of divisions might show up in a random sample of 25 voters?

"Evenly divided" means that $p = .5$, where $p$ is the proportion in the population who favor the proposal. So, from (6), $E(\hat{p}) = .5$. The standard deviation of $\hat{p}$, from (7), is

$$\sigma_{\hat{p}} = \sqrt{\frac{pq}{n}} = \sqrt{\frac{.5 \times .5}{25}} = .10.$$

Like sample statistics generally, sample proportions vary from sample to sample. In some samples, the proportion in favor will be less than the actual population proportion; in others, it will be greater. We interpret $\sigma_{\hat{p}} = .10$ to say that for samples of size 25, the sample proportions typically deviate from the true value of $p$ by about 10 percentage points when $p = \frac{1}{2}$.                                    ∎

When sampling is done *without replacement* from a finite population, the variance of the sample mean is complicated by the fact that the observations are not independent. In §3.5[4] we found [as (8)] that

$$\text{var}\left(\sum_1^n X\right) = n\sigma^2 \cdot \frac{N - n}{N - 1}. \tag{8}$$

---

[4]The derivation there for discrete populations is also valid in the continuous case.

Since $\bar{X} = \sum X_i / n$, it follows that var $\bar{X} = \frac{1}{n^2}$ var$(\sum X_i)$. Hence:

---

For random sampling without replacement from a finite population of size $N$ with mean $\mu$ and standard deviation $\sigma$,

$$E(\bar{X}) = \mu, \quad \sigma_{\bar{X}} = \frac{\sigma}{\sqrt{n}} \cdot \sqrt{\frac{N-n}{N-1}}. \tag{9}$$

---

Comparing (9) with (4) and (5), we see that although the means are the same, the s.d. of $\bar{X}$ is smaller under sampling without replacement (for $N > 1$).

In the case of a Bernoulli population, $\mu = p$ and $\sigma^2 = pq$, so the formulas (9) become

$$E(\hat{p}) = p, \quad \sigma_{\hat{p}} = \sqrt{\frac{pq}{n}} \sqrt{\frac{N-n}{N-1}}. \tag{10}$$

**Example 8.7c** | **Survey of a Small Population**
In the preceding example, suppose the sample of 25 voters had been drawn, at random but without replacement, from the 100 voters of a small town. Again assume an equal division on the issue of the levy. The expected value of $\hat{p}$ is still $p = .5$, but the standard deviation, from (10), is

$$\sigma_{\hat{p}} = \sqrt{\frac{.5 \times .5}{25}} \sqrt{\frac{100 - 25}{100 - 1}} \doteq .087,$$

reduced from the .10 of the preceding example by about 13% because of the nonreplacement feature. ∎

---

**8.8**                      **The Central Limit Theorem**

---

Finding sampling distributions can be difficult. Yet, for many of the statistics we commonly use, it turns out that when the sample size is large, the *normal* distribution gives good approximations to the actual distribution. This is the case in particular for the sample mean, and for a sample proportion (as a special case of a sample mean).

**Example 8.8a** | **A Sampling Experiment**
To simulate the sampling distribution of $\bar{X}$ in a particular case, we used a computer to generate 80 samples of size 100 from $\mathcal{U}(0, 1)$, and calculated the mean of each

sample, to obtain an empirical sampling distribution of $\overline{X}$. Figure 8-8 shows the c.d.f. of this distribution of a "sample" of 80 means—a step function that jumps $1/n$ at each sample value. Superimposed on this is the c.d.f. of a normal distribution whose mean and variance are those of $\overline{X}$:

$$\mu_{\overline{X}} = \mu_X = .5, \quad \sigma_{\overline{X}}^2 = \sigma_X^2/n = \frac{1}{12 \times 100} = (.0288)^2.$$

Were we to do a similar plot for the means of, say, 80,000 samples of size 100, the empirical c.d.f. would be much less rough, and the curves would be nearly indistinguishable. ∎

Consider, then, the distribution of the mean of a random sample from an arbitrary population. With only a mild restriction on "arbitrary": $\sigma_X < \infty$, it turns out (as the example suggests) that the distribution of $\overline{X}$ is approximately normal when $n$ is large. Moreover, quite remarkably, the limiting distribution does not depend on the population sampled! This striking result generalizes the DeMoivre-Laplace approximation of §4.5 in that the latter follows as a special case in which the population is Ber$(p)$. We'll defer a discussion of the proof until the next section.

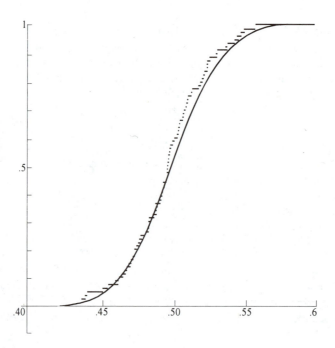

**Figure 8-8** C.d.f. of an empirical sampling distribution of $\overline{X}$

> **Central Limit Theorem:** For random samples of size $n$ and if $\sigma < \infty$,
>
> $$\lim_{n \to \infty} P \left( \frac{\overline{X} - \mu}{\sigma/\sqrt{n}} \le z \right) = \Phi(z). \tag{1}$$

In words, this theorem says that (as the sample size goes to infinity) the c.d.f. of the standardized variable $\overline{X}$ tends, at each point, to the standard normal c.d.f. If we set $c = \mu + z\sigma/\sqrt{n}$, equation (1) becomes

$$\lim_{n \to \infty} P(\overline{X} \le c) = \Phi \left( \frac{c - \mu}{\sigma/\sqrt{n}} \right). \tag{2}$$

In the case of a continuous population (as in the example), we can express the limiting relation in terms of *densities*. With a free interchange of limits, we should expect that the p.d.f. of $\overline{X}$ would tend to a normal p.d.f. as $n$ becomes infinite:

$$
\begin{aligned}
\lim_{n \to \infty} f_{\overline{X}}(u) &= \lim_{n \to \infty} \left\{ \lim_{\Delta u \to 0} \frac{F_{\overline{X}}(u + \Delta u) - F_{\overline{X}}(u)}{\Delta u} \right\} \\
&= \lim_{\Delta u \to 0} \left\{ \frac{1}{\Delta u} \left( \Phi \left\{ \frac{u + \Delta u - \mu}{\sigma/\sqrt{n}} \right\} - \Phi \left\{ \frac{u - \mu}{\sigma/\sqrt{n}} \right\} \right) \right\} \\
&= \Phi' \left( \frac{u - \mu}{\sigma/\sqrt{n}} \right)
\end{aligned}
$$

**Example 8.8b**

## Averaging Uniform Variates

We used a computer to find the actual sampling distribution of the means of random samples from $\mathcal{U}(0, 1)$ for various sample sizes. These were obtained (using the software package "S") by graphing the $n$-fold convolutions of uniform variates, described in §5.12 (page 222). Figure 8-9 shows the p.d.f.'s of $\overline{X}$ for $n = 1, 2, 5, 20$. Observe that the variance decreases, and the shape of the curve becomes more normal in appearance, as $n$ increases.

The shape becomes harder to see as the variance decreases further; it helps to look at the distribution on a standard scale. The population parameters are $EX = 1/2$, $\text{var} X = 1/12$, so the standardized sample mean is

$$Z_n = \frac{\overline{X} - \frac{1}{2}}{\sqrt{\frac{1}{12n}}}.$$

With the scale now stretched so that the variance is 1, the  graph of the p.d.f. for $n = 20$ will be almost indistinguishable from the graph of the standard normal density.  ∎

A convenient way of expressing the conclusion of the central limit theorem is to say that $\overline{X}$ is **asymptotically normal** with mean $\mu$ and variance $\sigma^2/n$, and we write $\overline{X} \approx \mathcal{N}(\mu,\ \sigma^2/n)$. We refer to $\mathcal{N}(\mu,\ \sigma^2/n)$ as the **asymptotic distribution** of $\overline{X}$.

The practical import of the central limit theorem is that when $n$ is large, we can use the normal distribution to find approximate probabilities for $\overline{X}$, *regardless of the shape of the population sampled,* provided only that the population variance is finite. That is, we can use the limiting value as an approximation to the desired probability:

$$P\left(\frac{\overline{X} - \mu}{\sigma/\sqrt{n}} \le z\right) \doteq \Phi(z).$$

Moreover, since the sample *sum* is a constant times the sample mean, it is also asymptotically normal. Indeed, the $Z$-score for the sum is precisely the same as the $Z$-score for the mean.

---

For a large random sample from $X$ with $\sigma_X < \infty$, we can use the standard normal table to find approximate probabilities for the sample *mean*:

$$P(\overline{X} < k) \doteq \Phi\left(\frac{k - \mu}{\sigma/\sqrt{n}}\right), \tag{3}$$

and for the sample *sum,*

$$P\left(\sum X_i < y\right) \doteq \Phi\left(\frac{y - n\mu}{\sigma\sqrt{n}}\right). \tag{4}$$

---

How large is "large"? Whether a sample size is large enough for a good approximation using (3) or (4) depends on the desired accuracy, on the population being sampled, and on whether we want to approximate $F(x)$ for $x$ in the tails of the distribution or toward the center. Roughly speaking, the closer the population is to being normal, the closer the distribution of $\overline{X}$ is to normal. Indeed, when the population itself *is* normal, the sample mean is normally distributed for any sample size! [This follows from Property (iv) in §6.1, page 234.]

If a population is nearly symmetric and has light tails (its p.d.f. tending to 0 rapidly as $x \to \pm\infty$), a sample size as small as 5 or so may give an approximation that is good enough for practical purposes. The uniform distribution is a case in point: It is symmetric and has no tails. However, if the population is markedly asymmetric, or if there is a substantial amount of probability far from the mean (as measured in standard deviations), then a very large sample may be needed for the distribution of $\overline{X}$ to be well approximated by the normal distribution.

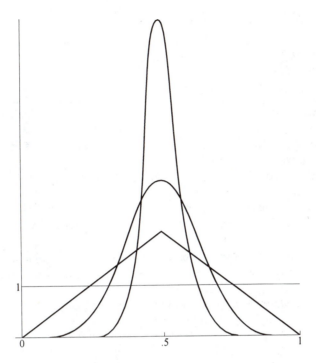

**Figure 8-9** Distribution of $\overline{X}$, uniform population

**Example 8.8c** | **An Elevator Load Limit**

We know of an elevator with this sign on the wall:

| Maximum number of passengers: |
| :---: |
| 16 |
| Maximum load: |
| 2500 lbs |

Suppose the mean weight of people who use the elevator is $\mu = 145$ lb, and the standard deviation is $\sigma = 30$ lb. The expected total weight of 16 people is 16 times the average of one: $n\mu = 2320$. However, many people weigh more than average, and the combined weight could exceed 2500 lb. What is the probability that the combined weight of 16 people selected at random from the population of riders exceeds 2500 lb?

We don't know the precise shape of the distribution of rider weights—it could be close to normal, or not. But at least the population variance is finite, since the population range is finite. Applying (4), we have

$$P\left(\sum X_i > 2500\right) \doteq 1 - \Phi\left(\frac{2500 - 2320}{\sqrt{16 \times 30^2}}\right) = \Phi(-1.5) \doteq .067.$$

You might speculate about the shape of the population distribution on the basis of your own experience. If the population of elevator riders includes both sexes, the population distribution is a mixture of the two distributions and could even be bimodal. Even so, combining as many as 16 observations is apt to result in a distribution for the sum that is close to normal.

Bear in mind that our calculation assumes a random sample. Since those riding the elevator in a particular instance are not infrequently "birds of a feather," the assumption of independence may not hold. (For example, visualize a group of 16 professional football linemen in a hotel elevator going up to an awards ceremony!) ∎

Consider now the special case of a Bernoulli population with parameter $p$, the population proportion of 1's, and a random sample of size $n$. The sample mean is the proportion of 1's in the sample: $\hat{p} = Y/n$, where $Y$ is the number of 1's in the sample. The standardized mean is also the standardized sum:

$$Z = \frac{Y - EY}{\sigma_Y} = \frac{Y - np}{\sqrt{npq}} = \frac{Y/n - p}{\sqrt{pq/n}} = \frac{\hat{p} - E(\hat{p})}{\sigma_{\hat{p}}}, \tag{5}$$

and the central limit theorem tells us that the c.d.f. of this standard score tends to the standard normal c.d.f.

---

For random samples of size $n$ from $\text{Ber}(p)$,

$$\frac{\hat{p} - p}{\sqrt{pq/n}} \approx \mathcal{N}(0, 1). \tag{6}$$

Thus, for large $n$, with $Y = n\hat{p}$,

$$P(Y \leq k) \doteq \Phi\left(\frac{k - np}{\sqrt{npq}}\right), \tag{7}$$

and

$$P(\hat{p} \leq c) \doteq \Phi\left(\frac{c - p}{\sqrt{pq/n}}\right). \tag{8}$$

---

This is the basis of the DeMoivre-Laplace approximation we gave in §4.5 for binomial probabilities when $n$ is large: For "large" $n$, we can use the normal table to find approximate probabilities for intervals of values of $\hat{p}$. In §4.5 we gave a rule of thumb for deciding if $n$ is large enough:   $npq > 5$. We also gave a "continuity correction," which puts an extra $+.5$ in the numerator of (7) for the purpose of improving the approximation.

**Example 8.8d** | **TV Ratings**

Suppose that 30% of the nation's TV viewers are watching a particular presidential press conference. Typically, a Nielsen survey will sample about 1000 viewers. Suppose the sampling is random. What is the probability that, at most, 28% of the viewers in the sample are watching the press conference? (Note: Although some elements of randomness are involved in their selection, Nielsen's samples are not random samples. Nielsen recognizes this, but nonetheless bases its assessments of sampling errors on formulas for random sampling.)

Because the population is very much larger than the sample size, we can use the formulas for sampling with replacement, (6) and (7) from §8.7. The population proportions are $p = .30$ and $q = .70$, so

$$E(\widehat{p}) = p = .30, \quad \sigma_{\widehat{p}} = \sqrt{\frac{pq}{n}} = \sqrt{\frac{.3 \times .7}{1000}} \doteq .0145.$$

And since $npq = 210$, we conclude that $\widehat{p}$ is approximately normally distributed. Figure 8-10 shows the approximating normal density function. We can use (8) and Table IIa in Appendix 1 to obtain, for instance,

$$P(\widehat{p} \leq .28) \doteq \Phi\left(\frac{.28 - .30}{.0145}\right) = \Phi(-1.38) \doteq .084.$$

This probability is the shaded area in Figure 8-10.

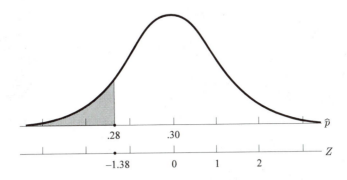

**Figure 8-10** Normal approximation of a binomial probability

We could also have calculated the desired probability using (7), expressed in terms of the sample sum $Y$, which is Bin(1000, 0.3). With $EY = np = 300$ and

$\text{var}Y = npq = 210$, we have

$$P(Y \leq 280) \doteq \Phi\left(\frac{280 - 300}{\sqrt{210}}\right) \doteq \Phi(-1.38),$$

the same result as before.

In these calculations we did not use the continuity correction from §4.5, because the difference is slight:

$$P(Y \leq 280) \doteq \Phi\left(\frac{280.5 - 300}{\sqrt{210}}\right) \doteq \Phi(-1.35).$$

But if you *do* use the continuity correction, make sure that the .5 is added to the *number* of successes and *not* to the proportion of successes. ■

**Example 8.8e** | ## The Continuity Correction

To explain why the correction seems to work, consider again the case $n = 8$, $p = .5$, first discussed in Example 4.5b. The mean and variance of $Y$, the number of successes in eight trials, are $EY = np = 4$, and $\sigma_Y^2 = npq = 2$. Table 8-1 gives values of $Y$, the corresponding $Z$-scores, the modified $Z$-scores with the extra $1/2$, the values of $\Phi(z)$ for each score, and the actual cumulative binomial probability (to four decimal places).

A study of the table reveals that the approximations that incorporate the continuity correction are closer to the binomial probabilities in the last column except at the extreme ends. A graph of the c.d.f. of $Y$, whose values at the integers are the cumulative binomial probabilities in the last column, is given in Figure 8-11, along with graphs of the approximating normal c.d.f.'s, with and without the correction.

**Table 8-1**

| $y$ | $\frac{y-4}{\sqrt{2}}$ | $\frac{y+.5-4}{\sqrt{2}}$ | $\Phi\left(\frac{y-4}{\sqrt{2}}\right)$ | $\Phi\left(\frac{y+.5-4}{\sqrt{2}}\right)$ | $\sum_{0}^{y}\binom{8}{k}.5^8$ |
|---|---|---|---|---|---|
| 0 | $-2.828$ | $-2.475$ | .002 | .007 | .0039 |
| 1 | $-2.121$ | $-1.768$ | .017 | .039 | .0352 |
| 2 | $-1.414$ | $-1.060$ | .079 | .145 | .1445 |
| 3 | $-.707$ | $-.354$ | .240 | .362 | .3633 |
| 4 | 0 | .354 | .500 | .638 | .6367 |
| 5 | .707 | 1.060 | .760 | .855 | .8555 |
| 6 | 1.414 | 1.768 | .921 | .961 | .9648 |
| 7 | 2.121 | 2.475 | .983 | .993 | .9961 |
| 8 | 2.828 | 3.182 | .998 | .999 | 1.0000 |

It is apparent in that figure (which is typical) that the height of the normal c.d.f. at a point halfway to the next integer is closer to the height of the binomial c.d.f. at the integer $k$. ∎

The continuity correction is not so crucial when we want the probability of an interval of values (such as $80 \leq Y \leq 92$), since the error that is being "corrected" tends to cancel.

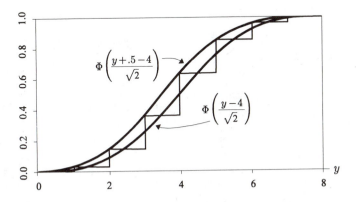

**Figure 8-11** A continuity-correction improvement

## 8.9    Using the Moment Generating Function

In §8.7 we saw how to find the first two moments (the mean and s.d.) of the distribution of the sample mean. In §8.8 we saw that these two moments allow us to find approximate probabilities of events such as $\overline{X} < k$ when the sample size is large. However, the sample size may be such that these approximations are not adequate, and the complete distribution of $\overline{X}$ is required. The m.g.f. is useful in finding the distribution of the sample mean.

It is easy to find the moment generating function of the mean, given the population p.f. or p.d.f.. First, we find the m.g.f. $\psi_X$ by summing or integrating $e^{tx}$ with respect to the population p.f. or p.d.f., respectively. Then,

$$\psi_{\overline{X}}(t) = E(e^{t\Sigma X/n}) = E(e^{(t/n)\Sigma X}).$$

This is just the m.g.f. of the sum $\sum X_i$, so

$$\psi_{\overline{X}}(t) = \psi_{\Sigma X}(t/n) = \prod \psi_{X_i}(t/n) = [\psi_X(t/n)]^n.$$

Here, we first used the independence of the $X_i$'s to write the m.g.f. of the sum as the product of the m.g.f.'s of the summands, and then used the fact that each of the observations has the population distribution, which means that all $n$ factors are the same, and we can write the product as the $n$th power.

---

The m.g.f. of the mean of a random sample of size $n$ is

$$\psi_{\overline{X}(t)} = [\psi_X(t/n)]^n,\tag{1}$$

where $\psi_X$ is the population m.g.f.

---

Calculating the p.d.f. or p.f. of the sample mean involves the several steps of the following scheme:

$$f_X \;\to\; \psi_X \;\to\; \psi_{\overline{X}} \;\to\; f_{\overline{X}}.$$

We've seen how to calculate $\psi_X$ from $f_X$ and from it, to find $\psi_{\overline{X}}$ using (1). Finding the p.d.f. of $\overline{X}$ from its m.g.f. is called *inverting* the transform that defines the m.g.f. There are integral expressions for the inverse, but it is often quite difficult to use them. Of course, if the m.g.f. of $\overline{X}$ happens to be a function we recognize as the m.g.f. of a familiar distribution, we're in business: The uniqueness theorem for m.g.f.'s says that (except for some pathological examples) there is one and only one distribution with a given m.g.f. So if we know a distribution whose m.g.f. is $\psi_{\overline{X}}$, we know that $\overline{X}$ has that distribution.

**Example 8.9a**

### Sample Mean for Exp($\lambda$)

The m.g.f. of Exp($\lambda$) is given by (4) of §6.2:

$$\psi_X(t) = \frac{\lambda}{\lambda - t}.$$

From (1), the m.g.f. of the mean of a random sample of size $n$ is then

$$\psi_{\overline{X}}(t) = [\psi_X(t/n)]^n = \left(\frac{\lambda}{\lambda - t/n}\right)^n.\tag{2}$$

Is this function of $t$ one that we have seen before as a m.g.f.? Not quite, but it is much like the m.g.f. of $Y \sim \text{Gam}(n, \lambda)$—different in that $t/n$ appears where there is just a $t$ in $\psi_Y$.

Looking back in §6.3, we see from (16) of that section that the m.g.f. (2) is $\psi_Y(t/n)$, where $Y \sim \text{Gam}(n, \lambda)$. But $\psi_Y(t/n) = \psi_{Y/n}(t)$, so $\overline{X}$ has the distribution of $Y/n$, whose p.d.f. is

$$f_{Y/n}(y) = n f_Y(ny) = \frac{(n\lambda)^n}{(n-1)!}\, y^{n-1} e^{-\lambda n y}, \quad y > 0.\tag{3}$$

This is the p.d.f. of $\overline{X}$. ∎

**Example 8.9b**

**Sample Mean for $\mathcal{N}(\mu, \sigma^2)$**

The m.g.f. of $X \sim \mathcal{N}(\mu, \sigma^2)$ is given by (9) of §6.1:

$$\psi_X(t) = \exp\left(\mu t + \frac{1}{2}\sigma^2 t^2\right). \tag{4}$$

Applying (1) we obtain the m.g.f. of the mean of a random sample:

$$\psi_{\overline{X}} = [\psi_X(t/n)]^n = \left(\exp\left\{\mu\frac{t}{n} + \frac{1}{2}\sigma^2\left(\frac{t}{n}\right)^2\right\}\right)^n = \exp\left\{\mu t + \frac{\sigma^2}{n}\frac{t^2}{2}\right\}.$$

And we recognize this, in view of (4), as the m.g.f. of $\mathcal{N}(\mu, \sigma^2/n)$. So when the population is normal, the mean of a random sample is exactly normal—for any sample size, as was claimed in the preceding section. ∎

A rigorous proof of the central limit theorem is not feasible at the mathematical level of this text, but we close this section with an outline of a proof using moment generating functions. (A similar, more general proof using characteristic functions covers cases in which we assume only $\sigma^2 < \infty$.)

Consider a sequence of independent, identically distributed random variables $X_1, X_2, \dots.$ Let $\mu$, $\sigma^2$, and $\psi$ denote the mean, variance, and m.g.f. (respectively) of the common distribution, and assume that $\psi_X(t)$ exists. The standardized mean of the first $n$ observations is the same as the standardized sum:

$$Z_n = \frac{\overline{X}_n - \mu}{\sigma/\sqrt{n}} = \frac{X_1 + \cdots + X_n - n\mu}{\sigma\sqrt{n}} = U_1 + \cdots + U_n,$$

where

$$U_i = \frac{X_i - \mu}{\sigma\sqrt{n}}.$$

Since $E(U_i) = 0$ and $\operatorname{var} U_i = 1/n$, the series expansion for the m.g.f. of each of the $U_i$ is

$$\psi_U(t) = 1 + \frac{t^2}{2n} + R,$$

where $R$ is the remainder term. It can be shown that as $n \to \infty$, this remainder tends to 0 faster than $1/n$, and that, therefore,

$$\psi_{Z_n}(t) = [\psi_U(t)]^n \doteq \left\{1 + \frac{t^2}{2n}\right\}^n \to e^{t^2/2}. \tag{5}$$

This is the m.g.f. of a standard normal variable, and since there is only one distribution with a given m.g.f., the distribution of $Z_n$ must be standard normal. [The conclusion also has used a "continuity theorem," which says that if $\psi_n$ is the m.g.f. of $Y_n$ and tends to $\psi$, the m.g.f. of $Y$, then the c.d.f. of $Y_n$ tends to the c.d.f. of $Y$ at every continuity point of $Y$.]

## 8.10          Normal Populations

The theory of inference for normal populations is mathematically quite tractable, and (as it turns out) it leads to procedures that work rather well for populations that, although not normal, are not too far from normal.

The sample mean and variance are *sufficient* for the mean and variance of a normal population, so it is useful to know their distributions. In standard and familiar notation, the (only) results we need are as follows:

---

Let $(X_1, ..., X_n)$ be a random sample from $\mathcal{N}(\mu, \sigma^2)$. For $n \geq 2$,

**(a)** $\overline{X} \sim \mathcal{N}(\mu, \sigma^2/n)$.

**(b)** $\frac{(n-1)S^2}{\sigma^2} \sim \text{chi}^2(n-1)$.

**(c)** $\overline{X}$ and $S^2$ are independent.

---

We showed (a) in Example 8.9b. We now show how (b) follows from (a) and (c), deferring discussion of (c). The parallel axis theorem for samples says

$$\sum (X_i - a)^2 = \sum (X_i - \overline{X})^2 + n(\overline{X} - a)^2.$$

Setting $a = \mu$ and dividing through both sides by $\sigma^2$, we obtain

$$\sum \left( \frac{X_i - \mu}{\sigma} \right)^2 = \sum \left( \frac{X_i - \overline{X}}{\sigma} \right)^2 + \left( \frac{\overline{X} - \mu}{\sigma/\sqrt{n}} \right)^2.$$

The left-hand side is the sum of squares of $n$ independent, standard normal random variables; it is therefore $\text{chi}^2(n)$. The second term on the right is the square of a single standard normal variable—it is $\text{chi}^2(1)$. For the moment granting (c), we see then that the first term on the right must be $\text{chi}^2(n-1)$, which establishes (b), since that first term is the ratio in (b).

As to the independence asserted in (c), consider first $n = 2$. In Example 5.9b we showed that $X_1 + X_2$ and $X_1 - X_2$ are uncorrelated. We'll see later that in the present context of normal populations, this implies that they are independent. And this in turn means that any function of $X_1 + X_2$ is independent of any function of $X_1 - X_2$. But $\overline{X}$ is clearly a function of $X_1 + X_2$, and $S^2 = (X_1 - X_2)^2/2$ [see Problem 8-44], a function of $X_1 - X_2$. But to show (c) for $n > 2$ is harder, and we'll not present the proof here. It can be done by induction on $n$, starting with $n = 2$. It can also be proved using multivariate generating functions.[5] [It is noteworthy that the independence claimed in (c) holds *only* when the population distribution is normal.[6]]

---

[5] See B. W. Lindgren, *Statistical Theory*, 4th Ed. (New York: Chapman & Hall, 1993), 212.

[6] See E. Lukacs and R. G. Laha, *Applications of Characteristic Functions* (New York: Hafner, 1964), 79.

## Problems

**8-33.**   Check the answers to Problem 8-18 using (9) of §8.7.

* **8-34.**   Consider weighings of the standard weight described in Example 8.1c, and assume that this standard weight is 10 gm. Many past weighings have shown that $\sigma = 6 \ \mu$g. For a new sample of 20 weighings,

 (a)   determine the mean and standard deviation of the sample mean.

 (b)   find the probability that the sample mean differs from the population mean by more than 2 $\mu$g.

**8-35.**   A population of males has mean height $\mu = 70$ inches and s.d. 3 inches. For a random sample of size 400, find

 (a)   the mean and s.d. of $\overline{X}$.

 (b)   $P(\overline{X} > 70.3)$.

 (c)   the probability that the sample mean differs from the population mean (either way) by more than .4 inch.

* **8-36.**   The mean and standard deviation of GRE scores in a particular year were 521 and 123, respectively. Find the probability that the mean score in a random sample of 100 students taking the GRE is less than 500.

**8-37.**   Boxes of detergent are filled by machine. The net weight of a box is a random variable with mean 28 oz and standard deviation .5 oz. Find

 (a)   the mean and s.d. of the total weight of a carton of 24 boxes.

 (b)   the probability that the carton weight exceeds 42.3 lb (16 oz = 1 lb).

* **8-38.**   In a carnival game (observed in Santa Barbara, California, in 1988), a player throws six darts at a board. The board has 396 squares (18 by 22), each with a number from 1 to 6, in no systematic order. We may assume that the number a dart hits has this distribution:

| $x$ | 1 | 2 | 3 | 4 | 5 | 6 |
|---|---|---|---|---|---|---|
| $f(x)$ | $\frac{57}{396}$ | $\frac{29}{396}$ | $\frac{142}{396}$ | $\frac{114}{396}$ | $\frac{27}{396}$ | $\frac{27}{396}$ |

(For instance, there are 142 squares numbered 3.)   Assuming six independent throws, let $Y$ denote the total score—a number from 6 to 36. You win a "large prize" if $Y \geq 29$ and a "small prize" if $Y \leq 14$. Find, approximately,

 (a)   the probability of winning a large prize.

 (b)   the probability of winning a small prize.

**8-39.**   Suppose 15% of the people in a certain population are left-handed. For a random sample of size 200, find

 (a)   the mean and s.d. of the proportion of left-handed people in the sample.

 (b)   the probability (approximate) that fewer than 10% in the sample are left-handed.

* **8-40.**   Suppose 25% of all families with TV's are watching a particular program. Find the probability that more than 26% of a random sample of 1000 TV viewers are watching it.

**8-41.**   Suppose people with type AB blood constitute only 5% of a certain population. Find the probability that fewer than two in a random sample of 100 individuals have this type of blood.

* **8-42.** A poultry farmer has an order for 30 black-winged chicks. Suppose that one-fourth of the chicks that hatch are black winged. How many chicks must hatch for the farmer to be 99.87% sure of having at least 30 black-winged chicks? [*Hint:* The equation $a + bn = c\sqrt{n}$ is quadratic in $\sqrt{n}$.]

**8-43.** Consider two independent random samples from the same normal population. Obtain the m.g.f. of the difference between the sample means (in terms of the parameters of the parent population) and deduce the distribution of the difference.

**8-44.** Show that when $n = 2$, the sample variance is $S^2 = (X_1 - X_2)^2/2$.

**8-45.** Find the m.g.f. of $\widehat{p}$, the proportion of successes in a random sample from Ber($p$). Show that as $n$ becomes infinite, the m.g.f. converges to $e^{pt}$, the m.g.f. of a singular distribution at $p$—one that concentrates all of the probability at a single value. [*Hint:* Using L'Hospital's rule, show that the log of the m.g.f. tends to $pt$ as $n$ becomes infinite.]

* **8-46.** Consider a random sample of size $n = 10$ from $\mathcal{N}(\mu, \sigma^2)$. Find the probability that the sample variance exceeds twice the population variance.

---

## 8.11     Updating Prior Probabilities via Likelihood

So far in this text, we have based inferences on probabilities calculated assuming a particular probability model. The motivation for statistical analyses generally is that the "target" population is not understood—there are aspects of it about which we are unsure. A basic premise in the mode of statistical inference termed *Bayesian* is that uncertainties in one's knowledge of a probability model can be represented by a probability distribution. What this distribution is depends on a person's beliefs, convictions, experience, hunches, etc. Since it depends on the person making the judgment, it is described as *personal* or *subjective*. We introduced this notion in §1.8, where we also mentioned the problem of eliciting a subjective probability and gave a simple example of how this can be done.

When there is uncertainty about a population parameter $\theta$, the Bayesian statistician treats $\theta$ as a random variable $\Theta$ whose distribution describes that uncertainty. Eliciting an entire distribution would appear to be more of a challenge than eliciting the probability of an event. However, with a suitable set of events of the form $[\Theta \leq \theta]$ and an elicitation of the probability of each as described in §1.8, one can sketch an approximate prior distribution for $\Theta$.[7]

Faced with a problem of inference or with a decision problem involving an unknown parameter $\theta$, the statistician can use his or her current probability distribution of $\theta$ to guide inferences and decisions.

A personal probability is subject to modification upon the acquisition of further information—the information supplied by experimental data. Suppose a distribution with p.f. or p.d.f. $g(\theta)$ describes one's present uncertainties about a probability model $f(x \mid \theta)$. Those uncertainties will change with the acquisition of data obtained by doing the experiment modeled by $f$. Bayes' theorem is the fundamental tool for a

---

[7]See D. A. Berry, *Basic Statistics, a Bayesian Perspective* (Belmont, CA: Duxbury Press, 1996).

learning process that involves changing one's probabilities *prior* to gathering the data into those *posterior* to the data. Bayes' theorem [(1) of §2.4] tells us that for any statement $H$ about $\theta$,

$$P(H \mid \text{data}) = \frac{P(\text{data} \mid H)P(H)}{P(\text{data})}. \tag{1}$$

Thus, Bayes' theorem provides the connection between (conditional) probabilities in the data space given a model, and conditional probabilities in the $\Theta$-space given some data.

The probability of $H$ given the data is called the **posterior probability** of $H$— posterior to the data. The (unconditional) probability of $H$, $P(H)$ in (1), is the **prior probability** of $H$. For given data, the probability of the data given $H$, $P(\text{data} \mid H)$, is the **likelihood of $H$**, as defined in §8.2. Since, for given data, the denominator in (1) is a constant, we often write (1) as a statement of proportionality:

$$P(H \mid \text{data}) \propto P(\text{data} \mid H)P(H). \tag{2}$$

In words, this says that the posterior is proportional to the likelihood times the prior, or

$$\text{posterior} \propto \text{likelihood} \times \text{prior}. \tag{3}$$

(It has become part of statistical parlance to shorten "prior distribution" to just "prior," and "posterior distribution" to "posterior.") Another way of expressing this relationship is in terms of odds:

$$\frac{P(H \mid \text{data})}{P(H^c \mid \text{data})} = \frac{P(\text{data} \mid H)}{P(\text{data} \mid H^c)} \cdot \frac{P(H)}{P(H^c)}. \tag{4}$$

In words:   To obtain the posterior odds, multiply the prior odds by the likelihood ratio.

We've written formulas (1)–(4) for the context of discrete distributions. In continuous cases we replace probabilities or p.f.'s by densities. If we denote the "data" by the random variable $Y$ (which could be a vector), and refer to a particular model by a value of $\theta$, we can write (2) as

$$h(\theta \mid Y) \propto L(\theta) \cdot g(\theta), \tag{5}$$

where $g$ is a p.f. or p.d.f. for the prior distribution of $\theta$, and $h$ is a p.f. or p.d.f. for its posterior distribution. And, of course, the likelihood function $L(\theta)$ is (proportional to) the p.f. or the p.d.f. of $Y$ given $\theta$.

We are now dealing with *two* random variables, $\Theta$ and $Y$, and their joint bivariate distribution. Then $\theta$ is a possible value of $\Theta$, $g(\theta)$ is the marginal p.f. or p.d.f. of $\Theta$, $h(\theta \mid y)$ is a conditional p.f. or p.d.f. given $Y = y$, and the $f(y \mid \theta)$ that defines $L(\theta)$ is a conditional p.f. or p.d.f. given $\Theta = \theta$. The marginal p.f. or p.d.f. of $Y$ is not exhibited in (5), since it is just a constant divisor—part of the proportionality constant.

**Example 8.11a** | **Disputed Paternity**

Consider calculating the probability of paternity using only ABO blood-group data. Suppose the mother is type O, the child is type B, and the alleged father is type AB. If the alleged father were type O or type A, he could not be the father. Suppose that he is not one of these two types. This is positive evidence of paternity—but not completely convincing, because there are many other men who are not excluded. Just how convincing is it?

There are two "states of nature" here: Either the alleged father is the father—call this $H$, or he is not: $H^c$. Appealing to Mendelian genetic theory, a geneticist comes up with these likelihoods, where the "data" is the fact that the alleged father's blood type is AB:

$$P(\text{data} \mid H) = .5, \quad P(\text{data} \mid H^c) = .1.$$

The likelihood ratio, of $H$ to $H^c$, is 5:1.

For a prior probability of $H$, blood banks always use $P(H) = 1/2$, and the corresponding odds are 1:1. Then, according to (4),

$$\frac{P(H \mid \text{data})}{P(H^c \mid \text{data})} = \frac{.5}{.1} \times 1 = 5.$$

This implies a probability of 5/6 that the alleged father is the father, and the blood bank would report this to the court in a paternity case. But in Example 2.4b we pointed out that taking $P(H)$ to be .5 has no basis and has silly implications. So, rather than accept the 5/6 figure, a juror (to make an informed decision) should listen to the other testimony and judge the possibility of lying by witnesses, and on this basis form an opinion about $H$ that defines his or her prior, $P(H)$. This prior can then be incorporated with the likelihood ratio using (3) or (4), to yield a corresponding posterior. ∎

A question: Could we use the posterior from one experiment as the prior to use with additional data? When those data come from an independent experiment, the answer is yes—but then the question arises: Might we get a different answer if we had waited with our calculations until all the data from both experiments were at hand, and then used these with the original prior?

To answer this question, suppose we have data $D_1$ with p.d.f. $f_1(D_1 \mid \theta)$ and a prior $g(\theta)$. The posterior p.d.f. is

$$h_1(\theta \mid D_1) \propto g(\theta) f_1(D_1 \mid \theta).$$

Using this function of $\theta$ as a prior in conjunction with new data $D_2$ with p.d.f. $f_2(D_2 \mid \theta)$, we obtain as the posterior

$$h_2(\theta \mid D_1 \text{ and } D_2) \propto g(\theta) f_1(D_1 \mid \theta) f_2(D_2 \mid \theta). \tag{6}$$

If the data $D_1$ and $D_2$ are independent, the p.d.f. of the combined data is

$$f(D_1, D_2) = f_1(D_1 \mid \theta) f_2(D_2 \mid \theta),$$

and the posterior for $\theta$ given the combined data would be the same as (6):

$$h(\theta \mid D_1 \text{ and } D_2) \propto g(\theta) f_1(D_1 \mid \theta) f_2(D_2 \mid \theta).$$

## 8.12   Some Conjugate Families

Sometimes a prior distribution can be approximated by one that is in a convenient family of distributions, one which combines neatly with the likelihood function to produce a posterior that is manageable. We'll find such families for two important population models—Bernoulli and normal.

The Bernoulli parameter $p$ is a number between 0 and 1, so a prior distribution must have the unit interval as its support. A family of distributions on this interval that includes a rather wide variety of individual distributions is the *beta family*, defined by densities of the form

$$f(p \mid r, s) \propto p^{r-1}(1-p)^{s-1}, \ 0 < p < 1. \tag{1}$$

The function on the right has a finite integral provided $r > 0$ and $s > 0$, expressible in terms of gamma functions:

$$B(r, s) = \int_0^1 p^{r-1}(1-p)^{s-1} dp = \frac{\Gamma(r)\Gamma(s)}{\Gamma(r+s)}. \tag{2}$$

This function of the parameters $r$ and $s$ is called the **beta function.** To show that it does give the value of the integral requires some manipulations with double integrals, which we shall forego.[8]

The integrand of (2) is nonnegative, so dividing through by $B(r, s)$ gives us a family of density functions.

---

Density function for the **beta distribution** family:

$$f(p \mid r, s) = \frac{1}{B(r, s)} p^{r-1}(1-p)^{s-1}, \ 0 < p < 1, \tag{3}$$

defined for $r > 0$, $s > 0$. If $X$ has this p.d.f., we write $X \sim \text{Beta}(r, s)$.

---

The beta family of distributions includes one that is uniform ($r = s = 1$), some that are skewed left, some skewed right, some quite peaked, and some more spread out. Figure 8-12 shows several beta densities.

---

[8]See B. W. Lindgren, *Statistical Theory*, 4th Ed. (New York: Chapman & Hall, 1993), 174.

The mean of a beta distribution is easy to find, since the integral that defines it is again of the form (2):

$$E(p) = \frac{1}{B(r,\,s)} \int_0^1 p \cdot p^{r-1}(1-p)^{s-1} dp = \frac{B(r+1,\,s)}{B(r,\,s)}.$$

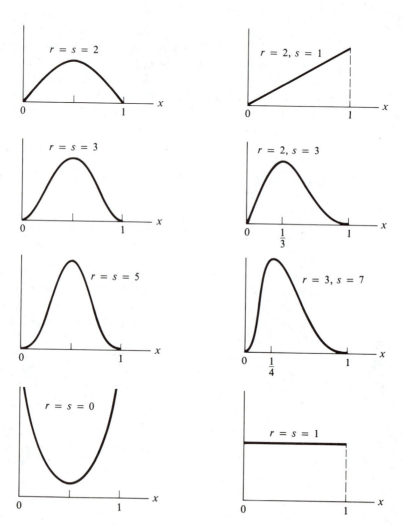

**Figure 8-12** Some beta densities

Using the expression (2) in terms of gamma functions, we obtain

$$E(p) = \frac{\Gamma(r+1)\Gamma(s)}{\Gamma(r+s+1)} \cdot \frac{\Gamma(r+s)}{\Gamma(r)\Gamma(s)} = \frac{r}{r+s}. \qquad (4)$$

[To achieve the final simple formula, we used the recursion formula for gamma functions: $\Gamma(\alpha + 1) = \alpha\Gamma(\alpha)$, from §6.3.] The expected square is found in similar fashion, and the variance follows (Problem 8-54).

**Example 8.12a** | **Bernoulli Populations**

Suppose the unknown population distribution is Ber$(p)$, and that one's prior for $p$ is Beta$(3, 7)$, defined by the p.d.f. $g(p) \propto p^2(1 - p)^6$, $0 < p < 1$. Suppose, further, that the Bernoulli experiment is performed five times, resulting in four successes and one failure. (The number of successes is a sufficient statistic.) The likelihood function for the observed result is $L(p) = p^4(1 - p)$. Then, according to (5) of §8.11,

$$h(p) \propto p^4(1 - p)p^2(1 - p)^6 = p^6(1 - p)^7.$$

So, the posterior distribution of $p$ is Beta$(7, 8)$. The prior mean is $3/10$, and the posterior mean is $7/15$. The result of four successes in five tries has shifted the distribution for $p$ somewhat to the right. The four successes in five tries have suggested that the prior assessment was perhaps too pessimistic. ∎

The normal family of populations is a two-parameter family. Prior distributions would have to be bivariate. To avoid going too far afield, we'll consider only the case of normal populations with known variance. In the next example we'll see that a normal prior distribution for the population mean fits nicely with the normal likelihood.

**Example 8.12b** | **A Normal Population**

Suppose we want to establish the precise weight of a particular block of metal whose nominal weight is 10 grams. Suppose the actual weight is 10 grams plus $\mu$ micrograms.

Using a precision balance known to be unbiased, giving the correct weight on average, we weigh the block five times. Experience suggests that as determined by the balance, measured weights for a given block are random, and (at least approximately) normally distributed, with standard deviation 6 micrograms. Let $X$ denote the amount (in micrograms) by which the measured weight exceeds the nominal weight, so that $X \sim \mathcal{N}(\mu, 36)$.

Suppose our prior for $\mu$ is given by a normal distribution with mean 0 and s.d. 4:

$$g(\mu) \propto e^{-\frac{1}{32}\mu^2}.$$

And suppose it turns out that the mean of our measurements, recorded as amounts in excess of the nominal weight, is $\overline{X} = 7.0$. Since $\sigma^2/n = 36/5$, the likelihood function (when $\overline{X} = 7$) is

$$L(\mu) = e^{-\frac{5}{72}(\mu-7)^2}.$$

The posterior density, from (5) of §8.11, is then

$$h(\mu \mid \overline{X} = 7) \propto e^{-\frac{5}{72}(\mu-7)^2} \cdot e^{-\frac{1}{32}\mu^2} \propto e^{-\frac{29}{288}\mu^2 + \frac{35}{36}\mu}.$$

Completing the square in $\mu$ (and dropping some factors not involving $\mu$), we obtain as the posterior p.d.f. for $\mu$

$$h(\mu \mid \overline{X} = 7) \propto e^{-\frac{29}{288}(\mu-140/29)^2}.$$

This is a normal p.d.f., with mean $140/29 \doteq 4.83$ and s.d. $\sqrt{144/29} \doteq 2.23$. The data have shifted the center of the distribution of $X$ from 0 to 4.83, and narrowed the distribution somewhat, from an s.d. of 4 to an s.d. of about 2.23. The graphs of the likelihood, prior, and posterior are shown in Figure 8-13. ∎

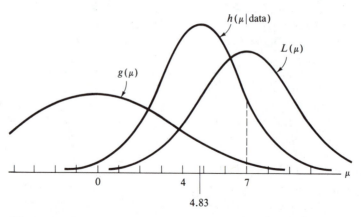

**Figure 8-13** Prior, likelihood, and posterior—Example 8.12b

Going through the same mathematical motions for the more general case: $\mu \sim \mathcal{N}(\nu, \tau^2)$, $X \sim \mathcal{N}(\mu, \sigma^2)$, and a random sample of size $n$, the result is a normal posterior with

$$E(\mu \mid \overline{X}) = \frac{1/\tau^2}{1/\tau^2 + n/\sigma^2}\, \nu + \frac{n/\sigma^2}{1/\tau^2 + n/\sigma^2}\, \overline{X}, \qquad (5)$$

and

$$\mathrm{var}(\mu \mid \overline{X}) = \frac{1}{1/\tau^2 + n/\sigma^2} = \frac{\sigma^2\tau^2}{\sigma^2 + n\tau^2}. \qquad (6)$$

In terms of *precisions*, defined as $\pi_{\text{pr}} = 1/\tau^2$, $\pi_{\text{data}} = n/\sigma^2$, we write the precision of the posterior as

$$\pi_{\text{po}} = \frac{1}{\text{var}\,(\mu \mid \overline{X})} = \pi_{\text{pr}} + \pi_{\text{data}}. \tag{7}$$

The posterior mean is a linear combination of prior mean and sample mean, weighted according to the precisions:

$$E(\mu \mid \overline{X}) = \frac{\pi_{\text{pr}}}{\pi_{\text{po}}} \nu + \frac{\pi_{\text{data}}}{\pi_{\text{po}}} \overline{X}. \tag{8}$$

We note, in particular, the case where $\pi_{\text{pr}} = 0$ ($\tau^2 = \infty$). Strictly speaking, the prior in this case is not a proper distribution. We can approach the notion of a constant density on $(-\infty, \infty)$ by taking the limit of a proper normal distribution as $\tau \to \infty$, which becomes flat. So we say that the improper "normal" prior with $\tau^2 = \infty$ is a flat prior.

Taking the limit of (7) and (8) as $\tau \to \infty$ (or $\pi_{\text{pr}} \to 0$), we see that the posterior mean tends to $\overline{X}$, and the posterior variance, to $\sigma^2/n$. And we obtain the *proper* posterior $\mathcal{N}(\overline{X}, \sigma^2/n)$ when we formally substitute a constant for "prior" in the expression "posterior $\propto$ likelihood $\times$ prior."

The normal family is said to be **conjugate** to the normal likelihood (when the population variance is known), because the posterior is in the same family as the prior. Similarly, the beta family is said to be conjugate to the Bernoulli likelihood. Another instance of a family of likelihoods with a conjugate family of priors is given in Problem 8-51.

---

## 8.13          Predictive Distributions

The inference to be presented in the next few chapters is primarily directed at the problems of estimating population parameters and testing specified population models. However, it is important in some situations to predict the next observation, or the next sequence of observations, based on data at hand.

Consider a random sample $\mathbf{X} = (X_1, \ldots, X_n)$ from $f(x \mid \theta)$ and the problem of predicting the next $m$ independent observations from the same population: $\mathbf{Y} = (Y_1, \ldots, Y_m)$. Given the data, the likelihood function is

$$L(\theta) = f(X_1 \mid \theta) \cdots f(X_n \mid \theta) \tag{1}$$

and the joint p.d.f. or p.f. of the new observations is

$$f^*(y_1, \ldots, y_m \mid \theta) = f(y_1 \mid \theta) \cdots f(y_m \mid \theta). \tag{2}$$

Given any particular distribution for the parameter $\theta$, we define the **Bayesian predictive distribution** of the new observations as the average of $f^*$ with respect to the distribution of $\theta$. With this we can also obtain a predictive distribution for any function of the new observations.

In the special case of a Bernoulli population, we might ask, "What is the probability that the next observation is a success?" Then $m = 1$, and the *conditional* probability of success at a single trial is the parameter $p$. The *unconditional* probability of success (see Problem 5-R17) is the average with respect to the current distribution of $p$:

$$E[P(\text{success} \mid p)] = E(p), \tag{3}$$

More generally, the probability of exactly $k$ successes in the next $m$ trials is

$$E\left\{ \binom{m}{k} p^k (1 - p)^{m-k} \right\}. \tag{4}$$

**Example 8.13a** | **Broken Windows**

The following situation arose in an actual legal case, although we have modified the specifics. Twenty windows in a high-rise office building broke in the first year after it was built. The question at issue is the number of these that were caused by a particular type $D$ of defect in the glass. If the breakages were caused by defect $D$, then the manufacturer would be at fault for the breakage and would be liable for replacing the thousands of windows in the building. If they were not caused by $D$, then the company that installed the windows would be liable.

Only four of the broken panes were available for analysis; assume these constitute a random sample of the 20 that broke. All four were found to have broken because of $D$. The question is, how many of the other 16 were caused by $D$?

The windows on the building were all from a particular lot. A glass expert testifies that the lot proportion $p$ of windows with defect $D$ among all windows with defects varies, having a $J$-shaped distribution with p.d.f.

$$g(p) \propto p^{-3/4}(1 - p)^{-1/4}, \; 0 < p < 1. \tag{5}$$

This is a beta density, shown in Figure 8-14, with mean $E(p) = 1/4$. In particular, $P(p > .5) \doteq .22$. [This is the area under $g(p)$, calculated with statistical software.] That is, $D$ is the predominant defect in only 22% of the lots.

Consider, then, 20 independent observations on $\text{Ber}(p)$, where $p$ has the prior distribution with p.d.f. given by (5). We have observed $U = 4$, where $U = X_1 + \cdots + X_4$, the number of 1's (defectives) among the first four observations. The likelihood function based on this is $p^4$. So, the posterior p.d.f. is

$$h(p \mid U = 4) \propto p^4 p^{-3/4}(1 - p)^{-1/4}, \; 0 < p < 1,$$

which is $Beta(4.25, .75)$, with mean $E(p \mid U = 4) = 4.25/5$. We'd like the predictive distribution of $Z = Y_1 + \cdots + Y_{16}$, the number of defectives among the 16 unavailable windows.

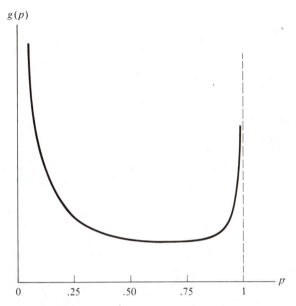

**Figure 8-14** Prior distribution for $p$—Example 8.13a

Given $p$, the p.f. of $Z$ is binomial:

$$f^*(z \mid p) = \binom{16}{z} p^z (1-p)^{16-z}, \quad z = 0, 1, ..., 16.$$

The predictive probability function for $Z$ is then

$$f(z) = E[f^*(z \mid p)] = \binom{16}{z} \frac{\int_0^1 p^{z+3.25}(1-p)^{15.75-z}dp}{\int_0^1 p^{3.25}(1-p)^{-.25}dp}.$$

The integrals in this fraction are beta functions, expressible in terms of gamma functions, and we found the values of $f$ by first calculating $f(16)$ and then using the fact that

$$f(k) = f(k+1) \cdot \frac{k+1}{k+4.25} \cdot \frac{15.75-k}{16-k}.$$

These predictive probabilities are given in Table 8-2 and shown in Figure 8-15. The table and figure are appropriate for presentation in court.

**Table 8-2**

| $z$ | $f(z)$ |
|---|---|
| 0 | .00008 |
| 1 | .00036 |
| 2 | .00096 |
| 3 | .00205 |
| 4 | .00378 |
| 5 | .00637 |
| 6 | .01006 |
| 7 | .01510 |
| 8 | .02184 |
| 9 | .03069 |
| 10 | .04217 |
| 11 | .05701 |
| 12 | .07626 |
| 13 | .10168 |
| 14 | .13667 |
| 15 | .19004 |
| 16 | .30486 |

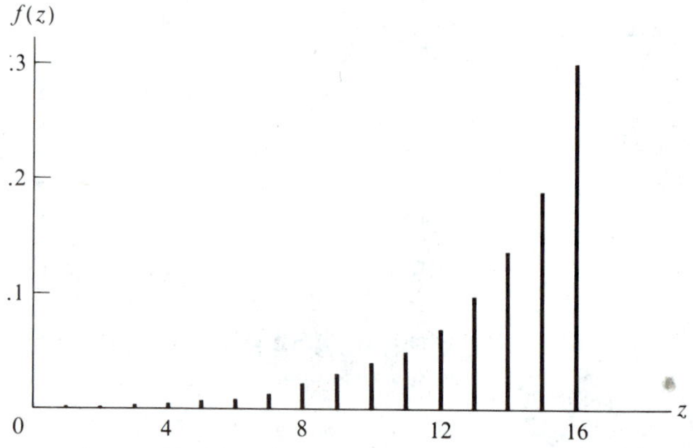

**Figure 8-15** Predictive distribution of $p$

The court may also be interested in unconditional mean $E(Z)$. Since, given $p$, $Z \sim \mathrm{Bin}(16, p)$, we see that $E(Z \mid p) = 16p$, so [by the iterated expectation formula, (6) of §5.11],

$$EZ = E[E(Z \mid p)] = 16E(p) = 13.6.$$

So, the expected number of the 20 broken windows caused by defect $D$, given $Y = 4$, is 17.6. ∎

## Problems

* **8-47.** Given a single observation $X$ from $\mathrm{Poi}(\lambda)$, find the posterior distribution for $\lambda$ when the prior p.d.f. is $g(\lambda) = 3e^{-3\lambda}$, $\lambda > 0$.

**8-48.** Suppose $X$ is geometric: $f(x \mid p) = p(1 - p)^{x-1}$, $x = 1, 2, \ldots$. Given that the prior for $p$ is $\mathrm{Beta}(r, s)$, find the posterior distribution for $p$ after obtaining a random sample of size $n$.

* **8-49.** In a segment of the TV show "The Odd Couple," Oscar was dubious of Felix's claim to have ESP. They conducted an experiment in which Oscar would draw a card at random from

four different cards and, without showing it, ask Felix to identify it. Let $p$ denote the probability that Felix correctly identifies the card that Oscar has drawn. Given that Felix succeeds six times in ten trials, find the posterior p.d.f. of $p$ when

    **(a)**   the prior for $p$ is $\mathcal{U}(0, 1)$.

    **(b)**   the prior p.d.f. is proportional to $p(1 - p)^4$.

    **(c)**   the prior p.d.f. is proportional to $p^5(1 - p)$.

**8-50.**   Repeat the calculations of the preceding problem, but with six successes out of ten trials replaced by ten out of ten.

**8-51.**   Consider a random sample of size $n$ from Exp($\lambda$), and assume a gamma prior for $\lambda$: $g(\lambda) \propto \lambda^{\alpha-1}e^{-\beta\lambda}$. Find the posterior p.d.f. for $\lambda$ in terms of $n$, $\alpha$, $\beta$, and $\overline{X}$.

**8-52.**   Box 1 contains one white and three black balls; box 2 contains one black and three white balls. A ball is selected from one of the boxes and found to be white. Find the (posterior) probability that it is from box 1, given that

    **\*(a)**   the boxes were equally likely to be chosen for the selection (prior to seeing the selected ball).

    **(b)**   the prior odds on box 1's being the one from which the ball will be drawn are 4 to 1.

**8-53.**   Show that if all of the prior probability for $\theta$ is concentrated at $\theta = \theta_0$, then (except for one case) the posterior distribution is the same as the prior. (That is, no amount of data can change the mind of a "know-it-all.")

**\* 8-54.**   Obtain a formula for the variance of a beta distribution.

**\* 8-55.**   Suppose a minicourse is directed at preparing students for a particular aptitude test, and that test scores are approximately $\mathcal{N}(\mu, 80^2)$. Assume that the prior for $\mu$ is $\mathcal{N}(520, 20^2)$. In a random sample of 50 students who have taken the course, the average score is found to be 513. Find the posterior distribution of the scores of those who have taken the course.

**8-56.**   A university professor (who was trying to convince a U.S. senator to vote against the MX missile program) posed this question: If an enemy missile has been fired successfully ten times in ten trials, what is the probability that in the next five firings, there are no failures? If one takes the sample proportion 10/10 as the value of $p$, the answer is 1. However, this is unrealistic. Suppose (prior to learning the data) we'd judge, from other evidence, that $p$ may be near 1 and adopt a beta prior for $p$ with p.d.f. $g(p) \propto p^4$, $0 < p < 1$. This peaks at $p = 1$ and has mean 5/6. Given this prior and the assumed data, find the predictive distribution of the number of failures in the next five firings.

# Review Problems

**8-R1.**   Given the single observation $X \sim$ Bin(4, $p$), suppose we observe $X = 3$. Let $L(p)$ denote the likelihood function, and find $L(\tfrac{1}{2})/L(\tfrac{1}{3})$.

**8-R2.**   Consider random samples of size $n$ from $f(x \mid \theta) = \theta x^{\theta-1}$, $0 < x < 1$.

    **(a)**   Find the joint p.d.f. of the sample observations.

    **(b)**   Find a (one-dimensional) sufficient statistic.

    **(c)**   Find the c.d.f. of the largest sample observation, $X_{(n)}$.

    **(d)**   Find $E[X_{(n)}]$.

**8-R3.**   The discrete observation $X$ has four possible values: 0, 1, 2, 3. Consider the four models given in the accompanying table of probabilities:

|   | $\theta_1$ | $\theta_2$ | $\theta_3$ | $\theta_4$ |
|---|---|---|---|---|
| 0 | .2 | .2 | .4 | .2 |
| 1 | .5 | .6 | .3 | .7 |
| 2 | .1 | .1 | .2 | .1 |
| 3 | .2 | .1 | .1 | 0 |

(a) Suppose we observe $X = 1$. Find the $\theta$ with the largest likelihood.

(b) For each $\theta$, calculate $P(X = i \mid X = 0 \text{ or } 2)$.

(c) Show that $T = |X - 1|$ is sufficient for $\theta$.

**8-R4.** Referring to the setting of the preceding problem, consider the prior $g(\theta_1) = .2$, $g(\theta_2) = .4$, $g(\theta_3) = .3$, $g(\theta_4) = .1$.

(a) Find the (unconditional) probability that $X = 0$.

(b) Find the posterior probability $h(\theta_3 \mid X = 0)$.

**8-R5.** Given that $X$ has the p.d.f. $\frac{x}{\theta^2} e^{-x/\theta}$ for $x > 0$, with $EX = 2\theta$, $\sigma_X^2 = 2\theta^2$ ($\theta > 0$), consider random samples of size $n$ and let $T = \overline{X}/2$. In terms of $\theta$,

(a) find the mean and standard deviation of $T$.

(b) give the asymptotic distribution of $T$.

**8-R6.** Consider again the population of the preceding problem, this time with the parameter $\lambda = 1/\theta$: $f(x \mid \lambda) = \lambda^2 x e^{-\lambda x}$, $x > 0$.

(a) Find the likelihood function $L(\lambda)$, given that in a random sample of size 10, the sample sum is 12.

(b) Given the prior p.d.f. $g(\lambda) = e^{-\lambda}$ ($\lambda > 0$), find the posterior p.d.f. of $\lambda$.

(c) Identify the distribution of the sample sum (given $\lambda$). [*Hint:* The population is of the gamma type.]

**8-R7.** Find $P(\overline{X} > .54)$, where $\overline{X}$ is the mean of a random sample of size 300 from a uniform population on the unit interval, $(0, 1)$.

**8-R8.** Consider a population in which the proportion of Hispanics is 10%. Find the probability that in a random sample of size 225, the proportion of Hispanics is less than 5%.

**8-R9.** Consider random samples from a Rayleigh population with density function $\frac{x}{\theta} \exp\left(-\frac{x^2}{2\theta}\right)$, $x > 0$. Given: $EX = \sqrt{\pi\theta/2}$, $\operatorname{var} X = 2\theta(1 - \pi/4)$.

(a) Find a sufficient statistic.

(b) Find $ET$, where $T = \sum X_i^2$.

(c) Find the distribution of $T$. [*Hint:* First show $X^2/\theta \sim \operatorname{Exp}(\frac{1}{2})$.]

**8-R10.** Consider independent random samples of size 400 from population 1 with mean $\mu_1$, and of size 100 from population 2 with mean $\mu_2$. Given that the population s.d.'s are $\sigma_1 = 3$, $\sigma_2 = 2$, find (approximately) the probability that the sample means will differ by more than .5

(a) if $\mu_1 = \mu_2$.   (b) if $\mu_1 - \mu_2 = .2$.

**8-R11.** Suppose $Y = \beta x + \epsilon$, where $\epsilon \sim \mathcal{N}(0, \sigma^2)$. For each of $n$ given values $x = x_i$, we observe a response $Y_i$.

(a) Find the likelihood function $L(\beta, \sigma^2)$.

(b) Find a pair of statistics that are sufficient for the two parameters.

**8-R12.** Refer to the expression for the p.d.f. of the $k$th smallest observation given as (1) in §8.6.

**(a)** Find the expected value of $X_{(k)}$ in the case of a random sample from $\mathcal{U}(0, 1)$. [*Hint:* The integral is a beta function.]

**(b)** The average spacing of the ordered observations in a random sample from $\mathcal{U}(0, 1)$ has some practical applications. Use the result in (a) to find the mean spacing, $E[X_{(k+1)} - X_{(k)}]$.

**8-R13.** In some environments, the length of life of electronic components follows an approximately exponential distribution. Consider a random sample of size $n$ from $f(x \mid \lambda) = \lambda e^{-\lambda x}$ $(x > 0)$.

**(a)** Write the likelihood function.

**(b)** Assume a gamma prior with p.d.f. $g(\lambda) \propto \lambda^{\alpha-1} e^{-\beta \lambda}$, $\lambda > 0$, and find $h(\lambda \mid \mathbf{x})$, the posterior p.d.f. of $\lambda$.

**(c)** Suppose we now take a new unit, with life-length $Y$. Find the predictive p.d.f. of $Y$: $f^*(y) = E_h[f(y \mid \lambda)]$. (It will be essentially the ratio of two gamma functions.)

**(d)** Find the mean of the predictive distribution in (c) and its limit as $n$ becomes infinite.

**8-R14.** The random variable $V_k = F(X_{(k)})$ is the area to the left of the $k$th smallest observation under the population p.d.f., $f(x) = F'(x)$.

**(a)** Assuming a random sample, show that the distribution of $V_k$ is that of the $k$th smallest in a random sample from $\mathcal{U}(0, 1)$.

**(b)** Use the result of (a) to find the average area under the population p.d.f. between successive ordered observations.

**8-R15.** The prior distribution with p.d.f. $g(p) \propto [p(1 - p)]^{-1/2}$, $0 < p < 1$, is known as *Jeffreys' prior* for the Bernoulli parameter $p$.

**(a)** Find the constant of proportionality.

**(b)** Find the (unconditional) probability of success, assuming Jeffreys' prior.

**(c)** Suppose you have adopted Jeffreys' prior and observe a success. What is the new probability of success at the next trial? That is, given that one trial results in success, find the predictive distribution of $p$ for the next trial.

---

## Chapter Perspective

Statistical inference is the process of drawing conclusions about populations on the basis of sample information. In this chapter we have seen that sample statistics are random variables—they have probability distributions. We have addressed various mathematical issues dealing with the distributions of sample statistics. In particular, we have studied how to derive (or simulate, if derivations are intractable) these "sampling distributions," how they depend on the various population characteristics, and how the reliability of sample information increases with the sample size.

The "central limit theorem" gives us a powerful tool for obtaining approximate sampling distributions, for use when the sample size is large. It also suggests why the distributions of so many variables in real-life situations are well approximated by a normal curve, being the combinations of many independent factors.

The notions of likelihood and sufficiency are helpful in seeing how best to extract the information from a sample that is pertinent to a given population characteristic.

The likelihood function summarizes what is useful in the sample and provides a powerful intuitive basis for designing inferential procedures. We'll see in later chapters that, broadly speaking, there is nothing lost when we base inferences on sufficient statistics.

The Bayesian approach to inference introduced in §8.11 is very different from the "classical" approach, in which parameters are treated as fixed constants, albeit unknown. The Bayesian approach permits one to do what users of statistics quite naturally want to do: namely, to attach probabilities to possible models. There is a price to pay, however—one must assume a prior distribution. We have seen how prior distributions are revised or "updated" on the basis of sample information. In subsequent chapters we'll describe ways to use a posterior distribution in various types of statistical inference.

# Estimation

A random sample from a population contains information about the various aspects of the population. Estimating population means, population proportions, and other sample parameters using sample information is an important practical problem.

**Example 9a**

## TV Ratings

Producers and advertisers are vitally interested in knowing what TV viewers watch. Surveying *all* viewers is impossible, but viewers can be *sampled*. TV ratings by Arbitron and Nielsen are based on samples of TV owners. These ratings are estimates of population proportions based on sample information. Television executives make decisions with far-reaching consequences on the basis of such estimates.

A Nielsen rating of 22 for a particular TV program means that 22% of viewers in the sample were watching that program. The sample proportion .22 is an estimate of the proportion of the population of TV owners who watched the program. This is a sensible estimate if the sample is reasonably representative of that population. ∎

**Example 9b**

## Auditing Accounts by Sampling

The book value of an account receivable of a large company may or may not agree with the actual value, owing to errors in the records. Auditors, faced with enormous numbers of accounts and transactions, may conduct an audit by taking a sample of accounts and examining the sampled accounts in detail. They use random sampling to estimate the error rate as well as the magnitudes of errors. ∎

An estimate based on a random sample is a statistic—a random variable. Different samples generally lead to different estimates. An estimate calculated from any one sample is apt to be wrong. The amount of the error is random.

**Example 9c**

## Variability of Sample Means

In Example 8.1c we referred to the very precise scales used by the National Bureau of Standards to calibrate standard weights. Despite the scale's high precision, different weighings of the same object give different results. The variation in these results can be modeled as a probability distribution. If it were possible to perform infinitely many weighings, their average would be the "true" weight—assuming that there is no systematic error caused by such factors as moisture on the scale. (Bureau scales are periodically calibrated to eliminate such biases.)

Suppose we want to calibrate a scale using a standard weight, a block of metal whose weight is rather precisely known. If we could weigh it an infinite number of times and average these weighings, this would tell us how to adjust the scale. We can't, so a finite number of weighings must suffice.

Suppose we weigh the block 20 times, observing the following differences between the observed weight and the nominal weight (in micrograms):

| 8.82 | 7.39 | 18.86 | 2.65 | 12.92 | 2.07 | 4.26 | $-3.05$ | 6.93 | $-15.40$ |
|------|------|-------|------|-------|------|------|---------|------|----------|
| 5.95 | 15.97 | $-6.78$ | $-3.80$ | 4.52 | 5.54 | 13.45 | $-5.17$ | 3.57 | 4.92 |

The mean of the 20 differences is 4.181. Actually, we generated the data artificially, sampling from a population with mean 5.00 and standard deviation 6.00. So the estimate 4.181 of the amount of needed adjustment is not correct. Whenever we base our estimate on a sample, the estimate is apt to be in error. The question is, how large is the error? ■

In this chapter we consider a number of natural estimates of population parameters. For example, the largest sample value $X_{(n)}$ estimates the largest population value, and a sample proportion $\widehat{p}$ estimates the corresponding population proportion $p$. And when a parameter to be estimated is a function of population moments, an obvious candidate for the estimate is the same function of the sample moments.

We begin by studying ways of assessing and reporting errors of estimation and showing how to determine the sample size required to keep the error within a specified level.

## 9.1     Errors in Estimation

We're going to be estimating various population parameters—means, proportions, variances, and so on. As in the preceding chapter, let $\theta$ denote a generic parameter. Let $T$ denote a statistic proposed for estimating $\theta$. When used for this purpose, the statistic $T$ is called an **estimator.** We call the value of $T$ calculated from a particular sample an **estimate** of $\theta$.

The value of an estimator varies from sample to sample. So, when we find the value of an estimator $T$ from a particular sample, it is too much to expect that $T$ exactly *equals* $\theta$. Yet, we can hope that it is *near* $\theta$ in some sense. We consider $T$ to be a good estimator of $\theta$ if its sampling distribution is concentrated near $\theta$.

**Example 9.1a**

## Estimating a Maximum

Suppose $X \sim \mathcal{U}(0, \theta)$, and consider estimating $\theta$ based on the observations in a random sample from $X$ of size $n$. The largest sample observation $Y = X_{(n)}$ is sufficient for $\theta$ (see Example 8.3d) and is an obvious candidate for estimating the largest population value $\theta$. In Example 8.4c we derived the c.d.f. of $Y$:

$$F_Y(y) = \left(\frac{y}{\theta}\right)^n, \quad 0 < y < \theta.$$

When $n = 100$, for example, the probability that $Y$ is within 2% of $\theta$ is

$$P(.98\theta < Y < \theta) = F_Y(\theta) - F_Y(.98\theta) = 1 - (.98)^{100} \doteq .867.$$

This large a probability in such a small interval results from the enormous concentration of probability near $\theta$. The standard deviation of $X_{(100)}$ is only about $(.01)\theta$. ∎

When we use $T$ to estimate $\theta$, the **error** is the difference $T - \theta$. This error varies from sample to sample because $T$ does. The **absolute error** of estimation is $|T - \theta|$. This is the distance between the estimated value $T$ and the actual value $\theta$. An obvious overall measure of how poorly an estimator $T$ performs is the **mean absolute error** $E|T - \theta|$, where the averaging is with respect to the sampling distribution of $T$. However, a more commonly used criterion is the **mean squared error:**

$$\text{m.s.e.}(T) = E[(T - \theta)^2]. \tag{1}$$

We use the parallel axis theorem [(8) in §3.3 and (3) in §5.6] to express the second moment on the right in terms of the variance:

$$E[(T - \theta)^2] = \operatorname{var} T + [ET - \theta]^2. \tag{2}$$

The quantity $ET - \theta$ is termed the **bias** in $T$ as an estimator of $\theta$. In words: the mean squared error is the variance of $T$ plus the square of its bias.

---

The **mean squared error** using $T$ as an estimator of $\theta$ is

$$\text{m.s.e.}(T) = E[(T - \theta)^2] = \operatorname{var} T + [b_T(\theta)]^2, \tag{3}$$

where $b_T$ is the **bias:**

$$b_T(\theta) = ET - \theta. \tag{4}$$

When $b_T(t) = 0$, $Y$ is said to be **unbiased.**

---

Since both the variance and the squared bias are nonnegative, the m.s.e. will be small only when *both* the bias and the variance of $T$ are small. When they are both small, most of the sampling distribution of $T$ is concentrated near $\theta$. However, unbiasedness is not in itself especially virtuous.

**Example 9.1b** | **Bias in Estimating $\sigma^2$**

In §7.5 we defined this measure of sample variability:

$$V = \frac{1}{n} \sum_{i=1}^{n} (X_i - \overline{X})^2. \tag{5}$$

According to the parallel axis theorem [(5) of §7.5, with $c = \mu$],

$$\frac{1}{n} \sum_{i=1}^{n} (X_i - \mu)^2 = V + (\overline{X} - \mu)^2.$$

Transposing the last term and taking expected values, we find

$$EV = \frac{1}{n} \sum_{i=1}^{n} E\big[(X_i - \mu)^2\big] - E\big[(\overline{X} - \mu)^2\big]. \tag{6}$$

But each $X_i$ has the population distribution, so $E[(X_i - \mu)^2] = \sigma^2$ for each $i$:

$$\frac{1}{n} \sum_{i=1}^{n} E\big[(X_i - \mu)^2\big] = \frac{1}{n}\big(\sigma^2 + \cdots \sigma^2\big) = \sigma^2.$$

And because $E\overline{X} = \mu$, the second term on the right of (6) is var $\overline{X}$, or $\sigma^2/n$. Thus, (6) becomes

$$EV = \sigma^2 - \frac{\sigma^2}{n} = \frac{n-1}{n}\sigma^2.$$

It follows that $V$ is biased in estimating $\sigma^2$; the bias is $b_V(\sigma^2) = -\sigma^2/n$.

On the other hand, the "sample variance" as defined by (6) of §7.5:

$$S^2 = \frac{n}{n-1}V = \frac{1}{n-1}\sum_{i=1}^{n}(X_i - \overline{X})^2,$$

is unbiased in estimating $\sigma^2$, because

$$E(S^2) = E\left(\frac{n}{n-1}V\right) = \frac{n}{n-1} \cdot \frac{n-1}{n}\sigma^2 = \sigma^2.$$

However, we shall show in §9.9 that in the case of a normal population, it is $V$ that has the smaller m.s.e., despite its bias.

Lest you become too enthralled with the unbiasedness of $S^2$, note that as an estimator of the more easily interpreted $\sigma$, the statistic $S$ is *biased*. For,

$$0 < \text{var } S = E(S^2) - (ES)^2 = \sigma^2 - (ES)^2$$

(unless $S$ is constant). And this says that $ES < \sigma$, so $b_S(\sigma) \neq 0$. ■

The units of m.s.e. are the units of $T^2$. This makes the m.s.e. hard to interpret. The square root of the m.s.e. has the same units as $T$ itself; we refer to this as the r.m.s. (*root-mean square*) error of estimation. It is interpreted as an "average" or "typical" amount of error. Of course, the estimate obtained from any particular sample can be off by more or by less than the typical error—in either direction.

---

The **r.m.s. error** of $T$ as an estimator of $\theta$ is

$$\text{r.m.s.e.}(T) = \sqrt{\text{m.s.e.}} = \sqrt{E[(T-\theta)^2]}. \qquad (7)$$

When $T$ is unbiased, r.m.s.e.$(T) = \sigma_T$.

---

To estimate a population mean $\mu$ from a random sample, we turn naturally to the sample mean $\overline{X}$. Its bias is 0, since $E\overline{X} = \mu$ [see (4) of §8.7], so the r.m.s. error of $\overline{X}$ is its standard deviation [(5) of §8.7]:

$$\text{r.m.s.e.}(\overline{X}) = \sigma_{\overline{X}} = \frac{\sigma}{\sqrt{n}}. \qquad (8)$$

This depends on a population parameter, but not on the one being estimated. Still, $\sigma$ is generally unknown, and when that is the case, we can only approximate the r.m.s. error (8). When $n$ is large, the sample distribution tends to be close to the population distribution, and the sample standard deviation $S$ will be close to $\sigma$, so we can approximate (8) by replacing $\sigma$ by $S$:

---

**Standard error** of $\overline{X}$ as an estimate of the population mean $\mu$:

$$\text{s.e.}(\overline{X}) = \frac{S}{\sqrt{n}}. \qquad (9)$$

---

Of course, in the uncommon situation in which the population variance is known, we'd simply use (8), the standard deviation of the estimator—which is what we really wanted in the first place—as the criterion. (Some statisticians use the term *standard error* for the standard deviation of the error and call the approximation (9) the *estimated standard error.*)

---

**Example 9.1c**   **Estimating a Mean Platelet Count**

The report cited in Example 7.2e gives blood platelet counts for 35 healthy males. The mean count per microliter was $\overline{X} = 234$, and the standard deviation was

$S = 45.05$. We assume random sampling. The standard error of $\overline{X}$ is

$$\text{s.e.}(\overline{X}) = \frac{S}{\sqrt{n}} = \frac{45.05}{\sqrt{35}} = 7.61.$$

The result would often be reported as $234 \pm 7.6$—the *point estimate* 234 plus or minus the standard error.     ■

Estimating a population proportion $p$ is a special case of estimating a mean. The corresponding sample mean is the proportion $\hat{p}$, a natural estimator of $p$. When we assume random sampling, $\hat{p}$ is sufficient for $p$. According to (6) of §8.7, $E(\hat{p}) = p$, so $\hat{p}$ is unbiased. Its r.m.s.e. is therefore its standard deviation, given by (7) of §8.7:

$$\text{r.m.s.e.}(\hat{p}) = \sqrt{\frac{p(1-p)}{n}}. \tag{10}$$

This depends on $p$, which is awkward in that we'd need to know the very parameter we're estimating in order to assess its performance! But there is a way around this.

When $n$ is large, the r.m.s. error is small, so $\hat{p}$ is likely to be close to $p$, and substituting $\hat{p}$ for $p$ in (10) would give us a good approximation to the r.m.s.e. And, actually, the function $\sqrt{p(1-p)}$ is only slowly changing with $p$, unless $p$ is close to 0 or 1; so $\sqrt{\hat{p}(1-\hat{p})}$ is a reasonably good approximation even if $\hat{p}$ isn't quite so close to $p$. Figure 9-1 shows a graph of $\sqrt{p(1-p)}$. With the replacement of $\hat{p}$ by $p$ in (10), we have the following definition:

---

**Standard error of $\hat{p}$** as an estimate of a population proportion $p$:

$$\text{s.e.}(\hat{p}) = \sqrt{\frac{\hat{p}(1-\hat{p})}{n}}. \tag{11}$$

---

Formula (11) for standard error of a proportion would be a special case of (9), the formula for the standard error of the mean, but for the fact that we use the divisor $n-1$ instead of $n$ in calculating $S^2$. The variance $V$ of the sample distribution (with divisor $n$) is $\hat{p}(1-\hat{p})$, so (11) is $\sqrt{V/n}$ instead of $S/\sqrt{n}$.

**Example 9.1d**  |  **TV Ratings**

A ratings company takes a random sample of 1000 TV owners to estimate the proportion $p$ who are watching a particular program. It finds that 290 of the viewers in the sample are watching the program. It then announces the program's rating as 29, taking the sample proportion $\hat{p}$ as an estimate of the population proportion $p$. The

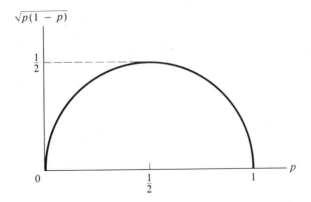

$\sqrt{p(1-p)}$

**Figure 9-1**

standard error (11) is

$$\text{s.e.}(\widehat{p}) = \sqrt{\frac{\widehat{p}(1-\widehat{p})}{n}} = \sqrt{\frac{.29 \times .71}{1000}} \doteq .0143.$$

This may be thought of as a typical error involved in estimating $p$ to be $\widehat{p}$. The estimate is often reported in the form $.29 \pm .014$.  ■

Having defined a measure of precision for estimation, we see next how specifying the desired precision can lead to a choice of sample size.

## 9.2        Determining Sample Size

In setting up a research study, the investigator must decide how much data is needed. Needed to do what? is a logical question. Perhaps what the investigator really wants to know is how much data is needed to achieve a specified accuracy. Ordinarily, the larger the sample, the smaller the standard error of estimate, but observations cost time and money, so one can't go on sampling indefinitely. If the researcher can give an *allowable* standard error of estimation, we can use this specification to determine a sample size large enough that the standard error will not exceed that amount.

Consider estimating a population mean when the population variance is known. To meet a specification that the standard deviation of the mean should not exceed $\epsilon$, we can choose $n$ such that

$$\sigma_{\bar{X}} = \frac{\sigma}{\sqrt{n}} \leq \epsilon, \text{ or } n \geq \frac{\sigma^2}{\epsilon^2}. \tag{1}$$

This will hold when $n$ is the next integer larger than $\sigma^2/\epsilon^2$, and this $n$ will be large enough to meet the specification.

**Example 9.2a**

## Calibrating a Scale

A standard weight is to be used in calibrating a scale, as in Example 8.1b. We require the standard deviation of $\overline{X}$ to be at most two micrograms. How many weighings will be needed, given that $\sigma = 6\,\mu$g? According to (1), we should take $n = \sigma^2/4 = 9$. ∎

Using (1) to determine a sample size requires that we know $\sigma$. In most applications, $\sigma$ is not known. The only experimental settings we can think of in which $\sigma$ is known but $\mu$ is not are those in which the error arises from using a measuring device whose variability is known from previous experiments. In such cases, the mean measurement $\mu$ will be unknown, depending on the object being measured. Even then, if we claimed to "know" $\sigma = 10$ but then found $S = 100$, we'd have to back down from our "knowledge."

There are various ways of getting around the problem of unknown $\sigma$. The investigator may be able to estimate $\sigma$ based on past experience; using this estimate in (1) gives an approximate sample size. If the investigator can give only an interval of possible values for $\sigma$, taking $\sigma$ in (1) to be the largest value in that interval leads to a sample size that is more than adequate for any other $\sigma$ in the interval.

This approach has the obvious disadvantage that the sample size found using (1) is only as reliable as the estimate of variance on which it is based. Indeed, the current circumstances may turn out to be quite different from those that produced the estimate.

A disadvantage of estimating $n$ based on an interval of $\sigma$-values is that the $n$-interval can turn out to be enormous—for instance, 10 to 1000. The conservative 1000 may cost more in time and resources than the investigator is willing to invest.

A second approach is more interesting. It is feasible only when data can be gathered in such a way that at least some intermediate results are available during the course of the experiment. The data available at any point provide an estimate of $\sigma$—namely, the current sample standard deviation $S$. Using $S$ in place of $\sigma$ in (1) yields an estimate of the required sample size. Such calculations can be done periodically—even after every observation!

**Example 9.2b**

## Sequential Adjustment of Sample Size

A device for measuring the amount of stretch or "drawer" in the human knee is used to determine the extent of damage to an injured anterior cruciate ligament. For such measurements to be useful, the distribution of drawers of normal knees must be known, at least approximately.

Although the entire distribution is of interest, we focus on the mean $\mu$. Suppose this is to be estimated with a standard error no greater than .2 mm. How many knees should we measure to meet this specification?

Table 9-1 lists the drawer measurements (in mm) of the normal knees of 34 patients whose other knee was thought to have an injured ligament.[1] Also shown, after each observation, is the current estimate of $\sigma$ (based on subjects measured up to that point), along with the corresponding estimate of required sample size [from (1)]:

---

[1] From the consulting practice of one of the authors.

$n = S^2/(.2)^2$, rounded up to an integer. Early estimates of sample size are unstable, but they eventually settle down.

The experiment could continue in this fashion beyond 34 patients to about 65; this would be the required sample size. The continual calculation of $S$ could then be abandoned, since the required sample size would have become pretty well stabilized. If feasible, one could check the value of the s.e. after 60 knees and continue sampling if it exceeds the specified .2 mm. ∎

**Table 9-1**

| Subject | Drawer | $\overline{X}$ | $S$ | $S^2/.04$ |
|---------|--------|------|------|-----------|
| 1  | 11 | 11.0 | —    | —   |
| 2  | 9  | 10.0 | 1.41 | 50  |
| 3  | 10 | 10.0 | 1.00 | 25  |
| 4  | 6  | 9.0  | 2.16 | 117 |
| 5  | 7  | 8.6  | 2.07 | 108 |
| 6  | 8  | 8.5  | 1.87 | 88  |
| 7  | 8  | 8.4  | 1.72 | 74  |
| 8  | 5  | 8.0  | 2.00 | 100 |
| 9  | 9  | 8.1  | 1.90 | 91  |
| 10 | 7  | 8.0  | 1.83 | 84  |
| 11 | 10 | 8.2  | 1.83 | 84  |
| 12 | 10 | 8.3  | 1.83 | 84  |
| 13 | 6  | 8.2  | 1.86 | 87  |
| 14 | 7  | 8.1  | 1.82 | 83  |
| 15 | 8  | 8.1  | 1.75 | 77  |
| 16 | 6  | 7.9  | 1.77 | 79  |
| 17 | 7  | 7.9  | 1.73 | 75  |
| 18 | 10 | 8.0  | 1.75 | 77  |
| 19 | 9  | 8.1  | 1.77 | 79  |
| 20 | 7  | 8.0  | 1.73 | 75  |
| 21 | 7  | 8.0  | 1.69 | 72  |
| 22 | 8  | 8.0  | 1.65 | 69  |
| 23 | 8  | 8.0  | 1.62 | 66  |
| 24 | 7  | 7.9  | 1.59 | 63  |
| 25 | 11 | 9.0  | 1.68 | 71  |
| 26 | 6  | 8.0  | 1.69 | 72  |
| 27 | 6  | 7.9  | 1.69 | 72  |
| 28 | 7  | 7.9  | 1.67 | 70  |
| 29 | 9  | 7.9  | 1.66 | 69  |
| 30 | 8  | 7.9  | 1.63 | 66  |
| 31 | 8  | 7.9  | 1.60 | 64  |
| 32 | 10 | 8.0  | 1.62 | 66  |
| 33 | 9  | 8.0  | 1.61 | 65  |
| 34 | 7  | 8.0  | 1.59 | 63  |

It may not be possible to incorporate the data into an estimate of $\sigma$ after each observation, as we've done in the example. In some cases it may be feasible only to collect all the data at once. Yet, something can be done if the data can be collected in even two batches: Take a pilot sample of $m$ observations to estimate $\sigma$, and a second sample of size $n - m$ with $n$ chosen to satisfy (1) using the estimated value of $\sigma$. However, $\sigma$ may be badly underestimated, especially if $n$ is small. (Overestimating $\sigma$ does not present as much of a problem.) With $\sigma$ underestimated, sampling might stop too soon. It is perhaps a good idea to take at least 20 observations in a pilot sample.

We consider next the estimation of a *population proportion,* using the sample proportion $\hat{p}$. Again in this case, the standard deviation of the estimator involves an unknown parameter. But this time the unknown parameter is the very one we're estimating:

$$\sigma_{\hat{p}} = \sqrt{\frac{p(1-p)}{n}}. \tag{2}$$

Even so, the situation is actually somewhat better than in the case of $\bar{X}$, because the s.d. (2) is *bounded* from above. The graph of $\sqrt{p(1-p)}$ in Figure 9-1 is a semicircle with center at $p = \frac{1}{2}$. The maximum value of (2) occurs at $p = \frac{1}{2}$, which means that

$$\sigma_{\hat{p}} \leq \sqrt{\frac{.5 \times .5}{n}} = \frac{.5}{\sqrt{n}}.$$

If we choose $n$ so that $.5/\sqrt{n} \leq \epsilon$, then $\sigma_{\hat{p}} \leq \epsilon$. Thus, to meet the specification that $\sigma_{\hat{p}} \leq \epsilon$, we take $n$ to be the smallest integer such that

$$n \geq \frac{.25}{\epsilon^2}. \tag{3}$$

With $n$ so determined, the standard error of $\hat{p}$ will not exceed $\epsilon$, whatever the true value of $p$. It actually will be substantially less than $\epsilon$ if $p$ is near 0 or 1.

In a particular application, we may know that $p$ is less than some $p_0 < .5$ or greater than some $p_0 > .5$. In either case, we can use $p_0$ in place of .5 as the "worst case" and get by with a smaller required sample size:

$$n \geq \frac{p_0(1-p_0)}{\epsilon^2}. \tag{4}$$

---

**Example 9.2c** | **A Political Poll**

Suppose we want to estimate the proportion of voters who favor a certain candidate with a standard error of at most .01. Setting $\epsilon = .01$ in (3), we obtain $n \geq .25/(.01)^2$, so 2500 is the smallest sample size guaranteed to achieve s.e. $\leq .01$: No matter what $\hat{p}$ turns out to be, its s.e. will not exceed .01 if $n = 2500$. However, if $\hat{p}$ is not equal to .5, the s.e. will be smaller than the specified .01. For instance, if $\hat{p} = .2$, the s.e. is $\sqrt{.2 \times .8/2500} = .008$.

Suppose we know that $p$ cannot be close to .5. For instance, suppose the candidate is running for president of the United States as a Socialist, and we are quite sure that the population proportion who will vote for that candidate is less than 10%.

With $p_0 = .1$ in (4), we find that $n \geq .1 \times .9/(.01)^2$ or 900 will guarantee the standard error to be at most .01. ∎

Even if little is known about $p$ in advance of sampling, it may be possible to save on sampling costs by estimating $\sqrt{p(1-p)}$ as data are collected, continually revising the estimate of required sample size accordingly.

## Problems

**\* 9-1.** A random survey of 1500 individuals shows that 630 watched a televised presidential news conference. Give an estimate of the population proportion who watched it, together with a standard error.

**9-2.** A TV show gets a rating of 27 (that is, 27%) based on a random sample of 1000 individuals. Find the standard error.

**\* 9-3.** A report on water pollution[2] included the following data on chemical oxygen demand (mg/l) of 18 water samples:

| | | | | | | | | |
|---|---|---|---|---|---|---|---|---|
| 580 | 674 | 512 | 540 | 616 | 298 | 960 | 570 | 640 |
| 588 | 556 | 588 | 582 | 844 | 574 | 420 | 696 | 620 |

Find the mean and the standard error of the mean.

**\* 9-4.** The statistic $X_{(n)}$, the largest observation in a random sample of size $n$, is proposed in Example 9.1a as an estimator for the parameter $\theta$ of a uniform distribution on the interval $(0, \theta)$.

    **(a)** Find the bias in $X_{(n)}$ as an estimator of $\theta$.

    **(b)** Find its mean squared error.

**9-5.** Laplace's rule of succession for an unknown probability $p$ of success estimates it to be $(Y + 1)/(n + 2)$, where $Y$ is the number of successes in $n$ independent trials.

    **(a)** Find the bias and mean squared error for this estimator.

    **(b)** For what range of values of $p$ is this estimator better than the sample proportion $Y/n$, according to the criterion of mean squared error?

**\* 9-6.** A newspaper poll reported that about 28% in a sample of U.S. voters (434 out of 1549) said they were Independents. Assuming random sampling, find the standard error of the sample proportion as an estimate of the population proportion of Independents.

**9-7.** A newspaper report says that a sample proportion of .43 has a "margin of error" of 5 percentage points (in estimating the population proportion). Interpret .05 as two standard errors and find the approximate sample size, assuming random sampling.

**\* 9-8.** Previous experience suggests that the s.d. for a certain kind of measurement $X$ is about 5.0. How large a sample should then be taken to estimate the population mean with

    **(a)** a standard error of 1.0?     **(b)** a standard error of .50?

**\* 9-9.** To estimate the proportion of TV viewers watching a certain special, how large a random sample is required so that the standard error is

---

[2]K. J. Shapland, "Industrial effluent treatability—A case study," *Water Pollution Control* 85 (1986), 77–80.

**(a)**   at most .01?      **(b)**   at most .02?

**9-10.**   In estimating the proportion $p$ in a population having a disease, how large a random sample is needed to be sure that the s.e. does not exceed .005,

   **(a)**   if you have no idea as to the value of $p$?
   **(b)**   if you assume that $p$ cannot exceed 5%?

**9-11.**   Let $\overline{X}_1$ and $\overline{X}_2$ denote the means of independent random samples from populations with the same mean $\mu$. Find the value of $\gamma$ for which the mixture $\gamma\overline{X}_1 + (1 - \gamma)\overline{X}_2$, where $0 \leq \gamma \leq 1$, has the smallest m.s.e., when

   **(a)**   the sample sizes are both $n$ and the population variances are $\sigma_1^2$ and $\sigma_2^2$.
   **(b)**   the population variances are the same, but the sample sizes are $n_1$ and $n_2$, respectively.

## 9.3          Consistency

The original notion of *consistency* was that an estimator, when applied to the whole population as a "sample," should yield the parameter being estimated. For infinite populations, real or conceptual, consistency has come to mean that the estimator (defined for each sample size) should approach the parameter being estimated, in some sense, as the sample size becomes infinite. The definition uses the notion of *convergence in probability*, and is as follows:

---

A sequence of estimators $\{T_n\}$ is **consistent** as an estimate of $\theta$ when

$$\lim_{n \to \infty} P(|T_n - \theta| \leq \epsilon) = 1 \qquad (1)$$

for every $\epsilon > 0$.

---

Equation (1) says that any interval centered at $\theta$, however small, will eventually capture an arbitrarily large fraction of the distribution of $T_n$. That is, sooner or later, most of the distribution crowds into any given tiny interval about $\theta$. This suggests that when $n$ is large, there is a high probability that the estimator will be close to the parameter being estimated.

Checking an estimator to see whether (1) holds is possible in some simple cases, but there is often an easier way. When the second moments involved exist, the condition (1) holds provided only that the mean squared error tends to 0 as $n$ becomes infinite. (We'll prove this at the end of the section.)

---

When $T_n$ has a finite variance, the sequence $\{T_n\}$ is consistent as an estimator for $\theta$ if it is *mean-square consistent*—that is, if

$$ET_n \to \theta \text{ and } \operatorname{var} T_n \to 0 \text{ as } n \to \infty.$$

---

**Example 9.3a** | **Lou Gehrig's Experiment**

A pencil-written notebook compiled by baseball Hall of Famer Lou Gehrig when he was a 17-year-old high school student was sold at auction in October, 1988, for $7,150.[3] It was compiled by Gehrig in 1920 for a statistics class at New York City's Commerce High School. After discussing rolling dice, he wrote, "If you want to find the average of any large number of items, take a small group (at random) and it will practically every time give you the average of the large group." He had rolled three dice 60 times and calculated the average to be 10.78. (The expected number of points showing is 10.5.)

He went on to say, "To find the average height of boys in the H. S. of C., I would take 1 class from every term, measure each boy's height, and I am almost sure that the average obtained would be the average height of the boys in the whole school." ∎

The consistency of the sample mean is a classical result in probability theory known as the **law of large numbers** or the **law of averages.** The special case of the sample proportion (which is a mean) was given in §4.7. The sample mean is unbiased, so its consistency can be established by showing that its variance ($\sigma^2/n$) tends to 0—which it does, provided that $\sigma^2 < \infty$. However, it is actually true that the sample mean is consistent, in the sense (1), even when the variance *is* infinite, provided that the population mean is finite. This fact is *Khintchine's theorem*, which we shall not prove.[4]

**Example 9.3b** | **Consistency of Sample Moments**

Consider estimating the population moment $E(X^k)$ using the corresponding sample moment, $Y_n = \frac{1}{n}\sum X_i^k$. The mean of $Y_n$ is $E(X^k)$, so $Y_n$ is unbiased. The variance is

$$\text{var } Y_n = \frac{\text{var } X^k}{n} = \frac{1}{n}\{E(X^{2k}) - [E(X^k)]^2\}.$$

This tends to 0 as $n$ becomes infinite if $E(X^{2k}) < \infty$, in which case $Y_n$ is mean-square consistent—and therefore consistent.

Even if $E(X^{2k})$ is *not* finite, $Y_n$ is consistent—as a result of Khintchine's theorem applied to the population variable $X^k$. Thus, sample moments about 0 are consistent estimators of the corresponding population moments. It can be shown, further, that any function of sample moments, including any central moment, is a consistent estimator of that function of the corresponding population moments.[5] ∎

**Example 9.3c** | **The Sample Maximum**

Suppose $X \sim \mathcal{U}(0, \theta)$. We have seen in Problem 8-28 that in the case of random samples, the largest sample observation $X_{(n)}$ (which is sufficient for $\theta$) has these

---

[3] *Sports Collectors' Digest* (Nov. 18, 1988), 54–60.

[4] See W. Feller, *An Introduction to Probability Theory and Its Applications* (New York: Wiley, 1968) 235.

[5] H. Cramér, *Mathematical Methods of Statistics* (Princeton NJ: Princeton Univ. Press, 1946), 367.

moments:

$$E(X_{(n)}) = \frac{n}{n+1}\,\theta, \text{ and } \operatorname{var} X_{(n)} = \frac{n\theta^2}{(n+1)^2(n+2)}.$$

The mean clearly tends to $\theta$, and the variance to 0, as $n$ becomes infinite. Thus, the largest sample observation is a consistent estimator for $\theta$.  ∎

Consistency is a "large-sample" property and offers little comfort when one is using a small sample for estimation. But in the not uncommon case of a large sample, consistency is a desirable property, because a consistent estimator has a good chance of coming close to the parameter it estimates.

We return now to the earlier claim that an estimator which is consistent in mean square is consistent in the sense of (1). To show this we use an inequality of Chebyshev's, which relates concentration of probability to variance.

---

**Chebyshev's inequality:** If $E(X^2) < \infty$, then for any real $\theta$ and $\epsilon > 0$,

$$P(|X - \theta| \geq \epsilon) \leq \frac{1}{\epsilon^2}\, E[(X - \theta)^2]. \tag{2}$$

---

To show (2), we use a special case of the result in Problem 5-R18(b):

$$E[h(X)] = E[h(X) \mid A] \cdot P(A) + E[h(X) \mid A^c] \cdot P(A^c). \tag{3}$$

Let $h(x) = (x - \theta)^2$, and take $A$ to be the set where $h(x) \geq \epsilon^2$. [Figure 9-2 shows $h(x)$ and the region $A$ as the set of $x$'s beyond $\theta \pm \epsilon$.] The second term on the right of (3) is nonnegative, so we can drop it and preserve the inequality. And since $h(x) \geq \epsilon^2$ on $A$,

$$E[h(X)] \geq E[h(X) \mid A] \cdot P(A) \geq \epsilon^2 P(A).$$

Dividing through by $\epsilon^2$ yields (2). [We have used a useful technique for showing a

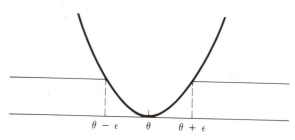

**Figure 9-2**

probability inequality: We replaced $h$ by a smaller function (shown in Figure 9-2), and then averaged.]

To show that consistency in mean square implies consistency in the sense (1), we apply Chebyshev's inequality with $X = Y_n$ and $\theta$ as the parameter to be estimated:

$$0 \leq P(|Y_n - \theta| \geq \epsilon) \leq \frac{1}{\epsilon^2} E[(Y_n - \theta)^2] = \frac{1}{\epsilon^2} \, \text{m.s.e.}(Y_n).$$

If, then, the m.s.e. tends to 0, so does $P(|Y_n - \theta| \geq \epsilon)$, as was to be proved.

---

## 9.4  Large-Sample Confidence Intervals

In Example 9.1c we reported an estimate of mean platelet count in the form $234 \pm 7.6$, where 7.6 is the standard error. Such a report is intended to convey that the point estimate of 234 could easily be in error by 7.6, or more. We could not say for sure that the population mean $\mu$ lies between $234 - 7.6$ and $234 + 7.6$. Yet, we are surely more confident that it is in that interval (226.4 to 241.6) than it is, say, in the interval 326.4 to 341.6. A natural interpretation of "confidence" is in terms of probability. But can we hope to calculate such a thing as $P(226.4 < \mu < 241.6)$?

So long as we view $\mu$ as a constant, it does not make sense to ask for the probability that it lies in a particular interval. In §8.11 we introduced the idea of representing uncertainty about $\mu$ using a probability distribution, and we'll return to this viewpoint in §9.11 and some later sections. Here, however, in the framework of "classical" statistical inference, we do not regard a parameter such as $\mu$ as a random variable. Nevertheless, there is an approach in which a probability is associated with an interval of $\mu$-values; but the interval is a *random* interval.

Until we actually obtain a sample and calculate its mean, the quantity $\overline{X}$ is a random variable. The interval with endpoints $\overline{X} \pm \text{s.e.}(\overline{X})$ is therefore a random interval—an interval centered at the random point $\overline{X}$. And it does make sense to talk about the probability that this random interval will cover any particular point, and $\mu$ in particular. Thus, we can define and calculate

$$P(\overline{X} - \text{s.e.} < \mu < \overline{X} + \text{s.e.}). \tag{1}$$

The inequality in this expression is satisfied if and only if $\overline{X}$ is within one standard error of $\mu$:

$$|\overline{X} - \mu| < \frac{S}{\sqrt{n}}, \quad \text{or} \quad \left| \frac{\overline{X} - \mu}{S/\sqrt{n}} \right| < 1. \tag{2}$$

In the second inequality, the ratio

$$Z = \frac{\overline{X} - \mu}{S/\sqrt{n}}$$

is approximately standard normal when $n$ is large, since then the denominator is practically the same as $\sigma/\sqrt{n}$, the standard deviation of the numerator. So (1) is approximately the probability that a standard normal variable falls between $-1$ and $+1$, or about 68%:

$$P(\overline{X} - \text{s.e.} < \mu < \overline{X} + \text{s.e.}) \doteq .68. \tag{3}$$

Keep in mind here that the $P$ refers not to $\mu$, but to the randomness in $\overline{X}$.

Each sample that might be drawn has its own $\overline{X}$-value and s.e., and hence, its own set of interval endpoints, $\overline{X} \pm$ s.e. The statement (3) says that just over two out of three possible samples result in an $\overline{X}$-value such that the interval with endpoints $\overline{X} \pm$ s.e. includes the true value of $\mu$. The process of obtaining a sample and constructing the interval from $\overline{X} -$ s.e. to $\overline{X} +$ s.e. has about two chances in three of producing an interval that covers the value $\mu$. We say that the probability .68 is a degree of *confidence* in the method we have used in constructing the interval. The interval is called a 68% **confidence interval.**

To increase our level of confidence, it seems quite apparent that using a wider interval, still centered at $\overline{X}$, would be the thing to try. If we repeat the development leading to (3), replacing the 1 by 2 in the second inequality of (2), we find

$$P(\overline{X} - 2\,\text{s.e.} < \mu < \overline{X} + 2\,\text{s.e.}) \doteq .9544, \tag{4}$$

where .9544 is the probability that a standard normal $Z$ falls between $-2$ and $+2$. The interval with endpoints $\overline{X} \pm 2\,\text{s.e.}$ is a 95.44% confidence interval for the mean $\mu$, and those endpoints are termed *95.44% confidence limits.*

We have given confidence limits for $\mu$ for situations in which $\sigma$ is unknown. In the less common situation in which $\sigma$ can be assumed known, we'd naturally use the s.d. of $\overline{X}$, or $\sigma/\sqrt{n}$, in place of $S/\sqrt{n}$.

To summarize, we give confidence limits for $\mu$, for the more commonly used confidence levels:

---

Approximate large-sample confidence limits for a population mean:

$$\overline{X} \pm k\,\frac{S}{\sqrt{n}} \ (\sigma \text{ unknown}), \ \text{or} \ \overline{X} \pm k\,\frac{\sigma}{\sqrt{n}} \ (\sigma \text{ known}), \tag{5}$$

where $k$ is determined by the desired confidence level (Table IId):

| Confidence level | $k$ |
|---|---|
| 68% | 1 |
| 90% | 1.645 |
| 95% | 1.96 |
| 99% | 2.58 |

---

Example **9.4a** | **Confidence Limits for Mean Platelet Count**
In Example 9.1c we found $\overline{X} = 234$ and s.e.$(\overline{X}) = 7.6$ for a sample of 35 blood platelet counts. Approximate 95% confidence limits for the mean count, from (5), are $234 \pm 1.96 \times 7.6$. The interval $219.1 < \mu < 248.9$ is a 95% confidence interval for $\mu$.

    Notice that we do **not** say:  the probability that $\mu$ is between 219.1 and 248.9 is 95%. Rather, we have used a procedure [obtain a sample, calculate its mean and variance, and calculate (5)] which has a 95% success rate in trapping the true value of $\mu$. That success rate is taken as a measure of confidence in the procedure. ■

    Large-sample confidence limits for a population *proportion* are constructed in the same way as for a population mean:  sample proportion plus and minus a multiple of the standard error. This is to be expected. When the population is Ber$(p)$, the population proportion and sample proportion are the population mean and sample mean, respectively.

---

Approximate large-sample confidence limits for a population proportion:

$$\widehat{p} \pm k\sqrt{\frac{\widehat{p}(1 - \widehat{p})}{n}} , \qquad (6)$$

where $k$ is given in Table IId.

---

Example **9.4b** | **Nielsen Ratings**
Nielsen ratings are based on samples of about 1000 TV owners. The Nielsen company says that the sampling error for a single telecast with rating of 22 is 1.3 percentage points (that is, .013). The actual sampling scheme is complicated, involving many stages and clusters, but Nielsen uses (6) to obtain confidence limits. With $\widehat{p} = .22$, the standard error is

$$\text{s.e.}(\widehat{p}) = \sqrt{\frac{.22 \times .78}{1000}} \doteq .013.$$

Taking $k = 1$ in (6), Nielsen reports $.22 \pm .013$ as 68% confidence limits.
    A Nielsen official was quoted in *TV Guide*[6] as saying, "The reason we use the 68% level is that it makes it very convenient for the user." The "convenience" is not clear, but (being narrower) the 68% interval suggests greater accuracy than a 95% interval—the latter is twice as wide: $(19.4, 24.6)$. In one survey, a show with a 19.4 rating was ranked 27th, while one with a 24.6 rating ranked 6th. Such differences in rankings have quite an impact when it comes to deciding the fate of a program. ■

---

[6]Vol. 26, No. 25 (1978).

In what follows, we'll encounter other estimators $T$ of a parameter $\theta$ that are approximately normal when $n$ is large. In any such case, if $ET = \theta$, we can use the same procedure to construct a confidence level for $\theta$. All that is needed is a formula for standard error:   The desired interval is $T \pm k \times$ s.e., with $k$ determined from Table IId as above.

## Problems

**9-12.**   Show that for estimating of a population proportion $p$, the sample proportion $\hat{p}$ is mean-square consistent (and hence consistent).

∗ **9-13.**   Consider the sample mean $\overline{X}$ as an estimate of $\mu$, the mean of a population with finite variance. For given $\epsilon > 0$, use the central limit theorem to approximate the probability outside the interval $(\mu - \epsilon, \mu + \epsilon)$, and show that it tends to 0 as $n$ becomes infinite.

**9-14.**   Show that the largest observation in a random sample from $\mathcal{U}(0, \theta)$ is a consistent estimator of $\theta$ by calculating $P(|X_{(n)} - \theta| > \epsilon)$.

**9-15.**   Show that for random samples from $\mathcal{N}(\mu, \sigma^2)$, the sample variance $S^2$ is mean-square consistent (and hence consistent).

**9-16.**   Given a random sample from a population with c.d.f. $F$, let $F_n$ denote the sample distribution function. Show that for each $x$, $F_n(x)$ is a consistent estimator of $F(x)$. [In fact, something much stronger is true:   Glivenko's theorem[7] states that $F_n(x)$ tends to $F(x)$ uniformly in $x$.]

∗ **9-17.**   A transportation study gave data on time from origin to destination for 201 buses on a particular bus route in Chicago.[8]   The sample average is $\overline{X} = 18.31$, and the standard deviation is $S = 2.111$. Find 95% confidence limits for the population mean time.

∗ **9-18.**   An opinion poll reports that 41% of 1500 who were polled think that the president is doing a good job. Give 68% and 95% confidence limits for the corresponding population proportion $p$.

**9-19.**   It can be shown that for random samples from $\mathcal{N}(\mu, \sigma^2)$, the sample median is asymptotically normal with mean $\mu$ and variance $\pi\sigma^2/2n$. Example 9c gave weights of a block of metal as deviations from the nominal weight:

| | | | | | | | | | |
|---|---|---|---|---|---|---|---|---|---|
| 8.82 | 7.39 | 18.86 | 2.65 | 12.92 | 2.07 | 4.26 | − 3.05 | 6.93 | − 15.40 |
| 5.95 | 15.97 | − 6.78 | − 3.80 | 4.52 | 5.54 | 13.45 | − 5.17 | 3.57 | 4.92 |

Assuming $\sigma = 6$, find 95% confidence limits for $\mu$ based on the sample median.

∗ **9-20.**   The spot-weld data in Example 7.5b have mean $\overline{X} = 404.3$ and standard deviation 18.9. Give 90% confidence limits for the population mean $\mu$.

**9-21.**   In a survey[9] of college students, it was found that at least one parent of 69 of the 224 students surveyed had had heart disease. Let $p$ denote the population proportion of students

---

[7]See C. R. Rao, *Linear Statistical Inference and its Applications*, 2nd Ed. (New York: Wiley, 1973) 42.

[8]A. Polus, "Modeling and measurements of bus service reliability," *Transportation Research* 12 (1978), 253–256.

[9]R. N. Tamragouri, "Cardiovascular risk factors and health knowledge among freshman college students..." *J. Amer. College Health* 34 (1986) 267–70.

with parents of whom at least one had had heart disease. Give 95% confidence limits for $p$. (The only relevant population is that of all students at that college; the incidence is apt to be different for adults without children in college and for different colleges.)

**9-22.** Suppose a research report is to include 20 confidence intervals, each at the .95 level. Although one would expect some relationships, suppose the intervals are based on independent statistics.

    **(a)** About how many of the intervals would you expect actually to include the true value of the parameter being estimated?

    **(b)** What is the probability that all 20 intervals will contain the true values of the parameters being estimated?

## 9.5       Small-Sample Confidence Intervals for $\mu$

In §9.4, when constructing confidence intervals for $\mu$, we assumed sample sizes to be large enough so that (i) the sample mean is approximately normal, and (ii) the sample s.d. is close to the population s.d. If the parent population is not too far from normal, the sample mean is apt to be close to normal even when $n$ is as small as 5 or so. Indeed, if the population is exactly normal, so is $\overline{X}$. However, the approximate score

$$T = \frac{\overline{X} - \mu}{S/\sqrt{n}} \tag{1}$$

that we used in developing the formula for confidence limits is likely not to be standard normal because of the unreliability of $S$ as estimating $\sigma$—which is what we really need in the denominator for a $Z$-score. That is, $T$ has greater variability than $Z$ because of the variability in $S$. So, when $n$ is not large, a confidence interval based on $T$ has to be wider, to take into account the extra variability.

    To construct a confidence interval for $\mu$, we need to know the distribution of the random variable $T$ defined by (1). This distribution was found nearly a century ago by W. S. Gosset for the case of *normal* populations. For reasons of company confidentiality (he worked for the Guiness Brewery), Gosset used the pseudonym *Student* in publishing his result, and the distribution now goes under the name "Student's $t$-distribution."

    The $t$-distribution is actually a *family* of distributions—symmetric distributions indexed by a parameter called the "number of *degrees of freedom*" or "d.f." We'll sometimes write $T \sim t(k)$ to mean that $T$ has a $t$-distribution with $k$ d.f. For the distribution of $T$ defined by (1), the number of degrees of freedom is $n - 1$. Tables IIIa and IIIb give percentiles and tail-probabilities for degrees of freedom 1 to 30. We'll derive the distribution in the next section, but first we illustrate its use.

    We again use the inequalities like (2) in §9.4 that lead us to the large-sample confidence limits, but now to find the probability of such inequalities we go to Table IIIa. Thus, for example, when $n = 20$, $T$ has 19 degrees of freedom, and the probability to the right of $T = 2$ is shown as .030 in Table IIIb. The probability to the left of $T = -2$ is also .030, since the $t$-distribution is symmetric about 0. Thus,

$P(-2 < T < 2) = 1 - .06$. Upon solving for $\mu$ as before, we obtain

$$P\left(\overline{X} - 2\,\frac{S}{\sqrt{n}} < \mu < \overline{X} + 2\,\frac{S}{\sqrt{n}}\right) = .94.$$

Thus, $\overline{X} \pm 2S/\sqrt{n}$ are 94% confidence limits for $\mu$—with the assumption that the population is normal. To get 95% limits we put .025 in each tail of the distribution and replace 2 by 2.09, the 97.5 percentile when d.f. = 19 (from Table IIIa).

The extra variability in $S$ when $n = 20$ has been taken into account by widening the confidence interval: For 95% confidence, the half-interval width for large samples is $1.96S/\sqrt{n}$, and for $n = 20$ it is $2.09S/\sqrt{n}$.

## Example 9.5a | Adjusting Limits for Unknown $\sigma$

In Example 9.4a we gave confidence limits for mean blood platelet count. We assumed that a sample size of 35 is large enough that the normal table can be used, even though we replaced $\sigma$ by $S$. If the population is normal, we can use the $t$-table to get more accurate limits. For 95% confidence, we need the 97.5 percentile of $t(34)$. This is not shown in the table but appears to be 2.03 (by a rough interpolation). Using 2.03 in place of 1.96, we find the confidence limits to be $234 \pm 2.03 \times 7.6$, or 218.6 and 249.4. The interval is about 4% wider than the interval we found in Example 9.4a. ∎

Suppose the population is *not* normal (the usual case), and that $n$ is not large and $\sigma$ is not known. Strictly speaking, the statistic $T$ given by (1) does not have a $t$-distribution. However, simulation studies show that the actual distribution of $T$ is well approximated by a $t$-distribution provided the population distribution is "not too far" from normal. The $t$-distribution does not give a good approximation if either or both tails of the distribution are heavy—that is, if there is too much probability far from the mean.

---

Approximate confidence limits for $\mu$ when $\sigma$ is unknown and the population is close to normal:

$$\overline{X} \pm k\,\frac{S}{\sqrt{n}}, \qquad (2)$$

where for confidence level $\gamma$, the multiplier $k$ is the $100(\frac{1+\gamma}{2})$ percentile of $t(n-1)$, from Table IIIa.

---

When should you use Table IIa, and when Tables IIIa and IIIb? The simplest rule is to use Tables IIIa and IIIb whenever $\sigma$ is unknown and the population distribution is not especially heavy tailed. If the sample size is beyond the range of sizes shown in Table IIIa, you will be led to the last row (d.f. = $\infty$), which gives the appropriate normal percentiles as multipliers. If $\sigma$ is known, then $T$ is never appropriate. Whether $Z$ is appropriate when $\sigma$ is known depends only on whether or not the sample is large enough that $\overline{X}$ is approximately normal. (If it is not, a different approach would be needed.)

Having seen how to modify large-sample confidence limits for a population mean when the sample size is small raises the question of whether there is a small-sample method for estimating a proportion. There is (see Problem 9-27), but when $n$ is small, the confidence interval turns out to be so wide that fine-tuning of the limits isn't worth the effort. For example, with a sample of 25 and a sample proportion of .32, the 95% limits given by (6) of §9.4 are .137 and .503. With the modification for small samples obtained in Problem 9-27, the 95% limits turn out to be .171 and .516. We'll explore this no further. However, lest there be confusion in this regard, you should *never* use the $t$-distribution in connection with sample proportions.

## 9.6   The Distribution of $T$

We now derive the distribution of the statistic $T$ given as (1) of the preceding section, under the assumption that the population being sampled is normal. We first write $T$, a ratio of independent variables, in the form

$$T = \frac{\sqrt{n}(\overline{X} - \mu)/\sigma}{S/\sigma} = \frac{Z}{\sqrt{\{(n-1)S^2/\sigma^2\} \div (n-1)}}. \tag{1}$$

Since $\overline{X} \sim \mathcal{N}(\mu, \sigma^2/n)$, the numerator is standard normal. And from §8.10 we know that the variable $(n-1)S^2/\sigma^2$ is distributed as $\text{chi}^2(n-1)$. So what we need is the distribution of the ratio of a standard normal $Z$ to the square root of a chi-square variable divided by its d.f., when these variables are independent.

In general, to find the p.d.f. of the ratio $W = U/V$, where $U$ and $V$ are independent and $V$ is nonnegative, we calculate the c.d.f. $P(U/V \leq w)$ and differentiate:

$$f_W(w) = \frac{d}{dw} P(U/V \leq w) = \frac{d}{dw} \int_0^\infty \int_{-\infty}^{vw} f_{U,V}(u, v)\, du\, dv.$$

Interchanging the differentiation and the first integration (which has constant limits), we then have to differentiate the inner integral with respect to $w$, which appears only in the upper limit. This is accomplished by substituting that upper limit for $u$ in the integrand and multiplying by $d(vw)/dw = v$:

$$f_W(w) = \int_0^\infty f_{U,V}(vw, v) \cdot v\, dv = \int_0^\infty v f_U(vw) f_V(v)\, dv. \tag{2}$$

In applying this in the case of (1), the "$U$" is the standard normal $Z$ in the numerator, and the "$V$" is the square root of a chi-square variable divided by its number of degrees of freedom: $V = \sqrt{Y/k}$, or $Y = kV^2$. Using the p.d.f. for $\text{chi}^2(k)$

from §6.4 and the formula for obtaining new densities from old under monotone transformations [(15) of §5.2], we find

$$f_V(v) = f_{\chi^2(k)}(kv^2) \cdot 2kv \propto v^{k-1}e^{-kv^2/2}, \, v > 0. \tag{3}$$

With $T$ playing the role of $W$ in (2), and substituting (3) for $f_V$ and the standard normal p.d.f. for $f_U$, we can write the p.d.f. of $T$ as

$$f_T(t) \propto \int_0^\infty v \cdot e^{-(vw)^2/2} v^{k-1} e^{-kv^2/2} \, dv. \tag{4}$$

or

$$f_T(t) \propto \int_0^\infty v^k e^{-(k+t^2)v^2/2} dv.$$

From (9) in §6.3, we have this integration formula:

$$\int_0^\infty v^{2\alpha-1} e^{-\lambda v^2/2} dv = \frac{2^{\alpha-1}}{\lambda^\alpha} \Gamma(\alpha).$$

With $\alpha = (k+1)/2$, and $\lambda = k + t^2$, we see, then, that

$$f_T(t) \propto \frac{1}{(k+t^2)^{(k+1)/2}} \propto \frac{1}{\left(1 + \frac{t^2}{k}\right)^{(k+1)/2}}. \tag{5}$$

The parameter $k$ is termed the number of degrees of freedom. For the variable $T$ given by (1), this number is $k = n - 1$.

---

When $T$ has the p.d.f.

$$f(t) = \propto \frac{1}{\left(1 + \frac{t^2}{k}\right)^{(k+1)/2}},$$

we say that $T$ has the $t$-distribution with $k$ degrees of freedom, and write $T \sim t(k)$.

    For a random sample of size $n$ from $\mathcal{N}(\mu, \sigma^2)$, the distribution of $T = \sqrt{n}(\bar{X} - \mu)/S$ is $t(n - 1)$.

---

Since we have a table of $t$-probabilities, we'll not need the $t$-p.d.f. *per se.* Observe that it is symmetric about 0; so when the mean exists, it is 0. When $n = 2$, the p.d.f. is that of a Cauchy distribution, which doesn't have a mean. And, of course, when $n = 1$, the p.d.f. is not defined since $S$ is not defined.

As $k$ tends to infinity, the p.d.f. of (4) tends to a standard normal p.d.f. This is to be expected: The $S/\sqrt{n}$ in the denominator of $T$ given by (1) tends to $\sigma/\sqrt{n}$, just what would be needed for $T$ to be a Z-score. The tendency to normality is seen in Table IIIa, where entries for $n = \infty$ are normal percentiles.

## 9.7      Pivotal Quantities

The ratio $T = \sqrt{n}(\overline{X} - \mu)/S$ that we used in §9.5 to construct confidence intervals for a population mean is approximately standard normal when the sample size is large. When the population being sampled is normal, $T$ has a $t$-distribution. These distributions (standard normal and $t$) do not involve population parameters. A function of a statistic and parameters whose distribution does not involve any unknown parameters is said to be a *pivotal quantity*, sometimes referred to simply as a "pivotal."

> A **pivotal** (quantity) is a function of a statistic and population parameters whose distribution does not depend on any population parameters.

A pivotal quantity can be used to construct confidence intervals. Faced with a problem of estimating $\theta$, suppose we can find a function $g(\mathbf{X}, \theta)$ which, as a random variable, has a distribution independent of $\theta$. Let $A$ and $B$ be numbers such that

$$P[A < g(\mathbf{X}, \theta) < B] = \gamma, \tag{1}$$

where $\gamma$ is a desired confidence level. Solving the inequality for $\theta$ will often produce (two-sided) confidence limits.

**Example 9.7a**

**Interval Estimate for $\sigma$**

We have seen that for a random sample of size $n$ from $\mathcal{N}(\mu, \sigma^2)$, the quantity $(n-1)S^2/\sigma^2$ has a chi-square distribution—which does not depend on $\mu$ or $\sigma^2$. So it is pivotal. To construct a 90% confidence interval, we look in Table Va for the 5th and 95th percentiles of chi$^2(n-1)$. For instance, if $n = 15$, we find $\chi^2_{.05} = 6.57$ and $\chi^2_{.95} = 23.7$ (in the row for 14 d.f.). Writing out (1) for this particular case, we have

$$P\left(6.57 < \frac{(n-1)S^2}{\sigma^2} < 23.7\right) = .90.$$

To solve the inequalities for $\sigma^2$, we invert (remembering that in inverting the fractions the inequalities are reversed):

$$\frac{1}{6.57} > \frac{\sigma^2}{(n-1)S^2} > \frac{1}{23.7},$$

and multiply through by $(n-1)S^2$:

$$P\left\{ \frac{(n-1)S^2}{23.7} < \sigma^2 < \frac{(n-1)S^2}{6.57} \right\} = .90.$$

The extremes in the extended inequality are sample statistics—the 90% confidence limits for $\sigma^2$. ∎

In some situations one might like a *one-sided* interval for a parameter $\theta$, defining a *confidence bound* for $\theta$. The confidence sets we have found, up to this point, have been two-sided (extending on both sides of the point estimate) simply because we split the probability $1 - \gamma$, putting half of it in one tail and the other half in the other tail of the distribution of the pivotal quantity. By putting it all in one tail *or* the other we get a one-sided confidence bound.

**Example 9.7b**   **A Lower Confidence Bound**

We have seen that the distribution of the largest observation $Y = X_{(n)}$ from a uniform distribution on $(0, \theta)$ has the p.d.f.

$$f(y) = \frac{n}{\theta^n} y^{n-1},\ 0 < y < \theta.$$

The quantity $U = Y/\theta$ then has the p.d.f.

$$f_U(u) = f_Y(\theta u) \cdot \theta = n u^{n-1},\ \ 0 < u < 1.$$

This is independent of $\theta$, so $U$ is pivotal.

Suppose we'd like a 90% lower confidence bound for $\theta$ when $n = 10$. The 90th percentile of $U$ is the number $u$ such that

$$F_U(u) = u^{10} = .90,$$

namely, $u = (.90)^{1/10} = .9895$. Thus,

$$P\left( \frac{X_{(10)}}{\theta} < .9895 \right) = .90.$$

Solving the inequality for $\theta$ we get $\theta > X_{(10)}/.9895 = 1.0106\, X_{(10)}$ as the desired lower confidence bound. ∎

## Problems

**∗ 9-23.** The water samples referred to in Problem 9-3 were also checked as to biological oxygen demand. The mean and s.d. of the 18 measurements are $\overline{X} = 89.14$ and $S = 25.19$. Give 95% and 99% confidence limits for $\mu$, assuming population near normality and random sampling.

**9-24.** Through careful examinations of sound and film records, it is possible to measure the distance at which a bat first detects an insect. Given the following data[10] (in cm): 62, 52, 68, 23, 34, 45, 27, 42, 83, 56, 40, construct a 90% confidence interval for the population mean distance (assuming near normality of the population and random sampling).

∗ **9-25.** Consider a random sample of size 15 from $\mathcal{N}(\mu, \sigma^2)$. Find
   **(a)** $P(|\overline{X} - \mu| < 2\sigma/\sqrt{n})$.    **(b)** $P(|\overline{X} - \mu| < 2S/\sqrt{n})$.

**9-26.** Find a 95% upper confidence interval for $\theta$ based on the largest observation in a random sample of size 10 from $\mathcal{U}(0, \theta)$. [This will be an interval of the form $X_{(n)} < \theta < kX_{(n)}$.]

**9-27.** Even for moderately large random samples from a Bernoulli population, the statistic $\hat{p}$ (sample proportion of successes) is approximately normal.
   **(a)** Give the $Z$-score and explain why it is (approximately) pivotal.
   ∗**(b)** For the $Z$ found in (a), solve the inequality $|Z| < 1.96$ for $p$. (Squaring each side of the inequality preserves the inequality, and the result is a quadratic inequality in $p$.)

**9-28.** In Problem 6-26 it was shown that for a random sample of size $n$ from $\text{Exp}(\lambda)$, the quantity $2\lambda\sum X_i$ is distributed as $\text{chi}^2(2n)$. Use this fact to obtain 90% confidence limits for $\lambda$, if $n = 10$ and $\sum X_i = 32$.

## 9.8    Estimating a Mean Difference

Comparing two populations on the basis of a sample from each is a common and important problem in many fields. For example, treated experimental units may respond differently than those that are not treated. When populations are compared in terms of mean response, the *difference* $\delta = \mu_1 - \mu_2$ is a relevant parameter. A natural estimate of $\delta$ is the corresponding difference between sample means.

We remarked at the end of §9.5 that to construct large-sample confidence intervals for a parameter $\theta$ using a statistic $Y$ that is approximately normal with mean $\theta$, we need only find an expression for standard error. The interval we want is then $Y \pm k \times$ s.e.

The difference between the means of two large, independent random samples can be shown to be approximately normal, as one would expect, since the mean of each sample is approximately normal, and because the difference between independent normal variables is normal. The moments of the mean difference derive from the population parameters:

$$E(\overline{X}_1 - \overline{X}_2) = \mu_1 - \mu_2, \tag{1}$$

and

$$\text{var}(\overline{X}_1 - \overline{X}_2) = \text{var}\,\overline{X}_1 + \text{var}\,\overline{X}_2 = \frac{\sigma_1^2}{n_1} + \frac{\sigma_2^2}{n_2}. \tag{2}$$

[10]Reported in Griffin et al., "The echolocation of flying insects by bats," *Animal Behavior* 8 (1960), 141–54.

This variance involves population parameters—variances, that are (usually) unknown. Replacing these with sample variances we obtain a formula for the standard error of the difference in sample means:

$$\text{s.e.}(\overline{X}_1 - \overline{X}_2) = \sqrt{\frac{S_1^2}{n_1} + \frac{S_2^2}{n_2}}. \tag{3}$$

**Example 9.8a** | **Two Methods of Filtration**

A water purification plant was equipped with a new filtration device. To test its effectiveness, 20 test runs were carried out[11] using one method of operation and 29 using a second method, to record the total water filtered (in cubic meters per square meter of filter). All runs lasted the same length of time.

The sample means were 202.0 for the first method and 278.3 for the second. The difference, 76.3, is a point estimate of the population mean difference. The published report also gives standard deviations: $S_1 = 74.35$ and $S_2 = 79.03$. The standard error (3) of the difference in means is thus

$$\text{s.e.} = \sqrt{\frac{74.35^2}{20} + \frac{79.03^2}{29}} \doteq 22.18.$$

So 95% confidence limits for the population mean difference $\delta$ are

$$76.3 \pm 1.96 \times 22.18,$$

and the confidence interval is $32.8 < \delta < 119.8$. ■

---

Confidence limits for $\delta = \mu_1 - \mu_2$, based on independent random samples of sizes $n_1$ and $n_2$ (large):

$$\overline{X}_1 - \overline{X}_2 \pm k \sqrt{\frac{S_1^2}{n_1} + \frac{S_2^2}{n_2}}, \tag{4}$$

where $k$ is found in Table IId.

---

A population proportion (in a Bernoulli model) is actually a mean, and estimating a difference in population proportions is estimating a difference in population means—a special case of what we have just been doing. Again with subscript $i$ referring to population $i$ (for $i = 1, 2$), we'll use the difference in sample proportions (i.e.,

---

[11]P. B. V. Vosloo, P. G. Williams, and R. G. Rademan,"Pilot and full-scale investigations on the use of combined dissolved-air flotation and filtration for water treatment," *Water Pollution Control* 85 (1986), 114–21.

sample means) as the estimator. We know that

$$E(\widehat{p}_1 - \widehat{p}_2) = p_1 - p_2,$$

and

$$\text{var}\,(\widehat{p}_1 - \widehat{p}_2) = \text{var}\,\widehat{p}_1 + \text{var}\,\widehat{p}_2 = \frac{p_1(1 - p_1)}{n_1} + \frac{p_2(1 - p_2)}{n_2}$$

The standard error of the difference in sample proportions is the square root of the variance, but with the unknown $p_i$'s replaced by sample $\widehat{p}_i$'s:

$$\text{s.e.}(\widehat{p}_1 - \widehat{p}_2) = \sqrt{\frac{\widehat{p}_1(1 - \widehat{p}_1)}{n_1} + \frac{\widehat{p}_2(1 - \widehat{p}_2)}{n_2}}. \tag{5}$$

And then we combine the point estimate $\widehat{p}_1 - \widehat{p}_2$ with a s.e. as before.

## Example 9.8a | A Gallup Poll

A 1987 Gallup survey (reported April 12, 1987) found "no evidence of growing public intolerance towards gays." The poll used a sample of 1015 adults from "scientifically selected localities across the nation." One question asked was this: "Do you think homosexual relations between consenting adults should or should not be legal?" In 1986 the percentage who said "should not" was 54; in the 1987 poll, it was 55. Is this "no evidence" of growing intolerance?

The estimate of the difference in proportions in the two years we take to be $.55 - .54 = .01$, or 1 percentage point. We know that $n_1 = 1015$, and we assume that the 1986 sample was the same, $n_2 = 1015$. It's rather clear that the samples were not taken as simple random samples, but we'll pretend that they were, and use (5) to find 95% confidence limits under that assumption:

$$.01 \pm k \sqrt{\frac{.55 \times .45}{1015} + \frac{.54 \times .46}{1015}} = .01 \pm 1.96 \times .022.$$

The desired interval estimate is thus $-.033 < p_1 - p_2 < .053$. In particular, the value $p_1 - p_2 = 0$ is well within this interval, so the Gallup statement of "no evidence" seems reasonable. ∎

---

Confidence limits for $p_1 - p_2$, based on proportions $\widehat{p}_1$ and $\widehat{p}_2$ in independent random samples of sizes $n_1$ and $n_2$:

$$\widehat{p}_1 - \widehat{p}_2 \pm k \sqrt{\frac{\widehat{p}_1(1 - \widehat{p}_1)}{n_1} + \frac{\widehat{p}_2(1 - \widehat{p}_2)}{n_2}}, \tag{6}$$

where $k$ is found in Table IId for a given confidence level.

## 9.9        **Estimating Variability**

We have needed point estimates of population standard deviations in judging the reliability of estimates of a population mean, as in setting confidence limits using $S$ in place of an unknown $\sigma$. Sometimes population variability is of interest for its own sake, as when it is necessary to control the variability of a production process to achieve product uniformity.

**Example 9.9a** | A circuit board in a large computer has many layers that must match up reasonably well, or else the board is defective. The board can be so bad that repairing it is impossible or prohibitively expensive. The process that matches layers can be absolutely perfect *on average,* and yet be so variable that no single board is worth keeping. Companies have experienced a rejection rate as high as 95% in the course of circuit board development.                                                                ■

Obvious estimators to try, in estimating $\sigma^2$ and $\sigma$, are the corresponding sample moments $S^2$ and $S$. We'll look first at the sample variance:

$$S^2 = \frac{1}{n-1}\sum_{i=1}^{n}(X_i - \overline{X})^2. \tag{1}$$

In Example 9.1b, we found that $S^2$ is unbiased: $E(S^2) = \sigma^2$. Its mean squared error is therefore its variance. This variance is a complicated function of the second and fourth population central moments, in which the principle term is inversely proportional to $n$.[12]

In the important special case of a normal population, we know the distribution of $(n-1)S^2/\sigma^2$ to be $\text{chi}^2(n-1)$, with variance $2(n-1)$. Thus,

$$\text{var } S^2 = \frac{\sigma^4}{(n-1)^2} \cdot 2(n-1) = \frac{2\sigma^4}{n-1} = \text{m.s.e.}(S^2).$$

But the second moment of the sample distribution:

$$V = \frac{1}{n}\sum_{i=1}^{n}(X_i - \overline{X})^2 = \frac{n-1}{n}\,S^2,$$

is biased (as we pointed out in Example 9.1b), having mean $(n-1)\sigma^2/n$. Its variance is

$$\text{var } V = \left(\frac{n-1}{n}\right)^2 \text{var } S^2 = \frac{2(n-1)\sigma^4}{n^2}.$$

---

[12]See H. Cramér, *Mathematical Methods of Statistics* (Princeton, NJ: Princeton Univ. Press, 1946), 348.

The mean squared error of $V$ is thus

$$\text{m.s.e.}(V) = \text{var } V + \text{bias}^2 = \frac{2(n-1)\sigma^4}{n^2} + \left(-\frac{\sigma^2}{n}\right)^2 = \frac{2n-1}{n^2}\sigma^4.$$

This is always *smaller* (despite its slight bias) than the m.s.e. of $S^2$.

Because the standard deviation $\sigma$ of a measurement $X$ has the same units as $X$, a confidence interval for $\sigma$ is perhaps better appreciated than an interval for $\sigma^2$. Of course, having found confidence limits for $\sigma^2$, all we need to do is take their square roots to get confidence limits for $\sigma$.

**Example 9.9b**

**A Confidence Interval for $\sigma$**

In Example 9.7a we showed how to use the pivotal quantity $(n-1)S^2/\sigma^2$ to construct a confidence interval for $\sigma^2$, under the assumption of a normal population. In Examples 7.2e and 9.1c we cited a report of blood platelet counts for 35 healthy males. The sample standard deviation was given as 45.05, so the 90% percent confidence interval for $\sigma^2$ is (interpolating in Table Va for 34 d.f.)

$$\frac{34 \times 45.05^2}{48.6} < \sigma^2 < \frac{34 \times 45.05^2}{21.7}.$$

Taking square roots, we get limits for $\sigma$:

$$37.7 < \sigma < 56.4. \qquad \blacksquare$$

Sometimes a different estimator of the standard deviation of a normal population is used, one based on the sample range, $R = X_{(n)} - X_{(1)}$. If we transform each observation to the corresponding $Z$-scores, the largest $Z$ will come from the largest $X$, and the smallest $Z$ from the smallest $X$. Thus,

$$R_Z = Z_{(n)} - Z_{(1)} = \frac{X_{(n)} - \mu}{\sigma} - \frac{X_{(1)} - \mu}{\sigma} = \frac{R_X}{\sigma}. \qquad (2)$$

The distribution of $R_Z$, based on the $n$ independent $Z$-scores, cannot depend on $\mu$ or $\sigma$, so $R_X/\sigma$ is pivotal.

Several percentiles of the distribution of $R_X/\sigma$ are given in Table XIII of the Appendix, along with the first two moments:

$$E(R_X/\sigma) = a_n, \quad \text{var}\,(R_X/\sigma) = b_n^2\sigma^2.$$

These assume a random sample from a normal population. The table entries are calculated numerically as values of integrals based on the joint p.d.f. of the smallest and largest observations for certain small sample sizes [(6) in §8.6]. We can use the entries in the table to get an unbiased estimate of $\sigma$ and to construct confidence intervals for $\sigma$.

Since $R_X = \sigma R_Z$ [from (2)], it follows that $ER_X = \sigma ER_Z = \sigma a_n$, in the notation of Table XIII. So, the statistic $R_X/a_n$ is unbiased:

$$E\left(\frac{R_X}{a_n}\right) = \sigma.$$

Its mean squared error is then its variance:

$$\text{var}\left(\frac{R_X}{a_n}\right) = \frac{1}{a_n^2}\,\text{var}\,(\sigma R_Z) = \frac{\sigma^2}{a_n^2}\,b_n^2.$$

We could define a standard error as the square root of this with the $\sigma$ replaced by the estimate $R_X/a_n$, with the caveat that the latter is not very accurate when $n$ is small.

**Example 9.9c**

## Tracheal Compliance

Measurements of tracheal compliance in six newborn lambs are given in a pediatric research study:[13] .029, .043, .022, .012, .020, .034. The sample range is $R = .031$, so a point estimate of $\sigma$ is $R/a_6 = .031/2.534 = .0122$.

Suppose we want a 90% confidence interval for the population s.d. We know that $W = R/\sigma$ is pivotal, and in Table XIII (for $n = 6$) we find: $W_{.05} = 1.25$, $W_{.95} = 4.03$. Thus

$$.90 = P\left(1.25 < \frac{R}{\sigma} < 4.03\right) = P\left(\frac{R}{4.03} < \sigma < \frac{R}{1.25}\right).$$

The sample range is $R = .031$, so the 90% confidence interval is (.0077, .0248).

Since there is an assumption of population normality, one should at least look at the data. In this case, a dot plot shows no evidence of nonnormality, but with only six observations, normality is hard to prove or disprove.   ∎

---

## Problems

* **9-29.** Blacks constitute 15% of the population in a certain city. Find the probability that the proportions of blacks in two independent random samples, each of size 500, will differ by more than 2 percentage points (either way).

* **9-30.** Aggregate with a low measure of thermal conductivity is desirable for paving. Two types of aggregate are tested for thermal conductivity, with 25 observations from $X$ (the lower cost type) with mean .485 and s.d. 0.180, and 25 observations from $Y$ (the higher cost type) with mean .372 and s.d. 0.160.

(a) Find the s.e. of the difference in sample means as an estimate of the difference in population mean conductivities.

(b) Obtain 99% confidence limits for the difference between the mean conductivities of the two types of aggregates.

---

[13]V. K. Bhutani, R. J. Koslo, and T. H. Shaffer, "The effect of tracheal smooth muscle tone on neonatal airway collapsibility," *Pediatric Research* 20 (1986), 492–95.

**9-31.** An experiment by Danish investigators[14] on the effectiveness of toothpaste involved 295 children who used fluoride toothpaste and 284 who used a nonfluoride toothpaste (colored and flavored to look and taste the same as the fluoride version). The mean number of cavities per child, developed over a 30-month period, was 10.88 for the fluoride sample and 13.41 for the nonfluoride sample. The sample s.d.'s were 6.36 and 7.20, respectively. Give an estimate of the difference in mean number of cavities, together with the standard error.

**✳ 9-32.** The following[15] are summary statistics on SAT-M scores, the mathematical part of the Scholastic Aptitude Test, of children of seventh-grade age who were found in a Johns Hopkins talent search. (To be included, the children had to be in the top 3% of any standard achievement test in verbal, mathematical, or overall intellectual ability.)

|  | $n$ | Mean | s.d. |
|---|---|---|---|
| *Boys* | 19,883 | 416 | 87 |
| *Girls* | 19,937 | 386 | 74 |

Assuming that the participants can be regarded as constituting independent random samples from some population of interest, determine 99% confidence limits for the population mean difference in boys' and girls' SAT-M scores.

**9-33.** In a large group of young couples, the s.d. of the husbands' ages is four years, and that of the wives' ages, three years. Let $D$ denote the age difference within a couple. Since $4^2 + 3^2 = 5^2$, you might expect to find the s.d. of $D$'s in the group to be about five years. Instead you find it to be two years. One explanation is that the discrepancy is the result of random variability. Give another explanation.

**✳ 9-34.** In a sample of 120 adults from the roster of a health maintenance organization in California, there were 39 smokers. In a sample of 100 adults from the roster of an HMO in Minnesota, there were 27 smokers. Find the standard error of the difference in sample proportions of smokers, as an estimate of the difference in population proportions.

**9-35.** An opinion poll surveyed a random sample of 1549 adults prior to the 1984 election. The sample included 638 Democrats and 441 Independents. Of the 638 Democrats, 43% felt that Walter Mondale would do a better job of maintaining prosperity than John Glenn; among the 441 Independents, 28% felt this way.

    **(a)** Give a 95% confidence interval for the population proportion of Democrats.

    **(b)** Assume that the 638 Democrats and 441 Independents are random samples of the Democrats and of the Independents in the population. Give 95% confidence limits for the difference in population proportions of Democrats and of Independents who felt that Mondale would do better.

**✳ 9-36.** Nine readings of an inlet oil temperature are as follows: 99, 93, 99, 97, 90, 96, 93, 88, 99. Find 95% confidence limits for the population s.d. $\sigma$, under the assumption that these readings constitute a random sample from a normal population,

[14] I. J. Moller, J. J. Holst, and E. Sorensen, "Caries reducing effect of a sodium monofluorophosphate dentifrice," *Brit. Dental Jour.* 124 (1968), 209–13.

[15] C. P. Benbow and J. C. Stanley, "Sex differences in mathematical reasoning ability: More facts," Science 213 (1983), 1029–31.

    **(a)**    based on the sample standard deviation.

    **(b)**    based on the sample range.

\* **9-37.**    Consider a random sample of size $n$ from $\mathcal{N}(0, \sigma^2)$. The statistic $\sum X_i^2$ is sufficient for $\sigma^2$. Show that $\sum X_i^2/\sigma^2$ is pivotal and obtain 95% confidence limits for $\sigma^2$ when $n = 15$.

**9-38.**    Cholesterol measurements of those in a random sample of 100 healthy males have mean 245.7 and s.d. 46.25. Assuming a normal population, find 95% confidence limits for

    **(a)**    the population mean.

    **(b)**    the population standard deviation.

**9-39.**    Verify the claim in §9.9 that the m.s.e. of the biased estimator $V$ is always smaller than the m.s.e. of $S^2$, in the case of a normal population.

---

## 9.10    Deriving Estimators

So far, we have considered estimators suggested by intuition. When intuition fails, or even when it doesn't, it is helpful to have some general procedures to follow that lead to estimators which are apt to be good in some sense. One such procedure is the **method of moments**, which develops more systematically what we have done in some special cases. The idea is simply that we estimate population moments using corresponding sample moments. The basis of this idea is that sample moments are *consistent* estimators of the corresponding population moments. This means that for large samples, at least, the estimates we obtain in this way should be close to the target values.

Suppose the population being studied is one of a family indexed by $k$ parameters. The procedure for obtaining estimators for those $k$ parameters involves three steps:

    **(i)**    Obtain formulas for the first $k$ population moments that involve the parameters to be estimated in terms of those parameters.

    **(ii)**    Solve the equations obtained in (i) for the parameters in terms of the $k$ population moments.

    **(iii)**    Replace the population moments by corresponding sample moments.

We need to clarify the terms "population moment" and "corresponding sample moment." We can use either moments about 0: $\mu_k' = E(X^k)$, or (for $k > 1$) central moments: $\mu_k = E[(X - \mu)^k]$. For the corresponding sample moment, we replace the expected value $E(\cdot)$ by the arithmetic mean $\frac{1}{n}\sum$, although in the case of the variance, the divisor $n$ could be replaced by $n - 1$.

---

Example **9.10a**  |  ### Estimating the Parameter of Exp $(\lambda)$

Consider the family of exponential distributions, with p.d.f.

$$f(x \mid \lambda) = \lambda e^{-\lambda x}, \, x > 0,$$

where $\lambda$ is a positive parameter. There is just the one parameter to be estimated, so in step (i) we write $EX = 1/\lambda$. In step (ii) we solve for $\lambda$ in terms of $EX$: $\lambda = 1/EX$. Finally, replacing $EX$ by $\overline{X}$, the corresponding sample moment, we have the method-of-moments estimator: $1/\overline{X}$.                          ∎

## Example 9.10b | Gamma Parameters

The gamma family is indexed by two parameters:

$$f(x \mid \alpha,\, \lambda) = \frac{\lambda^\alpha}{\Gamma(\alpha)}\, x^{\alpha-1} e^{-\lambda x}, \ x > 0,$$

and the first two moments are

$$EX = \alpha/\lambda, \ \text{var}\, X = \alpha/\lambda^2.$$

Solving these for $\alpha$ and $\lambda$, we find

$$\alpha = \frac{(EX)^2}{\text{var}\, X}, \ \lambda = \frac{EX}{\text{var}\, X}\, .$$

We then replace $EX$ by $\overline{X}$ and $\text{var}\, X$ by either $S^2$ or by $V = \frac{n-1}{n} S^2$ to obtain method-of-moments estimators of $\alpha$ and $\lambda$.                          ∎

In §8.2 we suggested that the plausibility of a model should depend on its likelihood, and that in comparing one model with another as possible explanations of data, most people would prefer the one with the larger likelihood. Carrying this to an extreme, we derive estimators by finding the model with the largest likelihood (for given data).

Consider first a one-dimensional parameter $\theta$. The likelihood $L(\theta)$ may have its largest value at a point where the tangent line is horizontal, and the maximizing value of $\theta$ can be found by setting the derivative equal to 0.

---

A **maximum likelihood estimate** of $\theta$ (m.l.e.) is any value $\widehat{\theta}$ that maximizes the likelihood function $L(\theta)$. When the maximum occurs at an interior point of the parameter space where $L'$ exists,

$$L'(\widehat{\theta}) = 0. \tag{1}$$

---

In the case of random samples, the likelihood function is a product, so the differentiation is easier if we work with its logarithm, which is a sum. The logarithm function is monotonically increasing, so a value of $\theta$ that maximizes $\log L(\theta)$ also maximizes $L(\theta)$.

**Example 9.10c** | **Mean Waiting Time**

For estimating the parameter $\lambda$ in the family of exponential distributions from a random sample, we start with the likelihood function:

$$L(\lambda) = \prod_{i=1}^{n} f(X_i \mid \lambda) = \lambda^n \exp\left(-\lambda \sum_{i=1}^{n} X_i\right) = \lambda^n \exp(-n\lambda\overline{X}), \ \ \lambda > 0,$$

and take logarithms:

$$\log L(\lambda) = n \log \lambda - n\lambda\overline{X}.$$

The derivatives are

$$\frac{d}{d\lambda} \log L = \frac{n}{\lambda} - n\overline{X}, \ \text{ and } \ \frac{d^2}{d\lambda^2} \log L = \frac{-n}{\lambda^2}.$$

The first derivative vanishes when $\lambda$ has the value $1/\overline{X}$, and the second derivative is negative there. So $\widehat{\lambda}$ is the (only) maximum likelihood estimator. [$L(\lambda)$ is 0 at both ends of its support interval, $(0, \infty)$.]    ∎

**Example 9.10d** | **Probability of Success**

Given $Y$ successes in independent trials of Ber($p$), the likelihood function is

$$L(p) = p^Y(1-p)^{n-Y}, \ 0 \le p \le 1.$$

If $Y$ is not 0 nor $n$, this bounded, nonnegative function of $p$ is 0 at the endpoints of the interval $[0, 1]$. So it has a maximum at an interior point, and the slope must be 0 at that maximum, the derivative being continuous. To find that point, we first take the logarithm (to make differentiation easier):

$$\log L(p) = Y \log p + (n - Y) \log (1 - p),$$

and then set the derivative with respect to $p$ equal to 0 and solve for $p$:

$$\frac{Y}{p} - \frac{n-Y}{1-p} = 0, \ Y - Yp = np - pY, \ \text{ or } \ p = \frac{Y}{n}.$$

So the m.l.e. is the relative frequency of success, $\widehat{p} = Y/n$.

If $Y = 0$ or $n$, the likelihood function is $(1 - p)^n$ or $p^n$, respectively. In these cases the maximum occurs at an endpoint of the interval $[0, 1]$. The slope of $L(p)$ is *not* zero at those endpoints, so we can't find the maximum by differentiating. It is clear, however, that when $Y = 0$, the likelihood $(1 - p)^n$ has its largest value at 0; and when $Y = n$, the likelihood $p^n$ has its largest value at 1. So even in these cases, the m.l.e. is $\widehat{p} = Y/n$.    ∎

When a distribution family is indexed by several parameters: $(\theta_1, ..., \theta_k)$, the values of these parameters that simultaneously maximize the likelihood function are termed **joint** maximum likelihood estimators. To find them, we use the fact that at any maximum point, the likelihood has its largest value with respect to a particular $\theta_j$ when the other $\theta$'s are fixed at their maximizing values. That is, the (partial) derivative with respect to each $\theta_j$ with the other $\theta$'s held fixed must vanish at a maximum point. These necessary conditions give us $k$ equations to solve simultaneously for the $k$ $\theta$'s. Their solution is a potential maximum point.

To make sure that the solution is a maximum point, one could look at a second derivative condition. However, it is sometimes clear that the solution must be a maximum point when the likelihood is bounded above and is a smooth function.

**Example 9.10e** | **Multinomial Samples**

In the preceding example we saw that the m.l.e. of the Bernoulli parameter $p$ is the sample proportion $\hat{p}$. More generally, consider now a categorical population with any finite number of categories. Suppose there are $k$ categories, and let $p_j$ denote the probability of category $j$, where $\sum p_j = 1$.

Random sampling of such a population is called *multinomial sampling*. The category frequencies $Y_j$ are clearly sufficient, in view of the likelihood function (from Example 8.2d):

$$L(p_1, ..., p_k) = p_1^{Y_1} \cdots p_k^{Y_k}, \ \ 0 \le p_j \le 1, \ \ j = 1, ..., k.$$

Suppose $Y_j > 0$ for all $j$. Substituting $p_k = 1 - p_1 - \cdots - p_{k-1}$ and taking logarithms, we get

$$\log L = \sum_{j=1}^{k-1} Y_j \log p_j + Y_k \log(1 - p_1 - \cdots - p_{k-1}).$$

The derivative with respect to each $p_j$ (with the other $p$'s held constant) must vanish at a maximum point:

$$\frac{Y_j}{p_j} - \frac{Y_k}{p_k} = 0 \ \text{ for } j = 1, ..., k-1. \tag{2}$$

This says that the $p$'s must be proportional to the $Y$'s. Since they must sum to 1, we divide the $Y$'s by their total, $\sum Y_j = n$. Thus, $\hat{p}_j = Y_j/n$, for $j = 1, ..., k$. When no frequency is 0, these relative frequencies do in fact maximize $L$ because $L$ is nonnegative, bounded above, and 0 on the boundary of the $(k-1)$-dimensional parameter space. So the maximum occurs on its interior—and there is only the one critical point.

It can be shown that the restriction $Y_j > 0$ can be dropped here, just as in the case of two categories (see Example 9.10d), so in general, $\hat{p}_j = Y_j/n$. ■

The way in which one happens to parameterize a family of distributions is not unique. For a given parameter $\theta$, any one-one function of $\theta$ gives another way of indexing the family members. Then the question arises, what relation does the m.l.e. of the new parameter have to the m.l.e. of the old?

**Example 9.10f**

**Exp $(1/\mu)$**

In Example 9.10a we considered the family $\text{Exp}(\lambda)$ and found that $\widehat{\lambda} = 1/\overline{X}$. This same family of distributions can be indexed equally well in terms of the population mean, $\mu = 1/\lambda$:

$$f(x \mid \mu) = \frac{1}{\mu} \exp\left(-\frac{x}{\mu}\right), \quad x > 0.$$

(The function $1/\lambda$ is a one-one function for $\lambda > 0$.)  For a random sample of size $n$,

$$\log L(\mu) = -n \log \mu - \frac{n\overline{X}}{\mu}.$$

The derivative with respect to $\mu$ is

$$\frac{\partial}{\partial \mu} \log L(\mu) = -\frac{n}{\mu} + \frac{n\overline{X}}{\mu^2},$$

which vanishes when $\mu$ is $\overline{X}$. The second derivative with respect to $\mu$ is

$$\frac{\partial^2}{\partial \mu^2} \log L(\mu) = \frac{n(\mu - 2\overline{X})}{\mu^3},$$

which is negative at $\mu = \overline{X}$, so $\overline{X}$ maximizes the likelihood. Thus $\widehat{\mu} = \overline{X}$, and this is $1/\widehat{\lambda}$. ∎

In this example we have seen that to find the m.l.e. of $\mu$, we could simply replace the $\lambda$ in $\mu = 1/\lambda$ by its m.l.e. This illustrates a property of m.l.e.'s that holds in general:

---

If $g(\theta)$ is a one-to-one function of $\theta$, and $\widehat{\theta}$ is a maximum likelihood estimator of $\theta$, then $g(\widehat{\theta})$ is a maximum likelihood estimator of $g(\theta)$.

---

To show this, we'll find the likelihood function $\widetilde{L}\,(\xi)$ of the new parameter $\xi = g(\theta)$, where $g$ is one-one and $\theta = g^{-1}(\xi)$. Let $x$ represent the data, with p.f. or p.d.f. $f$. Let

$$\widetilde{f}\,(x \mid \xi) = f[x \mid g^{-1}(\xi)] = f(x \mid \theta).$$

Then,

$$\widetilde{L}\,(\xi) = \widetilde{f}\,(x \mid \xi) = f[x \mid g^{-1}(\xi)] = L[g^{-1}(\xi)].$$

Suppose that $\widehat{\theta}$ maximizes $L(\theta)$, and define $\widehat{\xi} = g(\widehat{\theta})$, so that $\widehat{\theta} = g^{-1}(\widehat{\xi})$. But no value of $L$ can exceed $L(\widehat{\theta})$, so for any $\xi$, we have

$$\widetilde{L}(\xi) = L[g^{-1}(\xi)] \leq L(\widehat{\theta}) = L[g^{-1}(\widehat{\xi})] = \widetilde{L}(\widehat{\xi}).$$

This says that $\widehat{\xi}$ maximizes $\widetilde{L}(\xi)$, as claimed.

Suppose we want a maximum likelihood estimator for estimating some function of the parameters that index the members of a parametric family—a function that is not one-one. We simply define it to be that same function of the m.l.e.'s of those parameters. The rationale for this is what might be termed the *maximum likelihood principle*: A statistical procedure should be consistent with the assumption that the best explanation of a set of data is provided by the population model indexed by $\widehat{\theta}$, a value of $\theta$ that maximizes the likelihood function.

**Example 9.10g** | **Joint Parameter Estimates for $\mathcal{N}(\mu, \sigma^2)$**

Consider the normal family, $\mathcal{N}(\mu, \sigma^2)$. The likelihood function is

$$L(\mu, \sigma^2) = (\sigma^2)^{-n/2}\exp\left\{ -\frac{1}{2\sigma^2}\sum(x_i - \mu)^2 \right\}$$

$$= (\sigma^2)^{-n/2}\exp\left\{ -\frac{1}{2\sigma^2}[nV + n(\mu - \overline{X})^2] \right\},$$

with logarithm

$$\log L(\mu, \sigma^2) = -\frac{n}{2}\log\sigma^2 - \frac{1}{2\sigma^2}[nV + n(\mu - \overline{X})^2].$$

We set the derivatives with respect to $\mu$ and $\sigma^2$, respectively, equal to 0:

$$-\frac{2n}{2\sigma^2}(\mu - \overline{X}) = 0$$

$$-\frac{n}{2\sigma^2} + \frac{n}{2\sigma^4}[V + (\mu - \overline{X})]^2 = 0.$$

The first holds when $\mu$ is $\overline{X}$; substituting this into the second we find that the joint m.l.e.'s are $\overline{X}$ and $V$. (Rather than check a second derivative criterion, we note that we could find the maximum of $L$ by observing first that for any $\sigma^2$, the quantity in braces is smallest when $\mu = \overline{X}$; using this value for $\mu$ yields a function of the single variable $\sigma^2$ whose maximum is easily shown to occur when $\sigma^2$ has the value $V$.)

The joint m.l.e.'s of the parameters $(\mu, \sigma)$, a one-one transformation of $(\mu, \sigma^2)$, are $\overline{X}$ and $\sqrt{V}$. So, if we wanted to find the m.l.e. of $\sigma/\mu$ (for a normal population), we'd simply substitute: $\sqrt{V}/\overline{X}$.  ■

The method of maximum likelihood is in quite common use and often yields estimators that are those we think of intuitively. Moreover:

Maximum likelihood estimators have the following properties:

**(i)**    They are consistent.

**(ii)**   They are asymptotically normal.

**(iii)**  They are functions of sufficient statistics.

Proofs of (i)–(ii) are beyond our scope, and some conditions of regularity are required. (Theorems on consistency need to be carefully stated, because the likelihood can do some weird things in pathological cases.[16]) Property (iii) is almost obvious: If an estimator $T$ is sufficient, the likelihood function $L(\theta)$ depends on the data only through the value of $T$. So any value of $\theta$ that maximizes $L(\theta)$ must be a function of $T$.

Finally, as another approach to deriving estimators, we mention another connection with the notion of sufficiency—another partial justification for the term "sufficient." Roughly speaking, it is this: Given any estimator of $\theta$, if it is not sufficient, and there is a sufficient statistic $T$, we can use $T$ to derive an estimator that is at least as good as the given one:

If $U$ is unbiased as an estimator of $\theta$, and $T = t(\mathbf{X})$ is sufficient for $\theta$, then $V = E(U \mid T)$ is an unbiased estimator of $\theta$ whose variance is no larger than the variance of $U$. Moreover, $\operatorname{var} V = \operatorname{var} U$ if and only if $U$ is essentially a function of $T$.

This is known as the *Rao-Blackwell theorem.* The inequality $\operatorname{var} V \leq \operatorname{var} U$ is argued as follows: Since the conditional distribution of $\mathbf{X}$ given $T$ does not depend on $\theta$, the conditional mean $V$ is a statistic—a function of the sample observations alone. And it has the same mean as $U$:

$$EV = E[E(U \mid T)] = EU,$$

according to (6) of §5.11 (page 217). And by (7) of that same section,

$$\operatorname{var} U = \operatorname{var} V + E[\operatorname{var}(U \mid T)] \geq \operatorname{var} V,$$

as asserted. The occasions for actually deriving an estimator with this method are rare, and since the calculation of the conditional mean of $U$ can be tricky, we'll forego giving any examples.

---

[16]See E. Lehmann, *Theory of Point Estimation* (New York: Wiley, 1983) 410, 414.

## Problems

★ **9-40.** Given a random sample from $\mathcal{N}(0, 1/\theta)$ obtain an estimator of $\theta$ using
  (a)  the method of moments.
  (b)  the method of maximum likelihood.

★ **9-41.** Find the m.l.e. of $\sigma^2$ based on a random sample of size $n$ from $\mathcal{N}(0, \sigma^2)$
  (a)  by finding the likelihood function and maximizing.
  (b)  by using (b) of the preceding problem.

  **9-42.** Given a random sample of size $n$ from Geo($p$), find estimators for $p$
  (a)  using the method of moments.
  (b)  using the method of maximum likelihood.

★ **9-43.** Given a random sample of size $n$ from Poi($\lambda$), find the m.l.e. of $\lambda$.

★ **9-44.** Find the m.l.e.'s of the parameters $(\theta_1, \theta_2)$ of a uniform distribution on the interval $\theta_1 < X < \theta_2$, assuming a random sample of size $n$.

  **9-45.** Referring to the preceding problem, find an estimate of the parameters using the method of moments, given the sample (5, 7, 8, 12). (Compare these with the m.l.e.'s.)

★ **9-46.** Consider a random sample of pairs $(X, Y)$ from a bivariate population with joint p.d.f. $f(x, y) = \exp(-\theta x - y/\theta)$ for $x > 0$ and $y > 0$, where $\theta$ is a positive parameter. Find the maximum likelihood estimate. (Is the m.l.e. a function of the sufficient statistic you found in Problem 8-11?)

  **9-47.** Find the m.l.e. of $\theta$ based on a random sample of size $n$ from the truncated exponential distribution with p.d.f.

$$f(x \mid \theta) = \frac{e^{-x}}{1 - e^{-\theta}}, \quad 0 < x < \theta.$$

★ **9-48.** Find the joint m.l.e.'s of the parameters $\alpha$, $\beta$, and $\sigma^2$, given independent observations $Y_1, \ldots, Y_n$ and constants $c_1, \ldots, c_n$ such that

$$Y_i \sim \mathcal{N}(\alpha + \beta c_i, \sigma^2).$$

(This is a *regression* model, to be taken up in detail in §15.1.)

## 9.11   Bayes Estimators

In §8.11 we used a prior distribution to represent an individual's past experience, beliefs, and convictions about a population model with unknown features—usually a distribution for a parameter in a family of parametric models. We showed how Bayes' theorem can be used to modify or update a *prior* distribution to incorporate the information in new data, resulting in a *posterior* distribution.

In the problem of estimating a population parameter $\theta$, we need to ask how to use a distribution of $\theta$ to come up with an estimate of $\theta$. Of course, if there is no penalty for getting it wrong, we could announce any number as an estimate. If there is a penalty, a rational approach must take the nature of that penalty into account. Perhaps we could specify a function $\ell(\theta, a)$ whose value is the cost or loss incurred if we announce $a$ as the value of the parameter when the true value is $\theta$. The nature of this "loss function" will determine how we might best use the data to obtain an estimate.

It is reasonable to suppose that the bigger the error, the greater the loss—that loss is some monotonic function of $|\theta - a|$, the absolute error. We'll consider just two possible loss functions, the absolute error loss, and the squared-error loss: $(\theta - a)^2$. But, depending as they do on $\theta$, which we now think of as a random variable, these losses are random; so we'll judge an estimate according to the *average* loss—the average with respect to the current distribution of $\theta$, and define the **Bayes loss** as

$$B(a) = E_\theta[\ell(\theta - a)].$$

The number to choose as an estimate of $\theta$ is the value of $a$ that minimizes the Bayes loss, $B(a)$. We term such an estimate a **Bayes estimate.**

With absolute error loss, the Bayes loss is $B(a) = E_\theta|\theta - a|$; and this achieves a minimum value when $a$ is the *median* of the distribution of $\theta$. (This result is given in §5.6 as Property (ii) of the mean deviation.) With mean-square loss, $B(a)$ is $E_\theta[(\theta - a)^2]$, a second moment of $\theta$, minimized when $a$ is the *mean* of the distribution of $\theta$.

---

In estimating a parameter $\theta$, the Bayes estimate is

(i)   the *median* of the distribution of $\theta$, when $\ell(\theta, a) = |\theta - a|$,
(ii)  the *mean* of the distribution of $\theta$, when $\ell(\theta, a) = (\theta - a)^2$.

---

## Example 9.11a | Estimating the Bernoulli Parameter

In Example 8.11a we found the posterior distribution of $p$ based on a random sample from Ber$(p)$, in a particular case, assuming a beta distribution as the prior for $p$. More generally, suppose the prior is Beta$(r, s)$, and $Y$ is the number of successes in $n$ independent trials. The posterior p.d.f. is

$$h(p \mid Y) \propto L(p)g(p) = p^Y(1-p)^{n-Y}p^{r-1}(1-p)^{s-1}$$

$$= p^{Y+r-1}(1-p)^{n-Y+s-1}.$$

The mean of this beta posterior is

$$E(p \mid Y) = \frac{Y+r}{n+r+s},$$

and this is the Bayes estimate of $p$ under the assumption of a quadratic loss. It can also be written as

$$E(p \mid Y) = \left\{\frac{n}{n+r+s}\right\}\frac{Y}{n} + \left\{1 - \frac{n}{n+r+s}\right\}\frac{r}{r+s},$$

showing that it is a weighted average of the sample proportion and the prior mean. The larger the sample, the closer the estimate is to the sample proportion.

The special case $r = s = 1$ is that of a *uniform* prior, where in this case the posterior mean is

$$E(p \mid Y) = \frac{Y + 1}{n + 2}.$$

This estimate of $p$ has an extensive history and is known as *Laplace's rule of succession.*

It is noteworthy that if we set $g(p) \propto p^{-1}(1 - p)^{-1}$ and proceed formally, the posterior mean is $Y/n$, which is the maximum likelihood estimator. However, $g$ is not (and can't be made into) a proper density p.d.f. ∎

**Example 9.11b** | **Estimating a Normal Mean**

Suppose $X \sim \mathcal{N}(\mu, \sigma_0^2)$, where $\sigma_0^2$ is a known value of the population variance, and suppose we adopt a normal prior for the mean: $\mu \sim \mathcal{N}(\nu, \tau^2)$. With no data at hand, the Bayes estimate of $\mu$ is the prior mean $\nu$ (which is also the prior median), using either absolute error or mean-squared error loss.

Given a random sample of size $n$, with mean $\overline{X}$, the posterior distribution of $\mu$, as shown in §8.11, is *normal* with mean

$$E(\mu \mid \overline{X}) = \frac{\pi_{\text{pr}}}{\pi_{\text{po}}} \nu + \frac{\pi_{\text{data}}}{\pi_{\text{po}}} \overline{X}, \tag{1}$$

where "$\pi$" denotes precision, or reciprocal variance:

$$\pi_{\text{pr}} = \frac{1}{\tau^2}, \quad \pi_{\text{data}} = \frac{n}{\sigma_0^2}, \quad \pi_{\text{po}} = \pi_{\text{pr}} + \pi_{\text{data}}. \tag{2}$$

The posterior mean (1), which is also the posterior median, is the Bayes estimate of $\mu$, with either absolute error or mean-squared error loss. ∎

Turning to the notion of interval estimate, we are now in a position to attribute probability to a specific interval of possible values of a parameter $\theta$, since we have (either as prior or posterior) a probability distribution for $\theta$. With any given distribution for a parameter $\theta$, we can construct a probability interval for $\theta$, taking as upper and lower limits numbers surrounding $\theta$ such that the posterior probability between them is any specified amount (such as 95%).

If someone has found a confidence interval for $\theta$, we can find the probability of that interval in the posterior distribution. But this may or may not (usually it will not) agree with the confidence level.

**Example 9.11c** | **Interval Estimates for a Normal Mean**

In Example 9.1c we used the mean $\overline{X} = 234$ and s.e.$(\overline{X}) = 7.6$, given a sample of 35 blood platelet counts, to find an approximate 95% confidence interval for the mean count: $219.1 < \mu < 248.9$.

Suppose now we have a prior distribution for $\mu$ that is normal with mean $\nu = 250$ and s.d. $\tau = 5$. The prior precision is then $1/\tau^2 = .04$, and the precision of the data is $n/\sigma^2 = .0173$ (with $\sigma$ estimated as $S$). The posterior is then normal with precision

$$\pi_{po} = .0173 + .04 = .0573 = \frac{1}{4.18^2},$$

and mean

$$E(\mu \mid \text{data}) = \frac{.0173}{.0573}\overline{X} + \frac{.04}{.0573}\nu = 245.2.$$

The posterior probability of the interval $(219.1, 248.9)$ is

$$P(219.1 < \mu < 248.9) = \Phi\left(\frac{248.9 - 245.2}{4.18}\right) - \Phi\left(\frac{219.1 - 245.2}{4.18}\right)$$

$$\doteq \Phi(.89) - \Phi(-6.24) = .8133.$$

Of course, with a different prior, the interval would have a different probability. For instance, if $\nu = 240$ and $\tau = 5$, the probability of the above confidence interval would be about .995.

If we were to take a sequence of normal priors in which (for any fixed mean) the prior variance tends to infinity, we'd find that the posterior mean and variance would tend to the sample mean $\overline{X}$, and its variance to $\sigma^2/n$. This means that the posterior probability of the confidence interval tends to the confidence level. That is, in the limit, as the prior information becomes ever more diffuse, the popular interpretation of a confidence level (as the probability of the confidence interval) becomes correct.

With the above posterior for $\mu$: $\mathcal{N}(245.2, 4.18^2)$, we can define a 95% *probability interval* for $\mu$ centered at the mean, by the limits

$$245.2 \pm 1.96 \times 4.18, \ \text{ or } \ 237.0 < \mu < 253.4.$$

And we can now say (correctly) that $P(237.0 < \mu < 253.4) = .95.$   ■

---

## Problems

* **9-49.**  Obtain Bayes estimates of the Bernoulli parameter $p$ based on ten independent trials, assuming quadratic loss,
  - **(a)**  if the prior is $g(p) \propto p(1-p)^4$ and there are six successes.
  - **(b)**  if the prior is $g(p) \propto p^5(1-p)$ and there are six successes.

**9-50.**  Obtain the Bayes estimate of the Bernoulli parameter $p$ based on ten successes in ten independent trials if the prior is uniform, with the absolute error loss function.

**9-51.**  A U.S. Fish and Wildlife report[17] applies the Bayesian method to a census of breeding pairs of mallard ducks. The parameter of interest is the proportion $p$ of quarter sections in which breeding mallards are present. For the assumption of ignorance, the report uses the prior

---

[17]U.S. Department of the Interior, *Special Scientific Report—Wildlife*, No. 203, by D. H. Johnson.

$g(p) \propto p^{-.5}(1 - p)^{-.5}$, which is quite flat over the interval $.2 < p < .8$ but rises sharply to infinity at 0 and 1. A 1972 study[18] found mallards breeding in 79 of 130 randomly selected quarter sections. Assuming the binomial model to be applicable, find the Bayes estimate of $p$ when the loss is quadratic.

* **9-52.** Consider a random sample of size $n$ from a normal population with mean 0 and reciprocal variance $\theta$. (Thus, $\theta$ is the "precision" of each observation.)  Assume a quadratic loss function for estimating $\theta$.

   **(a)**   Given the prior $\text{Gam}(\alpha, \lambda)$ for $\theta$, find the (no-data) Bayes estimate of $\theta$.

   **(b)**   If the prior is $\text{Gam}(5, 5)$, the sample size is $n = 8$, and the observations $X$ are $(3, -1, 1, -1, -1, 0, 2, 0)$, find the Bayes estimate of $\theta$.

**9-53.**   Suppose you are estimating a normal mean $\mu$, and that your current distribution for $\mu$ is $\mathcal{N}(\mu_0, 1)$. Find the Bayes estimate of $\mu$ if the loss function $\ell(\mu, a)$ is 0 or 1 according as $|\mu - a| \le 1$ or $|\mu - a| > 1$.

* **9-54.**   Given a random sample of size 5 from $\text{Exp}(\lambda)$. Assuming quadratic loss, find the Bayes estimate of $\theta$ if the prior p.d.f. is $g(\lambda) = \lambda e^{-\lambda}$, $\lambda > 0$, and the sample sum is $\sum X_i = 10$.

**9-55.**   Some years ago, an extensive study[19] of times to failure gave data on failure times of 100 radar indicator tubes, with mean 297.5. Suppose we assume the time-to-failure distribution is $\text{Exp}(\lambda)$, and that (before examining the data) we had a gamma prior with $\mu_\lambda = \sigma_\lambda = .0005$. Find the Bayes estimate of $\lambda$, assuming quadratic loss.

## 9.12   Efficiency

When we use mean squared error as a criterion for accuracy of estimation, it is natural to choose between competing estimators in terms of m.s.e. In some cases, however, one estimator may have a smaller m.s.e. than another for some parameter values but not for others. For instance, the m.s.e. estimator $T \equiv 5$ of a population mean is $(5 - \mu)^2$. If $\mu = 5$, this is 0—smaller than the m.s.e. of the sample mean; but when $\mu$ is far from 5, the inequality will go the other way.

Consider a distribution family indexed by a single parameter $\theta$ and estimators $T$ and $T'$. When the *ratio* of their m.s.e.'s is independent of $\theta$, that ratio is called the *efficiency* of $T'$ relative to $T$:

$$\text{eff}(T', T) = \frac{\text{m.s.e.}(T)}{\text{m.s.e.}(T')}. \tag{1}$$

When this efficiency is less than 1, we think of $T$ as making more efficient use of the data than does $T'$.

**Example 9.12a** | **The Mean Is Not Always Best**

Consider a random sample of size $n$ from $\mathcal{U}(0, \theta)$. The population mean is $\mu = \theta/2$, and the variance $\theta^2/12$. The sample mean (as always) is an unbiased estimator of the

---

[18]"Population estimates of breeding birds in North Dakota," *Auk* 89(4), 766–88.

[19]From D. J. Davis, "An analysis of some failure data," *J. Amer. Stat. Assn.* 47 (1952), 147.

population mean, so the mean squared error is the variance:

$$\text{var}\,\overline{X} = \frac{\sigma^2}{n} = \frac{\theta^2}{12n}.$$

On the other hand, we know (see Problem 9-4) that the statistic

$$T = \frac{n+1}{2n}\,X_{(n)}$$

is unbiased for estimating $\mu$. The variance of $X_{(n)}$, from Example 8.4c, is

$$\text{var}\,X_{(n)} = \frac{n\theta^2}{(n+2)(n+1)^2},$$

so the mean squared error of $T$ is its variance:

$$\text{var}\,T = \left(\frac{n+1}{2n}\right)^2 \text{var}\,X_{(n)} = \frac{\theta^2}{4n(n+2)}.$$

For estimating $\mu$, the efficiency of $\overline{X}$ relative to $T$ is thus

$$\text{eff}(\overline{X},\,T) = \frac{\text{var}\,T}{\text{var}\,\overline{X}} = \frac{3}{n+2}.$$

So when $n > 1$, $T$ is more efficient than $\overline{X}$, and indeed, the larger the sample, the better it looks. ∎

One may ask, just how efficient can an estimator be? If there were a lower bound for the mean squared error, we could get a measure of an estimator's absolute efficiency by comparing its mean squared error with that lower bound. There isn't always such a bound, but in the class of *unbiased* estimators there is one:

---

### Cramér-Rao Inequality:

Let $T = g(\mathbf{X})$ be an unbiased estimator of $\theta$, where $\mathbf{X}$ has a p.f. or p.d.f. $f(\mathbf{x} \mid \theta)$ whose support does not depend on $\theta$. Under mild conditions of regularity,

$$\text{var}\,T \geq \frac{1}{E\left\{\left(\frac{\partial}{\partial\theta}\log f(\mathbf{X} \mid \theta)\right)^2\right\}}. \qquad (2)$$

---

The denominator in (2) is called the *information* in $\mathbf{X}$ about $\theta$, $I_{\mathbf{X}}(\theta \mid \mathbf{X})$. If $\mathbf{X}$ is a random sample from $f(x \mid \theta)$, the information in $\mathbf{X}$ is $n$ times the information in a single observation:

$$I(\theta \mid \mathbf{X}) = nI(\theta \mid X).$$

Thus, (2) can be written as

$$\operatorname{var} T \geq \frac{1}{n\, E\left\{\left(\frac{\partial}{\partial \theta} \log f(X \mid \theta)\right)^2\right\}}. \tag{3}$$

The inequality (2) [or (3)] is often referred to as the *information inequality*.

A formal proof of the Cramér-Rao inequality requires stating and using the "conditions of regularity." We give here an informal derivation using the fact (from §5.9 and §3.4) that a correlation coefficient cannot exceed 1 in magnitude. We'll write it out for the case of a continuous population variable; the discrete case is similar. Our manipulations involve interchanges of integral and derivative that call for the regularity conditions.

The correlation coefficient we use is that of an unbiased estimator $T$ and the variable $W$, defined as

$$W = \frac{\partial}{\partial \theta} \log f(\mathbf{X} \mid \theta) = \frac{f'(\mathbf{X} \mid \theta)}{f(\mathbf{X} \mid \theta)},$$

where here and in what follows a prime will denote differentiation with respect to the parameter $\theta$. The variable $W$ has mean 0, which we see as follows:

$$EW = \int \frac{f'(\mathbf{x} \mid \theta)}{f(\mathbf{x} \mid \theta)} f(\mathbf{x} \mid \theta)\, d\mathbf{x} = \frac{d}{d\theta} \int f(\mathbf{x} \mid \theta)\, d\mathbf{x} = \frac{d}{d\theta}(1) = 0.$$

(The integrals here are $n$-dimensional, over the whole space.) Because of this, the covariance of $W$ and $T = t(\mathbf{x})$ is just their expected product:

$$\operatorname{cov}(W, T) = E(WT) = \int \frac{f'(\mathbf{x} \mid \theta)}{f(\mathbf{x} \mid \theta)} t(\mathbf{x}) f(\mathbf{x} \mid \theta)\, d\mathbf{x}$$

$$= \frac{d}{d\theta} \int t(\mathbf{x})\, f(\mathbf{x} \mid \theta)\, d\mathbf{x} = \frac{d}{d\theta} ET = \frac{d\theta}{d\theta} = 1.$$

Now, using the fact that the correlation is bounded, we have

$$\rho_{W,T}^2 = \frac{\operatorname{cov}^2(W, T)}{(\operatorname{var} W)(\operatorname{var} T)} = \frac{1}{(\operatorname{var} W)(\operatorname{var} T)} \leq 1,$$

or, as claimed in (2),

$$\operatorname{var} T \geq \frac{1}{\operatorname{var} W}. \tag{4}$$

When $\mathbf{X}$ is a random sample from $f(x \mid \theta)$, the sample observations $X_i$ are independent, so

$$W = \frac{\partial}{\partial \theta} \log f(\mathbf{X} \mid \theta) = \frac{\partial}{\partial \theta} \log \prod f(X_i \mid \theta) = \sum \frac{\partial}{\partial \theta} \log f(X_i \mid \theta) = \sum W_i,$$

where

$$W_i = \frac{\partial}{\partial \theta} \log f(X_i \mid \theta).$$

The $W_i$ are independent and identically distributed with mean 0 and variance equal to the expected square. Hence,

$$\text{var } W = \text{var}\left(\sum W_i\right) = n \, E\left\{\left(\frac{\partial}{\partial \theta} \log f(X \mid \theta)\right)^2\right\}. \qquad (5)$$

When substituted in (4), this yields (3).

According to (2), no unbiased estimator can have a variance smaller than $1/I_{\mathbf{X}}(\theta)$. But can we find an estimator whose variance *equals* that lower bound? If so, we say that it is **efficient,** making fullest use of the data. The *efficiency* of an estimator is then defined as the ratio of the Cramér-Rao lower bound to the variance of the estimator.

---

The **efficiency** of an unbiased estimator $T$ in estimating $\theta$ is

$$\text{eff}(T) = \frac{1/I_{\mathbf{X}}(\theta)}{\text{var } T}. \qquad (6)$$

When its efficiency is 1, we say that $T$ is *efficient.*

---

**Example 9.12b**  **Exp $(1/\lambda)$**

Consider a random sample of size $n$ from an exponential population with the population mean $\mu$ as the parameter:

$$f(x \mid \mu) = \frac{1}{\mu} e^{-x/\mu}, \quad x > 0.$$

The likelihood, for one observation, is $f(X \mid \mu)$, and

$$\log L(\mu) = -\log \mu - \frac{X}{\mu}.$$

The partial derivative with respect to $\mu$ is

$$W = \frac{\partial}{\partial \mu} \log L(\mu) = -\frac{1}{\mu} + \frac{X}{\mu^2}.$$

Since $EX = \mu$, this has mean 0—as is generally the case. So the average of the square is the variance:

$$\text{var } W = \text{var}\left(-\frac{1}{\mu} + \frac{X}{\mu^2}\right) = \text{var}\left(\frac{X}{\mu^2}\right) = \frac{1}{\mu^4}\text{var } X = \frac{1}{\mu^2},$$

and this is $I(\mu)$ for a single observation. For a random sample of size $n$, the information is $n$ times this, so the lower bound in (5) is $\mu^2/n$.

The m.l.e. $\overline{X}$ is unbiased, and its variance is $\sigma^2/n = \mu^2/n$. This equals the lower bound, so $\overline{X}$ is efficient in this case. ∎

**Example 9.12c** | **The Exponential Family**

In Example 8.3e we introduced the exponential family of distributions. The p.d.f. of the one-parameter family is of the form

$$f(x \mid \theta) = B(\theta)h(x)\exp[Q(\theta)R(x)],$$

a form which includes the distribution of the preceding example as a special case. Given a random sample of size $n$, we have

$$W = \frac{\partial}{\partial\theta}\log f(\mathbf{X} \mid \theta) = n\frac{B'(\theta)}{B(\theta)} + Q'(\theta)\sum R(X_i).$$

So $W$ is a linear function of the statistic $Y = \sum R(X_i)$. This means that $\rho_{W,Y}^2 = 1$, and (2) is an equality. If $Y$ is unbiased, it is efficient. Of course, if it is not unbiased, but multiplying by a constant would produce an unbiased estimator of $\theta$, one can simply redefine $R$ and $Q$ so that $\sum R(X_i)$ is unbiased.

For instance, in the particular subfamily with p.d.f.

$$f(x \mid \theta) = \frac{x}{\theta^2}e^{-x/\theta}, \quad x > 0, \ \theta > 0,$$

the function $R(x)$ could be just $x$. But $\sum X_i$ is not unbiased; its mean is $n\mu$, where $\mu = E(X) = 2\theta$. If instead we take $R$ to be $x/(2n)$ and $Q$ to be $2n/\theta$, we'd have $\sum R(X_i) = \overline{X}/2$, which is unbiased—and hence, efficient. ∎

There is almost a kind of converse of what we found in the last example: Suppose $T$ is efficient in estimating $\theta$, a parameter that indexes a family of p.f.'s or p.d.f.'s, $\{f(x|\theta)\}$. Because $T$ is efficient, the correlation coefficient of the pair $(T, W)$ is $\pm 1$. Thus, $W$ is a linear function of $T$—one in which the slope and intercept constants can depend on $\theta$:

$$W = \frac{\partial}{\partial\theta}\log f(\mathbf{X} \mid \theta) = a(\theta) + b(\theta)T.$$

Integrating, we recover the log of $f(\mathbf{X} \mid \theta)$:

$$\log f(\mathbf{X} \mid \theta) = A(\theta) + B(\theta)T + K(\mathbf{X}). \tag{7}$$

[The functions $A$ and $B$ are indefinite integrals of $a$ and $b$, and the "constant" of integration $K$ can depend on $\mathbf{X}$ (though independent of $\theta$).]  And the likelihood function is

$$L(\theta) = \exp[A(\theta) + B(\theta)T].\tag{8}$$

From (7) and (8) we draw two conclusions about an estimator $T$ that is efficient: Since the likelihood function depends on $\mathbf{X}$ through the value of $T$, it follows that $T$ must be *sufficient*. Also, the distribution of $T$ is in the exponential family.

## Example 9.12d | A Parameter with No Efficient Estimator

Suppose $X \sim \text{Exp}(\lambda)$. For a random sample of size $n$, the sample mean $\bar{X}$ is sufficient, the information in the sample is $I(\lambda) = n/\lambda^2$ (see Problem 9-58), and the m.l.e. (Example 9.10a) is $1/\bar{X}$. Using Problem 9-61, we find the mean value of the m.l.e. to be

$$E\left(\frac{1}{\bar{X}}\right) = n E\left(\frac{1}{\sum X_i}\right) = n \int_0^\infty \frac{1}{u} f_{\Sigma X}(u)\, du = \frac{n}{n-1}\lambda.$$

So the estimator $Y = (n-1)/\sum X_i$ is unbiased for $\lambda$. Its variance (Problem 9-61) is $\lambda^2/(n-2)$, and no other unbiased estimator has a smaller variance! The information in the sample is $I(\lambda) = n/\lambda^2$, so the efficiency of $Y$ is

$$\text{eff}(Y) = \frac{1/I(\lambda)}{\text{var } Y} = \frac{\lambda^2/n}{\lambda^2/(n-2)} = 1 - \frac{2}{n}.$$

This is less than 1, so there is no efficient estimator. However, as $n$ becomes infinite, the efficiency of $Y$ tends to 1.  ∎

When the efficiency of an estimator approaches 1 as the sample size tends to infinity (as in the above example), it is said to be **asymptotically efficient.** The notion of asymptotic efficiency can be given a more general definition, one that applies to sequences of estimators that are biased and estimators whose variance is not finite. In regular cases, maximum likelihood estimators can be shown to be asymptotically efficient in this extended sense.[20]

## Example 9.12e | The Sample Median

The sample median and the sample mean are obvious candidates for estimating $\mu$ in $N(\mu, \sigma^2)$, since $\mu$ is both the population mean and median. When $\sigma^2$ is known, the sample mean is efficient for any $n$, so it is asymptotically efficient, with asymptotic variance $\sigma^2/n$. It can be shown that the sample median has asymptotic variance $\pi\sigma^2/(2n)$. The ratio of these asymptotic variances is the asymptotic efficiency of the median, or $2/\pi \doteq .637$. The interpretation of this is that in large samples, using the

---

[20]H. Cramér, *Mathematical Methods of Statistics* (Princeton, NJ:  Princeton Univ. Press, 1946), 500.

sample mean requires only about 64% as many observations to estimate $\mu$ with a specified precision as would be required using the sample median.

But don't read too much into this result. If the population is normal, the median is indeed less efficient than the mean. But if the population is *not* normal, the median can be much preferable—as, for example, when the population is normal except for an occasional observation that has a larger variance because it is recorded incorrectly.

∎

## Problems

**9-56.** Show that if $T_n$ is efficient for each $n$ as an estimator of $\theta$, it is also consistent.

**\* 9-57.** For a random sample of size $n$ from $\text{Ber}(p)$, show that the sample proportion $\hat{p}$ is an efficient estimator of the corresponding population proportion,
   (a) by calculating the information $I(p)$.
   (b) by using the fact that $\text{Ber}(p)$ is in the exponential family.

**9-58.** For a random sample of size $n$ from $\text{Exp}(\lambda)$, find the information $I(\lambda)$.

**9-59.** Show that the mean of a random sample from $\text{Poi}(\lambda)$ is an efficient estimator of $\lambda$.

**9-60.** The likelihood function for a single observation from $\mathcal{N}(0, \theta)$ is

$$L(\theta) = \theta^{-1/2} e^{-x^2/(2\theta)}$$

Show that, for a random sample of size $n$, the m.l.e. $\hat{\theta}$ is efficient,
   (a) by calculating the information $I(\theta)$.
   (b) by using the fact that $\mathcal{N}(0, \theta)$ is in the exponential family.

**\* 9-61.** Given that $V \sim \text{Gam}(n, \lambda)$, find the mean and variance of the random variable $1/V$, for use in Example 9-12d.

## Review Problems

**9-R1.** Find the sample size that will ensure a standard error of at most 1.2, when estimating a population mean, given that the population s.d. is $\sigma = 18$ and assuming a random sample.

**9-R2.** We want to estimate a population proportion using a random sample.
   (a) Find $n$ such that the standard error will not exceed .015.
   (b) With the sample size $n$ found in (a), suppose it turns out that $\hat{p} = .30$.
Find 95% confidence limits for $p$.

**9-R3.** Using independent random samples of size 100 from each of populations 1 and 2, find the standard error of $\overline{X}_1 - \overline{X}_2$ as an estimate of $\mu_1 - \mu_2$ (the population mean difference) when the sample s.d.'s are $S_1 = 9$ and $S_2 = 12$.

**9-R4.** Find the information $I(\theta)$ in one observation from a population with p.d.f. $f(x \mid \theta) = \theta x^{\theta-1}, 0 < x < 1$.

**9-R5.** Consider random samples from $\mathcal{N}(0, \theta)$. In Problem 8-9 it was seen that $T = \sum X_i^2/n$ is sufficient for $\theta$.
   (a) Find an efficient estimate of $\theta$.
   (b) Show that $\sum X_i^2/\theta$ is pivotal.

**9-R6.**   Consider random samples from Poi($\lambda$).

(a)   Find the information in one observation.

(b)   Find the information in $n$ observations.

(c)   Find the efficiency of $\overline{X}$.

(d)   Show consistency of $\overline{X}$ directly, rather than as a special case of a general theorem about m.l.e.'s.

**9-R7.**   For the setting of the preceding problem, suppose $n = 10$ and $\sum X_i = 9$, and we adopt a prior with p.d.f. $e^{-\lambda}$, $\lambda > 0$. Find the Bayes estimate of $\theta$,

(a)   assuming quadratic loss.

(b)   assuming absolute error loss. [*Hint:* Use (3) of §6.3 and interpolate in the Poisson tables.]

**9-R8.**   Let $T = \overline{X}/2$, where $\overline{X}$ is the mean of a random sample from a population with p.d.f. $\lambda^2 x e^{-\lambda x}$, $x > 0$ and $\lambda > 0$ (as in Problems 8-R5 and 8-R6, where we found the mean and s.d. of $T$).

(a)   Find the m.s.e. for $T$ as an estimate of $\theta = 1/\lambda$.

(b)   Find the m.l.e.'s of $\lambda$ and $\theta$.

(c)   Find the method-of-moments estimator of $\lambda$.

(d)   Suppose we observe $\sum X_i = 12$ for ten observations. Find the posterior mean of $\lambda$ when the prior p.d.f. is $\lambda e^{-\lambda}$, $\lambda > 0$. (Compare with the m.l.e.)

**9-R9.**   Given: $f(x \mid \theta) = \frac{x}{\theta} e^{-\frac{x^2}{2\theta}}$, $x > 0$, $EX = \sqrt{\pi\theta/2}$, and $\sigma_X^2 = 2\theta(1 - \pi/4)$ (as in Problem 8-R9.)

(a)   Find the m.l.e. of $\theta$.

(b)   Use the method of moments to obtain an estimator of $\theta$.

(c)   The given p.d.f. is that of $\sqrt{\theta(Z_1^2 + Z_2^2)}$, where $Z_1$ and $Z_2$ are standard normal and independent. Use this fact to find the distribution of $X^2/\theta$.

(d)   Find the mean squared error of the m.l.e.

**9-R10.**   Show that if $T$ is unbiased as an estimate of $\theta$, then

(a)   $\sqrt{T}$ is biased as an estimate of $\sqrt{\theta}$.

(b)   $T^2$ is biased as an estimate of $\theta^2$.

**9-R11.**   A dental study[21] found that the mean difference in the number of new cavities, after a trial period, between independent samples of children who used, respectively, Colgate's MFP fluoride toothpaste and a stannous fluoride toothpaste, was 2.41. The mean difference $\overline{D}$ can be assumed to be approximately normal with mean $\theta$ and variance 1.252 (as estimated from the sample data). Find a 95% confidence interval for $\theta$ and calculate its probability if the researchers' prior was

(a)   $\mathcal{N}(4, 4)$.     (b)   $\mathcal{N}(1, 16)$

---

## Chapter Perspective

We have studied the estimation of a population parameter $\theta$, mainly for cases in which $\theta$ is a single real parameter, and in terms of mean squared error as a criterion

---

[21]S. F. Frankl and J. E. Alman, *J. Oral Therapeutics Pharmacol.* 4 (1968), 443–49.

for accuracy. Using unbiased (or nearly unbiased) estimators based on large samples, we have assessed the accuracy of estimation in terms of mean squared error and its estimated square root, the standard error. For unbiased estimators, this is the estimated standard deviation of the estimator.

Properties of estimators that seem desirable are consistency—a large sample property—and efficiency, which is defined for any sample size. If an estimator is not a function of a sufficient statistic, one can find an estimator based on a sufficient statistic that is just as good (in terms of mean squared error).

Confidence intervals have provided a way of combining a point estimate with its standard error to give a range of parameter values in which, with a given degree of confidence, one thinks the true value lies. Confidence interval construction is closely related to the testing of hypotheses, as we'll see in the next chapter.

In presenting ways of "deriving" estimators, we have emphasized the most popular one, the method of maximum likelihood. Maximizing the likelihood is intuitive and leads to estimates that are consistent and sometimes efficient. It is mainly this method that we'll use later—in problems of regression, categorical data, and analysis of variance (Chapters 13–15) and in constructing "likelihood ratio" tests (Chapter 11).

The maximum likelihood technique is applicable when several parameters are jointly estimated. The notions of m.s.e. and efficiency can be extended to these multiparameter situations. For efficiency, one needs an extended definition of information. (See, for example, Rao [16].)

In Chapter 8 we saw how to incorporate the likelihood function (for given data) and a prior distribution for $\theta$ to produce a posterior distribution for $\theta$. In this chapter we have shown how to use that posterior to obtain an estimate of $\theta$ and how to construct an interval containing $\theta$ with a specified probability.

# Significance Testing

Sample data provide evidence concerning hypotheses about the population from which they are drawn. The following examples give some settings in which there are hypotheses about populations that need testing.

| Example **10a** | **Extrasensory Perception** |

In one segment of the TV show "The Odd Couple," Felix claimed to have ESP. Naturally, Oscar was skeptical and suggested testing the claim. Oscar would draw a card at random from four large cards, each with a different geometric figure on it, and without showing it, ask Felix to identify the card. They repeated this basic experiment several times.

At each such trial, an individual without ESP has one chance in four of correctly identifying the card. In ten trials, Felix made six correct identifications. Although he didn't claim to be perfect, 6 is rather more than 2.5, the average number correct if he does *not* have ESP. What does this prove?   Strictly speaking, it proves nothing, because it was possible for Felix to get six correct whether he used ESP or was only guessing. Nevertheless, his high score provides some evidence for ESP—which may or may not be convincing to you, depending on how you view ESP. (To be continued.) ∎

| Example **10b** | **Feeding Schedules and Blood Pressure** |

Falk et al.[1] report on a study of the effect of intermittent feeding on the blood pressure of rats. They fed eight rats intermittently over a period of weeks, and at the end of that time measured the rats' blood pressures. The results were as follows, in mm of Hg:

170, 168, 115, 181, 162, 199, 207, 162.

---

[1]J. Falk, M. Tang, and S. Forman, "Schedule-induced chronic hypertension," *Psychosomatic Medicine* 39 (1977), 252–63.

These numbers would mean little in the absence of data for comparison. So, blood pressures of seven rats that had been fed normally were also measured, with these results:

169, 155, 134, 152, 133, 108, 145.

(There was an eighth rat in this group, but its result is missing because it had died during the induction of ether anesthesia.)

The data are plotted as dot diagrams in Figure 10-1. It is apparent that the measurements in the first group tend to be higher than those in the second group, although there is one rather low reading in the first group. (It is possible that this "outlier" resulted from a misprint, a temporarily defective instrument, or a rat that was inadvertently exchanged with one in the second group; but without information of this nature, all we can do is proceed as though all measurements were equally valid.)

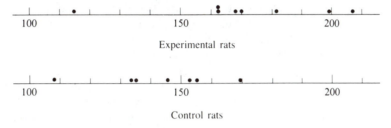

**Figure 10-1** Dot diagram for Example 10b

The question addressed in the study is whether intermittent feeding affects blood pressure. If it does, we might well see differences like those actually observed; but because of sampling variability, we would see differences even if the feeding schedule had *no* effect. The problem is to decide which explanation of the data is the more plausible—that feeding has an effect, or that it has no effect and that sampling variability is responsible for the discrepancy between the two samples. (To be continued.)

Example 10c

## Are Boys Better at Math?

Do boys have a greater natural proclivity toward mathematics than girls?[2] Some people think so, but how could such a claim be tested? First, of course, we'd need a way of measuring "proclivity." This is not easy, but suppose that it can be done, using a battery of tests administered at an early age. Since one can't possibly test every child, we must be content with testing those in a sample. Naturally, since the basic question is one of comparison, both boys and girls must be tested.

The claim is not that *all* boys have a greater proclivity toward mathematics than all girls. The question is one of *tendency,* ordinarily measured in terms of an *average.* To draw conclusions about the difference between averages for boys and for girls

[2]This question has been the subject of much research. For example, see L. H. Fox, L. Brody, and D. Tobin (Eds.), *Women and the Mathematical Mystique* (Baltimore: Johns Hopkins University Press, 1980).

generally, we could calculate the averages for a sample of boys and for a sample of girls. These sample means should tell us something about the difference (if any) between population means, especially when sample sizes are large.  ■

## 10.1        Hypotheses

Examples 10a–10c have some common elements. In each, there is a theory or claim and (at least implicitly) a countertheory or counterclaim. These theories and counter-theories are **hypotheses.** Each hypothesis can be expressed in terms of a probability model or a set of probability models. The model describes a population, real or conceptual, and it is assumed that sample observations are selected at random from this population.

Oscar's claim in Example 10a that Felix does not have ESP corresponds to a model in which the probability $p$ of a correct identification is 1/4. If Felix has ESP, then $p > 1/4$, and this defines a set of probability models.

In the feeding-schedule example, there are two populations—one of rats fed intermittently, the other of rats fed regularly. One hypothesis about these two populations is that their average blood pressures are the same—that how the rats are fed does not affect average blood pressure. A competing hypothesis, undoubtedly the one the investigators had in mind, is that the feeding schedule *is* a factor.

In Example 10c, one possible hypothesis is that boys and girls have the same natural proclivity for mathematics, on average. An alternative hypothesis, suggested by the original question, is that boys tend to have the greater proclivity.

The hypothesis of "no difference," in some sense, is usually called the **null hypothesis** and denoted by $H_0$. In Example 10a, Oscar's claim that Felix has no special ability is the null hypothesis. In Example 10b, the null hypothesis is that intermittent feeding has no effect on average blood pressure, or that the two populations have the same mean. In Example 10c, the null hypothesis is that on average there is no difference between the mathematical proclivities of boys and girls.

In each of the examples, the data are intended to provide evidence about a null hypothesis. Collecting data and using them as evidence about a null hypothesis is called **testing** the hypothesis.

Suppose the null hypothesis is not true. What, then, *is* true?  The **alternative** to $H_0$, denoted by $H_A$, is ordinarily either a particular probability distribution (for the population) or a family of distributions. For example, if the null hypothesis in Example 10a is $p = 1/4$, the alternative expressing the existence of ESP is that $p > 1/4$. When a researcher is conducting a test to establish a difference or an effect (that is, that $H_0$ is not true), the alternative is sometimes referred to as the *research hypothesis.*

Although it is common to have a rather well-defined set of probability distributions in mind for the alternative, there are settings in which the alternative is simply that $H_0$ is not true. An instance of this is the hypothesis that the stars are distributed at random; in this case, the alternative includes models that are not necessarily probability models.

In planning a test, it is important—before doing any sampling—to be clear as to the population of interest. If it is not possible to run an experiment on the population of interest, it may be possible to use another population and extrapolate the results. For instance, one might extrapolate from animals to humans, or from moderately sick or healthy people to very sick people.

How does one get experimental subjects? In hypothesis testing, we assume that the sampling is random. (We have not considered some reasonable but more sophisticated methods, such as stratified or clustered sampling schemes.) Obtaining random samples is usually very difficult and may be impossible. Often one must be content with a convenient sample; using this to carry out a statistical test assumes that it is something like a random sample, with a good chance of being "representative."

| 10.2 | **Assessing the Evidence** |
|---|---|

Samples vary from one sample to another. Because of this, the evidence in a particular sample can be misleading. There is no way to be absolutely sure we're not being misled by a sample, unless the sample is the entire population. However, properly designed tests using random samples have a good chance of leading to correct conclusions.

The essence of a statistical test of $H_0$ is a comparison of actual sample data with what might be expected when $H_0$ is true. Such comparisons are usually based on the value of a particularly relevant statistic, called the **test statistic.**

The sampling distribution of a statistic depends on what is assumed as the population distribution (see §8.4). The sampling distribution of a test statistic under $H_0$ is called its **null distribution.**

| Example 10.2a | **ESP (Continued)** |
|---|---|

We return to "The Odd Couple" (Example 10a). Let $p$ denote the probability that Felix correctly identifies a card. The null hypothesis $H_0$ is that Felix has no ESP and is only guessing or, in terms of the parameter $p$, that $p = 1/4$. A natural test statistic is $Y$, the number of correct identifications. (We know from §8.3 that $Y$ is sufficient for $p$.)

Suppose Oscar and Felix decided at the outset to conduct exactly ten trials. We assume that the trials are independent and that the probability $p$ is the same at each trial, so that $Y \sim \text{Bin}(10, p)$. Under $H_0$, the distribution of $Y$ is given in Table Ia of Appendix 1 as follows:

| $y$ | 0 | 1 | 2 | 3 | 4 | 5 | 6 | 7 | 8 | 9 | 10 |
|---|---|---|---|---|---|---|---|---|---|---|---|
| $f(y)$ | .056 | .188 | .282 | .250 | .146 | .058 | .016 | .003 | .000 | .000 | .000 |

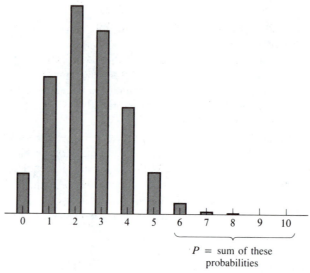

**Figure 10-2** Binomial probabilities: $n \doteq 10$, $p = \frac{1}{4}$

Figure 10-2 gives a graph of this distribution, whose mean value is 2.5 (that is, $np = 10 \times 1/4$), from §4.2.

Felix's score (six correct) is rather far from 2.5, out in a tail of the null distribution. In this sense, to get six correct is rather surprising when $H_0$ is true. A common way of indicating the discrepancy between what we expect (2.5) and what we've observed (6) is to give the tail-probability in the null distribution, beyond the value of $Y$ that we observed. In this example, the tail-probability is

$$P(Y \geq 6 \mid H_0) = .016 + .003 + .000 + .000 + .000 = .019.$$

The fact that this is small tells us that the observed $Y = 6$ is quite far from the expected value of $Y$ when $p = 1/4$. The sample results are rather inconsistent with the null hypothesis.

On the other hand, six correct is *not* very surprising if Felix really has some degree of ESP. The evidence appears to favor the conclusion that Felix has ESP. However, if you strongly disbelieve in the possibility of ESP, the evidence may not be convincing enough for you. You might attribute the evidence just to luck, or possibly to chicanery. (To be continued.) ■

As in the above example, it is common practice to describe the location of the observed test result in the null distribution of the test statistic by giving the tail-area or tail-probability beyond the observed value.[3]

---

[3]Summarizing sample results in terms of tail-probabilities is controversial. Many people feel that the sample actually observed is all that is relevant, so that in Example 10a the probabilities of 7, 8, 9, and 10 successes are irrelevant. In most of this chapter we present the predominant view in applied statistics; another view is presented in §10.6.

> The probability in the tail of the null distribution of the test statistic $Y$ at and beyond the observed value of $Y$ (in the direction corresponding to $H_A$) is called the **P-value.**

The smaller the $P$-value, the farther the observed value of the test statistic is from its expected value under $H_0$, and the harder it is to accept the observed discrepancy as just sampling variability. Thus, a very small $P$-value is taken as evidence against the null hypothesis.

How small must a $P$-value be to be convincing evidence against $H_0$? This is purely a subjective matter. Many statisticians take the following interpretations as benchmarks:

$P < .01$: Strong evidence against $H_0$.
$.01 < P < .05$: Moderate evidence against $H_0$.
$P > .10$: Little or no evidence against $H_0$.

There is a strong tradition that arbitrarily selects the values .01 and .05 as critical levels for $P$-values. It uses this language:   When $P < .05$, the result is called *statistically significant;* and when $P < .01$, the result is called *highly statistically significant.*[4]  In many research journals, a statistically significant result is indicated by attaching an asterisk ($*$) to a $P$-value less than .05, and a highly statistically significant result by attaching a double asterisk ($**$) to a $P$-value less than .01. There seems to be a common perception that the .01 and .05 critical levels have a theoretical basis, but they are in fact arbitrary.[5]

Clearly, stating whether $P < .05$ or $P > .05$ is not as informative as giving the $P$-value itself. Thus, both $P = .049$ and $P = .011$ are described as statistically significant, but the latter is stronger evidence against $H_0$. A report of statistical significance is quite like the red light on a dashboard of an automobile that tells you the water temperature has reached a certain arbitrary value; a thermometer that tells you the exact temperature is more useful.

Calculating a $P$-value from given data, or simply determining whether $P < .05$, is called a **significance test.** A $P$-value is sometimes called the *observed level of significance.*

One reason for the prevalence of significance tests is their championing by the great statistician and geneticist, Sir Ronald A. Fisher (1890–1962). Yet, even Fisher noted: "...in fact no scientific worker has a fixed level of significance at which from year to year, and in all circumstances, he rejects hypotheses; he rather gives his mind to each particular case in the light of his evidence and his ideas."[6]

---

[4]For an example of dissent, see J. Berger and T. Sellke, "Testing a point null hypothesis: The irreconcilability of $P$-values and evidence," *J. Amer. Stat. Assn.* 82 (1987), 112–22.

[5]R. A. Fisher introduced the practice of using 5% and 1% levels simply to overcome tabulation difficulties that no longer exist.

[6]R. A. Fisher, *Statistical Methods and Scientific Inference,* 3rd Ed. (Edinburgh: Oliver & Boyd, 1956), 45.

Although testing should not be reduced to a stepwise procedure that is carried out without thinking, a checklist can help. We list the main ingredients and then, following an example, discuss them in greater detail:

---

To carry out a test of significance:

1. Identify the *null* hypothesis, $H_0$.
2. Identify $H_A$, the *alternative* to $H_0$.
3. Specify or construct a test statistic $Y$, one that discriminates between $H_0$ and $H_A$.
4. Specify values of $Y$ that are "extreme" under $H_0$, in the direction of $H_A$, suggesting that $H_A$ better explains the data.
5. Obtain data and calculate the corresponding $P$-value. The smaller the $P$-value, the stronger the evidence against $H_0$.

---

**Example 10.2b** | **ESP (Conclusion)**

We return to the Odd Couple's experiment (Example 10.2a) and put our earlier analysis in the framework of the above five steps:

1. The null hypothesis is that Felix does not have ESP. In terms of $p$, the probability of a correct identification, we write $H_0$: $p = 1/4$.
2. The alternative to $H_0$, that Felix *has* ESP, is $H_A$: $p > 1/4$. Values of $p$ *less* than 1/4 are not included since Felix claims to do *better* than one who is choosing at random.
3. The test statistic $Y$ is the number of correct identifications in the ten trials.
4. Because $H_A$ specifies large values of $p$, we take $Y$-values in the *right* tail of its null distribution as extreme— more typical under $H_A$ than under $H_0$.
5. The $P$-value is the probability in the right tail, at and beyond the observed value $Y = 6$. In Example 10.2a we found this to be $P \doteq .019$, so six successes in ten trials are deemed "statistically significant." ∎

**Step 1**   The hypothesis $H_0$ is usually a very specific statement about the population or populations of interest. The null hypothesis in most applications is that there is no effect, no special ability, or no difference. Hypotheses are often expressed in terms of population parameters. For instance, one might test the hypothesis that the population mean is 0.

**Step 2**   The *alternative hypothesis* $H_A$ is a theory or claim that is different from $H_0$. This may be a hypothesis that an investigator either has a vested interest in establishing, or would particularly like to guard against.

**Step 3**   The test statistic $Y$ is especially designed to give evidence about $H_0$ and discriminate effectively between it and competing hypotheses. When $H_0$ is phrased in terms of a population parameter, the test statistic is often based on a sample estimate of that parameter.

**Step 4**   Values of the test statistic $Y$ that are in a tail of its null distribution are *extreme* when they tend to support $H_A$. Suppose $H_0$ specifies $\theta = \theta_0$, where $\theta$ is a parameter indexing the class of possible models. The alternatives $\theta < \theta_0$ and $\theta > \theta_0$ are *one-sided*. The alternative $\theta \neq \theta_0$ is *two-sided*. When values of the test statistic in *both* tails of its distribution are regarded as extreme, the *test* is two-sided; otherwise, one-sided.

**Step 5**   The $P$-value corresponding to an observed result is calculated according to what is considered extreme in Step 4. If large values of $Y$ are extreme and we observe $Y = y$, then the $P$-value is $P(Y \geq y \mid H_0)$. Figure 10-3(a) illustrates this in a typical continuous case. If small values are extreme, the $P$-value is $P(Y \leq y \mid H_0)$, as in Figure 10-3(b). If *both* large and small values of $Y$ are extreme, and we observe $Y = y$, we need to know more about the null distribution of $Y$. [In many practical settings this distribution is symmetric. When it is, many statisticians report *twice* the tail-area beyond $Y = y$ as a "two-sided" $P$-value. See Figure 10-3(c).]

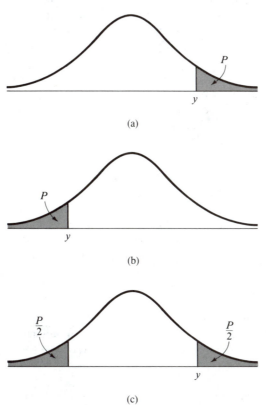

**Figure 10-3** $P$-values—three cases

When $H_A$ is of the form $\theta > \theta_0$, it may be that a more appropriate null hypothesis is $\theta \le \theta_0$ (rather than $\theta = \theta_0$). For instance, when a treatment is intended to increase an average response, the null hypothesis is really that the average does not increase it. The treatment would not be worthwhile if it should turn out that it decreased the average response. Even so, in such cases we may take the null hypothesis to be $\theta = \theta_0$ because we need a specific null distribution for calculation of a $P$-value. If the data cast doubt on $\theta = \theta_0$, they would cast even more doubt on any $\theta$ smaller than $\theta_0$. That is, testing $H_0$: $\theta = \theta_0$ tests $\theta \le \theta_0$ as well.

Whether one should use one-sided or two-sided $P$-values is a matter of controversy among statisticians.[7] Some argue that *all* $P$-values should be two-sided, reasoning that two-sided $P$-values are conservative. In our view, it would suffice to say what statistic was used and whether the reported $P$-value is one- or two-sided. The reader is then free to double or halve it, as may seem appropriate.

One reason for concern as to whether or not to double a $P$-value lies in the traditional language of significance testing. In that language, a $P$-value such as .04 indicates "significance," whereas the doubled value .08 does not! Clearly, whether or not $P$ is doubled can affect your conclusions if you adhere blindly to the standard benchmarks of .01 and .05.

A result that is not significant is sometimes taken as evidence that the null hypothesis is true. However, a result may be nonsignificant just because the sample was not large enough to reveal a difference that does exist—and one that may be important. (This has implications for the design of a study; sample sizes should be large enough that differences deemed important are likely to be detected. See §11.3.)

A common misinterpretation of "significance" is to think that a statistically significant result has great practical importance, or has proved something. However, a statistically significant deviation may have little or no practical significance. Moreover, a deviation that is important in the field of application may not show up as "statistically significant," simply because the sample was not large enough.

## Example 10.2c | Statistical Versus Practical Significance

A 1965 Alabama court case (*Swain vs. Alabama*) involved a claim of discrimination against blacks in the selection of grand juries.[8] According to census data, the population proportion of blacks was about 25%, but among those called to appear for jury duty the percentage of blacks was much smaller. Problem 10-16 calls for a test of $H_0$: $p = .25$; the result is a $P$-value of about $6 \times 10^{-10}$, which is so close to 0 as to be quite convincing evidence that the probability of a black's being called is *less* than 25%. However, the observed discrepancy did not strike the court as large enough for a *prima facie* case!

The author of the referenced article (a law professor) writes as follows: "That trivial differences can appear statistically significant only underscores the admonition that the $p$-value should not be considered in a vacuum. The courts are not likely to

---

[7]For a more detailed discussion of this controversy see J. Gibbons and J. Pratt, "$P$-values: Interpretation and methodology," *Amer. Statistician* 29 (1975), 20–25.

[8]Cited by D. Kaye in "Statistical evidence of discrimination," *J. Amer. Stat. Assn.* 77 (1982), 773–83.

lose sight of the question of practical significance and to shut their eyes to the possibility that the degree of discrimination is itself *de minimis.*"                                       ■

A *P*-value is a "probability," but *only* in a particular model, $H_0$—a model that may or may not be true (and is not true more often than true). It is better not to think of the *P*-value as a probability at all, since doing so only leads to confusion and misinterpretation. Rather, regard a *P*-value as a standard type of report of the degree of agreement between the null hypothesis model and the data—the smaller the *P*-value, the poorer the fit.

Studies often report observations on many variables; some of the *P*-values are likely to be less than .05 even if all of the corresponding hypotheses are true. (See Problem 10-2.) Even if only two independent *P*-values are calculated, the probability that at least one is less than .05 is $1 - (.95)^2$, or .0975. This is sometimes referred to as the overall "error rate." More generally, the overall error rate for $n$ independent tests (in terms of the level .05 as "significant") is $1 - (.95)^n$. Investigators who search their data will assuredly find at least one result that is statistically significant.

How does one properly summarize the results of several experiments, to draw an *overall* conclusion? In making simultaneous inferences and determining overall error rates, it is possible to adjust "critical" levels for individual *P*-values so that the overall critical level is .05. For example, the probability that at least one of two *P*-values is less than .0253 is $1 - .9747^2 = .050$. Although this kind of adjustment is recommended by many statisticians, the cure seems worse than the ailment. For instance, in Example 10b, why should the mere fact that heart rates were measured, as in fact they were, have any bearing on the conclusions about blood pressure? This is a severe limitation of significance testing. Scientists must decide subjectively how to weigh various experiments and their conclusions, and *P*-values are but one part of this process.

In the examples and problems that follow, we'll carry the analysis to the point of calculating a *P*-value, sometimes adding the words "strong evidence," or "some evidence," or "little or no evidence" against $H_0$, according to the benchmarks we've described above. We do not go farther and say which hypothesis you should regard as being true; this would take us into subjective interpretations and into considerations of the particular field of application.

We have given no guide to what is a "relevant" statistic, nor to what kinds of values of that statistic are "extreme"—explained better by $H_A$ than by $H_0$. These choices are often intuitively clear, based on the large sample tendencies of sample estimators of the parameters used in defining $H_0$. But there is another way, based on the intuition of likelihoods, that can lead to the choice of a test statistic and to deciding which of its values are extreme. This will be developed in more detail in the next chapter, but the essential idea is that we'd tend to favor any explanation of the data (a hypothesis) that is much more "likely" than the null hypothesis.

## Example **10.2d** | Testing by Comparing Likelihoods

Suppose we want to test the null hypothesis $\lambda = 1$ against the alternative $\lambda \neq 1$ with a random sample of size $n$ from $\text{Exp}(\lambda)$. The likelihood function can be written in terms of $\overline{X}$ (the sufficient statistic: see Problem 8-9b) as

$$L(\lambda) = \lambda^n \exp[-n\lambda\overline{X}].$$

The likelihood of $H_0$ is $L(1) = e^{-n\overline{X}}$, but there are values of $\lambda$ with greater likelihoods. The largest likelihood is achieved when $\lambda = \widehat{\lambda}$, the maximum likelihood estimator, $\widehat{\lambda} = 1/\overline{X}$: $L(\widehat{\lambda}) = (e\overline{X})^{-n}$. The ratio of these likelihoods is

$$\frac{L(1)}{L(\widehat{\lambda})} = (e\overline{X}e^{-\overline{X}})^n.$$

The likelihood of the alternative $\lambda = 1/\overline{X}$ is much greater than the likelihood of $\lambda = 1$ when $\overline{X}$ is either very small or very large—such values of $\overline{X}$ are thus "extreme" values of the test statistic $\overline{X}$ for testing $\lambda = 1$ vs. $\lambda \neq 1$. ∎

As in this example, we'll see in the chapters to follow that some algebraic manipulations with the likelihood ratio will often suggest a more familiar statistic ($\overline{X}$ in the example) to use as the test statistic. It will also suggest the kinds of values of that statistic that provide strong evidence against the null hypothesis. In the next chapter we'll develop this idea further. Of course, it may be that even some clever algebra does not serve to suggest a statistic that we know all about; in such a case, we can use the ratio $\Lambda$ itself as the test statistic—provided we can derive its sampling distribution.

## Problems

**∗ 10-1.** Referring to Oscar Madison's testing of Felix Unger for ESP (Example 10a), suppose there were 25 successes in 80 trials. Find an approximate $P$-value for testing $H_0$: $p = 1/4$.

**∗ 10-2.** Again referring to Example 10a, suppose 60 subjects are given Oscar's test for ESP, ten trials per subject. Suppose that, in fact, *none* of the subjects has ESP—all of them simply guess.

(a) For a single subject, what is the probability of obtaining a result that is statistically significant (that is, $P < .05$)?

(b) How many of the 60 subjects would you expect to yield statistically significant results?

(c) Find the probability that none of the results for the 60 subjects turns out to be statistically significant.

**∗ 10-3.** An observation $X$ has this triangular p.d.f., centered about $\theta$:

$$f(x \mid \theta) = 1 - |x - \theta|, \quad |x - \theta| < 1.$$

Given the single observation $X = .7$, find the $P$-value for testing the hypothesis $\theta = 0$ against the alternative $\theta \neq 0$.

**10-4.** Consider testing $H_0$: $p = 1/2$ against $H_A$: $p < 1/2$, where $p$ is a Bernoulli parameter. Ten independent Bernoulli trials resulted in this sample sequence: 0 0 1 0 0 0 0 0 0 1.

(a) Suppose this sequence is the result of carrying out ten trials where $n = 10$ was fixed in advance. In view of $H_A$, a small number of successes would be evidence against $H_0$. Find the $P$-value.

(b) Suppose the sequence resulted from sampling until two successes were obtained. In view of $H_A$, a large number of required trials would be taken as evidence against $H_0$. Find the $P$-value. [*Hint:* The number of trials required is negative binomial, and it may be easier to find the probability that the number required is nine or fewer.]

(c) Find the likelihood function in each case, (a) and (b). Is reporting $P$-values in accord with the likelihood principle enunciated in §8.2?

∗ **10-5.** In the comic strip "Hi and Lois," the twins Dot and Ditto once argued about the probability that a piece of buttered bread dropped on the floor would land buttered side down. Dot claimed it was .9, and Ditto that it was only .5. They conducted several trials, say $n$. Let $Y$ denote the number of times the bread landed buttered side down.

(a) Find the ratio of the two likelihoods, $L(.5)$ and $L(.9)$.

(b) Based on likelihoods, what kinds of $Y$-values are considered "extreme"?

**10-6.** Consider testing $H_0$: $\mu = 0$ against the alternative $H_A$: $\mu \neq 0$, where $\mu$ is the mean of a normal population with known variance, $\sigma^2 = 1$. Given a random sample of size $n$, the likelihood function (from Example 8.2c) is

$$L(\mu) = \exp\left\{\frac{n}{2}(\mu - \overline{X})^2\right\}.$$

(a) Find the ratio of the likelihood of $\mu = 0$ to the largest possible likelihood.

(b) Based on likelihoods, what kinds of values of $\overline{X}$ are "extreme"?

∗ **10-7.** Consider a random sample of size $n$ from $\mathcal{N}(\mu, \sigma^2)$ for testing the hypothesis $H_0$: $\sigma^2 = \sigma_0^2$, using the test statistic $Y = (n-1)S^2/\sigma_0^2$.

(a) Give the null distribution of $Y$.

(b) An experiment was conducted to check the variability in explosion times of detonators.[9] (When several are to explode together, the desired effect may not be achieved if the variability is too great.) Data for one run were reported as follows, expressed in milliseconds short of 2.7 seconds:

11, 23, 25, 9, 2, 6, −2, 2, −6, 8, 9, 19, 0, 2.

Find the $P$-value for a test of the hypothesis $H_0$: $\sigma \leq 7$ (milliseconds), assuming that the times are normally distributed.

[This test for a population variance is not (unfortunately) robust against the assumption of normality.]

## 10.3    One-Sample $Z$-tests

Many test statistics in common use have null distributions that are at least approximately *normal* when the sample size is large. If the null distribution of a test statistic $Y$ is approximately normal, the null distribution of the standard score

$$Z = \frac{Y - E(Y \mid H_0)}{\text{s.d.}(Y \mid H_0)} \tag{1}$$

---

[9]Reported in G. L. Tietjen and M. E. Johnson, "Exact statistical tolerance limits for sample variances," *Technometrics* 21 (1979), 107–10.

is approximately standard normal. The null hypothesis is ordinarily specific enough (for instance, $\mu = \mu_0$) that the mean $E(Y \mid H_0)$ is a specific number that we can calculate or look up in a table. However, the denominator can involve a parameter that may or may not be known.

If the quantity s.d.$(Y \mid H_0)$ is *known*, $Z$ is a statistic, and we can use it in place of $Y$ as a test statistic. But if s.d.$(Y \mid H_0)$ is not known, we can only approximate it using sample data, substituting the standard error of $Y$ for its standard deviation. In such cases, define

$$T = \frac{Y - E(Y \mid H_0)}{\text{s.e.}(Y \mid H_0)}. \tag{2}$$

When $n$ is large, the null distribution of $T$ is also approximately standard normal, and we could (and do) also call this statistic $Z$.

In both (1) and (2), the standard score measures the discrepancy between the observed $Y$ and $E(Y \mid H_0)$, the $Y$ we "expect" when $H_0$ is true. A $Z$- or $T$-score is unitless, being simply a number of s.d.'s [in (1)] or number of s.e.'s [in (2)] to the right of the expected value. [The units for the quantities in the numerator and denominator of (1) and of (2) cancel.]

Now, suppose $Y$ is approximately normal, and it turns out that $Z = 1$. This means that the observed $Y$ is about 1 standard deviation away from what it is expected to be under $H_0$. Such a deviation is typical when $H_0$ is true, so it is useless as evidence against $H_0$. The one-sided $P$-value for $Z = 1$ (when $H_A$ makes positive $Z$'s likely) is $1 - \Phi(1) = .1587$, and a $P$-value this large is usually not thought of as saying much on behalf of $H_A$. On the other hand, if $Z = 3$, say, then the observed $Y$ is 3 standard deviations from its mean value under $H_0$. The one-sided $P$-value is $.0013$, the area to the right of $Z = 3$. A value of $P$ this small is considered as very strong evidence against $H_0$ by most statisticians.

---

Significance test of $H_0$: $\mu = \mu_0$, based on the mean of a large, random sample: Calculate (according as $\sigma$ is known or unknown)

$$Z = \frac{\overline{X} - \mu_0}{\sigma/\sqrt{n}} \quad \text{or} \quad Z \doteq \frac{\overline{X} - \mu_0}{S/\sqrt{n}},$$

and find the $P$-value in Table IIa of Appendix 1:

$$P = \Phi(Z) \qquad \text{for } H_A\colon \mu < \mu_0,$$
$$P = 1 - \Phi(Z) \quad \text{for } H_A\colon \mu > \mu_0,$$
$$P = 2\,\Phi(-|Z|) \quad \text{for } H_A\colon \mu \neq \mu_0.$$

Example **10.3a** | **Average Speed Reduced?**

Suppose, over a long period of time in which the posted speed limit was 65 mph, the average speed along a certain stretch of highway was well established as 63.0 mph. After the speed limit dropped to 55 mph, the average speed of cars in a sample of size 100 was found to be 61.4 mph, and the s.d., 4.6 mph. Does this apparent reduction indicate a genuine reduction in mean speed, or could it be a manifestation of sampling variability? We test as follows:

1.   The null hypothesis is $H_0$: $\mu = 63.0$ (the new mean is not different from the old mean).
2.   The alternative is $H_A$: $\mu < 63.0$ (the new limit was intended to reduce the average speed).
3.   The value of the test statistic (2) is

$$Z = \frac{61.4 - 63.0}{4.6/\sqrt{100}} = -3.48.$$

4.   Since one would expect negative values of $Z$ under $H_A$, we take values of $Z$ in the left tail as "extreme."
5.   Because $n$ is large (100), the null distribution of (2) is approximately standard normal, and the $P$-value is the area under the standard normal p.d.f. to the *left* of $-3.48$:   $\Phi(-3.48) = .0002$.   The evidence against $H_0$ is strong, and the alternative explanation—that the mean speed has been reduced—is the more plausible. (Whether the lowering was *caused* by the new speed limit is another matter; the time periods were different, and there may be other factors involved.)

When such a result is described as statistically significant, it is apt to be reported in the media that "a significant reduction" has been achieved. This language suggests that the mean speed has been lowered by an amount that has important consequences. But, even if the 1.6 mph reduction is real, it may not be important. A statistically significant result signifies only that the sample result is far from what is expected under the null hypotheses, for samples of that size.

Rather than simply claiming statistical significance, it may be better to give an estimate of the mean reduction together with the standard error of estimation: $61.4 \pm .46$. (A 95% confidence interval has limits $61.4 \pm 1.96 \times .46$:   $60.5 < \mu < 62.3$, excluding not only 63 but values near 63 as well.)   ∎

We know from §8.8 that a sample proportion is approximately normal when the sample size is large. So a $Z$-score based on a sample proportion is an appropriate test statistic for a large-sample test of a specified value of the corresponding population proportion. The $Z$-score we use is a special case of (1), where $\overline{X}$ is the sample proportion, and $\mu$ is the population proportion. The population variance is $\sigma^2 = p(1 - p)$.

**Example 10.3b** | **Male Births and Timing of Conception**

A long-standing theory holds that a greater proportion of male births occur when conception is late in the woman's fertile period. One investigation, which studied thousands of births, reported 145 cases in which conception was reckoned as having occurred two days after ovulation.[10] Among these there were 95 male births, or a proportion $95/145 = .655$. The proportion of male births among the rest was .520. Is this evidence that the probability of a male child is higher when conception occurs late? Or is the observed difference in proportions attributable to sampling variations? To address this question, we may test the hypothesis $H_0$: $p = .520$, where $p$ is the proportion of male births among all those who conceive two days after ovulation. (To be continued.) ■

The mean and variance of a sample proportion $\hat{p}$, according to (6) and (7) of §8.7 are

$$E(\hat{p}) = p, \quad \operatorname{var}\hat{p} = \frac{p(1-p)}{n}. \tag{3}$$

To test the hypothesis $p = p_0$, we use the $Z$-score (1) with $Y$ replaced by $\hat{p}$, $E(Y \mid H_0)$ by $p_0$, and s.d.$(Y \mid H_0)$ by $\sqrt{p_0(1-p_0)/n}$:

$$Z = \frac{\hat{p} - p_0}{\sqrt{p_0(1-p_0)/n}}. \tag{4}$$

Exactly the same $Z$-score is obtained if, for $Y$, we use the sample *number* of "successes," which is $\operatorname{Bin}(n, p_0)$:

$$Z = \frac{Y - np_0}{\sqrt{np_0(1-p_0)}}. \tag{5}$$

[Dividing numerator and denominator of (5) by $n$ produces (4).]

**Example 10.3c** | **Male Births (Continued)**

Continuing Example 10.3b, we carry out a test of $H_0$: $p = .520$. The proportion of boys in the sample of size $n = 145$ is $\hat{p} = .655$:

1.  The null hypothesis is $H_0$: $p = .520$, where $p$ is the probability of a male child when conception occurs two days after ovulation.
2.  The alternative is the long-standing theory that the probability of a male child is greater when conception is late in the fertile period, $H_A$: $p > .52$.

---

[10]S. Harlap, "Gender of infants conceived on different days of the menstrual cycle," *New Engl. Jour. Med.* 300 (1979), 1445–48. Subjects of the study were Orthodox Jewish mothers, whose religious practices facilitated the documentation.

3.  The test statistic, calculated from $Y = 95$ or from $\hat{p} = Y/n = .655$, is

$$Z = \frac{95 - 145 \times .520}{\sqrt{145 \times .52 \times .48}} = \frac{.655 - .520}{\sqrt{.52 \times .48/145}} = 3.25.$$

4.  In view of $H_A$, large positive values of $Z$ are considered extreme.
5.  The one-sided $P$-value is the area under the standard normal p.d.f. to the right of $Z = 3.25$, or $1 - \Phi(3.25) = .0006$, strongly favoring $H_A$.  ∎

---

Large-sample test for $H_0$: $p = p_0$: From the given (random) sample, calculate the $Z$-score (4) or (5) and find the $P$-value in Table IIa:

$$P = \Phi(Z) \qquad \text{for } H_A: \mu < \mu_0,$$
$$P = 1 - \Phi(Z) \quad \text{for } H_A: \mu > \mu_0,$$
$$P = 2\,\Phi(-|Z|) \quad \text{for } H_A: \mu \neq \mu_0.$$

---

## Problems

**∗ 10-8.**  An educational testing organization has prepared a new version of an aptitude test. The average score on the old version, based on the scores of hundreds of thousands of individuals who took the test, is 500. The new test is tried out on a sample of 50 students selected at random from those eligible to take the test; their average score is 480, and their s.d. is 80. Does the evidence suggest that the new version is harder than the old?

**10-9.**  Suppose the mean weight of a rancher's cattle (live weight at the time of slaughter) has been 400 lb. A feed salesman has induced him to try a new supplement on 60 cattle. The mean weight of these is 405, with s.d. 50 lb. Is there evidence that the supplement is effective?

**∗ 10-10.**  For a number of years, the proportion of entering students who survived the first year in a certain engineering school was about 65%. The school then instituted various measures to improve the situation, including more diligent screening of admissions and more counseling help. In the first year after the change, there were 560 survivors in a class of 800, or 70%. Is this increase explainable as simply sampling variability?

**10-11.**  In a TV commercial, the makers of a medium-priced car $M$ reported that in a road test by 90 owners of a high-priced luxury car $C$, 57 preferred $M$ for overall ride. (Presumably, the makers of car $M$ felt that if you saw that a substantial majority of luxury car owners preferred car $M$ over their own cars, you'd at least want to try it out.) Is there evidence in the reported result that, in the population sampled, a majority of *all* drivers prefer car $M$?

**10-12.**  Someone has advanced the theory that people tend to postpone their death dates until after their birthdates.[11] If this be so, people are less likely to die in the month immediately

---

[11]"Deathday and birthday: An unexpected connection," in J. M. Tanur et al. (Eds.), *Statistics: A Guide to the Unknown* (San Francisco: Holden-Day, 1972), 52–66.

preceding their birthdates than in any other month. The skeptic's hypothesis is that all 12 months are equally likely.

   **(a)**   Data on 348 notables showed that 16 died in the month preceding their birthdates.[12] What evidence does this provide?

   **(b)**   A study of death dates of 1202 athletes showed that 91 died in the month preceding their birthdates.[13] Does this evidence support the theory?

$*$ **10-13.**   Can the average consumer distinguish between cheddar cheese that has been aged 9 months and cheddar cheese that has been aged 18 months? To answer this, 45 clerical employees of a cheese manufacturer, with no previous experience in taste testing, are given three samples of cheese, two aged 9 months and one aged 18 months. They are told that one piece is different from the others and asked to single out that piece. The result is that 25 of the 45 picked the odd piece correctly. Can one conclude that consumers generally can tell 9-month-old from 18-month-old cheeses by taste?

$*$ **10-14.**   Consider testing $H_0$: $\mu = 0$ against $H_A = \mu > 0$, where $\mu$ is the mean of $X \sim \mathcal{N}(\mu, 1)$.

   **(a)**   Suppose you use a sample of size $n = 1$ and observe $X = 2$. Find $P$.

   **(b)**   Suppose you use a sample of size $n = 400$ and find $\overline{X} = .1$. Find $P$.

**10-15.**   Consider a sample of size 100 from $\text{Exp}(\lambda)$, a distribution with p.d.f. $\lambda e^{-\lambda x}$ for $x > 0$. If $\overline{X} = 1.23$, find the $P$-value for a test of $\lambda = 1$ against $\lambda \neq 1$. [*Hint:* Use the approximate normality of the sample mean.]

$*$ **10-16.**   Example 10.2c cited a 1965 Alabama court case involving a claim of discrimination against blacks in the selection of grand juries. According to census data, the population proportion of blacks was about 25%, but there were only 177 blacks among 1050 called to appear for jury duty. Test the hypothesis that the 1050 constitute a random sample from a population in which the proportion of blacks is .25.

---

## 10.4   One-Sample $t$-tests

In a $Z$-test for $\mu = \mu_0$ (§10.3), the large sample size allowed us to assume (i) that the sample mean is approximately normal, and (ii) that the sample s.d. is a good approximation to the population s.d. As explained in §9.4, (ii) is more worrisome than (i) when the sample sizes are not large. For example, $\overline{X}$ can be fairly close to normal even when $n$ is as small as 5, depending on the population distribution. (Indeed, if the population itself is normal, then $\overline{X}$ is exactly normal for *every* $n$.) However, the sample standard deviation $S$ is quite unreliable as an estimator for $\sigma$ when $n$ is that small.

   In this section, we consider tests involving the population mean $\mu$ when $\sigma$ is unknown and $n$ is not large. To test $H_0$: $\mu = \mu_0$ we again measure the discrepancy

---

[12]From R. B. Morris (Ed.), *Four Hundred Notable Americans* (New York: Harper & Row, 1965).

[13]Data from R. Hickock, *Who Was Who in American Sports* (Hawthorne Books, 1971), reported in Larson & Stroup, *Statistics in the Real World* (New York: Macmillan, 1976).

between $\overline{X}$ and $\mu_0$ in terms of standard errors, using the ratio

$$T = \frac{\overline{X} - \mu_0}{S/\sqrt{n}}. \tag{1}$$

When the sampling is random and the population is normal, the null distribution of $T$ is $t(n-1)$, as we saw in §9.4. Tables IIIa and IIIb in Appendix 1 gives percentiles and tail-areas of the $t$-distribution.

Although a population encountered in practice may be nearly normal, you can never be certain that it is *exactly* normal. The statistic $T$ in (1) usually does not have a $t$-distribution when the population is nonnormal. However, as we said in §9.5, probabilities given in Tables IIIa and IIIb have been found to be nearly correct if the population is not far from normal—if it has a single hump, has tails that are not too heavy, and is not badly skewed. The $t$-test is said to be *robust* with respect to the assumption of population normality.

**Example 10.4a** | **Ventricular Premature Beats**

To study the effectiveness of a drug[14] in reducing ventricular premature beats (VPB's), a clinician administered the drug intravenously to ten patients. The following are reductions (VPB's per min) after a prescribed interval following administration of 2 mgs/kg:

0, 7, −2, 14, 15, 14, 6, 16, 19, 26,

with $\overline{X} = 11.5$, $S = 8.67$. Figure 10-4 shows a dot plot of the data.

**Figure 10-4**

1. The null hypothesis is that the mean reduction is $\mu = 0$.
2. The alternative is one-sided: $\mu > 0$.
3. The test statistic is

$$T = \frac{\overline{X} - \mu}{S/\sqrt{n}} = \frac{11.5 - 0}{8.67/\sqrt{10}} = 4.19.$$

4. In view of the alternative hypothesis, large values of $T$ are extreme.
5. Under $H_0$, $T \sim t(9)$, and from Table IIIb, $P \doteq .001$.

The result is "highly statistically significant"—the observed mean reduction is not easy to account for as only a phenomenon of random sampling. The evidence against $H_0$ is strong. ∎

---

[14]Private communication.

Test of $H_0$: $\mu = \mu_0$ when $\sigma^2$ is unknown:  Calculate

$$T = \frac{\overline{X} - \mu}{S/\sqrt{n}} \tag{2}$$

and find the $P$-value in Table IIIb, entering at $n - 1$ degrees of freedom (d.f.). If $n < 40$, the population should be not too nonnormal. (If $n > 40$, the table leads you to normal tail areas.)

Is the population close to normal in Example 10.4a?  Nonnormality is very hard to detect with a sample whose size is as small as 10, unless the nonnormality is quite pronounced. The plot in Figure 10-4 reveals neither skewness nor heavy tails that would bring the $t$-test into question. In Chapter 13, we'll consider some formal tests for normality. In §10.5 we'll take up a test that does not assume a normal population. The following example shows what nonnormality can do to a $t$-test.

## Example 10.4b    A Bad Use of $T$

The average normal heart rate for pigs is considered to be 114 beats/min, according to a 1984 research paper.[15]  The paper gives the heart rates of four pigs under anesthesia:  116, 85, 118, 118.

Suppose we test the null hypothesis $\mu = 114$ against the alternative that $\mu \neq 114$, where $\mu$ is the mean rate for pigs under anesthesia. Large values of $|T|$ are extreme. The mean and s.d. of the four given heart rates are 109.25 and $S = 16.194$, so

$$T = \frac{109.25 - 114}{16.19/\sqrt{4}} \doteq -.59.$$

The null distribution of $T$ is $t(3)$, and the observed value is not even close to being statistically significant.

However:  Suppose the second pig (the 85) had not been included in the sample. (Perhaps it did not qualify for some reason, or just had not been part of the sample.) This leaves only (116, 118, 118) as the sample, and now the mean is $\overline{X} = 117.33$ and the s.d., $S = 1.155$. The value of $T$ with this sample of size $n = 3$ (d.f. $= 2$) is 5.00. Leaving out the one observation *farthest* from $\mu_0 = 114$ makes the result significant (one-sided $P = .019$)!  This should seem strange, since that observation ought to *add* to the evidence against $H_0$. Numerically, what has happened is that even though the sample mean is now actually closer to 114 (albeit flipped to the other side), the sample s.d. has dramatically decreased, which makes the ratio defining $T$ large.

Although $n = 4$ is really too small to draw definite conclusions, the observation 85 suggests that the population distribution may have a long left tail. This would be

---

[15]"Xylazine-Ketamine-Oxymorphone:  An injectable anesthetic combination in swine," *J. Amer. Vet. Med. Assn.* (1984), 182–84.

an important violation of the assumption of normality, and the $t$-test would not be appropriate.                                                                      ■

The result in the preceding example reflects the fact that the standard deviation $S$ is much more dramatically affected by a large deviation than is $\overline{X}$. Consider $H_0$: $\mu = \mu_0$, and a particular sample of size $n$ for which $T$ is as large as you please, so that the $P$-value is small. Then take any single observation and let it get large. This increases $\overline{X}$, and one might expect an even smaller $P$-value. But $S$ also increases. The net result is that as the one observation grows, $T$ eventually starts to *decrease* and in fact tends to 1! So the (one-sided) $P$-value grows to about .16.

The moral here is the same as in the example:  Don't use the $t$-test when there are *outliers*—observations so unusual in size (either large or small) as to give them undue influence in the analysis. Outliers may be in error; if so, they should be deleted. If they are genuine, or if their legitimacy is in doubt, a $t$-test is not appropriate.

## Problems

* **10-17.**   Seven skulls found in a certain digging averaged 143.3 mm in width. The s.d. of the widths was $S = 5.62$. Use these statistics in a test of the hypothesis that these skulls belong to a race previously found nearby in which the average width is known to be 146 mm. (Assume a nearly normal population.)

**10-18.**   R. A. Fisher analyzed the results of testing two sleeping drugs on each of ten patients.[16]  The additional hours of sleep obtained when using drug A instead of drug B are as follows:

   1.2, 2.4, 1.3, 1.3, 0.0, 1.0, 1.8, 0.8, 4.6, 1.3

Test the hypothesis that the mean number of additional hours is 0, making necessary assumptions.

* **10-19.**   Female killdeers usually lay four eggs each spring. A scientist finds the following differences in weights of the first-hatched and last-hatched birds, in eight broods:

   .02, − .10, .06, .23, .14, .14, − .04, .29,

(first-hatched minus last-hatched). Is there evidence of a difference between average weights of first- and last-hatched eggs?  State your assumptions.

**10-20.**   Ten subjects were tested to determine the increase in reaction time after being given a certain drug. The mean increase was 10 milliseconds, and the s.d. 10.6 milliseconds. What does this say about the hypothesis of no average difference in before-drug and after-drug reaction times?

* **10-21.**   Example c in the Prologue (page 2) gave corneal thicknesses of the glaucomatous eye and the nonglaucomatous eye of eight patients. The differences in thickness are − 4, 0, 12, 18, − 4, − 2, 6, 16. Use a $t$-test for the hypothesis that the mean difference between the corneal thicknesses of glaucomatous and nonglaucomatous eyes is 0.

---

[16]Quoted in P. Meier, "Statistical analysis of clinical trials," in S. Shapiro and T. Louis (Eds.), *Clinical Trials: Issues and Approaches* (New York:  Marcel Dekker, 1983).

**10-22.** Six healthy young subjects were tested on a cycle ergometer.[17] The work rate was increased continuously and linearly, at a fast rate and at a slow rate. One of the measured variables was a difference in oxygen uptake at which the criterion for anaerobic threshold was met. The differences were

765, 650, 700, 550, 250, 20.

Assuming normality, test the hypothesis that the mean difference is 0.

---

| 10.5 | ## Some Nonparametric Tests |
|------|------|

The tests considered so far have limited applicability. They have assumed either very large samples or, when sample sizes are only moderate and $\sigma$ is unknown, populations that are not too different from normal populations. In this section we take up some tests concerning location that require fewer assumptions.

**Example 10.5a**

### Marijuana and Proficiency

A study on the effects of smoking marijuana included the following data.[18] Each of nine subjects was given a "digit substitution test," before and again 15 minutes after a marijuana smoking session. Changes in the number correct from the baseline number (that is, number correct after smoking minus number correct before) were reported as follows:

$$+5, -17, -7, -3, -7, -9, -6, +1, -3.$$

We could analyze these data using the $t$-test of the preceding section. But if we don't want to assume near-normality, the $t$-test is not appropriate. Instead, we focus on whether the predominance of negative changes in the data set indicates a real effect or is simply random variation.

Even if marijuana has no effect, the inherent variability in testing will produce changes from before to after measurements. Under the hypothesis of no effect, the probability of an increase is 1/2. That is, we'd expect about as many negative changes ($-$'s) as positive changes ($+$'s). Let $p = P(+)$:

1. The null hypothesis is that marijuana has no effect, so we test $H_0: p = 1/2$.
2. Undoubtedly, the research hypothesis was that scores tend to get worse after smoking marijuana, so the alternative is $H_A: p < 1/2$.
3. The test statistic is $Y$, the number of $+$'s in the sample sequence; with random sampling, the null distribution of $Y$ is Bin(9, .5).
4. Under $H_A$ we'd expect an unusually small number of $+$'s, so small values of $Y$ are extreme.

---

[17]R. L. Hughson and H. J. Green, "Blood acid-base and lactate relationships studied by ramp work tests," *Medicine and Science in Sports and Exercise* 14 (1982), 297–301. [Curiously, two of the six subjects were female (numbers 4 and 5), but the researchers made no attempt to take this into account in the analysis.]

[18]Reported in *Science 162* (1968), 1234.

5. With the given data, $Y = 2$. The $P$-value is

$$P(Y \leq 2 \mid p = .5) = \frac{1}{2^9} \left\{ \binom{9}{2} + \binom{9}{1} + \binom{9}{0} \right\} = \frac{46}{512} = .0898.$$

[We could also find this tail probability in Tables Ia and Ib with $n = 9$ and $p = .5$.] ∎

Following the idea in the example, we can test the hypothesis that the median $m$ of a continuous variable $X$ is $m_0$, using the number of $+$'s among the deviations $X_i - m_0$ in a random sample $(X_1, ..., X_n)$ as a test statistic. Let $Y$ denote that number of plus signs, and let $p = P(X > m_0) = P(+)$. The null hypothesis $H_0$: $m = m_0$ is equivalent to $H_0$: $p = 1/2$, and under $H_0$, $Y \sim \text{Bin}(n, .5)$. The test based on $Y$ is called a **sign test.**

When $X$ is continuous, $P(X = m_0) = 0$; but in practice, zero differences can occur because of round-off. Such observations carry no information about the hypothesis that positive and negative deviations are equally likely. So in applying a sign test, we ignore zeros and use the reduced sample consisting of the nonzero differences.

---

*Sign test* for $H_0$: $m = m_0$:

When there are $n$ nonzero differences $X_i - m_0$ in a random sample, let $Y$ denote the number of positive differences and $p = P(+)$. If the alternative is $m > m_0$, large values of $Y$ are extreme, and if we observe $Y = y$, the $P$-value is $P(Y \geq y)$. When $n \leq 12$, find $P$ in Tables Ia and Ib; for $n > 12$, find approximate $P$-values using Table IIa.

---

A useful property of the sign test is that the shape of the population distribution is immaterial, even when the sample size is small. The null distribution of the test statistic depends only on $n$. In particular, it does *not* depend on the population distribution—it is **distribution-free.** And because the distributions included in $H_0 \cup H_A$ do not constitute a parametric family, the sign test is a **nonparametric** test.

The sign test uses only the algebraic sign of the deviations from $m_0$, ignoring the magnitudes of the deviations. One might then expect it to be less sensitive (in some sense) than a test that exploits the sizes of the deviations. For instance, suppose the deviations are $24, -3, 33, 29, -1$. Three of these are positive, so $Y = 3$—not an unusual occurrence when $n = 5$ and $p = .5$. However, the sizes of the positive deviations suggest more of an imbalance than do their signs: Those that are negative are small in magnitude, and those that are positive are quite large.

Wilcoxon's **signed-rank** test is a refinement of the sign test that takes some account of the magnitudes of the deviations. To carry out the test, we first order the deviations $X_i - m_0$ *according to their magnitudes*, and then assign ranks $1, 2, ...$ from left to right. For the deviations of the preceding paragraph, the ordering produces $-1,\ -3, 24, 29, 33$. Rank 1 is assigned to $-1$, coming from the $X$ that is closest to $m_0$; rank 2 is assigned to $-3$, the second closest to $m_0$; and so on. The test statistic is the sum of the ranks of the negative deviations, or $1 + 2 = 3$.

---

**Signed-rank statistics** for testing $m = m_0$:

$$R_- = \text{sum of ranks of negative deviations } X_i - m_0,$$

$$R_+ = \text{sum of ranks of positive deviations } X_i - m_0,$$

when the deviations are ordered by magnitude and ranked from smallest to largest.

---

The idea in using the sum of the ranks $R_-$ is that when it is small, either there are very few negative observations, or the negative deviations are among those that are smallest in magnitude. We could just as effectively use the sum of the ranks of the positive deviations, $R_+$, but this would contain the same information since if you know one, you know the other:

$$R_- + R_+ = 1 + 2 + \cdots + n = \frac{1}{2}\, n(n + 1).$$

To calculate a $P$-value, we need to know the null distribution of $R_-$ (or of $R_+$). The possible values are 0, 1, 2, ..., $\frac{1}{2}\, n(n + 1)$. To find the distribution of $R_-$, we assume that the population distribution is *symmetric* about its median—in which case, of course, the median coincides with the mean. Any assumption about the population is restrictive, but this one is not nearly as restrictive as the assumption of normality.

The value of $R_-$ for a particular sample sequence *depends only on the pattern* of $+$ 's and $-$ 's, when the deviations are arranged according to magnitude. There are $2^n$ patterns of $+$ 's and $-$ 's in a sequence of length $n$. Under $H_0$ these are equally likely, a fact that we reason as follows: Symmetry implies that the deviation smallest in magnitude is as likely to be positive as it is negative; and given this deviation, the deviation second smallest in magnitude is as likely to be positive as negative—independent of whether the sign of the deviation smallest in magnitude is positive or negative. Continuing in this fashion, we find the probability of each sequence of signs to be $1/2^n$.

So, the patterns are equally likely, which means that to find the probability of a particular value of $R_-$, we need only count the number of patterns producing this value and divide by $2^n$. If $n$ is small, we simply list all patterns and determine the value of $R_-$ for each. For $n = 3$ there are eight, listed below, along with the implied table of the p.f., $f(r)$:

| Sequence | $R_-$ | | $R_-$ | $f(r)$ |
|----------|-------|--|-------|--------|
| $+ + +$  | 0 | | 0 | $1/8$ |
| $- + +$  | 1 | | 1 | $1/8$ |
| $+ - +$  | 2 | | 2 | $1/8$ |
| $+ + -$  | 3 | | 3 | $2/8$ |
| $- - +$  | 3 | | 4 | $1/8$ |
| $- + -$  | 4 | | 5 | $1/8$ |
| $+ - -$  | 5 | | 6 | $1/8$ |
| $- - -$  | 6 | | | |

If $n$ is large, it is only practical to list those sequences with the smallest values of $R_-$ and count to find the probabilities for these small values. Here we list beginnings of extreme sequences (all signs indicated by dots are $+$'s):

**Table 10-1**

| Sequence | $R_-$ |
|----------|-------|
| $+ + + + + + \cdots$ | 0 |
| $- + + + + + \cdots$ | 1 |
| $+ - + + + + \cdots$ | 2 |
| $+ + - + + + \cdots$ | 3 |
| $- - + + + + \cdots$ | 3 |
| $+ + + - + + \cdots$ | 4 |
| $- + - + + + \cdots$ | 4 |
| $+ + + + - + \cdots$ | 5 |
| $+ - - + + + \cdots$ | 5 |
| $- + + - + + \cdots$ | 5 |

Then, since we've listed all sequences with $R_- = 5$ or smaller, $f(5) = 3/2^n$, $f(4) = 2/2^n$, and so on. If the value of $R_-$ is not in the tail, the $P$-value is not apt to be significant, and its precise value is irrelevant.

This process can be tedious when $n$ is not small, but it has been done and the results tabled. Table VI gives the left-tail probabilities for $n \leq 15$. By symmetry, the null distribution of $R_+$ is the same as that of $R_-$, so the table entries can be used for

$R_+$ as well. Also, the probabilities in the right-hand tail of either distribution are found by symmetry:

$$P(R_- = c) = P(R_+ = c) = P\left\{R_- = \frac{1}{2}n(n+1) - c\right\}. \tag{1}$$

So the easiest way to use the table is to use, as the test statistic, whichever of $R_-$ and $R_+$ is the smaller.

As for the moments of the distribution, it is symmetric about its mean value, which must then be halfway from its smallest value 0 to its largest value, $\frac{1}{2}n(n+1)$:

$$E(R_-) = E(R_+) = \frac{1}{4}n(n+1). \tag{2}$$

The formula for variance is not so obvious. It is as follows:[19]

$$\text{var } R_- = \text{var } R_+ = \frac{1}{24}n(n+1)(2n+1). \tag{3}$$

With these moments, we can find approximate probabilities when $n > 15$, as it turns out that the signed-rank statistics are then approximately *normal*.

---

Signed-rank test for $H_0$: $m = m_0$ based on a random sample from a symmetric population:

For $H_A$: $m > m_0$, small values of $R_-$ are extreme.
For $H_A$: $m < m_0$, small values of $R_+$ are extreme.

For $n \leq 15$, find $P$-values in Table VI; for $n > 15$, use Table IIa and the $Z$-score

$$Z = \frac{R_- - \frac{1}{4}n(n+1)}{\sqrt{\frac{1}{24}n(n+1)(2n+1)}}. \tag{4}$$

---

The null distribution of the signed-rank statistic does not depend on the form of the population distribution, provided the latter is symmetric. Like the sign test, the signed-rank test is applicable with a minimum of assumptions about the nature of the population, and just the one table is needed for finding $P$-values.

Being distribution-free and not directed at parametric alternatives, the signed-rank test is another *nonparametric* test.

---

[19]See Lindgren [15], 509.

**Example 10.5b** | **Marijuana and Proficiency (Continued)**

In Example 10.5a, the changes in test score, arranged in order of increasing magnitude, are as follows:

$$+1, -3, -3, +5, -6, -7, -7, -9, -17.$$

There are two positive changes, $+1$ with rank 1 (smallest magnitude) and $+5$ with rank 4 (4th smallest magnitude). So $R_+ = 1 + 4 = 5$. Under $H_A$, we expect more negative scores, so *small* values of $R_+$ are extreme, and we take a left-tail probability as the $P$-value. (This is why we chose to calculate $R_+$ rather than $R_-$, since the table gives left-tail probabilities.) With $n = 9$ and $c = 5$ in Table VI, we find $P = P(R_+ \leq 5) = .020$. This is the sum of the probabilities for 0, 1, ..., 5, implied in the listing of Table 10-1:

$$\frac{1}{2^9}(1 + 1 + 1 + 2 + 2 + 3) = \frac{10}{512} = .0195$$

This is much smaller than the $P$-value found in Example 10.5a using the sign test (.0898), reflecting the greater sensitivity of the signed-rank test.

To get an idea as to how the large-sample approximation fares with a sample of this size, suppose we calculate the $P$-value from the $Z$-score (4). For this we need the mean and variance of $R_+$:

$$ER_+ = \frac{9 \cdot 10}{4} = 22.5, \quad \text{var } R_+ = \frac{9 \cdot 10 \cdot 19}{24} = 71.25.$$

Then

$$Z = \frac{5 - 22.5}{\sqrt{71.25}} = -2.07,$$

and $P \doteq \Phi(-2.07) = .019$ (from Table IIa). This is perhaps closer to the exact value (.020) than we could have expected for a sample size of only 9.

Suppose we had assumed (although the $-17$ is perhaps a bit out of line) that the population distribution could be close enough to normal for a $t$-test. The statistic is

$$T = \frac{\overline{X} - 0}{S/\sqrt{n}} = \frac{-5.111}{6.254/\sqrt{9}} = -2.45.$$

From Table IIIb (8 d.f.) we find $P = .020$, coincidentally about the same $P$ as what we obtained using the signed-rank test.

In an experiment such as this, where subjects are given a test before and after a treatment, one might worry that there is learning—that scores would be better the second time around (with no treatment). Since the subjects tended to do *worse* on the second test, a conclusion against the null hypothesis is conservative. ∎

The sign test, based on a cruder summary of data, is less sensitive than the signed-rank test. On the other hand, the sign test has advantages related to its simplicity. For one thing, calculations are easy. Moreover, it can be used when one can say whether a difference is positive or negative but would be reluctant to assign specific numerical values to the differences. (An example would be in judging the quality of fruit or wine, where a numerical scale might be difficult to apply.)

Sign test, signed-rank test, $t$-test—which should one use? If the population is normal, the $t$-test is best. If the population is not far from normal, the $t$-test works well—but only slightly better than the signed-rank test. If the population is quite different from normal, but symmetric, the signed-rank test is best. The sign test is the least sensitive, but if the population is actually far from $H_0$, even the sign test may provide strong evidence against it. And as mentioned before, the sign test can be used in some cases where the other two tests cannot.

## Problems

* **10-23.**   Redo Problem 10-19 using a signed-rank test.

**10-24.**   An alternative (but equivalent) version of the signed-rank test is based on *signed-ranks*. The signed-rank $S_i$ of the observation $X_i$ is its rank with a sign attached that is the sign of the original observation. The statistic is

$$W = \sum_1^n S_i = R_+ - R_- = \frac{n(n+1)}{2} - 2R_-.$$

(a)   Show that, knowing $n$ and any one of the three statistics $W$, $R_+$, $R_-$, one can calculate the other two.

(b)   Find the mean and variance of $W$.

(c)   Show that the $Z$-scores corresponding to $W$, $R_+$, $-R_-$ are the same.

* **10-25.**   A sprinkler system is designed so that the mean activation time is 25 seconds. A series of tests resulted in the following times of first activation:[20]

27, 41, 22, 27, 23, 35, 30, 33, 24, 27, 28, 22, 24.

Test the hypothesis that the mean time is 25 seconds against the alternative that it exceeds 25 seconds,

(a)   using the sign test.

(b)   using the signed-rank test. Find the $P$-value in Table VI and compare it with the approximate $P$-value found with a normal approximation.

(c)   using a $t$-test.

**10-26.**   Referring to Problem 10-21, test the hypothesis that the median difference is 0,

(a)   using the sign test.        (b)   using the signed-rank test.

* **10-27.**   A random sample of size $n = 8$ is drawn from a symmetric population with mean 0. Consider the sequence of signs of the observations in a sample sequence and the signed rank statistic $R_-$.

(a)   How many possible sign sequences are there?

(b)   Find $P(R_- \leq 4)$ by listing and counting the sign sequences for which $R_-$ is at most 4, and so verify the entry in Table VI.

---

[20]R. R. Burford, "Use of AFFF in sprinkler systems," *Fire Technology* (1976), 5.

**10-28.**   The study referred to in Example 10.4a also included these data, decreases in VPB's in the first 10 minutes after administration of a drug:

4, 5, 7, 3, 5, 12, 22, 7, 4, −20, 2, 15.

Plot these in a dot diagram (to see if you think the assumption of population normality is reasonable), and test $\mu = 0$ with an appropriate test.

* **10-29.**   Consider again the data of Problem 10-18.

   **(a)**   Make a dot plot. Is there reason to question the use of the $t$-test?
   **(b)**   Use a sign test for the hypothesis that the median difference is 0.
   **(c)**   Calculate the signed-rank statistic and test the hypothesis that the median difference is 0.

---

## 10.6                 Probability of the Null Hypothesis

We have mentioned the unfortunate tendency for researchers to view a $P$-value as the probability of the null hypothesis. It seems natural to think in terms of probabilities of hypotheses, but the significance testing presented so far makes no provision for this. In this "classical" approach, the symbols $P(H_0)$ and $P(H_0 \mid \text{data})$ do not have meaning.

In §8.11 we introduced the notion of treating a population parameter $\theta$ as a random variable, whose distribution represents the state of one's uncertainty about $\theta$. We showed there how Bayes' theorem allows us to update a prior distribution for $\theta$ in the light of data, to obtain a posterior distribution for $\theta$.

Consider a problem of testing $H_0\colon \theta \le \theta_0$ against $H_A\colon \theta > \theta_0$. With whatever one's current distribution may be, it makes sense to talk about and perhaps calculate $P(H_0) = P(\theta \le \theta_0)$. This probability is just the integral of the current p.d.f. of $\theta$ over the interval $(-\infty, \theta)$.

**Example 10.6a** | **Aptitude Scores**

Because of the role of college aptitude test scores in college entrance decisions, there are minicourses that purport to teach students how to take these tests. Suppose a particular aptitude test has been found to produce scores that are normally distributed with mean 500. If the minicourse directed at this test is effective, on average, the mean score $\mu$ of students who take the course is larger than 500, otherwise not.

We want to test $H_0\colon \mu \le 500$ against $H_A\colon \mu > 500$ and assume, for a prior, that $\mu \sim \mathcal{N}(520, 20^2)$. The prior probability of $H_0$ is then

$$P(H_0) = P(\mu \le 500) = \Phi\left(\frac{500 - 520}{20}\right) \doteq .16.$$

Suppose we take a random sample of 50 students who took the course and find that their test scores have mean $\overline{X} = 513$ and s.d. $S = 80$. To simplify things, assume that the population s.d. is known to be 80. The posterior distribution of $\mu$ is normal, with parameters defined in terms of the "precisions" defined in §8.12:

$$\pi_{pr} = \frac{1}{400}, \quad \pi_{data} = \frac{50}{6400} = \frac{1}{128}.$$

Using these in (8) of §8.11, we find

$$E(\mu \mid \overline{X} = 513) = \frac{\frac{1}{128} \times 513 + \frac{1}{400} \times 520}{\frac{1}{128} + \frac{1}{400}} = 514.7,$$

and

$$\text{var}\,(\mu \mid \overline{X} = 513) \;=\; \frac{1}{\frac{1}{128} + \frac{1}{400}} = (9.85)^2.$$

With these in hand, we can find the probability of the null hypothesis in the posterior distribution of $\mu$:

$$P(H_0 \mid \text{data}) = P(\mu \le 500 \mid \text{data}) = \Phi\left(\frac{500 - 514.7}{9.85}\right) \doteq \Phi(-1.49) \doteq .068.$$

Suppose we determine that to recommend the special training, we'd want to see an increase in mean score to 510 as the result of the training. With this in mind, we might want to calculate

$$P(\mu > 510 \mid \text{data}) = 1 - \Phi\left(\frac{510 - 514.7}{9.85}\right) \doteq \Phi(.48) \doteq .68. \qquad \blacksquare$$

The situation in the above example is one-sided. Consider a two-sided problem—testing $\theta = \theta_0$ against the alternative $\theta \ne \theta_0$. The "point hypothesis" $\theta = \theta_0$ is usually only a convenient idealization of the hypothesis that $\theta$ lies in some small interval including $H_0$. So as an approximation, suppose we assign a positive weight to $H_0$, regarding the distribution of $\theta$ as a mixture of a point distribution at $\theta_0$ and a continuous distribution over the alternative. The next example will illustrate this approach.

**Example 10.6b** | **Comparing Treatments**

Suppose treatment A is applied to one twin in each of ten sets of twins, and treatment B is applied to the other. Let $p$ denote the probability that treatment A produces a better outcome than treatment B (in a given set). The null hypothesis is $p = 1/2$—that there is no difference between the treatments. The alternative is $p \ne 1/2$. Let $Y$ denote the number of cases in which A wins, so that $Y \sim \text{Bin}(10, p)$.

Perhaps our prior feeling is that there is a good chance that the treatments are equally effective, and we assign a positive prior probability to the null hypothesis:

$$P(p = 1/2) = \gamma.$$

(If one treatment is a control and the other is a new cancer therapy, $\gamma$ is apt to be quite large, since there are countless new therapies, most of which are worthless.) On the other hand, there is a positive probability that one is more effective than the other;

suppose we decide on a symmetric p.d.f.

$$g(p \mid p \neq 1/2) = 6p(1 - p), \ 0 < p < 1,$$

as a prior p.d.f. for $p$ when $p \neq 1/2$.

If the experiment results in nine wins for treatment A and one for treatment B, the likelihood function is

$$L(p) = P(Y = 9 \mid p) = 10p^9(1 - p).$$

To apply Bayes' theorem, we need the unconditional probability of $Y = 9$:

$$P(Y = 9) = \gamma \times P(Y = 9 \mid p = .5) + (1 - \gamma) \times \int_0^1 P(Y = 9 \mid p \neq .5)g(p)dp$$

$$= \gamma \times 10\left(\frac{1}{2}\right)^9 \left(\frac{1}{2}\right) + (1 - \gamma) \times 10 \int_0^1 6p^{10}(1 - p)^2 \, dp$$

$$= 10\left\{\frac{\gamma}{1024} + \frac{1 - \gamma}{143}\right\}.$$

The probability of $H_0$ is then (as in our applications of Bayes' theorem in Chapter 2) the ratio of the first term to the sum:

$$P(H_0 \mid \text{data}) = \frac{143\gamma}{143\gamma + 1024(1 - \gamma)}.$$

Figure 10-5 shows this probability as a function of $\gamma$. When the prior probability of

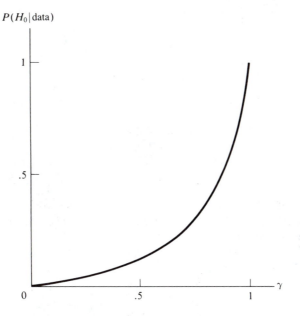

$P(H_0 \mid \text{data})$

**Figure 10-5** Posterior probability of $H_0$, Example 10.6b

$H_0$ is $\gamma = .5$, the probability of no treatment effect is reduced to .123 by the result of nine successes in ten trials; if the prior probability is $\gamma = .1$, the posterior probability of $H_0$ is only .015.                                                                    ∎

## Problems

**∗ 10-30.**   Consider the setting of Example 10.6a, with the data as given that Example ($\overline{X} = 513$ and $S = 80$).

(a)   Find the limiting posterior distribution of $\mu$ if the prior is assumed to be $\mathcal{N}(\nu, \tau^2)$ and $\tau^2 \to \infty$.

(b)   Find the probability of the null hypothesis in the posterior distribution found in (a).

(c)   Find the $P$-value for testing $\mu = 500$. [Compare with (b).]

**10-31.**   Using the same prior as in Example 10.6b, repeat the calculations for the case in which there are only three successes in ten trials to find the posterior probability of $H_0$ when the prior probability of $H_0$ is

(a)   $\gamma = .2$.       (b)   $\gamma = .5$.       (c)   $\gamma = .8$.

**∗ 10-32.**   Referring to Example 10.6b, suppose treatment B is "no treatment" (a control), and that the researcher expects to see an increase in mean response using treatment A. The appropriate null hypothesis is then $H_0$: $p \le .5$ and the alternative is $H_A$: $p > .5$. Find the probability of $H_0$ when the prior for $p$ is $g(p) \propto p(1 - p)$ and the sample result is as in the example: nine wins for A in ten trials.

## Review Problems

**10-R1.**   For the problem of testing $H_0$: $\mu \le 0$ against $H_A$: $\mu > 0$ using the mean of a random sample, and given $\overline{X} \sim \mathcal{N}(\mu, 4)$, we observe $\overline{X} = .75$.

(a)   Find the $P$-value.

(b)   Find $P(H_0)$ if the prior for $\mu$ is $\mathcal{N}(.5, 1)$.

**10-R2.**   In testing $\mu = 0$ with one observation from $\mathcal{N}(0, 1)$, suppose we obtain the result $X = 2$.

(a)   Find the $P$-value, if large values of $X$ are considered extreme.

(b)   Using the fact that $X^2 \sim \text{chi}^2(1)$, find the $P$-value if we use $Y = X^2$ as the test statistic and consider large values of $Y$ to be extreme. (Comment?)

**10-R3.**   Verify the entry for $c = 2$, $n = 6$, in Table VI.

**10-R4.**   Referring to Problem 10-18, which test could one use?

(a)   In 10-18 you did a $t$-test. Do the test again, this time without the 4.6.; why is $T$ now larger—without the observation *farthest* from 0?

(b)   Carry out the sign test and the signed-rank test, for the hypothesis that the median change is 0.

**10-R5.**   In an experiment to see whether coal dust would increase the mean heat flux of soil covered with coal dust, eight plots were so covered. Measured heat fluxes were reported as follows:[21]

24.9   35.4   18.4   37.7   34.7   32.5   34.7   28.0.

Given that the mean soil heat flux for plots covered with grass is 29.0, is the coal dust effective? [Do you think necessary assumptions are justified?]

**10-R6.**   Stand a quarter on its edge, with the face upright, and cause it to spin on the edge by flicking it with a finger. It will come to rest either heads or tails. Do this 25 times and, with the resulting number of heads, find a $P$-value appropriate to the hypothesis that the probability of heads is 1/2.

**10-R7.**   Suppose one tests an $H_0$ using the statistic $T$, regarding large values of $T$ as evidence against $H_0$. Given that the observed value is $T = t$, the $P$-value depends on $t$. Thus, $P$ depends on the value of $T$, and as such is a random variable. Show that $P$ is uniform on $(0, 1)$.

**10-R8.**   Problem 9-36 gave these nine readings of inlet oil temperature:

99, 93, 99, 97, 90, 96, 93, 88, 89.

Suppose the specifications call for a population s.d. not exceeding 3.00. Test the hypothesis that the specification is met, given $\overline{X} = 93.78$ and $S = 4.206$, and assuming a nearly normal population.

**10-R9.**   Problem 9-R11 referred to a dental study in which the mean difference in new cavities (after a trial period) between two types of toothpaste was found to be 2.41. In that problem you found posterior distributions for the population mean difference $\theta$ for two different priors, assuming that the s.d. of the difference is 1.252. Find the probability of the null hypothesis $\theta \le 0$ in each case.

**10-R10.**   Let $p$ denote the probability that a thumbtack falls with point up. If we adopt the prior $g(p) \propto p^2(1 - p)^4$, and then observe that six out of ten tacks fall with point up, find the posterior probability of the hypothesis $p \le .5$.

**10-R11.**   Consider testing $H_0$: $X \sim \text{Exp}(1)$ against $H_A$: $X \sim \text{Exp}(2)$, using a sample of size 8. The likelihood function is $L(\lambda) = \lambda^8 \exp\{ -\lambda \sum X_i\}$. What value of $\overline{X}$ would tell us that
  **(a)**   $\lambda = 1$ and $\lambda = 2$ are equally likely?
  **(b)**   $\lambda = 2$ is four times as likely as $\lambda = 1$?

## Chapter Perspective

Significance tests are extensively used in every field that applies statistical ideas. They are easily misinterpreted, so it is important to understand what they are and what they are not.

P-values serve to standardize reports of statistical analyses on a scale from 0 to 1. The closer a $P$-value is to 0, the stronger the evidence against the null hypothesis. But the interpretation of the $P$-value in a particular case is subjective: A value of $P$

[21]*Agric. & Forest Meteorology* (1988), 71–82.

that is convincing evidence to one person may be unconvincing to another. The classical framework provides no basis nor means for incorporating this subjectivity into the analysis.

The $P$-value corresponding to a given set of data is not always uniquely defined. It can depend on what alternative an investigator has in mind (e.g., one- or two-sided), and on how the decision was made as to when to stop sampling. (See Problem 10-13.) Moreover, comparing and combining $P$-values from various experiments is tricky and at best difficult.

The meaning of $P = .03$, say, is not clearly understood by users of statistics generally. A $P$-value is often incorrectly interpreted as the probability of $H_0$, although the framework of significance testing makes no provision for assigning probabilities to hypotheses. However, as discussed in the last section of the chapter, Bayesians do not use $P$-values; and for them it does make sense to talk about the probability of a hypothesis.

The point of view of this chapter is that experimental results are published or otherwise placed on file to be assimilated along with previous or subsequent evidence. In the next chapter we take up a type of inference that is essentially one of decision making, when an action must be taken based on experimental results. In such applications, the data are used as a guide in making decisions, and the problem can be formulated as a problem of testing a null hypothesis. One of two actions is taken, depending on the $P$-value.

A conclusion that $H_0$ is false or true, when based on the information in a sample, is subject to error—a point that is often overlooked in popular reporting. In the next chapter we consider the errors that can occur in drawing such conclusions.

# Tests as Decision Rules

In Chapter 10 we considered issues of statistical inference in scientific investigations. We were concerned generally with studies that add to scientific knowledge but that may not require an immediate choice of a specific action. Some situations, especially in business and industry, medicine, law, and public policy making, require more than a mere reporting of results. A decision must be made between two or more alternative courses of action.

Making the choice between two actions according to the results of a statistical analysis of data can be viewed as testing a null hypothesis $H_0$ against an alternative $H_A$. Thus, one action is preferred when $H_0$ is true, and the other when $H_A$ is true.

**Example 11a**

## Adjusting Lab Equipment

Hospital laboratories use various instruments to determine white cell count, sedimentation rate, and the like. Many states require daily checks of these instruments. For such checks, a laboratory makes measurements using a small sample from a standard blood supply, one whose characteristics are known rather precisely.

Measurements of hemoglobin with the lab's test equipment on any given blood sample will vary from one determination to another, and we regard a measurement as a random variable. If the mean of this variable is equal to the hemoglobin level of the standard blood sample (call this $H_0$), the equipment is properly adjusted. In this case, the appropriate action is to proceed with the day's tests. If the mean is not equal to the hemoglobin level of the blood sample (this is $H_A$), the equipment is out of adjustment. In this case, the appropriate action is to make an adjustment.

A particular standard supply used for checking the instrumentation has a hemoglobin level of 15.1. Let $Y$ denote the result of measuring the hemoglobin level of this standard supply on a particular day. The population of all possible such measurements has unknown mean $\mu$. If $Y$ is much different from 15.1, there is evidence that $\mu \neq 15.1$. The lab technician must have a rule for deciding whether the observed value of $Y$ is far enough from 15.1 to warrant a readjustment—to reject $H_0$: $\mu = 15.1$.

The decision rule used in one hospital lab is this: If the measured hemoglobin level is between 14.3 and 15.9, proceed with the day's lab work; otherwise, check the equipment and make necessary adjustments. Following this rule can result in a wrong decision—checking the equipment when it is working properly, or not checking when it is out of adjustment. (To be continued.) ∎

The difference between this example and those in Chapter 10 is rather subtle. Here, a rule is established *before* observing the data, a rule that will tell the technician what to do for every possible sample result. In the settings of Chapter 10, no immediate action is contemplated; one merely assesses the strength of the evidence against $H_0$ provided by a particular sample. [Even so, when researchers use the (arbitrary) convention that $P < .05$ means "statistical significance," they imply that one should act as if $H_0$ is false.]

## 11.1     Rejection Regions and Errors

As in previous chapters, we assume throughout that data are obtained by random sampling (in some cases without replacement). As in Chapter 10, we reduce the sample information to a *test statistic*. A test statistic should contain whatever information in the sample is relevant for deciding between $H_0$ and $H_A$. In the setting of this chapter, a **test** of $H_0$ is a *rule* that says which action to take according to the sample results.

### Example 11.1a  |  Adjusting a Filling Machine

Suppose "one-pound" bags (say, of sugar or flour) are filled by machine. The amount filled will actually vary from bag to bag, and the machine is supposedly adjusted so that the mean net weight in a bag is 16.0 oz. Perhaps the amount in a bag can be considered to be normally distributed with standard deviation $\sigma = .05$ oz (say).

Adjusting the machine can increase or decrease the average net weight. Because adjustments can change with time, they need to be checked periodically. We take the null hypothesis to be $H_0$: $\mu = 16.0$ (no adjustment needed), even though a setting that differs only slightly from 16.0 may be satisfactory in practice. The alternative is $H_A$: $\mu \neq 16.0$ (adjustment is required).

To check the adjustment, a technician takes a sample of $n$ bags and weighs the amount in each. Let $\overline{X}$ denote the sample average weight.

When $\mu = 16.0$, the sample mean $\overline{X}$ should be near 16.0. A large discrepancy between $\overline{X}$ and 16.0 suggests that $\mu$ is *not* close to 16.0, and that an adjustment is required. And the adjustment should be made whether $\mu$ has drifted *above* 16.0 (in which case too much is given away, accumulating over time), or *below* 16.0 (in which case the customer is shortchanged). We'll consider this type of decision rule:

$$\text{Reject } H_0 \text{ if } |\overline{X} - 16.0| > K, \text{ otherwise accept } H_0. \tag{1}$$

Each choice of the critical constant $K$ defines a rule; how to choose $K$ will be addressed later. For the moment, suppose $K = .05$. Then if $\overline{X} = 15.9$ (for instance), the

rule says to "reject $H_0$" (since 15.9 is farther than .05 from 16), which means: adjust
the setting.                                                                    ∎

Like the decision rule in the example, a test of $H_0$ versus $H_A$ is characterized by
the *rejection region*, sometimes called the *critical region*:

---

The **rejection region** $R$ of a test of $H_0$ against $H_A$ is the set of values of the
test statistic that call for rejecting $H_0$.

---

Because the decision is based on the value of a random variable (the test
statistic), using a particular set of values as the rejection region can lead to choosing
the wrong action. This is because the test statistic can fall in either the rejection region
or the acceptance region, no matter which hypothesis is governing. So one may take
the action appropriate under $H_0$ when in fact $H_0$ is false, or take the action appropriate under $H_A$ when $H_0$ is true. Since these kinds of errors involve different
consequences, we give them distinct labels:

---

**Type I error:**  Rejecting $H_0$ when $H_0$ is true.
**Type II error:**  Accepting $H_0$ when $H_0$ is false.

---

What are the chances that if we follow a particular rule, we make the wrong
decision?  Of course, if we *never* encounter $H_0$, we never make a type I error!  So to
answer the question, we'd need to know "how often" we might encounter $H_0$. This
means we'd have to have probabilities for $H_0$ and $H_A$. And in the classical approach
to testing, we don't conceive of these hypotheses as having probabilities. Thus,
instead of answering the question, we can only calculate the probabilities of taking the
wrong action *given $H_0$* and *given $H_A$*.

Whether a probability given $H_0$ or given $H_A$ is even defined depends on the form
of these hypotheses. If $H_0$ completely specifies a distribution for the test statistic, we
can calculate the probability of rejecting $H_0$ when $H_0$ is true. This probability is
called the *size* of the type I error, or the *significance level* of the test. It is quite
generally denoted by $\alpha$.

---

For a test with rejection region $R$, the **size** of the type I error, or **significance
level** of the test, is

$$\alpha = P(\text{reject } H_0|\, H_0) = P(R \mid H_0). \tag{2}$$

---

Example **11.1b** | **Adjusting the Amount (Continued)**
The rule suggested in Example 11.1a calls for an adjustment (that is, for rejecting $H_0$) if $\overline{X} > 16.05$ or $\overline{X} < 15.95$. Finding the probability of this critical region requires knowing the distribution of $\overline{X}$.

Suppose $n = 8$ and $\sigma = .06$, so that $\mu_{\overline{X}} = \mu$ and $\sigma_{\overline{X}} = .06/\sqrt{8} = .0212$. These parameters do not define a distribution for $\overline{X}$, but the central limit theorem suggests that $\overline{X} \approx \mathcal{N}(\mu, .0212^2)$. With this, we can calculate

$$\alpha = P(|\overline{X} - 16| > .05 \mid H_0) = 1 - P(15.95 < \overline{X} < 16.05 \mid \mu = 16)$$
$$= 1 - \Phi\left(\frac{16.05 - 16}{.0212}\right) + \Phi\left(\frac{15.95 - 16}{.0212}\right) = 2\Phi(-2.36) \doteq .018.$$

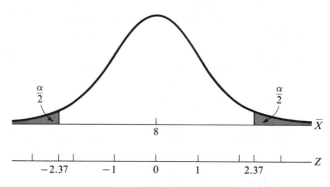

**Figure 11-1** Type I error size—Example 11.1b

Figure 11-1 shows the sampling distribution of $\overline{X}$ under $H_0$, with $\alpha$ indicated as the sum of two (equal) tail-areas.  ∎

If $H_A$ is specific enough for the probability of accepting $H_0$ to be well defined, we can find the probability of accepting $H_0$ when $H_A$ is true. It is called the **size of the type II error,** and denoted by $\beta$. The quantity $1 - \beta$, the probability of rejecting $H_0$ when it should be rejected, is called the **power** of the test, measuring the ability of the test to detect that $H_0$ is not true when it is not. But alternative hypotheses are often like the one in the preceding examples, stating only that a parameter $\theta$ is in some *range* of values.

The probability of erroneously accepting $H_0$ when $H_A$ is true depends on which model in $H_A$ is used in calculating the probability; that is, it depends on $\theta$. Given any particular value of $\theta$, we can find the corresponding size $\beta$.

For a test with rejection region $R$, the size of the type II error for a specific $\theta$ in $H_A$ is

$$\beta = P(\text{accept } H_0 \mid \theta) = P(R^c \mid \theta). \qquad (3)$$

The **power** of the test against that alternative is $1 - \beta$.

**Example 11.1c** | **Adjusting Lab Equipment (Continued)**

We return to the setting of Example 11a. The rule the lab uses for deciding whether to proceed with the day's testing or to make an adjustment is defined by the rejection region: $R = [Y > 15.9 \text{ or } Y < 14.3]$, where $Y$ is the result of a single measurement $(n = 1)$ on the standard blood supply. This operating procedure can be thought of as testing $H_0$: $\mu = 15.1$ against $H_A$: $\mu \neq 15.1$.

The lab assumes that $Y \sim \mathcal{N}(\mu, 0.16)$. With this we can calculate the size of the type I error, since specifying $\mu = 15.1$ defines the distribution of $Y$:

$$\alpha = P(R \mid H_0) = P(Y > 15.9 \text{ or } Y < 14.3 \mid \mu = 15.1)$$
$$= 1 - \Phi\left(\frac{15.9 - 15.1}{.40}\right) + \Phi\left(\frac{14.3 - 15.1}{.40}\right) = .0456.$$

But the probability of accepting $H_0$ given $H_A$ (a type II error) is *not* uniquely defined, since $H_A$ includes many values of $\mu$; it takes a particular value of $\mu$ to complete the description of the distribution of $Y$. We can, of course, find the size of the type II error for a particular $\mu$ in $H_A$. For example, when $\mu = 16.3$ (see Figure 11-2), we have

$$\beta = P(\text{accept } H_0 \mid \mu = 16.3) = P(14.3 < Y < 15.9 \mid \mu = 16.3)$$
$$= \Phi\left(\frac{15.9 - 16.3}{.40}\right) - \Phi\left(\frac{14.3 - 16.3}{.40}\right) \doteq .1587.$$

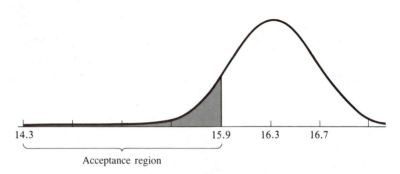

**Figure 11-2** Type II error size—Example 11.1c

The acceptance region for the test is the interval with endpoints that are two standard deviations on either side of the null hypothesis value 15.1. These endpoints are called "$2\sigma$-limits." State regulations actually call for using $3\sigma$ limits: $15.1 \pm 3 \times .40$, rejecting $H_0$ if $Y > 16.3$ or $Y < 13.9$. Using the rule implied by these limits, the significance level is smaller:

$$\alpha = P(|Z| > 3) = .0026.$$

But now, $\beta$'s are larger. Thus, for the particular alternative $\mu = 16.3$,

$$\beta = P(13.9 < Y < 16.3 \mid \mu = 16.3)$$
$$= \Phi\left(\frac{16.3 - 16.3}{.40}\right) - \Phi\left(\frac{13.9 - 16.3}{.40}\right) \doteq .50.$$

The lab's procedure is more stringent than required, as concerns type I errors. But observe that with the lab's rule, the probability of a type II error is smaller, which is more protective of the patient. ∎

In this example, changing the rejection region to reduce $\alpha$ increased the size of the type II error. This is typical. Indeed, if one critical region is contained in another: $R_1 \subset R_2$, the corresponding $\alpha$'s are in the order $\alpha_1 \leq \alpha_2$, and the $\beta$'s are in the opposite order: $\beta_2 \leq \beta_1$, since $(R_2)^c \subset (R_1)^c$.

Significance levels ($\alpha$'s) and $P$-values are both tail-probabilities in the null distribution of the test statistic. But a $P$-value is calculated from observed data, without regard to choices of action, whereas $\alpha$ is calculated for a given rule or test, before data are obtained. The $P$-value in a particular case can equal the preset $\alpha$; this happens when the observed value of the test statistic falls on the boundary of the rejection region. For this reason a $P$-value is sometimes referred to as the *observed significance level.*

## Example 11.1d | Adjusting the Amount (Continued)

In Example 11.1b we decided to reject $H_0$ when $|\overline{X} - 16| > .05$. Suppose we happen to observe $\overline{X} = 15.95$, a value on the boundary of the rejection region. The $P$-value is

$$P = 2 P(\overline{X} \leq 15.95 \mid H_0) = 2\Phi\left(\frac{15.95 - 16}{.0212}\right) = .018.$$

This is precisely the "$\alpha$" calculated in Example 11.1b—the observed significance level. ∎

## Problems

**∗ 11-1.** Suppose Oscar (in Example 10a) had decided to "accept" the hypothesis of ESP (that is, reject the null hypothesis, $p = .25$) if Felix could make seven or more correct identifications in ten trials. Find the size of the type I error for this decision rule.

∗ **11-2.** Consider the following two rejection regions: $R_1 = [|\overline{X} - 10| > .5]$, and $R_2 = [|\overline{X} - 10| > .8]$. Of these,

    **(a)** which has the larger $\alpha$?

    **(b)** which has the larger $\beta$?

∗ **11-3.** A random sample of size $n = 4$ from $\mathcal{N}(\mu, 1)$ is to be used in testing $H_0: \mu = 10$. For each of the rejection regions in the preceding problem,

    **(a)** find $\alpha$.

    **(b)** find $\beta$ for the alternative $\mu = 11$.

(It is instructive to sketch the distribution of $\overline{X}$ under $H_0$ and under the alternative $\mu = 11$, identifying the areas that represent the error sizes you have calculated.)

**11-4.** The hospital lab mentioned in Example 11a also checked its equipment for measuring white blood cell count, using standard supplies with low, average, and high counts. The s.d. of the measurements was assumed to be 400. For the "low" standard supply, the count was 3800; for this, an "upper control line" was drawn at 4600, and a "lower control line" was drawn at 3000. If the single measurement used fell outside these limits, they would stop and look for an assignable cause, making necessary adjustments. Assuming that the distribution of a single measurement is $\mathcal{N}(\mu, 400^2)$,

    **(a)** find the size of the type I error for the lab's operating rule.

    **(b)** find the probability that, following the rule, the lab would (incorrectly) fail to check the equipment if the mean of the measurements were in fact

        **(i)** $\mu = 5400$,     **(ii)** $\mu = 2800$.

∗ **11-5.** Consider this rule: Reject the hypothesis that $\mu = 10$ if the usual 90% confidence limits for $\mu$ based on the mean of a random sample of size 100 does not include the value $\mu = 10$. Assume $\sigma = 1$.

    **(a)** Express the rule as a rejection region for $\overline{X}$.

    **(b)** Find the $\alpha$ for the rule.

**11-6.** A panel of health experts will approve a certain drug for use (assuming no adverse side effects) if *two* studies comparing its effect with a placebo obtain statistically significant results—that is, if each of the two studies rejects the hypothesis of no treatment effect at the level $\alpha = .05$.

    **(a)** Find the probability that this rule serves to release a drug that has no effect whatever.

    **(b)** Suppose one wants the probability that an ineffective drug is released to be .01. What $\alpha$ should be prescribed (per study) to accomplish this?

∗ **11-7.** In Problem 10-5, twins Dot and Ditto argued about the probability that a piece of buttered bread dropped on the floor would land buttered side down. Dot claimed it was .9, and Ditto that it was .5. They conducted several trials (interrupted by their mother).

    **(a)** Suppose they had decided to carry out five trials and would reject $H_0: p = .5$ and accept $H_A: p = .9$ if $Y > K$, where $Y$ is the number of times the bread landed buttered side down in five trials. Calculate $\alpha$ and $\beta$ for this rule, for each value of $K$: $-1, 0, 1, 2, 3, 4, 5$. Plot the seven pairs $(\alpha, \beta)$ obtained in this way, taking note of the inverse nature of the relationship between $\alpha$ and $\beta$.

    **(b)** Consider the rule of rejecting $p = .5$ if $Y < K$. Plot the $(\alpha, \beta)$ pairs defined by the seven values of $K$ and explain why these rules are foolish.

## 11.2          The Power Function

In §11.1 we saw that the size of the type II error depends on the particular distribution in $H_A$ that is assumed. When the distribution of the test statistic is determined by a parameter $\theta$, the probability of rejecting $H_0$ is a function of $\theta$, the **power function,** $\pi(\theta)$. The term *power* is especially appropriate for $\theta \in H_A$, but $\pi(\theta)$ is defined for *all* $\theta$, including $\theta$'s in $H_0$. When $H_0$ is a point hypothesis: $\theta = \theta_0$, the power at $\theta_0$ is the significance level: $\pi(\theta_0) = \alpha$.

| Example 11.2a | **Adjusting Lab Equipment (Still More)** |
|---|---|

In Examples 11a and 11.1c we gave the rule used by a hospital lab for testing $H_0$: $\mu = 15.1$, defined by the rejection region

$$R = [Y < 14.3 \text{ or } Y > 15.9.]$$

The power function of the test is the probability that $Y$ falls in the rejection region when the population mean is $\mu$, as a function of $\mu$:

$$\pi(\mu) = P(\text{reject} \mid \mu) = P(R \mid \mu) = P(Y < 14.3 \text{ or } Y > 15.9 \mid \mu).$$

With the assumption in Example 11.1c that $Y \sim \mathcal{N}(\mu, 0.16)$, we find

$$\pi(\mu) = \Phi\left(\frac{14.3 - \mu}{.40}\right) + 1 - \Phi\left(\frac{15.9 - \mu}{.40}\right) = \Phi\left(\frac{14.3 - \mu}{.40}\right) + \Phi\left(\frac{\mu - 15.9}{.40}\right).$$

This is the sum of a decreasing function of $\mu$ (the next to last term) and an increasing function of $\mu$ (the last term). Its graph has the shape of an inverted bell, as sketched in Figure 11-3, with the minimum value $\alpha$ at $\mu = 15.1$:

$$\alpha = P(R \mid \mu = 15.1) = \pi(15.1) = 2\Phi(-2) = .0456.$$

As $\mu$ moves away from 15.1 in either direction, the power increases. Thus, the greater the misadjustment, the more likely it is that the technician will be led to take corrective action. However, the power function is flat near $\mu = 15.1$, so if the mean is only slightly out of adjustment, the test is not likely to call for a readjustment.   ■

The power function for a given test or rejection region does not depend on how $H_0$ and $H_A$ are defined. However, it does suggest the type of partition of $\theta$-values into a null and an alternative hypothesis for which the test is especially suited: Low power is desirable on $H_0$ and high power on $H_A$.

| Example 11.2b | **Power of a One-Sided Test** |
|---|---|

The campaign committee for a gubernatorial candidate commissions a poll to determine whether the candidate is already the choice of a majority of voters. If not, the committee will take action, spending more on TV time and newspaper ads. Let $p$ denote the population proportion who already favor the candidate, and $\hat{p}$ the corresponding proportion in a random sample of size $n = 400$. Consider this test: Buy more advertising if $\hat{p} < .54$, otherwise do not.

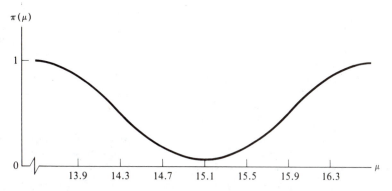

**Figure 11-3** Power function—Example 11.2a

Because $n$ is fairly large, we know that $\hat{p} \approx \mathcal{N}(p, \ pq/n)$, from §8.8. The power function of the test is

$$\pi(p) = P(\hat{p} < .54 \mid p) \doteq \Phi\left(\frac{.54 - p}{\sqrt{p(1-p)/400}}\right).$$

This is a decreasing function of $p$, shown in Figure 11-4. At $p = .50$,

$$\pi(.5) = \Phi\left(\frac{.54 - .50}{.5/20}\right) = .945.$$

At $p = .6$, the power is

$$\pi(.6) \doteq \Phi\left(\frac{.54 - .60}{\sqrt{.24/20}}\right) = .0071.$$

Thus, if only a bare majority are already in favor, there is a high probability that the

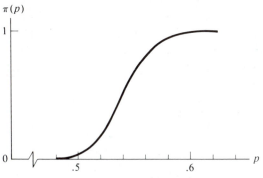

**Figure 11-4** Power function—Example 11.2b

poll result will call for action; if the majority is substantial (as for $p = .6$), the probability is small that action will be called for.

It was not necessary to say what $H_0$ and $H_A$ are in this situation, nor to define an "$\alpha$." What is important is the power function defined by the rule for taking action. ■

## 11.3   Choosing the Sample Size

In Chapter 10, and so far in this chapter, we have assumed a given, fixed sample size. This is appropriate when one has a particular data set to analyze. When setting up an experiment, one faces the question of how large a sample is needed. Economic considerations are usually a constraint on this choice. Aside from such constraints, the sample size should depend on how reliable we want our conclusions to be. In §9.2 we showed how, in estimating a parameter, one might determine a sample size from a specification of standard error. When testing hypotheses, one can determine the sample size by specifying allowable error sizes.

A finite amount of data seldom provides perfect information, so an experimenter who must make a decision based on those data must accept the possibility of error—both type I and type II. How large the sizes of these errors are allowed to be depends on the costs of wrong decisions and on the cost of gathering data. Here we'll assume that such considerations have led to a specification of error sizes.

Suppose that the distribution of the test statistic is completely determined by the value of a parameter $\theta$, and consider $H_0$: $\theta = \theta_0$. An investigator may be willing to tolerate $\alpha = .05$, but insists on $\beta = .20$ when $\theta = \theta_1$. We'll see that such restrictions on the power function serve to determine the sample size and the choice of a particular rejection region from the class of regions of a suitable type.

Consider testing hypotheses about the population mean $\mu$ using the sample mean $\overline{X}$ as the test statistic. Perhaps we can assume either that the population is normal or that the sample size is large enough that $\overline{X}$ is approximately normal. Then, if $\sigma$ is known, the power function for a rejection region of the form $\overline{X} > K$ is

$$\pi(\mu) = P(\overline{X} > K \mid \mu) \doteq \Phi\left(\frac{\mu - K}{\sigma/\sqrt{n}}\right).$$

As a function of $\mu$, this has the shape of a normal c.d.f. with mean $K$. The curve can be shifted to the right or left by increasing or decreasing $K$. For a given $K$, the slope of the curve at $K$ can be increased by increasing $n$:

$$\pi'(K) = \frac{1}{\sigma}\sqrt{\frac{n}{2\pi}}.$$

Requiring $\pi(\mu)$ to pass through a specified point is a condition on $n$ and $K$, and requiring it to pass through two specified points defines two simultaneous equations for $n$ and $K$. Solving this system of equations produces values of $n$ and $K$ that satisfy the given two requirements.

**Example 11.3a** | **Adjusting the Amount (Conclusion)**

We return once more to the setting of Example 11.1d, and the rejection region

$$R = [|\overline{X} - 16| > K].$$

Suppose, in deciding whether to adjust the machine or not, we want at most a 5% chance of calling for readjustment when none is needed (that is, when $\mu = 16$), and a 90% chance of being led to do the readjustment when the mean has shifted to either 16.1 oz or 15.9 oz. We translate these conditions in terms of power as follows:

$$\pi(16) = .05, \text{ and } \pi(15.9) = \pi(16.1) = .90. \tag{1}$$

We assume (as before) that $\sigma = .06$, and that the population shape is such that even a small $n$ would imply approximate normality of $\overline{X}$. The power function is then

$$\pi(\mu) = P(|\overline{X} - 16| > K \mid \mu) \doteq 1 - \Phi\left(\frac{16 + K - \mu}{.06/\sqrt{n}}\right) + \Phi\left(\frac{16 - K - \mu}{.06/\sqrt{n}}\right). \tag{2}$$

Imposing the first of the given conditions (1), we have

$$\pi(16) = 1 - \Phi\left(\frac{K\sqrt{n}}{.06}\right) + \Phi\left(\frac{-K\sqrt{n}}{.06}\right) = 2\Phi\left(\frac{-K\sqrt{n}}{.06}\right) = .05,$$

which means (according to Table IIa)

$$\frac{-K\sqrt{n}}{.06} = -1.96, \text{ or } K = \frac{.1176}{\sqrt{n}}. \tag{3}$$

With this $K$ and $\mu = 16.1$ in (2), we set $\pi(16.1) = .90$:

$$\pi(16.1) \doteq 1 - \Phi\left(\frac{.1176/\sqrt{n} - .1}{.06/\sqrt{n}}\right) + \Phi\left(\frac{-.1176/\sqrt{n} - .1}{.06/\sqrt{n}}\right) = .90.$$

Simplifying the fractions, we have

$$\Phi(1.96 - 1.67\sqrt{n}) - \Phi(-1.96 - 1.67\sqrt{n}) = .10.$$

We could obtain an approximate solution for $n$ by trial and error, but we see that the second term is negligible when $n > 1$. Then $1.96 - 1.67\sqrt{n} \doteq -1.28$, or $n \doteq 3.78$. We take the next larger integer for $n$ and substitute in (3) to get $K = .059$. Using the values $n = 4$ and $K = .059$ in (2), we find $\pi(16) \doteq .05$ and $\pi(16.1) = .915$. The latter is slightly greater than what was specified for the power at $\mu = 16.1$, because it was necessary to round $n$ to an integer. ∎

The issues we discussed in §9.2 regarding the selection of a sample size when $\sigma$ is not known apply in the present setting of testing hypotheses about $\mu$. In particular, taking a pilot sample or monitoring data as it accumulates can be effective in estimating $\sigma$.

## Problems

* **11-8.**  Find the power function for the rule given in Problem 11-1.

* **11-9.**  As in Problem 11-3, consider a random sample of size $n = 4$ from $\mathcal{N}(\mu, 1)$ to be used in testing $H_0$: $\mu = 10$. For each of the regions $R_1$ and $R_2$ in Problem 11-2, find the power function. (In the process of solving Problem 11-3, you would have actually calculated some points on the power curve.)

**11-10.**  Find and sketch the power function for the lab's operating rule in Problem 11-4. Suppose that instead of $2\sigma$-limits, the lab used the $3\sigma$-limits specified in the state regulations—rejecting $H_0$ if the observation $X$ fell outside the range $3800 \pm 1200$.

   **(a)**  What would this do to the power curve?

   **(b)**  Which rule is more protective of the patients?  (Explain.)

* **11-11.**  Consider testing $\theta = 0$ on the basis of a single observation of a Cauchy variable $X$ with median $\theta$: $f(x \mid \theta) \propto [1 + (x - \theta)^2]^{-1}$.

   **(a)**  Find $\alpha$ and the power function for the rejection region $X > 2$.

   **(b)**  Find $\alpha$ and the power function for the rejection region $|X| > 1.5$.

**11-12.**  Suppose a treatment is intended to increase a response. Consider the null hypothesis that the mean increase is $\mu = 0$. A study employing a sample of size $n$ is to be conducted to determine the treatment's effect. A treatment effect is to be inferred if $\overline{X} > K$. The test is to have $\alpha = .05$ and an 80% chance of detecting a mean increase of 2 units. What sample size $n$ and critical value $K$ should be used, given that $\sigma = 5$?

* **11-13.**  Problem 11-5 considered this rule:  Reject the hypothesis $\mu = 10$ if a 90% confidence interval for $\mu$ based on a random sample of size 100 does not include the value 10. Find the power function. (Again, assume $\sigma = 1$.)

## 11.4      Quality Control

**Example 11.4a**  |  **The Deming Medal**

In Japan, the Deming medal is the highest and most prestigious award that Japanese companies can earn for the advancement of precision and dependability of products. It is named for W. Edwards Deming (1900–1993), a statistician who gave assistance to Japanese industry after World War II. He is given credit for helping to improve the quality of their products and increase the productivity of their plants. Many Japanese feel that the practical methods he taught led to the spectacular strengthening of their competitive position in international markets. These methods include sampling inspection and acceptance sampling—statistical tools for monitoring product quality. ∎

The quality of manufactured items is important for consumers. It is equally important to the manufacturers, who may well find it unprofitable to sell items of poor quality in competitive markets. In controlling quality, it is important for a producer to monitor the quality of what is produced and for the customer to monitor the quality of what is received. One procedure aimed at quality control is the use of *control charts*. This technique is what was being used in the hospital lab that was the subject of Examples 11a, 11.1c, and 11.2a. Delving further into this area takes us too far afield, so in this section we'll just introduce another tool of the quality control engineer— **acceptance sampling.**[1]

Suppose that goods are shipped in lots. Both the manufacturer and the customer want to determine whether a lot is acceptable. Inspecting every item in a lot is costly, and complete inspection may not be totally reliable since errors can occur in the inspection process. In some cases, it may be necessary to destroy an item to assess reliability; in such cases complete sampling is out of the question. In any case, suppose the manufacturer decides to inspect the items in a sample taken from the lot to help decide whether or not to pass the lot. We'll deal here only with such "sampling inspection" when items can be classed as either *good* or *defective.*

The choice of action is between rejecting and accepting the lot. The decision rule used in practice is a natural one: Reject the lot if the sample has too many defectives, and otherwise accept the lot.

We assume that sampling is done at random, and without replacement. Our notation is consistent with that of Chapter 4:

$N$ = lot size,
$n$ = sample size,
$p$ = lot fraction defective,
$M = Np$ = number of defectives in the lot,
$Y$ = number of defectives in the sample.

For an acceptance region of the form $Y \leq c$, the critical value $c$ is called the **acceptance number.** We assume that $N$ is fixed, and take $p = M/N$ as the unknown parameter. Under sampling without replacement, $Y$ has a hypergeometric distribution (§4.3). The decision rule can be construed as a test of the hypothesis that $M \leq M_0$ against the alternative $M > M_0$ (for some $M_0$, which we don't need to specify). The power function of the rejection region $R = [Y > c]$ is

$$\pi(p) = P(R \mid p) = 1 - P(Y \leq c \mid p) = 1 - \sum_{k=0}^{c} \frac{\binom{Np}{k}\binom{N-Np}{n-k}}{\binom{N}{n}}. \qquad (1)$$

In industrial circles, the performance of a sampling plan is usually described by the **operating characteristic,** or **OC-curve,** defined as the probability that the lot is accepted (as a function of $p$):

$$\mathrm{OC}(p) = P(\text{lot is accepted} \mid p) = P(Y \leq c \mid p) = 1 - \pi(p). \qquad (2)$$

---

[1]For further reading on this and related topics, see A. J. Duncan, *Quality Control and Industrial Statistics*, 4th ed. (Homewood, IL: Irwin, 1974).

**Example 11.4b** | **An Industrial Sampling Plan**

Consider a sampling plan with $c = 1$, $n = 3$, and $N = 10$. The power function defined by these constants, as given by (1), is

$$\pi(p) = 1 - [P(Y = 0 \mid p) + P(Y = 1 \mid p)]$$

$$= 1 - \frac{\binom{10p}{0}\binom{10-10p}{3} + \binom{10p}{1}\binom{10-10p}{2}}{\binom{10}{3}}.$$

This, together with the OC-function, is given in the following table:

| $p$ | 0 | .1 | .2 | .3 | .4 | .5 | .6 | .7 | .8 | .9 | 1 |
|---|---|---|---|---|---|---|---|---|---|---|---|
| $\pi(p)$ | 0 | 0 | .07 | .18 | .33 | .50 | .67 | .82 | .93 | 1 | 1 |
| $OC(p)$ | 1 | 1 | .93 | .82 | .67 | .50 | .33 | .18 | .07 | 0 | 0 |

With this plan, the chance of rejecting the lot increases as $p$ increases—that is, as the lot quality worsens. (This is in fact a general characteristic of this type of sampling plan.) ∎

When the sample size is small in comparison with the population or lot size ($n \ll N$), the hypergeometric probabilities in (1) can be approximated by binomial probabilities:

$$\pi(p) \doteq 1 - \sum_{k=0}^{c} \binom{n}{k} p^k (1-p)^{n-k}. \tag{3}$$

These, in turn, can be approximated by Poisson probabilities when $p$ is small and $n$ is large, and by normal probabilities if $npq > 5$ (see §4.5 and §4.6). And in general, the power function and its binomial, Poisson, and normal approximations are increasing functions of $p$: The poorer the lot quality, the more likely it is that the lot will be rejected.

**Example 11.4c** | **Sampling a Large Lot**

Consider sampling 25 from a lot of 500 articles and the plan defined by the acceptance number $c = 2$. Because $n = 25$ is much smaller than $N = 500$, the binomial probabilities in (3) should provide sufficient accuracy. From (3),

$$\pi(p) \doteq 1 - \sum_{0}^{2} \binom{25}{k} p^k (1-p)^{25-k}$$

$$= 1 - (1-p)^{23}[(1-p)^2 + 25p(1-p) + 300p^2].$$

For any particular value of $p$, this is easy to evaluate on a hand calculator. When $p$ is small we could use a Poisson approximation with $m = np = 25p$:

$$\pi(p) = 1 - e^{-25p}[1 + 25p + (25p)^2/2].$$

The following table gives values of the power function for selected values of $p$. It also gives the binomial and Poisson approximations.

| $p$ | Hypergeometric | Binomial | Poisson |
|-----|----------------|----------|---------|
| .05 | .1224 | .1271 | .1315 |
| .10 | .4657 | .4629 | .4562 |
| .15 | .7534 | .7463 | .7229 |
| .20 | .9075 | .9018 | .8753 |

(Observe that the Poisson does not do well when $p$ is greater than .1.)  The difference between the power at $p = .05$ and $p = .20$ shows us that the test does a reasonably good job in discriminating between these two values of $p$.                                 ■

   In the preceding example we have analyzed the performance of a test, again without specifying a null hypothesis. Because the power function (1) increases with $p$, the sampling inspection procedure it describes is the appropriate type of test for a hypothesis of the form $H_0$: $p \leq p_0$ against the alternative $H_A$: $p > p_0$. Both $H_0$ and $H_A$ involve many values of $p$, so the $\alpha$ and $\beta$ are not uniquely defined. However, suppose the manufacturer specifies a fraction defective $p_0$ such that any lot with $p \leq p_0$ should be accepted. The power at $p_0$ is called the *producer's risk*:

$$\alpha = \max_{p \leq p_0} \pi(p) = \pi(p_0).$$

The consumer, on the other hand, may specify a fraction defective $p_1$ such that any lot with $p \geq p_1$ is unacceptable. The probability of acceptance when $p = p_1$ is called the *consumer's risk*:

$$\beta = \max_{p \geq p_1} [1 - \pi(p)] = 1 - \pi(p_1).$$

Specifying these risks (which may be part of a contract) amounts to requiring that the power function pass through the two points $(p_0, \alpha)$ and $(p_1, 1 - \beta)$. As in §11.3, these two conditions determine an acceptance number $c$ and a sample size $n$, which define the sampling plan. [See Problem 11-16(c).]

## Problems

∗ **11-14.**  A large shipment of disposable thermometers is to be accepted or rejected by a clinic according as a sample of ten thermometers includes no defectives or has at least one defective. Find and sketch the power function of this rule, as a function of $p$, the proportion of defectives in the shipment.

**11-15.**    An acceptance sampling plan for lots of size 20 is to accept a lot if a random selection of five from the lot includes at most one defective item. Find the power function of the plan.

∗ **11-16.**    Suppose a manufacturer of a battery-powered toy wants to buy batteries whose operating lives $X$ average at least 40 hours. The purchase agreement with a certain supplier specifies that for a shipment to be acceptable, a sample of 25 batteries taken at random from the shipment must satisfy the condition $T > -1.5$, where $T = 5(\overline{X} - 40)/S_X$.

(a)    Find $\alpha$ for this rule, taking $H_0$ to be $\mu = 40$.

(b)    Power calculations for the given rule require a distribution for $T$ when $\mu \neq 40$. This distribution (called *noncentral t*) is one that we have not studied. So instead, find the power function of the region

$$5(\overline{X} - 40)/S_X > 1.5$$

with the sample s.d. $S_X$ replaced by $\sigma = 4$, assumed known.

(c)    Suppose the manufacturer and supplier agree that a procedure with rejection region of the form $\overline{X} < K$ is acceptable to both if it has a 5% chance of rejecting a shipment in which $\mu = 42$ and a 1% chance of accepting a shipment in which $\mu = 39$. Again assuming $\sigma = 4$ is known, find the sample size and critical value $K$ so as to satisfy these criteria.

**11-17.**    Find the OC-function for a plan that accepts a lot if a random selection of 15 items from the lot includes at most one defective. (The lot size is much larger than 15.)

---

## 11.5    Most Powerful Tests

Finding good tests is straightforward when both the null and the alternative hypotheses are *simple*—complete specifications of a probability model, with no unknown parameter values.

Let $f$ denote the joint p.f. or p.d.f. of the sample observations, and consider testing

$$H_0: f = f_0 \text{ against } H_A: f = f_1,$$

where $f_0$ and $f_1$ involve no unknown parameters. A test is "good" if the sizes of both types of error are small. How do we find a rejection region that will have small error sizes of both types? One approach is to fix the size of $\alpha$ at some acceptable level and search for a rejection region whose $\beta$ is as small as possible (with that $\alpha$) or, equivalently, one whose power on the alternative is a maximum. In the continuous case, this calls for a region $R$ with specified $\alpha$:

$$\int_R f_0(\mathbf{x}) \, d\mathbf{x} = \alpha, \tag{1}$$

such that the power

$$1 - \beta = \int_R f_1(\mathbf{x}) \, d\mathbf{x} \tag{2}$$

is as large as possible. A clue as to how to construct such a rejection region $R$ from sample points **x** is found in the following example.

**Example 11.5a** | **Maximizing the Ratio of Value to Volume**

A device that has been used in TV game shows is to have someone with a shopping cart dash madly through a grocery store, filling the cart with as much loot as possible, taking at most one of each type of item. The usual constraint is one of time, but suppose you have a cart and enough time to reflect on how to fill it in the most efficient way. How would *you* go about filling the cart to get the most total value?

Surely you would not take a large box of cornflakes, or a 12-pack of cola, or a package of cotton puffs—these all take up too much room for what they're worth. Rather, you would take things that are expensive but don't take up much room, things that have a high cost per unit volume—raspberries out of season, swordfish, caviar, and the like. You might start by putting into your cart *first* those items that have the highest cost per unit volume. ■

Think of $f_0(\mathbf{x})$ as the volume of the "item" **x**, and $f_1(\mathbf{x})$ as its cost or value. The reasoning in the above example suggests that we get a region of greatest value (integral of $f_1$ over $R$) for a given volume (integral of $f_0$ over $R$) by putting into $R$ first those **x**'s which have the smallest volume per unit price (smallest ratio of $f_0$ to $f_1$). The ratio $f_0/f_1$ is a **likelihood ratio.** We keep putting points into $R$ until it gets filled up—that is, until the total volume (integral of $f_0$ over $R$) reaches $\alpha$, the capacity of the "cart" $R$. This suggests that the likelihood ratio should be a good test statistic, and that small values of the ratio are "extreme" (as suggested also in §10.2).

---

*Likelihood ratio statistic* for testing $f_0(\mathbf{X})$ versus $f_1(\mathbf{X})$:

$$\Lambda(\mathbf{X}) = \frac{f_0(\mathbf{X})}{f_1(\mathbf{X})}. \tag{3}$$

---

**Example 11.5b** | **Likelihood Ratio for Bin($p$)**

Consider these simple hypotheses about a Bernoulli population:

$$H_0\colon p = .50 \ \text{ versus } \ H_1\colon p = .25.$$

Suppose we conduct five independent trials and base a test on the sufficient statistic $Y$, the number of successes in those five trials. The possible values of $Y$ and corresponding probabilities under these two models (from Table Ia/Ib of the Appendix) are shown in the following table, along with a column of values of the likelihood ratio statistic $\Lambda$. (As is customary, our notation may drop the dependence of $\Lambda$ on $Y$ and refer to the likelihood ratio as $\Lambda$.)

| $y$ | $f_0(y)$ | $f_1(y)$ | $\Lambda(y)$ |
|-----|----------|----------|--------------|
| 0 | .0312 | .2373 | .13 |
| 1 | .1562 | .3955 | .40 |
| 2 | .3125 | .2637 | 1.2 |
| 3 | .3125 | .0879 | 3.6 |
| 4 | .1562 | .0146 | 10.6 |
| 5 | .0312 | .0010 | 31.2 |

From the column of values of $\Lambda$, it is clear that $\Lambda$ is an increasing function of $Y$—the values of $Y$ with the smallest $\Lambda$-values are those with the smallest $Y$-values. So in forming a rejection region, we put into it first $y = 0$; then the second smallest, $y = 1$, and so on, until the sum of the $f_0$'s for the next $Y$-value would exceed a specified $\alpha$. The resulting rejection region will have the smallest $\beta$ (greatest power) among all tests with the same $\alpha$. With $n = 5$, one could not hope that both $\alpha$ and $\beta$ would turn out to be small, but a rejection formed in this way is optimal. Some of these **most powerful** rejection regions of $Y$-values are as follows:

| $R$ | $\alpha$ | $\beta$ | Power |
|-----|----------|---------|-------|
| $\emptyset$ | 0 | 1 | 0 |
| {0} | .031 | .763 | .237 |
| {0,1} | .187 | .367 | .633 |
| {0,1,2} | .500 | .103 | .897 |

Notice, for instance, that the rejection regions $R = \{0, 1\}$ and $R = \{4, 5\}$ both have $\alpha$'s of .187; but the $\beta$ for the latter, which is not a likelihood ratio rejection region, is much higher: $\beta = .984$. (To be continued.)   ∎

The mathematical theorem states that rejection regions of the form $\Lambda < K$ are most powerful in the class of tests with the same $\alpha$. In the following, the "$f$" can be either a p.d.f. or a p.f., and $L_i = f_i(\mathbf{x})$:

**Neyman-Pearson lemma:**  For testing $f_0(\mathbf{x})$ against $f_1(\mathbf{x})$, a critical region of the form

$$\Lambda(\mathbf{x}) = \frac{f_0(\mathbf{x})}{f_1(\mathbf{x})} < K, \qquad (4)$$

where $K$ is a constant, has the greatest power (smallest $\beta$) in the class of tests with no larger $\alpha$.

We'll write out the proof only for the continuous case, where $f_i$ is a p.d.f. Consider the region $R$ as defined by (4), a region in the $n$-dimensional sample space of $\mathbf{X}$, and a competing region $S$ whose type I error size is no larger than that of $R$: $\alpha_S \leq \alpha_R$. Then, for any function $f$ (and in abbreviated notation),

$$\left( \int_R - \int_S \right) f = \int_{RS^c} f - \int_{R^c S} f, \qquad (5)$$

since the integral over the intersection $RS$ cancels. Applying (5) to $f_1$, we can calculate the difference in $\beta$'s:

$$\beta_S - \beta_R = \left( \int_R - \int_S \right) f_1 = \int_{RS^c} f_1 - \int_{R^c S} f_1.$$

According to (4), we know that $f_1 \geq f_0/K$ on $R$ and $-f_1 \geq -f_0/K$ on $R^c$. Hence, with another application of (2), we obtain

$$\beta_S - \beta_R \geq \frac{1}{K} \left( \int_{RS^c} f_0 - \int_{R^c S} f_0 \right) = \frac{1}{K} \left( \int_R - \int_S \right) f_0 = \frac{1}{K} (\alpha_R - \alpha_S).$$

In the competition, we allow only regions $S$ such that $\alpha_S \leq \alpha_R$, so the lemma is established.

A graphical representation of the error sizes of tests is helpful. Each proposed rejection region $R$ has an associated pair of error sizes, $(\alpha, \beta)$. Each such pair can be plotted as a point in a rectangular coordinate system. The collection of points corresponding to all rejection regions is contained in the unit square, since $\alpha$ and $\beta$ are probabilities.

**Example 11.5c** | **Neyman-Pearson Test for $\mathcal{N}(0, \theta)$**

Given a random sample from $X$ with p.d.f.

$$f(x \mid \theta) = \frac{1}{\sqrt{2\pi\theta}} \exp\left\{-\frac{1}{2\theta} x^2\right\},$$

the likelihood function is

$$L(\theta) = \theta^{-n/2} \exp\left\{-\frac{1}{2\theta}\sum X_i^2\right\}.$$

For the N-P test of $\theta_1$ against $\theta_2$, the critical region is of the form

$$\Lambda = \left(\frac{\theta_1}{\theta_2}\right)^{-n/2} \exp\left\{-\left(\frac{1}{2\theta_1} - \frac{1}{2\theta_2}\right)\sum X_i^2\right\} < K'.$$

If $\theta_1 < \theta_2$, this inequality holds for a given $K'$ if and only if, for some $K$,

$$\sum X_i^2 > K. \tag{6}$$

This rejection region is most powerful in the class of tests with no larger $\alpha$.

The distribution of $\sum X_i^2/\theta$ is chi$^2(n)$, and we can use this to find the error sizes for the test (6), given $K$:

$$\alpha = P\left(\sum X_i^2 > K \mid \theta_1\right) = 1 - F_{\chi^2(n)}(K/\theta_1),$$

$$\beta = P\left(\sum X_i^2 < K \mid \theta_2\right) = F_{\chi^2(n)}(K/\theta_2).$$

These are parametric equations for a curve in the $\alpha$-$\beta$ plane, the curve whose points (defined by values of $K$) are the representations of all of the most powerful tests. The curve is shown in Figure 11-5 for the particular case $n = 10$, $\theta_1 = 1$, and $\theta_2 = 2$. The $\alpha$-$\beta$ representations of all other tests must lie above that curve.

It is clear from the figure that not only do the Neyman-Pearson tests have the smallest $\beta$'s for a given $\alpha$, they also have the smallest $\alpha$ for a given $\beta$. According to the Neyman-Pearson lemma, there is no test whose $\alpha$-$\beta$ representation lies below or to the left of this curve. ∎

## 11.6 Randomized Tests

As we defined a decision rule or test in §11.1, it says (for given data), either "accept $H_0$" or "reject $H_0$", depending on the data. Consider instead a rule that says (for given data **x**), "reject $H_0$ with probability $\phi(\mathbf{x})$," where $0 \leq \phi(\mathbf{x}) \leq 1$. A test defined by $\phi$ in this way is a **randomized test**. A test defined by a rejection region $R$ is the special case for which $\phi(\mathbf{x}) = 1$ when $\mathbf{x} \in R$, and is 0 when $\mathbf{x} \notin R$.

The error sizes of the test $\phi$ are

$$\alpha = E_0[\phi(\mathbf{x})], \quad \beta = E_1[1 - \phi(\mathbf{x})], \tag{1}$$

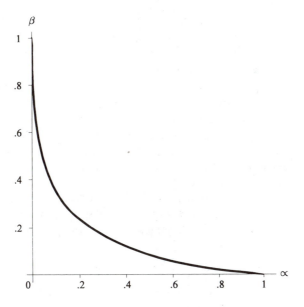

**Figure 11-5** Most powerful tests—Example 11.5c

where the subscript indicates whether the expected value assumes $H_0$ or $H_1$.

Given tests $\phi_1$ and $\phi_2$, the function $\phi = \gamma\phi_1 + (1 - \gamma)\phi_2$ also defines a test if $0 \leq \gamma \leq 1$. This is a *mixture* of $\phi_1$ and $\phi_2$, with error sizes

$$\alpha = \gamma\alpha_1 + (1 - \gamma)\alpha_2$$
$$\beta = \gamma\beta_1 + (1 - \gamma)\beta_2, \tag{2}$$

since expectations are linear operations. These equations are parametric equations of a straight line. The line contains $P_1 = (\alpha_1, \beta_1)$, when $\gamma = 1$, and $P_2 = (\alpha_2, \beta_2)$, when $\gamma = 0$. With the restriction $0 \leq \gamma \leq 1$, the point $(\alpha, \beta)$ lies on the line segment joining $P_1$ and $P_2$.

All of this means that the set of points representing all tests, randomized or not, is a convex set within the unit square, which always includes (0, 1) and (1, 0). [A convex set is one that includes every convex combination—every $(\alpha, \beta)$ of the form (2) for $0 \leq \gamma \leq 1$.] A somewhat more general form of the Neyman-Pearson lemma asserts that the likelihood ratio tests defined by (4) of the preceding section are most powerful in the class of all tests, including randomized tests, and that all tests represented by points on the lower left boundary of the convex set of all tests are most powerful.[2]

In the next example we'll be looking at rejection regions, one of which contains the other:  $R_1 \subset R_2$. In such a case, the mixture $\gamma R_1 + (1 - \gamma)R_2$ is a test of this

---

[2]For a proof, see Lindgren [15].

form: reject $H_0$ if $\mathbf{x} \in R_1$, reject $H_0$ with probability $\gamma$ if $\mathbf{x} \in R_2 R_1^c$, and accept $H_0$ if $\mathbf{x} \notin R_2$.

## Example 11.6a | Neyman-Pearson Test for Bin($p$)

We return to the problem of choosing between the two binomial models in Example 11.5b. Figure 11-6 shows the point $(\alpha, \beta)$ for each of the likelihood ratio tests (Neyman-Pearson tests) that we found in Example 11.5b.

We have also plotted the points representing tests of the form $\Lambda > K$, which are *worst* tests. (Reverse the argument for the best tests: Fill the shopping cart in Example 11.5a with voluminous items having little value.) All other possible tests, including randomized tests, are represented by points in the shaded region of the figure. This region is a convex set whose extreme or corner points are the best and worst (nonrandomized) tests. The points on the line segment joining consecutive corner points of the set of all tests are mixtures of those two corner points.

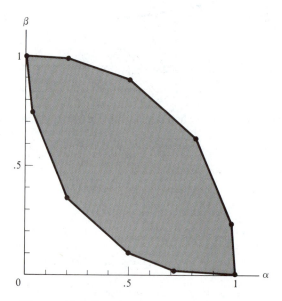

**Figure 11-6** Randomized tests—Example 11.6a

One such mixture is this: If $Y = 0$, reject $H_0$; if $Y > 1$, accept $H_0$; and if $Y = 1$, reject $H_0$ with probability $\gamma$, a number in the unit interval $(0, 1)$. The $\alpha$ for this mixture is

$$\alpha = P(\text{rej. } H_0 \mid H_0)$$
$$= P(\text{rej. } H_0 \mid Y = 0) \cdot P(Y = 0 \mid H_0) + P(\text{rej. } H_0 \mid Y = 1) \cdot P(Y = 1 \mid H_0)$$
$$= .0312 + \gamma \times .1563.$$

Depending on the value of $\gamma$, this $\alpha$ lies between .031 (the $\alpha$ of the rejection region $\{0\}$) and $.0312 + .1563 = .1875$ (the $\alpha$ of the rejection region $\{0, 1\}$). To define a test with $\alpha = .05$, for example, we solve $.05 = .0312 + .1563\gamma$ for $\gamma$, obtaining $\gamma = .0188/.1563 = .1203$. With this $\gamma$, the test has power

$$1 - \beta = P(\text{rej. } H_0 \mid H_1) = P(\text{rej. } H_0 \mid H_1) = .2373 + \gamma \times .3955 \doteq .285.$$

The $(\alpha, \beta)$-representation for this randomized test is $(.05, .715)$, a point on the line segment joining the points representing the regions $\{0\}$ and $\{0, 1\}$.  ■

Why do we need to consider such tests, and would anyone really (after getting some data) toss a "coin" in deciding what to do? People do not usually resort to extraneous games of chance to help with their statistical decisions. The only real need for a randomized test is to round out the theory. And there is a point to a randomized test only in the case of a test statistic with a discrete distribution. In such situations there may not be a rejection region whose $\alpha$ is exactly a specified size, such as .05.

The availability of randomized tests allows us to show that we may as well base our tests on *sufficient* statistics. We reason as follows: If $T$ is sufficient for $\theta$, and $\phi$ is any test (pure or randomized), define the test

$$\phi^*(t) = E[\phi(\mathbf{X}) \mid T = t].$$

(Since $T$ is sufficient, the distribution of $\mathbf{X}$ given $T = t$ does not depend on $\theta$, so $\phi^*$ is a function of $t$ alone and defines a test.)  The power function of $\phi^*$ is

$$\pi_{\phi^*}(\theta) = E\{E[\phi(\mathbf{X}) \mid T = t]\} = E[\phi(\mathbf{X})] = \pi_\phi(\theta).$$

Given any test $\phi$ and a *sufficient* statistic $T$, there is a test based on $T$ (only) with the same power function as that of $\phi$.

## 11.7  Uniformly Most Powerful Tests

A test that has the greatest power among tests with no larger $\alpha$, not only against a particular alternative but against *all* alternatives in $H_A$, is said to be **uniformly most powerful** (UMP). One way of constructing a UMP test for a simple null against a composite alternative hypothesis, say $H_0: \theta = \theta_0$ against the composite $H_A: \theta < \theta_0$, is first to select a particular $\theta_1 < \theta_0$ and construct a Neyman-Pearson test against $\theta = \theta_1$. If the form of this best test is the same for *every* $\theta_1 < \theta_0$, the test is uniformly most powerful for $\theta_0$ against $\theta < \theta_0$.

### Example 11.7a | UMP Test for Bin$(p)$

Suppose we test $H_0: p = .50$ based on $Y$, the number of successes (1's) in a random sample of size $n = 5$. We saw in Example 11.5b that rejecting $H_0$ for $Y < K$ is most powerful against $p = .25$ in the class of tests with $\alpha = .187$. But the same is true if we were to use an arbitrary $p < .50$: The likelihood ratio is

$$\Lambda = \frac{.5^Y .5^{5-Y}}{p^Y (1-p)^{5-Y}} \propto \left(\frac{1}{p} - 1\right)^Y,$$

an increasing function of $Y$ when $1/p > 2$, or $p < 1/2$. So $\Lambda$ is less than a constant when $Y$ is less than a constant. Thus, $Y < K$ is most powerful against *every* $p < .5$; it is UMP against $H_A$: $p < .5$. (To be continued.) ∎

When the *null* hypothesis is composite, $\alpha$ may not be uniquely defined. In this case, we define $\alpha^*$ to be the maximum (or supremum)[3] of the probabilities of rejection over all simple hypotheses in $H_0$. With $\alpha$ replaced by the extended size $\alpha^*$, we define uniformly most powerful tests as before.

Consider testing $\theta \leq \theta_0$ against $\theta > \theta_0$. Suppose we have found a region $R$ that is UMP for $\theta = \theta_0$ against $\theta > \theta_0$. If the power function is increasing, the maximum power on $H_0$ occurs *at $\theta_0$*, and this is the $\alpha^*$. Any competing test with extended size $\alpha^*$ is then included in the set of tests that compete *at $\theta_0$*, and among these, $R$ has the greatest power on $\theta > \theta_0$—it is UMP in the extended sense, $\theta \leq \theta_0$ against $\theta > \theta_0$.

**Example 11.7b** | **UMP Test for Bin($p$) (Continued)**

In the preceding example we showed that the rejection region $\{0, 1\}$ for the binomially distributed $Y$ is UMP for $p = .5$ against $p < .5$, in the class of tests with $\alpha = \pi(.5) = .187$. The power function is

$$\pi(p) = P(Y = 0 \text{ or } 1 \mid p) = (1-p)^5 + 5(1-p)^4 p = (1-p)^4(1+4p).$$

This is a decreasing function of $p$ on $(0, 1)$, so its largest value on $p \geq .5$ is at $p = .5$. So $\alpha^* = \pi(.5) = .187$. Since $\{Y \leq 1\}$ has the greatest power at any alternative $p < .5$ among tests with $\alpha \leq .187$, it has greatest power at any alternative among tests with $\alpha^* = .187$. (The latter class is a subclass of the former.) Thus, it is UMP for $p \geq .5$ against $p < .5$. ∎

In view of the foregoing, it is useful to know when a power function is monotonic. A general class of distributions for which the power function of the likelihood ratio test for $\theta = \theta_0$ against $\theta = \theta_1$ is monotonic in $\theta$ is the exponential family defined in Example 8.3e. [This family includes normal, Bernoulli, geometric, and exponential distributions, among others.][4]

---

[3]A set of numbers may not have a "largest" value, as for example the set of numbers strictly less than 1. But there will always be a least upper bound, or supremum. The least or smallest upper bound of the numbers less than 1 is 1.

[4]For a proof, see Lindgren [15].

Suppose $(X_1, ..., X_n)$ is a random sample from a population with p.d.f. $f(x \mid \theta) = B(\theta)h(x)e^{\theta x}$. The power function of the rejection region $\sum X_i > K$ is a nondecreasing function of $\theta$.

## Problems

* **11-18.** Consider testing $\mu = 0$ against $\mu = 1$, based on a random sample of size $n$ from $\mathcal{N}(\mu, 1)$.
  (a) Find the most powerful rejection regions in terms of $\overline{X}$.
  (b) Find $\alpha$ and $\beta$ as functions of the rejection boundary in (a), when $n = 4$.
  (c) Sketch a plot of $\beta$ versus $\alpha$ for the most powerful rejection regions.

* **11-19.** Find the uniformly most powerful tests for $\mu = 0$ against $\mu > 0$, based on a random sample of size $n$ from $\mathcal{N}(\mu, 1)$.

**11-20.** For a random sample of size $n$ from Exp($\lambda$),
  (a) find the most powerful tests for $\lambda = 1$ against $\lambda = 2$.
  (b) find the form of the uniformly most powerful tests for $H_0: \lambda = 1$ against $H_A: \lambda > 1$.

* **11-21.** Find the most powerful tests for $m = 1$ vs. $m = 2$ based on a random sample of size $n$ from Poi($m$).

**11-22.** The random variable $Z$ has one of the two discrete distributions shown below.
  (a) How many different rejection regions (for a single observation on $Z$) are possible?
  (b) Find the most powerful tests among those in (a).
  (c) Find the most powerful test of size $\alpha = .3$.
  (d) Find the most powerful (randomized) test of size $\alpha = .2$.

| $z$ | $z_1$ | $z_2$ | $z_3$ | $z_4$ | $z_5$ |
|-----|-------|-------|-------|-------|-------|
| $f_1(z)$ | .2 | .3 | .1 | .3 | .1 |
| $f_2(z)$ | .3 | .1 | .3 | .2 | .1 |

  (e) Construct an $\alpha$-$\beta$ plot representing the most powerful and least powerful tests (pure and randomized).

## 11.8    Likelihood Ratio Tests

In §11.5, when deciding between two simple hypotheses $\theta_0$ and $\theta_1$, we employed the likelihood ratio statistic:

$$\Lambda = \frac{L(\theta_0)}{L(\theta_1)}.$$

This statistic orders the sample points according to the value of $\Lambda$, and we showed that rejection regions of the form $\Lambda(\mathbf{X}) < K$ are most powerful. We now extend the idea of using likelihood comparisons to a general method of constructing tests for cases in which either the null or the alternative hypothesis (or both) is composite.

The idea of the **generalized likelihood ratio,** introduced in §10.2, is that we compare the best "explanation" of a given set of data among models in $H_0$ with the best explanation in $H_A$, where "best" means having maximum likelihood. A generalized likelihood ratio test often turns out to have good properties, especially when sample sizes are large. Indeed, in many situations the likelihood ratio method leads to rejection regions that we have already introduced on other intuitive grounds. But it also provides reasonable tests in cases where it is not so obvious how to proceed.

Consider a family of distributions indexed by $\theta$, and take as $H_0$ a subset of the parameter space. The alternative is the complement of that subset. Let $\widehat{\theta}$ denote the m.l.e. of $\theta$, and let $\widehat{\theta}_0$ denote the $\theta$ that maximizes the likelihood *within* $H_0$:

$$L(\widehat{\theta}) = \sup_{H_0 \cup H_A} L(\theta), \ \ L(\widehat{\theta}_0) = \sup_{H_0} L(\theta).$$

The generalized likelihood ratio is then

$$\Lambda = \frac{L(\widehat{\theta}_0)}{L(\widehat{\theta})}. \tag{1}$$

Clearly, $\Lambda \leq 1$, and $\Lambda = 1$ if $\widehat{\theta}_0 = \widehat{\theta}$. When $\Lambda$ is very small, the best explanation of the data to be found in $H_0$ is much worse than the best available explanation—the one in $H_A$. So we take regions of the form $\Lambda < K$ as rejection regions for the likelihood ratio statistic.

**Example 11.8a** | ## The $t$-Test as a Likelihood Ratio Test

Consider testing $H_0$: $\mu = \mu_0$ against $H_A$: $\mu \neq \mu_0$, where $\mu$ is the mean of a normal variable $X$ with unknown variance $\sigma^2$. The population distributions are indexed by the parameter pair $(\mu, \sigma^2)$, and $H_0$ restricts $\mu$ but leaves $\sigma^2$ as a free parameter. The likelihood function is

$$L(\mu, \sigma^2) = (\sigma^2)^{-n/2} \exp\left\{ -\frac{1}{2\sigma^2} \sum (X_i - \mu)^2 \right\}. \tag{2}$$

This has maximum value at $(\overline{X}, V)$, where $V$ is the mean squared deviation of the sample values about their mean:

$$V = \frac{1}{n} \sum (X_i - \overline{X})^2.$$

The maximum value of $L$ is

$$L(\overline{X}, V) = V^{-n/2} \exp\left\{ -\frac{1}{2V} \sum (X_i - \overline{X})^2 \right\} = V^{-n/2} e^{-n/2}. \tag{3}$$

And this maximum is the denominator of $\Lambda$.

To find the numerator of $\Lambda$, we need to maximize $L$ when $\mu$ is held fixed at $\mu_0$. The log of the likelihood, when $\mu = \mu_0$, is

$$\log L(\mu_0, \sigma^2) = -\frac{n}{2} \log \sigma^2 - \frac{1}{2\sigma^2} \sum (X_i - \mu_0)^2.$$

The derivative with respect to $\sigma^2$ is

$$\frac{\partial}{\partial \sigma^2} \log L(\mu_0, \sigma^2) = -\frac{n}{2\sigma^2} + \frac{1}{2\sigma^4} \sum (X_i - \mu_0)^2,$$

which vanishes when $\sigma^2$ has the value

$$V_0 = \frac{1}{n} \sum (X_i - \mu_0)^2. \tag{4}$$

The maximum value of $L$ in $H_0$ is

$$L(\mu_0, V_0) = V_0^{-n/2} \exp \left\{ -\frac{1}{2V_0} \sum (X_i - \mu_0)^2 \right\} = V_0^{-n/2} e^{-n/2}. \tag{5}$$

The likelihood ratio statistic is the ratio (5) divided by (3):

$$\Lambda = \left( \frac{V_0}{V} \right)^{-n/2}. \tag{6}$$

To simplify it, we first rewrite $V_0$ using the identity

$$(X_i - \mu_0)^2 = (X_i - \overline{X})^2 + 2(X_i - \overline{X})(\overline{X} - \mu_0) + (\overline{X} - \mu_0)^2.$$

Summing on $i$ and dividing by $n$, we see that

$$V_0 = \frac{1}{n} \sum (X_i - \overline{X})^2 + (\overline{X} - \mu_0)^2 = V + (\overline{X} - \mu_0)^2.$$

With this substituted in (6), we obtain

$$\Lambda = \left\{ 1 + \frac{(\overline{X} - \mu_0)^2}{V} \right\}^{-n/2}.$$

This ratio is small when the second term in the braces is large. That term is proportional to $T^2$, the test statistic we used for a two-sided test of $\mu = \mu_0$ [see (1) in §10.4]:

$$T^2 = \frac{(\overline{X} - \mu_0)^2}{V/(n-1)}.$$

So, the likelihood ratio rejection region $\Lambda < K$ is equivalent to a two-sided rejection region for $T$: $|T| > K$. ∎

In this example we have seen that $\Lambda$ was expressible in terms of a statistic whose null distribution is familiar. This is not always the case. The null distribution of $\Lambda$ can be quite complicated. However, when $H_0$ is formed by fixing the value of one or more components of a vector parameter $\theta$, it has been shown that as $n$ becomes infinite, the distribution of the statistic $-2 \log \Lambda$ is asymptotically chi-square under $H_0$.[5] The number of degrees of freedom is equal to the number of parameters assigned specific values under $H_0$. (In the preceding example, $H_0$ assigns a value to the one parameter $\mu$, so d.f. $= 1$.)  The next example illustrates how this fact can be used.

**Example 11.8b**

## A Two-Sided Test for $p = p_0$

Consider a random sample from Ber($p$) and the problem of testing $H_0$: $p = .5$ against the alternative $p \neq .5$.  Suppose we observe 100 independent trials and record 58 successes. The likelihood function (see Example 8.2b) is

$$L(p) = p^{58}(1 - p)^{42}.$$

This has its maximum value (over all $p$ in the unit interval) at $p = .58$, the sample relative frequency of successes. That maximum value is

$$L(\hat{p}) = (.58)^{58}(.42)^{42}.$$

Under $H_0$, $L$ has only one value:

$$L(.5) = (.5)^{58}(.5)^{42} = .5^{100}.$$

The test statistic (1) is

$$\Lambda = \frac{L(.5)}{L(\hat{p})} = \frac{.5^{100}}{(.58)^{58}(.42)^{42}},$$

and

$$-2 \log \Lambda = -2 \left[100 \log .5 - 58 \log .58 - 42 \log .42\right] \doteq 2.57.$$

The $P$-value, from Table Vb, is .109, which is not significant at the .05 level.

In a $Z$-test for this same problem, we'd calculate

$$Z = \frac{.585 - .50}{\sqrt{.5 \times .5/100}} = 1.70,$$

so $Z^2 = 2.89$. Under $H_0$, $Z \approx \mathcal{N}(0, 1)$, so $Z^2 \approx \text{chi}^2(1)$—the same asymptotic distribution as that of $-2 \log \Lambda$! (This is not just a happenstance, nor is the closeness of 2.89 to 2.57.)[6]  ∎

Here we have used likelihood ratio tests as rules for deciding between a null and an alternative hypothesis, but we saw in Chapter 10 that the likelihood ratio statistic $\Lambda$ is also useful in problems where a decision is not required, as in Example 10.2d: The smaller the value of $\Lambda$, the stronger the evidence against $H_0$. And when, as in the

---

[5]See Wilks [20], 419ff.
[6]See Lindgren [15], 367.

above example, $\Lambda$ is a monotonic function of some more familiar statistic whose distribution is known, we can find a $P$-value from that distribution. But in any case, if $n$ is large, we can find a $P$-value for $-2\log\Lambda$ in the chi-square table.

In the examples in which we have used it, the likelihood ratio approach has not given us new tests or new test statistics. And it, too, is based on intuition—this time, on the intuition of "likelihood." However, it is perhaps of interest that certain tests to which we were led on intuitive grounds are in fact likelihood ratio tests, or asymptotically equivalent to likelihood ratio tests; this means that they share some of the good large-sample properties of likelihood ratio tests.[7]  We'll see in Chapter 13 that the likelihood ratio statistic is quite useful in analyzing contingency tables.

## Problems

**∗ 11-23.**  Given a random sample from $\mathcal{N}(\mu, 1)$, find the likelihood ratio tests for $\mu = \mu_0$ against $\mu \neq \mu_0$, as rejection regions for the sample mean.

**11-24.**  Find the likelihood ratio rejection regions for testing $\sigma = 1$ against $\sigma = 2$, using a random sample of size $n$, when the population is normal with unknown mean.

**11-25.**  Find the likelihood ratio tests for $\lambda = \lambda_0$ against $\lambda \neq \lambda_0$, based on a random sample from $\text{Exp}(\lambda)$.

**∗ 11-26.**  A single observation $Z$ takes on one of four values. Its distribution is one of the three identified as $\theta_1$, $\theta_2$, and $\theta_3$ in the following table:

|            | $z_1$ | $z_2$ | $z_3$ | $z_4$ |
|------------|-------|-------|-------|-------|
| $\theta_1$ | .2    | .3    | .1    | .4    |
| $\theta_2$ | .6    | .1    | .1    | .2    |
| $\theta_3$ | .3    | 0     | .4    | .3    |

Find all the likelihood ratio rejection regions for testing $\theta = \theta_1$ against the alternative $H_A$: $\theta = \theta_2$ or $\theta_3$.

**11-27.**  Find the likelihood ratio test for $\theta = 0$ against $\theta \neq 0$ based on a single observation from a Cauchy distribution with median 0. (The likelihood function is $[1 + (X - \theta)^2]^{-1}$.)

**∗ 11-28.**  Consider a random sample of size $n$ from $\mathcal{N}(\mu, 1)$ for testing $H_0$: $\mu \leq 0$ against $H_A$: $\mu > 0$. The likelihood function [from (2) of §8.2] is

$$L(\mu) = \exp\left\{ -\frac{n}{2}(\mu - \overline{X})^2 \right\}.$$

**(a)**  Sketch $L(\mu)$ for a negative value of $\overline{X}$ and also for a positive value of $\overline{X}$. In each case, deduce from the graphs the location of the maximum of $L$ over $\mu \leq 0$, and the maximum over all $\mu$.

**(b)**  Calculate $\Lambda$ for an $\overline{X} < 0$ and for an $\overline{X} > 0$.

---

[7]See Wilks [20], 419ff.

**(c)** Show that the rejection region for $\Lambda$ of the form $\Lambda < K'$ is equivalent to a rejection region for $\overline{X}$ of the form $\overline{X} > K$.

**11-29.** The function $f(z \mid \theta)$ is shown in the accompanying table. We want to test $H_0 \colon \theta = 0$ against the composite hypothesis $0 < \theta < 1$.

| $Z$ | $z_1$ | $z_2$ | $z_3$ | $z_4$ |
|---|---|---|---|---|
| $\theta = 0$ | $1/12$ | $1/12$ | $1/6$ | $2/3$ |
| $0 < \theta < 1$ | $\theta/3$ | $1 - \theta/3$ | $1/2$ | $1/6$ |

**(a)** Give the likelihood ratio test with $\alpha = 1/6$ and find its power function.

**(b)** For the test with rejection region $R = \{z_3\}$, find $\alpha$ and the power function. How does this test compare with the likelihood ratio test in (a)?

[Moral: A test derived as a likelihood ratio test is not always the best test, except in the case of simple null hypothesis against a simple alternative.]

**11-30.** Find the likelihood ratio tests for $\mu = \mu_0$ against $\mu = \mu_1$, based on a random sample from $\mathcal{N}(\mu, \theta)$. Use the modified ratio

$$\Lambda^* = \frac{\sup_{\theta} L(\mu_0, \theta)}{\sup_{\theta} L(\mu_1, \theta)}.$$

For the special case $\Lambda^* < 1$, give the rejection region in terms of $\overline{X}$.

## 11.9    Bayesian Testing

In the classical approach presented so far, the choice of $\alpha$ is quite a subjective matter. The choice would generally hinge on the costs of wrong decisions, but there is no provision for introducing such costs except in an informal, subjective manner—as suggested in R. A. Fisher's statement quoted in Chapter 10, that one "gives his mind to each particular case in the light of his evidence and his ideas."

In the Bayesian approach, having assigned probabilities to hypotheses, we can determine the appropriate rejection region provided we assume a particular loss function $\ell(\theta, a)$, the loss incurred upon taking action $a$ when nature is in the state $\theta$.

Consider first the case of a simple null hypothesis $H_0$ and a simple alternative $H_1$. Let $f_0$ and $f_1$ denote the corresponding p.d.f.'s of a sample $\mathbf{X} = (X_1, \ldots, X_n)$. Suppose there is no loss for a correct choice of action, and for an incorrect choice let the losses be as follows:

$$A = \ell(H_0, \text{rej. } H_0), \quad \text{and} \quad B = \ell(H_1, \text{acc. } H_0).$$

To find the Bayes losses—the expected values of the losses with respect to the distribution of "$\theta$"—we need prior probabilities for the two competing hypotheses.

Suppose these are

$$g_0 = P(H_0), \text{ and } g_1 = P(H_1) = 1 - g_0.$$

The posterior probabilities are then

$$h(H_0 \mid \mathbf{X}) = \frac{g_0 f_0(\mathbf{X})}{g_0 f_0(\mathbf{X}) + g_1 f_1(\mathbf{X})}, \quad h(H_1 \mid \mathbf{X}) = \frac{g_1 f_1(\mathbf{X})}{g_0 f_0(\mathbf{X}) + g_1 f_1(\mathbf{X})}.$$

The Bayes losses (average losses with respect to the posterior) are

$$B(\text{rej.} \mid \mathbf{X}) = A \cdot h(H_0 \mid \mathbf{X}) + 0 \cdot h(H_1 \mid \mathbf{X}),$$

$$B(\text{acc.} \mid \mathbf{X}) = 0 \cdot h(H_0 \mid \mathbf{X}) + B \cdot h(H_1 \mid \mathbf{X}).$$

The ratio of the Bayes loss for rejecting and the Bayes loss for accepting $H_0$ is

$$\frac{B(\text{rej. } H_0 \mid \mathbf{X})}{B(\text{acc. } H_0 \mid \mathbf{X})} = \frac{A g_0 f_0(\mathbf{X})}{B g_1 f_1(\mathbf{X})}.$$

Rejecting $H_0$ is the better action when this ratio is less than 1, or when

$$\Lambda = \frac{f_0(\mathbf{X})}{f_1(\mathbf{X})} < K = \frac{B g_1}{A g_0}.$$

---

> For testing a simple $H_0$ ($f_0$) against a simple $H_1$ ($f_1$), a Bayes procedure is a Neyman-Pearson test—is most powerful; and a Neyman-Pearson rejection region $[f_0/f_1 < K]$ is a Bayes test for some choice of a prior.

---

Suppose next that $H_0$ and $H_A$ are composite hypotheses about a parameter $\theta$—that is, about sets of $\theta$-values. We assume no penalty for correct decisions; for incorrect decisions, suppose the losses are

$$\ell(H_0, \text{rej. } H_0) = a(\theta), \quad \ell(H_A, \text{acc. } H_0) = b(\theta),$$

where $a$ and $b$ are nonnegative functions. (An especially simple case is that in which these functions are constant.) When the current distribution for $\theta$ has the p.d.f. $g(\theta)$, the Bayes losses are

$$B(\text{acc. } H_0) = \int_{H_A} b(\theta) g(\theta) \, d\theta, \quad \text{and } B(\text{rej. } H_0) = \int_{H_0} a(\theta) g(\theta) \, d\theta.$$

## Example 11.9a | Adjustment of a Dispenser (Revisited)

We return to the filling machine of earlier examples, one that is adjusted so that the mean weight in a bag is 16.0 oz. The two actions are to let the process continue, and to stop the production line to adjust the level. If the process is allowed to continue, there is an obvious loss to the company if the mean weight is over 16.0; if the mean weight is under 16.0, customers are shortchanged, with a loss that would be more difficult to determine. If the process is stopped for readjustment, there is a fixed cost whether $\mu$ has changed either way or not changed. Suppose the loss function were determined to be

$$\ell(\mu, \text{continue}) = \begin{cases} 2(16 - \mu), & \mu \leq 16, \\ \mu - 16, & \mu > 16 \end{cases}$$

$$\ell(\mu, \text{adjust}) = a.$$

Suppose at this point one's prior for $\mu$ is $\mathcal{N}(16.0, .002)$. The bags in a sample of size 5 are checked, resulting in the statistic $\overline{X} = 16.40$. If we assume that $\sigma = .05$, the posterior is normal with mean 16.32 and precision 2500 (s.d. .02). The Bayes loss involved in accepting $H_0$ (no need to readjust) is

$$B(\text{accept}) = E[\ell(\mu, \text{accept})] = 2\int_{-\infty}^{16} (16 - \mu)h(\mu)\,d\mu + \int_{16}^{\infty} (\mu - 16)h(\mu)\,d\mu.$$

Evaluating the integrals is not a particularly pleasant prospect, but it can be done (at least numerically). This Bayes loss would be compared with the Bayes loss involved in stopping to make the adjustment: $E(a) = a$, to see which action to take. ■

## Problems

**∗ 11-31.**   Consider testing $H_0$: $p \leq .5$ against $p > .5$, based on a random sample from Ber($p$). We assume losses of 2 if we accept $H_0$ when $p > .5$, and 3 if we reject $H_0$ when $p \leq .5$, and 0 if we take the correct action. Suppose, as a prior, we assume $p \sim$ Beta$(1, 1)$, which is $\mathcal{U}(0, 1)$.
   (a)   Determine the no-data Bayes action.
   (b)   Determine the Bayes action, given nine successes in ten trials.

**11-32.**   To test $H_0$: $p = p_0$ against $H_A$: $p = p_1$, when $p_0 = .5$ and $p_1 = .9$, we carry out ten independent trials of Ber($p$). Our prior for $(p_0, p_1)$ is $(.7, .3)$—we feel that there are only three chances in ten that $p = .9$. Suppose it turns out that there are eight successes in our ten trials.
   (a)   Find the posterior distribution of $(p_0, p_1)$, given the data.
   (b)   If losses are assumed to be 5 for a type I error and 3 for a type II error (and 0 for a correct decision), which is the Bayes action—to reject or to accept the null hypothesis?

**∗ 11-33.**   Consider testing $p \leq .5$ against $p > .5$, where using a random sample from Ber($p$). Suppose losses are assumed to be $\ell(p, a) = 0$ if $a$ is the correct action, 3 if we reject $H_0$ when $p \leq .5$, and 2 if we accept $H_0$ when $p > .5$. Find the Bayes action if our prior for $p$ is Beta$(7, 3)$ and we observe three success in ten independent trials.

**11-34.**   Another approach to a Bayes test of $f_0(\mathbf{x})$ against $f_1(\mathbf{x})$: Given losses as in §11.6 [$A$, for rejecting $H_0$ when $H_0$ is true, $B$, for accepting $H_0$ when it is false], consider a proposed rejection region $R$. Define *risks* $L_0$ and $L_1$ for $R$ as average losses (averaging with respect to the distribution of **X**):

$$L_i(R) = \ell(H_i, \text{rej. } H_0)P_i(\text{rej. } H_0) + \ell(H_i, \text{acc. } H_0)P_i(\text{acc. } H_0), \quad i = 0, 1,$$

where $P_i$ means probability calculated using $f_i$. Thus,

$$L_0(R) = A\alpha + 0 \quad \text{and} \quad L_1 = 0 + B\beta.$$

The *Bayes risk* is the average of these with respect to the prior:

$$B(R) = g_0 L_0(r) + g_1 L_1(R) = (g_0 A)\alpha + (g_1 B)\beta.$$

Show geometrically that the region $R$ that achieves the minimum Bayes risk is represented in the plot of $(\alpha, \beta)$ as a point on the lower-left boundary of the set of all possible tests—and is therefore most powerful. [*Hint:* On the $(\alpha, \beta)$ plot, draw lines along which the Bayes risk is constant.]

## Review Problems

**11-R1.**  For testing $\mathcal{N}(0, 1)$ against $\mathcal{N}(0, 4)$, with a random sample of size $n$,
  (a)  find all most powerful tests.
  (b)  among the tests in (a), find one with $\alpha = .10$ when $n = 4$.
  (c)  find $\beta$ for the test in (b).
  (d)  find values of $n$ and $K$ such that $\alpha = \beta = .01$.

**11-R2.**  Assuming a random sample of size 5 from $\text{Poi}(\lambda)$, and the problem of testing $\lambda = 1$ against $\lambda = 2$,
  (a)  find the Neyman-Pearson rejection regions in terms of $\sum X_i$.
  (b)  find the type I and II error sizes for the rejection region $\sum X_i > 7$.

**11-R3.**  Suppose we use the rejection region $\bar{X} > 6$, in a test of $\mu = 5$, where $\mu$ is the mean of a normal distribution with $\sigma^2 = 1$, and $n = 4$. Find the power function of the test.

**11-R4.**  Consider a random sample of size $n = 10$ from $\mathcal{N}(\mu, \sigma^2)$, and the critical region $S_X^2 > 4$.
  (a)  Find $\alpha$, in testing $\sigma^2 = 2$.
  (b)  Find and plot the power function of the test.

**11-R5.**  Referring to the setup of Problem 11-26, suppose losses are as follows:

|          | $\theta_1$ | $\theta_2$ | $\theta_3$ |
|----------|:----:|:----:|:----:|
| *Rej.* $H_0$ | 4 | 0 | 0 |
| *Acc.* $H_0$ | 0 | 2 | 1 |

Given the prior  $(.2, .5, .3)$ for $(\theta_1, \theta_2, \theta_3)$,
  (a)  find the posterior, for each $z_i$.
  (b)  find the expected posterior losses if we reject $H_0$ and if we accept $H_0$, for each $z_i$, and from these deduce the Bayes decision procedure. (This is of the form "if $Z = z_1$, then ...; if $Z = z_2$, then ..., etc.)

**11-R6.** Construct the likelihood ratio $\Lambda$ for testing $\theta = 1$ against $\theta \neq 1$, where $\theta$ is the parameter of a Rayleigh distribution (see Problem 9-R9), and describe the region of values of the m.l.e. $\hat\theta$ that corresponds to $\Lambda < K$.

**11-R7.** For given constants $x_i$, let $Y \sim \mathcal{N}(\beta x_i, \sigma^2)$ (as in Problem 9-48, but with $\alpha = 0$). Find the likelihood ratio $\Lambda$ for testing $\beta = 0$ against $\beta \neq 0$, and show that $\Lambda < K'$ is equivalent to $(\hat\sigma^2/\hat\sigma_0^2) < K$, where $\hat\sigma_0^2 = \frac{1}{n}\Sigma Y_i^2$.

**11-R8.** A categorical variable has three possible values $A_i$ with corresponding probabilities $p_i$. To test the hypothesis that the categories are equally likely, we do the experiment 60 times in independent trials and observe the frequencies 15, 15, 30.

(a) Write the likelihood function and evaluate it when $p_i \equiv 1/3$, and also when the $p_i$'s are replaced by their m.l.e.'s (15/60, 15/60, and 30/60, respectively).

(b) Find the value of $-2 \log \Lambda$ and a corresponding $P$-value. [The number of parameters assigned specific values is 2, since $p_3 = 1 - p_1 - p_2$.]

## Chapter Perspective

In this chapter we have considered tests of hypotheses as decision rules—rules set up *prior* to the collection of data. The type of rejection region used is based on a consideration of alternative hypotheses, prior to analysis of the data. Specific rejection limits are chosen to satisfy specifications of error sizes or power characteristics.

Tests as presented in this chapter are closely related to confidence intervals. We have seen that one $\alpha$-level decision rule for testing $\theta = \theta_0$ is to reject $H_0$ if a confidence interval at the confidence level $1 - \alpha$ does not include the value $\theta_0$. It turns out that, although we did not go into it, there is actually a complete duality—a family of $\alpha$-level rejection regions, one for each $\theta_0$, implies a $100(1 - \alpha)\%$ confidence interval.[8]

The notion of the power of a test can be helpful even when (as in the preceding chapter) an immediate decision is *not* required, and we only report the strength of evidence against $H_0$. When planning an experiment, an investigator would want to use a sample of such a size that a deviation from $H_0$ by a practically important amount would not be missed for the lack of sufficient data, a notion that involves power.

The likelihood ratio approach to testing provides a systematic method of devising tests for null and alternative hypotheses that may be composite. This intuitive method gives optimal tests when both hypotheses are actually simple; and more generally, the method usually yields good tests—at least for large samples.

In studying $\alpha$'s, $\beta$'s, and power functions for the various decision rules, how does one choose the one to use? Comparing rules with respect to power depends on which model (among those in $H_0$ and $H_A$) is assumed, but in the classical framework there is no provision for specifying how often $H_0$ and how often $H_A$ would be encountered. About all one can do is somehow think about the various errors, their sizes and consequences, and subjectively specify desired power characteristics to define a rule.

---

[8]See Lindgren [15], 310.

Seeing how that rule operates in practice may call for a readjustment of the constants that define the decision rule.

The Bayesian approach (§11.9) is to determine costs of wrong decisions, combine these with a distribution for unknown parameters, to obtain the decision rule with minimum expected loss.

We have introduced hypothesis testing in settings where there is a single sample for testing hypotheses about the population from which it was drawn. The next chapter takes up the important problem of testing hypotheses about two populations, using a sample from each.

# Comparing Two Populations

Suppose a new treatment is to be evaluated. Researchers continually propose new treatments (drugs, surgical techniques, chemical baths, fertilizers, teaching methods, and so on), as well as new formulas (for making bread, detergents, alloys, and the like). These are aimed at making some aspect of life less painful, more pleasant, more productive, or more profitable.

Sometimes a "treatment" is simply a set of circumstances that exist and may have resulted in an effect, although no treatment has been deliberately applied. For example, in some localities, being black may affect the punishment for a crime; being female may have an effect on income; smoking may affect health; and so on.

To assess the effects of a treatment or of a difference in treatments, we'd want to compare the population of responses of treated individuals or experimental materials with those that are not treated (or are given a second kind of treatment). Generally speaking, we can compare populations only by comparing samples from them.

When comparing populations, the usual null hypothesis is that they are the same. Using samples to test the hypothesis that two populations are the same in every respect can be difficult, so we usually restrict consideration to a single population parameter. For example, we ask whether the two populations have the same mean, or have the same proportion of successes, or have the same variance. We'll employ the labels 1 and 2 to refer to the two populations—as subscripts on parameters ($\mu$, $p$, $\sigma^2$), sample sizes ($n$), and statistics ($\overline{X}$, $\hat{p}$, $S^2$), to identify them with their respective populations.

We'll take up particular two-sample tests in later sections but begin with a discussion of some general aspects of evaluating treatments.

## 12.1 Treatment Effects

The response to a treatment in some settings is categorical—for example, life or death. For the success-failure type of response (as when a patient lives or dies), the treatment **effect** is the change in probability of success. When there are more than two categories (such as little, moderate, or complete relief from pain), the treatment "effect" is not as easily described.

In other settings, the response to a treatment is a numerical characteristic, such as blood pressure, hardness, yield, and the like. Responses will usually be different for different individuals or different pieces of experimental material. (Even the same individual may respond differently at different times.) So the treatment may increase some responses and decrease others; but if it increases the *average* response, we say there is a treatment effect. The *effect* is the average amount of increase.

**Example 12.1a** | **Lowering Blood Pressure**

A biochemist designs a new drug, intended to lower blood pressure. Once it has been shown to be safe when used on animals and then healthy human volunteers, it is used on patients with high blood pressure. Patients' blood pressures change after taking the drug; most fall, but some rise. The question is: Does the drug lower blood pressure on average? ∎

In evaluating a treatment, one compares it with no treatment or with some standard treatment. There are two populations to consider: the **treatment population**—the population of responses of all individuals who might be treated, and the **control population**—the population of responses of individuals who are not treated or who are given the standard treatment.

**Example 12.1b** | **Surgical Controls**

The following is quoted from an article in an American Airlines flight magazine[1] by the late Dr. William Nolen, author and surgeon:

> Before coronary by-passes became popular, another operation, called internal mammary artery ligation, was used in the treatment of angina pectoris. The internal mammary artery consists of two branches, one on either side of the breastbone; one of these goes toward the heart. The inventor of the procedure, a surgeon named Glover, would ligate, or tie off, one branch of the artery so that more blood would be shunted toward the heart. About 50 percent of his angina patients got better. It was a simple operation and could even be done under a local anesthesia. However, another surgeon decided to test it with some sham operations. He made a skin incision but did not tie off the internal mammary. He achieved about the same results as Dr. Glover—about 45 to 50 percent of his patients got better.

Without a control group, the real effect of a treatment cannot be assessed. ∎

When the distribution of responses in the control population, is well known, it may be sufficient to sample only from the treatment population. Testing is then a one-sample problem and is treated using the methods discussed in Chapter 10. However, many statisticians recommend always using a control sample in addition to the treatment sample. It is a mistake to assume that the distribution of responses in the control population is completely known; the experimental units involved in the cur-

---

[1]*American Way*, August, 1983.

rent experiment are apt to be special in some important respect. The next example illustrates this point.

## Example 12.1c | Standard Treatment as a Control

Suppose one is planning a clinical trial designed to evaluate a new therapy for cardiac rhythm disorders. An important question is how it compares with quinidine sulfate, a very old, standard treatment. The null hypothesis is that the new therapy is no more effective than quinidine. The literature contains many studies evaluating the effectiveness of quinidine. It is tempting to assume that its average effectiveness is known to be the average in these studies.

The results for quinidine in these studies vary greatly—from 20% of the patients showing marked improvement to 65% showing the same level of improvement. These studies had different entry criteria; and even when the criteria are the same, two different clinicians may admit different types of patients. For example, one study may involve much sicker patients than another. It is difficult or impossible to determine which study group, if any, provides an appropriate comparison for the study being planned. The safest approach in a new study is to include a control group, assigning some patients (randomly) to the new treatment and the rest to quinidine.    ■

Randomizing some subjects to a treatment and the remainder to no treatment is the safest way to ascertain the effect of the treatment, but this may be impossible. For example, it could be considered unethical to assign patients who are terminally ill to a placebo when there is a treatment that might be effective. An alternative is to use the results for patients treated previously by the same clinicians. These are called *historical controls.* Another possibility is to cull patients from studies in the literature who match those in the current study—*literature controls.* The use of historical or literature controls is a controversial matter. Some statisticians claim that results based on such controls are invalid. In any case, they can serve an important function when randomized controls are not possible.

Studies that are planned before subjects are selected and treatments administered are called *prospective.* Studies that involve searching existing records of treatments and results are *retrospective.* A study may have prospective and retrospective aspects. For instance, two active treatments may be compared prospectively, but each may be compared to no treatment using historical controls.

An advantage of a prospective study is that subjects can be assigned to treatment or control independent of characteristics (such as severity of the disease) that may influence responses. The assignment should be made *randomly* so that these characteristics, including any that are unrecognized by the experimenter, tend to balance between treatment and control. A table of random digits can be used to make the assignment: Start at an arbitrary point in the table and assign each individual in sequence to treatment or control, depending on whether the next digit is even or odd. This does not guarantee equal sample sizes, but there are various ways of achieving equal sample sizes for the treatment and control groups if this is important. For example, assign each odd-numbered individual at random to treatment or control and assign the next (even-numbered) individual to the other group. (There may be selection bias if the investigator knows that the treatment is used on the odd-numbered individuals and can choose the next individual to be treated.)

### Example 12.1d │ Smoking and Health

Many studies have investigated the effect of smoking on health. Some of these have compared smoking histories of diseased persons with those of healthy persons; others have compared disease histories of smokers with those of nonsmokers. Prospective studies are impossible, since one could not *assign* some people to the treatment (smoking) and others to the control. Tobacco interests and others argue that individuals prone to disease may also be apt to continue smoking. They may concede that there is a relationship, but they point out that this is not enough to establish smoking as the *cause* of the disease.                                                        ■

An investigator needs to answer such questions as the following. What data are important? How expensive are the needed measurements? For a control population, do we want active controls (a standard treatment) or passive controls (no treatment)? What levels of treatment should be used—how large a dose? Or in the case of intermittent feeding (Example 10b), how intermittent? How long should a treatment last?

When the experimental units are humans, the design must avoid the possibility that results are confounded with a "placebo effect." A **placebo** is a nontherapeutic "treatment"—a *non*treatment that appears to the subject to be identical to a treatment—for example, a pill that contains no active ingredient but resembles the real thing. It is well established in medicine that people sometimes respond because they think they're being treated; this response is the placebo effect.

A 1980 newspaper article, dealing with the role of doctors' hands in healing—the "human touch"—stated:

> If you tell someone a drug will relieve pain—and that person believes it—the odds are very good that the pain will be relieved. Solid scientific studies show that placebos relieve pain by releasing endorphins.

### Example 12.1e │ Nonsurgery

In the article quoted in Example 12.1b, Dr. Nolen goes on to explain the success rate of the sham operation—a kind of placebo. "You might consider this an example of the placebo effect." Whether this is placebo effect is not clear. Many patients get better without *either* treatment or placebo. The placebo effect can be isolated only by using a control group of individuals who were given neither treatment nor placebo. (What is thought to be a placebo effect is sometimes a *regression effect*—see §15.7.)          ■

Using a placebo for the control group has become common practice in clinical trials. Of course, the subjects should not know who is getting the real thing; such trials are called **blind.** (The laws of many countries require that human subjects be informed that they *may* get placebo therapy.) It is also best that the clinician not know who is getting which treatment, lest the clinician subconsciously give better collateral treatment to those in one of the groups. Experiments are considered more reliable when they are **double blind,** with neither patient nor clinician aware of who is getting the actual treatment.

---

**12.2**          ## Large-Sample Comparison of Means

---

Suppose we have two large, independent random samples, one from population 1 and one from population 2. We'll use these to test $H_0$: $\mu_1 = \mu_2$. In terms of the single parameter $\delta = \mu_1 - \mu_2$ introduced in §9.8, this null hypothesis can be written as $H_0$: $\delta = 0$.

In §9.8 we estimated $\delta$ using the difference between the corresponding sample means: $\bar{D} = \bar{X}_1 - \bar{X}_2$. We used the fact that this difference is approximately normal when the sample sizes $n_1$ and $n_2$ are large. We exploit this approximate normality in a "$Z$-test" for $H_0$.

To standardize the statistic $\bar{D}$, we need its expected value and standard deviation. These (from §9.8) are

$$E\bar{D} = \delta, \ \text{s.d.}(\bar{D}) = \sqrt{\frac{\sigma_1^2}{n_1} + \frac{\sigma_2^2}{n_2}}. \tag{1}$$

If the sample mean difference is approximately normal, the $Z$-score

$$Z = \frac{\bar{D} - \delta}{\sqrt{\frac{\sigma_1^2}{n_1} + \frac{\sigma_2^2}{n_2}}} \tag{2}$$

is approximately *standard* normal. When the population variances $\sigma_i^2$ are known, this is a statistic, and we again have a $Z$-test. But those variances are usually not known. So in place of the s.d. of $\bar{D}$, we use the standard error:

$$\text{s.e.}(\bar{D}) = \sqrt{\frac{S_1^2}{n_1} + \frac{S_2^2}{n_2}} \tag{3}$$

in constructing an approximately standardized mean difference:

---

$Z$-test for testing $H_0$: $\delta = 0$. Calculate

$$Z = \frac{\bar{D} - 0}{\text{s.e.}(\bar{D})} = \frac{\bar{X}_1 - \bar{X}_2}{\sqrt{\frac{S_1^2}{n_1} + \frac{S_2^2}{n_2}}}. \tag{4}$$

When $n_1$ and $n_2$ are large, $Z$ is approximately normal; find $P$-values in Table IIa of Appendix 1.

---

The $Z$-score (4) is the distance between the sample mean difference $\bar{D}$ and the population mean difference under $H_0$, measured (as usual) in numbers of standard errors. How large the sample sizes have to be for (4) to be approximately standard normal depends on the shape of the population; in most cases, sample sizes of 325 or more would do. (The case of smaller sample sizes will be taken up in §12.4.)

| Example 12.2a | ### Platelet Counts in Cancer Diagnosis |

In the study referred to in Example 7.2e, investigators were interested in whether blood platelet count might be useful in cancer diagnosis. They obtained platelet counts of 153 male cancer patients and of 325 healthy males. The results were as follows:

|                  | $n$ | $\overline{X}$ | $S$ |
|------------------|-----|-----|------|
| *Cancer patients* | 153 | 395 | 170  |
| *Healthy males*   | 35  | 235 | 45.3 |

The investigators' null hypothesis was $\delta = 0$, although this could be questioned as the correct null hypothesis for deciding whether platelet count discriminates between healthy persons and cancer patients.

Since a difference either way would be of interest, perhaps a two-sided alternative is appropriate: $\delta \neq 0$. The $Z$-score (4) is

$$Z = \frac{(395 - 235) - 0}{\sqrt{170^2/153 + 45.3^2/35}} \doteq 10.17.$$

In view of the alternative, large values of $|Z|$ are "extreme." The $P$-value is the area under the standard normal curve outside $Z = \pm 10.17$. From Table IIa it is seen to be .0000 (actually, about $10^{-24}$).

So, the observed difference is highly statistically significant. The very small $P$ signifies that the evidence against $\delta = 0$ is overwhelming—which agrees with what is quite evident in the sample histograms in Figure 7-9.     ∎

In §11.3 we considered the problem of choosing the size of a single sample so that a deviation from $H_0$ deemed practically significant would be likely to be picked up as statistically significant. Here we have to choose *two* sample sizes, $n_1$ and $n_2$. As before, we need some knowledge of population variability; but now we also have to decide the relative sizes of $n_1$ and $n_2$. In many cases, if $n_1 + n_2$ is fixed as an even integer, then $n_1 = n_2$ provides the greatest power. The ratio $n_1/n_2$ is an important variable to be specified in advance, but in the next example we'll take sample sizes to be equal for simplicity.

**Example 12.2b** | **Hypertension and Sodium Levels**

Consider planning an experiment to determine whether hypertensive patients have higher sodium levels than normotensive patients. We want to be fairly confident that our test will detect an average difference of 5 mEq/L. We take this to mean that the test should be powerful, having a high probability of rejecting $H_0$ (no difference in mean levels), if the mean difference is $\delta = 5$. Past experience suggests that the population s.d.'s are about $\sigma = 5$. Suppose we use equal sample sizes and require (i) $\alpha = .05$, and (ii) $\pi(\delta) = .80$, when $\delta = 5$. These conditions will determine a critical value and sample size.

We take the rejection region to be of the form $\bar{D} > K$, where $\bar{D}$ is the difference between sample means. The condition (i) on $\alpha$ is as follows:

$$1 - \alpha = .95 = P(\bar{D} < K \mid \delta = 0) = \Phi \left( \frac{K - 0}{5\sqrt{1/n + 1/n}} \right),$$

which means that

$$\frac{K \sqrt{n}}{5\sqrt{2}} = 1.645,$$

the 95th percentile of the standard normal $Z$. So $K \sqrt{n} = 11.63$.

The requirement (ii), that the power at $\delta = 5$ be .80, is a second condition on $K$ and $n$:

$$.80 = P(\bar{D} > K \mid \delta = 5) = 1 - \Phi \left( \frac{K - 5}{5\sqrt{2/n}} \right).$$

The $Z$-score in parentheses must then be $-.8416$, the 20th percentile of $Z$:

$$(K - 5)\sqrt{n/2} = -4.208.$$

Substituting $K \sqrt{n} = 11.63$ and solving for $n$, we find $\sqrt{n} = 3.52$, or $n = 12.4$. Rounding to the nearest larger integer yields $n = 13$. Substituting this $n$ in the first condition: $K \sqrt{12.4} = 11.63$, yields $K = 3.30$.

Should it turn out that the data are not consistent with the assumption that $\sigma = 5$, we may find that we used larger (if $\sigma < 5$) or smaller ($\sigma > 5$) sample sizes than needed to satisfy requirements (i) and (ii).  ■

Power calculations are most relevant before the study. People sometimes criticize a study that has found a statistically significant difference on the grounds that the sample sizes were not large enough to detect a clinically significant difference. However, if a result has turned out to be statistically significant, the sample sizes *were* large enough to reject $H_0$.

## 12.3      Comparing Proportions

We turn next to the case of Bernoulli populations, comparing two population proportions (or probabilities of success), $p_1$ and $p_2$. The null hypothesis of no difference is $H_0$: $p_1 = p_2$ or, equivalently, with $\delta = p_1 - p_2$,

$$H_0: \delta = 0. \tag{1}$$

[This notation is consistent with our use of $\delta$ as the difference in population means, since with the Bernoulli coding $(0 - 1)$, the population mean is the population proportion.]

If we assume independent random samples of sizes $n_1$ and $n_2$, the maximum likelihood estimator of $\delta$ is the difference between sample proportions:

$$\widehat{\delta} = \widehat{p}_1 - \widehat{p}_2, \tag{2}$$

where $\widehat{p}_i = Y_i/n_i$, and $Y_i$ is the number of successes in sample $i$. We saw in §9.8 that the large sample distribution of $\widehat{\delta}$ is approximately normal. To transform $\widehat{\delta}$ into a $Z$-statistic for testing $H_0$, we need its mean and variance under $H_0$. The expected difference in sample proportions is the difference in population proportions: $E(\widehat{\delta}) = \delta$. Thus, in view of (1),

$$E(\widehat{\delta} \mid H_0) = 0. \tag{3}$$

Because the samples are independent, the variance of $\widehat{\delta}$ is the sum of the variances of $p_1$ and $p_2$ [see (7) in §8.7]:

$$\operatorname{var}\widehat{\delta} = \operatorname{var}\widehat{p}_1 + \operatorname{var}\widehat{p}_2 = \frac{p_1(1 - p_1)}{n_1} + \frac{p_2(1 - p_2)}{n_2}. \tag{4}$$

Under the null hypothesis, $p_1 = p_2$; call this common value $p$. Then,

$$\operatorname{var}\widehat{\delta} = \frac{p(1 - p)}{n_1} + \frac{p(1 - p)}{n_2} = p(1 - p)\left(\frac{1}{n_1} + \frac{1}{n_2}\right).$$

With this and (3), we can construct a $Z$-score:

$$Z = \frac{\widehat{\delta} - 0}{\text{s.d.}(\widehat{\delta} \mid H_0)} = \frac{\widehat{p}_1 - \widehat{p}_2}{\sqrt{p(1 - p)(1/n_1 + 1/n_2)}}. \tag{5}$$

Under $H_0$, this is approximately standard normal when $n_1$ and $n_2$ are large.

As just defined, $Z$ is not a statistic—the "$p$" is not known. It can be estimated from either of the two samples or, better still, by pooling the two samples. Under $H_0$, the two samples together constitute a single random sample of size $n_1 + n_2$ from a Bernoulli population with parameter $p$. The obvious (and maximum likelihood) estimator of the common $p$ is the proportion $\widehat{p}$ in the *combined* sample:

$$\widehat{p} = \frac{Y_1 + Y_2}{n_1 + n_2} = \frac{n_1\widehat{p}_1 + n_2\widehat{p}_2}{n_1 + n_2}. \tag{6}$$

Thus, $\widehat{p}$ is a weighted average of the two sample proportions, with weights proportional to the sample sizes.

Replacing $p$ in (5) by the estimate $\widehat{p}$ from (6), we obtain an approximate $Z$-score (now, a statistic) as the test statistic:

---

Test for the equality of population proportions $\widehat{p}_1$ and $\widehat{p}_2$ using large, independent random samples:  Calculate

$$Z = \frac{\widehat{p}_1 - \widehat{p}_2}{\sqrt{\widehat{p}(1 - \widehat{p})(1/n_1 + 1/n_2)}}, \tag{7}$$

where $\widehat{p}$ is the pooled estimate (6), and find the $P$-value in Table IIa.

---

**Example 12.3a** | ### Windowlessness

Some researchers in the field of behavior hypothesized that visual decor in windowless offices would be more nature-oriented than in windowed offices. Their study of the visual material used to decorate offices at the University of Washington found nature-dominated themes in 134 of 195 windowless offices and in 45 out of 82 windowed offices.[2]

Let $p_1$ denote the proportion of windowless offices with nature-dominated themes, and $p_2$ the corresponding proportion in windowed offices. With $\delta$ denoting the difference $p_1 - p_2$, the null hypothesis tested to be is $H_0$: $\delta = 0$. The "research hypothesis" is one-sided—$H_A$: $\delta > 0$.

The test statistic is the difference in sample proportions:

$$\widehat{p}_1 - \widehat{p}_2 = \frac{134}{195} - \frac{45}{82} \doteq .69 - .55 = .14.$$

Under $H_0$, $p_1 = p_2 = p$, and the maximum likelihood estimate of $p$ is

$$\widehat{p} = \frac{134 + 45}{195 + 82} = \frac{179}{277}.$$

Substituting in (7), we obtain

$$Z = \frac{\frac{134}{195} - \frac{45}{82}}{\sqrt{\frac{179}{277} \cdot \frac{98}{277}\left(\frac{1}{195} + \frac{1}{82}\right)}} \doteq \frac{.1384}{.06293} \doteq 2.20.$$

---

[2] J. Heerwager and G. Orians, "Adaptations to windowlessness," *Environment and Behavior 18* (1986), 623–39.

Since (in view of $H_A$) large values of $Z$ are extreme, the $P$-value is the area under the standard normal curve to the right of 2.20: $P = \Phi(2.2) = .0139$. The observed difference in proportions of .14 is hard to explain as sampling variability—there is moderate evidence against $H_0$ and supporting $H_A$. ∎

When sample sizes are small, or when one or the other of the sample proportions is small, the normal approximation for $Z$ is apt to be rather inaccurate. Although the null hypothesis does not specify $p$, we got around this by replacing it in the $Z$-score with the m.l.e. $\hat{p}$. This amounts to assuming that $Y = Y_1 + Y_2$ is given. Taking the same approach, we can find an exact conditional $P$-value, given $Y$, using the conditional distribution of (say) $Y_1$ given $Y = y$.

The joint p.f. of the pair $(Y_1, Y_2)$ is the product of binomial p.f.'s, given that the samples are independent:

$$f(y_1,\, y_2) = \binom{n_1}{y_1} p_1^{y_1}(1-p_1)^{n_1-y_1} \cdot \binom{n_2}{y_2} p_2^{y_2}(1-p_2)^{n_2-y_2}.$$

Under $H_0$: $p_1 = p_2 = p$, the distribution of $Y = Y_1 + Y_2$ is $\mathrm{Bin}(n_1 + n_2,\, p)$. When it is given that $Y = y$, the conditional distribution of $Y_1$ is hypergeometric, with p.f.

$$f_{Y_1}(y_1 \mid Y = y) = \frac{P(Y_1 = y_1,\, Y_2 = y - y_1)}{P(Y = y)} = \frac{\binom{n_1}{y_1}\binom{n_2}{y_2}}{\binom{n_1+n_2}{y}}.$$

(This result was obtained in Problem 4-20.) In the next example we see how to use this distribution for finding exact $P$-values. Doing so is referred to as *Fisher's exact test.*

**Example 12.3b** | **Pets as Therapy**

Example 2a referred to a study investigating the possibility that having a pet might be good therapy for patients with heart disease. The comparison was based on whether or not a patient was alive one year after a heart attack. The study included 92 patients hospitalized for a heart attack or other serious heart disease. It was found that 3 of the 53 patients who had a pet died within a year after release, whereas 11 of 39 who did not have a pet died within a year. Let $\hat{p}_i$ be the sample proportion, and $Y_i$ the number who died within a year ($i = 1, 2$).

We find an exact $P$-value using the fact that the conditional distribution of $Y_1$, given $Y = 3 + 11 = 14$, is hypergeometric. It is the tail-probability $P(Y_1 \leq 3)$ in that distribution:

$$P(Y_1 \leq 3 \mid Y = 14) = \frac{\binom{53}{3}\binom{39}{11} + \binom{53}{2}\binom{39}{12} + \binom{53}{1}\binom{39}{13} + \binom{53}{0}\binom{39}{14}}{\binom{92}{14}} \doteq .00358.$$

For comparison we'll also calculate a $P$-value using the normal approximation, given by (7):

$$Z = \frac{\frac{3}{53} - \frac{11}{39}}{\sqrt{\frac{14}{92} \cdot \frac{78}{92}\left(\frac{1}{53} + \frac{1}{39}\right)}} \doteq -2.98.$$

The corresponding $P$-value (from Table IIa) is .0014. In this case the approximation is poor because the observed relative frequency 3/53 is very small.

The observed difference in proportions offers strong evidence against the hypothesis of equality of population proportions. However, we cannot jump to the conclusion that having a pet is the *cause* of the higher survival proportion among pet owners, even though we reject $\delta = 0$. (This typifies a fundamental problem with retrospective studies.) There may be factors that lead a person to have a pet which are also factors that lead to survival. For example, a healthier person may feel more capable of caring for a pet. A prospective randomized trial is not out of the question here, but we know of none that has been conducted. ∎

In §13.5 we'll extend the problem of comparing proportions to that of comparing discrete populations with finitely many categories.

## Problems

∗ **12-1.**   A research study on fluoride in toothpaste was conducted using Colgate's "MFP" formula, and a leading stannous fluoride (SF) toothpaste. Data on the number of new cavities over a three-year period are summarized as follows:[3]

|       | $n$ | $\overline{X}$ | $S$ |
|-------|-----|------|------|
| MFP   | 208 | 19.98 | 10.6 |
| SF    | 201 | 22.39 | 11.96 |

What conclusion can be drawn concerning the mean difference in number of cavities between the MFP and SF populations generally?

**12-2.**   Two types of aggregate are tested for thermal conductivity, with 25 observations for each type, with these results:

|             | $\overline{X}$ | $S$ |
|-------------|------|------|
| Low cost    | .485 | .180 |
| Higher cost | .372 | .160 |

Before choosing the higher-cost type, a certain buyer wants to be convinced that it has mean conductivity at least .05 less than that of the low-cost type. Do the data provide such evidence? (Test $H_0: \mu_1 - \mu_2 = .05$ against the alternative $H_A: \mu_1 - \mu_2 > .05$.)

∗ **12-3.**   In a comparison of pain and activity levels for good and poor sleepers, researchers report these data on hours of activity:[4]

---

[3]S. F. Frankl and J. E. Alman, "Report of three-year clinical trials...," *J. Oral Therapeutics & Pharmacol.* 4 (1968), 443–49.

[4]I. Pilowsky, I. Crettenden, and M. Townley, "Sleep disturbance in pain clinic patients," *Pain* 26 (1985), 27–33.

|       | $n$ | $\overline{X}$ | $S$ |
|-------|-----|------|-----|
| *Good* | 28  | 10.7 | 4.8 |
| *Poor* | 70  | 8.6  | 4.8 |

Test the hypothesis of no average difference in hours of activity between good and poor sleepers.

* **12-4.** Poll A uses a random sample of size 1000 and reports that 42% are opposed to an amendment prohibiting abortions. Poll B uses the identically worded question in a random sample of size 1500, reporting 39% opposed. Assuming independent samples, test the hypothesis that the populations sampled have the same proportions of those who oppose the amendment.

**12-5.** Does the response rate on a questionnaire depend on whether the cover letter is a form letter or is semipersonal? In one study, it was found that 225 among 1022 receiving the form letter responded, and 325 of 1018 who received the semipersonal letter responded.[5] Test the hypothesis that the proportion of responders is the same for both types of cover letters.

* **12-6.** The proportion of smokers in the sample of male students listed in Table 7-1 is 8/56, and the proportion in the sample of female students listed in Table 7-5 is 5/48. Assuming these to be independent random samples, test the hypothesis that the proportion of smokers is the same for female students and male students (in the populations sampled).

**12-7.** Show that the pooled estimate of $p$ given by (6) of §12.3 is the m.l.e. of $p$ under $H_0$: $p_1 = p_2 = p$.

* **12-8.** A study found that of 238 individuals with coronary heart disease (CHD), 145 smoked cigarettes; of 476 in the same age group but with no CHD, 192 smoked cigarettes.[6] Test the hypothesis that the proportion of cigarette smokers is the same in the population of individuals with CHD as in the population with no CHD.

**12-9.** The *New York Times* (February 15, 1983) carried a report of a study which concluded that a medicine commonly used to treat middle-ear infections in young children is no more effective than a placebo. Of 278 children who took the drug, 25% had no ear infection after four weeks, whereas 24% of the 275 who received a placebo had no infection. (On the other hand, those who took the drug experienced "significantly more" side effects—mild sedation and weakness.) In view of the sample proportions, is the study's conclusion warranted?

* **12-10.** In a study on oral hygiene, 30 subjects used a test compound, and 34 subjects used a placebo.[7] Improvements in an oral hygiene index are as follows:

Test:     10, 15, 6, 10, 11, 3, 8, 8, 3, 13, 10, 9, 8, 9, 8,
          4, 10, 15, 11, 5, 14, 7, 8, 8, 2, 13, 6, 2, 7, 3.
Placebo:  5, 6, 4, 3, 3, 5, 6, 4, 4, 2, 0, 7, 0, 3, 2, 2, 3, 6,
          0, 3, − 1, 1 6, 6, 8, 2, 12, 24, 5, 3, 3, 13, 4, 3.

Test the hypothesis that the test compound is no more effective, on average, than the placebo.

---

[5] M. J. Matteson, "Type of transmittal letter and questionnaire color as two variables influencing response in a mail survey," *J. Applied Psych.* 59 (1974), 535–36.

[6] J. N. Morris et al., "Vigorous exercise in leisure-time and the incidence of coronary heart-disease," *The Lancet* (Feb. 17, 1973), 333–37.

[7] Zinner, Duany, and Chilton, *Pharmac. & Therapeutics in Dentistry I* (1970), 7–15.

**12-11.** In the magazine *Natural History* (September 1988), an article on the burrowing owl in the Columbia River basin reports that of 25 nests lined with dung, only two were lost to badgers, whereas 13 of 24 unlined nests were destroyed by these carnivores. Find the *P*-value for a test of the hypothesis that lining with dung does not affect destruction of owl nests by badgers,

    **(a)**   using a normal approximation.

    **(b)**   using an exact calculation based on the hypergeometric distribution.

---

## 12.4      Two-Sample *t*-Tests

---

In §12.2, we based a large-sample test of the hypothesis $\delta = 0$ (that is, $\mu_1 = \mu_2$) on the sample mean difference, $\bar{D} = \bar{X}_1 - \bar{X}_2$, using the statistic

$$\text{s.e.}(\bar{D}) = \sqrt{\frac{S_1^2}{n_1} + \frac{S_2^2}{n_2}} \tag{1}$$

as the standard error in the denominator of a "*Z*-score." However, when $n_1$ or $n_2$ is small, we encounter the same problem as in the one-sample case (§10.4): The variability in the sample standard deviations introduces more variability in the *Z*-score than is represented by the standard normal curve. As in the one-sample case, we can use the *t*-distribution to take this into account—under the assumption of normality of the populations.

A further assumption of equal population variances may be quite appropriate in many practical settings. Indeed, when the null hypothesis is that a treatment has no effect at all, the treatment population is identical with the control population, having the equal means and equal variances. And this is the case we'll consider in detail.

The *t*-test to be presented is directed at detecting a difference in means, and the alternative is either the two-sided $\delta \neq 0$, or one-sided ($\delta > 0$ or $\delta < 0$), depending on the context.

To estimate the variance of the sample mean difference, which is now

$$\text{var}\,\bar{D} = \frac{\sigma^2}{n_1} + \frac{\sigma^2}{n_2} = \sigma^2 \left( \frac{1}{n_1} + \frac{1}{n_2} \right), \tag{2}$$

we need an estimate of the common variance, $\sigma^2$. Either of the two sample variances provides such an estimate, but an estimate that is apt to be better than either of these is the **pooled variance,** a weighted average of the sample variances in which the weights are proportional to degrees of freedom:

$$S_p^2 = \frac{(n_1 - 1)S_1^2 + (n_2 - 1)S_2^2}{n_1 + n_2 - 2}. \tag{3}$$

The numerator of this fraction is simply the sum of the squared deviations of the observations in sample 1 about their mean, plus the sum of the squared deviations of the observations in sample 2 about their mean. [It is straightforward to show (Problem

12-20) that dividing the numerator by the combined sample size, $n_1 + n_2$, yields the *m.l.e.* of $\sigma^2$. But it is traditional and convenient to use (3) instead of the m.l.e.]

Note that the coefficients of the two sample variances in (3) are fractions that sum to 1. When the sample sizes are equal, these fractions are each 1/2, and the pooled variance is just the ordinary average of the two sample variances. If the sample sizes are not equal, the variance from the larger sample (presumably more reliable) is weighted more heavily.

Substituting $S_p^2$ for $\sigma^2$ in (2) and taking the square root gives us a formula for standard error for use in the denominator of a standard score, one which we now call $T$:

$$T = \frac{\overline{X}_1 - \overline{X}_2 - 0}{S_p\sqrt{1/n_1 + 1/n_2}}. \tag{4}$$

The reason for the name is that under $H_0$, $T \sim t(n_1 + n_2 - 2)$.

To see why this is so, observe that for random samples from normal populations with variance $\sigma^2$,

$$\frac{(n_1 - 1)S_1^2}{\sigma^2} \sim \text{chi}^2(n_1 - 1), \text{ and } \frac{(n_2 - 1)S_2^2}{\sigma^2} \sim \text{chi}^2(n_2 - 1).$$

And because we are assuming the two samples to be independent, the two sample variances are independent. But the sum of independent chi-square variables has a chi-square distribution with d.f. equal to the sum of the d.f.'s of the summands. From (3), then, we see that

$$\frac{(n_1 + n_2 - 2)S_p^2}{\sigma^2} \sim \text{chi}^2(n_1 + n_2 - 2).$$

In §9.6, we showed that the ratio of a standard normal variable to the positive square root of a chi-square variable divided by its d.f. has the $t$-distribution with that number of degrees of freedom. And dividing numerator and denominator of $T$ in (4) by $\sigma$ yields just such a ratio, which means that it does have the $t$-distribution as claimed.

As in the one-sample case, it turns out that the $t$-distribution gives a good approximation to the null distribution of $T$ even when the populations involved are not normal—provided they are not too nonnormal.

---

Test of $H_0$: $\mu_1 = \mu_2$, assuming nearly normal populations with equal variances: Calculate

$$T = \frac{\overline{X}_1 - \overline{X}_2}{S_p\sqrt{1/n_1 + 1/n_2}}, \tag{5}$$

where $S_p^2$ is the pooled variance (3). Find $P$-values in Table IIIb, with d.f. $= n_1 + n_2 - 2$.

Example **12.4a**

**Feeding Schedule and Blood Pressure (Continued)**

We return to the rat blood pressure data of Example 10b:

  Regular:      108, 133, 134, 145, 152, 155, 169
  Intermittent:  115, 162, 162, 168, 170, 181, 199, 207

The statistics necessary for calculating $T$ are as follows:

|        | $n$ | $\overline{X}$ | $S$ |
|--------|-----|--------|-------|
| *Reg.* | 7   | 142.29 | 19.61 |
| *Int.* | 8   | 170.50 | 27.99 |

The pooled variance, from (3), is

$$S_p^2 = \frac{6(19.61)^2 + 7(27.99)^2}{7 + 8 - 2} = 24.48^2,$$

and substituting in (5) we get

$$T = \frac{-28.21}{24.48\sqrt{1/7 + 1/8}} \doteq -2.23.$$

We find the one-sided $P$-value in Table IIIb (13 d.f.): $P = .022$.   ∎

When should one use $Z$, and when $T$? The simplest answer is that you should use $T$ whenever population variances cannot be assumed known. If the sample sizes are large enough, the $t$-table takes you automatically to the $Z$-approximation, in the row for d.f. $= \infty$. When population variances are known, $T$ is *not* appropriate. (And whether $Z$ is appropriate in this case hinges on whether the sample sizes are large enough that the difference in sample means is approximately normal.)

Example **12.4b**

**Feminism and Authoritarianism**

A study compared people's attitudes toward feminism with their degree of "authoritarianism."[8] Two samples were used, one consisting of 30 subjects who were rated high in authoritarianism, and a second sample of 31 subjects who were rated low. Each subject was given an 18-item test, designed to reveal attitudes on feminism, with scores reported on a scale from 18 to 90. High scores indicated pro-feminism. Summary statistics are as follows:

|        | $n$ | $\overline{X}$ | $S$ |
|--------|-----|------|------|
| *High* | 30  | 67.7 | 11.8 |
| *Low*  | 31  | 52.4 | 13.0 |

---

[8]G. Sarup, "Gender, authoritarianism, and attitude toward feminism," *Soc. Behav. Personality* 4 (1976), 57–64.

The null hypothesis in the study is that authoritarianism is not a factor in attitudes toward feminism. Under $H_0$, score of the high and low authoritarianism types are equal on average: $\delta = 0$. We'll take $\delta \neq 0$ as the alternative, so that large values of $|T|$ are extreme.

Using (3), we find the pooled variance to be $S_p^2 = (12.42)^2$. Then,

$$T = \frac{67.7 - 52.4}{12.42\sqrt{1/30 + 1/31}} = 4.81.$$

The $t$-distribution with 59 d.f. is close to normal. In Table IIIa of Appendix 1 we are driven to use the last row (d.f. $= \infty$), where we see that 4.81 is beyond the 99.5th percentile. So, $P < .005$. (Table IIIb shows $P = .000$ for 40 d.f.)

Alternatively, we could calculate $Z$ using (4) of §12.2, with

$$\frac{S_1^2}{n_1} + \frac{S_2^2}{n_2} = \frac{11.8^2}{30} + \frac{13.0^2}{31} = 10.09.$$

The result is $Z = 4.82$, very close to the value of $T$. (Ordinarily, $T$ and $Z$ will not agree exactly, since they are based on different estimates of $\sigma$.) ■

Although the assumption of equal variances under $H_0$ is often reasonable, it may not be correct. If one rejects $H_0$, it is not clear whether this is because $\delta$ is not 0, or because the variances are not equal. To handle the case of unequal population variances, various modifications of the basic two-sample $t$-test for equality of means have been proposed. One method uses the large sample "$Z$-score" of §12.2 in conjunction with the $t$-table and a modified number of degrees of freedom. The two-sample test for equal means in most statistical computer packages has such an option, but there is not a consensus as to what approach is best.

In the next section we consider a nonparametric alternative to the $t$-test, one that involves fewer assumptions.

## 12.5  Two-Sample Nonparametric Tests

The $t$-test is appropriate when populations are "nearly normal." We now take up a rank test for comparing the locations of two continuous populations that are unrestricted as to shape. The null hypothesis is that the two populations are the same. The alternative hypothesis of interest, as in the case of the two-sample $t$-test of §12.4, is that the populations differ in location—that one population is *shifted* to the right or to the left of the other.

Two samples that come from the same population tend to be sprinkled over the axis of possible values in roughly the same way, so the observations in the two samples will tend to be interspersed when the null hypothesis is true. Suppose we plot the observations from both populations on the same axis of values, using the symbol "1" for those from population 1 and the symbol "2" for those from population 2. This orders the $n_1 + n_2$ observations in the combined sample. If the populations are the same, the 1's and 2's will be interspersed in the combined order statistic. However, if population 1 is shifted to the right of population 2, the 1's will tend to be toward the right in the sequence. Thus, if sample 1 is (6, 8, 9, 11) and sample 2 is (10, 14, 17), the combined sample sequence, in numerical order, is (6, 8, 9, **10**, 11, **14**, **17**), with sample 2 observations indicated here in boldface type. The corresponding sequence of 1's and 2's is 1 1 1 2 1 2 2.

The next step is to assign a rank to each observation in the *combined* sample, rank 1 to the smallest, rank 2 to the second smallest, and so on. For the sequence 1 1 1 2 1 2 2, the 1-ranks are 1, 2, 3, 5, and the 2-ranks are 4, 6, and 7. If the observations from population 1 tend to be toward the right in the combined sample sequence, their average rank will be unusually large; if they tend to be toward the left, their average rank will be unusually small. So the average rank indicates the degree to which one population is shifted from the other. To avoid fractions, we use the *sum* of the ranks (which is proportional to the average) as our test statistic. This statistic was proposed by F. Wilcoxon.[9]

---

### Wilcoxon's Rank-Sum Statistic:

Given independent random samples of $n_1$ observations from population 1 and $n_2$ observations from population 2, combine them into a single ordered sequence and assign rank $j$ to the $j$th smallest. Then,

$$R_i = \text{sum of the ranks of the observations from population } i.$$

---

Since $R_1 + R_2$ is the sum of the first $n_1 + n_2$ positive integers, it is constant, depending only on $n_1 + n_2$:

$$R_1 + R_2 = 1 + 2 + \cdots + (n_1 + n_2) = \frac{1}{2}(n_1 + n_2)(n_1 + n_2 + 1).$$

This means that we need use just one (either one) of the rank sums as the test statistic.

As usual, we need to know the null distribution of the test statistic. Under $H_0$, the hypothesis that the two populations are the same, the two independent samples together constitute a single sample from the common population. The exchangeability of the $n_1 + n_2$ observations (under $H_0$) implies that all patterns of 1's and 2's are equally likely under $H_0$. To see how this fact leads to the null distribution of $R_1$, we consider an artificial example with small sample sizes.

## Example 12.5a | A Null Distribution for $R_1$

Consider independent random samples of sizes $n_1 = 2$ and $n_2 = 3$, and suppose the samples are as follows: (5, 10) and (2, 7, 9). Combining them into a single ordered

---

[9]Equivalent rank statistics were proposed more or less simultaneously by H. B. Mann and R. Whitney, and by J. B. S. Haldane and C. A. B. Smith.

sequence yields (2, **5**, 7, 9, **10**), where the observations from population 1 are shown in boldface type. With each observation replaced by its population of origin, the sequence becomes 2 1 2 2 1. This is one of $\binom{5}{2}$ sequences of two 1's and three 2's that could have occurred. All ten of them are shown in Table 12-1, along with the corresponding values of $R_1$ and $R_2$. (The one obtained from the given sample is in boldface.) Observe that in each case, $R_1 + R_2 = 15$, which is $\frac{1}{2}(5 \times 6)$.

Along with the table of sequences is the probability table for $R_1$, obtained using the fact that each sequence has probability 1/10. The null distribution of $R_1$ is symmetric, and this is true in general. The mean value is therefore the point of symmetry: $E(R_1 \mid H_0) = 6$.

For a test of the hypothesis that the two populations are identical, a one-sided $P$-value (for $R_1 = 7$) is

$$P = P(R_1 \geq 7 \mid H_0) = .2 + .1 + .1 = .4. \qquad \blacksquare$$

**Table 12-1**

| Sequence | $R_1$ | $R_2$ | | $r$ | $P(R_1 = r)$ |
|---|---|---|---|---|---|
| 2 2 2 1 1 | 9 | 6 | | 3 | .1 |
| 2 2 1 2 1 | 8 | 7 | | 4 | .1 |
| **2 1 2 2 1** | 7 | 8 | | 5 | .2 |
| 2 2 1 1 2 | 7 | 8 | | 6 | .2 |
| 2 1 2 1 2 | 6 | 9 | | 7 | .2 |
| 1 2 2 2 1 | 6 | 9 | | 8 | .1 |
| 1 2 2 1 2 | 5 | 10 | | 9 | .1 |
| 2 1 1 2 2 | 5 | 10 | | | |
| 1 2 1 2 2 | 4 | 11 | | | |
| 1 1 2 2 2 | 3 | 12 | | | |

In deriving the null distribution of $R_1$, we assumed nothing about the common population except that it is continuous. So, a test based on $R_1$ can be used in situations where a $t$-test is inappropriate.

In general, as in the example, the null distributions of $R_1$ do *not* depend on the shape of the common (but unspecified) distribution under $H_0$. The test statistic is said to be **distribution-free**. This means that a single table, with only the sample sizes as parameters, will suffice to give the null distribution we need for a test of $H_0$.

Counting sequences to determine the null distribution of a rank-sum statistic is, of course, much more tedious for larger sample sizes. However, tables are available, and Table VII of Appendix 1 gives tail-probabilities for the distribution of $R_1$ for sample sizes from 4 to 10. To use this particular table, labels 1 and 2 must be assigned so that

$n_1 \leq n_2$. The table gives only probabilities in the *left* tail, but these are sufficient because of the symmetry of the distribution about its mean.

When sample sizes exceed 10, we're in luck:  The null distribution of $R_1$ is asymptotically normal, so a normal approximation is possible. (Proof of this limit theorem is beyond our scope.) To approximate probabilities using a $Z$-score, we need formulas for the mean and variance of the distribution. We'll not derive these,[10] but you should check that they are correct for the case of Example 12.5a:

$$E(R_1) = \frac{1}{2}\, n_1(n_1 + n_2 + 1), \quad \operatorname{var} R_1 = \frac{1}{12}\, n_1 n_2 (n_1 + n_2 + 1). \tag{1}$$

Because we'd be approximating a discrete distribution with the continuous normal distribution, the approximation is improved using a "continuity correction," given in formula (2) below.

---

*Rank-sum test* for

$$H_0: \text{ Populations 1 and 2 are identical,}$$

based on independent random samples of sizes $n_1$ and $n_2$. Extreme values of $R_1$ are those in the right tail if the alternative is that population 1 is to the right of population 2, and in the left tail if the alternative is that population 1 is to the left of population 2. For $n_1 \leq n_2 \leq 10$, use Table VII to find $P$-values. For larger samples, use the following $Z$-score with Table IIa:

$$Z = \frac{R_1 + \frac{1}{2} - \frac{1}{2} n_1 (n_1 + n_2 + 1)}{\sqrt{\frac{1}{12} n_1 n_2 (n_1 + n_2 + 1)}}. \tag{2}$$

---

**Example 12.5b** | **Intermittent Feeding (Continued)**

Consider again the blood pressure data of Example 12.4a and earlier examples. Using the rank-sum test avoids assuming population normality. The sample sizes are 7 and 8, and (to use Table VII) we test $H_0$ using the rank sum for the *smaller* sample. The order statistic of the combined sample of 15 is

**108**, 115, **133, 134, 145, 152, 155**, 162, 162, 168, **169**, 170, 181, 199, 207,

where the sample observations from population 1 (regular feeding) are shown in boldface type. The corresponding sequence of population origins is thus

1 2 1 1 1 1 1 2 2 2 1 2 2 2 2.

The ranks of the observations from population 1 are 1, 3, 4, 5, 6, 7, 11, which sum to

---

[10]See Lindgren [15], 519–21.

37. With this value for $R_1$ we enter Table VII, in the block for sample sizes 7 and 8, and find .014 opposite $c = 37$. This is the one-sided $P$, and if the alternative is two-sided, some will double this to get $P = .028$.

What we have just found is an exact $P$-value; but let's see what we get using the normal approximation (2), since 7 and 8 are approaching the table limits of 10 for sample sizes. From (1), the mean and variance of $R_1$, are

$$E(R_1) = \frac{1}{2} \times 7 \times 16 = 56, \text{ and } \operatorname{var} R_1 = \frac{1}{12} \times 7 \times 8 \times 16 = \frac{224}{3}.$$

Then, with the continuity correction,

$$Z = \frac{37 + .5 - 56}{\sqrt{224/3}} = -2.14.$$

The two-sided $P$-value is $P = 2\,\Phi(-2.14) = .032$, not terribly far from the above exact value of .028. (From the $t$-test in Example 12.4a, $P = .044$.) ∎

Rank tests are appropriate when the assumptions needed for a $t$-test are in question, but they can be used even when a $t$-test seems appropriate. The matter of which test is preferred in such a case has received extensive study. The $t$-test is only slightly better when the populations are normal, the case for which the $t$-test is designed. But when the populations are nonnormal, the rank-sum test can be very much better than the $t$-test. Indeed, a $t$-test may be worthless if the populations are sufficiently far from normal. (See our comments after Example 10.4b.) Most researchers use $t$-tests (this is what they learned), but rank tests are better all-purpose procedures.

When data suggest population nonnormality, a transformation may make the data appear more like data from a normal population. Commonly used transformations are logarithms and powers—the square root, for example. One might select the transformation that makes the data look most like data from a normal population and carry out a $t$-test using the transformed data.

A **rank transformation** is an all-purpose transformation in which each observation is replaced by its rank in the combined order statistic.[11] We then apply the ordinary two-sample $t$-test using the two sets of ranks. This $t$-test on ranks is in fact approximately equivalent to the rank-sum test. One reason to use the $T$-score based on ranks is that it uses the commonly understood $t$-test and conveniently available $t$-table, but does not require the assumption of normality of the original data.

## Example 12.5c | Intermittent Feeding (Once More)

Replacing the sample observations in Example 12.5b by their respective ranks in the combined order statistic, we obtain these samples and sample statistics:

---

[11]See W. J. Conover and R. L. Iman, "Rank transformations as a bridge between parametric and nonparametric statistics," *American Statistician* 35 (1981), 124–29.

| | Ranks | $\overline{X}$ | $S$ |
|---|---|---|---|
| Regular | 1, 3, 4, 5, 6, 7, 11 | 5.286 | 3.200 |
| Intermit. | 2, 8, 9, 10, 12, 13, 14, 15 | 10.375 | 4.173 |

The pooled variance is

$$S_p^2 = \frac{6 \times 3.200^2 + 7 \times 4.173^2}{6 + 7} = (3.755)^2,$$

and so,

$$T = \frac{5.286 - 10.375}{3.755\sqrt{1/7 + 1/8}} \doteq -2.62.$$

From Table IIIb (13 d.f.) we find the one-sided $P$-value to be about .01.   ∎

## Problems

**✱ 12-12.** Measurements on anteroposterior chest diameter for 25 male pulmonary emphysema patients and 16 normal males yielded the following:[12]

| | $\overline{X}$ | $S$ |
|---|---|---|
| Normal | 20.2 | 2.0 |
| Emphysema | 23.0 | 2.4 |

Test the hypothesis that there is no difference between average chest diameters of normal males and males with emphysema, assuming independent random samples. (The study is actually flawed, because it confounds the difference of interest with both age and weight: The mean age in the normal sample was 30.9 years, and in the other sample 56 years; the mean weight in the normal sample was 74.5 kg, and in the other 62.1 kg. Taking weights into consideration suggests that the effect of emphysema is even greater, since the emphysema patients weigh less than normal but have larger chests.)

**12-13.** Students in a statistics class measured their pulse rates:

Smokers:      62, 66, 90, 92, 66, 70, 68, 70, 78, 100, 88, 62.
Nonsmokers:   64, 58, 64, 74, 84, 68, 62, 76, 80, 68, 60, 62, 72, 70, 74, 66.

Test the hypothesis of no difference between pulse rate distributions for smokers and non-smokers (among statistics students and at the particular school),

(a) using a $t$-test.     (b) using a rank test.

---

[12]Kilburn and Asmundsson, "Anteroposterior chest diameter in emphysema," *Archiv. Internal Med.* 123 (1969) 379–82.

* **12-14.**   The following are summary statistics of plasma clearances (in l/sec) of a drug given intravenously to eight smokers and four nonsmokers:

|           | $n$ | $\overline{X}$ | $S$  |
|-----------|-----|------|------|
| *Smoker*    | 8   | 12.12 | 4.18 |
| *Nonsmoker* | 4   | 15.14 | 5.06 |

Test the hypothesis of no difference between clearances of smokers and nonsmokers against a two-sided alternative.

**12-15.**   Write out the sequences of four 1's and four 2's in which the 1's have rank-sums less than 14. Assume the 70 possible patterns to be equally likely, to verify the entry in Table VII of Appendix 1 for $n_1 = n_2 = 4$ and $c = 13$.

* **12-16.**   Platelet counts (in 1000's per $mm^3$) for 10 normal males and for 14 patients with a recent thrombosis were obtained as follows:[13]

Normal:   257, 185, 231, 220, 141, 237, 199, 295, 276, 319.
Patient:   597, 415, 264, 403, 681, 188, 364, 169, 426, 388, 364, 294, 368, 466.

Apply the rank-sum test for the hypothesis that the population distributions are the same against the alternative that the counts for thrombosis patients tend to be higher.

**12-17.**   A study addressed the relationship between the attitudes of children toward their fathers and their birth order.[14]   Fifteen firstborn and 15 second-born males (independent samples) were given a questionnaire dealing with these attitudes. The scores are as follows:

Firstborn:      40, 41, 44, 49, 53, 53, 54, 54, 56, 61, 62, 64, 65, 67, 67.
Second-born:   23, 25, 38, 43, 44, 47, 49, 54, 55, 58, 58, 60, 66, 66, 72.

(A large score means that the child identifies with and supports the role of the father in the family.)  Apply the rank-sum test for the hypothesis of no difference between attitudes of first- and second-born children against a two-sided alternative.

* **12-18.**   For independent random samples of sizes $n_1 = 8$ and $n_2 = 10$, compare the (exact) $P$-value given in Table VII for $R_1 = 57$ with the normal approximation given by (2) of §12.5.

**12-19.**   Example 9.8a dealt with two methods of operation of a new filtration device. The data that were summarized in the means and variances given in that example are as follows:

Method 1:   308 365 221 172 81 277 286 248 243 119
            243 216 196 182 157 194 148 182 105 98
Method 2:   221 249 271 187 99 222 161 379 307 294
            305 325 280 308 311 258 203 400 354 296
            253 415 344 329 282 356 323 248 91

Test for the hypothesis of no difference between the two methods using the rank-sum test.

---

[13]From a consulting project.
[14]A. Roost, "A *Q*-sort analysis of family ordinal position," PhD. thesis, U. of Minnesota (1975).

∗ **12-20.** Obtain the m.l.e. of the common variance $\sigma^2$, for independent random samples from normal populations, and give its relationship to the pooled variance of §12.4.

**12-21.** In a study[15] on the effect of cod-liver oil, seven pigs were fed a diet containing large amounts of cod-liver oil. Eleven pigs were fed normally. The extent of artery blockage was determined for several arteries of each pig. For the right coronary artery, the results (measured in percentages) were:

|          | $n$ | $\overline{X}$ | s.d. |
| -------- | --- | ----- | ----- |
| *Oil fed* | 7  | 12.75 | 13.77 |
| *Control* | 11 | 53.46 | 22.80 |

Test $H_0$: $\delta = 0$, where $\delta$ is the population mean difference in blockage.

## 12.6    Paired Comparisons

*Pairing* observations is one way of reducing sampling variability. Thus, when one of the twins in a pair is treated and the other not, their genetic similarity suggests that a difference in response is more apt to be the result of the treatment than it would be if the individuals were unrelated. It is common practice in animal experimentation to use littermates for this reason—the control animal is then as much like the treatment animal as possible. When twins or other littermates are not available, subjects can sometimes be artificially paired by matching individuals who are similar in aspects that may be related to their responses.

Paired data occur naturally when an individual serves as both treatment and control. "Before and after" experiments, as in Examples 10.5a and 10.5b, are of this type. In those examples, we used one-sample tests, based on the single sample of differences. Indeed, we gave only the *changes,* not the individual "before and after" measurements. Two-sample tests are inappropriate because the pairing makes the two sets of data dependent.

Example **12.6a** | **Newborn Lambs**

A study of six newborn lambs was conducted to determine whether specific tracheal muscle stimulation could increase the strength of the trachea.[16] (According to previous research, this might reduce the chance of airway collapse.) Tracheal compliances before and after smooth muscle stimulation are as follows, for each lamb:

---

[15]B. H. Weiner et al., "Inhibition of atherosclerosis by cod-liver oil in a hyperlipidemic swine model," *New Engl. J. Med.* 315 (1986) 841–46.

[16]B. Bhutani, R. Koslo, and T. Shaffer, "The effect of tracheal smooth muscle tone on neonatal airway collapsibility," *Pediatric Research* 20 (1986), 492–95.

| Lamb # | Before | After | Difference |
|--------|--------|-------|------------|
| 1 | .029 | .020 | + .009 |
| 2 | .043 | .011 | + .032 |
| 3 | .022 | .008 | + .014 |
| 4 | .012 | .005 | + .007 |
| 5 | .020 | .009 | + .011 |
| 6 | .034 | .009 | + .025 |
| Mean | .0267 | .0103 | .0164 |
| S.D. | .011 | .0051 | .0099 |

You will often find summary statistics given in this form; do not be misled! The standard deviations shown for "before" and "after" measurements are of no use. In fact, *we use only the column of differences*. (Of course, the mean difference *is* the difference of the means: $.0267 - .0103 = .0164$, but there is not a simple relationship that would enable you to find the s.d. of the differences from the other two s.d.'s shown.)

As before, let $\delta$ denote the (population) mean difference. For testing $\delta = 0$ against $\delta > 0$, we may use a one-sample test from Chapter 10. To review, we'll carry out the work for the sign test, the signed-rank test, the $t$-test, and the $t$-test on ranks:

The sign test simply takes note that all of the six signs of the differences are positive; this leads to the one-sided $P$-value $(.5)^6 \doteq .0156$. The signed rank statistic $R_-$ is 0, since there are no negative observations; the $P$-value is thus the same as for the sign test, or .0156. (Table VI gives the $P$-value to three decimal places as .016.)

The $T$-statistic is calculated using the mean and s.d. of the differences:

$$T = \frac{\overline{D} - E(\overline{D} \mid \delta = 0)}{\text{s.e.}(\overline{D})} = \frac{.0164 - 0}{.0099/\sqrt{6}} \doteq 4.06.$$

The one-sided $P$-value, from Table IIIb (5 d.f.), is $P \doteq .005$.

We recommend the $t$-test on ranks; it is sensitive to the data, yet not overly influenced by outliers. The ranks are shown in the table below. The mean difference in ranks is 5.5, and the s.d. of the difference is 2.0. With these we find $T = 6.74$ (with 5 d.f.), and $P < .002$ (from Table IIIb).

| Lamb # | Before | After | Difference |
|--------|--------|-------|------------|
| 1 | 10 | 7.5 | + 2.5 |
| 2 | 12 | 5 | + 7 |
| 3 | 9 | 2 | + 7 |
| 4 | 6 | 1 | + 5 |
| 5 | 7.5 | 3.5 | + 4 |
| 6 | 11 | 3.5 | + 7.5 |

At first glance, paired data may seem to consist of two samples. However, when you see a column giving "subject number" or "pair number" or "batch number" or (in the above example) "lamb number," this is a clue that the data are apt to be paired; check out this possibility, before proceeding with a two-sample test that assumes independent samples. The pair differences may not be shown explicitly, but you'll need them to carry out the appropriate paired-$t$-test.

If you happen to calculate a two-sample $T$-statistic in a paired-data setting by mistake, you are apt to end up with a smaller $T$ and a larger $P$-value than if you had proceeded correctly. This is because the denominator of $T$ will include variability between subjects, which would tend to be eliminated in working with differences. However, in Example 12.6a, the two-sample $T$-statistic is $T = 3.13$, and the corresponding $P$-value (10 d.f.) is again about .01. Although the paired-$T$ is the appropriate statistic, pairing has not gained much because the before and after measurements are not highly correlated. With the increase in d.f., the $P$-value from the two-sample statistic is about the same as with the one-sample statistic. Pairing is most effective in reducing error when the loss in degrees of freedom is more than compensated by the elimination of individual differences.

Pairing to reduce sampling error is a special case of a design technique called *blocking*, which we'll encounter in §14.4.

## Problems

* **12-22.**   In a diabetes study, glycosylated hemoglobin was measured for each individual in four pairs of twins.[17]   One twin in each pair had diabetes and one did not.

| Twin pair no. | 1 | 2 | 3 | 4 |
|---|---|---|---|---|
| Diabetic twin | 9.6 | 8.5 | 8.9 | 7.0 |
| Nondiabetic twin | 4.9 | 5.1 | 5.2 | 5.8 |

(The normal range for glycosylated hemoglobin is between 4 and 6.)  Test the hypothesis of 0 mean difference between diabetics and nondiabetics.

**12-23.**   Darwin did an experiment to learn whether self- or cross-pollinated seeds would produce more vigorous seeds (as indicated by plants of greater height). He obtained the following data, giving mature heights (in eighths of inches) for pairs of plants of genus *Zea mays* (Indian corn). Plants in a pair were matched genetically and assigned to self- and cross-pollination at random, planted at opposite sides of a pot and separated by a plane of glass parallel to the light rays.[18]

[17]W. A. Kaye et al., "Acquired defect in Interleukin-2 production ...," *New Engl. J. Med.* 315 (1986), 920–24.

[18]*Effect of Cross- and Self-Fertilization in the Vegetable Kingdom* (New York: Appleton, 1902).

| Pair | 1 | 2 | 3 | 4 | 5 | 6 | 7 | 8 | 9 | 10 | 11 | 12 | 13 | 14 | 15 |
|------|---|---|---|---|---|---|---|---|---|----|----|----|----|----|----|
| Cross | 188 | 96 | 168 | 176 | 153 | 172 | 177 | 163 | 146 | 171 | 186 | 168 | 177 | 184 | 96 |
| Self | 139 | 163 | 160 | 160 | 147 | 149 | 149 | 122 | 132 | 144 | 130 | 144 | 102 | 124 | 144 |

Assuming a two-sided alternative, test the hypothesis of no difference between self- and cross-pollinated seeds, using

(a)   the sign test.     (b)   the signed-rank test.     (c)   a $t$-test.

* **12-24.**   Two pigs are selected from each of ten litters. One of each pair is fed a test diet, and the other a standard diet.  Test the hypothesis of no difference against the alternative that the test diet results in greater average weight gains, given these gains (over a given period) under the test and standard diets:

| Pair | 1 | 2 | 3 | 4 | 5 | 6 | 7 | 8 | 9 | 10 |
|------|---|---|---|---|---|---|---|---|---|----|
| Test | 36.0 | 32.7 | 39.2 | 37.6 | 32.0 | 40.2 | 34.4 | 30.7 | 36.4 | 37.2 |
| Standard | 35.2 | 30.0 | 36.5 | 38.1 | 29.4 | 36.0 | 31.3 | 31.6 | 31.1 | 34.0 |

**12-25.**   Using the data of the preceding problem to illustrate the point discussed in the next to last paragraph of §12.6,

(a)   calculate the correlation coefficient of weight gains.

(b)   proceed with the test you would use if the results for the test diet were independent of the results for the standard diet. Observe that eliminating litter-to-litter differences by pairing more than compensates for the loss of sensitivity associated with fewer degrees of freedom.

* **12-26.**   A prospective study was designed to see if honey in a diet would increase hemoglobin. Six pairs of twins in a children's home were the subjects. For a period of six weeks, each child was given a cup of milk at 9 P.M. One of the twins (always the same one) in each pair received a tablespoon of honey dissolved in the milk. The following increases in hemoglobin were recorded:

| Twin pair | 1 | 2 | 3 | 4 | 5 | 6 |
|-----------|---|---|---|---|---|---|
| Honey | 19 | 12 | 9 | 17 | 24 | 24 |
| No honey | 14 | 8 | 4 | 4 | 11 | 11 |

(a)   Apply the sign test and give a one-sided $P$-value.

(b)   Apply the signed-rank test and give a one-sided $P$-value. Explain why this is the same as the $P$-value in (a).

(c)   Apply a $t$-test on the ranks of the observations in the combined sample.

## 12.7          The $F$-Distribution

For carrying out the test of equal population variances to be presented in the next section, we need to interpose a development of yet another sampling distribution. This same distribution will also be needed in Chapters 14 and 15, where certain test statistics are ratios of variance estimates.

The family of **$F$-distributions** is indexed by two parameters, both non-negative integers. The p.d.f. is

$$f(x \mid k,\, m) \propto \frac{x^{k/2-1}}{(1 + kx/m)^{(k+m)/2}}, \quad x > 0. \tag{1}$$

(The constant of proportionality is what is needed to make the area under the graph of $f$ equal to 1. It is a function of the parameters $k$ and $m$—which we'll not need to know.)

When a random variable $X$ has the distribution defined by the p.d.f. (1), we write $X \sim F(k,\, m)$. The parameters $k$ and $m$ are called *numerator degrees of freedom* and *denominator degrees of freedom,* respectively—for reasons that will emerge. Table VIII of Appendix 1 gives tail probabilities and certain commonly used percentiles of the distribution for selected d.f.'s.

The $t$- and $F$-distributions are related:  Consider the distribution of $T^2$, where $T \sim t(m)$. We use the transformation formula (18) of §5.2 to write the p.d.f. of $T^2$:

$$f_{T^2}(x) = f_T(\sqrt{x}) \cdot \frac{d\sqrt{x}}{dx} \propto \frac{x^{-1/2}}{(1 + x/m)^{(m+1)/2}}, \quad x > 0.$$

This is a special case of (1) where $k = 1$. Thus, $T^2 \sim F(1,\, m)$.

The $F$-distribution arises in inference problems as the distribution of the ratio of independent chi-square distributions, each divided by its degrees of freedom. To show this, we first recall [from (2) of §9.6, where we derived the $t$-density] a formula for the p.d.f. of a ratio of the form $W = U/V$, where $U$ and $V$ are independent variables, and $V$ is supported on the positive real line:

$$f_{U/V}(w) = \int_0^\infty v\, f_U(vw)\, f_V(v)\, dv. \tag{2}$$

Now suppose in particular that $U$ and $V$ have independent chi-square distributions, with $k$ and $m$ degrees of freedom, respectively. Substituting the chi-square p.d.f.'s in (2), we obtain

$$f_{U/V}(w) = \frac{1}{2^{(k+m)/2}\Gamma(\frac{k}{2})\Gamma(\frac{m}{2})} \int_0^\infty v\, [(vw)^{k/2-1} e^{-vw/2}][v^{m/2-1} e^{-v/2}]\, dv.$$

To exhibit the dependence on $w$ more clearly, we factor out the powers of $w$:

$$f_W(w) \propto w^{k/2-1} \int_0^\infty v^{m/2} v^{k/2-1} e^{-v(w+1)/2} dv, \quad w > 0,$$

and make the change of variable $y = (w+1)v$:

$$f_W(w) \propto \frac{w^{k/2-1}}{(w+1)^{(k+m)/2}} \int_0^\infty y^{(k+m-2)/2} e^{-y/2} dy, \quad w > 0.$$

The integral expression does not depend on $w$, and we can lump it together with the constant of integration, and write

$$f_W(w) \propto \frac{w^{k/2-1}}{(w+1)^{(k+m)/2}}, \quad w > 0. \tag{3}$$

Next, we define the random variable $F$ as the ratio of independent chi-square variable, each divided by its number of degrees of freedom. It is related to the above $W$ as follows:

$$F = \frac{U/k}{V/m} = \frac{m}{k} W.$$

The p.d.f. of this ratio, a constant times $W$, follows from (3) when we apply the transformation formula (16) from §5.2:

$$f_F(z) \propto \frac{z^{k/2-1}}{(1+kz/m)^{(k+m)/2}}, \quad z > 0, \tag{4}$$

which is just the $F$-density given by (1).

Notice that the reciprocal of $F$ is also a ratio of independent chi-square variables each divided by degrees of freedom:

$$\frac{1}{F} \sim F(m, k).$$

This means that we can find lower percentiles from upper percentiles. For instance, the 5th percentile of $F(k, m)$ is equal to the reciprocal of the 95th percentile of an $F$-distribution with degrees of freedom reversed. For,

$$.05 = P(F < \lambda) = P\left(\frac{1}{F} > \frac{1}{\lambda}\right) = 1 - P\left(\frac{1}{F} < \frac{1}{\lambda}\right) = 1 - .95,$$

so $1/\lambda$ is the 95th percentile of $F(m, k)$.

**12.8**          ## Comparing Variances

We said in Chapter 10 that the occasions for testing hypotheses about the variance of a single population are rare. The hypothesis that two populations have the same variance is more common. In particular, we assumed equality of variances in the two-sample $t$-test of §12.4, so a collateral test of equality may be in order.

We now present a test for the equality of the variances of two *normal* populations. (In a two-sample $t$-test, we assume population normality and equal variances.) The null hypothesis to be tested is

$$H_0: \ \sigma_1^2 = \sigma_2^2 = \sigma^2. \tag{1}$$

The alternative can be either two-sided ($\sigma_1^2 \neq \sigma_2^2$) or one-sided (for example, $\sigma_1^2 > \sigma_2^2$).

The basis for the test is a random sample from each population, assumed to be *independent* samples. As before, let $S_i^2$ denote the variance and $n_i$ the size of the sample from population $i$, $i = 1, 2$. Since these sample variances are consistent estimates of the corresponding population variances, the difference $S_1^2 - S_2^2$ is sensitive to any discrepancy between the population variances when the $n$'s are sufficiently large. However, even under the null hypothesis, the distribution of the difference depends on the common variance. On the other hand, the *ratio* $S_1^2/S_2^2$ has a null distribution that is independent of the common variance, as we'll see shortly. In view of this, we phrase the null and alternative hypotheses in terms of ratios:

$$H_0: \ \frac{\sigma_1^2}{\sigma_2^2} = 1, \quad H_A: \ \frac{\sigma_1^2}{\sigma_2^2} \neq 1.$$

Given independent random samples from the two populations, the test statistic we'll use is the corresponding ratio of sample variances:

$$F = \frac{S_1^2}{S_2^2}. \tag{2}$$

For the above alternative $H_A$, we take either vary large or very small values of $F$ as extreme.

As usual, we need the *null* distribution of the test statistic. Under $H_0$, each population is normal with variance $\sigma^2$, and so (from §8.10),

$$Y_i = \frac{(n_i - 1)S_i^2}{\sigma^2} \sim \text{chi}^2(n_i - 1), \ i = 1, 2.$$

Since the samples are assumed independent, the $Y_i$'s are independent, and we can write the ratio $F$ in terms of them as follows:

$$F = \frac{S_1^2}{S_2^2} = \frac{S_1^2/\sigma^2}{S_2^2/\sigma^2} = \frac{Y_1/(n_1 - 1)}{Y_2/(n_2 - 1)}. \tag{3}$$

In the preceding section we showed that such a ratio (independent chi-square variables, each divided by degrees of freedom) has an $F$-distribution.

To test $H_0$: $\sigma_1^2 = \sigma_2^2$, using independent random samples from normal populations, calculate $F = S_1^2/S_2^2$ and use Table VIIIa/b to find $P$-values or determine critical values: Under $H_0$,

$$F \sim F(n_1 - 1, n_2 - 1). \tag{4}$$

Large values of $F$ are extreme for the alternative $\sigma_1^2 > \sigma_2^2$; both large and small values of $F$ are extreme for the alternative $\sigma_1^2 \neq \sigma_2^2$.

**Example 12.8a** | **Intermittent Feeding**

In Examples 12.5b and 12.5c, we gave blood pressures for two independent samples of rats. The sample variances are $S_2^2 = 783.7$ and $S_1^2 = 384.6$. Their ratio is $S_2^2/S_1^2 = 783.7/384.6 \doteq 2.04$. Table VIIIa does not include the case $n_2 - 1 = 7$, but in Table VIIIb we find that the 95th percentile of $F(7, 6)$ is about 4.2, substantially larger than 2.04. So the ratio is *not* "statistically significant," even at the 10% level (in a two-sided test).   ∎

Unfortunately, it has been found that the $F$-test for equal variances is *not robust* with respect to the assumption of normality of the population distributions.

## Problems

\* **12-27.**   For the data in Problem 12-19, the standard deviations (given in Example 9.8a) are $S_1 = 74.35$ and $S_2 = 79.03$. (And $n_1 = 20$, $n_2 = 29$.) Test the hypothesis that the two procedures are the same with regard to the variance of the amount filtered.

**12-28.**   Test the hypothesis that the variances of pulse rate are the same for smokers and non-smokers, using the data in Problem 12-13.

**12-29.**   Test the hypothesis that the population variances are equal, given the data in Problem 12-21, where $S_1 = 13.77$, $S_2 = 22.80$, $n_1 = 7$, $n_2 = 11$.

\* **12-30.**   The p.d.f. of $\text{chi}^2(m)$ is

$$f(v \mid m) = \frac{1}{2^{m/2}\Gamma(m/2)} \, v^{m/2-1} e^{-v/2}, \, v > 0.$$

**(a)**   Use this to find $E(1/V)$ when $V \sim \text{chi}^2(m)$.

**(b)**   Find $E(U/V)$ when $U$ and $V$ are independent, and $U \sim \text{chi}^2(k)$ and $V \sim \text{chi}^2(m)$. [*Hint:* Write $U/V$ as $U \cdot (1/V)$ and use (a).]

**(c)**   Use the result in (b) to find the mean value of the $F$-distribution.

**12-31.**   Trace through the steps leading to (3) in §12.7, starting at the point where the chi-square p.d.f.'s are substituted in (2); but now include all the constants that were dropped, to obtain the constant of proportionality in (3) as well as the constant of proportionality for the $F$-density in (4).

**12-32.**   Using the p.d.f. from the preceding problem (that is, including the constant of proportionality), find the mean value of $W$ directly, as the integral of $w f_W(w)$ over $(0, \infty)$.

## Review Problems

**12-R1.**   In a seven-year study on aspirin as stroke prevention, 416 men who had had at least one TIA were followed for a period of time. Of these, 200 were in the treatment (aspirin) group, and 216 used a placebo or sulfinpyrazone. At the end of the period, 29 of the aspirin group had strokes or died, and 56 of the control group had strokes or died. Test the hypothesis that the probability of stroke or death is the same for both "treatments."

**12-R2.**   The observations in independent random samples from populations 1 and 2 were ordered together, to make a combined sample of size 10. This sample sequence, coded by population of origin, is (1, 1, 1, 2, 1, 1, 2, 2, 2, 2).

  **(a)**   Find the (one-sided) $P$-value using the appropriate table.

  **(b)**   List all sequences of five 1's and five 2's that are as extreme as or more extreme than the one observed. Explain how this confirms the table entry.

**12-R3.**   Consider independent random samples of size $n_1$ from population 1 and size $n_2$ from population 2, each normal with $\sigma = 1$. Let $\delta$ denote the population mean difference: $\mu_1 - \mu_2$.

  **(a)**   What is the sampling distribution of $\widehat{\delta} = \overline{X}_1 - \overline{X}_2$, the sample mean difference?

  **(b)**   Find the power of the test with rejection region $|\widehat{\delta}| > 1$ against the alternative $\widehat{\delta} = 2$, when the sample sizes are $n_1 = n_2 = 8$.

**12-R4.**   Each of two political polls used random samples of size 1000, to determine voter sentiment on a certain proposition. Poll A found 500 favoring the proposition and Poll B found 480 in favor. Test the hypothesis that the populations polled were the same (or at least, had the same proportion in favor).

**12-R5.**   Amounts of chlorine in glacial moraines are related to their ages. In one study, accumulations of chlorine in samples from two moraines were measured (in parts per million): From "Older Tahoe": 73, 75, 49, 76, 115, and from "Younger Tahoe:" 74, 31, 64, 74, 100, 38, 90.[19]   Are the amounts in the two moraines different on average?

**12-R6.**   Using the data from Problem 12-13, test the hypothesis of no difference in mean pulse rate using a $t$-test on ranks.

**12-R7.**   To test a new method for measuring concentration of plutonium-239, each of ten sample solution media of various concentrations was split into two parts, one measured using the new method, and one using the old.[20]   It was thought that the new method would tend to give a higher average reading.

| Sample | 1 | 2 | 3 | 4 | 5 | 6 | 7 | 8 | 9 | 10 |
|---|---|---|---|---|---|---|---|---|---|---|
| New method | 3.78 | 3.58 | 3.77 | 3.82 | 3.67 | 3.66 | 3.48 | 3.63 | 3.88 | 3.53 |
| Old method | 3.35 | 3.60 | 3.41 | 3.69 | 3.48 | 3.50 | 3.33 | 3.64 | 3.65 | 3.64 |

Test the hypothesis that there is no difference, on average, between the two methods of measurement using

---

[19]F. M. Phillips et al., "Cosmogenic chlorine—a chronology for glacial deposits at Bloody Canyon,...," *Science* 248 (1990), 1529–32.

[20]N. Chaudhuri and V. Natarajan, "Analysis of uranium and plutonium ...," *Nuclear Tracks and Radiation Measurements* (June 1982), 109–13.

(a) a $t$-statistic.    (b) a rank statistic.    (c) the sign test.

[Because the data are paired, the two-sample $t$ is not appropriate; however, if you calculate it by mistake, you'll find $t \doteq 2.56$. Typically in such situations, the two-sample $t$ will be quite a bit smaller than the one-sample $t$. What is there about the data here that might explain the near equality of the two $t$'s?]

**12-R8.** Example 10c raised the question "Are boys better at math?" One study gives SAT-M (math) scores for 3674 intellectually gifted 7th and 8th graders, all of whom had been accelerated at least one grade level.[21] The scores are summarized as follows:

|       | Number | Mean | S.D. |
|-------|--------|------|------|
| Boys  | 2046   | 436  | 87   |
| Girls | 1628   | 404  | 77   |

Test the hypothesis that the corresponding population means are equal.

**12-R9.** An Associated Press report was headlined in one newspaper, "Link found between male homosexuality, fingerprints." It reported on a study investigating a theory that sexual orientation is determined before birth.[22] Hall and Kimura compared the number of tiny ridges on the fingertips of 66 gay men (G) with the fingerprint patterns of 182 heterosexual men (H). Thirty percent of G's showed more ridges on their left hands than on their right, while only 14% of the H's showed the same pattern. "This certainly suggests sexual orientation is somehow determined by prenatal events," said Kimura, who continued, "What we found is a statistically significant difference between groups of heterosexual and homosexual men." Check the latter claim.

(A UCLA neurobiologist called the study "another suggestion that there's a biological component to sexuality." But he said he had some trouble making the connection between ridges on fingers and sexual orientation. It's unlikely that one gene, that one hormone, that one environmental experience—or that one fingerprint—is going to be the explanation for everything. The AP report continues: "[Kimura] speculates that there is a link between finger ridge patterns and the development of the nervous system. Many of the gay men with more ridges on their left hands were also left-handed. Research has established that there is a higher incidence of left-handedness among gays than in the general population.")

**12-R10.** Consider comparing the parameters $p_1$ and $p_2$ of two Bernoulli populations. Suppose your prior for $(p_1, p_2)$ is uniform on the unit square.

(a) Find the probability that $p_1 > p_2$.

(b) Suppose taking independent samples of size 1 from each population we obtain a success from population 1 and a failure from population 2. In view of this, update your probability that $p_1 > p_2$.

**12-R11.** The accompanying data gives forced vital capacities after a period of use of an aerosol spray containing amiloride, and also after an equal period of use of a spray without the drug (vehicle only), for each of 14 patients suffering from cystic fibrosis, with these results:[23]

---

[21]C. P. Benlow and J. C. Stanby, "Sex differences in mathematical ability: Fact or artifact?" *Science* 210 (1980), 1262–64.

[22]D. Kimura and J. Hall, in the December 1994 issue of *Behavioral Neuroscience*.

[23]M. R. Knowles et al., "A pilot study of aerosolized amiloride...," *New England J. of Medicine*, 322 (1990), 1189–94.

| Patient | Vehicle | Drug |
|---------|---------|------|
| 1 | 2925 | 2760 |
| 2 | 4190 | 4490 |
| 3 | 5067 | 5617 |
| 4 | 2588 | 2549 |
| 5 | 3934 | 3810 |
| 6 | 3952 | 3985 |
| 7 | 2547 | 2392 |
| 8 | 4108 | 3880 |
| 9 | 2646 | 2732 |
| 10 | 3635 | 3758 |
| 11 | 2890 | 2960 |
| 12 | 3125 | 3387 |
| 13 | 3805 | 4048 |
| 14 | 1741 | 1787 |

Is the drug effective? (Larger capacity indicates more favorable response.)

**12-R12.**   Referring to the birthweight data in Problem 7-R10, test the hypothesis that there is no difference in mean birthweight of babies whose mothers had never smoked and babies whose mothers had smoked.

## Chapter Perspective

Using samples to compare two populations with respect to location is a common statistical problem. In particular, determining whether a treatment is effective is such a problem. "Two-sample" tests assume independent random samples, one from each population. However, when a subject serves both as treatment and as control, or when treatment and control are assigned at random in a pair of like subjects, we test the hypothesis of no mean treatment effect by applying a *one*-sample test to the sample of differences, using the methods of Chapter 10.

In this chapter we have given both parametric and nonparametric tests. The latter can be used when it is not appropriate to assume a particular class of parametric models. But they can be used with little loss of sensitivity, even if one is comfortable assuming a particular parametric class of models. Rank transformation tests are non-parametric tests in which one uses the $t$-test (a parametric test) applied to ranks.

The two-sample $t$-test, like the one-sample $t$-test, is fairly robust with respect to the underlying assumption of population normality. The $F$-test for equal variances, on the other hand, is not robust in this regard, so it is not recommended—and if used, should be used with caution. One purpose in presenting this $F$-test is to introduce the $F$-distribution.

The $F$-distribution will be used extensively in both Chapters 14 and 15 for inference based on an "analysis of variance." There, we'll extend the $t$-test for comparing two populations to a test for comparing several populations. The test statistic is a ratio of variance estimates whose null distribution is $F$; and in the special case of two populations, this $F$-ratio is the square of the two-sample $t$-statistic. These $F$-tests share the robustness of $t$ (with respect to population normality), which they generalize.

# Goodness of Fit

Tests for **goodness of fit** check the consistency of a set of data and a proposed model. The various tests presented in the preceding chapters are actually tests of fit. In formulating those tests, however, we usually phrased both null and alternative hypotheses in terms of a population parameter such as the population mean. Tests were designed to be sensitive to a particular type of parametric alternative such as $\mu > \mu_0$. In this chapter we take up some tests that are not directed at such specific kinds of alternatives. Rather, the alternative hypothesis $H_A$ is simply that the null hypothesis is not true. These are nonparametric tests.

We begin with the testing of a completely defined distribution as the null hypothesis, first in the discrete case and then in the continuous case. We then address the commonly occurring question of whether a data set fits, not a specific model, but a *type* of model—a class of probability distributions indexed by a small number of parameters. For instance, we want to know whether a population is normal (or binomial, or exponential, etc.) without specifying a particular member of the family. We'll show how to adapt the test of fit of a completely specified distribution to this more general type of null hypothesis. Finally, we use tests of fit for hypotheses about bivariate categorical populations based on contingency table data.

## 13.1    Fitting a Distribution with Two Categories

As in earlier chapters, let $p$ denote the probability or population proportion of "successes," and $1 - p$ the probability or proportion of "failures." In §10.3 we introduced a large-sample test for a hypothesis of the form $H_0$: $p = p_0$. The test was based on the sample relative frequency of successes ($\hat{p}$), and employed this $Z$-score:

$$Z = \frac{\hat{p} - p_0}{\sqrt{\frac{p_0(1-p_0)}{n}}}. \tag{1}$$

We now want to generalize (1) to the case of a distribution with more than two categories.

To see how to do this, we'll recast the statistic $Z^2$ in a form that treats the categories symmetrically. For this purpose we change the notation:  Let the two categories be designated 1 and 2, and their probabilities $p_1$ and $p_2$. That is, $p_1$ plays the role of "$p$," and $p_2$ plays the role of "$q$" or $1 - p$. Let $Y_i$ denote the frequency of category $i$ in a sample of size $n$:

| Category | Probability | Frequency |
|----------|-------------|-----------|
| 1 | $p_1$ | $Y_1$ |
| 2 | $p_2$ | $Y_2$ |
| Sums: | 1 | $n$ |

In §10.3 we called the specific value of $p$ to be tested, $p_0$. We replace $p_0$ by $\pi_1$ and $1 - p_0$ by $\pi_2$, so that the null hypothesis to be tested is $H_0: p_1 = \pi_1$ (or, equivalently, $p_2 = \pi_2$). In this notation, the test statistic (1) becomes

$$Z = \frac{\frac{Y_1}{n} - \pi_1}{\sqrt{\frac{\pi_1 \pi_2}{n}}}. \tag{2}$$

For the two-sided alternative $p_1 \neq \pi_1$, we'd take large values of $|Z|$, or large values of $Z^2$, as extreme. Squaring each side of (2) we have

$$Z^2 = \frac{\left(\frac{Y_1}{n} - \pi_1\right)^2}{\frac{\pi_1 \pi_2}{n}} = \frac{(Y_1 - n\pi_1)^2}{n\pi_1 \pi_2}. \tag{3}$$

Next, we use the identity

$$\frac{1}{n\pi_1 \pi_2} = \frac{\pi_1 + \pi_2}{n\pi_1 \pi_2} = \frac{1}{n\pi_1} + \frac{1}{n\pi_2},$$

to rewrite (3) as

$$Z^2 = \frac{(Y_1 - n\pi_1)^2}{n\pi_1} + \frac{(Y_1 - n\pi_1)^2}{n\pi_2}. \tag{4}$$

But then, since

$$(Y_1 - n\pi_1)^2 = [n - Y_2 - n(1 - \pi_2)]^2 = (Y_2 - n\pi_2)^2,$$

we can substitute this in the second term of (4) to express $Z^2$ in a form that is easily generalized to more than two categories:

$$Z^2 = \frac{(Y_1 - n\pi_1)^2}{n\pi_1} + \frac{(Y_2 - n\pi_2)^2}{n\pi_2}. \tag{5}$$

The large-sample distribution of $Z$ is approximately $\mathcal{N}(0, 1)$, so the large-sample distribution of $Z^2$ is approximately chi$^2(1)$.

**Example 13.1a** | **Male Births (Revisited)**

In Example 10.3c we tested $p = .520$, where $p$ is the probability of a male birth. The proportion of boys in a sample of size $n = 145$ was $\hat{p} = 95/145$.

Here we'll calculate $Z^2$ using (5). To set the pattern for calculating an extension of (5) to several categories, we give the null hypothesis and the sample frequencies in the following table form:

| Category | $\pi_i$ | $Y_i$ | $145\pi_i$ | $Y_i - 145\pi_i$ |
|---|---|---|---|---|
| 1 | .520 | 95 | 75.4 | 19.6 |
| 2 | .480 | 50 | 69.6 | $-19.6$ |
| Sums: | 1 | 145 | 145 | 0 |

Substituting in (5) yields $Z^2 = 10.6 = 3.25^2$. (In our earlier $Z$-test, we had found that $Z = 3.26$.) The two-sided $P$-value from Table II is

$$P(Z > 3.25 \text{ or } Z < -3.25) = 2\Phi(-3.25) \doteq .0012.$$

We could also use Table Vb (chi-square table) to find the $P$-value corresponding to $Z^2 = 10.6$:

$$P(Z^2 > 10.6) = P[\chi^2(1) > 10.6] \doteq .001,$$

which differs from .0012 only because of round-off in the table.    ∎

## 13.2    The Chi-Square Test

Consider a discrete population with $k$ categories. We label these 1, 2, ..., $k$ and denote the corresponding probabilities by $p_1$, $p_2$, ..., $p_k$. We want to test the null hypothesis that these category probabilities are $\pi_1$, $\pi_2$, ..., $\pi_k$ (respectively), specific positive numbers that sum to 1:

$$H_0: p_j = \pi_j, \ j = 1, 2, ..., k.$$

Suppose we obtain a random sample of size $n$ from such a population, for the purpose of testing $H_0$. Let $Y_j$ denote the frequency of category $j$. These sample frequencies are sufficient (see §8.3), and from §4.8 we know that $Y_j \sim \text{Bin}(n, p_j)$. So the expected frequency of category $j$, under $H_0$, is $n\pi_j$. The following table gives the category probabilities under $H_0$, the expected frequencies under $H_0$, and the observed frequencies:

| Category | Probability under $H_0$ | Expected frequency | Observed frequency |
|:---:|:---:|:---:|:---:|
| 1 | $\pi_1$ | $n\pi_1$ | $Y_1$ |
| 2 | $\pi_2$ | $n\pi_2$ | $Y_2$ |
| $\vdots$ | $\vdots$ | $\vdots$ | $\vdots$ |
| $k$ | $\pi_k$ | $n\pi_k$ | $Y_k$ |
| Sums: | 1 | $n$ | $n$ |

In general, we could not expect the observed sample frequencies to be equal to the expected frequencies, even when $H_0$ is the correct model. But if there are large differences, perhaps the poor "fit" is evidence that $H_0$ is *not* the correct model.

The **chi-square statistic**[1] is an overall measure of discrepancy between the observed frequencies and the expected frequencies under $H_0$. It is a natural generalization of the statistic $Z^2$ given as (5) of the preceding section:

$$\chi^2 = \sum_1^k \frac{(Y_j - n\pi_j)^2}{n\pi_j}.\tag{1}$$

This weighted sum of squared differences is equal to 0 if and only if every observed frequency is equal to the corresponding expected frequency under $H_0$, that is, if the fit is perfect. Otherwise $\chi^2$ is positive. If it is large, the fit of the data to the model given as $H_0$ is poor. So we take large values of $\chi^2$ as extreme—as evidence against $H_0$. This means that $P$-values are areas in the right tail of the null distribution of $\chi^2$. But what *is* that distribution, and how large is "large"?

The null distribution of $\chi^2$ is actually quite complicated and depends on the particular model being tested. However, when $n$ is large, the distribution of $\chi^2$ under $H_0$ is approximately one of the chi-square family introduced in §6.5, namely, $\text{chi}^2(k-1)$—independent of the particular model we call $H_0$. In the next example we show how to use $\chi^2$ in testing a genetic theory.

**Example 13.2a** | ## Mendel's Pea Experiment

In Example 8.2d we gave Mendel's famous pea data, obtained by crossing round yellow pea plants with wrinkled green pea plants, obtaining plants bearing peas in one of four categories. According to Mendel's theory, the expected frequencies of these characteristics should be in the proportion 9:3:3:1. This defines the probabilities shown in the table that follows, along with the sample frequencies $Y_j$ (from Mendel's experiment) and expected frequencies:

---

[1]The English statistician Karl Pearson proposed this statistic and gave its large-sample distribution in about 1900.

| $j$ | Type | $\pi_j$ | $Y_j$ | $556\pi_j$ | $Y_j - 556\pi_j$ |
|-----|------|---------|-------|------------|------------------|
| 1 | RY | 9/16 | 315 | 312.75 | 2.25 |
| 2 | RG | 3/16 | 108 | 104.25 | 3.75 |
| 3 | WY | 3/16 | 101 | 104.25 | $-3.25$ |
| 4 | WG | 1/16 | 32 | 34.75 | $-2.75$ |
| Sums: | | 1 | 556 | 556 | 0 |

The null hypothesis is that the probabilities $\pi_j$ give the true proportions. The fit of the model is not perfect, since the differences between observed and expected frequencies are not 0. (But observe that they sum to 0, as they always will.) Substituting in (1), we find the value of the test statistic:

$$\chi^2 = \frac{2.25^2}{312.75} + \frac{3.75^2}{104.25} + \frac{(-3.25)^2}{104.25} + \frac{(-2.75)^2}{34.75} \doteq .47.$$

So $\chi^2$ is positive—the fit is not perfect. But is this just sampling variability, or is the model of the null hypothesis wrong?

Under $H_0$, $\chi^2 \sim \text{chi}^2(3)$, since there are four categories ($k = 4$). The graph of this distribution is shown in Figure 13-1. The calculated value of $\chi^2$ turns out to be in the *left* tail of the null distribution, whereas it is *large* values of $\chi^2$ that are extreme. The data are not inconsistent with $H_0$.

Indeed, the data fit the model *too* well! It is a bit surprising to have obtained a value as small as .47 when $H_0$ is true, but surely we cannot deny $H_0$ when the fit is so close to perfect—almost too good to be true. (Some have suggested that the gardener, having learned what Mendel was expecting, may have fudged the data to improve the fit; but this is pure speculation.)  ■

---

Chi-square test for $H_0$: $p_j = \pi_j$, $j = 1, 2, ..., k$: Given frequencies $Y_j$ in a random sample of size $n$, calculate

$$\chi^2 = \sum_1^k \frac{(Y_j - n\pi_j)^2}{n\pi_j}. \tag{2}$$

When $n$ is large, find the $P$-value as a right-tail area of $\text{chi}^2(k-1)$ in Table Vb of the Appendix.

---

One naturally wonders how large $n$ must be for the chi-square approximation to work well. This depends on the cell probabilities. A conservative rule of thumb is that the expected frequencies $n\pi_j$ should all be about 5 or larger. However, it has been found that when there are many cells, the approximation is good enough even if one or two expected frequencies are as small as 1. Cells with small expected frequencies

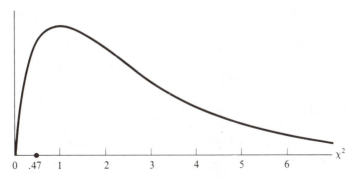

**Figure 13-1** Null distribution of $\chi^2$—Example 13.2a

can be combined to improve the approximation; but by combining them, we'd be testing a slightly different model—an approximation to the null hypothesis model.

The chi-square statistic can be used to test a distribution *type* without specifying a particular distribution of that type—for instance, to test the hypothesis that a population distribution is binomial, or Poisson, or normal. Such null hypotheses are *families* of distributions indexed by unknown parameters. It is necessary to specify the values of these parameters to define category probabilities. However, if we replace the parameters by sample estimates, we can calculate estimated mean frequencies for use as "expected frequencies" in the chi-square statistic (2).

What are appropriate estimates of the unknown parameters, and what does using them in place of the parameters do to the null distribution of $\chi^2$? With regard to the second question, using the sample to help define the null distribution seems like cheating, for it amounts to tailoring the null hypothesis to make it more like the sample. The fit then appears to be better than it really is. Since $\chi^2$ is apt to be smaller, this suggests that its true null distribution is shifted to the left.

With regard to the first question, we'll usually replace unknown parameters by their maximum likelihood estimates (§9.10). For the examples of this chapter, we'll use the sample mean, sample variance, and sample proportion as estimates of the corresponding population parameters. We let $\widehat{p}_j$ denote the estimated probability of category $j$, calculated using m.l.e.'s of any unknown parameters. It can be shown[2] that under $H_0$, the large-sample distribution of  the resulting value of the chi-square statistic

$$\chi^2 = \sum_1^k \frac{(Y_j - n\widehat{p}_j)^2}{n\widehat{p}_j},\tag{3}$$

is again approximately chi-square. However, the number of degrees of freedom must be reduced to $k - 1 - r$, where $r$ is the number of estimated parameters. Indeed, the

---

[2]See H. Cramér [5], 417ff.  This result assumes that unknown parameters are replaced by m.l.e.'s or by estimators asymptotically equivalent to m.l.e.'s.

distribution of $\text{chi}^2(k-1-r)$ is shifted to the left as compared with that of $\text{chi}^2(k-1)$. So any particular percentile of $\chi^2$ is decreased accordingly.

**Example 13.2b** | **Boy-Girl Ratio in Families**

Does the sex of successive children in a family behave like independent Bernoulli trials, with a fixed probability $p$ of having a boy? If so, the number of boys in a family of given size is binomially distributed. Consider families with eight children. The table below summarizes data on the numbers of boys in each of 1000 families with eight children.[3] We'll test the composite null hypothesis that the distribution of the number of boys among the eight children in a family is binomial, with unspecified $p$ (but with $n = 8$):

$$H_0:\ p_j = \binom{8}{j} p^j (1-p)^{8-j}, \quad j = 0, 1, ..., 8, \tag{4}$$

where $p_j$ is the probability of $j$ boys in a family. The number of boys among the 8000 children in these families is $\sum j Y_j = 4040$, so $\hat{p} = 4040/8000 = .505$. The category probabilities (3) are thus estimated to be

$$\hat{p}_j = \binom{8}{j} (.505)^j (.495)^{8-j}, \quad j = 0, 1, ..., 8.$$

Multiplying by 1000 yields the expected frequencies $n\hat{p}_j$.

| $j$ | $Y_j$ | $jY_j$ | $\hat{p}_j$ | $n\hat{p}_j$ |
|---|---|---|---|---|
| 0 | 10 | 0 | .0036 | 3.6 |
| 1 | 34 | 34 | .0294 | 29.4 |
| 2 | 111 | 222 | .1050 | 105.0 |
| 3 | 215 | 645 | .2143 | 214.3 |
| 4 | 239 | 956 | .2733 | 273.3 |
| 5 | 227 | 1135 | .2231 | 223.1 |
| 6 | 115 | 690 | .1138 | 113.8 |
| 7 | 34 | 238 | .0332 | 33.2 |
| 8 | 15 | 120 | .0042 | 4.2 |

Substituting the expected frequencies into (2), we find that $\chi^2 \doteq 50$. We estimated the one parameter $p$, so the number of degrees of freedom of $\chi^2$ is $k-1-r = 9-1-1 = 7$. The $P$-value is much smaller than .002—strong evidence against the hypothesis that the distribution is binomial.

---

[3]The data are artificial but follow the tendency observed in actual data, which are so extensive as to be quite convincing that the binomial model is not correct.

(The distribution apparently differs from binomial in a very special way: There are more families than expected in which one or the other of the sexes predominates. Data such as these could arise if the sex distribution within a family is binomial but different families have different $p$'s. Estimating the distribution of $p$'s from data such as the above is an interesting but difficult problem.) ∎

---

To test a discrete distribution whose probabilities $p_j$ depend on $r$ parameters: $p_j(\theta_1, ..., \theta_r)$, obtain maximum likelihood estimators $\widehat{\theta}_j$ and calculate $\chi^2$ with $p_j$ replaced by $p_j(\widehat{\theta}_1, ..., \widehat{\theta}_r) = \widehat{p}_j$:

$$\chi^2 = \sum_1^k \frac{(Y_j - n\widehat{p}_j)^2}{n\widehat{p}_j}.$$

Find the $P$-value as a right-tail area of $\mathrm{chi}^2(k - 1 - r)$.

---

The chi-square test of fit is for a *discrete* distribution with a finite number of categories. It can be adapted to test a discrete distribution with a countably infinite number of categories by lumping all the categories from some point on into one. It can also be adapted to test a continuous distribution by partitioning the axis of values into a finite number of class intervals (as when data are summarized) and using these as categories of a discrete approximation to the continuous model. The latter procedure is not entirely satisfactory, in part because it only tests the discrete approximation, but also because the chi-square statistic does not take any account of the ordering that is an essential aspect of a numerical variable—whether discrete or continuous.

## Problems

* **13-1.** Peanut M&M's, prior to the introduction of a safe red coloring agent, came in four colors: brown, orange, yellow, and green. We counted the colors in one 2-pound bag and found 106 brown, 105 orange, 62 yellow, and 87 green. Assuming this to be a random sample, test the manufacturer's claim that colors were being mixed in equal proportions.

**13-2.** We tossed a die 150 times and found these frequencies of the six sides: 28, 25, 30, 26, 22, 19. Use these data to test the hypothesis that (in the way we did the tossing) the six sides were equally likely.

* **13-3.** Suppose your friend believes that when two coins are tossed together, the numbers of heads showing, 0, 1, or 2, are equally likely. You decide to test this hypothesis by tossing two coins 100 times, obtaining 23 0's, 52 1's, 25 2's.

   **(a)** Carry out a chi-square test of the equal likelihood of 0, 1, 2.

   **(b)** Carry out a chi-square test of the hypothesis that obtaining two heads is just as probable as obtaining none, but half as probable as obtaining one.

**13-4.**   In the 1987 season, Gary Gaetti (then of the Minnesota Twins baseball team) played in 80 games in which he was at bat four times. He got four hits in 1 game, three hits in 2 games, two hits in 16, one hit in 35, and none in 26 games. Test the hypothesis that the number of hits in four at-bats has a binomial distribution (for this player).

**★ 13-5.**   An industrial plant compiled the following summary of accidents per month over a period of 32 months:  No accidents in 22 months, one in 6 months, two in 2 months, and three in 2 months. Test the hypothesis that the number of accidents per month follows a Poisson distribution. (Since $\chi^2$ is for testing a model with finitely many categories, combine the values 3, 4, ... into a single category "3 or more" and test the resulting finite model.)

**13-6.**   Example 7.5b (page 287) gives a frequency distribution of 50 spot-weld strengths. The mean and s.d. are found in that example to be $\overline{X} = 404.3$ and $S = 18.89$. Test the hypothesis of normality using a value of $\chi^2$ based on the class intervals in the frequency table.

**★ 13-7.**   Some years ago an extensive study of times to failure reported data on a variety of units and components.[4]  In each of the following cases, test the hypothesis that the time to failure follows an exponential distribution using a chi-square test. (To approximate the sample mean, use the class marks shown in the table.)

(a)   Fifth bus motor failures:

| Miles (1000's) | Failures | Class Mark |
|---|---|---|
| 0–20 | 29 | 8 |
| 20–40 | 27 | 27 |
| 40–60 | 14 | 45 |
| 60–80 | 8 | 68 |
| 80–up | 7 | 120 |

(b)   A type of radar indicator tube:

| Hours | Failures | Class Mark |
|---|---|---|
| 0–100 | 29 | 36 |
| 100–200 | 22 | 136 |
| 200–300 | 12 | 237 |
| 300–400 | 10 | 340 |
| 400–600 | 10 | 475 |
| 600–800 | 9 | 685 |
| 800–up | 8 | 1100 |

**13-8.**   The formula for the chi-square statistic is given in (1) of §13.2 in terms of *frequencies*. Show that if one uses *proportions* in place of frequencies (a common student error), the result is $1/n$ times the correct value of $\chi^2$.

---

[4]From D. J. Davis, "An analysis of some failure data," *J. Amer. Stat. Assn. 47* (1952), 147.

## 13.3   Tests Based on Distribution Functions

In this section we consider testing a specific *continuous* probability distribution in terms of the population c.d.f. $F(x)$:

$$H_0: \ F(x) = F_0(x) \ \text{ for all } x, \text{ against } H_A: \ F(x) \neq F_0(x) \ \text{ for some } x,$$

where $F_0$ is a completely defined c.d.f.

Although we have indicated how a chi-square test might be used, various tests have been designed especially for continuous distributions. Some of these compare $F_0(x)$ with the c.d.f. of the sample distribution:

---

**Sample distribution function:**

$$F_n(x) = \frac{\text{Number of observations } \leq x}{n}.$$ (1)

---

The sample d.f. characterizes the discrete distribution that assigns probability $1/n$ to each sample value. (We referred to this distribution at the beginning of §7.5.) Thus, $F_n$ is a step function, jumping an amount $1/n$ at each observation and staying constant between successive observations. If there are $k$ observations with the same value, then the jump at that value is $k/n$. (To be sure, when we assume that the observation $X$ has a continuous distribution, repeated values should not occur. However, data are necessarily recorded on the discrete scale defined by some round-off scheme.)

There are various ways of measuring the discrepancy between $F_n$ and $F_0$, some of them based on the absolute differences $|F_n(x) - F_0(x)|$, the (vertical) distances between $F_n$ and $F_0$. The *Cramér-von Mises* statistic measures "discrepancy" as the average squared difference—"average" meaning integral with respect to the weighting defined by $F_0$. The **Kolmogorov-Smirnov statistic** measures discrepancy as the maximum[5] absolute difference:

$$D_n = \max_x |F_n(x) - F_0(x)|.$$ (2)

Since $F_n$ is a step function, and $F_0$ is a continuous function, the largest vertical distance between them occurs at one of the jump points—at one of the sample values, say at the $i$th smallest, $X_{(i)}$. Then, $D_n$ is either

$$a_i = |F_0(X_{(i)}) - F_n(X_{(i)})| \ \text{ or } \ b_i = |F_0(X_{(i)}) - F_n(X_{(i-1)})|,$$ (3)

---

[5]To be technically correct, *maximum* should read *supremum*, or least upper bound. (See footnote 3 in Section 11.7, p. 501.)

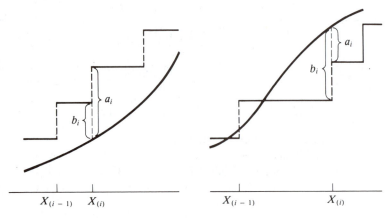

**Figure 13-2**

as is evident in the graphs shown in Figure 13-2.

There are recursion formulas for calculating tail-probabilities that have been used to construct Table IXb of the Appendix. This table gives "critical" values of $D_n$ for $\alpha = .01, .05, .10, .15,$ and $.20$, and for sample sizes 1 to 20 and 25, 30, 35. For $n > 35$, an approximation based on the known asymptotic distribution of $D_n$ can be used:

$$P(D_n > c) \doteq 2e^{-2nc^2}. \tag{4}$$

**Example 13.3a** | **Hemoglobin Level**

In setting control limits for a control chart such as that described in Example 11a, it is usually assumed that observations are normally distributed. In the chart of that example, it was assumed further that $\sigma = .40$. On ten consecutive days, the following measurements of hemoglobin were recorded, using a standard blood sample with hemoglobin level 15.5:

15.36, 14.24, 15.69, 15.07, 16.89, 15.21, 15.09, 15.52, 16.28, 15.56.

Provided that the test equipment did not change over the ten-day period, these can perhaps be considered to constitute a random sample from (under the null hypothesis) $\mathcal{N}(15.5, .40^2)$. Let $F_0$ denote the c.d.f. of this normal distribution. This and the sample d.f. are shown in Figure 13-3.

To find $D_n$ we could simply estimate it from the graph. Or, since this maximum distance occurs at a jump point, we can find $D_n$ by looking at Table 13-1, which shows the data, the corresponding $Z$-scores, the values of $F_0$ and $F_n$ at the data points, and the quantities $a_n$ and $b_n$ given by (3). The largest distance is $D_n = .174$, shown outlined in the table. The question is, how surprising is this value of $D_n$ if $H_0$ is true? In Table IXb we see that .174 is not in the tail of the null distribution of $D_{10}$. So there is no reason to question that the population c.d.f. is $F_0$—on the basis of the particular sample we used. (If we used the graph in Table IXa, we'd see that $10D_{10} = 1.74$ would have a $P$-value of at least .7.)

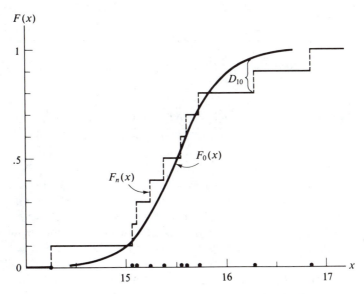

**Figure 13-3** $F_0$ and $F_n$ for Example 13.3a

**Table 13-1**

| $X_{(i)}$ | $Z_{(i)}$ | $F_0(X_{(i)})$ | $F_n(X_{(i)})$ | $a_i$ | $b_i$ |
|-----------|-----------|----------------|----------------|-------|-------|
| 14.24 | $-3.15$ | .001 | .1 | .099 | .001 |
| 15.07 | $-1.08$ | .140 | .2 | .060 | .040 |
| 15.09 | $-1.02$ | .154 | .3 | .146 | .046 |
| 15.21 | $-.72$ | .236 | .4 | .164 | .064 |
| 15.36 | $-.35$ | .363 | .5 | .137 | .037 |
| 15.52 | .05 | .520 | .6 | .080 | .020 |
| 15.56 | .15 | .560 | .7 | .140 | .040 |
| 15.69 | .48 | .684 | .8 | .116 | .016 |
| 16.28 | 1.95 | .974 | .9 | .074 | .174 |
| 16.89 | 3.47 | 1.000 | 1.0 | .000 | .100 |

The sample size is too small for a "large-sample" approximation of the $P$-value, but to illustrate its use we substitute the value .174 in (4):

$$P \doteq 2 \exp\{-20(.174)^2\} = 1.09.$$

[Although we didn't find the exact $P$-value (above), we did see that it is larger than .7, and it could be close to 1—but not larger than 1!]                                    ■

---

*Kolmogorov-Smirnov test* for a continuous c.d.f. $F_0(x)$:  Calculate

$$D_n = \max_x |F_n(x) - F_0(x)|,$$

where $F_n(x)$ is the sample c.d.f., and find the $P$-value or critical value in Table IX. For large $n$, an approximate $P$-value for $D_n = c$ is $2\exp(-2nc^2)$.

---

The K-S test has its good and bad points. On the good side, the small-sample distribution is known, and the test does not require a large sample. Also, it is distribution-free: The same table is used *no matter what $F_0$ is being tested*. On the bad side, it is designed with no very specific kind of alternative in mind. So we could not expect it to be as sensitive as the $Z$-test or the $t$-test in detecting a shift in mean value—an alternative for which those tests are especially designed. Thus, in the above example, if the alternative to the model $\mathcal{N}(15.5, .16)$ were $\mathcal{N}(\mu, .16)$ with $\mu \neq 15.5$, a $Z$-test would be better.

When $H_0$ involves unknown parameters, we could calculate a value for $D_n$ by using sample estimates in place of those parameters—just as we did in the case of the chi-square test (§13.2). As in that case, this replacement adjusts the model we're testing so as to be closer to the data, which means that the null distribution of $D_n$ is not what it is when the model being tested is completely specified. We need a new table. And, unfortunately, we'd need a new table for each problem. The effect of using parameter estimates in the test statistic is not known in general, but it has been studied empirically in the case of a test for normality. The result will be presented in the next section.

Although the sampling distribution of $D_n$ on which Table IX is based assumes that the population distribution is continuous, it can be used when the population is discrete. Doing so is conservative in the sense that the actual $P$-value does not exceed that given by the table, but it could be smaller.

**Example 13.3b** | **Testing Binomiality**

We return to the situation in Example 13.2b, where we tested the hypothesis that the distribution of the number of boys in a family with eight children is binomial. Here we'll use the data of that example to test the hypothesis that the distribution is binomial with $p = 1/2$.

The number of boys is a numerical random variable. So the K-S test, which involves the ordering on the number scale, may be more appropriate than the chi-square test, which ignores that ordering. But the variable is discrete. The following table gives the sample d.f., derived from the data in the earlier example, along with the population c.d.f. of the null hypothesis.

| $x$ | $F_n(x)$ | $F_0(x)$ | $|F_n(x) - F_0(x)|$ |
|---|---|---|---|
| 0 | .010 | .004 | .006 |
| 1 | .044 | .035 | .009 |
| 2 | .155 | .145 | .011 |
| 3 | .370 | .363 | .007 |
| 4 | .609 | .637 | .028 |
| 5 | .836 | .856 | .020 |
| 6 | .951 | .965 | .014 |
| 7 | .985 | .996 | .011 |
| 8 | 1.000 | 1.000 | 0.000 |

In this case, since both c.d.f.'s jump at the same points, there is no need (as in the usual application of the K-S test) to look at what we called "$b_n$" in (3). Only the differences *at* each $x$ are relevant. In the table, we see that the maximum difference (a true maximum) is .028, and the tail-area in the limiting distribution of the K-S statistic under $H_0$ is

$$P(D_n > .028) \doteq 2\exp\left[-2000(.028)^2\right] = 2e^{-1.57} \doteq .42.$$

This is not "significant." But the chi-square test carried out in Example 13.2b resulted in a $P$-value less than .01. How can this be? Observe that the differences between c.d.f.'s near 0 and near 8 are small; but it is the heavy tails that led to the low $P$-value using $\chi^2$, and (since the c.d.f.'s necessarily start out together and end together) heavy tails are not easily picked up using $D_n$. ∎

The idea behind the one-sample statistic we have been considering—using a single sample to test a proposed model—can be adapted to comparing two sample distributions, for testing the hypothesis that two population c.d.f.'s are exactly the same. Thus, we'll test

$$H_0\colon F(x) \equiv G(x) \ \text{ vs. } \ H_A\colon F(x) \neq G(x) \ \text{ for some } x,$$

using a statistic which is the maximum distance between the d.f.'s of random samples, one from $F(x)$ and one from $G(x)$.

Given independent random samples from continuous populations, we define the two-sample K-S statistic[6] as the maximum distance between the sample d.f.'s. The null distribution has been tabulated for small sample sizes. Critical values are given in Table X, which also includes asymptotic formulas for critical values and $P$-values.

---

[6]This two-sample version was actually proposed and studied by Smirnov; the test is sometimes referred to as the Smirnov test.

> Two-sample K-S statistic for testing for $F(x) \equiv G(x)$, based on independent random samples of sizes $m$ and $n$ with c.d.f.'s $F_m$ and $G_n$:
>
> $$D = \max_{x} |F_m(x) - G_n(x)|.$$
>
> Critical values and $P$-values are given in Table X.

The sizes of the jumps in the sample c.d.f.'s $F_m$ and $G_n$ are multiples of $1/m$ and $1/n$, respectively. This means that the difference between the c.d.f.'s will always be a multiple of the least common multiple of $m$ and $n$. And this is why the table of critical values for given significance levels for $m = 7$ and $n = 8$, for instance, are multiples of 56.

## Example 13.3c | Intermittent Feeding and Blood Pressure

Several earlier examples (12.4a) dealt with blood pressure data for rats in an experimental group that were fed intermittently and rats in a control group that were fed normally:

Normal feeding:       169  155  134  152  133  108  145
Intermittent feeding:   170  168  115  181  162  199  207  162

Figure 13-4 gives the sample d.f.'s. There we see that $D = \frac{6}{7} - \frac{1}{8} = \frac{41}{56}$:

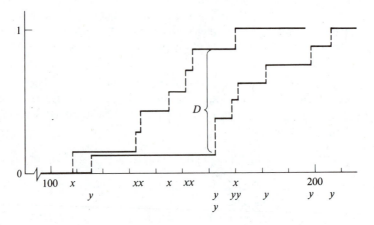

**Figure 13-4** Sample d.f.'s—Example 13.3c

The critical value for a test at level $\alpha = .01$ is shown in Table Xb as 42/56, so the $P$-value is slightly larger than .01.

(The sample sizes are not very "large," but we can try the large-sample approximation of $P$ using Case 2 of Table Xa, with

$$y = \sqrt{\frac{mn}{m+n}} \times D = \sqrt{\frac{56}{15}} \times \frac{41}{56} = 1.41.$$

The result is $P = .038$.)

For comparison, we note that the two-sample $t$-test (Example 12.4a) and the rank-sum test (Example 12.5b) yielded $P$-values of .044 and .028, respectively. These tests were designed to pick up differences in location, whereas the K-S test is a general-purpose test that is apt to be less sensitive in detecting specific types of alternatives. In the present example, however, it happens to provide about the same degree of evidence against the hypothesis of identical populations as do the location tests.                                                                         ∎

A comparison of two distributions can be carried out using a Q-Q plot. When the samples are of the same size, this is just a plot of the pairs of corresponding ordered observations (knowing which is equivalent to knowing the sample d.f.'s).

**Example 13.3d** | **Intermittent Feeding**

Returning to the data of the preceding example, suppose we have investigated the "outlier" 115 in the intermittent sample and determined that there was reason to delete it from the sample. The samples are then of the same size, and in Figure 13-5 we have plotted the pairs of corresponding ordered values, which are corresponding quantiles:

$$(108, 162), (133, 162), (134, 168), (145, 170), (152, 181), (155, 199), (169, 207).$$

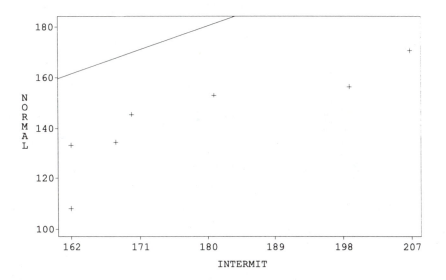

**Figure 13-5** Q-Q plot for Example 13.3d

On the graph, we have drawn a portion of the line $x = y$ and observe that the intermittent scores are consistently higher. The data points are not far from being on a line ($r = .86$), so there is not much here to suggest a difference in population distributions except in location.  ■

## Example 13.3e | A Sampling Experiment

We used a computer to generate independent samples of size 100 from populations on $(0, 1)$. One sample is from $\mathcal{U}(0, 1)$; the other is from a triangular distribution (the distribution of the average of two independent uniform random variables). We ordered the samples and paired corresponding ordered observations. The Q-Q plot in Figure 13-6 shows rather convincingly that the populations sampled are different, which is not surprising in view of the sample sizes we used.

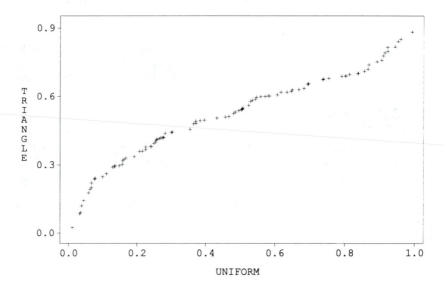

**Figure 13-6** Q-Q plot for Example 13.3e

We also carried out the two-sample Kolmogorov-Smirnov test, which was available in our software only for grouped data. Using class intervals of size .05 to obtain frequency distributions, we then called for a K-S test. The printout is shown in Figure 13-7. The observed maximum distance is .25, and from Table Xa we obtain $P = .004$. (The printout gives $P = .0039$).  ■

## 13.4    Testing Normality

The assumption of population normality is fundamental to the $t$-tests of Chapters 10 and 12, as well as to various tests yet to come. Normality can be tested using the chi-square statistic, as in Problem 13-6. However, a test using the K-S measure of dis-

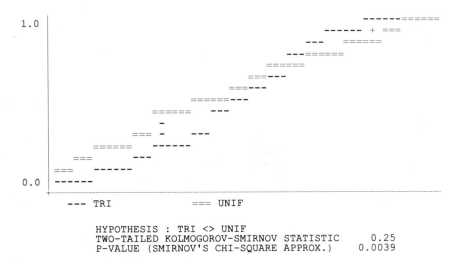

HYPOTHESIS : TRI <> UNIF
TWO-TAILED KOLMOGOROV-SMIRNOV STATISTIC        0.25
P-VALUE (SMIRNOV'S CHI-SQUARE APPROX.)      0.0039

**Figure 13-7** Computer printout for Example 13.3e

crepancy, with unknown mean and variance estimated from sample data, has been found to be generally more powerful against a wider class of alternatives.

For this test of normality, we define an $F_0$ for use in the usual K-S statistic as the c.d.f. of $\mathcal{N}(\overline{X}, S^2)$. The null distribution of the resulting test statistic is not known analytically, but extensive simulation, such as described in §8.5, has led to Table XI of the Appendix, due to Lilliefors. It gives some percentiles of $D_n$ that are accurate enough for practical purposes.[7]

**Example 13.4a** | **Sea Pollution**

A study of the effects of dumping wastes into the sea reports the following measurements of zinc concentration (in mg/kg) at 13 points within one mile of a dump site:[8]

13.5, 23.8, 23.3, 20.9, 23.8, 29.0, 20.9, 24.4, 16.4, 18.3, 17.6, 25.4, 23.3.

The mean and s.d. are $\overline{X} = 21.6$ and $S = 4.21$, and to test normality, we use these as the mean and s.d. of a normal distribution, defining

$$F_0(x) = \Phi\left(\frac{x - 21.6}{4.21}\right).$$

[7]H. W. Lilliefors, "On the Kolmogorov-Smirnov test for normality with mean and variance unknown," *J. Amer. Stat. Assoc.* 64 (1967), 399–402. Slight adjustments of his tabled values based on more extensive sampling are given in G. Dallal and L. Wilkinson, "An analytic approximation to the distribution of Lilliefors' test statistic for normality," *Amer. Statistician* 40 (1986), 294–96.

[8]K. J. Borwell, D. M. G. Kingston, and J. Webster, "Sludge disposal at sea—the Lothian experience," *Water Pollution Control* 85 (1986), 269–76.

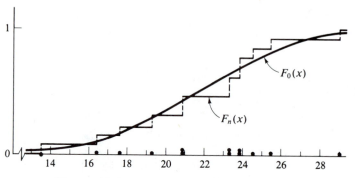

**Figure 13-8** $F_0$ and $F_n$ for Example 13.4a

Figure 13-8 shows this $F_0$ and the sample d.f. $F_n$ $(n = 13)$.

The largest discrepancy between $F_n$ and $F_0$ can be read approximately from such a graph. We can calculate it more precisely if we make a table like Table 13-1 (page 531), with $a_i$ and $b_i$ defined as in that table:

| $X_{(i)}$ | $F_0(X_{(i)})$ | $F_n(X_{(i)})$ | $a_i$ | $b_i$ |
|---|---|---|---|---|
| 13.5 | .0272 | 1/13 | .050 | .027 |
| 16.4 | .1084 | 2/13 | .045 | .031 |
| 17.6 | .1711 | 3/13 | .059 | .017 |
| 18.3 | .2165 | 4/13 | .091 | .014 |
| 20.9 | .4340 | 6/13 | .028 | .126 |
| 23.3 | .6568 | 8/13 | .041 | .195 |
| 23.8 | .6992 | 10/13 | .070 | .084 |
| 24.4 | .7470 | 11/13 | .099 | .022 |
| 25.4 | .8166 | 12/13 | .106 | .030 |
| 29.0 | .9606 | 13/13 | .039 | .038 |

As an example of the calculations, the entry under $F_0$ for $X_{(4)}$ is

$$\Phi\left(\frac{18.3 - 21.6}{4.21}\right) = \Phi(-.784) = .2165.$$

And for that entry, $a_4 = |.2165 - 4/13|$ and $b_4 = |.2165 - 3/13|$. (One could, of course, use the graph to see about where the maximum occurs and calculate only that portion of the above table in that vicinity.)

The value $D_n = .195$ is outlined in the above table. Table XI shows the area in the null distribution beyond $D_{13} = .193$ to be .20, so we conclude that $P > .20$. The sample sizes are small, but the data do not call the hypothesis of normality into serious question.  ∎

Yet another kind of test for normality is due to Shapiro and Wilk. It is based on a comparison of ordered sample values with their expected locations under the hypothesis of normality. This is done as follows.

Let $Z_{(i)}$ denote the $i$th smallest in a random sample from a *standard* normal distribution, and define $m_i = E[Z_{(i)}]$. These constants $m_i$ are called **normal scores** or **rankits.** They are given in Table XII of the Appendix for sample sizes 2 to 20. Now consider the pairs

$$(X_{(1)},\ m_1),\ (X_{(2)},\ m_2),\ ...,\ (X_{(n)},\ m_n),$$

where (as before) $X_{(i)}$ is the $i$th smallest value among the observations $X_i$ in a random sample from the population whose normality is to be tested. A scatter plot of the pairs $(X_{(i)},\ m_i)$ is called a **normal-scores plot** or a **rankit plot.**

If the population is normal, with mean $\mu$ and s.d. $\sigma$, then the $Z$-scores

$$Z_{(i)} = \frac{X_{(i)} - \mu}{\sigma}$$

are ordered observations from a standard normal distribution, and

$$E[X_{(i)}] = \sigma\, E[Z_{(i)}] + \mu = \sigma m_i + \mu.$$

This says that the expected values of the ordered sample values and the rankits $m_i$ are *linearly* related.

So, if the population really is normal, the correlation of the sample observations with the $m_i$ should be near 1. The actual correlation $r$ will generally not equal 1— because of sampling variability. But it should be high. We'll use the quantity $W = r^2$ as the test statistic and take an unexpectedly small value of $r^2$ as indicating a departure from normality of the population. The statistic $W$, although not exactly equal to the statistic proposed by Shapiro and Wilk, is a good approximation to it, and we'll refer to $W$ as the *Shapiro-Wilk* (or Wilk-Shapiro) statistic.

With some experience in interpreting such a plot, one can make an informal judgment as to population normality according as the plotted pairs are close to lying on a straight line, or not. For a formal test, we'd need to know the distribution of $r^2$. This is not known analytically, but tables have been given, developed in simulation studies. One such is given in the Appendix as Table XIV.[9]

**Example 13.4b** | **Sea Pollution (Continued)**

Consider once more the 13 zinc concentrations given in Example 13.4a. The (ordered) normal scores or rankits for $n = 13$, from Table XII, are paired with the

---

[9]The table is taken from S. Weisberg's *Multreg User's Manual* as adapted from S. Weisberg, "An empirical comparison of $W$ and $W'$," *Biometrika* 61 (1974), 645–46, and S. S. Shapiro and R. S. Francia, "An approximate analysis of variance test for normality," *J. Amer. Stat. Assoc.* 67 (1972), 215–16.

ordered sample observations as follows:

$(13.5, -1.668)$   $(16.4, -1.164)$   $(17.6, -.850)$   $(18.3, -.603)$
$(20.9, -.388)$   $(20.9, -.190)$   $(23.3, 0)$   $(23.3, .190)$   $(23.8, .388)$
$(23.8, .603)$   $(24.4, .850)$   $(25.4, 1.164)$   $(29.0, 1.668)$

The rankit plot is shown in Figure 13-9.

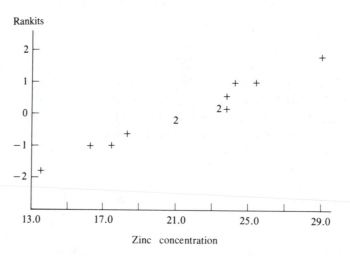

**Figure 13-9** Rankit plot for Example 13.4b

The sample points lie very nearly on a straight line, so there appears to be little evidence against normality. The correlation coefficient is $r = .979$, and $r^2 = .959$. We see in Table XIV that this is not in the left half of the null distribution of $W$. (For $n = 10$, the median is .94, and for $n = 15$ the median is about .95.) The $P$-value is larger than .5, and this is consistent with what we see in the rankit plot—little or no evidence against population normality.

This is not to be interpreted as "proving" normality, since the data would also give no evidence against other population distributions not far from normality. But it gives one a measure of confidence in subsequently using a test that assumes population normality or near normality.                                                      ■

Some researchers use a test of normality to decide whether or not a $t$-test is valid. Using such a preliminary test with the same data as is used in the $t$-test can make the calculated $P$-value incorrect. However, it is usually a good idea to make at least an informal analysis with a rankit plot to see that there is not an obvious and drastic departure from normality. When there is, it may be appropriate to transform the observations so that the transformed data are more nearly like data from a normal population. Alternatively, one can use a nonparametric procedure that does not assume population normality. (See §10.5 and 12.5.)

Yet another way of looking into normality is to plot sample quantiles (as defined in Chapter 7) against corresponding quantiles of a normal distribution. If the population sampled is actually normal, we'd expect the quantiles to match if the sample and population parameters agree, and to fall more or less on a straight line if we use any normal population. Because the quantiles of $\mathcal{N}(0, 1)$ are in fact quite close to the normal scores, a Q-Q plot and the normal scores or rankit plot are nearly indistinguishable, as we'll see in the next example. Some statistical packages include rankit plots; others use Q-Q plots.

**Example 13.4c** | **Sea Pollution (The End)**

The graph shown in Figure 13-10 is a plot of zinc concentration quantiles (in particular, the sample observations) against the corresponding normal quantiles. (It was produced using the software StataQuest, which does not provide for rankit plots.) With this plot, there is not a formal test—one that produces a $P$-value for the hypothesis of normality. However, the Q-Q plot is quite close to the rankit plot of the preceding example.                                                                      ■

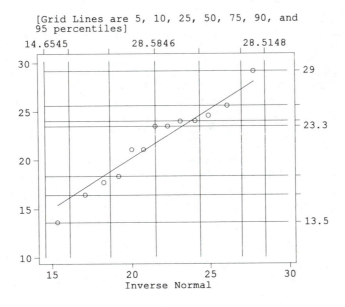

**Figure 13-10** Q-Q plot for Example 13.4c

## Problems

* **13-9.** The following are successive entries taken from a published table of "Gaussian (normal) deviates:"

.275  −.514  .982  −.071  −1.975  .027  .016  −.584  .669  .070

The table entries were generated using a computer program, and are supposed to be usable as a random sample from a standard normal population. Use the K-S statistic to test the hypothesis that they constitute a random sample from a standard normal population.

**13-10.**   The following are successive sequences of three digits taken from a table of "random digits." We have inserted decimal points to make them numbers on the interval (0, 1):

| .633 | .871 | .026 | .698 | .276 | .597 | .465 | .136 | .828 | .979 |
|------|------|------|------|------|------|------|------|------|------|
| .095 | .216 | .973 | .165 | .677 | .914 | .740 | .245 | .119 | .612 |

Test the hypothesis that the population sampled is $\mathcal{U}(0, 1)$.

\* **13-11.**   In Problem 13-3 are data on the number of heads showing in each of 100 tosses of two coins: 23 0's, 52 1's, and 25 2's. Using the K-S statistic, test each of these hypotheses:
   **(a)**  $f(0) = f(1) = f(2) = 1/3$.     **(b)**  $f(0) = f(2) = 1/4$.

**13-12.**   Use the K-S statistic to test the hypothesis that the following is a random sample from an exponential population with mean 1:

   .890  .038  .014  .297  .077  1.98  1.94  .150  1.39  1.77

[The population actually sampled was $\mathcal{U}(0, 2)$.]

**13-13.**   Use the K-S statistic to test the hypothesis that the data in Problem 13-5 (page 528) are from Poi(1/2).

\* **13-14.**   Apply the two-sample K-S test for the hypothesis of no difference between attitudes of firstborn and second-born children on role of the father against a two-sided alternative, using the data in Problem 12-17. The data are repeated here:
   1st:    40, 41, 44, 49, 53, 53, 54, 54, 56, 61, 62, 64, 65, 67, 67
   2nd:    23, 25, 38, 43, 44, 47, 49, 54, 55, 58, 58, 60, 66, 66, 72

**13-15.**   The study cited in Example 13.4a included these measurements of zinc concentration at various locations in an outer zone and an inner zone around an ocean dump site:
   Inner:    13.5 23.8 23.3 20.9 23.8 20.0 24.4 16.4 18.3 17.6 25.4 23.3
   Outer:    26.4 20.6 19.8 15.0 16.8 20.4 23.4 21.5

Use the two-sample K-S statistic to test the hypothesis of no difference in distributions of concentrations in the two zones.

\* **13-16.**   A pollution study gives measures of biological oxygen demand (BOD) and chemical oxygen demand (COD) for 18 samples of the effluent from a factory manufacturing detergents and related materials:[10]
   BOD:    881  104  74  118  74  46  59  96  78
               59  78  101  128  144  82  77  111  87
   COD:    580  674  512  540  616  298  960  570  640
               588  556  588  582  844  574  420  696  620

Make rankit plots for BOD and COD amounts and test each for normality.

---

[10]K. J. Shapland, "Industrial effluent treatability—a case study," *Water Pollution Control* 85 (1986), 75–80.

| **13.5** | **Testing Independence** |

A common problem in applied statistics is deciding whether two variables are related. In this section, we show how the chi-square statistic can be adapted to test the independence of two categorical variables.

**Example 13.5a** | **Arrowhead Breakage**

Seventy-seven fractured arrowheads were found in central Tennessee.[11] The investigator considered the null hypothesis that the location of fracture and the cause of fracture are unrelated. A cross-classification of the 77 arrowheads yielded the following frequencies, laid out in a contingency table:

|  |  | *Location* | | | |
|---|---|---|---|---|---|
|  |  | *Base* | *Middle* | *Tip* | |
| *Cause* | *Fire* | 21 | 8 | 18 | 47 |
|  | *Other* | 15 | 11 | 4 | 30 |
|  |  | 36 | 19 | 22 | 77 |

The proportions fractured by fire are different for the different locations, and the statistical question is whether these sample differences are explainable as random variations or are telling us that the true proportions caused by fire depend on location of the fracture. (To be continued.) ∎

The example calls for a test of the hypothesis of independence in a discrete bivariate population. Consider discrete variables $X$ with $r$ categories and $Y$ with $c$ categories. According to §2.5, $X$ and $Y$ are independent if and only if their joint probabilities factor into products of marginal probabilities. Denote the probability of the pair $(i, j)$ by $p_{ij}$, where $i$ is a category of $X$ and $j$ is a category of $Y$. Marginal probabilities are the row and column sums in the two-way table of joint probabilities:

$$P(X = i) = p_{i1} + \cdots + p_{ic} = p_{i+},$$
$$P(Y = j) = p_{1j} + \cdots + p_{rj} = p_{+j}. \tag{1}$$

(A "+" in a subscript will indicate that a summation has been performed over the corresponding index.) In this notation, the hypothesis of independence is

$$H_0: \ p_{ij} = p_{i+}p_{+j} \ \text{for all } i \text{ and } j. \tag{2}$$

The alternative is that this $H_0$ is false—that $X$ and $Y$ are not independent.

---

[11]Reported by J. L. Hofman in "Eva projectile point breakage at Cave Spring: Pattern recognition and interpretive possibilities," *Midcont. J. Arch.* 11 (1986), 79–95.

The null hypothesis gives the $rc$ cell probabilities in the joint distribution as functions of the $r-1$ parameters $p_{i+}$ and the $c-1$ parameters $p_{+j}$. We may then apply the chi-square statistic (2) of §13.2 to the case of a discrete distribution whose $rc$ category probabilities depend on unknown parameters. The number of degrees of freedom of $\chi^2$ is $rc-1$ minus the number of estimated parameters:

$$\text{d.f.} = rc - 1 - (r-1+c-1) = (r-1)(c-1).$$

Consider a random sample of $n$ observations on $(X, Y)$. Let $n_{ij}$ denote the frequency of the cell $(i, j)$. And, just as for the $p$'s, let a "+" in a subscript indicate a summation over that subscript. Thus, $n_{1+}$ is the number of observations in which $i = 1$:

$$n_{1+} = n_{11} + \cdots + n_{1c}.$$

The joint maximum likelihood estimates of the cell probabilities are the corresponding cell relative frequencies:

$$\widehat{p}_{ij} = \frac{n_{ij}}{n},$$

where $\sum n_{ij} = n$, the sample size. It follows that the m.l.e.'s of the marginal probabilities are

$$\widehat{p}_{i+} = \frac{n_{i+}}{n}, \quad \widehat{p}_{+j} = \frac{n_{+j}}{n}. \tag{3}$$

And under $H_0$, the m.l.e. of the cell expectation $np_{ij} = np_{i+}p_{+j}$ is

$$n\widehat{p}_{i+}\widehat{p}_{+j} = \frac{n_{i+}n_{+j}}{n} = e_{ij}. \tag{4}$$

[The "$e$" is for (estimated) expected frequency.]

## Example 13.5b  Arrowhead Breakage (Continued)

Returning to the data of Example 13.5a, we calculate two of the expected cell frequencies:

$$e_{11} = \frac{36 \times 47}{77} = 21.97, \quad e_{12} = \frac{19 \times 47}{77} = 11.60.$$

The other four $e_{ij}$ follow from these two by subtraction from marginal totals. The complete table of expected frequencies is as follows:

|  |  | Base | Middle | Tip |  |
|---|---|---|---|---|---|
| Cause | Fire | 21.97 | 11.60 | 13.43 | 47 |
|  | Other | 14.03 | 7.40 | 8.57 | 30 |
|  |  | 36 | 19 | 22 | 77 |

Location

Observe that in this table, as in the table of joint probabilities for independent variables, the rows are proportional and the columns are proportional. The observed frequencies in the contingency table do not have this property. If they did, the fit would be perfect. But in the present case, the chi-square statistic is

$$\chi^2 = \frac{(21-21.97)^2}{21.97} + \frac{(8-11.60)^2}{11.60} + \frac{(18-13.43)^2}{13.43} + \\ \frac{(15-14.03)^2}{14.03} + \frac{(11-7.40)^2}{7.40} + \frac{(4-8.57)^2}{8.57} = 6.97.$$

The number of degrees of freedom in this instance ($r = 2$ and $c = 3$) is d.f. $= 6 - 1 - 3 = (2-1)(3-1) = 2$. This is just the number of $e_{ij}$'s we had to calculate using (4) before obtaining the others by subtraction. From Table Vb we find $P = .030$, moderate evidence against $H_0$. ■

(In carrying out the calculation of $\chi^2$ by hand, you will wonder how much to round off. In the example, in the table of expected frequencies, we kept two decimal places. If we had rounded to one decimal place, we'd have found $\chi^2 = 7.025$, not far from 6.97. Rounding to the nearest integer would produce $\chi^2 = 8.44$, which is appreciably in error. Rounding expected frequencies to the nearest tenth will usually give sufficient accuracy.)

---

Chi-square test for independence in an $r \times c$ contingency table: Calculate

$$\chi^2 = \sum_i \sum_j \frac{(n_{ij} - e_{ij})^2}{e_{ij}}, \qquad (5)$$

where

$$e_{ij} = \frac{n_{i+}n_{+j}}{n}. \qquad (6)$$

Under $H_0$, $\chi^2$ has a chi-square distribution with $(r-1)(c-1)$ d.f.

---

The sampling scheme used in Example 13.5a was to take a random sample of individuals from a population and record two characteristics or variables for each population member (each arrowhead). An alternative scheme is to control or fix one variable, say $x$. It is possible to control a variable when it defines identifiable subpopulations, such as males and females. Or, suppose individuals are to be given different cold remedies; the remedy is a controlled variable. In contrast with the uncontrolled case, we can specify a sample size for each category of the controlled variable, and this fixes the marginal frequencies for that variable in the contingency table.

A null hypothesis corresponding to "no difference among subpopulations" is that the probability distributions in those subpopulations are the same. This is the hypothesis of **homogeneity.** In Example 2.5a we saw that when $X$ and $Y$ are independent categorical variables, the conditional distributions in the subpopulations defined by categories of one variable are identical. And in general, independence implies homogeneity of conditional distributions. It is also true that homogeneity implies independence, if we introduce a distribution on the categories of the controlled variable. So it is perhaps not surprising that we test homogeneity and independence in exactly the same way—using the same estimated expected frequencies and the same chi-square statistic.

| Example 13.5c | ### Behavior Modification |

A study of opinions on the usefulness of "behavior modification" reported the opinions of psychologists and of psychiatrists in this summary of data:[12]

|  | Never | Occasionally | Often | Always |  |
|---|---|---|---|---|---|
| Psychologist | 12 | 26 | 29 | 18 | 85 |
| Psychiatrist | 5 | 13 | 7 | 0 | 25 |
|  | 17 | 39 | 36 | 18 | 110 |

The psychologists and psychiatrists are two identified populations, and samples were selected from each. The null hypothesis is that the psychologists and psychiatrists divide in the same way in their opinions as to the degree of usefulness of behavior modification.

To apply the chi-square test, we need expected cell frequencies, and we calculate three of these (so d.f. $= 3$) using (4) and the others by subtraction:

|  | Never | Occasionally | Often | Always |  |
|---|---|---|---|---|---|
| Psychologist | 13.14 | 30.14 | 27.82 | 13.90 | 85 |
| Psychiatrist | 3.86 | 8.86 | 8.18 | 4.10 | 25 |
|  | 17 | 39 | 36 | 18 | 110 |

Using (5), we find $\chi^2 = 8.45$, and reference to Table Vb yields $P = .037$.

The authors reported that 280 questionnaires were sent out. There were 138 responses, but only 110 were usable. Such a low response rate really invalidates the study when, as was likely the case, the responses were voluntary. (A rule of thumb says that the response rate should be at least 80%.) If these are to be thought of as random samples, they are from populations of those who would respond, but not necessarily from populations of psychiatrists and psychologists generally.  ∎

---

[12]G. P. Koocher and B. M. Pedulla, "Current practices in child psychotherapy," *Profess. Psychol. 8* (1977), 275–86.

The preceding example is an instance of the many situations in which the categories of one or both of the variables are naturally ordered. However, the chi-square test does not take order into account. For instance, if the frequencies for "never" and "always" were reversed, $\chi^2$ would be unchanged. It might be better to use a test based on location. This can be done by coding the ordered categories, perhaps by rank, and applying a t-test (say).

**Example 13.5d** | **Behavior Modification—Another Approach**

We'll consider again the data of the preceding example, ignoring problems with the response rate. Suppose we adopt a coding by rank:

$$\text{Never} = 1, \quad \text{Occasionally} = 2, \quad \text{Often} = 3, \quad \text{Always} = 4.$$

The data can then be summarized in two frequency tables:

| Code $(x)$ | 1 | 2 | 3 | 4 | $n$ | Mean | S.d. |
|---|---|---|---|---|---|---|---|
| Psychologists | 12 | 26 | 29 | 18 | 85 | 2.62 | .976 |
| Psychiatrists | 5 | 13 | 7 | 0 | 25 | 2.08 | .702 |

Applying the $Z$-test from §12.2, we obtain

$$Z = \frac{2.62 - 2.08}{\sqrt{.976^2/85 + .702^2/25}} \doteq 3.09.$$

The two-sided $P$-value is .002, indicating strong evidence against equal average ranks—indeed, much stronger than the evidence that the two populations differ in some unspecified way. (Recall that with $\chi^2$, $P = .037$.)

Because this approach considers the ordering explicitly, we prefer it to a chi-square test but recognize that it depends on the assignment of numbers to categories. For instance, if "always" = 5 instead of 4, then $Z = 3.78$, whereas if "always" is combined with "often" and given the value 3, then $Z = 2.06$. ∎

We have considered here only the case of two categorical variables and the corresponding two-dimensional contingency table. A considerable amount of research has been directed at understanding higher-dimensional contingency tables. For an introduction to this area, see Fienberg et al. [9].

## Problems

∗ **13-17.** A study reports the following cross-classification of perceptions by 28 school superintendents and 43 school social workers as to the extent of social workers' involvement in the school entry process.[13]

---

[13]R. Constable and E. Montgomery, "Perception of the school social worker's role," *Social Work in Education* 7 (1985), 244–57.

|        | Perceived by | |
|--------|:---:|:---:|
|        | *Superintendents* | *Social workers* |
| *Much*   | 15% | 47% |
| *Little* | 85% | 53% |

Convert the percentages to frequencies and test for homogeneity of superintendents and social workers with respect to perception,

   **(a)**   using the chi-square test.

   **(b)**   using a $Z$-test.

∗ **13-18.**   The sizes of a person's left and right feet are seldom exactly the same. The following table cross-classifies 127 right-handed individuals according to sex and according to which shoe size is larger:[14]

|          | $L > R$ | $L < R$ | $L = R$ |
|----------|:---:|:---:|:---:|
| *Males*   | 2  | 28 | 10 |
| *Females* | 55 | 14 | 18 |

Test for independence of relative shoe size and sex in the population.

**13-19.**   An investigation of the relationship between sex and field of study used questionnaires to classify individuals as either masculine (M), feminine (F), androgynous (A), or undifferentiated (U).[15] The data were reported as shown in the following cross-classification:

|           | $M$ | $F$ | $A$ | $U$ |
|-----------|:---:|:---:|:---:|:---:|
| *Nonsci.*   | 8  | 6  | 2 | 14 |
| *Biol. Sci.* | 11 | 10 | 2 | 7  |
| *Phys. Sci.* | 4  | 9  | 8 | 9  |

Test the independence of these two variables.

**13-20.**   Various research studies have dealt with the relationship between church attendance and truancy from school. A 1986 study used data from a sample of teenagers aged 13–18 throughout the United States in 1975.[16]

---

[14]J. Levy and M. M. Levy, "Human lateralization from head to foot: Sex-related factors," *Science* 200 (1978), 1291.

[15]R. Baker, "Masculinity, femininity, and androgyny among male and female science and non-science majors," *School Science and Math* 84 (1984), 459–67.

[16]D. M. Sloane and R. H. Potvin, "Religion and delinquency: Cutting through the maze," *Social Forces* 65 (1986), 87–105.

|        |           | Truancy: | | |
|--------|-----------|-------|------------|----------|
|        |           | *Never* | *Occasional* | *Frequent* |
| | *Rare/never* | 91 | 68 | 136 |
| *Church:* | *Occasional* | 140 | 78 | 119 |
| | *Frequent* | 296 | 106 | 90 |

Test the hypothesis of independence of truancy and church attendance.

**∗ 13-21.** A survey of college students reports these data on smoking and cardiovascular disease (CVD) in their parents.[17] Test for independence.

|        | Smokers: | | | |
|--------|--------|--------|------|--------|
|        | *Father* | *Mother* | *Both* | *Neither* |
| *CVD* | 18 | 6 | 22 | 23 |
| *No CVD* | 35 | 27 | 18 | 75 |

("Father" means father only, and "Mother" means mother only.)

**13-22.** The data in the preceding problem can be used in testing an interesting hypothesis not involving CVD: Construct a $2 \times 2$ contingency table in which both variables (one for Father and one for Mother) have the categories $S$ (smoke) and $N$ (not smoke). Test the hypothesis that the smoking habits of spouses are independent.

**∗ 13-23.** One theory for the rise of Japanese companies in world markets centers on Japanese managers' ability to foster worker productivity. Seventy-two managers from Japan and 65 from the United States were asked this question: "How would you attempt to convince a consistently tardy employee to report to work on time?"[18] There were several categories of response, but we repeat here only the totals for those involving punishment and those not:

|        | *American* | *Japanese* |
|--------|----------|----------|
| *Punishment* | 48 | 29 |
| *Nonpunishment* | 17 | 43 |

**(a)** Use $\chi^2$ to test the hypothesis of homogeneity.

**(b)** Test homogeneity using a $Z$-test based on sample proportions.

**13-24.** Show that in testing homogeneity with a $2 \times 2$ contingency table, the value of Pearson's $\chi^2$ is equal to the square of the $Z$-statistic used when comparing two population proportions.

---

[17]"Cardiovascular risk factors and health knowledge among freshman college students ...," *J. Amer. College Health* 34 (1986), 267–70.

[18]R. Hirokawa and A. Miyahara, "A comparison of influence strategies utilized by managers in American and Japanese organizations," *Comm. Quart.* (1986), 250–65.

* **13-25.**   The possibility of a link between coffee drinking and coronary heart disease (CHD) was investigated in a study that reported the following data.[19] Test for independence using $\chi^2$.

|          | Cups per day | | | |
|----------|------|------|------|------|
|          | 0    | 1–2  | 3–4  | $\geq 5$ |
| *CHD*    | 4    | 17   | 17   | 9    |
| *No CHD* | 185  | 457  | 226  | 125  |

## 13.6          A Likelihood Ratio Test for Goodness of Fit

The likelihood ratio method introduced in §11.6 is applicable in testing goodness of fit. We consider first the testing of a completely defined model for a categorical variable with finitely many categories, as in §13.2. In the notation of earlier sections, the frequency of category $j$ in a random sample of size $n$ is $Y_j$, for $j = 1, 2, ..., k$. The likelihood function is

$$L(\mathbf{p}) = p_1^{Y_1} \cdots p_k^{Y_k}, \tag{1}$$

where $p_j$ is the probability of category $j$. To construct a likelihood ratio test of the hypothesis $\mathbf{p} = \boldsymbol{\pi}$, for a model with given constants $\pi_j$ as probability of category $j$, we need the maximum of the likelihood (1) over all $\mathbf{p}$ in the unit hypercube (all $p_j$ on the interval $[0, 1]$). We saw in Example 9.10e that (1) has its maximum value when each $p_j$ is replaced by the corresponding sample relative frequency, $Y_j/n$, which we call $\widehat{p}_j$. That maximum value is

$$L(\widehat{\mathbf{p}}) = \sup L(\mathbf{p}) = \left(\frac{Y_1}{n}\right)^{Y_1} \cdots \left(\frac{Y_k}{n}\right)^{Y_k}. \tag{2}$$

The likelihood ratio is thus

$$\Lambda = \frac{L(\boldsymbol{\pi})}{L(\widehat{\mathbf{p}})} = \prod_{j=1}^{k} \frac{\pi_j^{Y_j}}{(Y_j/n)^{Y_j}}.$$

For a large sample test, we use the fact that $-2 \log \Lambda$ is approximately $\text{chi}^2(k-1)$, where the d.f. $k-1$ is the number of free parameters assigned values by $H_0$.

## Example 13.6a | A Fair Die

Problem 13-2 gave the frequencies 28, 25, 30, 26, 22, 19 as the results of 150 tosses of a die. To test the hypothesis that the six sides are equally likely, we calculate the

---

[19]A. Z. LaCroix et al., "Coffee consumption and the incidence of coronary heart disease," *New Engl. J. Med.* 315 (1986), 977–82.

likelihood ratio:

$$\Lambda = \frac{\left(\frac{1}{6}\right)^{150}}{\left(\frac{28}{150}\right)^{28}\left(\frac{25}{150}\right)^{25}\left(\frac{30}{150}\right)^{30}\left(\frac{26}{150}\right)^{26}\left(\frac{22}{150}\right)^{22}\left(\frac{19}{150}\right)^{19}}.$$

The (natural) logarithm is

$$\log \Lambda = 150(\log 150 - \log 6) - 28 \log 28 - 25 \log 25 - 30 \log 30$$

$$- 26 \log 26 - 22 \log 22 - 19 \log 19 = -1.636.$$

So $-2 \log \Lambda = 3.27$, and the $P$-value (from the chi-square table, 5 d.f.) exceeds .50. The data are not inconsistent with the null hypothesis.

The null distribution of the Pearson's $\chi^2$ is also chi$^2$(5), and the value of $\chi^2$ (see Problem 13-2) is 3.2. Indeed, it can be shown that not only do the statistics $-2 \log \Lambda$ and $\chi^2$ have the same asymptotic distribution, they are asymptotically equal under $H_0$.[20]    ∎

We can also use $\Lambda$ to test a *composite* null hypothesis, one of the form

$$H_0: \ p_1 = \pi_1(\theta), \ ..., \ p_k = \pi_k(\theta),$$

where $\theta$ is an unknown parameter. The likelihood function under $H_0$ is

$$L(\theta) = [\pi_1(\theta)]^{Y_1} \cdots [\pi_k(\theta)]^{Y_k}, \tag{3}$$

which has its maximum at $\widehat{\theta}$, the m.l.e. of $\theta$. The likelihood ratio is then

$$\Lambda = \frac{L(\widehat{\theta})}{L(\widehat{\mathbf{p}})} = \prod_{j=1}^{k} \frac{[\pi_j(\widehat{\theta})]^{Y_j}}{(Y_j/n)^{Y_j}}. \tag{4}$$

If $\theta$ is a vector parameter with $m$ components, the number of degrees of freedom for the asymptotic chi-square distribution of $-2 \log \Lambda$ is $k - 1 - m$.

| Example **13.6b** | **Kirby Puckett** |

Consider the hypothesis that a baseball player's successive times at bat in a game are independent Bernoulli trials, with fixed $p$—the player's "true" probability of getting a hit. The number of hits in four at-bats is then Bin(4, $p$) for some $p$. The following table gives the distribution of the number of hits by Kirby Puckett of the Minnesota Twins in the 1987 regular season, in the 78 games in which he had four official times at bat:

| *Number of hits* | 0 | 1 | 2 | 3 | 4 |
|---|---|---|---|---|---|
| *Frequency* | 19 | 26 | 26 | 4 | 3 |

---

[20]See Lindgren [15], 366.

The null hypothesis is

$$H_0: \; p_j = \binom{4}{j} p^j (1-p)^{4-j}, \;\; j = 0, \, 1, \, 2, \, 3, \, 4.$$

The alternative is that this is not correct—that the number of hits is not binomial for any $p$. The likelihood function is

$$L(p_0, \, ..., \, p_4) = p_0^{19} p_1^{26} p_2^{26} p_3^4 p_4^3,$$

where $p_0 + \cdots + p_4 = 1$. Under $H_0 \cup H_A$, this has its maximum value when $p_j$ is $Y_j/78 = \widehat{p}_j$:

$$\sup L(\mathbf{p}) = L(\widehat{\mathbf{p}}) = \left(\frac{19}{78}\right)^{19} \left(\frac{26}{78}\right)^{26} \left(\frac{26}{78}\right)^{26} \left(\frac{4}{78}\right)^4 \left(\frac{3}{78}\right)^3.$$

Under $H_0$, the likelihood function is

$$L_0(p) = [(1-p)^4]^{19}[p(1-p)^3]^{26}[p^2(1-p)^2]^{26}[p^3(1-p)]^4[p^4]^3$$
$$= p^{102}(1-p)^{210}.$$

This has its maximum value at $\widehat{p} = 102/312 \doteq .3269$ (number of hits divided by number of times at bat—Puckett's batting average in the 78 games). So,

$$\sup L_0(p) = (.3269)^{102}(.6731)^{210},$$

and

$$\log \Lambda = \log[L_0(\widehat{p})] - \log[L(\widehat{\mathbf{p}})] \doteq -3.409.$$

Then $-2 \log \Lambda \doteq 6.82$ ($5 - 1 - 1$ d.f.), and $P \doteq .078$ (Table Vb).
   [Applying the chi-square test of §13.2 yields $\chi^2 \doteq 8.4$ and $P \doteq .038$. Both $P$-values are small and of the same order of magnitude. But they are based on different approximations and will not generally agree. Perhaps the small (estimated) expected frequency for four hits (about .89) makes the chi-square the less reliable.]    ∎

The likelihood ratio statistic can also be applied to the data in an $r \times c$ contingency table, to test independence. The likelihood function, given cell frequencies $n_{ij}$, is

$$L(p_{11}, \, ... \, p_{rc}) = \prod_{i,j} p_{ij}^{n_{ij}}. \tag{5}$$

The joint m.l.e.'s of the $p_{ij}$'s are the cell relative frequencies: $\widehat{p}_{ij} = n_{ij}/n$, where $n$ is the sample size.
   Under $H_0$: $p_{ij} = p_{i+} p_{+j}$, there are $(r-1) + (c-1)$ free parameters, and their joint m.l.e.'s are the corresponding marginal relative frequencies. So the cell

probabilities are estimated (under $H_0$) to be

$$\widehat{p}_{ij0} = \frac{n_{i+}}{n} \cdot \frac{n_{+j}}{n} = \frac{n_{i+}n_{+j}}{n^2}.$$

The likelihood ratio is thus

$$\Lambda = \frac{\prod\limits_{i,j}\left(\frac{n_{i+}n_{+j}}{n^2}\right)^{n_{ij}}}{\prod\limits_{i,j}\left(\frac{n_{ij}}{n}\right)^{n_{ij}}} = \frac{\prod\limits_{i}(n_{i+})^{n_{i\cdot}}\prod(n_{+j})^{n_{\cdot j}}}{n^n\prod\limits_{i}\prod\limits_{j}n_{ij}^{n_{ij}}}. \tag{6}$$

Taking logarithms, we obtain the following:

---

Large-sample likelihood ratio statistic for testing independence:

$$-2\log\Lambda = 2n\log n + 2\sum_{1}^{r}\sum_{1}^{c}n_{ij}\log n_{ij}$$

$$-2\sum_{1}^{r}n_{i+}\log n_{i+} - 2\sum_{1}^{c}n_{+j}\log n_{+j}.$$

Under independence, $-2\log\Lambda \approx \text{chi}^2[(r-1)(c-1)]$.

---

[With $rc$ cells and $r + c - 2$ parameters to be estimated, the asymptotic distribution has $rc - 1 - (r + c - 2) = (r-1)(c-1)$ degrees of freedom.]

**Example 13.6c** | ## Coffee and Heart Disease

In Problem 13-25 we used $\chi^2$ to test for independence in this table:

|        | 0   | 1–2 | 3–4 | ≥ 5 |      |
|--------|-----|-----|-----|-----|------|
|        |     | *Cups per day* |  |  |      |
| CHD    | 4   | 17  | 17  | 9   | 47   |
| No CHD | 185 | 457 | 226 | 125 | 993  |
|        | 189 | 474 | 243 | 134 | 1040 |

Using the likelihood ratio statistic, we have

$$-\log\Lambda = 1040\log 1040 + 4\log 4 + 17\log 17 + \cdots + 125\log 125$$

$$- 189\log 189 - \cdots - 47\log 47 - 993\log 993 = 4.22$$

So $-2\log\Lambda = 8.44$, and with 3 d.f., $P = .038$.

Again we have a case in which the variable "cups per day" has ordered categories. As in Example 13.5d, we can code these categories and test for a difference in means. Suppose we adopt the coding 0 for 0 cups, 1.5 for 1–2 cups, 3.5 for 3–4 cups, and 6 for 5 or more cups. For the coded data, the summary statistics are as follows:

|  | $n$ | Mean | SD |
|---|---|---|---|
| CHD | 47 | 2.957 | 1.862 |
| No CHD | 993 | 2.242 | 1.825 |

The $Z$-statistic is

$$Z = \frac{2.96 - 2.24}{\sqrt{1.86^2/47 + 1.83^2/993}} \doteq 2.6.$$

The two-sided $P$-value is .01; the evidence against independence is somewhat stronger with this analysis. ∎

## 13.7    Testing Homogeneity Using Paired Data

### Example 13.7a | Tonsillectomies and Hodgkin's Disease

A study involved 85 patients with Hodgkin's disease.[21] Each of these had a normal sibling (one who did not have the disease). In 26 of the pairs, both individuals had had tonsillectomies (T); in 37 pairs, both individuals had not had tonsillectomies (N); in 15 pairs, only the normal individual had had a tonsillectomy; in 7 pairs, only the one with Hodgkin's disease had had a tonsillectomy:

|  |  | Normal | |
|---|---|---|---|
|  |  | T | N |
| Patient | T | 26 | 15 |
|  | N | 7 | 37 |

A goal of the study was to determine whether there was a link between the disease and having had a tonsillectomy. Is the proportion of those who had tonsillectomies the same among those with Hodgkin's disease as among those who don't have it? (To be continued.) ∎

Consider a response variable with two categories, which we observe for each individual in $n$ randomly selected matched pairs. In each pair, one is a "treatment" subject and the other a "control." We adopt the 0–1 coding for the variable of interest

---

[21]S. Johnson and R. Johnson, "Tonsillectomy history in Hodgkin's disease," *New Engl. J. Med.* 287 (1972), 1122–25.

and also for treatment control. Each pair will then be in one of these four cells: $(0, 0)$, $(0, 1)$, $(1, 0)$, $(1, 1)$. Given a random sample of $n$ pairs, each pair will fall in one of these categories, so the category frequencies are multinomial. The frequencies and corresponding probabilities are denoted as follows:

|  | Data: 0 | 1 |  |
|---|---|---|---|
| 0 | $n_{00}$ | $n_{01}$ | $n_{0+}$ |
| 1 | $n_{10}$ | $n_{11}$ | $n_{1+}$ |
|  | $n_{+0}$ | $n_{+1}$ | $n$ |

|  | Model: 0 | 1 |  |
|---|---|---|---|
| 0 | $p_{00}$ | $p_{01}$ | $p_{0+}$ |
| 1 | $p_{10}$ | $p_{11}$ | $p_{1+}$ |
|  | $p_{+0}$ | $p_{+1}$ | 1 |

The hypothesis to be tested is $H_0$: $p_{1+} = p_{+1}$. We can't use a standard comparison of proportions (as in §12.3) because we don't have independent samples—the data are paired.

The null hypothesis is equivalent to $p_{10} = p_{01}$. Intuition suggests that there is no information about this hypothesis in the main-diagonal frequencies, $n_{00}$ and $n_{11}$. Consider the conditional probability of $(0, 1)$, given the off-diagonal total:

$$p_* = \frac{p_{01}}{p_{01} + p_{10}}.$$

In terms of $p_*$, the null hypothesis is $p_* = 1/2$. The $Z$-score for testing this hypothesis based on $n_{01}$ successes in $n_{01} + n_{10}$ trials (§10.3) is

$$Z = \frac{n_{01} - \frac{1}{2}(n_{01} + n_{10})}{\frac{1}{2}\sqrt{(n_{01} + n_{10})}} = \frac{n_{01} - n_{10}}{\sqrt{n_{01} + n_{10}}}.$$

In a two-sided test, large values of $|Z|$ or of $Z^2$ are extreme:

$$Z^2 = \frac{(n_{01} - n_{10})^2}{n_{01} + n_{10}}. \tag{1}$$

The asymptotic null distribution of $Z^2$ is $\text{chi}^2(1)$, and the test based on (1) using the chi-square table is called **McNemar's test.**

The test is in fact equivalent to a chi-square test: The four cell probabilities $p_{ij}$ are functions of two parameters, $p_{11}$ and $p_*$. The likelihood function under $H_0$ is a multinomial probability:

$$L(p_{11}, p_*) = p_{11}^{n_{11}} p_*^{n_{10}} p_*^{n_{01}} (1 - 2p_* - p_{11})^{n_{00}}. \tag{2}$$

Maximizing (after taking logs on both sides), we find these m.l.e.'s:

$$\widehat{p}_{11} = \frac{n_{11}}{n}, \quad \widehat{p}_* = \frac{n_{01} + n_{10}}{2n}, \quad \widehat{p}_{00} = \frac{n_{00}}{n}. \tag{3}$$

Using these estimates to form estimates of expected cell frequencies, one can show that Pearson's chi-square statistic is identical with McNemar's statistic. The next example illustrates the equivalence.

**Example 13.7b** | **Tonsillectomies and Hodgkin's Disease (Continued)**

Consider again the data in Example 13.7a. Substituting $n_{01} = 15$ and $n_{10} = 7$ into (1), we find $Z^2 = 2.91$. Table IIa of the Appendix gives $P \doteq .09$. (This can also be read from Table Vb, with 1 d.f.)

To calculate Pearson's chi-square statistic, we arrange the cells in a column, as in §13.2, along with observed frequencies, estimates of category probabilities under $H_0$ [using the m.l.e.'s (3)]:

| Category | $f$ | $\widehat{p}$ | $n\widehat{p}$ |
|----------|-----|---------------|----------------|
| 00 | 26 | 26/85 | 26 |
| 01 | 15 | 11/85 | 11 |
| 10 | 7 | 11/85 | 11 |
| 11 | 37 | 37/85 | 37 |

(Observe the perfect fit in the main diagonal cells, where we anticipated no information to be available.)  The test statistic is

$$\chi^2 = \frac{(7 - 11)^2}{11} + \frac{(15 - 11)^2}{11} = 2.91.$$

This is identical with what we found as $Z^2$, above.                                    ∎

We can also use a likelihood ratio test for $p_{01} = p_{10}$: Under $H_0$, the likelihood function (2) has a maximum when we substitute the m.l.e.'s (3). In the unrestricted model, the m.l.e.'s of the cell probabilities are the corresponding cell relative frequencies. In the likelihood ratio statistic, the factors that involve $n_{00}$ and $n_{11}$ cancel, leaving

$$\Lambda = \frac{\left(\frac{n_{01}+n_{10}}{2n}\right)^{n_{01}+n_{10}}}{\left(\frac{n_{01}}{n}\right)^{n_{01}} \left(\frac{n_{10}}{n}\right)^{n_{10}}}. \tag{4}$$

**Example 13.7c** | **Tonsillectomies (Conclusion)**

Substituting the data from the preceding examples into (3), we obtain the estimates

$$\widehat{p}_{11} = \frac{37}{85}, \quad \widehat{p}_{00} = \frac{26}{85}, \quad \widehat{p}_* = \frac{7 + 15}{2 \times 85} = \frac{11}{85}.$$

Then,

$$\Lambda = \frac{11^{22}}{7^7 15^{15}} = .22573,$$

and $-2 \log \Lambda \doteq 2.98$. The three parameters in the general likelihood were restricted by $H_0$ to two, so d.f. $= 1$, for the approximating chi-square distribution. In Table Vb we find $P \doteq .08$.                                                                                      ■

As in these examples, McNemar's test and the likelihood ratio test will usually give similar results. We gave the latter as yet another illustration of the likelihood ratio method, a method that is especially useful in analyzing higher-dimensional contingency tables.

## Problems

**∗ 13-26.**   Test the manufacturer's claim in Problem 13-1 using $\Lambda$.

**13-27.**   Use the likelihood ratio method with the data in Problem 13-3 to test the hypothesis in Problem 13-3(a)—that the probabilities are 1/3, 1/3, and 1/3.

**∗ 13-28.**   For the data of Problem 13-23, calculate $-2 \log \Lambda$, where $\Lambda$ is the likelihood ratio statistic for testing independence. Find the approximate $P$-value.

**13-29.**   Four instructors, each teaching a section of the same statistics course, assigned final grades as follows:

| Instructor | 1 | 2 | 3 | 4 |
|------------|-----|-----|-----|-----|
| A | 13 | 8 | 31 | 11 |
| B | 21 | 20 | 45 | 18 |
| C | 9 | 35 | 20 | 20 |
| D | 2 | 16 | 8 | 4 |

Are the patterns of grade assignment different for the four instructors? Test using the likelihood ratio.

**∗ 13-30.**   In the context of $2 \times 2$ contingency tables, consider the null hypothesis defined by the following table of joint probabilities:

|       | $B_1$ | $B_2$ |
|-------|-------|-------|
| $A_1$ | $p^2$ | $p(1-p)$ |
| $A_2$ | $p(1-p)$ | $(1-p)^2$ |

Under this hypothesis, $A$ and $B$ are independent and identically distributed.

    **(a)**   Find the m.l.e.'s of the parameters in $H_0$ in terms of cell frequencies.

    **(b)**   Using the data in Example 13.7b, test $H_0$ against the hypothesis that $A$ and $B$ are merely exchangeable, using the likelihood ratio method.

    **(c)**   Use the likelihood ratio statistic with the data of Example 13.7b to test $H_0$ against the hypothesis that the cell probabilities are unrestricted (except that their sum is 1). Observe the relationship among the values of $-2 \log \Lambda$ here, in (b), and in Example 13.7b.

**13-31.** A multicenter trial was conducted to assess the possibility of preventing blindness in premature babies by temporarily freezing the whites of their eyes.[22] In the study, 135 babies with disease in both eyes were randomized—half of them had their left eyes treated and the other half, their right eyes. The outcomes in both eyes were the same in all but 40 babies. In 34 of these, the outcome in the treated eye was favorable while that in the untreated eye was unfavorable, and in the other six, the outcome in the treated eye was unfavorable while that in the untreated eye was favorable. Use McNemar's test to judge whether the treatment is effective.

---

## Review Problems

**13-R1.** Test the hypothesis that a tetrahedral die (four sides, equally likely), if 72 tosses result in 12 1's, 16 2's, 20 3's, and 24 4's.

**13-R2.** Test the hypothesis that observations in the following random sample are from $\mathcal{U}(0, 1)$:

.88 .64 .60 .95 .35 .64 .82 .24 .79 .70.

**13-R3.** Forty students reported on their parents' smoking habits: neither parent smoked in 24 cases, the father but not the mother in 3, and the mother but not the father in 10. Test the hypothesis that the proportions of smokers is the same for mothers as for fathers.

**13-R4.** Polls A and B, each based on random samples of size 1000, reported these frequencies, for how an individual would vote on a proposition that is to appear in a coming election:

| | | |
|---|---|---|
| Yes | 500 | 520 |
| No | 300 | 320 |
| Undecided | 200 | 160 |

Test the hypothesis that the proportions in the corresponding population categories were the same in the populations sampled,

(a)  using the $\chi^2$ statistic.
(b)  using the likelihood ratio statistic.

**13-R5.** In a random sample of size 100, responses were distributed as shown in the following table, which also shows category probabilities (depending on a parameter $\theta$) defining the null hypothesis:

| | | |
|---|---|---|
| A | $\theta^2$ | 32 |
| B | $\theta(1 - \theta)$ | 16 |
| C | $\theta(1 - \theta)$ | 20 |
| D | $(1 - \theta)^2$ | 32 |

---

[22]"Multicenter trial of cryotherapy for retinopathy of prematurity," *Archives of Ophthalmology* 106 (1988), 471–79.

**(a)** Find the likelihood function.

**(b)** Obtain the maximum likelihood estimate of $\theta$.

**(c)** Test the null hypothesis using $\chi^2$.

**13-R6.** Consider again the Fisher data on sleeping drugs from Problem 10-18:

1.2, 2.4, 1.3, 1.3, 0.0, 1.0, 1.8, 0.8, 4.6, 1.3.

Construct a rankit plot and calculate the value of the Shapiro-Wilk statistic $W$. Conclusion?

**13-R7.** Air Force lore says that fighter pilots are more likely to sire daughters than sons. A study investigating the possibility that high $G$-force exposure may be a factor found the following data, for the variables "sex of offspring" and "$G$-force exposure level":[23]

|        | Low | High |
|--------|-----|------|
| Male   | 295 | 66   |
| Female | 287 | 100  |

Test the hypothesis of no relationship between these variables.

**13-R8.** The following are residual flame times (sec) for strips of treated night wear for children:[24]

9.85, 9.94, 9.88, 9.93, 9.85, 9.95, 9.75, 9.75, 9.95, 9.77,
9.83, 9.93, 9.67, 9.92, 9.92, 9.87, 9.74, 9.89, 9.67, 9.99.

**(a)** Obtain a rankit plot and find the value of the Shapiro-Wilk statistic for a test of population normality.
[Rankits are given in Table XII for $n = 20$, but if you have access to computer software, do this problem on the computer. But don't bother to enter the "9."]

**(b)** Explain why it is not necessary to include the "9." in doing (a).

**13-R9.** Fifty patients in a U.S. Air Force hospital with active duodenal ulcerations were assigned at random to either a treatment regimen involving an antacid or to one involving a placebo made of starch, lactose, sodium, saccharine, and flavoring.[25] After 30 days of treatment, a "healing" was recorded if the ulcer was reduced in size by two-thirds. Results are as follows:

|         | Healing | No healing |
|---------|---------|------------|
| Antacid | 24      | 3          |
| Placebo | 17      | 6          |

Test for independence using $\chi^2$, and using the likelihood ratio.

---

[23]B. B. Little et al., "Pilot and astronaut offspring: Possible G-force effects on human sex ratio," *Aviation Space & Environmental Medicine* 58 (1987), 707–9.

[24]"An introduction to some precision and accuracy of measurement problems," *J. of Testing and Evaluation* (1982), 132–40.

[25]Reported in an article by D. Hollander and J. Harlan in *J. Amer. Med. Assn.* 226 (1973), 1181–85.

## Chapter Perspective

Seeing how well data fit a proposed model is the idea behind most statistical tests, including the $Z$- and $t$-tests of the preceding chapters. But "goodness-of-fit tests" are special in that they are not directed at particular classes of alternatives (parametric or nonparametric). So they are less sensitive to special classes of alternatives than are tests specifically designed to detect them. A goodness-of-fit test may detect gross deviations from a null hypothesis, but the lack of sensitivity of such tests limits their usefulness. In spite of this, we recommend normal-scores (rankit) plots and the Shapiro-Wilk test for normality because they are sensitive to the kind of nonnormality for which $t$-tests are most inappropriate.

Goodness-of-fit tests, like many of the tests we've encountered, are tests of very precise null hypotheses. Making a null hypothesis precise allows for calculating a $P$-value. A small $P$-value is evidence against the null hypothesis, but a large $P$-value should not be taken as evidence that $H_0$ is true. For instance, in a test of normality, failure to reject the hypothesis does not mean that the population is normal; there are many other distributions or types of distributions that would fit at least as well, and the normal distribution plays no special role. A goodness-of-fit test can easily mislead.

# Analysis of Variance

In this chapter we introduce tests for treatment effects, including tests that also take into account other factors that may affect responses. Suppose there are $k$ treatments or levels of treatment; each defines a population of responses. The hypothesis that the population means are equal is the hypothesis that there is no treatment effect. The analytical method we use in such problems is called the **analysis of variance** (ANOVA).

The analysis of variance is a method for comparing the means of several populations using ratios of variance estimates. A **one-way** ANOVA (§14.1 and 14.2) generalizes the two-sample $t$-test to the case of more than two treatments. Identifying important mean differences involves multiple comparisons of means, which we take up in §14.3. A **two-way** ANOVA (§14.4) generalizes paired $t$-tests.

This is but a brief introduction to the extensive field of experimental designs and their analyses.[1] We'll only study the simplest kinds of models, making the simplifying assumptions we made in developing $t$-tests—independent, normally distributed responses, and homoscedasticity (equal variances).

## 14.1      One-Way ANOVA: The Method

In §12.1 we discussed the running of an experiment to compare a treatment with a control, or to compare two treatments. Here we consider the case of more than two treatments. In designing an experiment to compare their effects, we first identify a population of individuals who are potential subjects for treatment. Then we randomly select **experimental units**—individuals or animals or pieces of experimental material—from that population. In a **completely randomized design,** we then randomly assign one of the treatments to each experimental unit. If there are $k$ treatments, we get $k$ samples, one from each treatment population.

---

[1] For further study in this area, see the references (page 721) for various texts on experimental design and linear models.

In an extension of the two-sample $t$-test of §12.4, our test for the equality of $k$ population means assumes $k$ independent random samples, one from each treatment population. As in the two-sample case, we assume (as the simplest case) equality of the population variances.

**Example 14.1a**  | **Dial Coatings**

To compare the effectiveness of three different types of phosphorescent coating of instrument dials ($k = 3$), each type of coating was applied to eight dials. After a period of illumination by an ultraviolet light, the number of minutes each dial glowed was recorded. The following are summary statistics of the results:

| Coating type | Sample size | Sample mean | Sample s.d. |
|:---:|:---:|:---:|:---:|
| 1 | 8 | 57.1 | 4.250 |
| 2 | 8 | 59.175 | 3.185 |
| 3 | 8 | 68.45 | 4.098 |

The sample means are different. But we know that sample means are variable and that differences can arise solely because of sampling variability. The question is whether the differences in the sample means are larger than would be typical if they were only the result of sampling variability. If so, an alternative explanation is that the population means are different. (To be continued.) ∎

Some notation: Let $n_i$ denote the size of the sample from population $i$, $i = 1, 2,$ ..., $k$, and $n = n_1 + \cdots + n_k$. Let $X_{ij}$ denote the $j$th observation in the $i$th sample, $j = 1, ..., n_i$. In summary, with some additional notation:

| Treatment | Mean | Variance | Sample | Mean | Variance |
|:---:|:---:|:---:|:---:|:---:|:---:|
| 1 | $\mu_1$ | $\sigma^2$ | $X_{11}, ..., X_{1n_1}$ | $\overline{X}_1$ | $S_1^2$ |
| ⋮ | ⋮ | ⋮ | ⋮ | ⋮ | ⋮ |
| $k$ | $\mu_k$ | $\sigma^2$ | $X_{k1}, ..., X_{kn_k}$ | $\overline{X}_k$ | $S_k^2$ |

The parameter $\sigma^2$, which measures the variability in the individual observations $X_{ij}$, is referred to as the **error variance.** The null hypothesis to be tested is that there are no treatment differences on average:

$$H_0: \mu_1 = \cdots = \mu_k = \mu, \tag{1}$$

where $\mu$ is unknown. The alternative is that $H_0$ is not true in some way.

We denote by $\overline{X}$ the mean of the combined sample—the set of all $n$ observations. It is called the **grand mean:**

$$\overline{X} = \frac{1}{n}\sum_{1}^{k}\sum_{1}^{n_i}X_{ij} = \frac{1}{n}\sum_{1}^{k}n_i\overline{X}_i. \tag{2}$$

When there are large differences among the several sample means, their deviations from the grand mean $\overline{X}$ are large. The squares of these deviations are the basis of the statistic we'll use to test (1). When $H_0$ is true, the combined sample (the set of all $n$ observations) is a random sample of size $n$ from $\mathcal{N}(\mu, \sigma^2)$. The sum of squares of their deviations from the grand mean is called the **total sum of squares:**

$$\text{SSTot} = \sum_{i=1}^{k}\sum_{j=1}^{n_i}(X_{ij} - \overline{X})^2. \tag{3}$$

This is large when there is a lot of variability among the $X_{ij}$. Some of that variability is sampling variability, which would be present even under $H_0$. But some may be owing to differences in population means. To sort out these sources of variability, we'll decompose the total sum of squares (3) into two parts. One we attribute to differences among population means, and the other to sampling variability.

The decomposition is as follows. First, we express the deviation of $X_{ij}$ from $\overline{X}$ as the sum of its deviation from the mean of its sample plus the deviation of that sample mean from the grand mean:

$$X_{ij} - \overline{X} = (X_{ij} - \overline{X}_i) + (\overline{X}_i - \overline{X}).$$

Squaring both sides, we get

$$(X_{ij} - \overline{X})^2 = (X_{ij} - \overline{X}_i)^2 + (\overline{X}_i - \overline{X})^2 + 2\,(X_{ij} - \overline{X}_i)(\overline{X}_i - \overline{X}).$$

Adding these over $i$ and over $j$ produces SSTot. We carry out that summation in two steps, first over $j$ (that is, within the $i$th sample):

$$\sum_{j=1}^{n_i}(X_{ij} - \overline{X})^2 = \sum_{j=1}^{n_i}(X_{ij} - \overline{X}_i)^2 + \sum_{j=1}^{n_i}(\overline{X}_i - \overline{X})^2 + 2\sum_{j=1}^{n_i}(X_{ij} - \overline{X}_i)(\overline{X}_i - \overline{X})$$

$$= \sum_{j=1}^{n_i}(X_{ij} - \overline{X}_i)^2 + n_i(\overline{X}_i - \overline{X})^2 + 2\,(\overline{X}_i - \overline{X})\sum_{j=1}^{n_i}(X_{ij} - \overline{X}_i).$$

The last sum on $j$ is 0, being the sum of the deviations of the $i$th sample observations about their mean. So when we sum both sides on $i$, we obtain

$$\text{SSTot} = \sum_{i=1}^{k}\sum_{j=1}^{n_i}(X_{ij} - \overline{X}_i)^2 + \sum_{i=1}^{k}n_i(\overline{X}_i - \overline{X})^2. \tag{4}$$

[From this point on, it will be understood that double sums will be over the indices indicated in the first term on the right—that is, over all the observations. Single sums such as the second term on the right will be over $i$, the sample number—that is, over the $k$ samples.]

The first term on the right of (4) is based on differences *within* each sample—a measure of variability attributed to sampling variation, called the **error sum of squares:**

$$\text{SSE} = \sum\sum (X_{ij} - \overline{X}_i)^2. \tag{5}$$

The second term, based on differences *among* the sample means, is 0 if the means are all equal and large if they are quite different. It is the **treatment sum of squares:**

$$\text{SSTr} = \sum n_i(\overline{X}_i - \overline{X})^2. \tag{6}$$

Thus, (4) can be written

$$\text{SSTot} = \text{SSTr} + \text{SSE}. \tag{7}$$

It is this decomposition of SSTot that embodies the idea of the "analysis of variance." If SSTr is small, the variation among the observations is considered to be primarily sampling variability. If SSTr is a substantial fraction of SSTot, this will be evidence of a "treatment effect"—of differences among treatment means.

The error sum of squares (5) is an extension to $k$ samples of the numerator of a "pooled variance" (§12.4). As we did for $k = 2$, we can express it in terms of the $k$ sample variances:

$$\text{SSE} = \sum_i \left\{ \sum_j (X_{ij} - \overline{X}_i)^2 \right\} = \sum_i (n_i - 1)S_i^2. \tag{8}$$

And since each sample variance $S_i^2$ is an unbiased estimate of $\sigma^2$,

$$E(\text{SSE}) = \sum E[(n_i - 1)S_i^2] = \sigma^2 \sum(n_i - 1) = (n - k)\sigma^2.$$

This tells us that the **error mean square:**

$$\text{MSE} = \frac{\text{SSE}}{n - k} \tag{9}$$

is an unbiased estimate of $\sigma^2$. And this is so *whether $H_0$ holds or not!*

The treatment sum of squares (6) is a weighted sum of the squared deviations of the sample means from the grand mean. In Problem 14-9 you are asked to show that

$$E(\text{SSTr}) = (k - 1)\sigma^2 + \sum n_i(\mu_i - \mu)^2. \tag{10}$$

Under $H_0$, when each $\mu_i = \mu$, the second term on the right is 0. This means that if there is no treatment effect, the **treatment mean square:**

$$\text{MSTr} = \frac{\text{SSTr}}{k - 1} \tag{11}$$

is an unbiased estimate of $\sigma^2$. So when $H_0$ is *true*, MSTr is expected to be about the same size as MSE. On the other hand, when $H_0$ is *false*—when there is a treatment effect, then MSTr tends to be larger than MSE. Thus, the *ratio* of MSTr to MSE is a statistic that should be near 1 under $H_0$, but will tend to be large when there is a treatment effect. This ratio is our test statistic:

$$F = \frac{\text{SSTr}/(k-1)}{\text{SSE}/(n-k)} = \frac{\text{MSTr}}{\text{MSE}}. \tag{12}$$

Large values of $F$ will be considered extreme and taken as evidence against the null hypothesis.

The divisors in the $F$-ratio (12) are termed "degrees of freedom" of the corresponding sums of squares. The total sum of squares, being the numerator of the variance of all the observations considered as a single sample, is said to have $n-1$ degrees of freedom, just as we used the term earlier for a sample variance. And we note that $n - 1 = (k - 1) + (n - k)$—the degrees of freedom add just as the corresponding sums of squares in (7) add.

**Example 14.1b**  **Dial Coatings (Continued)**

We return to the data of Example 14.1a. For these data, $k = 3$, the sample sizes are equal: $n_1 = n_2 = n_3 = 8$, and $n = 24$. The grand mean is

$$\overline{X} = \frac{1}{n}\sum n_i \overline{X}_i = \frac{8}{24}\,(57.1 + 59.175 + 68.45) = 61.575.$$

The treatment sum of squares is

$$\begin{aligned}\text{SSTr} &= \sum n_i(\overline{X}_i - \overline{X})^2 \\ &= 8[(57.1 - \overline{X})^2 + (59.175 - \overline{X})^2 + (68.45 - \overline{X})^2] = 584.41.\end{aligned}$$

Is this unusually large—larger than expected under $H_0$? To answer this, we find the error sum of squares:

$$\text{SSE} = \sum_i (n_i - 1)S_i^2 = 7[4.250^2 + 3.185^2 + 4.098^2] = 315.0.$$

It is common practice to enter these quantities in an "ANOVA table":

| Source | SS | d.f. | MS | F |
|--------|------|------|-------|------|
| Treatment | 584.4 | 2 | 292.2 | 19.5 |
| Error | 315.0 | 21 | 15.0 | — |
| Total | 899.41 | 23 | — | |

The estimate MSTr of $\sigma^2$ is almost 20 times the estimate MSE. This suggests that a treatment effect has inflated MSTr.   ∎

In the next section we'll see that, under the assumption of normal populations, the null distribution of the $F$-ratio (12) is an $F$-distribution. The parameters of that distribution are the degrees of freedom of the sums of squares that are the basis of the ratio.

---

Test for the equality of means of $k$ normal populations with equal variances: Calculate the $F$-ratio

$$F = \frac{\text{MSTr}}{\text{MSE}} = \frac{\sum n_i(\bar{X}_i - \bar{X})^2/(k-1)}{\sum\sum(X_{ij} - \bar{X}_i)^2/(n-k)}. \tag{13}$$

Large values of $F$ are extreme, and critical values and $P$-values are given in Table VIIIa/b of the Appendix, $k-1$ numerator d.f. and $n-k$ denominator d.f.

---

**Example 14.1c** | **Dial Coatings (Conclusion)**

In the ANOVA table of the preceding example, we saw that the treatment mean square is almost 20 times the error mean square:  $F = 19.5$. When the treatment means are all equal (that is, under $H_0$), the $F$-ratio is distributed as $F(k-1, n-k)$. In the case of the three types of dial coating, $k-1 = 2$, and with 24 observations in all, $n-k = 24-3 = 21$. To find the $P$-value we look in Table VIIIa under $F(2, 21)$: $P \doteq .000$, very strong evidence against $H_0$.   ∎

In the next section we'll show that the two-sided, two-sample $t$-test of §12.4 is a special case of the $F$-test. Like the $t$-test, the $F$-test is based on the assumption of normal populations. But, also like the $t$-test, the $F$-test is fairly robust with respect to this assumption. If there are no unusually large or unusually small observations, the $P$-value given by the $F$-distribution won't be far off. The $F$-test is also not very sensitive to the assumption of equal population variances.

In the above examples, we have strong evidence that the means under the various treatments are different. We saw that $\bar{X}_3 > \bar{X}_2 > \bar{X}_1$, but can we conclude that $\mu_3 > \mu_2 > \mu_1$? Or that just $\mu_3 > \mu_2$? These kinds of questions are important to the experimenter. Answering them is not easy. A number of methods for multiple comparisons have been proposed and are in use. We'll take up two of them in §14.3.

---

## Problems

**∗ 14-1.**   Calculate the $F$-statistic for testing the hypothesis of equal means, given the following data:

| Treatment | Sample |
|-----------|--------|
| A | 2, 6, 4 |
| B | 12, 9, 7, 4, 8 |
| C | 4, 6, 8, 2 |

**14-2.**  Find the within-samples estimate of error variance for these data:

| Treatment | Sample |
|-----------|--------|
| A | 4, 4, 4 |
| B | 6, 6, 6, 6, 6 |
| C | 7, 7 |

What is the implication of your result, for testing for a treatment effect?

**∗ 14-3.**  Three plots each of five varieties of clover were planted at the Rosemount Experiment Station in Minnesota, with yields (tons per acre):

| Variety | Yield | Mean | s.d. |
|---------|-------|------|------|
| Spanish | 2.79, 2.26, 3.09 | 2.713 | .420 |
| Evergreen | 1.93, 2.07, 2.45 | 2.150 | .269 |
| Commercial yellow | 2.76, 2.34, 1.87 | 2.323 | .445 |
| Madrid | 2.31, 2.30, 2.49 | 2.367 | .107 |
| Wisconsin A46 | 2.39, 2.05. 2.68 | 2.373 | .315 |

**(a)**  Construct the ANOVA table and find the $F$-statistic for testing the hypothesis of no difference in mean yield among the five varieties.

**(b)**  Construct a 90% confidence interval for the mean difference in yield between the Spanish and Evergreen varieties. [*Hint:*  Use the estimate of $\sigma^2$ from all five varieties—the MSE.]

**14-4.**  In some large computers, printed circuit boards have several layers. Each layer has many small holes. A substantial clearance between holes in different layers is necessary to avoid shorting. For six holes selected from each of five locations on a board, the clearances were measured as follows (in thousandths of an inch):

| Location | Clearance | Mean | s.d. |
|----------|-----------|------|------|
| A | 6.9, 6.9, 7.5 | 7.10 | .346 |
| B | 10.1, 10.1, 7.3, 7.3, 9.1, 9.9 | 8.97 | 1.343 |
| C | 5.0, 8.0, 6.4, 6.8, 7.2, 8.8 | 7.03 | 1.317 |
| D | 8.8, 8.0, 7.3, 8.5, 7.8, 8.9 | 8.22 | .624 |
| E | 7.8, 8.8, 8.9, 7.2, 8.1, 7.1 | 7.98 | .768 |

(Three holes at location A were accidentally ground through in preparing the board for measurement.) Test the hypothesis of no difference in mean clearance at the various locations on the board.

* **14-5.**   Use a one-way ANOVA to redo Problem 12-13. (Compare your answer with that given for Problem 12-13.)

* **14-6.**   The following data were obtained (in part) to examine the variation in strength between and within bobbins for a type of worsted yarn.[2] Test the hypothesis of no average difference in strengths among bobbins.

| Bobbin | Data | | | | Mean | $S^2$ |
|--------|------|------|------|------|------|-------|
| 1 | 18.2 | 16.8 | 18.1 | 17.0 | 17.52 | .5292 |
| 2 | 17.2 | 18.5 | 15.0 | 16.2 | 16.72 | 2.209 |
| 3 | 15.3 | 15.9 | 14.5 | 14.2 | 14.97 | .5958 |
| 4 | 15.6 | 16.0 | 15.2 | 14.9 | 15.42 | .2292 |
| 5 | 19.2 | 18.0 | 17.0 | 16.9 | 17.77 | 1.149 |
| 6 | 16.2 | 15.9 | 14.9 | 15.5 | 15.62 | .3158 |

**14-7.**   The following are amounts filled by a 24-head machine for filling bottles of vegetable oil:[3]

| Group | Amount | | | | |
|-------|--------|-------|-------|-------|-------|
| 1 | 15.70 | 15.68 | 15.64 | 15.60 | |
| 2 | 15.69 | 15.71 | | | |
| 3 | 15.75 | 15.82 | 15.75 | 15.71 | 15.84 |
| 4 | 15.61 | 15.66 | 15.59 | | |
| 5 | 15.65 | 15.60 | | | |

Test the hypothesis of no difference among the group means. (The five groups of bottles reflect, in part, head variability.)

## 14.2     One-Way ANOVA:  The Theory

We now give the theoretical basis and some further motivation for the method presented in §14.1. We assume that the $X_{ij}$ are independent, with common variance

---

[2]From E. J. Snell, *Applied Statistics* (New York: Chapman and Hall, 1987), 104.

[3]Reported in W. H. Swallow and S. R. Searle, "Minimum variance quadratic unbiased estimation of variance components," *Technometrics 20* (1978), 265–72.

$\sigma^2$. As in §14.1, let $\mu_i$ denote the mean of the $i$th population, and $\epsilon_{ij}$ the random error component $X_{ij} - \mu_i$. We define $\mu$ by analogy with our definition of $\overline{X}$ [(1), §14.1] as a weighted average[4] of population means:

$$\mu = \frac{1}{n} \sum_{i=1}^{k} n_i \mu_i. \tag{1}$$

Under the null hypothesis that all $\mu_i$ are equal, $\mu$ is their common value.

The mean of the $i$th sample is an unbiased estimate of the $i$th population mean:

$$E(\overline{X}_i) = \frac{1}{n_i} \sum_{j=1}^{n_i} E(X_{ij}) = \mu_i.$$

And the grand mean $\overline{X}$ is an unbiased estimate of $\mu$:

$$E(\overline{X}) = \frac{1}{n} \sum_i \sum_j E(X_{ij}) = \frac{1}{n} \sum_i \sum_j \mu_i = \frac{1}{n} \sum_i n_i \mu_i = \mu.$$

So, the expected deviation of $\overline{X}_i$ from the grand mean is

$$E(\overline{X}_i - \overline{X}) = \mu_i - \mu = \tau_i.$$

We call this mean difference $\tau_i$ the $i$th **treatment effect.** (It is actually a differential effect—the amount by which the average response to treatment $i$ differs from the overall average response.) Under the null hypothesis that the treatment means are all the same, the $i$th treatment effect is 0:

$$H_0: \ \tau_i = 0 \ \text{ for all } i.$$

To complete the description of the model, we assume (as in §14.1) that the random errors are *normally distributed.* This, in turn, means that the observations are normally distributed, with means $EX_{ij} = \mu + \tau_i$ and constant variance $\sigma^2$.

---

Model for a one-way ANOVA:

$$X_{ij} = \mu + \tau_i + \epsilon_{ij}, \tag{2}$$

where $\sum n_i \tau_i = 0$, and the errors $\epsilon_{ij}$ are *independent*, with

$$\epsilon_{ij} \sim \mathcal{N}(0, \sigma^2). \tag{3}$$

The null hypothesis is $H_0: \ \tau_i \equiv 0$, and the alternative is $H_A$: not all $\tau_i$ are 0.

---

[4]It may seem odd that this definition of $\mu$ depends on the sample sizes $n_j$. This is for convenience; it does not affect our testing of $H_0$ versus $H_A$. (See also Problem 14–13.)

Next, we'll find the joint m.l.e.'s of the model parameters, $\mu_i$ and $\sigma^2$. The p.d.f. of $X_{ij}$, with $\theta = \sigma^2$ for convenience in differentiating, is

$$f(x \mid \mu_i, \theta) \propto \frac{1}{\sqrt{\theta}} \exp\left\{ -\frac{1}{2\theta}(x - \mu_i)^2 \right\},$$

and the likelihood function is the product

$$L(\mu_1, \ldots, \mu_k; \theta) = \prod_i \prod_j \theta^{-1/2} \exp\left\{ -\frac{1}{2\theta}(X_{ij} - \mu_i)^2 \right\}$$

$$= \theta^{-n/2} \exp\left\{ -\frac{1}{2\theta}\sum_i \sum_j (X_{ij} - \mu_i)^2 \right\},$$

with logarithm

$$\log L = -\frac{n}{2}\log \theta - \frac{1}{2\theta}\sum_i \sum_j (X_{ij} - \mu_i)^2. \tag{4}$$

Clearly, for any given $\theta$, (4) is maximized when we minimize the double sum. To do this, we differentiate it with respect to each of the $k$ $\mu$'s and set the derivatives equal to 0 simultaneously, to find the minimum. With respect to $\mu_1$, the derivative is 0 except for the term in the outer sum in which $i = 1$; the derivative of this term is

$$\sum_{j=1}^{n_1} (X_{1j} - \mu_1) = n_1 \overline{X}_1 - n_1 \mu_1.$$

This vanishes when $\mu_1$ has the value $\overline{X}_1$: $\widehat{\mu}_1 = \overline{X}_1$. Similarly,

$$\widehat{\mu}_i = \overline{X}_i \quad \text{for every } i. \tag{5}$$

That these choices of the $\mu_i$ do in fact minimize the double sum follows from the fact that the minimum exists (the sum is bounded below by 0), and there is only the one point where the derivatives all vanish.

Then, having minimized the double sum, we can choose the $\theta$ that maximizes log $L$ by differentiating with respect to $\theta$ (or $\sigma^2$):

$$\frac{\partial}{\partial \theta}\log L = -\frac{n}{2\theta} + \frac{1}{2\theta^2}\sum_i \sum_j (X_{ij} - \widehat{\mu}_i)^2.$$

This vanishes when $\theta$ has the value

$$\widehat{\sigma}^2 = \frac{1}{n}\sum \sum (X_{ij} - \overline{X}_i)^2 = \frac{\text{SSE}}{n}. \tag{6}$$

Since the second derivative of log $L$ is always negative, this value of $\theta$ and the $\widehat{\mu}_i$'s given by (5), together, do maximize the likelihood.

From (5) we deduce the m.l.e.'s of the parameters $\mu$ and $\tau$:

$$\widehat{\mu} = \frac{1}{n}\sum n_i\widehat{\mu}_i = \frac{1}{n}\sum n_i\overline{X}_i = \overline{X}, \quad \text{and} \quad \widehat{\tau}_i = \widehat{\mu}_i - \widehat{\mu} = \overline{X}_i - \overline{X}. \qquad (7)$$

So, the differences we identified in §14.1 as attributed to treatment effect are the m.l.e.'s of the treatment effects $\tau_i$.

We observe that the m.l.e. of $\sigma^2$ is not quite the MSE, since the denominator of (6) is $n$ rather than $n - k$. But it is quite natural to base an estimate of the error variance on SSE, the numerator of (6), since $X_{ij} - \overline{X}_i$ is an estimate of $X_{ij} - \mu_i = \epsilon_{ij}$, and $\sigma^2$ is the average square of $\epsilon_{ij}$ (the mean is 0). The divisor $n - k$ makes the MSE unbiased. (It is not that we need an unbiased estimator, but the usual tables are constructed so as to be used with ratios of unbiased variance estimates. And this is a convenience rather than a necessity—they could have been made differently.)

We show next that the null distribution of the test ratio $F$ defined in §14.1 does indeed have an $F$-distribution, as claimed. The $F$-ratio for testing equality of means is

$$F = \frac{\text{MSTr}}{\text{MSE}} = \frac{\text{SSTr}/(k-1)}{\text{SSE}/(n-k)} = \frac{\sum \frac{n_i(\overline{X}_i - \overline{X})^2}{\sigma^2} \Big/ (k-1)}{\sum\sum \frac{(X_{ij} - \overline{X}_i)^2}{\sigma^2} \Big/ (n-k)}. \qquad (8)$$

To show that this is $F(k-1, n-k)$, we'll show that the numerator and denominator are independent and that each is a chi-square variable divided by its number of degrees of freedom. This is how we characterized the $F$-distribution in §12.7.

We can write the double sum in the denominator of the above $F$-ratio, using (8) of the preceding section, as

$$\sum\sum \frac{(X_{ij} - \overline{X}_i)^2}{\sigma^2} = \sum_i \frac{(n_i - 1)S_i^2}{\sigma^2}. \qquad (9)$$

This is a sum of the $k$ independent chi-square variables

$$\frac{(n_i - 1)S_i^2}{\sigma^2} \sim \text{chi}^2(n_i - 1),$$

(See §8.10, page 350.)   So the sum (9) is distributed as $\text{chi}^2(n - k)$, because $\sum(n_i - 1) = n - k$. And this distribution, incidentally, is correct whether or not the population means are equal—that is, whether $H_0$ holds or not.

To get at the distribution of the sum in the numerator of (8), we go back to the decomposition (4) of §14.1:

$$\sum\sum(X_{ij} - \overline{X})^2 = \sum_{i=1}^{k}\sum_{j=1}^{n_i}(X_{ij} - \overline{X}_i)^2 + \sum_{i=1}^{k}n_i(\overline{X}_i - \overline{X})^2, \qquad (10)$$

which is SSTot = SSE + SSTr, and divide through by $\sigma^2$:

$$\frac{\text{SSTot}}{\sigma^2} = \sum \frac{(n_i - 1)S_i^2}{\sigma^2} + \sum \frac{n_i(\overline{X}_i - \overline{X})^2}{\sigma^2}. \tag{11}$$

Now suppose that the means $\mu_i$ *are* all equal, so that the observations $X_{ij}$ can be thought of as constituting a single sample of size $n$ from $\mathcal{N}(\mu, \sigma^2)$. The variance of this combined sample is

$$S^2 = \frac{1}{n-1} \sum \sum (X_{ij} - \overline{X})^2 = \frac{\text{SSTot}}{n-1},$$

and from §8.10 we know that

$$\frac{\text{SSTot}}{\sigma^2} = \frac{(n-1)S^2}{\sigma^2} \sim \text{chi}^2(n-1).$$

The two terms on the right of (11) are *independent,* because the first is a function of sample variances, and the second is a function of sample means. And we know that the mean and variance of a sample from a normal population are independent. Then, according to §6.4 (see the box on page 248), since the l.h.s. is $\text{chi}^2(n-1)$, and the first term on the right is $\text{chi}^2(n-k)$, it follows that the second term must have a chi-square distribution with $n - 1 - (n - k)$ or $k - 1$ degrees of freedom.

So $F$, given by (8), is a ratio of independent chi-square variables, each divided by its d.f., and this is the definition of the $F$-distribution.

The next example carries out another one-way ANOVA, this time for the special case in which there are just two treatments—or a treatment and a control. It will turn out that the $F$-statistic is just the square of the two-sample $t$-statistic introduced in §12.4. This is generally true when $k = 2$, as we'll show after the example.

## Example 14.2a | Intermittent Feeding (Revisited)

In Example 12.4a and others, we used $T$ to analyze rat blood pressure data:

Regular:       108, 133, 134, 145, 152, 155, 169
Intermittent:  115, 162, 162, 168, 170, 181, 199, 207

When $k = 2$, the MSE [see (9) and (5) of §14.1, page 564] is just what we termed the pooled variance in §12.4:

$$\text{MSE} = S_p^2 = \frac{6 \times 19.61^2 + 7 \times 27.99^2}{7 + 8 - 2} = \frac{7791.39}{13} = 599.34.$$

The sample means are $\overline{X}_R = 142.29$, $\overline{X}_I = 170.5$, and $\overline{X} = 2360/15 = 157.33$. The treatment sum of squares is

$$\text{SSTr} = 7(142.29 - 157.33)^2 + 8(170.5 - 157.33)^2 = 2971.9.$$

The ANOVA table is thus:

| Source | SS | d.f. | MS | F |
|--------|------|------|--------|------|
| Treatment | 2971.9 | 1 | 2971.9 | 4.96 |
| Error | 7791.4 | 13 | 599.34 | — |

The $P$-value would be found in the table for $F(1, 13)$, but we haven't provided this table, for the reason to be explained below. However, we can turn to Table VIIIa and see that $.01 < P < .05$. In Example 12.4a, we found that $|T| = 2.226$, with two-sided $P$-value

$$P = P(|T| > 2.226) = 2\,P(T > 2.226) = .044,$$

Notice that $T^2 = 2.226^2 = 4.96 = F$. This is not just a coincidence, as we'll show next. Moreover, the two-sided $P$-value based on $T$ is equal to the $P$-value based on $F$: The $F$-test for equal means is the same as the two-sided $t$-test for equal means when there are just the two populations. ∎

When there are just two treatments ($k = 2$), the deviations of the sample means from the grand means can be expressed in terms of the difference in means:

$$\overline{X}_1 - \overline{X} = \overline{X}_1 - \frac{n_1\overline{X}_1 + n_2\overline{X}_2}{n_1 + n_2} = \frac{n_2}{n_1 + n_2}(\overline{X}_1 - \overline{X}_2),$$

with a similar expression for $\overline{X}_2 - \overline{X}$. Then,

$$\text{SSTr} = \sum_1^2 n_i(\overline{X}_i - \overline{X})^2 = n_1\left\{\frac{n_2}{n_1 + n_2}(\overline{X}_1 - \overline{X}_2)\right\}^2 + n_2\left\{\frac{n_1}{n_1 + n_2}(\overline{X}_2 - \overline{X}_1)\right\}^2$$

$$= \frac{n_1 n_2^2 + n_1^2 n_2}{(n_1 + n_2)^2}(\overline{X}_1 - \overline{X}_2)^2 = \frac{n_1 n_2}{n_1 + n_2}(\overline{X}_1 - \overline{X}_2)^2 = \frac{(\overline{X}_1 - \overline{X}_2)^2}{1/n_1 + 1/n_2}.$$

The MSE, for every $k$ including $k = 2$, is the pooled variance; so the $F$-ratio (8), when $k = 2$, is

$$F = \frac{\text{MSTr}}{\text{MSE}} = \left\{\frac{(\overline{X}_1 - \overline{X}_2)}{S_p\sqrt{1/n_1 + 1/n_2}}\right\}^2 = T^2.$$

This is the square of the two-sample $t$-statistic. Thus, the square of the two-sample $t$-statistic not only has the same distribution as the $F$-ratio (when $k = 2$), they are equal; so the one-sided $P$-value using $F$ is equal to the two-sided $P$-value using $T$. [It is for this reason that we have not included a table of $P$-values for $F(1, \nu)$—we need only look in the $t$-table and double the one-sided $P$.]

A final theoretical note: The $F$-test for equality of means is a likelihood ratio test. We'll show this by deriving the latter. We have already found the likelihood function and the parameter values that maximize it. The maximum value is

$$L(\overline{X}_1, \ldots, \overline{X}_k; \widehat{\theta}) = (\widehat{\theta})^{-n/2} \exp\left\{-\frac{1}{2\widehat{\theta}}\sum\sum(X_{ij} - \overline{X}_i)^2\right\} = (\widehat{\theta}e)^{-n/2},$$

since (in the denominator of the exponent)

$$\widehat{\theta} = \widehat{\sigma}^2 = \frac{1}{n}\sum\sum(X_{ij} - \overline{X}_i)^2.$$

Under $H_0$, the likelihood function involves only the two parameters $\mu$ and $\theta$:

$$L_0(\mu, \theta) = L(\mu, \ldots, \mu; \theta) = \theta^{-n/2}\exp\left\{-\frac{1}{2\theta}\sum_i\sum_j(X_{ij} - \mu)^2\right\}.$$

And under $H_0$, the $n$ observations in the $k$ samples constitute a sample of size $n$ from $\mathcal{N}(\mu, \sigma^2)$; so the m.l.e. of $\mu$ is $\overline{X}$, the grand mean. For the same reason, the m.l.e. of the error variance is just the "$V$" (second sample moment, as in §7.5) of that combined sample:

$$\widehat{\theta}_0 = \frac{1}{n}\sum\sum(X_{ij} - \overline{X})^2 = \frac{\text{SStot}}{n}.$$

The maximum value of $L_0$ is therefore

$$L_0(\widehat{\mu}, \widehat{\theta}_0) = L_0\left\{\overline{X}, \frac{\text{SStot}}{n}\right\} = (\widehat{\theta}_0 e)^{-n/2},$$

and

$$\Lambda = \frac{(\widehat{\theta}_0 e)^{-n/2}}{(\widehat{\theta}e)^{-n/2}} = \left(\frac{\widehat{\theta}_0}{\widehat{\theta}}\right)^{-n/2}.$$

This will be small when the ratio in parentheses is large:

$$\frac{\widehat{\theta}_0}{\widehat{\theta}} = \frac{\text{SStot}}{\text{SSE}} = \frac{\text{SSE} + \text{SSTr}}{\text{SSE}} = 1 + \frac{\text{SSTr}}{\text{SSE}} = 1 + \left(\frac{k-1}{n-k}\right)F.$$

And this is large when $F$ is large, so a critical region of the form $F > K$ is a likelihood ratio critical test.

## Problems

* **14-8.**  In the setting and notation of §14.1 and §14.2, calculate
     **(a)**  $E(\overline{X}_i - \overline{X})$.      **(b)**  $E(X_{ij} - \overline{X})$.

**14-9.**  In the setting and notation of §14.1 and §14.2, show (10) of §14.1:

$$E\left[\sum n_i(\overline{X}_i - \overline{X})^2\right] = (k-1)\sigma^2 + \sum n_i\tau_i^2.$$

[*Hint:* Start with $\overline{X}_i - \mu_i = (\overline{X}_i - \overline{X}) + (\overline{X} - \mu) + (\mu - \mu_i)$ and expand the square of this trinomial, multiply by $n_i$, sum, and take expected values.]

**14-10.**   Show that MSE in §14.1 and §14.2 reduces to the pooled variance $S_p^2$ of §12.4 when $k = 2$.

**14-11.**   For a one-way ANOVA, show the following:
  (a)  $\sum\sum(X_{ij} - \overline{X})^2 = \sum\sum X_{ij}^2 - \frac{1}{n}\left(\sum\sum X_{ij}\right)^2$.
  (b)  SSTr $= \sum n_i \overline{X}_i^2 - \frac{1}{n}\left(\sum\sum X_{ij}\right)^2$.

**14-12.**   Carry out the details of the maximization of the likelihood function $L_0$, the likelihood function under the hypothesis of equal means.

**14-13.**   Suppose we had defined $\mu$ as the *unweighted* average, $\sum\mu_i/k$. Then define $\alpha_i = \mu_i - \mu$, so that the model is $X_{ij} = \mu + \alpha_i + \epsilon_{ij}$, where as before, $\epsilon_{ij} \sim \mathcal{N}(0, \sigma^2)$. Show that the m.l.e. of this $\mu$ is the unweighted average of the sample means. [This would be the same as the m.l.e. of the $\mu$ defined by (1) of §14.2 if the sample sizes were all the same. In general, it is not the $\overline{X}$ used in forming SSTot, and not the $\overline{X}$ in the treatment sum of squares that appears in the natural $F$-ratio.]

## 14.3        Multiple Comparisons

The $F$-test we've given for equality of $k$ population means may suggest that the means are not all the same, but it does not tell us which are different. A particular ordering of the sample means suggests the same ordering of the population means, but this ordering may be wrong because of sampling variability. One could do $t$-tests for the equality of the means in each pair of treatments, but the question of significance in such multiple testing is problematical, as the following example shows.

**Example 14.3a**   We generated ten independent random samples of size 100 from $\mathcal{N}(20, 1)$. The summary statistics are as follows:

| Treatment # | Mean | s.d. |
|:---:|:---:|:---:|
| 1 | 19.83 | 1.0305 |
| 2 | 20.27 | .9588 |
| 3 | 20.08 | 1.0045 |
| 4 | 19.97 | .9086 |
| 5 | 19.89 | .9764 |
| 6 | 19.78 | .9814 |
| 7 | 20.01 | 1.0170 |
| 8 | 20.04 | .9846 |
| 9 | 20.11 | .9243 |
| 10 | 20.06 | 1.0071 |

There are $\binom{10}{2} = 45$ possible comparisons of two means. Applying the two-sample $t$-test to each such pair, we find differences in sample means that are significant at $\alpha = .05$ in seven of the 45 pairs: Treatment 6 with 2, 6 with 9, 1 with 2, 1 with 9, 5 with 2, 3 with 6, and 2 with 4. This is despite the fact that in every case the population means were identical!     ∎

The samples in the preceding example are not particularly unusual; when there are 45 opportunities to find significant differences, it is not surprising to find some! The expected number of pairs of sample means that are significantly different when the population means are actually equal is $45 \times .05$ or 2.25; the variance is quite large because the comparisons are highly correlated.

We saw in Chapter 11 that constructing a confidence interval at the confidence level $1 - \alpha$ is equivalent to testing a hypothesis at the level $\alpha$. In the remainder of this section, we focus on confidence limits, with the obvious implication of a corresponding test.

Consider now $k$ populations, where population $i$ is $N(\mu_i, \sigma^2)$, and (in the notation of the preceding sections) the ratio

$$T = \frac{\overline{X}_i - \overline{X}_j - (\mu_i - \mu_j)}{S_p\sqrt{1/n_i + 1/n_j}}, \tag{1}$$

for some pair $(i, j)$. In §12.4, the $S_p$ in the denominator of $T$ was a pooled estimate of $\sigma$ based on the two samples; here we have more information about $\sigma$ and take the pooled variance to be the MSE of the preceding sections:

$$S_p^2 = \text{MSE} = \frac{1}{n-k}\sum(n_i - 1)S_i^2.$$

Then $(n - k)S_p^2/\sigma^2 \sim \text{chi}^2(n - k)$, and the distribution of $T$ as given by (1) is $t(n - k)$. So $T$ is pivotal (see §9.5); and with an appropriate percentile of its distribution, we can construct a confidence interval for each difference $\mu_i - \mu_j$. But what is the appropriate percentile if we do this for *all* such differences?

Suppose we want $m$ confidence intervals at the overall level $1 - \alpha$. Let $t(m, \alpha)$ denote the value of $t(n - k)$ with area $\alpha/(2m)$ to its right:

$$P\{|T| > t(m, \alpha)\} = \frac{\alpha}{m}. \tag{2}$$

Let $E_{ij}$ denote the event that the confidence interval formed from (1) using the multiplier $t(m, \alpha)$ covers the difference $\mu_i - \mu_j$—that is, that

$$\overline{X}_i - \overline{X}_j - t(m, \alpha) \times \text{s.e.} < \mu_i - \mu_j < \overline{X}_i - \overline{X}_j + t(m, \alpha) \times \text{s.e.}, \tag{3}$$

where

$$\text{s.e.} = S_p\sqrt{\frac{1}{n_i} + \frac{1}{n_j}}.$$

Because of how we chose $t(m, \alpha)$, the probability of $E_{ij}$ is $1 - \alpha/m$. Then, the Bonferroni inequality [Problem 1-56] implies that

$$P(\bigcap E_{ij}) \geq 1 - \sum P(E_{ij}^c) = 1 - m(\alpha/m) = 1 - \alpha.$$

That is, the probability that *all* of the $m$ confidence intervals will cover the corresponding true mean differences is at least $1 - \alpha$. For simultaneous tests of all $\binom{k}{2}$ hypotheses of the form $\mu_i = \mu_j$, we construct confidence intervals for the corresponding differences in means as given by (3) and reject any for which the interval does not include 0. The probability is at most $\alpha$ that we'd incorrectly reject one or more of the hypotheses when all the means are actually equal.

**Example 14.3b** | **Dial Coatings (Revisited)**

Consider the data of Examples 14.1a and 14.1b. There were three populations ($k = 3$) and samples of size 8 from each. The mean square for error was found to be 15.0, with 21 d.f.

We might be interested in the three differences of the two means: $m = 3$. To obtain simultaneous confidence limits for the three differences at level .95, we set $\alpha = .05$, so that $\alpha/m = .05/3 \doteq .0167$. In Table IIIb of Appendix 1 we see that for $n = 21$, the value of $T$ with .0083 in each tail is about 2.6:

$$P(|T| > 2.6) = 2 \times .0083 = .0166.$$

So, for confidence limits, we go either way from the point estimate of a difference by the amount

$$2.6\sqrt{15.0\left(\frac{1}{8} + \frac{1}{8}\right)} \doteq 5.04.$$

The sample means were 57.1, 59.2, 68.5, for populations 1, 2, 3, respectively. So the three sets of confidence limits are as follows:

For $\mu_2 - \mu_1$: $2.1 \pm 5.04$.
For $\mu_3 - \mu_2$: $9.3 \pm 5.04$.
For $\mu_3 - \mu_1$: $11.4 \pm 5.04$.

The procedure used to obtain these limits has at least a 95% success rate. Since the first interval includes the value 0, the difference $\overline{X}_2 - \overline{X}_1$ is declared not significant; the other two sample mean differences are significant. ∎

A method of multiple comparisons due to H. Scheffé deals with all possible **contrasts**. A contrast is a linear combination of population means, $\sum a_i \mu_i$, in which the coefficients sum to 0: $\sum a_i = 0$. For instance, $\mu_1 - \mu_3$ is a contrast, and $\mu_1 + \mu_2 - 2\mu_3$ is another. The latter, in effect, compares the average mean of populations 1 and 2 with the mean of population 3. Similarly, the contrast $3(\mu_1 + \mu_2) - 2(\mu_3 + \mu_4 + \mu_5)$ compares the average of the means of populations 1 and 2 with the average of the means of populations 3, 4, and 5.

Scheffé's method constructs confidence limits for all possible contrasts of a set of $k$ means. It is based on the fact that the null distribution of the ratio

$$F = \frac{\sum n_i (\overline{X}_i - \overline{X} - \tau_i)^2}{(k-1)S_p^2} \tag{4}$$

is $F(k-1, n-k)$. When $\tau = 0$, this $F$-ratio is the test ratio in the one-way ANOVA [(8) of §14.1]. So if we choose a number $f$ such that $P(F < f)$ is the desired overall confidence level $1 - \alpha$, then

$$P\left(\sum n_i(\overline{X}_i - \overline{X} - \tau_i)^2 \le (k-1)fS_p^2\right) = 1 - \alpha.$$

Some further algebra[5] shows that the probability is at least $1 - \alpha$ that for every contrast $\sum a_i \mu_i$,

$$\sum a_i \overline{X}_i - S_p \sqrt{b} \le \sum a_i \mu_i \le \sum a_i \overline{X}_i + S_p \sqrt{b}, \tag{5}$$

where

$$b = (k-1)f \sum \frac{a_i^2}{n_i}. \tag{6}$$

The inequalities (5) define confidence intervals, and the probability is at least $1 - \alpha$ that they simultaneously cover the respective contrasts.

**Example 14.3c** | **Dial Coatings (Last Time)**

Consider again the data of the preceding example. The distribution of (4) is $F(2, 21)$, and for an overall confidence level of .95 we use $f$ defined as the 95th percentile: $f = 3.47$. For the contrasts $\mu_i - \mu_j$: $a_i = 1$, $a_j = -1$, and the other $a$'s are 0. The sample sizes $n_i$ are all 8. From (6) we have

$$b = 2 \times 3.47 \times (1/8 + 1/8) = 1.735.$$

The pooled variance (MSE), as in the preceding example, is 15.0, so

$$S_p\sqrt{b} = \sqrt{15 \times 1.735} \doteq 5.10.$$

The confidence intervals for the three contrasts considered in Example 14.3a are $2.1 \pm 5.10$, $9.3 \pm 5.10$, $11.4 \pm 5.10$. Each is a bit wider than the corresponding interval obtained using the Bonferroni inequality in Example 14.3a. But the Scheffé procedure gives simultaneous confidence intervals, not just for the three pairwise comparisons, but for all possible contrasts; the extra width allows for this.  ∎

For a detailed presentation of these and some other methods of multiple comparison, see, for instance, Chapter 11 of *An Introduction to Statistical Methods and Data Analysis* by Lyman Ott (Boston: Duxbury Press, 1984).

---

[5]See Lindgren [15], 522–23.

## Problems

**\* 14-14.** A study of the effects of exercise reports the following summary statistics for physical fitness scores.[6] The "treatment" subjects were participants in an exercise program; control subjects were volunteers for the program who were unable to attend; subjects in the other two groups were similar to those in the first two groups in age and other characteristics.

| Group | $n$ | Mean | s.d. |
|---|---|---|---|
| Treatment | 10 | 291.9 | 38.17 |
| Control | 5 | 309.0 | 32.07 |
| Joggers | 11 | 366.9 | 41.19 |
| Sedentary | 10 | 226.1 | 63.53 |

   **(a)**  Test the hypothesis of no average difference among the types of subjects.
   **(b)**  Apply the Bonferroni method for comparing the six pairs of means.
   **(c)**  Apply the Scheffé method for multiple comparisons.

**14-15.** Apply the Bonferroni and Scheffé methods for multiple comparisons to the data in Problem 14-6, using .05 as the overall $\alpha$.

**\* 14-16.** Compare the ten pairs of means for the data in Problem 14-4 using $\alpha = .10$ and
   **(a)**  the Bonferroni method.
   **(b)**  the Scheffé method, given $F_{.90}(4, 22) = 2.5$.

## 14.4      Two-Way ANOVA

In comparing two treatments in §12.6, we explained how matching pairs of experimental subjects can help to eliminate the effect of irrelevant individual differences. In such a design, subjects are paired according to variables that may have an effect on the measured response. Then one member of the pair is assigned to each treatment, at random. This matched-pairs design is a special case of a **randomized block design** for comparing $r$ treatments. In this design, we group $n = rc$ subjects or pieces of experimental material into $c$ matched **blocks** of size $r$ and randomly assign one member of each block to each treatment. The data consist of a rectangular array of $rc$ responses $X_{ij}$—the response to treatment level $i$ of the individual in block $j$ receiving that treatment level (for $i = 1, 2, ..., r$ and $j = 1, 2, ..., c$).

**Example 14.4a**   |   **Abrasion Resistance**
A wear-testing machine is used to determine resistance to abrasion. It has three weighted brushes under which samples of fabrics are fixed. Resistance is measured in

---

[6]Ph.D. dissertation by D. Lobstein, Purdue Univ., 1983, as recorded in Moore and McCabe's *Introduction to the Practice of Statistics*, 2nd Ed. (New York: Freeman, 1993).

terms of loss of weight after a specified number of cycles. Four different fabrics were tested and the weight loss measured at each brush position. The data are as follows:[7]

| | | *Brush* | | |
|---|---|---|---|---|
| | | 1 | 2 | 3 |
| *Fabric* | *A* | 1.93 | 2.38 | 2.20 |
| | *B* | 2.55 | 2.72 | 2.75 |
| | *C* | 2.40 | 2.68 | 2.31 |
| | *D* | 2.33 | 2.40 | 2.28 |

The main interest is in fabric differences, but measured wear may also depend on the brush or its position. It would be a mistake to ignore the brush and compare the fabrics using a one-way ANOVA if there is appreciable variation due to brush or brush position. In a one-way analysis, this variation would simply be part of "experimental error" and inflate the sum of squares in the denominator of the $F$-ratio. Real fabric differences would be less apt to show up than they would if the brush effect could be removed. (To be continued.) ■

As in a one-way ANOVA, we assume independent observations, normally distributed with constant variance:

$$X_{ij} = \mu_{ij} + \epsilon_{ij}, \tag{1}$$

where $\epsilon_{ij} \sim \mathcal{N}(0, \sigma^2)$. The mean response $\mu_{ij}$ depends on the treatment level $i$ and the block index $j$. Define

$$\theta_i = \frac{1}{c}\sum_j E(X_{ij}), \quad \phi_j = \frac{1}{r}\sum_i E(X_{ij}), \quad \mu = \frac{1}{rc}\sum_i\sum_j E(X_{ij}),$$

and

$$\tau_i = \theta_i - \mu, \quad \beta_j = \phi_j - \mu.$$

The $\tau$'s and $\beta$'s defined in this way satisfy

$$\sum_i \tau_i = \sum_j \beta_j = 0.$$

(See Problem 14-R6.) Thus, $\tau_i$ and $\beta_j$ are, respectively, treatment and block "effects"—differences of the mean treatment response and the mean block response from the overall mean response.

In some cases, treatment differences and block differences may contribute to a response additively: $\mu_{ij} = \mu + \tau_i + \beta_j$. But when the structure is not that simple, we

---

[7]Data from a consulting project.

define

$$\psi_{ij} = \mu_{ij} - \mu - \tau_i - \beta_j \tag{2}$$

and term this the *interaction*. With only one observation for each $ij$-combination, we can't get at this interaction (if present) and are forced to treat it as part of the "error," $\epsilon_{ij}$.

---

Additive model for a two-way ANOVA:

$$X_{ij} = \mu + \tau_i + \beta_j + \epsilon_{ij}, \tag{3}$$

where the $\epsilon_{ij}$ are independent and normally distributed with mean 0 and constant variance $\sigma^2$, and $\sum \tau_i = \sum \beta_j = 0$.

---

The null hypothesis to be tested is that the means of the responses at the different treatment levels are the same:

$$H_0: \quad \tau_i = 0, \quad \text{all } i.$$

Now, observe the symmetry of the mathematical problem: We could equally well test $\beta_j = 0$ for all $j$. Indeed, this same model (3) is applicable when the "treatment" is thought of as one factor, and the "block" as another factor affecting responses. For instance, in an agricultural experiment, yield may depend on seed type and on fertilizer type, and the researcher would be interested in both—not merely using seed type as a blocking variable. So what we'll develop for the randomized block design can be used also to investigate the effect on responses of two factors, say factor A and factor B. A "treatment" then really consists of a combination of seed type and fertilizer type. In such settings one might have enough resources to obtain more than one measurement of response for each combination of factor types. In such a case it is possible to exploit the replication of responses to check on the possibility of "interaction," as represented by (2). We'll leave this to more intensive developments of the material and present only the case of one observation per "cell" or factor combination, referring to the factors as treatment and block.

Maximum likelihood estimates of the parameters of the model (3) are fairly straightforward to obtain. (See Problem 14-23.) They are as follows:

$$\hat{\mu} = \overline{X} = \frac{1}{rc} \sum_i \sum_j X_{ij}, \quad \hat{\tau}_i = \overline{X}_{i\cdot} - \overline{X}, \quad \hat{\beta}_j = \overline{X}_{\cdot j} - \overline{X}, \tag{4}$$

where

$$\overline{X}_{i\cdot} = \frac{1}{c} \sum_{j=1}^{c} X_{ij}, \quad \overline{X}_{\cdot j} = \frac{1}{r} \sum_{i=1}^{r} X_{ij}.$$

The "error" involved in a particular response [from (3)] is $X_{ij} - \mu - \tau_i - \beta_j$, which we could estimate to be

$$\widehat{\epsilon}_{ij} = X_{ij} - \widehat{\mu} - \widehat{\tau}_i - \widehat{\beta}_j = X_{ij} - \overline{X}_{i\cdot} - \overline{X}_{\cdot j} + \overline{X}. \tag{5}$$

We write the deviation of each response $X_{ij}$ from the grand mean $\overline{X}$ in a way that exhibits the treatment component and the blocking component:

$$X_{ij} - \overline{X} = (\overline{X}_{i\cdot} - \overline{X}) + (\overline{X}_{\cdot j} - \overline{X}) + (X_{ij} - \overline{X}_{i\cdot} - \overline{X}_{\cdot j} + \overline{X}), \tag{6}$$

or

$$X_{ij} - \widehat{\mu} = \widehat{\tau}_i + \widehat{\beta}_j + \widehat{\epsilon}_{ij}.$$

Squaring (6) as the trinomial on the right-hand side, and summing over both $i$ and $j$, we find that all cross-product terms drop out upon summing, leaving only sums of squares:

$$\sum\sum(X_{ij} - \overline{X})^2 = \sum\sum(\overline{X}_{i\cdot} - \overline{X})^2 + \sum\sum(\overline{X}_{\cdot j} - \overline{X})^2 \tag{7}$$
$$+ \sum\sum(X_{ij} - \overline{X}_{i\cdot} - \overline{X}_{\cdot j} + \overline{X})^2.$$

Define SS's as follows:

$$\text{SSTot} = \sum\sum(X_{ij} - \overline{X})^2.$$

$$\text{SSTr} = \sum\sum(\overline{X}_{i\cdot} - \overline{X})^2 = \sum_i c(\overline{X}_{i\cdot} - \overline{X})^2 = \sum_i c\widehat{\tau}_i^2.$$

$$\text{SSB} = \sum\sum(\overline{X}_{\cdot j} - \overline{X})^2 = \sum_j r(\overline{X}_{\cdot j} - \overline{X})^2 = \sum_j r\widehat{\beta}_j^2.$$

$$\text{SSE} = \sum\sum(X_{ij} - \overline{X}_{i\cdot} - \overline{X}_{\cdot j} + \overline{X})^2 = \sum\sum\epsilon_{ij}^2.$$

In terms of these, we can write (7) as

$$\text{SSTot} = \text{SSTr} + \text{SSB} + \text{SSE}. \tag{8}$$

This decomposition of the variability about the grand mean into components identifiable as belonging to treatment, to block, and to error is the "analysis of variance" for the two-way model.

When there is a treatment effect, SSTr will tend to be large; when there is a block effect, SSB will tend to be large. However, SSE is based on the residual when the overall mean, treatment, and block effects are removed [see (5)]; it measures "error" and should be independent of treatment and block effects. A large ratio of SSTr to SSE would then indicate a treatment effect.

Under the hypothesis of no treatment or block effect, the $rc$ observations constitute a random sample from $\mathcal{N}(0, \sigma^2)$, so that

$$\frac{\text{SSTot}}{\sigma^2} = \sum \sum \frac{(X_{ij} - \overline{X})^2}{\sigma^2} \sim \text{chi}^2(rc - 1).$$

Dividing (8) through by $\sigma^2$, we have

$$\frac{\text{SSTot}}{\sigma^2} = \frac{\text{SSTr}}{\sigma^2} + \frac{\text{SSB}}{\sigma^2} + \frac{\text{SSE}}{\sigma^2}. \tag{9}$$

It can be shown that (when $\tau_i \equiv 0$ and $\beta_j \equiv 0$) the terms on the right are independent, chi-square variables, with d.f.'s $r - 1$, $c - 1$, and $(r - 1)(c - 1)$, respectively.[8]

When there is a treatment effect (not all $\tau_i = 0$), the treatment sum of squares will tend to be inflated, and the ratio of SSTr to the error sum of squares SSE (which is not affected by a treatment effect) will be unusually large. The ratio we'll use is the ratio of mean squares, MSTr/MSE, since this ratio has a distribution for which we have tables—an $F$-distribution. The MSE is an unbiased estimate of the error variance—whether or not the $\tau_i$'s are 0; and under the hypothesis of no treatment effect, the MSTr is also an unbiased estimate of the error variance. (This is shown in much the same way as a similar claim was made for a one-way ANOVA, and shown in Problem 14-9.) So the test ratio tends to be near 1 when $H_0$ is true, and large values are taken as "extreme," suggesting a treatment effect.

---

$F$-ratio for testing $H_0$: $\tau_i = 0$, $i = 1, \dots, r$:

$$F_{\text{Tr}} = \frac{\text{SSTr}/(r - 1)}{\text{SSE}/[(r - 1)(c - 1)]}. \tag{10}$$

Large values of $F$ are extreme. Find $P$ in Table VIIIa of Appendix I, with $[(r - 1), (r - 1)(c - 1)]$ d.f.

For testing no block effect, use

$$F_{\text{B}} = \frac{\text{SSB}/(c - 1)}{\text{SSE}/[(r - 1)(c - 1)]}, \tag{11}$$

an $F$-statistic with $[(c - 1), (r - 1)(c - 1)]$ d.f.

---

**Example 14.4b** | **Abrasion Resistance (Continued)**

We return to the data of Example 14.4a, repeated here with an added 6 column of sums:

---

[8]See Lindgren [15], 530.

|        |   | *Brush* |      |      |        |
|--------|---|------|------|------|--------|
|        |   | 1    | 2    | 3    | *Sum*  |
|        | A | 1.93 | 2.38 | 2.20 | 6.51   |
| *Fabric* | B | 2.55 | 2.72 | 2.75 | 8.02   |
|        | C | 2.40 | 2.68 | 2.31 | 7.39   |
|        | D | 2.33 | 2.40 | 2.28 | 7.01   |

The ANOVA table was obtained using statistical software:[9]

```
SOURCE              DF        SS          MS         F         P
--------------      ----      --------    --------   ------    -------
FABRIC (A)           3        0.40549     0.13516    9.43      0.0109
BRUSH (B)            2        0.12162     0.06081    4.24      0.0710
A*B                  6        0.08598     0.01433
--------------      ----      --------    --------
TOTAL               11        0.61309
GRAND AVERAGE        1       69.7454
```

If such software is not readily available, hand computation is feasible with this amount of data. Problem 14-22 gives some calculating formulas for the various SS's. The constant $C$ in those formulas is the square of the sum of all the data, divided by $rc$:  $C = 69.7459$. The formula for SSTr, which uses the row sums shown above, yields  SSTr $= 70.1509 - 69.7459 = .405$. The total sum of squares is just the numerator of the variance of the combined sample of the 12 observations, and subtracting SSTr and SSB from it yields SSE. The $P$-value for the hypothesis of no fabric differences, from Table VIIIa, is the tail-area beyond 9.43 (3 and 6 d.f.), or about .011.

The $F$-ratio for the blocking variable is not quite "significant," with $P \doteq .071$. But there is a hint of an effect, and one way to see if it matters is to ignore it and treat the data as consisting of four independent random samples of size 3. The one-way ANOVA table is easy to get, since it follows from the above two-way table by simply combining the brush and error rows to get a new error row:

```
SOURCE              DF        SS          MS         F         P
--------------      ----      --------    --------   ------    -------
BETWEEN              3        0.40549     0.13516    5.21      0.0276
WITHIN               8        0.20760     0.02595
TOTAL               11        0.61309
```

With 3 and 8 d.f., this $F$ yields a $P$-value of .028. There is still evidence against $H_0$, but the case is not as strong as it was when we took "brush" into account in a two-way analysis. ∎

---

[9]Obtained using *Statistix* 4.0  (Atlanta: Analytical Software).

Some final comments: It can be shown (Problem 14-25) that the $F$-tests using (10) and (11) are likelihood ratio tests. And when $r = 2$, the $F$-test using (10) is equivalent to a two-sided paired $t$-test (Problem 14-24).

The above analysis does not take into account any interaction—the possibility that treatment $i$ has a different effect in block $j$ from what it has in block $j'$ (or that factor A affects the response differently when paired with different levels of factor B). If one can obtain more than one response for at least one factor level combination, there will be a within-cells estimate of error variance available for checking interaction. For the model and the analysis of data, see any text on design of experiments (e.g., Kuehl [11]).

## Problems

**\* 14-17.** Construct the ANOVA table and calculate $F$-ratios for this set of artificial data on two factors, each at three levels:

|        |   | B: |    |    |
|--------|---|-----|-----|-----|
|        |   | 1   | 2   | 3   |
|        | 1 | 3   | 5   | 4   |
| A:     | 2 | 11  | 10  | 12  |
|        | 3 | 16  | 21  | 17  |

**14-18.** Find the SSE for the following set of artificial data on a treatment at four levels with three blocks:

|     |   | Tr: |    |   |    |
|-----|---|------|-----|-----|-----|
|     |   | 1    | 2   | 3   | 4   |
|     | 1 | 16   | 12  | 9   | 11  |
| B:  | 2 | 11   | 7   | 4   | 6   |
|     | 3 | 15   | 11  | 8   | 10  |

(a) What do the $F$-ratios suggest?

(b) Give estimates of the treatment and block effects.

**\* 14-19.** The whiteness of clothes after washing may depend on both water temperature and brand of detergent. With three water temperatures and three brands of detergent, suppose whiteness measurements are obtained as follows:

|        |        | Detergent |    |    |
|--------|--------|-----------|-----|-----|
|        |        | A         | B   | C   |
|        | Warm   | 45        | 47  | 55  |
| Temp.  | Cold   | 36        | 41  | 55  |
|        | Hot    | 42        | 47  | 46  |

Give the ANOVA table, the $F$-statistic, and the $P$-value, for testing
   (a)   the hypothesis of no difference among types of detergent.
   (b)   the hypothesis of no water temperature effect.

**14-20.**   A study of methods of freezing meat loaf investigated the effect of oven position on drip loss.[10] Three batches of eight loaves each (the "blocks") were baked and analyzed. The drip losses were as follows:

*Oven position*

|        | 1     | 2     | 3     | 4     | 5     | 6     | 7     | 8     | *Means* |
|--------|-------|-------|-------|-------|-------|-------|-------|-------|---------|
| 1      | 7.33  | 3.22  | 3.28  | 6.44  | 3.83  | 3.28  | 5.06  | 4.44  | 4.610   |
| 2      | 8.11  | 3.72  | 5.11  | 5.78  | 6.50  | 5.11  | 5.11  | 4.28  | 5.465   |
| 3      | 8.06  | 4.28  | 4.56  | 8.61  | 7.72  | 5.56  | 7.83  | 6.33  | 6.619   |
| *Means* | 7.833 | 3.740 | 4.317 | 6.943 | 6.017 | 4.650 | 6.000 | 5.017 | 5.565   |

   (a)   Test the hypothesis of no oven position effect, given SSE $= 9.290$.
   (b)   Test the hypothesis of no batch effect.

\* **14-21.**   In the preceding problem, suppose the data had been analyzed as consisting of eight random samples of size 3, ignoring the possibility of a batch effect. Combine the error and batch sums of squares to obtain an error sum of squares in a one-way ANOVA, to test the hypothesis of no oven-position effect. (Was the blocking effective?)

**14-22.**   Obtain these formulas (convenient in hand calculations) for the two-way classification:

$$\mathrm{SSTr} = \frac{1}{c}\sum_i \left\{\sum_j X_{ij}\right\}^2 - C, \quad \mathrm{SSB} = \frac{1}{r}\sum_j \left\{\sum_i X_{ij}\right\}^2 - C,$$

where

$$C = \frac{1}{rc}\left\{\sum\sum x_{ij}\right\}^2$$

**14-23.**   Derive the m.l.e.'s of the parameters $\tau_i$, $\beta_j$, and $\sigma^2$ given as (4) in §14.4 (page 581).

**14-24.**   Derive the tests based on the $F$-ratios (10) and (11) on page 583 as likelihood ratio tests.

**14-25.**   Show that when there are just two treatments, the $F$-test for no treatment effect is equivalent to a two-sided paired-$t$ test.

## Review Problems

**14-R1.**   Using telephoto lenses, researchers photographed pairs of unacquainted men on park benches in San Francisco, California; Tangiers, Morocco; and Seville, Spain.[11] The men in a

---

[10]B. Bobeng and B. David, described by T. J. Ryan, Jr. et al., in *Minitab: A Student Handbook* (North Scituate, MA: Duxbury Press, 1976).

[11]A. Mazur, "Interpersonal spacing on public benches in contact vs. noncontact cultures," *J. Soc. Psych.* 101 (1977), 53–58.

pair were not of different racial groups and were not tourists. The data were obtained by making measurements on enlargements, with reference to the known bench lengths. Summary statistics are as follows:

|  | n (pairs) | $\overline{X}$ (in.) | s.d. (in.) |
|---|---|---|---|
| San Francisco | 35 | 28 | 10 |
| Tangiers | 25 | 35 | 10 |
| Seville | 22 | 33 | 7 |

Test the hypothesis of no difference in population mean spacing.

**14-R2.** In one-way ANOVA, under what circumstance is the grand mean the ordinary average of the sample means?

**14-R3.** Explain why the Scheffé method of multiple comparisons is apt to result in wider confidence intervals than the Bonferroni method.

**14-R4.** Suppose you calculate sums of squares using the formulas in Problem 14-22 and find one to be negative. Is this possible? What could account for it?

**14-R5.** With $\tau_i$ defined as in §14.1, show that $\sum n_i \tau_i = 0$, as claimed.

**14-R6.** With $\tau_i$ and $\beta_j$ defined as in §14.4, show that $\sum \tau_i = \sum \beta_j = 0$, as asserted there.

**14-R7.** A study reports these summary statistics for the force on the hand at impact, in a one-handed backhand drive in handball:[12]

|  | n | $\overline{X}$ | s.d. |
|---|---|---|---|
| Advanced player | 16 | 40.3 | 11.3 |
| Intermediate | 8 | 21.4 | 8.3 |

Analyze, for the null hypothesis of no difference between advanced and intermediate players, using

**(a)** an analysis of variance.

**(b)** a two-sample $t$-test.

**14-R8.** Testing for interaction in a two-factor design requires replicate observations in some cells. Suppose there are $m$ observations in each of the $rc$ cells of a two-way table: $X_{ijk}$, where $k = 1, ..., m$, and let $\overline{X}_{jk.}$ denote the mean of the $ij$-cell. Under the usual assumptions of normal errors with equal variances, show that

$$SSE = \sum_i \sum_j \sum_k (X_{ijk} - \overline{X}_{ij.})^2 \sim \sigma^2 \times \text{chi}^2[rc(m-1)],$$

and if $\mu_i = \mu$,

$$SSTotal = \sum_i \sum_j \sum_k (X_{ijk} - \overline{X})^2 \sim \sigma^2 \times \text{chi}^2(rcm - 1).$$

---

[12]*Int. J. of Sport Biomechanisms*, (1991), 282–92.

## Chapter Perspective

We have seen how a suitable decomposition of total variability can focus on the various factors affecting a response, for two simple experimental designs, which generalized the settings of the two-sample $t$-test and the paired $t$-test of Chapter 12. Such designs can be further generalized in many ways. The basic ANOVA technique can be extended to deal with these more complicated designs. We refer the student to courses and textbooks in this extensively developed area of applied statistics.

The underlying probability structure we have assumed in this chapter is a special case of what is termed a *linear model*—one in which a response depends on parameters identifying the various factors that affect the response. The next chapter takes up linear models in which the factors of interest are numerical variables. In the analysis of data based on these "regression" models, we'll again encounter an ANOVA.

The analyses of this chapter are fairly robust against the hypothesis of normality. However, when normality cannot be assumed, one can use nonparametric tests for analyzing the data in a completely randomized design (Kruskal-Wallis test) and in a randomized block design (Friedman test). See, for instance, Rice [15].

# Regression

**Example 15a**

The height $(Y)$ that a cake rises depends on the oven temperature $(x_1)$ and on the amounts of the various ingredients—the amount of flour $(x_2)$, the amount of soda $(x_3)$, and so on. Thus, for some function $g$, we might expect that

$$Y = g(x_1, x_2, ..., x_k).$$

Because of variations in the ingredient materials and factors not taken into account, the height $Y$ is not completely determined by the $x$'s. It includes a *random error*. We'd want to know the importance of each of the explanatory variables $x_j$ and to be able to predict $Y$, given the values of the $x$'s. The randomness makes the problems of analysis and prediction statistical ones. An obvious approach is to gather some data— to bake several cakes using various combinations of values of these variables, and from the results, to draw conclusions about the underlying relationship as defined by $g$. ∎

Regression models are used to describe how a random numerical response $Y$ depends on one or more explanatory or predictor variables, $x_1$, $x_2$, ..., $x_k$. As suggested in the example, we might expect that the relationship is functional, described mathematically by $Y = g(x_1, x_2, ..., x_k)$. But in practice, such relationships are usually corrupted by factors that we can only account for as random "error." So our model will have to include a catch-all term to take account of the error component. We assume that it is an additive contribution to the observed response $Y$:

$$Y = g(x_1, x_2, ..., x_k) + \epsilon,$$

where $\epsilon$ will be described in terms of some probability model.

We assume that $E(\epsilon) = 0$, incorporating any constant into the $g$, so that the function $g$ is the mean value of $Y$, called the **regression function.** The regression function is usually assumed to have a specific form or type, involving one or more unknown parameters. The type to be considered in this chapter is linear:

$$Y = \beta_0 + \beta_1 x_1 + \cdots + \beta_k x_k,$$

although certain nonlinear functions can be subsumed within this type of relationship. A *regression analysis* involves inferences about the parameters $\beta_j$, based on data.

We begin by studying the special case of a single predictor or explanatory variable.

## 15.1    Models with a Single Explanatory Variable

We consider two types of sampling schemes. Both types result in pairs $(X, Y)$ in which $Y$ is the **response,** and $X$ is the **explanatory** or **predictor variable.** The schemes are analogous to those we considered in §13.5 for sampling bivariate categorical populations. In one scheme, $X$ can be *controlled*—set at a particular value. To stress that $X$ is nonrandom, we use the lower case $x$ for this variable. The variable $Y = Y_x$ is the response when the controlled variable has the value $x$. For instance, a reactor yield $(Y)$ may depend on temperature $(x)$; sales $(Y)$ on amount of advertising $(x)$; change in blood pressure $(Y)$ on dosage of an antihypertensive drug $(x)$; and so on.

The relationship between response and controlled variable may be completely predictable, but more often than not, there are unavoidable, uncontrolled factors that result in different responses for the same $x$. Such factors include measurement errors and changes in experimental conditions such as temperature, humidity, power surges, vibrations, and so on. Their effects on the response are thought of as random components of the measured response. The aim of a statistical analysis is to sort out the underlying relationship as defined by $g(x)$ from the "noise," the random component of the observed $Y$-values.

In such contexts the *data* consist of pairs $(x_i, Y_i)$, where the $x_i$'s are selected, fixed values of $x$, and the $Y_i$'s are the corresponding responses.

Example **15.1a** | ## A Welding Experiment

The data in the accompanying table are from the report of an inertia welding experiment.[1] The controlled variable $x$ is the angular velocity of a rotating part, and the response $Y$ is the breaking strength of the weld. Seven welds were made, using four different velocities of rotation:

| Velocity $(x)$ $(10^2 \text{ ft/min})$ | Breaking strength $(Y)$ $(ksi)$ |
|---|---|
| 2.00 | 89 |
| 2.5 | 97, 91 |
| 2.75 | 98 |
| 3.00 | 100, 104, 97 |

[1] Reported in G. E. P. Box, W. G. Hunter and J. S. Hunter, *Statistics for Experimenters* (New York: Wiley, 1978), 473.

The data are plotted in Figure 15-1. The plot suggests an underlying relationship between velocity and breaking strength that is linear, with deviations from linearity thought of as random errors. ∎

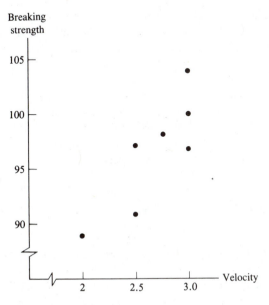

**Figure 15-1** Scatter plot—strength vs. velocity

Let $x$ denote a fixed value of the controlled variable, and $Y_x$ the corresponding response. We assume the response to be the sum of a deterministic function $g(x)$ and a random error $\epsilon$:

$$Y_x = g(x) + \epsilon, \tag{1}$$

where $E(\epsilon) = 0$, so that $EY_x = g(x)$, the regression function. The function $g$ specifies what we think of as the underlying or true relation between $Y$ and $x$, and the model includes the random error $\epsilon$ as an additive component of the $Y$ we observe. Researchers are interested in $g$ and will collect data (as in the above example)—perhaps a sample of $n$ pairs $(x_i, Y_i)$. But inferring anything about $g$ from the data is complicated by the presence of the error component $\epsilon$.

We said at the outset that there are two sampling schemes. In the second, we select $n$ experimental subjects or materials at random and make two measurements $(X, Y)$ on each. Both $X$ and $Y$ are random variables, and the basic model is a bivariate distribution. In such settings, the goal is usually to be able to predict the value of one variable, given a value of the other. The variable to be predicted will be called $Y$, and the predictor variable $X$.

**Example 15.1b** | **GPA vs. SAT Scores**

Many high school students take college aptitude tests such as the SAT (Scholastic Aptitude Test) in their junior or senior year. Such tests are intended to indicate a student's readiness for college and to predict how well the student will do in college. Let $X$ denote a student's SAT score, and $Y$ the student's GPA (grade point average) at the end of the first year in college. Both $X$ and $Y$ are random when the student is selected at random. They are correlated (as one would hope, and experience shows), but not perfectly so. (Typically, $\rho$ is about .6.) The relationship between the variables can aid in the prediction of GPA for a student whose SAT score is known, but any such prediction is subject to error when $|\rho| < 1$.          ∎

To predict $Y$ given $X = x^*$, we'll use the mean of the subpopulation of $Y$-values in which $X$ has the value $x^*$. In the setting where both $X$ and $Y$ are random, this mean is a *conditional* mean, which we write as $E(Y \mid x)$. In the setting where $X$ is controlled, the mean of $Y$ varies with $x$ and in this sense is conditional. So, we still write it as $E(Y \mid x)$. The choice of the mean as the parameter of the distribution of $Y \mid x$ relevant to prediction will be explained in §15.5.

In either sampling framework, the function $g(x) = E(Y \mid x)$ is called the regression function of $Y$ on $x$, and an observation $Y$ at $X = x$ has the structure of (1): $Y_x = g(x) + \epsilon$. In some applications, the distribution of the random error $\epsilon$ might depend on $x$, but we consider only the simple case in which $\epsilon$ is normally distributed with mean 0 and a variance independent of $x$. Figure 15-2 indicates this model schematically:  At each of several values of the explanatory variable $x$, the response follows a distribution centered at $g(x)$ with a standard deviation that is the same for all $x$.

---

Regression Model:

$$Y_x = g(x) + \epsilon,$$

where $\epsilon \sim \mathcal{N}(0, \sigma^2)$. Observed responses $Y_x$ are assumed independent.

---

The error distribution is usually assumed to be *normal*, as in the figure. The regression function is usually taken to be one of a parametric family of functions—trigonometric $(\alpha + x \sin \beta)$, exponential $(\alpha e^{\beta x})$, etc. The simplest form of regression function is *linear*:  $\alpha + \beta x$. The model parameters $(\alpha, \beta, ...)$ are unknown constants, and data points $(x_i, Y_i)$ can be used to estimate these parameters and to predict the value of $Y$ at other values of $x$.

In the next two sections, we consider only a simple linear regression:

$$Y_x = \alpha + \beta x + \epsilon,$$

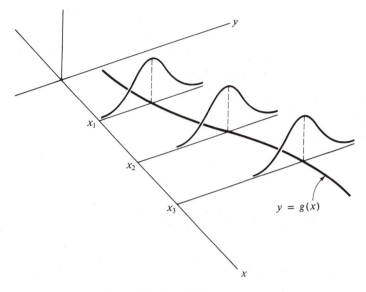

**Figure 15-2**

and independent, normally distributed errors $\epsilon$ with mean 0 and constant variance. The assumption of a linear regression function may be valid only to a limited range of values of $x$; and even there, a linear function may only be an approximation to the true regression function.

## 15.2   Least Squares

We turn now to finding estimates of the slope $\beta$ and intercept $\alpha$ of a linear regression function $\alpha + \beta x$, based on $n$ data pairs $(x_i, Y_i)$. We find such estimates by finding a linear function whose graph comes as nearly as possible to passing through all the data points. A look at Figure 15-1 should convince you that it may not be possible for a line to pass through all data points. The method of least squares, introduced by K. F. Gauss in the early nineteenth century, is to locate the line so that the sum of the squared "residuals" is as small as possible. The **residual** is the amount by which a particular $Y_x$ differs from the value of the regression function at that same value of $x$. For fitting a straight line $y = a + bx$, the residual at the point $x$ is

$$Y_x - (a + bx). \tag{1}$$

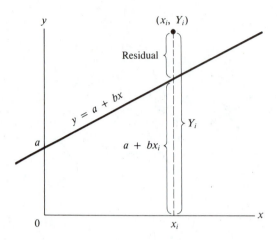

**Figure 15-3** Residual at $x_i$

The residual is shown for one of the data points $x_i$ in Figure 15-3. The sum of squared residuals, which we want to minimize, is

$$\sum_{1}^{n}\{Y_i - a - bx_i\}^2. \tag{2}$$

The values of $a$ and $b$ that minimize (2) define the **least-squares line.**

**Example 15.2a**   We'll illustrate the least-squares process with numbers chosen to make the arithmetic simple. Consider these three data points: $(0, 1)$, $(1, 0)$, $(2, 2)$. For the linear function $a + bx$, the residuals are

$$1 - (a + 0b), \ \ 0 - (a + b), \ \ 2 - (a + 2b),$$

respectively. The sum of their squares (2) is

$$(1 - 2a + a^2) + (a^2 + 2ab + b^2) + (4 + a^2 + 4b^2 - 4a + 4ab - 8b)$$
$$= 3a^2 - 6a + 5b^2 - 8b + 6ab + 5 \equiv R(a, b).$$

To find the $a$ and $b$ that minimize this $R(a, b)$, we can use either algebra or calculus. According to the method of calculus, we set the derivative with respect to $a$ (holding $b$ fixed) equal to 0, because at the maximum point the function $R$ must be largest as $a$ varies when $b$ is held fixed. Likewise, we set the derivative with respect to $b$ (holding $a$ fixed) equal to 0:

$$\begin{cases} 6a + 6b - 6 = 0 \\ 6a + 10b - 8 = 0. \end{cases}$$

Solving these equations simultaneously yields $a = b = 1/2$ as the only critical point—the only point where the surface $z = R(a, b)$ has a horizontal tangent plane. This point must minimize $R$, because the sum of squared residuals is bounded below (by 0) and so *has* a minimum. The least-squares line is therefore $y = (1 + x)/2$, shown along with the three data points in Figure 15-4.

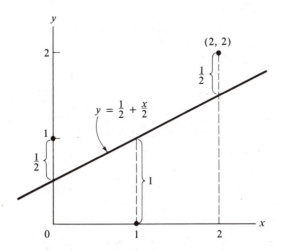

**Figure 15-4** Least-squares line for Example 15.2a

The residuals for the three data points are easy to calculate here—or they can be read from the graph: $1/2, -1, 1/2$. (Notice that the residuals sum to 0, as is always the case according to Problem 15-8.) The sum of their squares is $1/4 + 1 + 1/4 = 3/2$, and this is the minimum value of $R(a, b)$. The sum of the squared residuals about any other line would be larger. ∎

The detailed calculation of the last example was tedious, even with simple arithmetic. To derive general formulas for the least-squares coefficients, we use an algebraic method. As a by-product, the derivation will yield a convenient formula for the smallest sum of squared residuals.

To minimize (2), we first rewrite each residual thus:

$$Y_i - a - bx_i = (Y_i - \overline{Y}) + (\overline{Y} - a - b\overline{x}) - b(x_i - \overline{x}).$$

After squaring this as a trinomial, we sum on $i$:

$$\sum [Y_i - (a + bx_i)]^2 = \sum (Y_i - \overline{Y})^2 + n(\overline{Y} - a - b\overline{x})^2$$
$$+ b^2 \sum (x_i - \overline{x})^2 - 2b \sum (x_i - \overline{x})(Y_i - \overline{Y}). \qquad (3)$$

(The other cross-product terms vanish because the sum of the deviations of the $x$'s about their mean and the sum of the deviations of the $Y$'s about their mean are both 0.) Some notation will prove helpful. The sum of squared deviations of any set of numbers $x$ about their mean $\bar{x}$ is

$$\text{SS}_{xx} = \sum (x - \bar{x})^2 = \sum x^2 - \frac{1}{n} \left( \sum x \right)^2, \tag{4}$$

and the sum of products of deviations of $x$'s and $y$'s, each about their respective means, is

$$\text{SS}_{xy} = \sum (x - \bar{x})(y - \bar{y}) = \sum xy - \frac{1}{n} \left( \sum x \right) \left( \sum y \right). \tag{5}$$

(Here and in what follows, sums extend from $i = 1$ to $i = n$, the number of data points.) In this "SS" notation, we rewrite (3) as

$$R(a, b) = \text{SS}_{YY} + n(\overline{Y} - a - b\bar{x})^2 + b^2 \text{SS}_{xx} - 2b\text{SS}_{xY}. \tag{6}$$

The last two terms start out as the square of a binomial, which we complete as follows:

$$b^2 \text{SS}_{xx} - 2b\text{SS}_{xY} = \text{SS}_{xx} \left\{ b^2 - 2b \frac{\text{SS}_{xY}}{\text{SS}_{xx}} + \left( \frac{\text{SS}_{xY}}{\text{SS}_{xx}} \right)^2 \right\} - \frac{\text{SS}_{xY}^2}{\text{SS}_{xx}}$$

$$= \text{SS}_{xx} \left\{ b - \frac{\text{SS}_{xY}}{\text{SS}_{xx}} \right\}^2 - r^2 \text{SS}_{YY},$$

where $r$ is the correlation coefficient of the $n$ pairs $(x_i, y_i)$:

$$r = \frac{\text{SS}_{xy}}{\sqrt{\text{SS}_{xx}\text{SS}_{yy}}}. \tag{7}$$

We can now rewrite (6) as

$$R(a, b) = \text{SS}_{YY} + n(\overline{Y} - a - b\bar{x})^2 + \text{SS}_{xx} \left\{ b - \frac{\text{SS}_{xY}}{\text{SS}_{xx}} \right\}^2 - r^2 \text{SS}_{YY}. \tag{8}$$

Clearly, $R$ is minimized when the second and third terms on the right are 0:

$$b = \frac{\text{SS}_{xY}}{\text{SS}_{xx}}, \quad a = \overline{Y} - b\bar{x}. \tag{9}$$

We'll denote these minimizing values by $\widehat{\beta}$ and $\widehat{\alpha}$, because they will be our estimates of the regression parameters $\beta$ and $\alpha$. Indeed, with the assumption of normal errors, they will be seen to be m.l.e.'s. And with these in place of $a$ and $b$ in $R(a, b)$, we obtain its minimum value: $\text{SS}_{YY}(1 - r^2)$.

The least-squares line (empirical regression line) is given by

$$y = \widehat{\alpha} + \widehat{\beta} x, \tag{10}$$

where

$$\widehat{\beta} = \frac{SS_{xY}}{SS_{xx}}, \quad \widehat{\alpha} = \overline{Y} - \widehat{\beta}\overline{x}. \tag{11}$$

The sum of squares about the least-squares line is

$$SSRes = SS_{YY}(1 - r^2). \tag{12}$$

Comparing $r$ as defined in (7) and $\widehat{\beta}$ in (11), we see that these quantities are related:

$$\widehat{\beta} = r\sqrt{\frac{SS_{yy}}{SS_{xx}}} = \frac{S_y}{S_x} r. \tag{13}$$

In particular, then, $r = 0$ is equivalent to $\widehat{\beta} = 0$. Also, if $S_y = S_x$, then $\widehat{\beta} = r$. Notice that in (12) we have another demonstration of the fact that $r^2 \le 1$, since $SSRes \ge 0$. It also follows from (12) that $r = \pm 1$ if and only if $SSRes$ vanishes, which can only happen if all residuals are 0—that is, if all the data points lie on a straight line.

**Example 15.2b**  Consider again the three data points of Example 15.2a. A table including squares and products is convenient for hand computation:

| $x$ | $y$ | $x^2$ | $xy$ | $y^2$ |
|---|---|---|---|---|
| 1 | 0 | 1 | 0 | 0 |
| 0 | 1 | 0 | 0 | 1 |
| 2 | 2 | 4 | 4 | 4 |
| Sums   3 | 3 | 5 | 4 | 5 |

With these sums in (4) and (5), we obtain

$$SS_{xy} = 4 - \frac{3 \times 3}{3} = 1, \quad SS_{xx} = 5 - \frac{3^2}{3} = 2 = SS_{yy},$$

and from (11),

$$\widehat{\beta} = \frac{1}{2} = r, \quad \widehat{\alpha} = 1 - \frac{1}{2} = \frac{1}{2}.$$

As with the calculation in the earlier example, we find $y = \frac{1}{2}(1 + x)$ as the equation of the least-squares line. To find the sum of squared residuals about the least-squares line, we substitute into (12):  $2(1 - 1/4) = 3/2$. This is what we found earlier by squaring the residuals and summing.  ∎

## Example 15.2c | A Dose-Response Curve

A study in surgery examined the increase ($Y$) in pancreatic intraductal pressure (PIP) in response to doses of a potent cholinesterase inhibitor ($x$).[2]  Six different doses were administered to one dog, with the results shown in the following table:

| x (dose) | y (pressure) | xy | $x^2$ |
|---------|-------------|------|------|
| 0 | 14.6 | 0 | 0 |
| 5 | 24.5 | 122.5 | 25 |
| 10 | 21.8 | 218 | 100 |
| 15 | 34.5 | 517.5 | 225 |
| 20 | 35.1 | 702 | 400 |
| 25 | 43.0 | 1075 | 625 |
| 75 | 173.5 | 2635 | 1375 |

The means and SS's are as follows:

$$\bar{x} = 12.5, \quad \bar{y} = 28.92, \quad SS_{xx} = 437.5, \quad SS_{yy} = 542.868, \quad SS_{xy} = 466.25.$$

Substitution into (11) yields $\hat{\beta} = 1.065$, $\hat{\alpha} = 15.6$, so the least-squares line, shown in the scatter plot of Figure 15-5, is

$$y = 15.6 + 1.065x.$$

The correlation coefficient [from (7) of sec 15.2] is

$$r = \frac{466.25}{\sqrt{437.5 \times 542.868}} = \sqrt{.9153}.$$

With this we can find the residual sum of squares using (12) of §15.2:

$$SSRes = 542.868(1 - .9153) = 45.98.$$

We have assumed that the dose-response curve is linear—that the increment in PIP is the same for each unit increment in dosage, regardless of the dose level. This assumption is not correct for some dose levels, because PIP is bounded. However, it may be reasonable to assume linearity over the range of doses considered. In practice,

---

[2]T. D. Dressel et al., "Sensitivity of the canine pancreatic intraductal pressure to subclinical reduction in cholinesterase activity," *Annals of Surgery* 190, No. 1 (1979).

one should check the assumption of linearity, at least by an informal inspection of the scatter plot. A more formal way is to introduce, for instance, a quadratic term in the regression function, to see if the resulting model gives an appreciably better fit—and better predictions. ∎

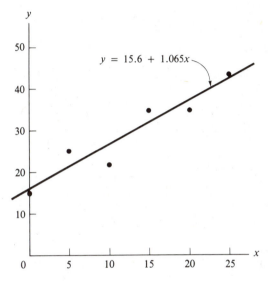

**Figure 15-5** Least-squares line for
Example 15.2c

## Problems

* **15-1.**   Given these data points:   $(0, 1), (1, 0), (1, 1), (0, 0), (2, 2)$,
   **(a)**   determine the least-squares line.
   **(b)**   find the sum of the squared residuals about the least-squares line
      **(i)**   directly.      **(ii)**   using (12) of §15.2.

* **15-2.**   Decrease in surface tension of liquid copper $(y)$ appears to be a linear function of the logarithm of the percentage of sulfur $(x)$. Given the following data, find the least-squares line $(y$ on $x)$ and the correlation coefficient.[3]

| Log percent sulfur | $-3.38$ | $-2.38$ | $-1.20$ | $-.92$ | $-.49$ | $-.19$ |
| --- | --- | --- | --- | --- | --- | --- |
| Decrease (deg/cm) | 308 | 426 | 590 | 624 | 649 | 727 |

---

[3]Modified from data of Baer and Kellogg, *Journal of Metals* 5 (1953), 634–48.

**15-3.** The following data are ages in years ($x$) and annual trunk diameter growth increment in mm ($y$) for a sample of trees:[4]

| | | |
|---|---|---|
| (17, 2.20) | (50, .75) | (40, 1.70) |
| (30, .85) | (93, .70) | (23, 1.25) |
| (54, .50) | (55, 1.00) | (46.5, .75) |

Make a scatter plot and find

**(a)** the least-squares line, and superimpose its graph on your scatter plot.
**(b)** the correlation coefficient.
**(c)** the residual sum of squares.

\* **15-4.** In an investigation of the damage to residential structures caused by blasting, the following data on frequency ($x$, in cps) and particle displacement ($y$, in inches) were obtained, for 13 blasts.[5] Find the least-squares regression line of $Y$ on $\log x$.

| $x$ | $y$ |
|---|---|
| 2.5 | .390, .250, .200 |
| 3.0 | .360 |
| 3.5 | .250 |
| 9.3 | .077 |
| 9.9 | .140 |
| 11.0 | .180, .093, .080 |
| 16.0 | .052 |
| 25.0 | .051, .150 |

**15-5.** Use calculus to minimize (2) of §15.2, to find the least-squares line. (First differentiate with respect to $a$, holding $b$ fixed, then with respect to $b$, holding $a$ fixed. Set these derivatives equal to 0 and solve simultaneously.)

\* **15-6.** Apply the method of least squares to find the *parabola* $y = a + bx + cx^2$ that best fits these four points: (0, 4), (1, 0), (2, 0), (3, 3).

**15-7.** Suppose there are $m$ observations at each $x_i$, for $i = 1, 2, ..., k$.
**(a)** Show that the least-squares line for the $mk$ data points can be found by applying (11) of §15.2 to the $k$ pairs $(x_i, \overline{Y}_i)$, where $\overline{Y}_i$ is the mean of the observations made at $x = x_i$.
**(b)** Explain why the method in (a) does not work when the numbers of observations at the various $x_i$ are not all equal.

**15-8.** Show that the sum of the residuals about the least-squares line is 0.

---

[4]"Pendunculate oak woodland severe environment," *J. Ecology* (1978), 707–40.
[5]H. Nicholls, C. Johnson, and W. Duvall, "Blasting vibrations and their effects on structures," U.S. Dept. of the Interior, Bureau of Mines *Bulletin 656* (1971), 17, Fig. 3.3.

## 15.3          Distribution of the Least-Squares Estimators

The least-squares coefficients $\hat{\alpha}$ and $\hat{\beta}$ [(11) in §15.2] depend on the data, so they vary from sample to sample. They are *random variables*—functions of the random responses $Y_i$. In order to understand what they have to tell us about the true regression coefficients $\alpha$ and $\beta$—how to use them in estimation and testing—we need to know *how* they vary from sample to sample. In this section, we derive their sampling distributions.

The distributions of $\hat{\alpha}$ and $\hat{\beta}$ follow from our assumptions about the random error component $\epsilon$. These assumptions, as given at the end of §15.1, are that the errors are independent and $\mathcal{N}(0, \sigma^2)$. They imply that for each $x$,

$$EY_x = \alpha + \beta x, \ \ \operatorname{var} Y_x = \sigma^2.$$

The data points are $(x_1, Y_1), \ldots, (x_n, Y_n)$, where the $Y_i$'s are independent.

In our derivations, we'll find it convenient to express $SS_{xY}$ in a form that exhibits its linear dependence on the $Y$'s:

$$SS_{xY} = \sum (x_i - \overline{x})(Y_i - \overline{Y}) = \sum (x_i - \overline{x}) Y_i. \tag{1}$$

What we left out in going from the first sum to the second is $\overline{Y} \sum (x_i - \overline{x}) = 0$. Similarly, we have

$$SS_{xx} = \sum (x_i - \overline{x})(x_i - \overline{x}) = \sum (x_i - \overline{x}) x_i. \tag{2}$$

The slope of the least-squares line is then

$$\hat{\beta} = \frac{SS_{xY}}{SS_{xx}} = \sum \frac{(x_i - \overline{x})}{SS_{xx}} Y_i. \tag{3}$$

Since $\hat{\beta}$ is a linear function of the independent variables $Y_i$, it is easy to find its mean and variance:

$$E\hat{\beta} = \sum \frac{(x_i - \overline{x})}{SS_{xx}} EY_i = \sum \frac{(x_i - \overline{x})}{SS_{xx}} (\alpha + \beta x_i)$$

$$= \frac{\alpha}{SS_{xx}} \sum (x_i - \overline{x}) + \frac{\beta}{SS_{xx}} \sum (x_i - \overline{x}) x_i = \beta.$$

So $\hat{\beta}$ is an unbiased estimator of $\beta$. Problem 15-9 asks you to show that $\hat{\alpha}$ is an unbiased estimate of $\alpha$.

The mean squared error of an unbiased estimator is its variance:

$$\operatorname{var} \hat{\beta} = \sum \left\{ \frac{x_i - \overline{x}}{SS_{xx}} \right\}^2 \operatorname{var} Y_i = \frac{\sigma^2}{(SS_{xx})^2} \sum (x_i - \overline{x})^2 = \frac{\sigma^2}{SS_{xx}}.$$

[Here we have used (4) of §5.10, page 212.]

The intercept estimate $\widehat{\alpha}$ is also a linear function of the $Y_i$'s. But we can get its variance from the variance of $\widehat{\beta}$:

$$\text{var }\widehat{\alpha} = \text{var}(\overline{Y} - \widehat{\beta}\,\overline{x}) = \text{var }\overline{Y} + \overline{x}^2 \text{ var }\widehat{\beta} - 2\overline{x}\text{ cov}(\overline{Y}, \widehat{\beta}).$$

Problem 15-11 asks you to show that the covariance of $\overline{Y}$ and $\widehat{\beta}$ is 0. Using this and the fact that var $\overline{Y} = \sigma^2/n$ produces the variance of $\widehat{\alpha}$.

---

When $Y_1, ..., Y_n$ are independent with mean $\alpha + \beta x_i$ and constant variance $\sigma^2$, the least-squares estimates $\widehat{\alpha}$ and $\widehat{\beta}$ are *unbiased,* and

$$\text{var }\widehat{\alpha} = \sigma^2 \left( \frac{1}{n} + \frac{\overline{x}^2}{\text{SS}_{xx}} \right), \quad \text{var }\widehat{\beta} = \frac{\sigma^2}{\text{SS}_{xx}}. \tag{4}$$

---

From (4) we see that the amount of variability in the estimator of the regression line's slope depends on (i) the amount of variability in the errors (or in the responses $Y_i$), and (ii) on the degree to which the $x_i$'s are spread out. When $x$ is a controlled variable, an investigator can design the experiment—choose the $x_i$'s—with the aim of minimizing the variability of the slope. If responses are observed at only one value of $x$, then $\text{SS}_{xx} = 0$—there is no hope of a reliable estimate of slope when only one $x$-value is used. The estimate of slope is most accurate when the $x_i$'s are chosen, half at one end and half at the other end of the interval over which it is feasible to collect data (and where the assumption of linearity is reasonable).

In order to get at the distribution of the least-squares coefficients, and to draw inferences and make predictions for the true regression coefficients, we need to refer to the error distribution, which we have assumed normal. Thus,

$$Y_x \sim \mathcal{N}(\alpha + \beta x, \sigma^2), \tag{5}$$

with p.d.f.

$$f(y \mid \alpha, \beta; \sigma^2) = \frac{1}{\sigma\sqrt{2\pi}} \exp\left\{ -\frac{1}{2\sigma^2}(Y_i - \alpha - \beta x_i)^2 \right\}.$$

With this assumption we can show that the least-squares estimators $\widehat{\alpha}$ and $\widehat{\beta}$ are *maximum likelihood estimators.* Because errors are assumed independent, the likelihood function is proportional to the product of the p.d.f.'s of the individual observations. Thus,

$$L(\alpha, \beta; \sigma^2) = \prod \sigma^{-1} \exp\left\{ -\frac{1}{2\sigma^2}(Y_i - \alpha - \beta x_i)^2 \right\}$$

$$= \sigma^{-n} \exp\left\{ -\frac{1}{2\sigma^2}\sum(Y_i - \alpha - \beta x_i)^2 \right\}. \tag{6}$$

Although there are three parameters to vary, it is clear that at a maximum point, the sum in the exponent (which does not involve $\sigma$) must have its minimum value. And minimizing this sum is precisely what led us to the least-squares estimators. So the m.l.e.'s of the parameters $\alpha$ and $\beta$ are their least-squares estimators.

To find the value of $\sigma^2$ that, along with $\hat{\alpha}$ and $\hat{\beta}$, maximizes the likelihood function, we differentiate $\log L$ with respect to $\sigma^2$:

$$\frac{\partial \log L}{\partial \sigma^2} = -\frac{n}{2\sigma^2} + \frac{1}{2\sigma^4}\sum(Y_i - \alpha - \beta x_i)^2.$$

Setting this equal to 0 and substituting the values $\hat{\alpha}$ and $\hat{\beta}$ we found from minimizing the sum of squared residuals, we obtain

$$\hat{\sigma}^2 = \frac{1}{n}\sum(Y_i - \hat{\alpha} - \hat{\beta}x_i)^2. \tag{7}$$

This is an average of the squared residuals about the empirical regression line, but it is not quite the MSRes of §15.2, since the latter uses the divisor $n-2$.

With the assumption of independent, normally distributed errors $\epsilon_i$, we can find the sampling distributions of the estimators $\hat{\alpha}$, $\hat{\beta}$, and $\hat{\sigma}^2$. The slope $\hat{\beta}$ and intercept $\hat{\alpha}$ are linear functions of the $Y_i$'s—linear combinations of independent normal variables. This means that these estimators are *normally* distributed. (See Property iv in §6.1, page 234.)

Finding the distribution of $\hat{\sigma}^2$ requires more work, and we can only indicate the approach and give the result. We start with the identity

$$Y_i - \alpha - \beta x_i = (Y_i - \hat{\alpha} - \hat{\beta}x_i) + (\hat{\alpha} - \alpha) + (\hat{\beta} - \beta)x_i.$$

Substituting

$$\hat{\alpha} - \alpha = \overline{Y} - \hat{\beta}\,\overline{x} - \alpha = (\overline{Y} - \alpha - \beta\overline{x}) - (\hat{\beta} - \beta)\overline{x}$$

in the right-hand side yields

$$Y_i - \alpha - \beta x_i = (Y_i - \hat{\alpha} - \hat{\beta}x_i) + (\overline{Y} - \alpha - \beta\overline{x}) + (\hat{\beta} - \beta)(x_i - \overline{x}).$$

Next we square the r.h.s. as a trinomial, sum the result on $i$, and divide by $\sigma^2$:

$$\sum\left(\frac{Y_i - \alpha - \beta x_i}{\sigma}\right)^2 = \sum\left(\frac{Y_i - \hat{\alpha} - \hat{\beta}x_i}{\sigma}\right)^2 \tag{8}$$
$$+ \frac{(\overline{Y} - \alpha - \beta\overline{x})^2}{\sigma^2/n} + \frac{(\hat{\beta} - \beta)^2}{\sigma^2}\sum(x_i - \overline{x})^2.$$

(The cross-product terms in the square have all vanished. See Problem 15-13.)

The left-hand side of (8) is the sum of squares of $n$ independent, standard normal variables—is chi$^2(n)$. Each of the last two terms on the right is the square of a standard normal variable—is chi$^2(1)$. At this point we appeal to a general decomposition theorem[6] to conclude that the first term on the right is $\text{chi}^2(n-2)$ and that the three terms on the right are independent.

---

If $Y_1, ..., Y_n$ are independent, and $Y_i \sim \mathcal{N}(\alpha + \beta x_i, \sigma^2)$, then the m.l.e.'s $\widehat{\alpha}$ and $\widehat{\beta}$ are least-squares estimators, and

$$\widehat{\alpha} \sim \mathcal{N}\left(\alpha, \sigma^2\left\{\frac{1}{n} + \frac{\overline{x}^2}{\text{SS}_{xx}}\right\}\right) \text{ and } \widehat{\beta} \sim \mathcal{N}\left(\beta, \frac{\sigma^2}{\text{SS}_{xx}}\right). \qquad (9)$$

Also, $\widehat{\sigma}^2$ and $\widehat{\alpha} + \widehat{\beta}x$ are independent, and

$$\frac{n\widehat{\sigma}^2}{\sigma^2} = \frac{\text{SSRes}}{\sigma^2} \sim \text{chi}^2(n-2). \qquad (10)$$

---

The m.l.e. $\widehat{\sigma}^2$ given by (7) is the ordinary average of the squared residuals about the least-squares line. In practice, the **mean squared residual** (MSRes or MSE) is defined as the unbiased estimator of $\sigma^2$ obtained by dividing SSRes by its d.f.:

$$\text{MSRes} = \frac{\text{SSRes}}{n-2} = \frac{n}{n-2} \widehat{\sigma}^2 \equiv S_\epsilon^2. \qquad (11)$$

The unbiasedness follows from (10), since the mean of $\text{chi}^2(n-2)$ is the number of degrees of freedom, $n-2$.

## Problems

**15-9.**   Show that $\widehat{\alpha}$, the $y$-intercept of the least-squares line, is an unbiased estimator of $\alpha$.

**∗ 15-10.**   When we use $\widehat{\alpha}$ as an estimator of the intercept $\alpha$ in the linear regression function,
   **(a)**   obtain a formula for the standard error.
   **(b)**   give the form of a confidence interval for $\alpha$.

**15-11.**   Show, under the assumptions of the preceding sections, that $\overline{Y}$ and the slope estimator $\widehat{\beta}$ are uncorrelated.

**15-12.**   Use the result of the preceding problem to show that

$$\text{cov}(\widehat{\alpha}, \widehat{\beta}) = -\frac{\sigma^2 \overline{x}}{\text{SS}_{xx}}.$$

**15-13.**   Show that the cross-product terms in the expansion (8) of §15.3 are all 0, as claimed.

**15-14.**   Suppose we use the least-squares line obtained using $n$ data points [(10) of §15.2] to estimate the mean response at $x = x^*$, when we substitute $x^*$ for $x$: $\widehat{Y} = \widehat{\alpha} + \widehat{\beta}x^*$.
   **(a)**   Show that $\widehat{Y}$ is an unbiased estimator of the mean at $x^*$: $\alpha + \beta x^*$.

---

[6]See Lindgren [15], 441, Theorem 16.

**(b)**   Show the following:

$$\text{var}\,\widehat{Y} = \sigma^2 \left( \frac{1}{n} + \frac{(x^* - \bar{x})^2}{\text{SS}_{xx}} \right). \tag{12}$$

## 15.4           Inference for the Regression Parameters

In applications it is usually the slope $\beta$ that is of special interest. It gives the increase in the mean response $Y$ when $x$ is increased by one unit. (In particular, when $\beta = 0$, the response $Y$ does not depend on $x$.)

The reliability of $\widehat{\beta}$ as an estimator of $\beta$ is judged by the size of its standard deviation:

$$\sigma_{\widehat{\beta}} = \frac{\sigma}{\sqrt{\text{SS}_{xx}}}. \tag{1}$$

This depends on $\sigma$, which is ordinarily unknown. An estimated s.d. of $\widehat{\beta}$ is obtained by replacing the unknown $\sigma$ by an estimator of $\sigma$. The estimator we use is $S_\epsilon$, the square root of the MSE. The **standard error** of $\beta$ is thus

$$\text{s.e.}(\widehat{\beta}) = \frac{S_\epsilon}{\sqrt{\text{SS}_{xx}}}. \tag{2}$$

With this we can construct confidence limits for $\beta$, much as we did for a population mean.

According to (9) of §15.3, the quantity

$$Z = \frac{\widehat{\beta} - \beta}{\sigma / \sqrt{\text{SS}_{xx}}}$$

is standard normal. From (10) of that section, we see that the quantity

$$\frac{S_\epsilon}{\sigma} = \sqrt{\frac{\text{SSRes}}{\sigma^2} \Big/ (n - 2)}$$

is the square root of a chi-square variable divided by its d.f. So, the ratio

$$T = \frac{\widehat{\beta} - \beta}{\text{s.e.}(\widehat{\beta})} = \frac{\widehat{\beta} - \beta}{S_\epsilon / \sqrt{\text{SS}_{xx}}} = \frac{Z}{S_\epsilon / \sigma}$$

is the ratio of a standard normal variable to the square root of a chi-square variable divided by its degrees of freedom. According to §9.6, this ratio has a $t$-distribution with $n - 2$ degrees of freedom. So it is pivotal. Just as in the case of estimating a mean, we now find confidence limits for $\beta$:

$$\widehat{\beta} \pm t^* \times \text{s.e.}(\widehat{\beta}) = \widehat{\beta} \pm t^* \frac{S_\epsilon}{\sqrt{\text{SS}_{xx}}}, \tag{3}$$

where $t^*$ is the percentile of $t(n-2)$ appropriate for the given confidence level.

Example 15.4a | **Dose Response (Continued)**
The estimate of the slope of the regression line was found in Example 15.2c to be $\widehat{\beta} = 1.065$. From that example, we have SSRes $= 45.98$ and $n = 6$, so

$$S_\epsilon^2 = \text{MSRes} = \frac{45.98}{4} = 11.495, \quad \text{SS}_{xx} = 437.5,$$

and, from (2),

$$\text{s.e.}(\widehat{\beta}) = \sqrt{\frac{11.495}{437.5}} = .1621.$$

In view of the small sample size ($n = 6$) and the need to estimate $\sigma$, we turn to the $t$-table for the appropriate multiplier of standard error. For 95% confidence, we use the multiplier 2.78, the 97.5th percentile of the $t$-distribution with 4 d.f. from Table IIIa. The confidence limits are

$$\widehat{\beta} \pm 2.78[\text{s.e.}(\widehat{\beta})] = 1.065 \pm 2.78 \times .1621,$$

or $.614 < \beta < 1.516$. Using the $t$-table assumes that the error distribution is normal or close to normal. In practice, some consideration should be given to checking this assumption—or at least worrying about it.  ∎

We turn now to the problem of testing $H_0$: $\beta = \beta_0$ against $\beta \neq \beta_0$. The obvious test would find strong evidence against $H_0$ in a set of data if the estimate $\widehat{\beta}$ is far from $\beta_0$, or if

$$T = \frac{\widehat{\beta} - \beta_0}{\text{s.e.}(\widehat{\beta})} \tag{4}$$

is large in magnitude. Testing the particular value $\beta_0 = 0$ is a test of the hypothesis that the slope of the regression line is 0. When $\beta = 0$, the distribution of responses is the same at each $x$. This would mean that there is no point to knowing $x$ when predicting $Y$. Statistical software for regression problems will ordinarily include in the printout, along with the value of $\widehat{\beta}$, the standard error of $\widehat{\beta}$ and the ratio of $\widehat{\beta}$ to its standard error; this ratio is the value of $T$ for testing $\beta = 0$.

Example 15.4b | **Dose Response (More)**
The $t$-statistic for testing $\beta = 0$ is

$$T = \frac{1.065}{.1621} = 6.57,$$

so the evidence against $\beta = 0$ is strong. (But the evidence against $\beta = 0$ is usually strong—the investigator's choice of $x$ as a predictor or controlled variable is apt to be well founded.)  ∎

The test described above is in fact a likelihood ratio test, as we now show. The likelihood function [(6) of §15.3] is

$$L(\alpha, \beta, \sigma^2) = \sigma^{-n}\exp\left\{-\frac{1}{2\sigma^2}\sum(Y_i - \alpha - \beta x_i)^2\right\},$$

whose maximum value over all parameter combinations is achieved when we substitute the maximum likelihood estimates:

$$\max_{(\alpha,\beta,\sigma^2)} L(\alpha, \beta, \sigma^2) = L(\widehat{\alpha}, \widehat{\beta}, \widehat{\sigma}^2) = (\widehat{\sigma}^2)^{-n/2}e^{-n/2}.$$

Under $H_0$ $(\beta = \beta_0)$, the likelihood function is $L_0(\alpha, \sigma^2) = L(\alpha, \beta_0; \sigma^2)$, and the m.l.e. of $\alpha$ is $\overline{Y} - \beta_0\overline{x}$. The value of $\sigma^2$ that goes with this is

$$\widehat{\sigma}_0^2 = \frac{1}{n}\sum(Y_i - \widehat{\alpha}_0 - \beta_0 x_i)^2 = \frac{1}{n}\sum[(Y_i - \overline{Y}) - \beta_0(x_i - \overline{x})]^2.$$

Adding and subtracting $\widehat{\beta}x_i$ inside the brackets on the right, we have

$$n\widehat{\sigma}_0^2 = \sum[(Y_i - \widehat{\alpha} - \widehat{\beta}x_i) + (\widehat{\beta} - \beta_0)(x_i - \overline{x})]^2 = \text{SSRes} + (\widehat{\beta} - \beta_0)^2\text{SS}_{xx}.$$

(Upon summation, the cross-products have conveniently vanished after summing, by virtue of the equations you obtain for finding least-squares estimators in Problem 15-5.) The maximum value of $L_0$ is

$$L_0(\widehat{\alpha}_0, \widehat{\sigma}_0^2) = (\widehat{\sigma}_0^2)^{-n/2}e^{-n/2}.$$

So,

$$\Lambda^{-2/n} = \frac{n\widehat{\sigma}_0^2}{n\widehat{\sigma}^2} = \frac{\text{SSRes} + (\widehat{\beta} - \beta_0)^2\,\text{SS}_{xx}}{\text{SSRes}} = 1 + T^2,$$

where

$$T = \frac{\widehat{\beta} - \beta_0}{\text{s.e.}(\widehat{\beta})}. \tag{5}$$

The likelihood ratio rejection region $\Lambda < K$ is thus equivalent to a region of the form $T^2 > K'$. Large values of $|T|$ are taken as evidence against $\beta = \beta_0$.

In §15.2 we gave [as equation (12)] an expression for the sum of the squared residuals in terms of $r$ and $\text{SS}_{YY}$:

$$\text{SSRes} = \text{SS}_{YY}(1 - r^2) = \text{SS}_{YY} - r^2\,\text{SS}_{YY}.$$

Transposing the last term, we rewrite this as

$$\text{SSTotal} = \text{SS}_{YY} = \text{SSRes} + \text{SSReg},$$

where

$$\text{SSReg} = r^2\,\text{SS}_{YY},$$

called the *regression sum of squares*. This is another instance of *analysis of variance*—a decomposition of the total sum of squares into two terms, one attributed to "error" (SSRes, sometimes called SSE) and the other attributed to "regression." That is, the variation of the responses $Y_i$ is explained in part as the presence of random error and in part as the effect of the regression of $Y$ on $x$: When $\beta$ is positive (say), the regression line has an upward slope, which means that the $Y$'s toward the right tend to be higher than those to the left; the $Y$'s are spread out more than they would be if $\beta$ were 0.

The test statistic $T^2$ for testing $\beta = 0$ can be written in terms of the sums of squares in the ANOVA, if we use (13) of §15.2 to relate $\widehat{\beta}$ and the correlation coefficient:

$$T^2 = \frac{\widehat{\beta}^2}{S_\epsilon^2/\mathrm{SS}_{xx}} = \frac{r^2\mathrm{SS}_{YY}/\mathrm{SS}_{xx}}{\mathrm{MSRes}/\mathrm{SS}_{xx}} = \frac{\mathrm{SSReg}}{\mathrm{MSRes}} = F. \qquad (6)$$

We call this $F$ because (when $\beta = 0$) it is the ratio of independent chi-square variables, each divided by its d.f.: $F \sim F(1, n-2)$.

**Example 15.4c**

## Dose Response (The End)

In Example 15.2c, we found $r^2 = .915$ for the dose response data. We then attribute 91.5% of the variation in responses to regression and 8.5% to random error. The various sums of squares and corresponding d.f.'s are usually given in computer printouts as an ANOVA table, which we give here, drawing upon our earlier calculations of SS's:

| Source     | SS     | df | MS     | F     |
|------------|--------|----|--------|-------|
| Regression | 496.89 | 1  | 496.89 | 43.22 |
| Residual   | 45.98  | 4  | 11.495 | —     |
| Total      | 542.87 | 5  | —      |       |

The value of $F$ is the square of the value we found for $T$: $43.22 = 6.57^2$, and the $P$-value can be found either in an $F$-table or a $t$-table. ■

So far, we have thought of $x$ as a controlled variable. Suppose we have a random sample of pairs $(X, Y)$ from a bivariate population with correlation $\rho$. Suppose further that the conditional mean given $X = x$ (that is, the regression function) is linear in $x$: $E(Y \mid x) = \alpha + \beta x$, and that the conditional variance is constant: $\mathrm{var}(Y \mid x) = \sigma^2$. The distribution of the $Y$'s, given the values of the $X$'s, is then exactly the same as it is when $x$ is a controlled variable. So, conditional on the values of the $X$'s in the sample, the estimators $\widehat{\alpha}$, $\widehat{\beta}$, and $\widehat{\sigma}^2$ have the same distributions as before.

In a bivariate population, testing $\beta = 0$ is equivalent to testing $\rho = 0$. So it is natural to express the test statistic $T$ in terms of the sample correlation coefficient $r$.

This we can do using (6):

$$T^2 = \frac{r^2\,\mathrm{SS}_{YY}}{(1-r^2)\,\mathrm{SS}_{YY}/(n-2)}, \quad \text{or } T = \sqrt{n-2}\,\frac{r}{\sqrt{1-r^2}}.$$

In §15.7 we'll define the *bivariate normal* population and see that in that model, the conditional distributions of $Y$ given $x$ are normal. So, in that case, when $\beta = 0$, the *conditional* distribution of $T$ is $t(n-2)$, given the $x$'s. But since the $x$'s are not involved in the distribution, the *unconditional* distribution of $T$ is then also $t(n-2)$ and the $t$-table can be used with the statistic $T$ to test $\rho = 0$.

## Problems

**∗ 15-15.**   Find a $P$-value for testing the hypothesis $\beta = 0$ against $\beta \neq 0$ in the setting and with the data of Problem 15-3.

**15-16.**   Give 90% confidence limits for the slope of the regression line in Problem 15-4.

**15-17.**   For any $x$, let $\widehat{Y} = \widehat{\alpha} + \widehat{\beta}x$. In Problem 15-14 you found the mean and variance of $\widehat{Y}$. Form the $Z$-score for $\widehat{Y}$, substitute $\widehat{\sigma}$ for $\sigma$, and call the new ratio $T$. Show (assuming the normal model) that its distribution is $t(n-2)$.

**15-18.**   Use the result of the preceding problem to obtain confidence limits for $\alpha + \beta x$, the mean response at $x$ (given a data set), assuming that responses are normally distributed.

**∗ 15-19.**   With the data in Problem 15-3, we want to estimate the mean response at $x = 60$ using $\widehat{Y} = \widehat{\alpha} + \widehat{\beta}x$ as the estimator, as in Problem 15-18. Find
  **(a)**   the standard error of $\widehat{Y}$ (the denominator of the $T$ you constructed in Problem 15-17).
  **(b)**   90% confidence limits for $\alpha + 60\beta$, the mean response at $x = 60$.

**15-20.**   Construct the ANOVA table
  **(a)**   for the regression analysis in Problem 15-15.
  **(b)**   for the regression analysis in Problem 15-4.

**15-21.**   Verify the expression for the maximum of the likelihood $L_0$ under $\beta = \beta_0$ on page 607 (§15.4).

**∗ 15-22.**   Consider the regression model with independent normal errors and constant variance, but with regression function $g(x) = \beta x$. This is appropriate when it is clear that the response $Y$ must be 0 when $x = 0$.
  **(a)**   Obtain the least-squares line.
  **(b)**   Find the distribution of $\widehat{\beta}$, the slope of the least-squares line.
  **(c)**   Show the following:

$$\sum (Y_i - \beta_0 x)^2 = \sum (Y_i - \widehat{\beta}x_i)^2 + (\widehat{\beta} - \beta_0)^2 \sum x_i^2.$$

  **(d)**   Find the likelihood ratio test for $\beta = \beta_0$ versus $\beta \neq \beta_0$, using (c) to express the test ratio in terms of the difference $\widehat{\beta} - \beta_0$.

**15-23.**   Thinking that students who sit toward the front of a class tend to do better on exams than those who sit toward the back, we once collected data on the exam score ($Y$) and the number of the row ($x$) in which a student usually sat. For the 65 students in one class, it turned out that $r = -.1047$. Assuming normality of $Y$ given $x$, test $H_0$: $\rho = 0$ against $\rho < 0$ (we had thought high scores would be associated with low row numbers).

---

## 15.5    Predicting $Y$ from $x$

Suppose you want to predict the demand for a new product or the response to a drug. You have information $X$ that may aid in the prediction. In the first instance, $X$ may come from a market survey; in the second, it may come from a dose-response clinical trial. How should you use the information, and how helpful is it?

First, consider a random demand or response $Y$ in the *absence* of correlative information $X$. The error in predicting that $Y$ will be $c$ is the difference $Y - c$. The absolute error is $|Y - c|$, a random variable. In choosing the predicted value $c$, we'd like to make the absolute error small in some average sense. In the spirit of least squares, we choose $c$ to minimize the **mean squared prediction error:**

$$\text{m.s.p.e.} = E[(Y - c)^2].$$

The m.s.p.e. is a second moment of $Y$ about $c$. We saw in §3.3 and §5.5 that the second moment of a random variable about its mean value is the smallest second moment. So to minimize the mean squared error in predicting $Y$, we use $c = E(Y)$. With this choice of $c$, the m.s.p.e. is the variance of $Y$, and the root-mean-square (r.m.s.) prediction error is $\sigma_Y$.

Now, when $X$ and $Y$ are *related*, and it becomes known that $X = x$, the appropriate distribution for predicting $Y$ is *conditional*—the distribution of $Y \mid x$. Applying the conclusion of the preceding paragraph, we use the *conditional mean* $E(Y \mid x)$ as our predicted value of $Y$ when we know that $X = x$. This function of $x$ is the regression function of $Y$ on $x$. When we use it to predict, the m.s.p.e. is the conditional variance, $\text{var}(Y \mid x)$.

The regression function may be *linear* in $x$. Suppose it is:

$$E(Y \mid x) = \alpha + \beta x. \tag{1}$$

Using iterated expectations (§5.9), we have

$$EY = E[E(Y \mid X)] = \alpha + \beta E(X), \text{ or } \mu_Y = \alpha + \beta \mu_X.$$

Substituting $\alpha = \mu_Y - \beta \mu_X$ in (1), we find

$$E(Y \mid x) = \mu_Y + \beta(x - \mu_X).$$

From this, we can find the conditional mean of the product $XY$:

$$E(XY \mid X = x) = E(xY \mid X = x) = x\,E(Y \mid x) = \alpha x + \beta x^2.$$

Then, treating "$x$" as random and using iterated expectations, we find that

$$E(XY) = EE(XY \mid X) = E(\alpha X + \beta X^2) = \alpha \mu_X + \beta E(X^2).$$

To find the covariance, we subtract the product of the means:

$$\sigma_{X,Y} = \alpha\mu_X + \beta E(X^2) - \mu_X(\alpha + \beta\mu_X) = \beta\sigma_X^2,$$

which yields

$$\beta = \frac{\sigma_{X,Y}}{\sigma_X^2}.$$

So, the regression line is

$$y = \mu_Y + \frac{\sigma_{X,Y}}{\sigma_X^2}(x - \mu_X), \tag{2}$$

or

$$\frac{y - \mu_Y}{\sigma_Y} = \rho\frac{x - \mu_X}{\sigma_X}. \tag{3}$$

Thus, the $Z$-score for $y$ is the fraction $\rho$ of the $Z$-score for the corresponding $x$.

Suppose next that the regression function is linear *and* the conditional variance is constant. The value of that constant [see (7) in §5.11] is:

$$\text{var}(Y \mid x) = E[\text{var}(Y \mid x)] = \sigma_Y^2 - \text{var}[E(Y \mid X)]$$

$$= \sigma_Y^2 - \text{var}(\beta X) = \sigma_Y^2 - \beta^2\sigma_X^2 = \sigma_Y^2(1 - \rho^2). \tag{4}$$

This says that the m.s.p.e. using the linear predictor (1) is $(1 - \rho^2)$ times $\sigma_Y^2$, the m.s.p.e. we'd incur if we ignored the value of $X$. So, using the value of $X$ reduces the r.m.s. prediction error by the factor $\sqrt{1 - \rho^2}$.

For instance, if $\rho = \sqrt{3/4} \doteq .866$, the r.m.s.p.e. can be cut in half by using the regression function to predict $Y$. But if $\rho = 0$, the r.m.s. prediction error is $\sigma_Y$, whether we know $X$ or not.

**Example 15.5a** | **Predicting Achievement from Aptitude**

In educational testing, a typical correlation between an aptitude score $(X)$ and an achievement score $(Y)$ is about .6. The reduction in r.m.s. prediction error using (1) is about 20%: $\sqrt{1 - .36} = .80$.

Suppose $\rho = .6$, $\mu_X = 22$, $\sigma_X = 2.4$, $\mu_Y = 540$, and $\sigma_Y = 50$. For any given $x$, the achievement would be predicted by (2) to be

$$y = 540 + .6 \times \frac{50}{2.4}(x - 22) = 265 + 12.5x.$$

The predicted achievement score for an individual with an aptitude score of 28 is $265 + 12.5 \times 28 = 615$. The m.s.p.e., from (4), is $.64\sigma_Y^2$, and the r.m.s.p.e. is $.8\sigma_Y = 40$. Without knowing the aptitude score, or simply ignoring it, one would predict the achievement score of 540 (that is, $\mu_Y$), and the r.m.s.p.e. would be $\sigma_Y = 50$. ∎

In general, the regression function may not be linear. Suppose it's not. The regression function is still the best predictor in the sense of mean squared error. But perhaps a linear predictor could serve almost as well—what is the best *linear* predictor? Consider $a + bX$ as a predictor. The best choice of $a$ and $b$, in terms of mean square, minimizes

$$\text{m.s.p.e.} = E\{[Y - (a + bX)]^2\}. \tag{5}$$

Expanding the square and averaging with respect to the joint distribution of $X$ and $Y$ and rearranging terms [as we did in obtaining (8) of §15.2], we end up with

$$\text{m.s.p.e.} = (a - [\mu_Y - b\mu_X])^2 + \sigma_X^2 \left(b - \frac{\sigma_{X,Y}}{\sigma_X^2}\right)^2 + \sigma_Y^2(1 - \rho^2). \tag{6}$$

Clearly, this is smallest if

$$a = \mu_Y - b\mu_X \quad \text{and} \quad b = \frac{\sigma_{X,Y}}{\sigma_X^2}. \tag{7}$$

So, the best *straight-line* predictor of $Y$, given $X = x$, is

$$y = \mu_Y + \frac{\sigma_{X,Y}}{\sigma_X^2}(x - \mu_X). \tag{8}$$

With this, the m.s.p.e. is $\sigma_Y^2(1 - \rho^2)$.

**Example 15.5b** | **A Nonlinear Regression Function**

Consider a uniform distribution on the triangle bounded by $y = 0$ and by $y = 1 - |x|$, shown in Figure 15-6. It is apparent from symmetry that $\rho = 0$, so the best *linear* predicting function is a  horizontal line:  $y = \mu_Y$, or $y = 1/3$. The corresponding m.s.p.e. is $\sigma_Y^2 = 1/18$.

The best predictor (mean-square sense) of any type is the conditional mean, $E(Y \mid x)$. The conditional distribution of $Y$, given $x$, is *uniform* on the interval $0 < y < 1 - |x|$. Thus, the conditional mean is $\frac{1}{2}(1 - |x|)$, and the conditional variance is $\frac{1}{12}(1 - |x|)^2$. Figure 15-6 shows the predicting function as the "curve" halfway between $y = 0$ and $y = 1 - |x|$, which is not linear in $x$. The m.s.p.e. is the conditional variance, which is not constant. ∎

We turn now to the problem of predicting $Y$ when the model parameters are unknown, using data in the form of $n$ pairs $(x_i, Y_i)$. We can use the data to determine a least-squares line, and the obvious estimator of the best linear predicting function (8) is $\widehat{\alpha} + \widehat{\beta}x$, with $\widehat{\alpha}$ and $\widehat{\beta}$ defined as in §15.2. If we use this function, which is not $\alpha + \beta x$ (assumed unknown), how accurate will the prediction be?

As in §15.3, we assume that responses $Y$ are independent, and that the distribution of $Y$ for a fixed $x$ is normal, with constant variance and mean depending linearly on $x$:

$$Y \mid x \sim \mathcal{N}(\alpha + \beta x, \sigma^2).$$

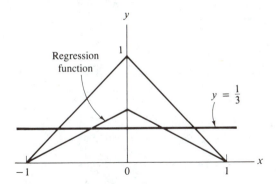

**Figure 15-6**

Given a set of $n$ data pairs, we calculate the least-squares coefficients and use $\widehat{\alpha} + \widehat{\beta} x_0$ as the predicted value of $Y$ for any given $X = x_0$. The prediction error is $Y - (\widehat{\alpha} + \widehat{\beta} x_0)$. In this expression for error, the random variables are $Y$, $\widehat{\alpha}$, and $\widehat{\beta}$. And $Y$, the (new) response at $x_0$, is independent of the coefficients $\widehat{\alpha}$ and $\widehat{\beta}$, which were based on responses at $x_1$, ..., $x_n$. The mean prediction error is 0:

$$E[Y - (\widehat{\alpha} + \widehat{\beta} x_0)] = \alpha + \beta x_0 - (\alpha + \beta x_0) = 0.$$

The mean squared prediction error is therefore the variance:

$$\text{m.s.p.e.} = E\{[Y - (\widehat{\alpha} + \widehat{\beta} x_0)]^2\} = \text{var}[Y - (\widehat{\alpha} + \widehat{\beta} x_0)]$$
$$= \text{var}\, Y + \text{var}\,(\widehat{\alpha} + \widehat{\beta} x_0).$$

The second term on the right is given by (12) in Problem 15-14 (page 600). The first term is just $\sigma^2$, the error variance for a single observation. Combining these, we have this:

---

In using $\widehat{\alpha} + \widehat{\beta} x_0$ to predict the response $Y$ when $x = x_0$, the mean squared prediction error is

$$\text{m.s.p.e.} = \sigma^2 \left\{ 1 + \frac{1}{n} + \frac{(x_0 - \overline{x})^2}{\text{SS}_{xx}} \right\}. \tag{9}$$

The error variance $\sigma^2$ can be approximated using $S_\epsilon^2 = \text{MSE}$.

---

Example **15.5c** | **An Arthropodal Thermometer**

The frequency of chirping of a cricket is thought to be related to temperature. This suggests the possibility that temperature can be estimated (predicted) from the chirp frequency. The following data give frequency-temperature pairs, observed for the striped ground cricket:[7]

| Chirps/sec $(x)$ | 20 | 16 | 20 | 18 | 17 | 16 | 15 | 17 | 15 | 16 |
|---|---|---|---|---|---|---|---|---|---|---|
| Temperature, °F | 89 | 72 | 93 | 84 | 81 | 75 | 70 | 82 | 69 | 83 |

The scatter plot is shown in Figure 15-7.

To find the correlation and least-squares line, we need these statistics:

$$\overline{X} = 17, \ \overline{Y} = 79.8, \ \mathrm{SS}_{xx} = 30, \ \mathrm{SS}_{yy} = 589.6, \ \mathrm{SS}_{xy} = 122.$$

Substituting in (11) and (7) and (12) of §15.2, we obtain

$$\widehat{\beta} = \frac{122}{30} \doteq 4.067, \ \widehat{\alpha} = 79.8 - \frac{122}{30} \times 17 \doteq 10.67,$$

and

$$r = \frac{122}{\sqrt{30 \times 589.6}} \doteq .917, \ \mathrm{SSRes} = 589.6 \times (1 - .917^2) \doteq 93.47.$$

So, the empirical regression line is $y = 10.67 + 4.067x$.

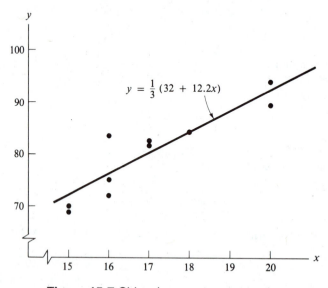

$$y = \tfrac{1}{3}(32 + 12.2x)$$

**Figure 15-7** Chirps/sec vs. temperature

---

[7]G. W. Pierce, *The Songs of Insects,* (Cambridge, MA: Harvard Univ. Press, 1949), 12–21.

For a chirp frequency of 19/sec, we predict the temperature to be

$$10.67 + 4.067 \times 19 \doteq 87.9.$$

With $\sigma^2$ estimated to be $S_\epsilon^2 = \text{SSRes}/8 \doteq 11.68$, the m.s.p.e. (9) is

$$11.68 \left\{ 1 + \frac{1}{10} + \frac{(19-17)^2}{30} \right\} \doteq 14.4.$$

The r.m.s. error is about 3.8°. Had we ignored $x$ and predicted the temperature to be $\overline{Y} = 79.8$, the r.m.s.p.e. would be $S_Y = 8.09$. Using the value of $x$ in a prediction does not eliminate prediction error, but substantially reduces it.   ∎

The term involving $(x_0 - \overline{x})^2$ in the m.s.p.e. (9) is large when $Y$ is predicted at a value of $x_0$ that is far from the mean of the $x$'s in the data set. This is the result of the uncertainty in estimating the slope parameter $\beta$. Moreover, the true regression function may be nonlinear outside the range of $x$-values in the data set, so using a linear predictor beyond the range where one has sample information is doubly risky.

## Problems

**\* 15-24.**   The director of graduate studies in a statistics department finds that for 25 entering students with GRE scores, the correlation of GRE with end-of-first-year GPA is .75. For GPA, the mean is 3.4 and the s.d. is .22. For GRE, the mean is 700 and the s.d. is 30.
   **(a)**   Find the least-squares line for predicting GPA from GRE.
   **(b)**   What GPA is predicted for an incoming student with GRE = 780?
   **(c)**   Find the r.m.s. prediction error for the prediction in (b).
   **(d)**   Compare $Z$-scores for the GRE and GPA in (b).

**15-25.**   Given the data of Problem 15-4, predict the particle displacement $(Y)$ for a frequency of 6.0 and give the r.m.s. prediction error.

**\* 15-26.**   Given the joint p.d.f. $f(x, y) = 24xy$ in the triangle bounded by $x = 0$, $y = 0$, and $x + y = 1$,
   **(a)**   find $E(Y \mid x)$, the regression function of $Y$ on $X$.
   **(b)**   find the best *linear* predictor of $Y$ given $X = x$.

**15-27.**   Given the joint p.d.f. $f(x, y) = e^{-y}$ for $0 < x < y$,
   **(a)**   find $E(Y \mid x)$, the regression function of $Y$ on $X$.
   **(b)**   find the best *linear* predictor of $Y$ given $X = x$.

**15-28.**   It has been found that the correlation between first and second exam scores in a certain course is .90. The mean scores are 65, and the s.d.'s are both 12. Predict the second exam score of a student whose first exam score is
   **(a)**   45.      **(b)**   92.      **(c)**   unknown.

**\* 15-29.**   A research paper on water management suggests that a linear relation between the logarithm or runoff volume $(V)$ and logarithm of peak discharge $(D)$, reporting the data in the table that follows.[8]

---

[8]V. P. Singh and H. Ainian, "An empirical relation between volume and peak of direct runoff," *Water Resources Bulletin* 22 (1986), 725–30.

**(a)**   Using statistical computer software, find the correlation coefficient.

**(b)**   Obtain the (empirical) regression line of $\log D$ on $\log V$.

**(c)**   Use the line in (b) to predict the peak discharge for a runoff volume of 1000 and give an estimate of the r.m.s. prediction error.

| $\log V$ | $\log D$ | $\log V$ | $\log D$ | $\log V$ | $\log D$ |
|----------|----------|----------|----------|----------|----------|
| 8.1440   | .6981    | 6.0638   | $-4.1997$ | 8.1291   | $-.0758$ |
| 7.0049   | $-.9916$ | 7.8002   | $-3.2702$ | 7.2513   | $-.8393$ |
| 7.5132   | $-.2256$ | 6.7639   | $-4.1352$ | 8.0938   | .3148    |
| 8.1693   | .4625    | 4.6634   | $-6.2146$ | 7.5099   | $-.1458$ |
| 7.7227   | .2624    | 5.6595   | $-4.6356$ | 7.3447   | $-.5327$ |

**15-30.**   Using batting averages during the first half $(x)$ and during the second half $(y)$ for most of the regulars in the American League East in the 1984 season, we found the following summary statistics: $\overline{X} = .2714$, $S_x = .03403$, $\overline{Y} = .2702$, $S_y = .03944$, $n = 66$, $r = .4719$.

**(a)**   Find the least-squares line in a regression of second-half averages on first-half averages.

**(b)**   Find SSRes.

**(c)**   Find $S_\epsilon$, the estimate of the error s.d., $\sigma$.

**(d)**   Use the line in (a) to predict the second-half batting average for a batter whose first-half batting average is (i) .300, (ii) .200.

**(e)**   Find the (approximate) r.m.s. prediction error for the predictions in (d).

**15-31.**   Suppose $(X, Y)$ is uniform on the triangle with vertices $(0, 0)$, $(0, 1)$, and $(1, 1)$. Find the least-squares regression line of $Y$ on $X$.

**15-32.**   Verify the equivalence of equations (5) and (6) in §15.5.

## 15.6   Residuals

Having made the rather stringent assumption of normally distributed errors with a fixed variance, an analyst should at least wonder about the validity of these assumptions, in each particular case, since the various $t$ and $F$ distributions that arise depend on them. Although a sample from the error distribution is not available for checking the assumptions, the residuals about the fitted regression function constitute an approximate sample of errors. These  are often used as a basis for a check.

When there are not many data points, you can calculate the residual for each case by hand, as the observed value of the dependent variable minus the corresponding fitted value: $y - \hat{y}$. Computer software packages include options to produce residuals and *standardized* (or "studentized") residuals—the residual divided by its standard error. The latter are more comparable on the same scale, since the residuals near the extremes of the data set tend to be somewhat less variable, the regression function being pulled, in a sense, toward those cases.

Example **15.6a**  | **A High Temperature Experiment**

Tukey, in his *Exploratory Data Analysis*, gives data on equilibrium splitting of plutonium tribromide by water (gases at high temperature).[9]  For each given value of absolute temperature $T$, the equilibrium constant $K$ was observed. The data are given in Table 15-1. The correlation coefficient is $r = .952$, suggesting that the relationship could be linear—at least over the range of temperatures used.  The least-squares line (from a computer printout) is

$$K = .4528 - .0004829\,T$$

The plot in Figure 15-8 shows the data points and the fitted line.

**Table 15-1**

| Case | $T$ | $K$ | Case | $T$ | $K$ |
|------|-----|------|------|-----|------|
| 1 | 911 | .0153 | 7 | 875 | .0247 |
| 2 | 914 | .0156 | 8 | 883 | .0243 |
| 3 | 919 | .0149 | 9 | 815 | .0704 |
| 4 | 920 | .0163 | 10 | 817 | .0502 |
| 5 | 882 | .0246 | 11 | 816 | .0692 |
| 6 | 876 | .0282 | | | |

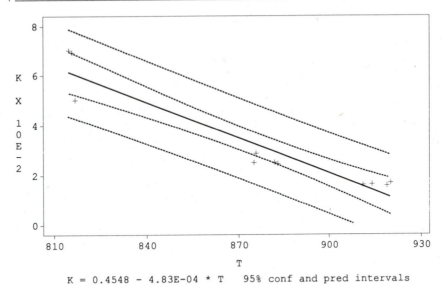

K = 0.4548 - 4.83E-04 * T    95% conf and pred intervals

**Figure 15-8** Equilibrium constant vs. temperature

---

[9]I. Shift and N. R. Davidson, "Equilibrium in the vapor-phase hydrolysis of plutonium tribromide" (1949), in *The Transuranium Elements*, Seaborg, Katz, and Manning, Eds.,  National Nuclear Energy Series IV-14B (New York:  McGraw-Hill, 1957), 831–40.

The plot also shows the loci of 95% confidence limits for $\alpha + \beta t$ (the inner pair of dotted curves) and of the 95% prediction limits for predicting the value of $K$ when the predictor $T$ has the value $t$ (the outer dotted curves—see (9) in §15.5).

We note that the data points in the middle are below the fitted line, and those to the right are above it. This somewhat systematic behavior shows up even more clearly in the plot of standardized residuals in Figure 15-9.

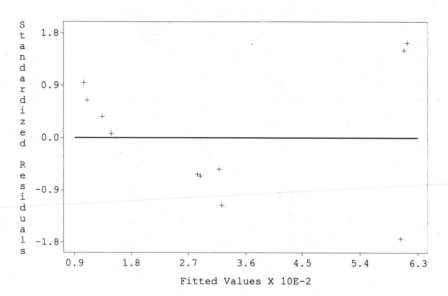

**Figure 15-9** Standardized residual plot for $K$ vs. $T$

A nonlinear function is suggested. A quadratic function might fit better over the range covered, but a parabolic graph could not be correct over a wide range of temperatures. As temperature increases, the equilibrium constant seems to be leveling off, suggesting the possibility of an exponential approach to some limiting value, a function of the form $K = ae^{bT}$, where then $\log K$ is a linear function of $T$. However, the investigators regressed $\log K$ against $U = 1/T$. Our computer printed out the following coefficient table:

| PREDICTOR VARIABLES | COEFFICIENT | STD ERROR | STUDENT'S T | P |
|---|---|---|---|---|
| CONSTANT | -15.4930 | 0.64890 | -23.88 | 0.0000 |
| U | 10382.0 | 566.153 | 18.34 | 0.0000 |

| R-SQUARED | 0.9739 | RESID. MEAN SQUARE (MSE) | 0.00999 |
|---|---|---|---|

The value of $R^2$ (which is $r^2$) is higher, and the least-squares line (pictured in Figure 15-10) appears to be a reasonably good fit.

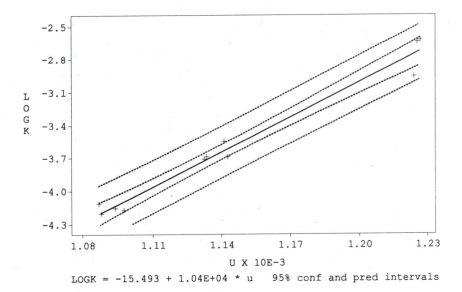

LOGK = -15.493 + 1.04E+04 * u    95% conf and pred intervals

**Figure 15-10** Log $K$ vs. $1/T$

The new residual plot is shown in Figure 15-11 and is more like what we expect to see as a random sample from the error population.

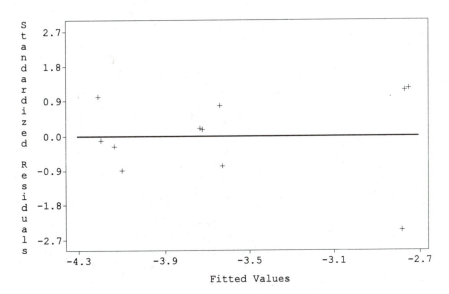

**Figure 15-11** Residual plot for log $K$ vs. $1/T$

Case 10, the lowest one at the far right, might be considered an outlier, being more than 2 s.e.'s from the mean. At any rate, the investigators deleted it, after they investigated and found that it was based on a run of only 16 hours as against 40 hours for the others. (They felt that equilibrium might not have been reached.) They also deleted cases 4 and 6, with reasoning that was not quite so convincing.

The cases are clustered in groups—four just over 900°, three just over 800°, and four clustered at a point in between. The deleted cases are one from each group. The equilibrium constant in case 10 is quite different from the other two in the group, but we don't know whether it was looked into just for that reason, or because the residual is over 2. (There was a time when it was common practice in the physical sciences to make triplicate measurements and discard the one that was quite different from the other two. This is an extremely poor tactic, deluding investigators into a wrong notion of the amount of variability present, and leading to wrong conclusions and "results" that can't be reproduced by others.)  ■

Analysis of residuals has become a standard tool for diagnostics in applied regression analysis, and we leave further consideration to books in the references.

## Problems

**15-33.** A study of isobaric heat capacity of alcohols included measurements of temperature ($T$, in $°K$) and heat capacity ($C$) for liquid isopropyl alcohol:[10]

| $T$ | $C$ |
|-----|--------|
| 200 | 1.8554 |
| 220 | 1.9302 |
| 240 | 2.0318 |
| 260 | 2.1715 |
| 280 | 2.3612 |
| 300 | 2.5942 |
| 320 | 2.8904 |
| 330 | 3.0401 |

**(a)** Calculate $r$ and obtain the least-squares regression line.

**(b)** Predict the heat capacity when $T = 250$, and find the standard error of prediction (estimated standard deviation of the prediction error).

**(c)** If you have done the above using computer software, obtain a plot of the standardized residuals against $T$. These are estimates of the standardized i.i.d. normal errors. Do you see anything noteworthy?

---

[10]Y. M. Nagiev, et al., "The thermal properties of monatomic alcohols," translation in *High Temperatures* 32 (1994), 864–85.

**15-34.** Thirteen specimens of Cu-Ni alloys, each with a known content of iron ($x$), were tested for corrosion in salt water for 60 days. Corrosion was measured by weight loss ($Y$) in mg/decimeter$^2$/day, and recorded as follows:[11]

| Iron, x | Loss, Y |
|---------|---------|
| .01 | 127.6, 130.1, 128.0 |
| .48 | 124.0, 122.0 |
| .71 | 113,1, 110.8 |
| .95 | 103.9 |
| 1.19 | 101.5 |
| 1.44 | 91.4, 92.3 |
| 1.96 | 83.7, 86.2 |

Use computer software to obtain a least-squares line and check the residuals from a residual plot.

## 15.7   Bivariate Normal Distributions

In some applications, the bivariate distribution of $(X, Y)$ has a bell-shaped density that is well approximated by a **bivariate normal** p.d.f.

> The random pair $(X, Y)$ is said to have a *bivariate normal distribution* if and only if every linear combination $aX + bY$ has a univariate normal distribution.

This definition has some immediate and important consequences. The linear combination in which $a = 1$ and $b = 0$, which is $X$ itself, is univariate normal. Similarly, $Y$ is univariate normal.

> The marginal distributions of a bivariate normal distribution are univariate normal.

Next, consider the linear transformation from $(X, Y)$ to $(U, V)$ given by

$$\begin{cases} U = aX + bY \\ V = cX + dY. \end{cases}$$

---

[11]Quoted in N. R. Draper and W. L. Smith, *Applied Regression Analysis* (New York: John Wiley & Sons, 1966), 37.

A linear combination of $U$ and $V$ is also a linear combination of $X$ and $Y$:

$$\alpha U + \beta V = \alpha(aX + bY) + \beta(cX + dY) = (\alpha a + \beta c)X + (\alpha b + \beta d)Y.$$

So, $U$ and $V$ are bivariate normal.

> If $(U, V)$ is obtained from a bivariate normal pair $(X, Y)$ by a linear trans-
> formation, then $(U, V)$ has a bivariate normal distribution.

A convenient tool for studying bivariate normal distributions is the bivariate moment generating function. The m.g.f. of $(X, Y)$ is

$$\psi(s, t) = E(e^{sX+tY}). \tag{1}$$

This function of two variables yields the univariate marginal m.g.f.'s when we set one or the other argument equal to 0:

$$\psi(s, 0) = E(e^{sX}) = \psi_X(s), \quad \psi(0, t) = E(e^{tY}) = \psi_Y(t).$$

When $X$ and $Y$ are *independent,* their joint m.g.f. factors into the product of the marginal m.g.f.'s:

$$\psi_{X,Y}(s, t) = E(e^{sX+Yt}) = E(e^{sX})E(e^{tY}) = \psi_X(s)\psi_Y(t).$$

And the converse is also true: If the joint m.g.f. factors into the product of a function of $s$ alone and a function of $t$ alone, then the marginal variables are independent. For, suppose

$$\psi_{X,Y}(s, t) = g(s)h(t).$$

Then,

$$\psi_X(s) = g(s)h(0) \text{ and } \psi_Y(t) = g(0)h(t).$$

Now set $s = 0$:

$$\psi_X(0) = g(0)h(0) = 1,$$

which implies

$$\psi_X(s)\psi_Y(t) = g(0)h(0)g(s)h(t) = g(s)h(t) = \psi_{X,Y}(s, t).$$

The left-hand side is the m.g.f. of independent $X$ and $Y$, and the uniqueness theorem for m.g.f.'s then implies that the joint distribution with m.g.f. $\psi_{X,Y}$ is that of independent variables.

We're now ready to find the m.g.f. of a bivariate normal distribution. From the definition, the variable $U = sX + tY$ is univariate normal. Hence

$$\psi_{X,Y}(s, t) = E(e^{sX+Yt}) = E(e^U) = \psi_U(1) = \exp\left\{\mu_U + \frac{1}{2}\sigma_U^2\right\}. \qquad (2)$$

But we know how to find the mean and variance of the sum $U = X + Y$:

$$\mu_U = s\mu_X + t\mu_Y, \quad \sigma_U^2 = s^2\sigma_X^2 + t^2\sigma_Y^2 + 2st\sigma_{X,Y}.$$

Substituting these in the rightmost member of (2), we obtain

$$\psi_{X,Y}(s, t) = \exp\left\{s\mu_X + t\mu_Y + \frac{1}{2}[s^2\sigma_X^2 + t^2\sigma_Y^2 + 2st\sigma_{X,Y}]\right\}. \qquad (3)$$

An interesting result, which we can extract from (3), is that when the variables $X$ and $Y$ are uncorrelated ($\rho = 0$), the m.g.f. is the product of a function of $s$ and a function of $t$—the criterion for independence:

---

Suppose $X$ and $Y$ are bivariate normal; then they are independent if and only if they are uncorrelated.

---

We use this fact to get at the p.d.f. of a general bivariate normal distribution. First we show that we can always find a linear transformation that transforms a bivariate normal pair into a pair of *independent* normals: Define new variables $U$ and $V$ as the result of a *rotation* through the angle $\theta$:

$$\begin{cases} U = (\cos\theta)X - (\sin\theta)Y \\ V = (\sin\theta)X + (\cos\theta)Y. \end{cases}$$

The covariance of $U$ and $V$ [see Problem 15-45] will vanish provided only that we choose $\theta$ so that

$$\cot 2\theta = \frac{\operatorname{var} Y - \operatorname{var} X}{2\operatorname{cov}(X, Y)}.$$

This means that we can account for all bivariate normal distributions by taking all possible linear transformations of a pair $(U, V)$ in which the $U$ and $V$ are independent, standard normal variables.

The joint p.d.f. of independent, standard normal variables $U$ and $V$ is the product of the marginal p.d.f.'s:

$$f_{U,V}(u, v) = \frac{1}{2\pi}\exp\left\{-\frac{1}{2}u^2 - \frac{1}{2}v^2\right\}.$$

The p.d.f. of $X$ and $Y$ obtained by a general nonsingular linear transformation of $(U, V)$ is obtained using a formula similar to that used in a linear transformation of a single variable. To keep from going too far afield, we'll simply give the result. It essentially substitutes the linear functions of $x$ and $y$ for the $u$ and $v$; the new exponent will be quadratic in $x$ and $y$. And when the dust settles, one finds that the general bivariate normal p.d.f. has the form

$$f(x, y) = \frac{1}{2\pi\sigma_X\sigma_Y\sqrt{1-\rho^2}}\, e^{-\frac{1}{2}Q}, \tag{4}$$

where $Q$ is a "quadratic form" in $x$ and $y$:

$$Q = \frac{1}{1-\rho^2}\left\{\left(\frac{x-\mu_X}{\sigma_X}\right)^2 - 2\rho\left(\frac{x-\mu_X}{\sigma_X}\right)\left(\frac{y-\mu_Y}{\sigma_Y}\right) + \left(\frac{y-\mu_Y}{\sigma_Y}\right)^2\right\}. \tag{5}$$

From this we see again that if the variables $X$ and $Y$ are uncorrelated ($\rho = 0$), they are independent, because the p.d.f. then factors into the product of a function of $x$ and a function of $y$. The graph of (4) is a bell-shaped surface, such as shown in Figure 15-12.

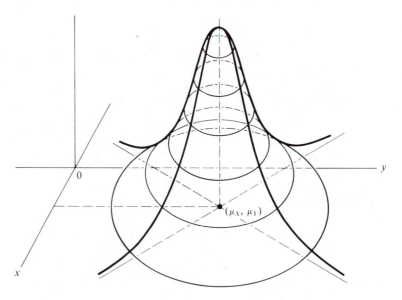

**Figure 15-12** A bivariate normal density

A level curve (locus of points where the p.d.f. is constant) consists of points where $Q$ is constant, and the graph of an equation of the form $Q = k$ is an ellipse. And these are circles only when $\rho = 0$, that is, when the variables are independent.

[Those who remember some analytic geometry will recall that the conic section $Q = k$ is an ellipse when the discriminant of $Q$, which here is $(\rho^2 - 1)$ divided by $\sigma_X^2 \sigma_Y^2$, is negative. Since $\rho^2 \leq 1$, the discriminant is negative except when $\rho^2 = 1$. And we know that when $\rho^2 = 1$, the variables $X$ and $Y$ are linearly related, so the distribution is concentrated on a line, and the density is not defined.]

Because the conditional p.d.f.'s, given a value of one variable or the other, are proportional to the joint p.d.f., and the joint p.d.f. is *quadratic* in each of the variables $x$ and $y$, it follows that the conditional p.d.f.'s are proportional to an exponential with a quadratic exponent—are *normal* p.d.f.'s.

---

When the distribution of $(X, Y)$ is bivariate normal, the conditional distributions (of $X \mid y$ and of $Y \mid x$) are univariate normal.

---

Being normal, the conditional distributions are completely specified by their means and variances. The following example shows how we can find them, in a special case.

**Example 15.7a**  Suppose the joint p.d.f. of $(X, Y)$ is the following exponential function with a quadratic exponent:

$$f(x, y) \propto \exp[-x^2 + xy - 2y^2 + 5x - 6y].$$

Given that $Y = y_0$, the conditional p.d.f. of $X$ is proportional to $f(x, y_0)$:

$$f(x \mid y_0) \propto \exp[-\{x^2 - (y_0 + 5)x\}].$$

(The constant of proportionality can involve $y_0$, a constant.) We complete the square in $x$ by inserting $(y_0 + 5)^2/4$ inside the braces:

$$x^2 - (y_0 + 5)x = \left\{ x - \frac{1}{2}(y_0 + 5) \right\}^2 + \text{term in } y_0.$$

The additive term in $y_0$ becomes a constant multiplier, when it's in the exponent. Thus, the conditional p.d.f. can be written

$$f(x \mid y_0) \propto \exp\left( -\frac{1}{2 \times \frac{1}{2}} \left\{ x - \frac{1}{2}(y_0 + 5) \right\}^2 \right).$$

This is a normal p.d.f. with mean $(y_0 + 5)/2$ and variance $1/2$. Thus, the regression function of $X$ on $Y$ is $(y + 5)/2$, and the conditional variance is $1/2$—a constant.  ■

The technique of this last example can be applied to the general bivariate normal p.d.f. (4) to yield the following general formulas for conditional mean and variance:

If $X$ and $Y$ are bivariate normal, the conditional distribution of $Y$ given $X = x$ is *normal*. The regression function of $Y$ on $x$ is linear in the predictor variable $x$:

$$E(Y \mid x) = \mu_Y + \rho \frac{\sigma_Y}{\sigma_X}(x - \mu_X). \tag{6}$$

The conditional variance is constant:

$$\text{var}(Y \mid x) = \sigma_Y^2(1 - \rho^2). \tag{7}$$

So, the best predictor of $Y$ in the sense of mean square—the conditional mean—is a linear function of the predictor variable $x$. Thus, the best linear predictor is the best predictor when the population is bivariate normal.

## 15.8    The Regression Effect

### Example 15.8a | Training Teachers

A psychology professor, D. Kahneman of the University of British Columbia, related this anecdote.[12] He was teaching a course in psychology of training to air force flight instructors at Hebrew University in the 1960s. He cited studies showing that in teaching, rewards are more effective than punishment. One of his students objected, saying, "I've often praised people warmly for beautifully executed maneuvers, and the next time they almost always do worse. And I've screamed at people for badly executed maneuvers, and by and large the next time they improve. Don't tell me that reward works and punishment doesn't. My experience contradicts it." Other students agreed.

Professor Kahneman said, "I suddenly realized that this was an example of the statistical principle of regression to the mean, and that nobody else had ever seen this before. I think this was one of the most exciting moments of my career." "Once you become sensitized to it," he remembered saying that day, "you see regression everywhere." Elaborating further on this ubiquity, he pointed out that great movies have disappointing sequels, and disastrous presidents have better successors. ∎

### Example 15.8b | Regression in Baseball

Astute baseball analysts have noticed the regression effect in American baseball. They call it the *law of competitive balance*. It works like this. Teams or players who do extremely well one year *tend* to do worse in the following year—better than the rest of the league but worse than their own previous high. For example, a "Rookie of the

---

[12]Quoted in *Discover* (June, 1985).

Year" will usually do worse in the second year. Conversely, teams or players that do very poorly one year tend to do better the next, although usually worse than average.

The reason for this is really quite simple. To do very well in a particular year requires some combination of two things: skill and luck. The fact that a particular team did well suggests that it had both. In the following year, the skill may still be there (barring extensive changes in personnel), but the luck is apt not to be—at least not to the same degree. So the team will *tend* to do well (the skill part), but not as well (the luck part) as before.

If you want to predict a player's batting average (BA) next year, average his previous year's BA with that year's BA of the entire league. This assumes that skill and luck contribute to BA in equal measure; they don't, but the formula works quite well. In particular, it does much better than using the player's previous year's BA as a predictor. ∎

The nineteenth-century geneticist Francis Galton measured the sizes of the seeds of mother and daughter sweet pea plants. He observed that the sizes of daughter plants seemed to revert (or regress) to the mean. From scatter plots of the heights of fathers and sons, he again noted a regression to average: Sons of tall men tended to be tall, but not as tall as their fathers, while sons of short men tended to be short, but not as short as their fathers.

In §15.5, we saw that when the regression function is linear (as it is, in particular, in the bivariate normal case), the equation of the regression line is

$$\frac{y - \mu_Y}{\sigma_Y} = \rho \, \frac{x - \mu_X}{\sigma_X}, \ \text{ or } \ z_Y = \rho z_X,$$

where $z_X$ and $z_Y$ are the standard scores corresponding to $x$ and $y$. Suppose we observe a value of $X$ that is two s.d.'s above average: $Z_X = 2$. The best prediction for the corresponding $Y$ (in the mean-square sense) is a value of $Y$ that is *less* than two s.d.'s above average by the factor $\rho$. Similarly, if we observe an $X$ that is three s.d.'s below average, the best prediction for $Y$ is a value that is *less* than three s.d.'s below average, by the factor $\rho$. So even though there may be as many tall sons as tall fathers, a father who is very tall will have a son who, on average, is not quite so tall.

Figure 15-13 shows a contour curve for a bivariate normal distribution with $\rho = 1/2$, drawn on $z_x$-$z_y$ axes—that is, axes with standard ($Z$-) scales. The equation of each contour curve is

$$z_x^2 - z_x z_y + z_y^2 = c^2,$$

where $c^2$ is a positive constant. [See (4) and (5) in §15.6, with $\rho = 1/2$.] Since a rotation through 45 degrees results in an equation of the form $u^2 + 3v^2$ equal to a constant, the major axis of the ellipse is the 45-degree line, $z_x = z_y$. However, the regression line, the line we use for predicting, $z_y = z_x/2$, is a line that bisects the vertical cords of the ellipse.

When $|\rho| = 1$, of course, there is no regression effect, and when $|\rho|$ is close to 1, the effect is slight.

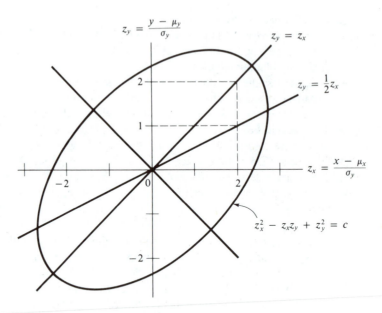

**Figure 15-13**

| Example 15.8c | **Regression in Golf** |
|---|---|

First- and second-round scores in the 1988 LPGA championship match in Mason, Ohio, are shown in the scatter diagram of Figure 15-14. The correlation between first- and second-round scores is about .52. It is evident that among those with very low first-round scores, some got high and some got low second-round scores, the average being even a bit above the overall average. So the second-round score of a player with a low first-round score should increase.

The regression effect is shown somewhat more clearly in the plot of Figure 15-15, in which the *change* in score from first to second round is plotted against first-round scores. The correlation here is about $-.58$. Golfers who had higher-than-average scores for the first round tended to go down in the second; those whose first-round scores were lower than average tended to go up in the second round.    ■

The regression effect (as Dr. Kahneman asserted) shows up practically everywhere, but in the case of clinical trials, its misinterpretation is particularly unfortunate. A common criterion for entry in a clinical trial is that the patient be "very sick." Thus, in a trial of an antihypertensive drug, only those with "high" blood pressure are included. Blood pressure measurements are quite variable within each individual. Just as in baseball, where a result is a combination of skill and luck, a diagnosis of high blood pressure is a combination of a real propensity to have high blood pressure and a random component. Those selected tend to have a high random component at entry, so their blood pressures tend to decrease—even if the drug has no effect whatever! The regression effect can easily be confused with a treatment effect.

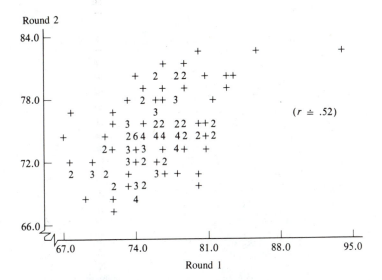

**Figure 15-14** Round 2 vs. Round 1

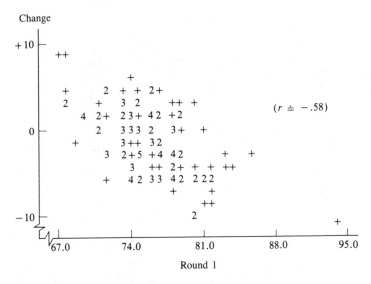

**Figure 15-15** Change vs. Round 1

One solution is to use separate measurements for admission to the trial and for data analysis. Another is to compare the drug group with a control group only, paying no heed to the apparent effect in the drug group alone.

## Problems

* **15-35.** Find the marginal distributions of $X$ and $Y$, given that their joint p.d.f. is proportional to

$$\exp\left\{-\frac{1}{4}x^2 - \frac{1}{8}y^2 + x - \frac{1}{2}y\right\}.$$

**15-36.** Given the joint p.d.f.

$$f(x, y) \propto \exp\left\{-\frac{1}{4}(x - 2y + 1)^2 - \frac{1}{2}y^2\right\},$$

  **(a)** find the marginal p.d.f. of $Y$.
  **(b)** find the conditional mean, $E(X \mid y)$.
  **(c)** find the conditional variance, $\text{var}(X \mid y)$.

* **15-37.** Write the p.d.f. of a bivariate normal distribution for $(X, Y)$ with $EX = 0$, $EY = 4$, $\text{var }X = 1$, $\text{var }Y = 9$, and $\text{cov}(X, Y) = 2$.

**15-38.** For the distribution in the preceding problem, find $E(X \mid y)$ and the conditional variance, $\text{var}(X \mid y)$.

* **15-39.** For the distribution with p.d.f. proportional to

$$\exp\left\{-\frac{1}{2}x^2 + xy - y^2 - x - 2y\right\},$$

  **(a)** find the regression functions (of $Y$ on $X$, and of $X$ on $Y$).
  **(b)** find the means, variances, covariance, and correlation.

* **15-40.** Given that $U$ and $V$ are independent, standard normal variables, find the joint distribution of $X = U + 2V$ and $Y = 3U - V$.

* **15-41.** Given this m.g.f. of $(X, Y)$: $\psi(s, t) = \exp[4s^2 - 4st + 9t^2 - 8s + 6t]$,
  **(a)** find the means and variances and the correlation $\rho$.
  **(b)** find the regression function of $X$ on $Y$.

**15-42.** For $(X, Y)$ as defined in Problem 15-37, find the joint distribution of $U = X - 2Y$ and $V = 3X + Y$.

* **15-43.** In a large data set, we find $r = .7$ and use the least-squares line to predict the $Y$-value for an individual whose $X$-score is $\overline{X} - 2S_X$. Will the predicted value be equal to, greater than, or less than $\overline{Y} - 2S_Y$?

**15-44.** Find a data set that you think may show the regression effect. (Sports data are particularly accessible.) Make a scatter plot of $(x, y)$ and a second scatter plot of $(x, y - x)$. Comment on the strength of the effect for your data set—no calculations are necessary.

**15-45.** Refer to Problem 15-28, page 615. Will your answers change if it is assumed that the joint distribution is bivariate normal?

**15-46.** Show what is claimed on page 623 about rotating to get independence.

## 15.9      Multiple Regression

**Example 15.9a**    **Moisture in Flour**

A study reports the following data on water absorption in wheat flower $(y)$, flour protein percentage $(x_1)$ and starch damage $(x_2)$:[13]

| $y$ | $x_1$ | $x_2$ | $y$ | $x_1$ | $x_2$ | $y$ | $x_1$ | $x_2$ | $y$ | $x_1$ | $x_2$ |
|---|---|---|---|---|---|---|---|---|---|---|---|
| 30.9 | 8.5 | 2 | 47.6 | 12.0 | 32 | 47.0 | 12.9 | 24 | 48.3 | 12.1 | 34 |
| 32.7 | 8.9 | 3 | 47.2 | 12.5 | 31 | 46.8 | 12.0 | 25 | 48.6 | 11.3 | 35 |
| 36.7 | 10.6 | 3 | 44.0 | 10.9 | 28 | 45.9 | 12.9 | 28 | 50.2 | 11.1 | 40 |
| 41.9 | 10.2 | 20 | 47.7 | 12.2 | 36 | 48.8 | 13.1 | 28 | 49.6 | 11.5 | 45 |
| 40.9 | 9.8 | 22 | 43.9 | 11.9 | 28 | 46.2 | 11.4 | 32 | 53.2 | 11.6 | 50 |
| 42.9 | 10.8 | 20 | 46.8 | 11.3 | 30 | 47.8 | 13.2 | 28 | 54.3 | 11.7 | 55 |
| 46.3 | 11.6 | 31 | 46.2 | 13.0 | 27 | 49.2 | 11.6 | 35 | 55.8 | 11.7 | 57 |

Figure 15-16 shows plots of $y$ vs. $x_1$ (circles) and $y$ vs. $x_2$ (squares). Both predictors appear to be closely related to the response $y$—linear correlations between $y$ and $x_1$ and between $y$ and $x_2$ are .69 and .95, respectively. Perhaps it would be useful to use both $x_1$ and $x_2$ in predicting $y$. (To be continued.) ■

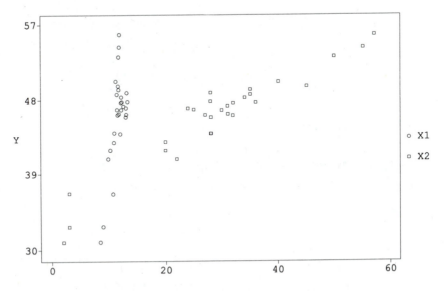

**Figure 15-16** Scatter diagrams for Example 15.9a

[13]"An ultracentrifuge flour absorption method," *Cereal Chemistry* (1978), 96–101.

In a simple extension of the linear regression in §15.1 to the case of two predictors, we assume a regression model of the form

$$g(x_1, x_2) = \beta_0 + \beta_1 x_1 + \beta_2 x_2 + \epsilon, \tag{1}$$

where $\epsilon$ is the "error." The error is usually assumed to be normally distributed with mean 0 and variance independent of the predictors. As when there is a single predictor variable, $\beta_i$ is the amount of increase in the mean response $y$ when $x_i$ is increased by one unit.

Since we are using a subscript to index the parameters $\beta$ and predictors $x$, a second subscript is needed to index observations. Data consist of triples: $(x_{1j}, x_{2j}, y_j)$ for $j = 1, ..., n$, represented graphically as a point in three dimensions. Each data point is referred to as a *case*. The regression function $g$ in (1) is represented graphically as a plane, but data points usually are not on this plane because of the error component $\epsilon$.

The method of least squares is again applicable to provide estimates of the coefficients $\beta_i$, finding the plane that "best" fits the $n$ data points. The residual of the $j$th data point about the regression plane is $y_j - g(x_{1j}, x_{2j})$, and the sum of squared residuals is thus

$$\sum_j (y_j - \beta_0 - \beta_1 x_{1j} - \beta_2 x_{2j})^2.$$

Equating its derivatives with respect to each $\beta_i$ in turn and setting them equal to 0 yields these "normal equations" of the least-squares process:

$$\sum y_j = n\beta_0 + \left(\sum x_{1j}\right)\beta_1 + \left(\sum x_{2j}\right)\beta_2$$

$$\sum x_{1j} y_j = \left(\sum x_{1j}\right)\beta_0 + \left(\sum x_{1j}^2\right)\beta_1 + \left(\sum x_{1j} x_{2j}\right)\beta_2$$

$$\sum x_{2j} y_j = \left(\sum x_{2j}\right)\beta_0 + \left(\sum x_{2j} x_{1j}\right)\beta_1 + \left(\sum x_{2j}^2\right)\beta_2.$$

There is ordinarily a unique solution $(\hat{\beta}_0, \hat{\beta}_1, \hat{\beta}_2)$, and these $\hat{\beta}$'s are the least-squares estimates.

**Example 15.9b** | **Moisture in Flour (Continued)**

For the data in Example 15.9a, calculation of the sums involved in the normal equations yields these equations:

$$1287.4 = 28\beta_0 + 322.3\beta_1 + 829.0\beta_2$$

$$14940 = 322.3\beta_0 + 3746.4\beta_1 + 9746.6\beta_2$$

$$40016 = 829\beta_0 + 9746.6\beta_2 + 29327\beta_2.$$

The unique solution is $\widehat{\beta}_0 = 19.44$, $\widehat{\beta}_1 = 1.442$, $\widehat{\beta}_2 = .3356$, so the least-squares plane is

$$y = 19.44 + 1.442x_1 + .3356x_2.$$

Carrying out the arithmetic is laborious (as you might suspect), but computers do it easily with statistical software. They will also produce the predicted response $\widehat{y}$ for given values of the predictors and compare them with the observed responses at the $x_1$-$x_2$ combinations in the data to produce the residuals, $\widehat{y} - y$. Figure 15-17 shows a plot of these residuals[14] against rankits (normal scores—see §13.4), and the Wilk-Shapiro statistic for checking normality: $W = .9541$.   ∎

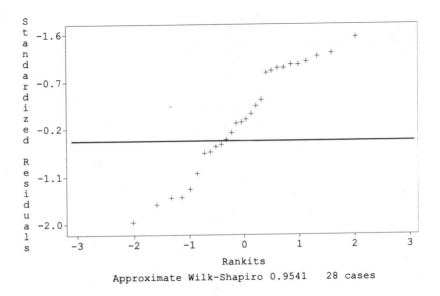

**Figure 15-17** Rankit plot of standardized residuals for Example 15.9b

When there are $k$ predictor variables, there are $k + 1$ parameters $\beta_j$. The method of least squares leads to $k + 1$ equations in these $\beta$'s. Solving them by hand calculations is at best tedious and (for most of us) almost prohibitive. Statistical software will usually do the job without strain. However, in some applications more than 100 predictors are used, and special software or a larger computer may be required in such cases.

In doing the algebra of least squares for multiple regression, it usually does not matter that some predictor variables may depend functionally on others, as long as the dependence is not exactly linear. Thus, one can take certain types of interaction into

---

[14]The plot was produced using the PC software "*Statistix.*" The residuals shown have been "studentized"—divided by standard errors, the latter being calculated by a formula we'll not go into.

account by including a predictor that is a product of two others. For instance, a regression function of this form can be accommodated:

$$\beta_0 + \beta_1 x_1 + \beta_2 x_2 + \beta_3 x_1^2 + \beta_4 x_1 x_2.$$

But if one predictor is a linear function of others ($x_2 = \beta_1 x_1 + \beta_2 x_3$, for instance), it is clear that the $\beta$'s are not uniquely determined. The least-squares equations are indeterminate. The computer program will balk. It may also balk if there are relations among the $x$'s that are close to being linear.

Whenever one includes several predictors in a regression function, it is natural to ask whether they are all needed, or all contribute to the accuracy of predictions. An ANOVA is particularly useful in considering a hierarchy of regression functions. To illustrate, assume that the regression function depends linearly on two predictors: $g(x_1, x_2) = \beta_0 + \beta_1 x_1 + \beta_2 x_2$. Testing the hypothesis $\beta_2 = 0$ against the alternative $\beta_2 \neq 0$ is a way of determining whether or not $x_2$ makes a useful contribution to the model. It tests whether the full model provides an appreciably better fit than a model with $x_1$ as the sole predictor.

As we did in the simple linear regression we considered first, suppose we assume that the errors are independent normal variables with constant variance $\sigma^2$. Let $\widehat{\beta}_{00}$ and $\widehat{\beta}_{10}$ denote the m.l.e.'s (which are also least-squares estimates in this case) of $\beta_0$ and $\beta_1$ when $\beta_2 = 0$. Given the $n$ cases $(x_{1j}, x_{2j}, y_j)$, the likelihood ratio statistic for testing $H_0$: $\beta_2 = 0$ against the alternative $H_A$: $\beta_2 \neq 0$ is

$$\Lambda = \left(\frac{\widehat{\sigma}_1^2}{\widehat{\sigma}_2^2}\right)^{-n/2}, \tag{2}$$

where

$$n\widehat{\sigma}_1^2 = \sum (y_j - \widehat{\beta}_{00} - \widehat{\beta}_{10} x_{1j})^2 = \text{SS1}$$

and

$$n\widehat{\sigma}_2^2 = \sum (y_j - \widehat{\beta}_0 - \widehat{\beta}_1 x_{1j} - \widehat{\beta}_2 x_{2j})^2 = \text{SS2}.$$

The quantity SS1 is the residual sum of squares for testing $\beta_2 = 0$, and SS2 is a residual sum of squares that we identify as "error" after fitting the regression function of the full model. The likelihood ratio test is thus defined by the equivalent critical regions

$$\Lambda < C'', \quad \frac{\text{SS1}}{\text{SS2}} > C', \quad \frac{\text{SS1} - \text{SS2}}{\text{SS2}} > C.$$

We use the notation $\text{SSTot} = \text{SS}_{yy}$ for *total sum of squares,* and base an ANOVA on the identity

$$\text{SSTot} = (\text{SSTot} - \text{SS1}) + (\text{SS1} - \text{SS2}) + \text{SS2}. \tag{3}$$

When the terms in this equation are each divided by $\sigma^2$, their distributions are chi-square, with d.f.'s $n - 1$, $1$, $1$, and $n - 3$, respectively, and the terms on the right are independent.[15]   Under $H_0$, the test ratio

$$F = \frac{(\text{SS1} - \text{SS2})/1}{\text{SS2}/(n-3)} \tag{4}$$

has an $F$-distribution with $(1, n - 3)$ d.f. Its square root, which is distributed as $t(n - 3)$, is the $t$-statistic (for testing $H_0$) found in computer printouts.

An $F$-ratio for testing the hypothesis that $\beta_1 = \beta_2 = 0$ is called an *overall F* in many statistical computer packages. It addresses the question of whether the part of the regression function that follows the constant term is useful, as a whole, in predicting $y$. Combining the first two terms on the right-hand side of (3), we obtain

$$\text{SSTot} = \text{SSReg} + \text{SSRes},$$

where $\text{SSReg} = \text{SSTot} - \text{SS2}$, and $\text{SSRes} = \text{SS2}$. The test statistic is

$$\text{Overall } F = \frac{\text{SSReg}/2}{\text{SSRes}/(n-3)}, \tag{5}$$

whose null distribution is $F(2, n - 3)$. The ratios

$$R^2 = \frac{\text{SSReg}}{\text{SSTot}} \quad \text{and} \quad 1 - R^2 = \frac{\text{SSRes}}{\text{SSTot}}$$

are fractions of the total variation of the $Y$'s about their mean that are attributed, respectively, to regression and to error. The ratio $R^2$ is called the **coefficient of determination.**

A large $R^2$ corresponds to a large $F$. But adding terms to the regression function, no matter how useless they may be, will always increase $R^2$. But it will also reduce the degrees of freedom for error estimation, and predictions may be less accurate. So the model with the largest $R^2$ is not necessarily best. A more useful indicator of a good predictor is usually found on computer printouts:

$$\text{Adjusted } R^2 = 1 - \frac{(n-1)\text{SSRes}}{(n-p-1)\text{SSTot}}, \tag{6}$$

where $p = k + 1$ is the number of parameters. Adding a term to the regression function will result in a closer fit, but at the cost of a more complicated model and the loss of a degree of freedom from SSRes. Choosing a model with a large "adjusted $R^2$" is apt to result in better predictions.

---

**Example 15.9c** | **Timber Volume from Diameter and Height**

To estimate the volume of wood in a mature tree, it is desirable to know how that volume is related to dimensions that can be measured before felling the tree. Trees of various heights and diameters were cut, and the diameter (in inches) at 4.5 feet above

---

[15]See Lindgren [15], 441.

the ground, the height (in feet, presumably obtained using trigonometry), and the volume of wood were recorded. These data for a sample of size 16 were reported:[16]

**Table 15-2**

| D | H | V |
|---|---|---|
| 8.3 | 70 | 10.3 |
| 8.8 | 63 | 10.2 |
| 10.7 | 81 | 18.8 |
| 11.0 | 66 | 15.6 |
| 11.1 | 80 | 22.6 |
| 11.3 | 79 | 24.2 |
| 11.4 | 76 | 21.4 |
| 12.0 | 75 | 19.1 |
| 12.9 | 85 | 33.8 |
| 13.7 | 71 | 25.7 |
| 14.0 | 78 | 34.5 |
| 14.5 | 74 | 36.3 |
| 16.3 | 77 | 42.6 |
| 17.5 | 82 | 55.7 |
| 18.0 | 80 | 51.5 |
| 20.6 | 87 | 77.0 |

The correlation between volume and diameter is quite high (.97), and between volume and height it is smaller (.70). We obtained a regression of volume on height and diameter, with computer printout as follows:

```
VARIABLES   COEFFICIENT   STD ERROR   STUDENT'S T      P       VIF
---------   -----------   ---------   -----------    ------    ---
CONSTANT      -64.2162     12.6126      -5.09        0.0002
D               4.63955     0.37421     12.40        0.0000    1.6
H               0.44339     0.19606      2.26        0.0415    1.6

R-SQUARED                0.9598    RESID. MEAN SQUARE (MSE)    15.3112
ADJUSTED R-SQUARED       0.9536    STANDARD DEVIATION          3.91295
```

---

[16]From the book *Forest Mensurations* (State College, PA.: Pennsylvania Valley Publishers, 1953).

| SOURCE | DF | SS | MS | F | P |
|--------|----|-----|-----|-----|-----|
| REGRESSION | 2 | 4755.14 | 2377.57 | 155.28 | 0.0000 |
| RESIDUAL | 13 | 199.046 | 15.3112 | | |
| TOTAL | 15 | 4954.19 | | | |

The error or residual sum of squares is shown as SSRes $= 199.046$, and SSReg for the numerator of the $F$-statistic (4) is SS1 $-$ SS2 $= 4755.14$:

$$F = \frac{4755.14/2}{199.046/13} = 155.28.$$

The $t$-statistic shown opposite "$D$," like all such $t$-values in most statistical software, is for testing the hypothesis that the coefficient in that line is 0 when the other coefficients are *not* assumed to be 0. The idea is that possibly that predictor can be left out *provided* that the others are left in the model. The $t$-table (with $n - 3 = 13$ d.f.) shows that the $P$-value is practically 0. On this basis it would be folly to leave out $D$ as a predictor.

The "overall $F$" in the printout (155.28) is the ratio (5), for testing the hypothesis that *both* $\beta_1$ and $\beta_2$ are 0. The printout also gives $R^2 = .9598$:

$$R^2 = \frac{\text{SSReg}}{\text{SSTot}} = 1 - \frac{\text{SSRes}}{\text{SSTot}} = 1 - \frac{4755.14}{4954.19} \doteq .9598,$$

as well as

$$\text{Adjusted } R^2 = 1 - \frac{(16 - 1)\text{SSRes}}{(16 - 2 - 1)\text{SSTot}} = 1 - \frac{15 \times 199.046}{13 \times 4954.19} = .9536.$$

The residual plot is shown in Figure 15-18. It isn't quite what we expect a random sample of errors to look like, dipping down somewhat in the middle.

What would be desired in this instance is a good predicting relationship, and the equation obtained above should do fairly well. However, since there is some geometry to be considered, we note that if the tree were a perfect cylinder, the volume should depend on the *square* of the diameter and the height. A regression of volume on diameter, diameter squared, and height yielded the following:

| PREDICTOR VARIABLES | COEFFICIENT | STD ERROR | STUDENT'S T | P | VIF |
|---------------------|-------------|-----------|-------------|-----|-----|
| CONSTANT | -23.8286 | 12.5197 | -1.90 | 0.0813 | |
| D | -1.91492 | 1.54788 | -1.24 | 0.2397 | 64.3 |
| DSQ | 0.22804 | 0.05318 | 4.29 | 0.0011 | 62.5 |
| H | 0.49468 | 0.12879 | 3.84 | 0.0023 | 1.6 |

| | | |
|---|---|---|
| R-SQUARED | 0.9841 | RESID. MEAN SQUARE (MSE)    6.54956 |
| ADJUSTED R-SQUARED | 0.9802 | STANDARD DEVIATION    2.55921 |

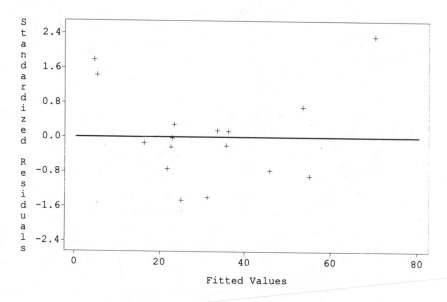

**Figure 15-18** Residuals: $V$ vs. $H$ and $D$

This adjusted $R^2$ is higher, and the residual plot in Figure 15-19 looks better, but perhaps we should try the formula for the volume of a circular cylinder:

$$V \propto hd^2$$

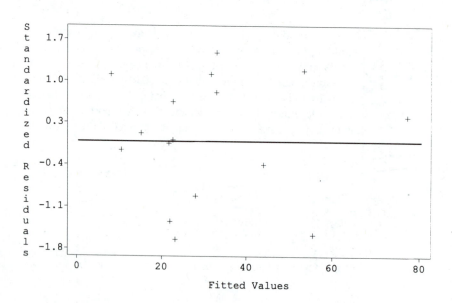

**Figure 15-19** Residuals: $V$ vs. $D$, $D^2$, $H$

So we regressed $V$ on the single predictor $HV^2$ and found that the adjusted $R^2$ is a bit higher yet:

```
PREDICTOR
VARIABLES    COEFFICIENT   STD ERROR   STUDENT'S T      P
---------    -----------   ---------   -----------    ------
CONSTANT       0.61317      1.22927        0.50        0.6257
HXDSQ          0.00209      7.292E-05     28.65        0.0000

R-SQUARED                0.9832    RESID. MEAN SQUARE (MSE)   5.93430
ADJUSTED R-SQUARED       0.9820    STANDARD DEVIATION         2.43604
```

Since we now have a single predictor, we can show the regression function, together with the loci of the 95% confidence limits for the mean volume and the 95% prediction limits for the volume, in Figure 15-20. The residual plot is quite usual, much like (though different from) the one shown in Figure 15-19.

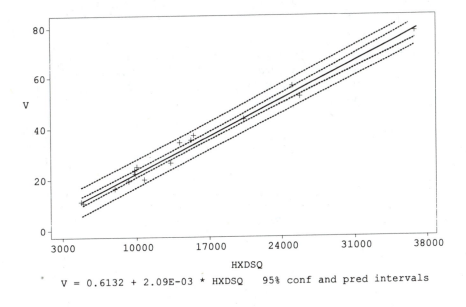

$$V = 0.6132 + 2.09E\text{-}03 * HXDSQ \quad 95\% \text{ conf and pred intervals}$$

**Figure 15-20** Least-squares line—$V$ vs. $HD^2$

Just to hint at the computing power now available, we have used our computer software to obtain the "best" fitting models with 1, 2, ... of the available predictor variables (taken to be $D$, $D^2$, $H$, $HD^2$). These are shown in the printout of Table 15-3 (obtained in a fraction of a second). What is best is a matter of judgment, often guided by the value of the adjusted $R^2$ and the constant $C_P$ (which we don't define, but should be close to the value of $p = 1 + k$, where $k$ is the number of predictors). Although some might choose a different "best" model, the one in Figure 15-20 is not bad and is appealingly simple and intuitive. ∎

## Table 15-3

BEST SUBSET REGRESSION MODELS FOR V

UNFORCED INDEPENDENT VARIABLES:    (A)D   (B)DSQ   (C)H   (D)HXDSQ
3 "BEST" MODELS FROM EACH SUBSET SIZE LISTED.

| P | CP | ADJUSTED<br>R SQUARE | R SQUARE | RESID SS | MODEL VARIABLES |
|----|------|--------|--------|---------|-----------------|
| 1 | 891.1 | 0.0000 | 0.0000 | 4954.19 | INTERCEPT ONLY |
| 2 | 3.2 | 0.9820 | 0.9832 | 83.0802 | D |
| 2 | 20.3 | 0.9617 | 0.9643 | 177.017 | B |
| 2 | 38.7 | 0.9400 | 0.9440 | 277.355 | A |
| 3 | 2.2 | 0.9844 | 0.9865 | 67.0226 | C D |
| 3 | 3.6 | 0.9827 | 0.9850 | 74.4396 | B D |
| 3 | 5.2 | 0.9807 | 0.9832 | 82.9899 | A D |
| 4 | 3.2 | 0.9845 | 0.9876 | 61.4045 | A B D |
| 4 | 4.2 | 0.9832 | 0.9865 | 66.6886 | A C D |
| 4 | 4.2 | 0.9831 | 0.9865 | 67.0065 | B C D |
| 5 | 5.0 | 0.9834 | 0.9878 | 60.2126 | A B C D |

The mathematics of linear multiple regression models is much simplified with the use of matrix algebra. Because we do not assume linear algebra as a prerequisite, we only go so far as to set up the least-squares problem in matrix notation. With $k$ predictor variables, let

$$\mathbf{Y} = \begin{bmatrix} Y_1 \\ \vdots \\ Y_n \end{bmatrix}, \ \beta = \begin{bmatrix} \beta_0 \\ \vdots \\ \beta_k \end{bmatrix}, \ \epsilon = \begin{bmatrix} \epsilon_1 \\ \vdots \\ \epsilon_n \end{bmatrix}.$$

The design matrix $\mathbf{X}$ is the following $n \times (k+1)$ matrix:

$$\mathbf{X} = \begin{bmatrix} 1 & x_{11} & \cdot & \cdot & \cdot & x_{k1} \\ 1 & x_{12} & \cdot & \cdot & \cdot & x_{21} \\ \vdots & \vdots & & & & \vdots \\ 1 & x_{1n} & \cdot & \cdot & \cdot & x_{kn} \end{bmatrix}.$$

Together, $\mathbf{X}$ and $\mathbf{Y}$ constitute the data. The model is the matrix equation

$$\mathbf{Y} = \mathbf{X}\beta + \epsilon,$$

together with the assumption that $\epsilon$ is multivariate normal with independent components and var $\epsilon_i = \sigma^2$.

According to the least-squares criterion, we choose $\beta$ to minimize the *quadratic form*

$$(\mathbf{Y} - \mathbf{X}\beta)'(\mathbf{Y} - \mathbf{X}\beta).$$

The solution, when $\mathbf{X}'\mathbf{X} \neq 0$, can be shown to be

$$\hat{\beta} = (\mathbf{X}'\mathbf{X})^{-1}\mathbf{X}'\mathbf{Y}.$$

One interested in pursuing this approach is referred to the texts on linear models or regression in the references.

## Problems

* **15-47.** The following data are power plant cost ($C$), date of construction permit ($D$), and power plant net capacity ($S$) for nine light water reactor power plants in the United States, outside the Northeast region.[17]

| Plant | C | D | S |
|-------|--------|-------|------|
| 1 | 452.99 | 67.33 | 1065 |
| 2 | 443.22 | 69.33 | 1065 |
| 3 | 412.18 | 68.42 | 530 |
| 4 | 289.66 | 68.42 | 530 |
| 5 | 567.79 | 68.75 | 913 |
| 6 | 621.45 | 69.67 | 786 |
| 7 | 473.64 | 70.42 | 538 |
| 8 | 697.14 | 71.80 | 1130 |
| 9 | 288.48 | 67.17 | 821 |

**(a)** Using a computer, find the least-squares regression of $C$ on $D$ and $S$.
**(b)** Find the overall $F$-ratio and corresponding $P$-value.
**(c)** Find $R^2$.
**(d)** Test the hypothesis that $\beta_S = 0$ (with the other $\beta$'s in the model).

**15-48.** In studies of the reaction rate of a synthetase of a bovine lens, it has seemed that the reciprocal of that rate ($Y$) is linearly related to the reciprocal of the substrate concentration ($x$). The following data were obtained:[18]

---

[17]From W. E. Mooz, "Cost analysis of light water reactor power plants," Report R-2304-DOE, Rand Corporation, Santa Monica, CA.
[18]From a consulting file.

| $x$ | $Y$ |
|-----|-----|
| 24 | .429, .444 |
| 20 | .293, .293 |
| 16 | .251, .268 |
| 12 | .207, .216 |
| 8 | .239, .218 |
| 6 | .156, .167 |

(a)   Test the hypothesis that the coefficient of $x^2$ in a quadratic regression function is 0.

(b)   Assuming linearity of the regression function, and combining SS's for a new SSE, obtain estimates of the coefficients in a linear function, together with standard errors.

**\* 15-49.**   Derive the likelihood ratio statistic for testing $\beta_2 = 0$ in the regression function (1) of §15.9, as given by (2).

**15-50.**   Using the data of Problem 15-4,

(a)   fit a quadratic regression function: $y = \alpha + \beta x + \gamma x^2$.

(b)   test the hypothesis $\gamma = 0$.

**15-51.**   The data in the following table give body weight $(x)$ and metabolic clearance rate per body weight $(y)$ for 14 cattle. Use statistical software to investigate the possibility of nonlinearity of the regression function, introducing a quadratic term: $x_1 = x$, and $x_2 = x^2$.

| $x$ | $y$ |
|-----|-----|
| 110 | 234, 198, 173 |
| 230 | 174, 149, 124 |
| 360 | 115, 130, 102, 95 |
| 505 | 122, 112, 98, 96 |

## Review Problems

**15-R1.**   Referring to the data in Problem 15-34 (corrosion vs. iron content),

(a)   Use a statistical hand calculator to verify the equation of the empirical (least-squares) regression line: $y = 129.8 - 24.02\,x$, and statistics $r = -.9847$ and $S_Y^2 = 283.0514$.

(b)   Calculate $S_x^2$. [Keep in mind that there are 13 $x$'s.]

(c)   Give 90% confidence limits for the slope.

(d)   What fraction of the total variation (SSTotal) is attributed to error?

**15-R2.**   Data on setting time and percent lithium chloride for a certain type of cement are repeated here from Example 7.4a:

| % Lithium | Setting time |
|-----------|--------------|
| 0 | 16110 |
| .0005 | 7110 |
| .001 | 1550 |
| .005 | 680 |
| .01 | 340 |
| .05 | (set while mixing) |

Transform the data to the logarithm of percent lithium ($x$) and logarithm of setting time ($y$), and
   **(a)**   fit a straight line to the transformed data using least squares. [You'll have to ignore the first case (log 0 is undefined) and last case (no $y$).]
   **(b)**   obtain the residuals $y_i - \hat{y}$ and plot these against $x$.
   **(c)**   calculate the Shapiro-Wilk statistic (§13.4) for the residuals in (b).

**15-R3.**   The article cited in the preceding problem also gives data on compressive strength ($Y$) as a function of time (hrs):

| $t$ | .5 | 1 | 1.5 | 2 | 2.5 | 3 | 3.5 |
|-----|-----|-----|-----|-----|-----|-----|-----|
| $Y$ | .68 | 2.6 | 4.9 | 10.1 | 14.79 | 18.33 | 23.75 |

| $t$ | 4 | 4.5 | 5 | 6 | 10 | 15 | 20 |
|-----|-----|-----|-----|-----|-----|-----|-----|
| $Y$ | 26.95 | 27.86 | 31.25 | 36.46 | 37.50 | 39.67 | 44.06 |

The correlation coefficient is $r = .82$.
   **(a)**   Plot the data. Would it be useful to fit a line?
   **(b)**   Enter the first 11 cases in a computer or statistical calculator and again calculate the correlation coefficient.
   **(c)**   Fit a least-squares line to those 11 cases.
   **(d)**   Using the line in (c), predict the compressive strength at 5.5 hrs. (Would you use this line to predict the strength at 12 hrs?)
   **(e)**   If you are using statistical software, obtain and plot the standardized residuals, checking normality with the Shapiro-Wilk statistic. (If your software does not calculate this, you can read out the residuals and find the rankits in Table XII.)

**15-R4.**   (Use computer software.) Referring to Problem 7-R5 (page 302) and assuming the simple linear regression model of §15.1–15.3,
   **(a)**   predict the count of the number of bears when the wind velocity is 15 and give the standard error.
   **(b)**   obtain and comment on the residual plot.

**15-R5.** Show directly: $\sum(Y_i - \hat{\alpha} - \hat{\beta}x_i)^2 = SS_{YY}(1 - r^2)$, which we derived indirectly as (12) of §15.2. [*Hint:* $Y_i - \hat{\alpha} - \hat{\beta}x_i = (Y_i - \overline{Y}) - \hat{\beta}(x_i - \overline{x})$.]

**15-R6.** Use what you found in Problem 15-5 to explain why the sum of the residuals about the empirical regression $y = \hat{\alpha} + \hat{\beta}x$ line must be 0.

**15-R7.** Carry out the differentiation of the sum of squared residuals (described near the beginning of §15.8) that led to the three equations for the least-squares estimates of the $\beta$'s in the regression function $y = \beta_0 + \beta_1 x_1 + \beta_2 x_2$.

**15-R8.** Explain why the estimators $\beta_i$ in the preceding problem are normally distributed when the errors are assumed to be independent and normal. [*Hint:* The inverse of a linear transformation is a linear transformation; that is, the solutions of the three equations are linear functions of the left-hand sides.]

**15-R9.** Another set of capacity vs. temperature data from the study referred to in Problem 15-33 is the following, for isobutyl alcohol in its crystal phase. Use computer software to study the relationship along the lines of Example 15.8c.

| T | C |
|---|---|
| 10 | .0225 |
| 20 | .1345 |
| 30 | .2621 |
| 40 | .3720 |
| 50 | .4660 |
| 60 | .5490 |
| 70 | .6217 |
| 80 | .7018 |
| 100 | .8373 |
| 120 | .9576 |
| 140 | 1.0740 |
| 160 | 1.1859 |

**15-R10.** Problem 12-R7 gave the following measurements of concentration of plutonium-239 in each of ten sample solution media using two methods of measurement:

| Sample | 1 | 2 | 3 | 4 | 5 | 6 | 7 | 8 | 9 | 10 |
|---|---|---|---|---|---|---|---|---|---|---|
| New method | 3.78 | 3.58 | 3.77 | 3.82 | 3.67 | 3.66 | 3.48 | 3.63 | 3.88 | 3.53 |
| Old method | 3.35 | 3.60 | 3.41 | 3.69 | 3.48 | 3.50 | 3.33 | 3.64 | 3.65 | 3.64 |

Are the methods correlated? (Test $\rho = 0$.)

## Chapter Perspective

The regression models we've assumed so far are sometimes overly restrictive. For example, it may be necessary to allow for the error variance to depend on $x$. Also, errors may be correlated, and the assumption of *normally* distributed random errors may not be reasonable. When normality is questionable, inferences based on $t$- and $F$-distributions are also questionable.

Many regression models fall in the category of *linear models*. The term *linear* refers not to the dependence of the response on the predictor variable, but rather to its dependence on the model parameters. With these models, the equations that result from minimizing the sum of squared residuals are linear and can be solved by methods of linear algebra to yield parameter estimates.

When the appropriate regression function $g$ involves parameters in a nonlinear way, transformations of response and/or predictor variables may produce a linear model. Transformations are also sometimes used in attempting to make the error distribution closer to normal.

## References & Further Readings

1. Berger, J. O., *Statistical Decision Theory and Bayesian Analysis,* 2nd Ed., New York: Springer Verlag, 1985.
2. Bickel, P., and K. Doksum, *Mathematical Statistics,* Oakland, CA: Holden-Day, 1977.
3. Casella, G., and R. L. Berger, *Statistical Inference,* Belmont, CA: Brooks-Cole Publ. Co., 1990.
4. Cleveland, W. S., *The Elements of Graphing Data,* Pacific Grove, CA: Wadsworth Advanced Books, 1985.
5. Cramér, H., *Mathematical Methods of Statistics,* Princeton, NJ: Princeton University Press, 1946.
6. Draper, N., and H. Smith, *Applied Regression Analysis,* New York: Wiley, 1981.
7. Feller, W., *An Introduction to Probability Theory and Its Applications,* Vol. 1, 3rd Ed., New York: Wiley, 1968.
8. Fienberg, S., Y. Bishop, and P. Holland, *Categorical Data Analysis,* Cambridge, MA: M. I. T. Press, 1975.
9. Fisher, R. A., *The Design of Experiments,* 9th Ed., New York: Hafner Press, 1971.
10. Fisher, R. A., *Statistical Methods and Scientific Inference,* 3rd Ed., New York: Hafner Press, 1973.
11. Hollander, M., and D. A. Wolfe, *Nonparametric Statistical Methods,* New York: Wiley, 1973.
12. Kuehl, R., *Statistical Principles of Research Design and Analysis,* Belmont, CA: Duxbury Press, 1994.
13. Lehmann, E., *Theory of Point Estimation,* 2nd Ed., New York: Wiley, 1983.
14. Lehmann, E., *Nonparametrics: Statistical Methods Based on Ranks,* San Francisco, CA: Holden-Day, 1975.
15. Lindgren, B. W., *Statistical Theory,* 4th Ed., New York: Chapman & Hall, 1993.
16. Rao, C. R., *Linear Statistical Inference and Its Applications,* 2nd Ed., New York: Wiley, 1973.

17. Rice, J. A., *Mathematical Statistics and Data Analysis,* 2nd Ed., Belmont, CA: Duxbury Press, 1995.

18. Snedecor, G. W., and W. G. Cochran, *Statistical Methods,* 9th Ed., Ames, IA: Iowa State Univ. Press, 1980.

19. Thompson, S. K., *Sampling,* New York: Wiley, 1992.

20. Tukey, J., *Exploratory Data Analysis,* Reading, MA: Addison-Wesley, 1977.

21. Weisberg, S., *Applied Regression Analysis,* 2nd Ed., New York: Wiley, 1985.

22. Wilks, S. S., *Mathematical Statistics,* New York: Wiley, 1962.

# Tables

**Table Ia** Binomial Probabilities

$$P(k \text{ successes in } n \text{ trials}) = \binom{n}{k} p^k (1-p)^{n-k}$$

| n | k | .01 | .05 | .10 | .15 | 1/6 | .20 | .25 | .30 | 1/3 | .35 | .40 | .45 | .50 |
|---|---|-----|-----|-----|-----|-----|-----|-----|-----|-----|-----|-----|-----|-----|
| 5 | 0 | .9510 | .7738 | .5905 | .4437 | .4019 | .3277 | .2373 | .1681 | .1317 | .1160 | .0778 | .0503 | .0312 |
|   | 1 | .0480 | .2036 | .3280 | .3915 | .4019 | .4096 | .3955 | .3601 | .3292 | .3124 | .2592 | .2059 | .1562 |
|   | 2 | .0010 | .0214 | .0729 | .1382 | .1608 | .2048 | .2637 | .3087 | .3292 | .3364 | .3456 | .3369 | .3125 |
|   | 3 | .0000 | .0011 | .0081 | .0244 | .0322 | .0512 | .0879 | .1323 | .1646 | .1811 | .2304 | .2757 | .3125 |
|   | 4 | .0000 | .0000 | .0004 | .0022 | .0032 | .0064 | .0146 | .0283 | .0412 | .0488 | .0768 | .1128 | .1562 |
|   | 5 | .0000 | .0000 | .0000 | .0001 | .0001 | .0003 | .0010 | .0024 | .0041 | .0053 | .0102 | .0185 | .0312 |
| 6 | 0 | .9415 | .7351 | .5314 | .3771 | .3349 | .2621 | .1780 | .1176 | .0878 | .0754 | .0467 | .0277 | .0156 |
|   | 1 | .0571 | .2321 | .3543 | .3993 | .4019 | .3932 | .3560 | .3025 | .2634 | .2437 | .1866 | .1359 | .0938 |
|   | 2 | .0014 | .0305 | .0984 | .1762 | .2009 | .2458 | .2966 | .3241 | .3293 | .3280 | .3110 | .2780 | .2344 |
|   | 3 | .0000 | .0021 | .0146 | .0415 | .0536 | .0819 | .1318 | .1852 | .2195 | .2355 | .2765 | .3032 | .3125 |
|   | 4 | .0000 | .0001 | .0012 | .0055 | .0080 | .0154 | .0330 | .0595 | .0823 | .0951 | .1382 | .1861 | .2344 |
|   | 5 | .0000 | .0000 | .0001 | .0004 | .0006 | .0015 | .0044 | .0102 | .0165 | .0205 | .0369 | .0609 | .0938 |
|   | 6 | .0000 | .0000 | .0000 | .0000 | .0000 | .0001 | .0002 | .0007 | .0014 | .0018 | .0041 | .0083 | .0156 |
| 7 | 0 | .9321 | .6983 | .4783 | .3206 | .2791 | .2097 | .1335 | .0824 | .0585 | .0490 | .0280 | .0152 | .0078 |
|   | 1 | .0659 | .2573 | .3720 | .3960 | .3907 | .3670 | .3115 | .2471 | .2049 | .1848 | .1306 | .0872 | .0547 |
|   | 2 | .0020 | .0406 | .1240 | .2097 | .2344 | .2753 | .3115 | .3177 | .3073 | .2985 | .2613 | .2140 | .1641 |
|   | 3 | .0000 | .0036 | .0230 | .0617 | .0781 | .1147 | .1730 | .2269 | .2561 | .2679 | .2903 | .2918 | .2734 |
|   | 4 | .0000 | .0002 | .0026 | .0109 | .0156 | .0287 | .0577 | .0972 | .1280 | .1442 | .1935 | .2388 | .2734 |
|   | 5 | .0000 | .0000 | .0002 | .0012 | .0018 | .0043 | .0115 | .0250 | .0384 | .0466 | .0774 | .1172 | .1641 |
|   | 6 | .0000 | .0000 | .0000 | .0001 | .0001 | .0004 | .0013 | .0036 | .0064 | .0084 | .0172 | .0320 | .0547 |
|   | 7 | .0000 | .0000 | .0000 | .0000 | .0000 | .0000 | .0001 | .0002 | .0005 | .0006 | .0016 | .0037 | .0078 |
| 8 | 0 | .9227 | .6634 | .4305 | .2725 | .2326 | .1678 | .1001 | .0576 | .0390 | .0319 | .0168 | .0084 | .0039 |
|   | 1 | .0746 | .2793 | .3826 | .3847 | .3721 | .3355 | .2670 | .1977 | .1561 | .1373 | .0896 | .0548 | .0312 |
|   | 2 | .0026 | .0515 | .1488 | .2376 | .2605 | .2936 | .3115 | .2965 | .2731 | .2587 | .2090 | .1569 | .1094 |
|   | 3 | .0001 | .0054 | .0331 | .0839 | .1042 | .1468 | .2076 | .2541 | .2731 | .2786 | .2787 | .2568 | .2187 |
|   | 4 | .0000 | .0004 | .0046 | .0185 | .0260 | .0459 | .0865 | .1361 | .1707 | .1875 | .2322 | .2627 | .2734 |
|   | 5 | .0000 | .0000 | .0004 | .0026 | .0042 | .0092 | .0231 | .0467 | .0683 | .0808 | .1239 | .1719 | .2187 |
|   | 6 | .0000 | .0000 | .0000 | .0002 | .0004 | .0011 | .0038 | .0100 | .0171 | .0217 | .0413 | .0703 | .1094 |
|   | 7 | .0000 | .0000 | .0000 | .0000 | .0000 | .0001 | .0004 | .0012 | .0024 | .0033 | .0079 | .0164 | .0312 |
|   | 8 | .0000 | .0000 | .0000 | .0000 | .0000 | .0000 | .0000 | .0001 | .0002 | .0002 | .0007 | .0017 | .0039 |
| 9 | 0 | .9135 | .6302 | .3874 | .2316 | .1938 | .1342 | .0751 | .0404 | .0260 | .0207 | .0101 | .0046 | .0020 |
|   | 1 | .0830 | .2985 | .3874 | .3679 | .3489 | .3020 | .2253 | .1556 | .1171 | .1004 | .0605 | .0339 | .0176 |
|   | 2 | .0034 | .0629 | .1722 | .2597 | .2791 | .3020 | .3003 | .2668 | .2341 | .2162 | .1612 | .1110 | .0703 |
|   | 3 | .0001 | .0077 | .0446 | .1069 | .1302 | .1762 | .2336 | .2668 | .2731 | .2716 | .2508 | .2119 | .1641 |
|   | 4 | .0000 | .0006 | .0074 | .0283 | .0391 | .0661 | .1168 | .1715 | .2048 | .2194 | .2508 | .2600 | .2461 |
|   | 5 | .0000 | .0000 | .0008 | .0050 | .0078 | .0165 | .0389 | .0735 | .1024 | .1181 | .1672 | .2128 | .2461 |
|   | 6 | .0000 | .0000 | .0001 | .0006 | .0013 | .0028 | .0087 | .0210 | .0341 | .0424 | .0743 | .1160 | .1641 |
|   | 7 | .0000 | .0000 | .0000 | .0000 | .0001 | .0003 | .0012 | .0039 | .0073 | .0098 | .0212 | .0407 | .0703 |
|   | 8 | .0000 | .0000 | .0000 | .0000 | .0000 | .0000 | .0001 | .0004 | .0009 | .0013 | .0035 | .0083 | .0176 |
|   | 9 | .0000 | .0000 | .0000 | .0000 | .0000 | .0000 | .0000 | .0000 | .0001 | .0001 | .0003 | .0008 | .0020 |

**Table Ia** (continued)

| n | k | .01 | .05 | .10 | .15 | 1/6 | .20 | .25 | .30 | 1/3 | .35 | .40 | .45 | .50 |
|---|---|-----|-----|-----|-----|-----|-----|-----|-----|-----|-----|-----|-----|-----|
| | | | | | | | | $p$ | | | | | | |
| 10 | 0 | .9044 | .5987 | .3487 | .1969 | .1615 | .1074 | .0563 | .0282 | .0173 | .0135 | .0060 | .0025 | .0010 |
| | 1 | .0914 | .3151 | .3874 | .3474 | .3230 | .2684 | .1877 | .1211 | .0867 | .0725 | .0403 | .0207 | .0098 |
| | 2 | .0042 | .0746 | .1937 | .2759 | .2907 | .3020 | .2816 | .2335 | .1951 | .1757 | .1209 | .0763 | .0439 |
| | 3 | .0001 | .0105 | .0574 | .1298 | .1550 | .2013 | .2503 | .2668 | .2601 | .2522 | .2150 | .1665 | .1172 |
| | 4 | .0000 | .0010 | .0112 | .0401 | .0543 | .0881 | .1460 | .2001 | .2276 | .2377 | .2508 | .2384 | .2051 |
| | 5 | .0000 | .0001 | .0015 | .0085 | .0130 | .0264 | .0584 | .1029 | .1366 | .1536 | .2007 | .2340 | .2461 |
| | 6 | .0000 | .0000 | .0001 | .0012 | .0022 | .0055 | .0162 | .0368 | .0569 | .0689 | .1115 | .1596 | .2051 |
| | 7 | .0000 | .0000 | .0000 | .0001 | .0002 | .0008 | .0031 | .0090 | .0163 | .0212 | .0425 | .0746 | .1172 |
| | 8 | .0000 | .0000 | .0000 | .0000 | .0000 | .0001 | .0004 | .0014 | .0030 | .0043 | .0106 | .0229 | .0439 |
| | 9 | .0000 | .0000 | .0000 | .0000 | .0000 | .0000 | .0000 | .0001 | .0003 | .0005 | .0016 | .0042 | .0098 |
| | 10 | .0000 | .0000 | .0000 | .0000 | .0000 | .0000 | .0000 | .0000 | .0000 | .0000 | .0001 | .0003 | .0010 |
| 11 | 0 | .8953 | .5688 | .3138 | .1673 | .1346 | .0859 | .0422 | .0198 | .0116 | .0088 | .0036 | .0014 | .0005 |
| | 1 | .0995 | .3293 | .3835 | .3248 | .2961 | .2362 | .1549 | .0932 | .0636 | .0518 | .0266 | .0125 | .0054 |
| | 2 | .0050 | .0867 | .2131 | .2866 | .2961 | .2953 | .2581 | .1998 | .1590 | .1395 | .0887 | .0513 | .0054 |
| | 3 | .0002 | .0137 | .0710 | .1517 | .1777 | .2215 | .2581 | .2568 | .2384 | .2254 | .1774 | .1259 | .0806 |
| | 4 | .0000 | .0014 | .0158 | .0536 | .0711 | .1107 | .1721 | .2201 | .2384 | .2428 | .2365 | .2060 | .1611 |
| | 5 | .0000 | .0001 | .0025 | .0132 | .0199 | .0388 | .0803 | .1321 | .1669 | .1830 | .2207 | .2360 | .2256 |
| | 6 | .0000 | .0000 | .0003 | .0023 | .0040 | .0097 | .0268 | .0566 | .0835 | .0985 | .1471 | .1931 | .2256 |
| | 7 | .0000 | .0000 | .0000 | .0003 | .0006 | .0017 | .0064 | .0173 | .0298 | .0379 | .0701 | .1128 | .1611 |
| | 8 | .0000 | .0000 | .0000 | .0000 | .0001 | .0002 | .0011 | .0037 | .0075 | .0102 | .0234 | .0462 | .0806 |
| | 9 | .0000 | .0000 | .0000 | .0000 | .0000 | .0000 | .0001 | .0005 | .0012 | .0018 | .0052 | .0126 | .0269 |
| | 10 | | .0000 | .0000 | .0000 | .0000 | .0000 | .0000 | .0000 | .0001 | .0002 | .0007 | .0021 | .0054 |
| | 11 | | .0000 | .0000 | .0000 | .0000 | .0000 | .0000 | .0000 | .0000 | .0000 | .0000 | .0000 | .0005 |
| 12 | 0 | .8864 | .5404 | .2824 | .1322 | .1122 | .0687 | .0317 | .0138 | .0077 | .0057 | .0022 | .0008 | .0002 |
| | 1 | .1074 | .3413 | .3766 | .3012 | .2692 | .2062 | .1267 | .0712 | .0462 | .0368 | .0174 | .0075 | .0029 |
| | 2 | .0060 | .0988 | .2301 | .2924 | .2961 | .2835 | .2323 | .1678 | .1272 | .1088 | .0639 | .0339 | .0161 |
| | 3 | .0002 | .0173 | .0852 | .1720 | .1974 | .2362 | .2581 | .2397 | .2120 | .1954 | .1419 | .0923 | .0537 |
| | 4 | .0000 | .0021 | .0213 | .0683 | .0888 | .1329 | .1936 | .2311 | .2384 | .2367 | .2128 | .1700 | .1208 |
| | 5 | .0000 | .0002 | .0038 | .0193 | .0284 | .0532 | .1032 | .1585 | .1908 | .2039 | .2270 | .2225 | .1934 |
| | 6 | .0000 | .0000 | .0005 | .0040 | .0066 | .0155 | .0401 | .0792 | .1113 | .1281 | .1655 | .2124 | .2256 |
| | 7 | .0000 | .0000 | .0000 | .0006 | .0011 | .0033 | .0115 | .0291 | .0477 | .0591 | .1009 | .1489 | .1934 |
| | 8 | .0000 | .0000 | .0000 | .0001 | .0001 | .0005 | .0024 | .0078 | .0149 | .0199 | .0420 | .0762 | .1208 |
| | 9 | .0000 | .0000 | .0000 | .0000 | .0000 | .0001 | .0004 | .0015 | .0033 | .0048 | .0125 | .0277 | .0537 |
| | 10 | .0000 | .0000 | .0000 | .0000 | .0000 | .0000 | .0000 | .0002 | .0005 | .0008 | .0025 | .0068 | .0161 |
| | 11 | .0000 | .0000 | .0000 | .0000 | .0000 | .0000 | .0000 | .0000 | .0000 | .0001 | .0003 | .0010 | .0029 |
| | 12 | .0000 | .0000 | .0000 | .0000 | .0000 | .0000 | .0000 | .0000 | .0000 | .0000 | .0000 | .0001 | .0002 |

NOTE: For $p > .5$, reverse the roles of $p$ and $q = 1 - p$. (For example, the probability of $k$ successes when $p = .7$ is found as the entry for $n - k$ under $p = .3$.)

**Table Ib** Cumulative Binomial Probabilities

$$P(\text{at least } k \text{ successes in } n \text{ trials}) = \sum_{i=k}^{n} \binom{n}{i} p^i (1-p)^{n-i}$$

| n | k | .01 | .05 | .10 | .15 | 1/6 | .20 | .25 | .30 | 1/3 | .35 | .40 | .45 | .50 |
|---|---|-----|-----|-----|-----|-----|-----|-----|-----|-----|-----|-----|-----|-----|
| 5 | 1 | .0490 | .2262 | .4095 | .5563 | .5981 | .6723 | .7627 | .8319 | .8683 | .8840 | .9222 | .9497 | .9688 |
|   | 2 | .0010 | .0226 | .0815 | .1648 | .1962 | .2627 | .3672 | .4718 | .5391 | .5716 | .6630 | .7538 | .8125 |
|   | 3 | .0000 | .0012 | .0086 | .0266 | .0355 | .0579 | .1035 | .1631 | .2099 | .2352 | .3174 | .4069 | .5000 |
|   | 4 | .0000 | .0000 | .0005 | .0022 | .0033 | .0067 | .0156 | .0308 | .0453 | .0540 | .0870 | .1312 | .1875 |
|   | 5 | .0000 | .0000 | .0000 | .0001 | .0001 | .0003 | .0010 | .0024 | .0041 | .0053 | .0102 | .0185 | .0313 |
| 6 | 1 | .0585 | .2649 | .4686 | .6229 | .6651 | .7379 | .8220 | .8824 | .9122 | .9246 | .9533 | .9723 | .9844 |
|   | 2 | .0015 | .0328 | .1143 | .2235 | .2632 | .3446 | .4661 | .5798 | .6488 | .6809 | .7667 | .8364 | .8906 |
|   | 3 | .0000 | .0022 | .0159 | .0473 | .0623 | .0989 | .1694 | .2557 | .3196 | .3529 | .4557 | .5585 | .6563 |
|   | 4 | .0000 | .0001 | .0013 | .0059 | .0087 | .0170 | .0379 | .0705 | .1001 | .1174 | .1792 | .2553 | .3438 |
|   | 5 | .0000 | .0000 | .0001 | .0004 | .0007 | .0016 | .0046 | .0109 | .0178 | .0223 | .0410 | .0692 | .1094 |
|   | 6 | .0000 | .0000 | .0000 | .0000 | .0000 | .0001 | .0002 | .0007 | .0014 | .0018 | .0041 | .0083 | .0156 |
| 7 | 1 | .0679 | .3017 | .5217 | .6794 | .7209 | .7903 | .8665 | .9176 | .9415 | .9510 | .9720 | .9848 | .9922 |
|   | 2 | .0020 | .0444 | .1497 | .2834 | .3302 | .4233 | .5551 | .6706 | .7366 | .7662 | .8414 | .8976 | .9375 |
|   | 3 | .0000 | .0038 | .0257 | .0738 | .0958 | .1480 | .2436 | .3529 | .4294 | .4677 | .5801 | .6836 | .7734 |
|   | 4 | .0000 | .0002 | .0027 | .0121 | .0176 | .0333 | .0706 | .1260 | .1733 | .1998 | .2898 | .3917 | .5000 |
|   | 5 | .0000 | .0000 | .0002 | .0012 | .0020 | .0047 | .0129 | .0288 | .0453 | .0556 | .0963 | .1529 | .2266 |
|   | 6 | .0000 | .0000 | .0000 | .0001 | .0001 | .0004 | .0013 | .0038 | .0069 | .0090 | .0188 | .0357 | .0625 |
|   | 7 | .0000 | .0000 | .0000 | .0000 | .0000 | .0000 | .0001 | .0002 | .0005 | .0006 | .0016 | .0037 | .0078 |
| 8 | 1 | .0773 | .3366 | .5695 | .7275 | .7674 | .8322 | .8999 | .9424 | .9610 | .9681 | .9832 | .9916 | .9961 |
|   | 2 | .0027 | .0572 | .1869 | .3428 | .3953 | .4967 | .6329 | .7447 | .8049 | .8309 | .8936 | .9368 | .9648 |
|   | 3 | .0001 | .0058 | .0381 | .1052 | .1348 | .2031 | .3215 | .4482 | .5318 | .5722 | .6846 | .7799 | .8555 |
|   | 4 | .0000 | .0004 | .0050 | .0214 | .0307 | .0563 | .1138 | .1941 | .2587 | .2936 | .4059 | .5230 | .6367 |
|   | 5 | .0000 | .0000 | .0004 | .0029 | .0046 | .0104 | .0273 | .0580 | .0879 | .1061 | .1737 | .2604 | .3633 |
|   | 6 | .0000 | .0000 | .0000 | .0002 | .0004 | .0012 | .0042 | .0113 | .0197 | .0253 | .0498 | .0885 | .1445 |
|   | 7 | .0000 | .0000 | .0000 | .0000 | .0000 | .0001 | .0004 | .0013 | .0026 | .0036 | .0085 | .0181 | .0352 |
|   | 8 | .0000 | .0000 | .0000 | .0000 | .0000 | .0000 | .0000 | .0001 | .0002 | .0002 | .0007 | .0017 | .0039 |
| 9 | 1 | .0865 | .3698 | .6126 | .7684 | .8062 | .8658 | .9249 | .9596 | .9740 | .9793 | .9899 | .9954 | .9980 |
|   | 2 | .0034 | .0712 | .2252 | .4005 | .4573 | .5638 | .6997 | .8040 | .8569 | .8789 | .9295 | .9615 | .9805 |
|   | 3 | .0001 | .0084 | .0530 | .1409 | .1783 | .2618 | .3993 | .5372 | .6228 | .6627 | .7682 | .8505 | .9102 |
|   | 4 | .0000 | .0006 | .0083 | .0339 | .0480 | .0856 | .1657 | .2703 | .3497 | .3911 | .5174 | .6386 | .7461 |
|   | 5 | .0000 | .0000 | .0009 | .0056 | .0090 | .0196 | .0489 | .0988 | .1448 | .1717 | .2666 | .3786 | .5000 |
|   | 6 | .0000 | .0000 | .0001 | .0006 | .0011 | .0031 | .0100 | .0253 | .0424 | .0536 | .0994 | .1658 | .2539 |
|   | 7 | .0000 | .0000 | .0000 | .0000 | .0001 | .0003 | .0013 | .0043 | .0083 | .0112 | .0250 | .0498 | .0898 |
|   | 8 | .0000 | .0000 | .0000 | .0000 | .0000 | .0000 | .0001 | .0004 | .0010 | .0014 | .0038 | .0091 | .0195 |
|   | 9 | .0000 | .0000 | .0000 | .0000 | .0000 | .0000 | .0000 | .0000 | .0001 | .0001 | .0003 | .0008 | .0020 |

**Table Ib** (*continued*)

| n | k | .01 | .05 | .10 | .15 | 1/6 | .20 | .25 | .30 | 1/3 | .35 | .40 | .45 | .50 |
|---|---|-----|-----|-----|-----|-----|-----|-----|-----|-----|-----|-----|-----|-----|
| 10 | 1 | .0956 | .4013 | .6513 | .8031 | .8385 | .8926 | .9437 | .9718 | .9827 | .9865 | .9940 | .9975 | .9990 |
| | 2 | .0043 | .0861 | .2639 | .4557 | .5155 | .6242 | .7560 | .8507 | .8960 | .9140 | .9536 | .9767 | .9893 |
| | 3 | .0001 | .0115 | .0702 | .1798 | .2248 | .3222 | .4744 | .6172 | .7009 | .7384 | .8327 | .9004 | .9453 |
| | 4 | .0000 | .0010 | .0128 | .0500 | .0697 | .1209 | .2241 | .3504 | .4407 | .4862 | .6177 | .7340 | .8281 |
| | 5 | .0000 | .0001 | .0016 | .0099 | .0155 | .0328 | .0781 | .1503 | .2131 | .2485 | .3669 | .4956 | .6230 |
| | 6 | .0000 | .0000 | .0001 | .0014 | .0024 | .0064 | .0197 | .0473 | .0766 | .0949 | .1662 | .2616 | .3770 |
| | 7 | .0000 | .0000 | .0000 | .0001 | .0003 | .0009 | .0035 | .0106 | .0197 | .0260 | .0548 | .1020 | .1719 |
| | 8 | .0000 | .0000 | .0000 | .0000 | .0000 | .0001 | .0004 | .0016 | .0034 | .0048 | .0123 | .0274 | .0547 |
| | 9 | .0000 | .0000 | .0000 | .0000 | .0000 | .0000 | .0000 | .0001 | .0004 | .0005 | .0017 | .0045 | .0107 |
| | 10 | .0000 | .0000 | .0000 | .0000 | .0000 | .0000 | .0000 | .0000 | .0000 | .0000 | .0001 | .0003 | .0010 |
| 11 | 1 | .1047 | .4312 | .6862 | .8327 | .8654 | .9141 | .9578 | .9802 | .9884 | .9912 | .9964 | .9986 | .9995 |
| | 2 | .0052 | .1019 | .3026 | .5078 | .5693 | .6779 | .8029 | .8870 | .9249 | .9394 | .9698 | .9861 | .9941 |
| | 3 | .0002 | .0152 | .0896 | .2212 | .2732 | .3826 | .5448 | .6873 | .7659 | .7999 | .8811 | .9348 | .9673 |
| | 4 | .0000 | .0016 | .0185 | .0694 | .0956 | .1611 | .2867 | .4304 | .5274 | .5745 | .7037 | .8089 | .8867 |
| | 5 | .0000 | .0001 | .0028 | .0159 | .0245 | .0504 | .1146 | .2103 | .2890 | .3317 | .4672 | .6029 | .7256 |
| | 6 | .0000 | .0000 | .0003 | .0027 | .0046 | .0117 | .0343 | .0782 | .1221 | .1487 | .2465 | .3669 | .5000 |
| | 7 | .0000 | .0000 | .0000 | .0003 | .0006 | .0020 | .0076 | .0216 | .0386 | .0501 | .0994 | .1738 | .2744 |
| | 8 | .0000 | .0000 | .0000 | .0000 | .0001 | .0002 | .0012 | .0043 | .0088 | .0122 | .0293 | .0610 | .1133 |
| | 9 | .0000 | .0000 | .0000 | .0000 | .0000 | .0000 | .0001 | .0006 | .0014 | .0020 | .0059 | .0148 | .0327 |
| | 10 | .0000 | .0000 | .0000 | .0000 | .0000 | .0000 | .0000 | .0000 | .0001 | .0002 | .0007 | .0022 | .0059 |
| | 11 | .0000 | .0000 | .0000 | .0000 | .0000 | .0000 | .0000 | .0000 | .0000 | .0000 | .0000 | .0002 | .0005 |
| 12 | 1 | .1136 | .4596 | .7176 | .8578 | .8878 | .9313 | .9683 | .9862 | .9923 | .9943 | .9978 | .9992 | .9998 |
| | 2 | .0062 | .1184 | .3410 | .5565 | .6187 | .7251 | .8416 | .9150 | .9460 | .9576 | .9804 | .9917 | .9968 |
| | 3 | .0002 | .0196 | .1109 | .2642 | .3226 | .4417 | .6093 | .7472 | .8189 | .8487 | .9166 | .9579 | .9807 |
| | 4 | .0000 | .0022 | .0256 | .0922 | .1252 | .2054 | .3512 | .5075 | .6069 | .6533 | .7747 | .8655 | .9270 |
| | 5 | .0000 | .0002 | .0043 | .0239 | .0364 | .0726 | .1576 | .2763 | .3685 | .4167 | .5618 | .6956 | .8062 |
| | 6 | .0000 | .0000 | .0005 | .0046 | .0079 | .0194 | .0544 | .1178 | .1777 | .2127 | .3348 | .4731 | .6128 |
| | 7 | .0000 | .0000 | .0000 | .0007 | .0013 | .0039 | .0143 | .0386 | .0664 | .0846 | .1582 | .2607 | .3872 |
| | 8 | .0000 | .0000 | .0000 | .0001 | .0002 | .0006 | .0028 | .0095 | .0188 | .0255 | .0573 | .1117 | .1938 |
| | 9 | .0000 | .0000 | .0000 | .0000 | .0000 | .0001 | .0004 | .0017 | .0039 | .0056 | .0153 | .0356 | .0730 |
| | 10 | .0000 | .0000 | .0000 | .0000 | .0000 | .0000 | .0000 | .0002 | .0005 | .0008 | .0028 | .0079 | .0193 |
| | 11 | .0000 | .0000 | .0000 | .0000 | .0000 | .0000 | .0000 | .0000 | .0000 | .0001 | .0003 | .0011 | .0032 |
| | 12 | .0000 | .0000 | .0000 | .0000 | .0000 | .0000 | .0000 | .0000 | .0000 | .0000 | .0000 | .0001 | .0002 |

**Table IIa** Standard Normal c.d.f.

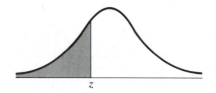

| $z$ | 0 | 1 | 2 | 3 | 4 | 5 | 6 | 7 | 8 | 9 |
|---|---|---|---|---|---|---|---|---|---|---|
| $-3.$ | .0013 | .0010 | .0007 | .0005 | .0003 | .0002 | .0002 | .0001 | .0001 | .0000 |
| $-2.9$ | .0019 | .0018 | .0017 | .0017 | .0016 | .0016 | .0015 | .0015 | .0014 | .0014 |
| $-2.8$ | .0026 | .0025 | .0024 | .0023 | .0023 | .0022 | .0021 | .0021 | .0020 | .0019 |
| $-2.7$ | .0035 | .0034 | .0033 | .0032 | .0031 | .0030 | .0029 | .0028 | .0027 | .0026 |
| $-2.6$ | .0047 | .0045 | .0044 | .0043 | .0041 | .0040 | .0039 | .0038 | .0037 | .0036 |
| $-2.5$ | .0062 | .0060 | .0059 | .0057 | .0055 | .0054 | .0052 | .0051 | .0049 | .0048 |
| $-2.4$ | .0082 | .0080 | .0078 | .0075 | .0073 | .0071 | .0069 | .0068 | .0066 | .0064 |
| $-2.3$ | .0107 | .0104 | .0102 | .0099 | .0096 | .0094 | .0091 | .0089 | .0087 | .0084 |
| $-2.2$ | .0139 | .0136 | .0132 | .0129 | .0126 | .0122 | .0119 | .0116 | .0113 | .0110 |
| $-2.1$ | .0179 | .0174 | .0170 | .0166 | .0162 | .0158 | .0154 | .0150 | .0146 | .0143 |
| $-2.0$ | .0228 | .0222 | .0217 | .0212 | .0207 | .0202 | .0197 | .0192 | .0188 | .0183 |
| $-1.9$ | .0287 | .0281 | .0274 | .0268 | .0262 | .0256 | .0250 | .0244 | .0238 | .0233 |
| $-1.8$ | .0359 | .0352 | .0344 | .0336 | .0329 | .0322 | .0314 | .0307 | .0300 | .0294 |
| $-1.7$ | .0446 | .0436 | .0427 | .0418 | .0409 | .0401 | .0392 | .0384 | .0375 | .0367 |
| $-1.6$ | .0548 | .0537 | .0526 | .0516 | .0505 | .0495 | .0485 | .0475 | .0465 | .0455 |
| $-1.5$ | .0668 | .0655 | .0643 | .0630 | .0618 | .0606 | .0594 | .0582 | .0570 | .0559 |
| $-1.4$ | .0808 | .0793 | .0778 | .0764 | .0749 | .0735 | .0722 | .0708 | .0694 | .0681 |
| $-1.3$ | .0968 | .0951 | .0934 | .0918 | .0901 | .0885 | .0869 | .0853 | .0838 | .0823 |
| $-1.2$ | .1151 | .1131 | .1112 | .1093 | .1075 | .1056 | .1038 | .1020 | .1003 | .0985 |
| $-1.1$ | .1357 | .1335 | .1314 | .1292 | .1271 | .1251 | .1230 | .1210 | .1190 | .1170 |
| $-1.0$ | .1587 | .1562 | .1539 | .1515 | .1492 | .1469 | .1446 | .1423 | .1401 | .1379 |
| $-.9$ | .1841 | .1814 | .1788 | .1762 | .1736 | .1711 | .1685 | .1660 | .1635 | .1611 |
| $-.8$ | .2119 | .2090 | .2061 | .2033 | .2005 | .1977 | .1949 | .1922 | .1894 | .1867 |
| $-.7$ | .2420 | .2389 | .2358 | .2327 | .2297 | .2266 | .2236 | .2206 | .2177 | .2148 |
| $-.6$ | .2743 | .2709 | .2676 | .2643 | .2611 | .2578 | .2546 | .2514 | .2483 | .2451 |
| $-.5$ | .3085 | .3050 | .3015 | .2981 | .2946 | .2912 | .2877 | .2843 | .2810 | .2776 |
| $-.4$ | .3446 | .3409 | .3372 | .3336 | .3300 | .3264 | .3228 | .3192 | .3156 | .3121 |
| $-.3$ | .3821 | .3783 | .3745 | .3707 | .3669 | .3632 | .3594 | .3557 | .3520 | .3483 |
| $-.2$ | .4207 | .4168 | .4129 | .4090 | .4052 | .4013 | .3974 | .3936 | .3897 | .3859 |
| $-.1$ | .4602 | .4562 | .4522 | .4483 | .4443 | .4404 | .4364 | .4325 | .4286 | .4247 |
| $-.0$ | .5000 | .4960 | .4920 | .4880 | .4840 | .4801 | .4761 | .4721 | .4681 | .4641 |

**Table IIa** (*continued*)

| z | 0 | 1 | 2 | 3 | 4 | 5 | 6 | 7 | 8 | 9 |
|---|---|---|---|---|---|---|---|---|---|---|
| .0 | .5000 | .5040 | .5080 | .5120 | .5160 | .5199 | .5239 | .5279 | .5319 | .5359 |
| .1 | .5398 | .5438 | .5478 | .5517 | .5557 | .5596 | .5636 | .5675 | .5714 | .5753 |
| .2 | .5793 | .5832 | .5871 | .5910 | .5948 | .5987 | .6026 | .6064 | .6103 | .6141 |
| .3 | .6179 | .6217 | .6255 | .6293 | .6331 | .6368 | .6406 | .6443 | .6480 | .6517 |
| .4 | .6554 | .6591 | .6628 | .6664 | .6700 | .6736 | .6772 | .6808 | .6844 | .6879 |
| .5 | .6915 | .6950 | .6985 | .7019 | .7054 | .7088 | .7123 | .7157 | .7190 | .7224 |
| .6 | .7257 | .7291 | .7324 | .7357 | .7389 | .7422 | .7454 | .7486 | .7517 | .7549 |
| .7 | .7580 | .7611 | .7642 | .7673 | .7703 | .7734 | .7764 | .7794 | .7823 | .7852 |
| .8 | .7881 | .7910 | .7939 | .7967 | .7995 | .8023 | .8051 | .8078 | .8106 | .8133 |
| .9 | .8159 | .8186 | .8212 | .8238 | .8264 | .8289 | .8315 | .8340 | .8365 | .8389 |
| 1.0 | .8413 | .8438 | .8461 | .8485 | .8508 | .8531 | .8554 | .8577 | .8599 | .8621 |
| 1.1 | .8643 | .8665 | .8686 | .8708 | .8729 | .8749 | .8770 | .8790 | .8810 | .8830 |
| 1.2 | .8849 | .8869 | .8888 | .8907 | .8925 | .8944 | .8962 | .8980 | .8997 | .9015 |
| 1.3 | .9032 | .9049 | .9066 | .9082 | .9099 | .9115 | .9131 | .9147 | .9162 | .9177 |
| 1.4 | .9192 | .9207 | .9222 | .9236 | .9251 | .9265 | .9278 | .9292 | .9306 | .9319 |
| 1.5 | .9332 | .9345 | .9357 | .9370 | .9382 | .9384 | .9406 | .9418 | .9430 | .9441 |
| 1.6 | .9452 | .9463 | .9474 | .9484 | .9495 | .9505 | .9515 | .9525 | .9535 | .9545 |
| 1.7 | .9554 | .9564 | .9573 | .9582 | .9591 | .9599 | .9608 | .9616 | .9625 | .9633 |
| 1.8 | .9641 | .9648 | .9656 | .9664 | .9671 | .9678 | .9686 | .9693 | .9700 | .9706 |
| 1.9 | .9713 | .9719 | .9726 | .9732 | .9738 | .9744 | .9750 | .9756 | .9762 | .9767 |
| 2.0 | .9772 | .9778 | .9783 | .9788 | .9793 | .9798 | .9803 | .9808 | .9812 | .9817 |
| 2.1 | .9821 | .9826 | .9830 | .9834 | .9838 | .9842 | .9846 | .9850 | .9854 | .9857 |
| 2.2 | .9861 | .9864 | .9868 | .9871 | .9874 | .9878 | .9881 | .9884 | .9887 | .9890 |
| 2.3 | .9893 | .9896 | .9898 | .9901 | .9904 | .9906 | .9909 | .9911 | .9913 | .9916 |
| 2.4 | .9918 | .9920 | .9922 | .9925 | .9927 | .9929 | .9931 | .9932 | .9934 | .9936 |
| 2.5 | .9938 | .9940 | .9941 | .9943 | .9945 | .9946 | .9948 | .9949 | .9951 | .9952 |
| 2.6 | .9953 | .9955 | .9956 | .9957 | .9959 | .9960 | .9961 | .9962 | .9963 | .9964 |
| 2.7 | .9965 | .9966 | .9967 | .9968 | .9969 | .9970 | .9971 | .9972 | .9973 | .9974 |
| 2.8 | .9974 | .9975 | .9976 | .9977 | .9977 | .9978 | .9979 | .9979 | .9980 | .9981 |
| 2.9 | .9981 | .9982 | .9982 | .9983 | .9984 | .9984 | .9985 | .9985 | .9986 | .9986 |
| 3. | .9987 | .9990 | .9993 | .9995 | .9997 | .9998 | .9998 | .9999 | .9999 | 1.0000 |

Notes: 1. Enter table at $Z$, read out $P(Z \leq z)$, the shaded area.

2. For a general normal $X$, enter table at $z = (x - \mu)/\sigma$ to read $P(X \leq x)$.

3. Entries opposite 3 are for 3.0, 3.1, 3.2 . . . , 3.9.

4. For $z \geq 4$, $P(Z > z) = P(Z < -z) \doteq \dfrac{1}{\sqrt{2\pi}} e^{-z^2/2}$.

**Table IIb** Standard Normal Percentiles

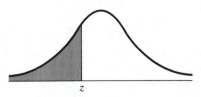

| $P(Z \leq z)$ | $z$ |
|---|---|
| .001 | −3.0902 |
| .005 | −2.5758 |
| .01 | −2.3263 |
| .02 | −2.0537 |
| .03 | −1.8808 |
| .04 | −1.7507 |
| .05 | −1.6449 |
| .10 | −1.2816 |
| .15 | −1.0364 |
| .20 | −.8416 |
| .25 | −.6745 |
| .30 | −.5244 |
| .35 | −.3853 |
| .40 | −.2533 |
| .45 | −.1257 |
| .50 | 0 |
| .55 | .1257 |
| .60 | .2533 |
| .65 | .3853 |
| .70 | .5244 |
| .75 | .6745 |
| .80 | .8416 |
| .85 | 1.0364 |
| .90 | 1.2816 |
| .95 | 1.6449 |
| .96 | 1.7507 |
| .97 | 1.8808 |
| .98 | 2.0537 |
| .99 | 2.3263 |
| .995 | 2.5758 |
| .999 | 3.0902 |

**Table IIc** Two-Tailed Probabilities for the Standard Normal Distribution

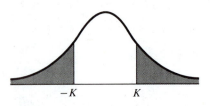

| $P(|Z| > K)$ | $K$ |
|---|---|
| .001 | 3.2905 |
| .002 | 3.0902 |
| .005 | 2.8070 |
| .01 | 2.5758 |
| .02 | 2.3263 |
| .03 | 2.1701 |
| .04 | 2.0537 |
| .05 | 1.9600 |
| .06 | 1.8808 |
| .08 | 1.7507 |
| .10 | 1.6449 |
| .15 | 1.4395 |
| .20 | 1.2816 |
| .30 | 1.0364 |

**Table IId** Multipliers for Large-Sample Confidence Intervals

| Confidence Level | Multiplier |
|---|---|
| .68 | 1 |
| .80 | 1.28 |
| .90 | 1.645 |
| .95 | 1.96 |
| .99 | 2.58 |

**Table IIIa** Percentiles of the $t$-Distribution

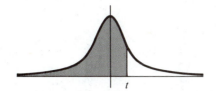

| Degrees of Freedom | $p$ | | | | | | | | |
|---|---|---|---|---|---|---|---|---|---|
| | .60 | .70 | .80 | .85 | .90 | .95 | .975 | .99 | .995 |
| 1 | .325 | .727 | 1.38 | 1.96 | 3.08 | 6.31 | 12.7 | 31.8 | 63.7 |
| 2 | .289 | .617 | 1.06 | 1.39 | 1.89 | 2.92 | 4.30 | 6.96 | 9.92 |
| 3 | .277 | .584 | .978 | 1.25 | 1.64 | 2.35 | 3.18 | 4.54 | 5.84 |
| 4 | .271 | .569 | .941 | 1.19 | 1.53 | 2.13 | 2.78 | 3.75 | 4.60 |
| 5 | .267 | .559 | .920 | 1.16 | 1.48 | 2.01 | 2.57 | 3.36 | 4.03 |
| 6 | .265 | .553 | .906 | 1.13 | 1.44 | 1.94 | 2.45 | 3.14 | 3.71 |
| 7 | .263 | .549 | .896 | 1.12 | 1.42 | 1.90 | 2.36 | 3.00 | 3.50 |
| 8 | .262 | .546 | .889 | 1.11 | 1.40 | 1.86 | 2.31 | 2.90 | 3.36 |
| 9 | .261 | .543 | .883 | 1.10 | 1.38 | 1.83 | 2.26 | 2.82 | 3.25 |
| 10 | .260 | .542 | .879 | 1.09 | 1.37 | 1.81 | 2.23 | 2.76 | 3.17 |
| 11 | .260 | .540 | .876 | 1.09 | 1.36 | 1.80 | 2.20 | 2.72 | 3.11 |
| 12 | .259 | .539 | .873 | 1.08 | 1.36 | 1.78 | 2.18 | 2.68 | 3.06 |
| 13 | .259 | .538 | .870 | 1.08 | 1.35 | 1.77 | 2.16 | 2.65 | 3.01 |
| 14 | .258 | .537 | .868 | 1.08 | 1.34 | 1.76 | 2.14 | 2.62 | 2.98 |
| 15 | .258 | .536 | .866 | 1.07 | 1.34 | 1.75 | 2.13 | 2.60 | 2.95 |
| 16 | .258 | .535 | .865 | 1.07 | 1.34 | 1.75 | 2.12 | 2.58 | 2.92 |
| 17 | .257 | .534 | .863 | 1.07 | 1.33 | 1.74 | 2.11 | 2.57 | 2.90 |
| 18 | .257 | .534 | .862 | 1.07 | 1.33 | 1.73 | 2.10 | 2.55 | 2.88 |
| 19 | .257 | .533 | .861 | 1.07 | 1.33 | 1.73 | 2.09 | 2.54 | 2.86 |
| 20 | .257 | .533 | .860 | 1.07 | 1.32 | 1.72 | 2.09 | 2.53 | 2.84 |
| 21 | .257 | .532 | .859 | 1.06 | 1.32 | 1.72 | 2.08 | 2.52 | 2.83 |
| 22 | .256 | .532 | .858 | 1.06 | 1.32 | 1.72 | 2.07 | 2.51 | 2.82 |
| 23 | .256 | .532 | .858 | 1.06 | 1.32 | 1.71 | 2.07 | 2.50 | 2.81 |
| 24 | .256 | .531 | .857 | 1.06 | 1.32 | 1.71 | 2.06 | 2.49 | 2.80 |
| 25 | .256 | .531 | .856 | 1.06 | 1.32 | 1.71 | 2.06 | 2.48 | 2.79 |
| 26 | .256 | .531 | .856 | 1.06 | 1.32 | 1.71 | 2.06 | 2.48 | 2.78 |
| 27 | .256 | .531 | .855 | 1.06 | 1.31 | 1.70 | 2.05 | 2.47 | 2.77 |
| 28 | .256 | .530 | .855 | 1.06 | 1.31 | 1.70 | 2.05 | 2.47 | 2.76 |
| 29 | .256 | .530 | .854 | 1.06 | 1.31 | 1.70 | 2.04 | 2.46 | 2.76 |
| 30 | .256 | .530 | .854 | 1.05 | 1.31 | 1.70 | 2.04 | 2.46 | 2.75 |
| 40 | .255 | .529 | .851 | 1.05 | 1.30 | 1.68 | 2.02 | 2.42 | 2.70 |
| ∞ | .253 | .524 | .842 | 1.04 | 1.28 | 1.64 | 1.96 | 2.33 | 2.58 |

Notes: 1. The area to the right of the table entry is $1 - p$.
2. The distribution is symmetric. For example, for 10 degrees of freedom, $P(-1.37 < t < 1.37) = .9 - .1 = .8$.

**Table IIIb** Tail-Probabilities of Student's $t$-Distribution

| $t$ | | | | | | | Degrees of freedom | | | | | | | |
|-----|------|------|------|------|------|------|------|------|------|------|------|------|------|------|
| | 4 | 5 | 6 | 7 | 8 | 9 | 10 | 11 | 12 | 13 | 14 | 15 | 16 | 17 |
| 1.0 | .186 | .181 | .178 | .175 | .173 | .172 | .170 | .169 | .169 | .168 | .167 | .167 | .166 | .166 |
| 1.1 | .166 | .160 | .157 | .154 | .152 | .150 | .149 | .147 | .146 | .146 | .145 | .144 | .144 | .143 |
| 1.2 | .145 | .142 | .138 | .135 | .132 | .130 | .129 | .128 | .127 | .126 | .125 | .124 | .124 | .123 |
| 1.3 | .131 | .125 | .121 | .117 | .115 | .113 | .111 | .110 | .109 | .108 | .107 | .107 | .106 | .105 |
| 1.4 | .117 | .110 | .105 | .102 | .100 | .098 | .096 | .095 | .093 | .092 | .092 | .091 | .090 | .090 |
| 1.5 | .104 | .097 | .092 | .089 | .086 | .084 | .082 | .081 | .080 | .079 | .078 | .077 | .077 | .076 |
| 1.6 | .092 | .085 | .080 | .077 | .074 | .072 | .070 | .069 | .068 | .067 | .066 | .065 | .065 | .064 |
| 1.7 | .082 | .075 | .070 | .066 | .064 | .062 | .060 | .059 | .057 | .056 | .056 | .055 | .054 | .054 |
| 1.8 | .073 | .066 | .061 | .057 | .055 | .053 | .051 | .050 | .049 | .048 | .047 | .046 | .045 | .045 |
| 1.9 | .065 | .058 | .053 | .050 | .047 | .045 | .043 | .042 | .041 | .040 | .039 | .038 | .038 | .037 |
| 2.0 | .058 | .051 | .046 | .043 | .040 | .038 | .037 | .035 | .034 | .033 | .033 | .032 | .031 | .031 |
| 2.1 | .052 | .045 | .040 | .037 | .034 | .033 | .031 | .030 | .029 | .028 | .027 | .027 | .026 | .025 |
| 2.2 | .046 | .040 | .035 | .032 | .029 | .028 | .026 | .025 | .024 | .023 | .023 | .022 | .021 | .021 |
| 2.3 | .042 | .035 | .031 | .027 | .025 | .023 | .022 | .021 | .020 | .019 | .019 | .018 | .018 | .017 |
| 2.4 | .037 | .031 | .027 | .024 | .022 | .020 | .019 | .018 | .017 | .016 | .015 | .015 | .014 | .014 |
| 2.5 | .033 | .027 | .023 | .020 | .018 | .017 | .016 | .015 | .014 | .013 | .013 | .012 | .012 | .011 |
| 2.6 | .030 | .024 | .020 | .018 | .016 | .014 | .013 | .012 | .012 | .011 | .010 | .010 | .010 | .009 |
| 2.7 | .027 | .021 | .018 | .015 | .014 | .012 | .011 | .010 | .010 | .009 | .009 | .008 | .008 | .008 |
| 2.8 | .025 | .019 | .016 | .013 | .012 | .010 | .009 | .009 | .008 | .008 | .007 | .007 | .006 | .006 |
| 2.9 | .022 | .017 | .014 | .011 | .010 | .009 | .008 | .007 | .007 | .006 | .006 | .005 | .005 | .005 |
| 3.0 | .020 | .015 | .012 | .010 | .009 | .007 | .007 | .006 | .006 | .005 | .005 | .004 | .004 | .004 |
| 3.1 | .018 | .013 | .011 | .009 | .007 | .006 | .006 | .005 | .005 | .004 | .004 | .004 | .003 | .003 |
| 3.2 | .017 | .012 | .009 | .008 | .006 | .005 | .005 | .004 | .004 | .003 | .003 | .003 | .003 | .003 |
| 3.3 | .015 | .011 | .008 | .007 | .005 | .005 | .004 | .004 | .003 | .003 | .003 | .002 | .002 | .002 |
| 3.4 | .014 | .010 | .007 | .006 | .005 | .004 | .003 | .003 | .002 | .002 | .002 | .002 | .002 | .002 |
| 3.5 | .013 | .009 | .006 | .005 | .004 | .003 | .003 | .002 | .002 | .002 | .002 | .002 | .001 | .001 |
| 3.6 | .012 | .008 | .006 | .004 | .003 | .003 | .002 | .002 | .002 | .002 | .001 | .001 | .001 | .001 |
| 3.7 | .011 | .007 | .005 | .004 | .003 | .002 | .002 | .002 | .002 | .001 | .001 | .001 | .001 | .001 |
| 3.8 | .010 | .006 | .004 | .003 | .003 | .002 | .002 | .001 | .001 | .001 | .001 | .001 | .001 | .001 |
| 3.9 | .009 | .006 | .004 | .003 | .002 | .002 | .001 | .001 | .001 | .001 | .001 | .001 | .001 | .001 |
| 4.0 | .008 | .005 | .004 | .003 | .002 | .002 | .001 | .001 | .001 | .001 | .001 | .001 | .001 | .000 |
| 4.5 | .006 | .003 | .002 | .001 | .001 | .001 | .001 | .000 | .000 | .000 | .000 | .000 | .000 | .000 |
| 5.0 | .004 | .002 | .001 | .001 | .000 | .000 | .000 | .000 | .000 | .000 | .000 | .000 | .000 | .000 |

**Table IIIb** (*continued*)

| t | 18 | 19 | 20 | 21 | 22 | 23 | 24 | 25 | 26 | 27 | 28 | 29 | 30 | 35 | 40 |
|---|----|----|----|----|----|----|----|----|----|----|----|----|----|----|----|
| | | | | | | | Degrees of freedom | | | | | | | | |
| 1.0 | .165 | .165 | .165 | .164 | .164 | .164 | .164 | .163 | .163 | .163 | .163 | .163 | .163 | .162 | .162 |
| 1.1 | .143 | .143 | .142 | .142 | .142 | .141 | .141 | .141 | 1.41 | .141 | .140 | .140 | .140 | .139 | .139 |
| 1.2 | .123 | .122 | .122 | .122 | .121 | .121 | .121 | .121 | .120 | .120 | .120 | .120 | .120 | .119 | .119 |
| 1.3 | .105 | .105 | .104 | .104 | .104 | .103 | .103 | .103 | .103 | .102 | .102 | .102 | .102 | .101 | .101 |
| 1.4 | .089 | .089 | .088 | .088 | .088 | .087 | .087 | .087 | .087 | .086 | .086 | .086 | .086 | .085 | .085 |
| 1.5 | .075 | .075 | .075 | .074 | .074 | .074 | .073 | .073 | .073 | .073 | .072 | .072 | .072 | .071 | .071 |
| 1.6 | .064 | .063 | .063 | .062 | .062 | .062 | .061 | .061 | .061 | .061 | .060 | .060 | .060 | .059 | .059 |
| 1.7 | .053 | .053 | .052 | .052 | .052 | .051 | .051 | .051 | .051 | .050 | .050 | .050 | .050 | .049 | .048 |
| 1.8 | .044 | .044 | .043 | .043 | .043 | .042 | .042 | .042 | .042 | .042 | .041 | .041 | .041 | .040 | .040 |
| 1.9 | .037 | .036 | .036 | .036 | .035 | .035 | .035 | .035 | .034 | .034 | .034 | .034 | .034 | .033 | .032 |
| 2.0 | .030 | .030 | .030 | .029 | .029 | .029 | .028 | .028 | .028 | .028 | .028 | .027 | .027 | .027 | .026 |
| 2.1 | .025 | .025 | .024 | .024 | .024 | .023 | .023 | .023 | .023 | .023 | .022 | .022 | .022 | .022 | .021 |
| 2.2 | .021 | .020 | .020 | .020 | .019 | .019 | .019 | .019 | .018 | .018 | .018 | .018 | .018 | .017 | .017 |
| 2.3 | .017 | .016 | .016 | .016 | .016 | .015 | .015 | .015 | .015 | .015 | .015 | .014 | .014 | .014 | .013 |
| 2.4 | .014 | .013 | .013 | .013 | .013 | .012 | .012 | .012 | .012 | .012 | .012 | .012 | .011 | .011 | .011 |
| 2.5 | .011 | .011 | .011 | .010 | .010 | .010 | .010 | .010 | .010 | .009 | .009 | .009 | .009 | .009 | .008 |
| 2.6 | .009 | .009 | .009 | .008 | .008 | .008 | .008 | .008 | .008 | .007 | .007 | .007 | .007 | .007 | .006 |
| 2.7 | .007 | .007 | .007 | .007 | .007 | .006 | .006 | .006 | .006 | .006 | .006 | .006 | .006 | .005 | .005 |
| 2.8 | .006 | .006 | .006 | .005 | .005 | .005 | .005 | .005 | .005 | .005 | .005 | .004 | .004 | .004 | .004 |
| 2.9 | .005 | .005 | .004 | .004 | .004 | .004 | .004 | .004 | .004 | .004 | .004 | .004 | .003 | .003 | .003 |
| 3.0 | .004 | .004 | .004 | .003 | .003 | .003 | .003 | .003 | .003 | .003 | .003 | .003 | .003 | .002 | .002 |
| 3.1 | .003 | .003 | .003 | .003 | .003 | .003 | .002 | .002 | .002 | .002 | .002 | .002 | .002 | .002 | .002 |
| 3.2 | .002 | .002 | .002 | .002 | .002 | .002 | .002 | .002 | .002 | .002 | .002 | .002 | .002 | .001 | .001 |
| 3.3 | .002 | .002 | .001 | .001 | .001 | .001 | .001 | .001 | .001 | .001 | .001 | .001 | .001 | .001 | .001 |
| 3.4 | .001 | .001 | .001 | .001 | .001 | .001 | .001 | .001 | .001 | .001 | .001 | .001 | .001 | .001 | .001 |
| 3.5 | .001 | .001 | .001 | .001 | .001 | .001 | .001 | .001 | .001 | .001 | .001 | .001 | .001 | .000 | .000 |
| 3.6 | .001 | .001 | .001 | .001 | .001 | .001 | .001 | .001 | .001 | .000 | .000 | .000 | .000 | .000 | .000 |
| 3.7 | .001 | .001 | .001 | .001 | .001 | .000 | .000 | .000 | .000 | .000 | .000 | .000 | .000 | .000 | .000 |
| 3.8 | .001 | .001 | .000 | .000 | .000 | .000 | .000 | .000 | .000 | .000 | .000 | .000 | .000 | .000 | .000 |
| 3.9 | .001 | .000 | .000 | .000 | .000 | .000 | .000 | .000 | .000 | .000 | .000 | .000 | .000 | .000 | .000 |
| 4.0 | .000 | .000 | .000 | .000 | .000 | .000 | .000 | .000 | .000 | .000 | .000 | .000 | .000 | .000 | .000 |
| 4.5 | .000 | .000 | .000 | .000 | .000 | .000 | .000 | .000 | .000 | .000 | .000 | .000 | .000 | .000 | .000 |
| 5.0 | .000 | .000 | .000 | .000 | .000 | .000 | .000 | .000 | .000 | .000 | .000 | .000 | .000 | .000 | .000 |

**Table IV** Cumulative Poisson Probabilities

$$P(X \le c \mid m) = \sum_0^c \frac{m^k}{k!} e^{-m}$$

*m* (expected value)

| $c$ | .02 | .04 | .06 | .08 | .10 | .15 | .20 | .25 | .30 | .35 | .40 |
|---|---|---|---|---|---|---|---|---|---|---|---|
| 0 | .980 | .961 | .942 | .923 | .905 | .861 | .819 | .779 | .741 | .705 | .670 |
| 1 | 1.000 | .999 | .998 | .997 | .995 | .990 | .982 | .974 | .963 | .951 | .938 |
| 2 | | 1.000 | 1.000 | 1.000 | 1.000 | .999 | .999 | .998 | .996 | .994 | .992 |
| 3 | | | | | | 1.000 | 1.000 | 1.000 | 1.000 | 1.000 | .999 |
| 4 | | | | | | | | | | | 1.000 |

| $c$ | .45 | .50 | .55 | .60 | .65 | .70 | .75 | .80 | .85 | .90 | .95 |
|---|---|---|---|---|---|---|---|---|---|---|---|
| 0 | .638 | .607 | .577 | .549 | .522 | .497 | .472 | .449 | .427 | .407 | .387 |
| 1 | .925 | .910 | .894 | .878 | .861 | .844 | .827 | .809 | .791 | .772 | .754 |
| 2 | .989 | .986 | .982 | .977 | .972 | .966 | .959 | .953 | .945 | .937 | .929 |
| 3 | .999 | .998 | .998 | .997 | .996 | .994 | .993 | .991 | .989 | .987 | .984 |
| 4 | 1.000 | 1.000 | 1.000 | 1.000 | .999 | .999 | .999 | .999 | .998 | .998 | .997 |
| 5 | | | | | 1.000 | 1.000 | 1.000 | 1.000 | 1.000 | 1.000 | 1.000 |

| $c$ | 1.0 | 1.1 | 1.2 | 1.3 | 1.4 | 1.5 | 1.6 | 1.7 | 1.8 | 1.9 | 2.0 |
|---|---|---|---|---|---|---|---|---|---|---|---|
| 0 | .368 | .333 | .301 | .273 | .247 | .223 | .202 | .183 | .165 | .150 | .135 |
| 1 | .736 | .699 | .663 | .627 | .592 | .558 | .525 | .493 | .463 | .434 | .406 |
| 2 | .920 | .900 | .879 | .857 | .833 | .809 | .783 | .757 | .731 | .704 | .677 |
| 3 | .981 | .974 | .966 | .957 | .946 | .934 | .921 | .907 | .891 | .875 | .857 |
| 4 | .996 | .996 | .992 | .989 | .986 | .981 | .976 | .970 | .964 | .956 | .947 |
| 5 | .999 | .999 | .998 | .998 | .997 | .996 | .994 | .992 | .990 | .987 | .983 |
| 6 | 1.000 | 1.000 | 1.000 | 1.000 | .999 | .999 | .999 | .998 | .997 | .997 | .995 |
| 7 | | | | | 1.000 | 1.000 | 1.000 | 1.000 | .999 | .999 | .999 |
| 8 | | | | | | | | | .000 | 1.000 | 1.000 |

| $c$ | 2.2 | 2.4 | 2.6 | 2.8 | 3.0 | 3.2 | 3.4 | 3.6 | 3.8 | 4.0 | 4.2 |
|---|---|---|---|---|---|---|---|---|---|---|---|
| 0 | .111 | .091 | .074 | .061 | .050 | .041 | .033 | .027 | .022 | .018 | .015 |
| 1 | .355 | .308 | .267 | .231 | .199 | .171 | .147 | .126 | .107 | .092 | .078 |
| 2 | .623 | .570 | .518 | .469 | .423 | .380 | .340 | .303 | .269 | .238 | .210 |
| 3 | .819 | .779 | .736 | .692 | .647 | .603 | .558 | .515 | .473 | .433 | .395 |
| 4 | .928 | .904 | .877 | .848 | .815 | .781 | .744 | .706 | .668 | .629 | .590 |
| 5 | .975 | .964 | .951 | .935 | .916 | .895 | .871 | .844 | .816 | .785 | .753 |
| 6 | .993 | .988 | .983 | .976 | .966 | .955 | .942 | .927 | .909 | .889 | .867 |
| 7 | .998 | .997 | .995 | .992 | .988 | .983 | .977 | .969 | .960 | .949 | .936 |
| 8 | 1.000 | .999 | .999 | .998 | .996 | .994 | .992 | .988 | .984 | .979 | .972 |
| 9 | | 1.000 | 1.000 | .999 | .999 | .998 | .997 | .996 | .994 | .992 | .989 |
| 10 | | | | 1.000 | 1.000 | 1.000 | .999 | .999 | .998 | .997 | .996 |
| 11 | | | | | | | 1.000 | 1.000 | .999 | .999 | .999 |
| 12 | | | | | | | | | 1.000 | 1.000 | 1.000 |

**Table IV** (*continued*)

| | | | | | | $m$ | | | | | |
|---|---|---|---|---|---|---|---|---|---|---|---|
| $c$ | 4.4 | 4.6 | 4.8 | 5.0 | 5.2 | 5.4 | 5.6 | 5.8 | 6.0 | 6.2 | 6.4 |
| 0 | .012 | .010 | .008 | .007 | .006 | .005 | .004 | .003 | .002 | .002 | .002 |
| 1 | .066 | .056 | .048 | .040 | .034 | .029 | .024 | .021 | .017 | .015 | .012 |
| 2 | .185 | .163 | .143 | .125 | .109 | .095 | .082 | .072 | .062 | .054 | .046 |
| 3 | .359 | .326 | .294 | .265 | .238 | .213 | .191 | .170 | .151 | .134 | .119 |
| 4 | .551 | .513 | .476 | .440 | .406 | .373 | .342 | .313 | .285 | .259 | .235 |
| 5 | .720 | .686 | .651 | .616 | .581 | .546 | .512 | .478 | .446 | .414 | .384 |
| 6 | .844 | .818 | .791 | .762 | .732 | .702 | .670 | .638 | .606 | .574 | .542 |
| 7 | .921 | .905 | .887 | .867 | .845 | .822 | .797 | .771 | .744 | .716 | .687 |
| 8 | .964 | .955 | .944 | .932 | .918 | .903 | .886 | .867 | .847 | .826 | .803 |
| 9 | .985 | .980 | .975 | .968 | .960 | .951 | .941 | .929 | .916 | .902 | .886 |
| 10 | .994 | .992 | .990 | .986 | .982 | .977 | .972 | .965 | .957 | .949 | .939 |
| 11 | .998 | .997 | .996 | .995 | .993 | .990 | .988 | .984 | .980 | .975 | .969 |
| 12 | .999 | .999 | .999 | .998 | .997 | .996 | .995 | .993 | .991 | .989 | .986 |
| 13 | 1.000 | 1.000 | 1.000 | .999 | .999 | .999 | .998 | .997 | .996 | .995 | .994 |
| 14 | | | | 1.000 | 1.000 | 1.000 | .999 | .999 | .999 | .998 | .997 |
| 15 | | | | | | | 1.000 | 1.000 | .999 | .999 | .999 |
| 16 | | | | | | | | | 1.000 | 1.000 | 1.000 |

| | | | | | | $m$ | | | | | |
|---|---|---|---|---|---|---|---|---|---|---|---|
| $c$ | 6.6 | 6.8 | 7.0 | 7.2 | 7.4 | 7.6 | 7.8 | 8.0 | 8.5 | 9.0 | 9.5 |
| 0 | .001 | .001 | .001 | .001 | .001 | .001 | .000 | .000 | .000 | .000 | .000 |
| 1 | .010 | .009 | .007 | .006 | .005 | .004 | .004 | .003 | .002 | .001 | .001 |
| 2 | .040 | .034 | .030 | .025 | .022 | .019 | .016 | .014 | .009 | .006 | .004 |
| 3 | .105 | .093 | .082 | .072 | .063 | .055 | .048 | .042 | .030 | .021 | .015 |
| 4 | .213 | .192 | .173 | .156 | .140 | .125 | .112 | .100 | .074 | .055 | .040 |
| 5 | .355 | .327 | .301 | .276 | .253 | .231 | .210 | .191 | .150 | .116 | .089 |
| 6 | .511 | .480 | .450 | .420 | .392 | .365 | .338 | .313 | .256 | .207 | .165 |
| 7 | .658 | .628 | .599 | .569 | .539 | .510 | .481 | .453 | .386 | .324 | .269 |
| 8 | .780 | .755 | .729 | .703 | .676 | .648 | .620 | .593 | .523 | .456 | .392 |
| 9 | .869 | .850 | .830 | .810 | .788 | .765 | .741 | .717 | .653 | .587 | .522 |
| 10 | .927 | .915 | .901 | .887 | .871 | .854 | .835 | .816 | .763 | .706 | .645 |
| 11 | .963 | .955 | .947 | .937 | .926 | .915 | .902 | .888 | .849 | .803 | .752 |
| 12 | .982 | .978 | .973 | .967 | .961 | .954 | .945 | .936 | .909 | .876 | .836 |
| 13 | .992 | .990 | .987 | .984 | .980 | .976 | .971 | .966 | .949 | .926 | .898 |
| 14 | .997 | .996 | .994 | .993 | .991 | .989 | .986 | .983 | .973 | .959 | .940 |
| 15 | .999 | .998 | .998 | .997 | .996 | .995 | .993 | .992 | .986 | .978 | .967 |
| 16 | .999 | .999 | .999 | .999 | .998 | .998 | .997 | .996 | .993 | .989 | .982 |
| 17 | 1.000 | 1.000 | 1.000 | .999 | .999 | .999 | .999 | .998 | .997 | .995 | .991 |
| 18 | | | | 1.000 | 1.000 | 1.000 | 1.000 | .999 | .999 | .998 | .996 |
| 19 | | | | | | | | 1.000 | .999 | .999 | .998 |
| 20 | | | | | | | | | 1.000 | 1.000 | .999 |
| 21 | | | | | | | | | | | 1.000 |

**Table IV** Cumulative Poisson Probabilities (*continued*)

| c | 10.0 | 10.5 | 11.0 | 11.5 | 12.0 | m 12.5 | 13.0 | 13.5 | 14.0 | 14.5 | 15.0 |
|---|------|------|------|------|------|------|------|------|------|------|------|
| 2 | .003 | .002 | .001 | .001 | .001 | .000 | | | | | |
| 3 | .010 | .007 | .005 | .003 | .002 | .002 | .001 | .001 | .000 | | |
| 4 | .029 | .021 | .015 | .011 | .008 | .005 | .004 | .003 | .002 | .001 | .001 |
| 5 | .067 | .050 | .038 | .028 | .020 | .015 | .011 | .008 | .006 | .004 | .003 |
| 6 | .130 | .102 | .079 | .060 | .046 | .035 | .026 | .019 | .014 | .010 | .008 |
| 7 | .220 | .179 | .143 | .114 | .090 | .070 | .054 | .041 | .032 | .024 | .018 |
| 8 | .333 | .279 | .232 | .191 | .155 | .125 | .100 | .079 | .062 | .048 | .037 |
| 9 | .458 | .397 | .341 | .289 | .242 | .201 | .166 | .135 | .109 | .088 | .070 |
| 10 | .583 | .521 | .460 | .402 | .347 | .297 | .252 | .211 | .176 | .145 | .118 |
| 11 | .697 | .629 | .579 | .520 | .462 | .406 | .353 | .304 | .260 | .220 | .185 |
| 12 | .792 | .742 | .689 | .633 | .576 | .519 | .463 | .409 | .358 | .311 | .268 |
| 13 | .864 | .825 | .781 | .733 | .682 | .628 | .573 | .518 | .464 | .413 | .363 |
| 14 | .917 | .888 | .854 | .815 | .772 | .725 | .675 | .623 | .570 | .518 | .466 |
| 15 | .951 | .932 | .907 | .878 | .844 | .806 | .764 | .718 | .669 | .619 | .568 |
| 16 | .973 | .960 | .944 | .924 | .899 | .869 | .835 | .798 | .756 | .711 | .664 |
| 17 | .986 | .978 | .968 | .954 | .937 | .916 | .890 | .861 | .827 | .790 | .749 |
| 18 | .993 | .988 | .982 | .974 | .963 | .948 | .930 | .908 | .883 | .853 | .819 |
| 19 | .997 | .994 | .991 | .986 | .979 | .969 | .957 | .942 | .923 | .901 | .875 |
| 20 | .998 | .997 | .995 | .992 | .988 | .983 | .975 | .965 | .952 | .936 | .917 |
| 21 | .999 | .999 | .998 | .996 | .994 | .991 | .986 | .980 | .971 | .960 | .947 |
| 22 | 1.000 | .999 | .999 | .998 | .997 | .995 | .992 | .989 | .983 | .976 | .967 |
| 23 | | 1.000 | 1.000 | .999 | .999 | .998 | .996 | .994 | .991 | .986 | .981 |
| 24 | | | | 1.000 | .999 | .999 | .998 | .997 | .995 | .992 | .989 |
| 25 | | | | | 1.000 | .999 | .999 | .998 | .997 | .996 | .994 |
| 26 | | | | | | 1.000 | 1.000 | .999 | .999 | .998 | .997 |
| 27 | | | | | | | | 1.000 | .999 | .999 | .998 |
| 28 | | | | | | | | | 1.000 | .999 | .999 |
| 29 | | | | | | | | | | 1.000 | 1.000 |

**Table Va** Percentiles of the Chi-Square Distribution

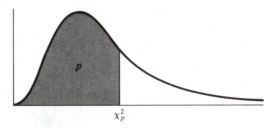

$\chi_p^2$

| Degrees of freedom | *p* | | | | | | | | | | | | |
|---|---|---|---|---|---|---|---|---|---|---|---|---|---|
| | .01 | .025 | .05 | .10 | .20 | .30 | .50 | .70 | .80 | .90 | .95 | .975 | .99 |
| 1 | .000 | .001 | .004 | .016 | .064 | .148 | .455 | 1.07 | 1.64 | 2.71 | 3.84 | 5.02 | 6.63 |
| 2 | .020 | .051 | .103 | .211 | .446 | .713 | 1.39 | 2.41 | 3.22 | 4.61 | 5.99 | 7.38 | 9.21 |
| 3 | .115 | .216 | .352 | .584 | 1.01 | 1.42 | 2.37 | 3.66 | 4.64 | 6.25 | 7.81 | 9.35 | 11.3 |
| 4 | .297 | .484 | .711 | 1.06 | 1.65 | 2.19 | 3.36 | 4.88 | 5.99 | 7.78 | 9.49 | 11.1 | 13.3 |
| 5 | .554 | .831 | 1.15 | 1.61 | 2.34 | 3.00 | 4.35 | 6.06 | 7.29 | 9.24 | 11.1 | 12.8 | 15.1 |
| 6 | .872 | 1.24 | 1.64 | 2.20 | 3.07 | 3.83 | 5.35 | 7.23 | 8.56 | 10.6 | 12.6 | 14.4 | 16.8 |
| 7 | 1.24 | 1.69 | 2.17 | 2.83 | 3.82 | 4.67 | 6.35 | 8.38 | 9.80 | 12.0 | 14.1 | 16.0 | 18.5 |
| 8 | 1.65 | 2.18 | 2.73 | 3.49 | 4.59 | 5.53 | 7.34 | 9.52 | 11.0 | 13.4 | 15.5 | 17.5 | 20.1 |
| 9 | 2.09 | 2.70 | 3.33 | 4.17 | 5.38 | 6.39 | 8.34 | 10.7 | 12.2 | 14.7 | 16.9 | 19.0 | 21.7 |
| 10 | 2.56 | 3.25 | 3.94 | 4.87 | 6.18 | 7.27 | 9.34 | 11.8 | 13.4 | 16.0 | 18.3 | 20.5 | 23.2 |
| 11 | 3.05 | 3.82 | 4.57 | 5.58 | 6.99 | 8.15 | 10.3 | 12.9 | 14.6 | 17.3 | 19.7 | 21.9 | 24.7 |
| 12 | 3.57 | 4.40 | 5.23 | 6.30 | 7.81 | 9.03 | 11.3 | 14.0 | 15.8 | 18.5 | 21.0 | 23.3 | 26.2 |
| 13 | 4.11 | 5.01 | 5.89 | 7.04 | 8.63 | 9.93 | 12.3 | 15.1 | 17.0 | 19.8 | 22.4 | 24.7 | 27.7 |
| 14 | 4.66 | 5.63 | 6.57 | 7.79 | 9.47 | 10.8 | 13.3 | 16.2 | 18.2 | 21.1 | 23.7 | 26.1 | 29.1 |
| 15 | 5.23 | 6.26 | 7.26 | 8.55 | 10.3 | 11.7 | 14.3 | 17.3 | 19.3 | 22.3 | 25.0 | 27.5 | 30.6 |
| 16 | 5.81 | 6.91 | 7.96 | 9.31 | 11.2 | 12.6 | 15.3 | 18.4 | 20.5 | 23.5 | 26.3 | 28.8 | 32.0 |
| 17 | 6.41 | 7.56 | 8.67 | 10.1 | 12.0 | 13.5 | 16.3 | 19.5 | 21.6 | 24.8 | 27.6 | 30.2 | 33.4 |
| 18 | 7.01 | 8.23 | 9.39 | 10.9 | 12.9 | 14.4 | 17.3 | 20.6 | 22.8 | 26.0 | 28.9 | 31.5 | 34.8 |
| 19 | 7.63 | 8.91 | 10.1 | 11.7 | 13.7 | 15.4 | 18.3 | 21.7 | 23.9 | 27.2 | 30.1 | 32.9 | 36.2 |
| 20 | 8.26 | 9.59 | 10.9 | 12.4 | 14.6 | 16.3 | 19.3 | 22.8 | 25.0 | 28.4 | 31.4 | 34.2 | 37.6 |
| 21 | 8.90 | 10.3 | 11.6 | 13.2 | 15.4 | 17.2 | 20.3 | 23.9 | 26.2 | 29.6 | 32.7 | 35.5 | 38.9 |
| 22 | 9.54 | 11.0 | 12.3 | 14.0 | 16.3 | 18.1 | 21.3 | 24.9 | 27.3 | 30.8 | 33.9 | 36.8 | 40.3 |
| 23 | 10.2 | 11.7 | 13.1 | 14.8 | 17.2 | 19.0 | 22.3 | 26.0 | 28.4 | 32.0 | 35.2 | 38.1 | 41.6 |
| 24 | 10.9 | 12.4 | 13.8 | 15.7 | 18.1 | 19.9 | 23.3 | 27.1 | 29.6 | 33.2 | 36.4 | 39.4 | 43.0 |
| 25 | 11.5 | 13.1 | 14.6 | 16.5 | 18.9 | 20.9 | 24.3 | 28.2 | 30.7 | 34.4 | 37.7 | 40.6 | 44.3 |
| 26 | 12.2 | 13.8 | 15.4 | 17.3 | 19.8 | 21.8 | 25.3 | 29.2 | 31.8 | 35.6 | 38.9 | 41.9 | 45.6 |
| 27 | 12.9 | 14.6 | 16.2 | 18.1 | 20.7 | 22.7 | 26.3 | 30.3 | 32.9 | 36.7 | 40.1 | 43.2 | 47.0 |
| 28 | 13.6 | 15.3 | 16.9 | 18.9 | 21.6 | 23.6 | 27.3 | 31.4 | 34.0 | 37.9 | 41.3 | 44.5 | 48.3 |
| 29 | 14.3 | 16.0 | 17.7 | 19.8 | 22.5 | 24.6 | 28.3 | 32.5 | 35.1 | 39.1 | 42.6 | 45.7 | 49.6 |
| 30 | 15.0 | 16.8 | 18.5 | 20.6 | 23.4 | 25.5 | 29.3 | 33.5 | 36.2 | 40.3 | 43.8 | 47.0 | 50.9 |
| 40 | 22.1 | 24.4 | 26.5 | 29.0 | 32.4 | 35.0 | 39.3 | 44.2 | 47.3 | 51.8 | 55.8 | 59.3 | 63.7 |
| 50 | 29.7 | 32.3 | 34.8 | 37.7 | 41.5 | 44.4 | 49.3 | 54.7 | 58.2 | 63.2 | 67.5 | 71.4 | 76.2 |
| 60 | 37.5 | 40.5 | 43.2 | 46.5 | 50.7 | 53.9 | 59.3 | 65.2 | 69.0 | 74.4 | 79.1 | 83.3 | 88.4 |

Note: For degrees of freedom $k > 30$, use $\chi_p^2 = \frac{1}{2}(z_p + \sqrt{2k-1})^2$, where $z_p$ is the corresponding percentile of the standard normal distribution.

**Table Vb** Tail-Areas of the Chi-Square Distribution

| | d.f. | | | | d.f. | | | | d.f. | | |
|---|---|---|---|---|---|---|---|---|---|---|---|
| $\chi^2$ | 1 | 2 | 3 | $\chi^2$ | 4 | 5 | 6 | $\chi^2$ | 7 | 8 | 9 |
| 3.0 | .083 | .223 | .392 | 8.0 | .092 | .156 | .238 | 13.0 | .072 | .112 | .163 |
| 3.2 | .074 | .202 | .362 | 8.2 | .085 | .146 | .224 | 13.2 | .067 | .105 | .154 |
| 3.4 | .065 | .183 | .334 | 8.4 | .078 | .136 | .210 | 13.4 | .063 | .099 | .145 |
| 3.6 | .058 | .165 | .308 | 8.6 | .072 | .126 | .197 | 13.6 | .059 | .093 | .137 |
| 3.8 | .051 | .150 | .284 | 8.8 | .066 | .117 | .185 | 13.8 | .055 | .087 | .130 |
| 4.0 | .045 | .135 | .261 | 9.0 | .061 | .109 | .174 | 14.0 | .051 | .082 | .122 |
| 4.2 | .040 | .122 | .241 | 9.2 | .056 | .101 | .163 | 14.2 | .048 | .077 | .115 |
| 4.4 | .036 | .111 | .221 | 9.4 | .052 | .094 | .152 | 14.4 | .045 | .072 | .109 |
| 4.6 | .032 | .105 | .204 | 9.6 | .048 | .087 | .143 | 14.6 | .041 | .067 | .103 |
| 4.8 | .028 | .091 | .187 | 9.8 | .044 | .081 | .133 | 14.8 | .039 | .063 | .097 |
| 5.0 | .025 | .082 | .172 | 10.0 | .040 | .075 | .125 | 15.0 | .036 | .059 | .091 |
| 5.2 | .023 | .074 | .158 | 10.2 | .037 | .070 | .116 | 15.2 | .034 | .055 | .086 |
| 5.4 | .020 | .067 | .145 | 10.4 | .034 | .065 | .109 | 15.4 | .031 | .052 | .081 |
| 5.6 | .018 | .061 | .133 | 10.6 | .031 | .060 | .102 | 15.6 | .029 | .048 | .076 |
| 5.8 | .016 | .055 | .122 | 10.8 | .029 | .055 | .095 | 15.8 | .027 | .045 | .071 |
| 6.0 | .014 | .050 | .111 | 11.0 | .027 | .051 | .088 | 16.0 | .025 | .042 | .067 |
| 6.2 | .013 | .045 | .102 | 11.2 | .024 | .048 | .082 | 16.2 | .023 | .040 | .063 |
| 6.4 | .011 | .041 | .094 | 11.4 | .022 | .044 | .077 | 16.4 | .022 | .037 | .059 |
| 6.6 | .010 | .037 | .086 | 11.6 | .021 | .041 | .072 | 16.6 | .020 | .035 | .055 |
| 6.8 | .009 | .033 | .079 | 11.8 | .019 | .038 | .067 | 16.8 | .019 | .032 | .052 |
| 7.0 | .008 | .030 | .072 | 12.0 | .017 | .035 | .062 | 17.0 | .017 | .030 | .049 |
| 7.2 | .007 | .027 | .066 | 12.2 | .016 | .032 | .058 | 17.2 | .016 | .028 | .046 |
| 7.4 | .007 | .025 | .060 | 12.4 | .015 | .030 | .054 | 17.4 | .015 | .026 | .043 |
| 7.6 | .006 | .022 | .055 | 12.6 | .013 | .027 | .050 | 17.6 | .014 | .024 | .040 |
| 7.8 | .005 | .020 | .050 | 12.8 | .012 | .025 | .046 | 17.8 | .013 | .023 | .038 |
| 8.0 | .005 | .018 | .046 | 13.0 | .011 | .023 | .043 | 18.0 | .012 | .021 | .035 |
| 8.2 | .004 | .017 | .042 | 13.2 | .010 | .022 | .040 | 18.2 | .011 | .020 | .033 |
| 8.4 | .004 | .015 | .038 | 13.4 | .009 | .020 | .037 | 18.4 | .010 | .018 | .031 |
| 8.6 | .003 | .014 | .035 | 13.6 | .009 | .018 | .034 | 18.6 | .010 | .017 | .029 |
| 8.8 | .003 | .012 | .032 | 13.8 | .008 | .017 | .032 | 18.8 | .009 | .016 | .027 |
| 9.0 | .003 | .011 | .029 | 14.0 | .007 | .016 | .030 | 19.0 | .008 | .015 | .025 |
| 9.2 | .002 | .010 | .027 | 14.2 | .007 | .014 | .027 | 19.2 | .008 | .014 | .024 |
| 9.4 | .002 | .009 | .024 | 14.4 | .006 | .013 | .025 | 19.4 | .007 | .013 | .022 |
| 9.6 | .002 | .008 | .022 | 14.6 | .006 | .012 | .024 | 19.6 | .006 | .012 | .021 |
| 9.8 | .002 | .007 | .020 | 14.8 | .005 | .011 | .022 | 19.8 | .006 | .011 | .019 |
| 10.0 | .001 | .007 | .019 | 15.0 | .005 | .010 | .020 | 20.0 | .006 | .010 | .018 |
| 10.2 | .001 | .006 | .017 | 15.2 | .004 | .010 | .019 | 20.2 | .005 | .010 | .017 |
| 10.4 | .001 | .006 | .015 | 15.4 | .004 | .009 | .017 | 20.4 | .005 | .009 | .016 |
| 10.6 | .001 | .005 | .014 | 15.6 | .004 | .008 | .016 | 20.6 | .004 | .008 | .015 |
| 10.8 | .001 | .005 | .013 | 15.8 | .003 | .007 | .015 | 20.8 | .004 | .008 | .014 |
| 11.0 | .001 | .004 | .012 | 16.0 | .003 | .007 | .014 | 21.0 | .004 | .007 | .013 |

**Table Vb** (*continued*)

| $\chi^2$ | d.f. 10 | 11 | 12 | $\chi^2$ | d.f. 13 | 14 | 15 | $\chi^2$ | d.f. 16 | 17 | 18 |
|---|---|---|---|---|---|---|---|---|---|---|---|
| 17.0 | .074 | .108 | .150 | 21.0 | .073 | .102 | .137 | 25.0 | .070 | .095 | .125 |
| 17.2 | .070 | .102 | .142 | 21.2 | .071 | .097 | .131 | 25.2 | .066 | .090 | .120 |
| 17.4 | .066 | .097 | .135 | 21.4 | .065 | .092 | .125 | 25.4 | .063 | .086 | .114 |
| 17.6 | .062 | .091 | .128 | 21.6 | .062 | .087 | .119 | 25.6 | .060 | .082 | .109 |
| 17.8 | .058 | .086 | .122 | 21.8 | .059 | .083 | .113 | 25.9 | .057 | .078 | .104 |
| 18.0 | .055 | .082 | .116 | 22.0 | .055 | .079 | .108 | 26.0 | .054 | .074 | .100 |
| 18.2 | .052 | .077 | .110 | 22.2 | .052 | .075 | .103 | 26.2 | .051 | .071 | .095 |
| 18.4 | .049 | .073 | .104 | 22.4 | .049 | .071 | .098 | 26.4 | .049 | .067 | .091 |
| 18.6 | .046 | .069 | .099 | 22.6 | .047 | .067 | .093 | 26.6 | .046 | .064 | .087 |
| 18.8 | .043 | .065 | .093 | 22.8 | .044 | .064 | .088 | 26.8 | .044 | .061 | .083 |
| 19.0 | .040 | .061 | .089 | 23.0 | .042 | .060 | .084 | 27.0 | .041 | .058 | .079 |
| 19.2 | .038 | .058 | .084 | 23.2 | .039 | .057 | .080 | 27.2 | .039 | .055 | .075 |
| 19.4 | .035 | .054 | .079 | 23.4 | .037 | .054 | .076 | 27.4 | .037 | .052 | .072 |
| 19.6 | .033 | .051 | .075 | 23.6 | .035 | .051 | .072 | 27.6 | .035 | .050 | .068 |
| 19.8 | .031 | .048 | .071 | 23.8 | .033 | .048 | .069 | 27.8 | .033 | .047 | .065 |
| 20.0 | .029 | .045 | .067 | 24.0 | .031 | .046 | .065 | 28.0 | .032 | .045 | .062 |
| 20.2 | .027 | .043 | .063 | 24.2 | .029 | .043 | .062 | 28.2 | .030 | .043 | .059 |
| 20.4 | .026 | .040 | .060 | 24.4 | .028 | .041 | .059 | 28.4 | .028 | .040 | .056 |
| 20.6 | .024 | .038 | .057 | 24.6 | .026 | .039 | .056 | 28.6 | .027 | .038 | .053 |
| 20.8 | .023 | .036 | .053 | 24.8 | .025 | .037 | .053 | 28.8 | .025 | .036 | .051 |
| 21.0 | .021 | .033 | .050 | 25.0 | .023 | .035 | .050 | 29.0 | .024 | .035 | .048 |
| 21.2 | .020 | .031 | .048 | 25.2 | .022 | .033 | .047 | 29.2 | .023 | .033 | .046 |
| 21.4 | .018 | .029 | .045 | 25.4 | .020 | .031 | .045 | 29.4 | .021 | .031 | .044 |
| 21.6 | .017 | .028 | .042 | 25.6 | .019 | .029 | .042 | 29.6 | .020 | .029 | .042 |
| 21.8 | .016 | .026 | .040 | 25.8 | .018 | .026 | .040 | 29.8 | .019 | .028 | .039 |
| 22.0 | .014 | .024 | .038 | 26.0 | .017 | .026 | .038 | 30.0 | .018 | .026 | .037 |
| 22.2 | .013 | .023 | .035 | 26.2 | .016 | .024 | .036 | 30.2 | .017 | .025 | .036 |
| 22.4 | .012 | .021 | .033 | 26.4 | .015 | .023 | .034 | 30.4 | .016 | .024 | .034 |
| 22.6 | .012 | .020 | .031 | 26.6 | .014 | .022 | .032 | 30.6 | .015 | .022 | .032 |
| 22.8 | .011 | .019 | .029 | 26.8 | .013 | .020 | .030 | 30.8 | .014 | .021 | .030 |
| 23.0 | .010 | .018 | .028 | 27.0 | .012 | .019 | .029 | 31.0 | .013 | .020 | .029 |
| 23.2 | .009 | .017 | .026 | 27.2 | .012 | .018 | .027 | 31.2 | .013 | .019 | .027 |
| 23.4 | .009 | .016 | .025 | 27.4 | .011 | .017 | .026 | 31.4 | .012 | .018 | .026 |
| 23.6 | .008 | .015 | .023 | 27.6 | .010 | .016 | .024 | 31.6 | .011 | .017 | .025 |
| 23.8 | .008 | .014 | .022 | 27.8 | .010 | .015 | .023 | 31.8 | .011 | .016 | .023 |
| 24.0 | .004 | .013 | .020 | 28.0 | .009 | .014 | .022 | 32.0 | .010 | .015 | .022 |
| 24.2 | .004 | .012 | .019 | 28.2 | .008 | .013 | .020 | 32.2 | .009 | .014 | .021 |
| 24.4 | .004 | .011 | .018 | 28.4 | .008 | .013 | .019 | 32.4 | .009 | .013 | .020 |
| 24.6 | .003 | .010 | .017 | 28.6 | .007 | .012 | .018 | 32.6 | .008 | .013 | .019 |
| 24.8 | .003 | .010 | .016 | 28.8 | .007 | .011 | .017 | 32.8 | .008 | .012 | .018 |
| 25.0 | .003 | .009 | .015 | 29.0 | .007 | .010 | .016 | 33.0 | .007 | .011 | .017 |

**Table VI** Tail-Probabilities of the One-Sample Wilcoxon Signed-Rank Statistic

| c \ n | 4 | 5 | 6 | 7 | 8 | 9 | 10 | 11 | 12 | 13 | 14 | 15 | n \ c |
|---|---|---|---|---|---|---|---|---|---|---|---|---|---|
| 0 | .062 | .031 | .016 | .008 | .004 | .002 | .001 | .000 | .000 | .000 | .000 | .000 | 0 |
| 1 | .125 | .062 | .031 | .016 | .008 | .004 | .002 | .001 | .000 | .000 | .000 | .000 | 1 |
| 2 | .188 | .094 | .047 | .023 | .012 | .006 | .003 | .001 | .001 | .000 | .000 | .000 | 2 |
| 3 | .312 | .156 | .078 | .039 | .020 | .010 | .005 | .002 | .001 | .001 | .000 | .000 | 3 |
| 4 |  | .219 | .109 | .055 | .027 | .014 | .007 | .003 | .002 | .001 | .000 | .000 | 4 |
| 5 |  |  | .156 | .078 | .039 | .020 | .010 | .005 | .002 | .001 | .001 | .000 | 5 |
| 6 |  |  | .219 | .109 | .055 | .027 | .014 | .007 | .003 | .002 | .001 | .000 | 6 |
| 7 |  |  |  | .148 | .074 | .037 | .019 | .009 | .005 | .002 | .001 | .001 | 7 |
| 8 |  |  |  | .188 | .098 | .049 | .024 | .012 | .006 | .003 | .002 | .001 | 8 |
| 9 |  |  |  | .234 | .125 | .064 | .032 | .016 | .008 | .004 | .002 | .001 | 9 |
| 10 |  |  |  |  | .156 | .082 | .042 | .021 | .010 | .005 | .003 | .001 | 10 |
| 11 |  |  |  |  | .191 | .102 | .053 | .027 | .013 | .007 | .003 | .002 | 11 |
| 12 |  |  |  |  | .230 | .125 | .065 | .034 | .017 | .009 | .004 | .002 | 12 |
| 13 |  |  |  |  |  | .150 | .080 | .042 | .021 | .011 | .005 | .003 | 13 |
| 14 |  |  |  |  |  | .180 | .097 | .051 | .026 | .013 | .007 | .003 | 14 |
| 15 |  |  |  |  |  | .213 | .116 | .062 | .032 | .016 | .008 | .004 | 15 |
| 16 |  |  |  |  |  |  | .138 | .074 | .039 | .020 | .010 | .005 | 16 |
| 17 |  |  |  |  |  |  | .161 | .087 | .046 | .024 | .012 | .006 | 17 |
| 18 |  |  |  |  |  |  | .188 | .103 | .055 | .029 | .015 | .008 | 18 |
| 19 |  |  |  |  |  |  | .216 | .120 | .065 | .034 | .018 | .009 | 19 |
| 20 |  |  |  |  |  |  |  | .139 | .076 | .040 | .021 | .011 | 20 |
| 21 |  |  |  |  |  |  |  | .160 | .088 | .047 | .025 | .013 | 21 |
| 22 |  |  |  |  |  |  |  | .183 | .102 | .055 | .029 | .015 | 22 |
| 23 |  |  |  |  |  |  |  | .207 | .117 | .064 | .034 | .018 | 23 |
| 24 |  |  |  |  |  |  |  |  | .113 | .073 | .039 | .021 | 24 |
| 25 |  |  |  |  |  |  |  |  | .151 | .084 | .045 | .024 | 25 |
| 26 |  |  |  |  |  |  |  |  | .170 | .095 | .052 | .028 | 26 |
| 27 |  |  |  |  |  |  |  |  | .190 | .108 | .059 | .032 | 27 |
| 28 |  |  |  |  |  |  |  |  | .212 | .122 | .068 | .036 | 28 |
| 29 |  |  |  |  |  |  |  |  |  | .137 | .077 | .042 | 29 |
| 30 |  |  |  |  |  |  |  |  |  | .153 | .086 | .047 | 30 |
| 31 |  |  |  |  |  |  |  |  |  | .170 | .097 | .053 | 31 |
| 32 |  |  |  |  |  |  |  |  |  | .188 | .108 | .060 | 32 |
| 33 |  |  |  |  |  |  |  |  |  | .207 | .121 | .068 | 33 |
| 34 |  |  |  |  |  |  |  |  |  |  | .134 | .076 | 34 |
| 35 |  |  |  |  |  |  |  |  |  |  | .148 | .084 | 35 |
| 36 |  |  |  |  |  |  |  |  |  |  | .163 | .094 | 36 |
| 37 |  |  |  |  |  |  |  |  |  |  | .179 | .104 | 37 |
| 38 |  |  |  |  |  |  |  |  |  |  | .196 | .115 | 38 |
| $\dfrac{n(n+1)}{2}$ | 10 | 15 | 21 | 28 | 36 | 45 | 55 | 66 | 78 | 91 | 105 | 120 | |

Notes: 1. Table entries are $P(R_+ \le c) = P[R_+ \ge \frac{1}{2}n(n+1) - c]$.
2. $R_- = \frac{1}{2}n(n+1) - R_+$.
3. For $n > 15$, use a normal approximation (see Section 10.5).

**Table VII** Tail-Probabilities of the Two-Sample Wilcoxon Rank-Sum Statistic

| m / n / c | 4 / 4 | 4 / 5 | 4 / 6 | 4 / 7 | 4 / 8 | 4 / 9 | 4 / 10 |
|---|---|---|---|---|---|---|---|
| 10 | .014 | .008 | .005 | .003 | .002 | .001 | .001 |
| 11 | .029 | .016 | .010 | .006 | .004 | .003 | .002 |
| 12 | .057 | .032 | .019 | .012 | .008 | .006 | .004 |
| 13 | .100 | .056 | .033 | .021 | .014 | .010 | .007 |
| 14 | .171 | .096 | .057 | .036 | .024 | .017 | .012 |
| 15 | .243 | .143 | .086 | .055 | .036 | .025 | .018 |
| 16 | | .206 | .129 | .082 | .055 | .038 | .027 |
| 17 | | | .176 | .115 | .077 | .053 | .038 |
| 18 | | | | .158 | .107 | .074 | .053 |
| 19 | | | | .206 | .141 | .099 | .071 |
| 20 | | | | | .184 | .130 | .094 |
| 21 | | | | | | .165 | .120 |
| M | 36 | 40 | 44 | 48 | 52 | 56 | 60 |

| m / n / c | 5 / 5 | 5 / 6 | 5 / 7 | 5 / 8 | 5 / 9 | 5 / 10 |
|---|---|---|---|---|---|---|
| 15 | .004 | .002 | .001 | .001 | .000 | .000 |
| 16 | .008 | .004 | .003 | .002 | .001 | .001 |
| 17 | .016 | .009 | .005 | .003 | .002 | .001 |
| 18 | .028 | .015 | .009 | .005 | .003 | .002 |
| 19 | .048 | .026 | .015 | .009 | .006 | .004 |
| 20 | .075 | .041 | .024 | .015 | .009 | .006 |
| 21 | .111 | .063 | .037 | .023 | .014 | .010 |
| 22 | .155 | .089 | .053 | .033 | .021 | .014 |
| 23 | .210 | .123 | .074 | .047 | .030 | .020 |
| 24 | | .165 | .101 | .064 | .041 | .028 |
| 25 | | .214 | .134 | .085 | .056 | .038 |
| 26 | | | .172 | .111 | .073 | .050 |
| 27 | | | .216 | .142 | .095 | .065 |
| 28 | | | | .177 | .120 | .082 |
| 28 | | | | | .149 | .103 |
| M | 55 | 60 | 65 | 70 | 75 | 80 |

Notes: 1. *m* is the size of the smaller sample.
2. The entry opposite *c* is the cumulative tail probability:

$$P(R \le c) = P(R \ge M - c),$$

where $R$ is the rank sum for the smaller sample, and $M$ is the sum of the minimum and maximum values of $R$. [Note that $\frac{M}{2} = $ m.v.$(R)$.]

3. For $m$ or $n > 10$, use a normal approximation (see Section 12.5).

**Table VII** Tail-Probabilities of the Two-Sample Wilcoxon Rank-Sum Statistic (*continued*)

| c \ m,n | 6,6 | 6,7 | 6,8 | 6,9 | 6,10 |
|---|---|---|---|---|---|
| 24 | .008 | .004 | .002 | .001 | .001 |
| 25 | .013 | .007 | .004 | .002 | .001 |
| 26 | .021 | .011 | .006 | .004 | .002 |
| 27 | .032 | .017 | .010 | .006 | .004 |
| 28 | .047 | .026 | .015 | .009 | .005 |
| 29 | .066 | .037 | .021 | .013 | .008 |
| 30 | .090 | .051 | .030 | .018 | .011 |
| 31 | .120 | .069 | .041 | .025 | .016 |
| 32 | .155 | .090 | .054 | .033 | .021 |
| 33 | .197 | .117 | .071 | .044 | .028 |
| 34 | | .147 | .091 | .057 | .036 |
| 35 | | .183 | .114 | .072 | .047 |
| 36 | | | .141 | .091 | .059 |
| 37 | | | .172 | .112 | .074 |
| 38 | | | | 136 | .090 |
| 39 | | | | .164 | .110 |
| M | 78 | 84 | 90 | 96 | 102 |

| c \ m,n | 7,7 | 7,8 | 7,9 | 7,10 |
|---|---|---|---|---|
| 34 | .009 | .005 | .003 | .002 |
| 35 | .013 | .007 | .004 | .002 |
| 36 | .019 | .010 | .006 | .003 |
| 37 | .027 | .014 | .008 | .005 |
| 38 | .036 | .020 | .011 | .007 |
| 39 | .049 | .027 | .016 | .009 |
| 40 | .064 | .036 | .001 | .012 |
| 41 | .082 | .047 | .027 | .017 |
| 42 | .104 | .060 | .036 | .022 |
| 43 | .130 | .076 | .045 | .028 |
| 44 | .159 | .095 | .057 | .035 |
| 45 | .191 | .116 | .071 | .044 |
| 46 | | .140 | .087 | .054 |
| 47 | | .168 | .105 | .067 |
| 48 | | .198 | .126 | .081 |
| 49 | | | .150 | .097 |
| 50 | | | .176 | .115 |
| M | 105 | 112 | 119 | 126 |

| c \ m,n | 8,8 | 8,9 | 8,10 |
|---|---|---|---|
| 46 | .010 | .006 | .003 |
| 47 | .014 | .008 | .004 |
| 48 | .019 | .010 | .006 |
| 49 | .025 | .014 | .008 |
| 50 | .032 | .018 | .010 |
| 51 | .041 | .023 | .013 |
| 52 | .052 | .030 | .017 |
| 53 | .065 | .037 | .022 |
| 54 | .080 | .046 | .027 |
| 55 | .097 | .057 | .034 |
| 56 | .117 | .069 | .042 |
| 57 | .139 | .084 | .051 |
| 58 | .164 | .100 | .061 |
| 59 | .191 | .118 | .073 |
| 60 | | .138 | .086 |
| 61 | | .161 | .102 |
| M | 136 | 144 | 152 |

| c \ m,n | 9,9 | 9,10 |
|---|---|---|
| 59 | .009 | .005 |
| 60 | .012 | .007 |
| 61 | .016 | .009 |
| 62 | .020 | .011 |
| 63 | .025 | .014 |
| 64 | .031 | .017 |
| 65 | .039 | .022 |
| 66 | .047 | .027 |
| 67 | .057 | .033 |
| 68 | .068 | .039 |
| 69 | .081 | .047 |
| 70 | .095 | .056 |
| 71 | .111 | .067 |
| 72 | .129 | .078 |
| 73 | .149 | .091 |
| 74 | .170 | .106 |
| M | 171 | 180 |

| c \ m,n | 10,10 |
|---|---|
| 74 | .009 |
| 75 | .012 |
| 76 | .014 |
| 77 | .018 |
| 78 | .022 |
| 79 | .026 |
| 80 | .032 |
| 81 | .038 |
| 82 | .045 |
| 83 | .053 |
| 84 | .062 |
| 85 | .072 |
| 86 | .083 |
| 87 | .095 |
| 88 | .109 |
| 89 | .124 |
| M | 210 |

**Table VIIIa** Tail-Probabilities of the $F$-Distribution

2 d.f. in numerator

| $F$ | 5 | 6 | 7 | 8 | 9 | 10 | 11 | 12 | 13 | 14 | 15 | 20 | 30 | 40 |
|------|------|------|------|------|------|------|------|------|------|------|------|------|------|------|
| | | | | | | *Denominator degrees of freedom* | | | | | | | | |
| 2.0 | .230 | .216 | .206 | .198 | .191 | .186 | .182 | .178 | .175 | .172 | .170 | .162 | .153 | .149 |
| 2.2 | .206 | .192 | .181 | .173 | .167 | .162 | .157 | .153 | .150 | .148 | .145 | .137 | .128 | .124 |
| 2.4 | .186 | .171 | .161 | .153 | .146 | .141 | .136 | .133 | .130 | .127 | .125 | .116 | .108 | .104 |
| 2.6 | .168 | .154 | .143 | .135 | .128 | .123 | .119 | .115 | .112 | .110 | .107 | .099 | .091 | .087 |
| 2.8 | .153 | .138 | .128 | .120 | .113 | .108 | .104 | .100 | .099 | .095 | .093 | .085 | .077 | .073 |
| 3.0 | .139 | .125 | .115 | .107 | .100 | .095 | .091 | .088 | .085 | .082 | .080 | .073 | .065 | .061 |
| 3.2 | .127 | .113 | .103 | .095 | .089 | .084 | .080 | .077 | .074 | .072 | .070 | .062 | .055 | .051 |
| 3.4 | .117 | .103 | .093 | .085 | .079 | .075 | .071 | .068 | .065 | .063 | .061 | .054 | .047 | .043 |
| 3.6 | .108 | .094 | .084 | .077 | .071 | .066 | .063 | .060 | .057 | .055 | .053 | .046 | .040 | .037 |
| 3.8 | .099 | .086 | .076 | .069 | .064 | .059 | .056 | .053 | .050 | .048 | .046 | .040 | .034 | .031 |
| 4.0 | .092 | .079 | .069 | .063 | .057 | .053 | .049 | .047 | .044 | .042 | .041 | .035 | .029 | .026 |
| 4.2 | .085 | .072 | .063 | .057 | .051 | .047 | .044 | .041 | .039 | .037 | .036 | .030 | .025 | .022 |
| 4.4 | .079 | .067 | .058 | .051 | .046 | .043 | .039 | .037 | .035 | .033 | .031 | .026 | .021 | .019 |
| 4.6 | .074 | .062 | .053 | .047 | .042 | .038 | .035 | .033 | .031 | .029 | .028 | .023 | .018 | .016 |
| 4.8 | .069 | .057 | .049 | .043 | .038 | .035 | .032 | .029 | .027 | .026 | .024 | .020 | .016 | .014 |
| 5.0 | .064 | .053 | .045 | .039 | .035 | .031 | .029 | .026 | .025 | .023 | .022 | .017 | .013 | .012 |
| 5.2 | .060 | .049 | .041 | .036 | .032 | .028 | .026 | .024 | .022 | .020 | .019 | .015 | .012 | .010 |
| 5.4 | .056 | .046 | .038 | .033 | .029 | .026 | .023 | .021 | .020 | .018 | .017 | .013 | .010 | .008 |
| 5.6 | .053 | .042 | .035 | .030 | .026 | .023 | .021 | .019 | .018 | .016 | .015 | .012 | .009 | .007 |
| 5.8 | .050 | .040 | .033 | .028 | .024 | .021 | .019 | .017 | .016 | .014 | .013 | .012 | .009 | .006 |
| 6.0 | .047 | .037 | .030 | .026 | .022 | .019 | .017 | .016 | .014 | .013 | .012 | .009 | .006 | .005 |
| 6.2 | .044 | .035 | .028 | .024 | .020 | .018 | .016 | .014 | .013 | .012 | .011 | .008 | .006 | .005 |
| 6.4 | .042 | .033 | .026 | .022 | .019 | .016 | .014 | .013 | .012 | .011 | .010 | .007 | .005 | .004 |
| 6.6 | .040 | .031 | .024 | .020 | .017 | .015 | .013 | .012 | .011 | .010 | .009 | .006 | .004 | .003 |
| 6.8 | .037 | .029 | .023 | .019 | .016 | .014 | .012 | .011 | .010 | .009 | .008 | .006 | .004 | .003 |
| 7.0 | .036 | .027 | .021 | .017 | .015 | .013 | .011 | .010 | .009 | .008 | .007 | .005 | .003 | .002 |
| 7.2 | .034 | .025 | .020 | .016 | .014 | .012 | .010 | .009 | .008 | .007 | .006 | .004 | .003 | .002 |
| 7.4 | .032 | .024 | .019 | .015 | .013 | .011 | .009 | .008 | .007 | .006 | .006 | .004 | .002 | .002 |
| 7.6 | .030 | .023 | .018 | .014 | .012 | .010 | .008 | .007 | .007 | .006 | .005 | .004 | .002 | .002 |
| 7.8 | .029 | .021 | .017 | .013 | .011 | .009 | .008 | .007 | .006 | .005 | .005 | .003 | .002 | .001 |
| 8.0 | .028 | .020 | .016 | .012 | .010 | .008 | .007 | .006 | .005 | .005 | .004 | .003 | .002 | .001 |
| 8.2 | .026 | .019 | .015 | .012 | .009 | .008 | .007 | .006 | .005 | .004 | .004 | .003 | .001 | .001 |
| 8.4 | .025 | .018 | .014 | .011 | .009 | .007 | .006 | .005 | .005 | .004 | .004 | .002 | .001 | .001 |
| 8.6 | .024 | .017 | .013 | .010 | .008 | .007 | .006 | .005 | .004 | .004 | .003 | .002 | .001 | .001 |
| 8.8 | .023 | .016 | .012 | .010 | .008 | .006 | .005 | .004 | .004 | .003 | .003 | .002 | .001 | .001 |
| 9.0 | .022 | .016 | .012 | .009 | .007 | .006 | .005 | .004 | .004 | .003 | .003 | .002 | .001 | .001 |
| 9.5 | .020 | .014 | .010 | .008 | .006 | .005 | .004 | .003 | .003 | .002 | .002 | .001 | .001 | .000 |
| 10.0 | .018 | .012 | .009 | .007 | .005 | .004 | .003 | .003 | .002 | .002 | .002 | .001 | .000 | .000 |

**Table VIIIa** Tail-Probabilities of the $F$-Distribution (*continued*)

2 d.f. in numerator

| $F$ | 5 | 6 | 7 | 8 | 9 | Denominator degrees of freedom 10 | 11 | 12 | 13 | 14 | 15 | 20 | 30 | 40 |
|---|---|---|---|---|---|---|---|---|---|---|---|---|---|---|
| 10.5 | .016 | .011 | .008 | .006 | .004 | .003 | .003 | .002 | .002 | .002 | .001 | .001 | .000 | .000 |
| 11.0 | .015 | .010 | .007 | .005 | .004 | .003 | .002 | .002 | .002 | .001 | .001 | .001 | .000 | .000 |
| 11.5 | .013 | .009 | .006 | .005 | .004 | .004 | .003 | .003 | .002 | .001 | .001 | .000 | .000 | .000 |
| 12.0 | .012 | .008 | .005 | .004 | .003 | .002 | .002 | .001 | .001 | .001 | .001 | .000 | .000 | .000 |
| 13.0 | .010 | .007 | .004 | .003 | .002 | .002 | .001 | .001 | .001 | .001 | .001 | .000 | .000 | .000 |
| 15.0 | .008 | .005 | .003 | .002 | .001 | .001 | .001 | .001 | .000 | .000 | .000 | .000 | .000 | .000 |
| 20.0 | .004 | .002 | .001 | .001 | .000 | .000 | .000 | .000 | .000 | .000 | .000 | .000 | .000 | .000 |

3 d.f. in numerator

| $F$ | 5 | 6 | 7 | 8 | 9 | 10 | 11 | 12 | 13 | 14 | 15 | 20 | 30 | 40 |
|---|---|---|---|---|---|---|---|---|---|---|---|---|---|---|
| 2.0 | .233 | .216 | .203 | .193 | .185 | .178 | .172 | .168 | .164 | .160 | .157 | .146 | .135 | .129 |
| 2.2 | .206 | .189 | .176 | .166 | .158 | .151 | .146 | .141 | .137 | .133 | .130 | .120 | .109 | .103 |
| 2.4 | .184 | .166 | .153 | .143 | .135 | .129 | .123 | .119 | .115 | .111 | .108 | .098 | .087 | .082 |
| 2.6 | .158 | .147 | .134 | .124 | .117 | .110 | .105 | .100 | .097 | .093 | .091 | .081 | .070 | .065 |
| 2.8 | .148 | .131 | .118 | .109 | .101 | .095 | .090 | .085 | .082 | .079 | .076 | .066 | .057 | .052 |
| 3.0 | .134 | .117 | .105 | .095 | .085 | .082 | .077 | .073 | .069 | .066 | .064 | .055 | .046 | .042 |
| 3.2 | .121 | .105 | .093 | .084 | .077 | .071 | .066 | .062 | .059 | .056 | .054 | .045 | .037 | .033 |
| 3.4 | .110 | .094 | .083 | .074 | .067 | .062 | .057 | .053 | .050 | .048 | .046 | .038 | .030 | .027 |
| 3.6 | .101 | .085 | .074 | .065 | .059 | .054 | .050 | .046 | .043 | .041 | .039 | .032 | .025 | .022 |
| 3.8 | .092 | .077 | .066 | .058 | .052 | .047 | .043 | .040 | .037 | .035 | .033 | .026 | .020 | .017 |
| 4.0 | .085 | .070 | .060 | .052 | .046 | .041 | .038 | .035 | .032 | .030 | .028 | .022 | .017 | .014 |
| 4.2 | .078 | .064 | .054 | .046 | .041 | .036 | .033 | .030 | .028 | .026 | .024 | .019 | .014 | .011 |
| 4.4 | .072 | .058 | .049 | .042 | .036 | .032 | .029 | .026 | .024 | .022 | .021 | .016 | .011 | .009 |
| 4.6 | .067 | .053 | .044 | .037 | .032 | .029 | .025 | .023 | .021 | .019 | .018 | .013 | .009 | .007 |
| 4.8 | .062 | .049 | .040 | .034 | .029 | .025 | .022 | .020 | .018 | .017 | .015 | .011 | .008 | .006 |
| 5.0 | .058 | .045 | .037 | .031 | .026 | .023 | .020 | .018 | .016 | .015 | .013 | .010 | .006 | .005 |
| 5.2 | .054 | .042 | .034 | .028 | .023 | .020 | .018 | .016 | .014 | .013 | .012 | .008 | .005 | .004 |
| 5.4 | .050 | .039 | .031 | .025 | .021 | .018 | .016 | .014 | .012 | .011 | .010 | .007 | .004 | .003 |
| 5.6 | .047 | .036 | .028 | .023 | .019 | .016 | .014 | .012 | .011 | .010 | .009 | .006 | .004 | .003 |
| 5.8 | .044 | .033 | .026 | .021 | .017 | .015 | .013 | .011 | .110 | .009 | .008 | .005 | .003 | .002 |
| 6.0 | .041 | .031 | .024 | .019 | .016 | .014 | .011 | .010 | .009 | .008 | .007 | .004 | .002 | .002 |
| 6.2 | .039 | .029 | .022 | .018 | .014 | .012 | .010 | .009 | .008 | .007 | .006 | .004 | .002 | .001 |
| 6.4 | .036 | .027 | .020 | .016 | .013 | .011 | .009 | .008 | .007 | .006 | .005 | .003 | .002 | .001 |
| 6.6 | .034 | .025 | .019 | .015 | .012 | .010 | .008 | .007 | .006 | .005 | .005 | .003 | .001 | .001 |
| 6.8 | .032 | .023 | .018 | .014 | .011 | .009 | .007 | .006 | .005 | .005 | .004 | .002 | .001 | .001 |

(*continues*)

**Table VIIIa** (*continued*)

3 d.f. in numerator

| | | | | | | Denominator degrees of freedom | | | | | | | | |
|---|---|---|---|---|---|---|---|---|---|---|---|---|---|---|
| $F$ | 5 | 6 | 7 | 8 | 9 | 10 | 11 | 12 | 13 | 14 | 15 | 20 | 30 | 40 |
| 7.0 | .031 | .022 | .016 | .013 | .010 | .008 | .007 | .006 | .005 | .004 | .004 | .002 | .001 | .001 |
| 7.2 | .029 | .021 | .015 | .012 | .009 | .007 | .006 | .005 | .004 | .004 | .003 | .002 | .001 | .001 |
| 7.4 | .028 | .019 | .014 | .011 | .008 | .007 | .005 | .004 | .004 | .003 | .003 | .002 | .001 | .000 |
| 7.6 | .026 | .018 | .013 | .010 | .008 | .006 | .005 | .004 | .003 | .003 | .003 | .001 | .001 | .001 |
| 7.8 | .025 | .017 | .012 | .009 | .007 | .006 | .005 | .004 | .003 | .003 | .002 | .001 | .001 | .000 |
| 8.0 | .024 | .016 | .012 | .009 | .007 | .005 | .004 | .003 | .003 | .002 | .002 | .001 | .001 | .000 |
| 8.2 | .022 | .015 | .011 | .008 | .006 | .005 | .004 | .003 | .003 | .002 | .002 | .001 | .001 | .000 |
| 8.4 | .021 | .014 | .010 | .007 | .006 | .005 | .004 | .003 | .003 | .002 | .002 | .001 | .000 | .001 |
| 8.6 | .020 | .014 | .010 | .007 | .005 | .004 | .003 | .003 | .002 | .002 | .001 | .001 | .000 | .000 |
| 8.8 | .019 | .013 | .009 | .006 | .005 | .004 | .003 | .002 | .002 | .002 | .001 | .001 | .000 | .000 |
| 9.0 | .019 | .012 | .008 | .006 | .005 | .003 | .003 | .002 | .002 | .001 | .001 | .001 | .000 | .000 |
| 9.5 | .017 | .011 | .007 | .005 | .004 | .003 | .002 | .002 | .001 | .001 | .001 | .000 | .000 | .000 |
| 10.0 | .015 | .009 | .006 | .004 | .003 | .002 | .002 | .001 | .001 | .001 | .001 | .000 | .000 | .000 |
| 10.5 | .013 | .008 | .006 | .004 | .003 | .002 | .001 | .001 | .001 | .001 | .001 | .000 | .000 | .000 |
| 11.0 | .012 | .007 | .005 | .003 | .002 | .002 | .001 | .001 | .001 | .001 | .000 | .000 | .000 | .000 |
| 11.5 | .011 | .007 | .004 | .003 | .002 | .001 | .001 | .001 | .001 | .000 | .000 | .000 | .000 | .000 |
| 12.0 | .010 | .006 | .003 | .002 | .002 | .001 | .001 | .001 | .000 | .000 | .000 | .000 | .000 | .000 |
| 13.0 | .008 | .005 | .003 | .002 | .001 | .001 | .001 | .000 | .000 | .000 | .000 | .000 | .000 | .000 |
| 15.0 | .006 | .003 | .002 | .001 | .001 | .000 | .000 | .000 | .000 | .000 | .000 | .000 | .000 | .000 |
| 20.0 | .003 | .002 | .001 | .000 | .000 | .000 | .000 | .000 | .000 | .000 | .000 | .000 | .000 | .000 |

4 d.f. in numerator

| | | | | | | | | | | | | | | |
|---|---|---|---|---|---|---|---|---|---|---|---|---|---|---|
| 2.0 | .233 | .214 | .199 | .188 | .178 | .171 | .164 | .159 | .154 | .150 | .146 | .133 | .120 | .113 |
| 2.2 | .205 | .185 | .171 | .159 | .150 | .142 | .136 | .130 | .126 | .122 | .118 | .106 | .093 | .086 |
| 2.4 | .181 | .162 | .147 | .136 | .127 | .119 | .113 | .108 | .103 | .099 | .096 | .074 | .072 | .066 |
| 2.6 | .161 | .142 | .128 | .117 | .108 | .100 | .094 | .089 | .085 | .082 | .078 | .067 | .056 | .050 |
| 2.8 | .144 | .125 | .111 | .100 | .092 | .085 | .079 | .075 | .071 | .067 | .064 | .054 | .044 | .039 |
| 3.0 | .134 | .111 | .097 | .087 | .079 | .072 | .067 | .063 | .059 | .056 | .053 | .043 | .034 | .030 |
| 3.2 | .117 | .099 | .086 | .076 | .068 | .062 | .057 | .053 | .049 | .046 | .044 | .035 | .027 | .023 |
| 3.4 | .106 | .088 | .076 | .066 | .059 | .053 | .048 | .046 | .041 | .038 | .036 | .028 | .021 | .017 |
| 3.6 | .096 | .079 | .067 | .058 | .051 | .046 | .041 | .038 | .035 | .032 | .030 | .023 | .016 | .013 |
| 3.8 | .088 | .071 | .060 | .051 | .045 | .040 | .035 | .032 | .029 | .027 | .025 | .019 | .013 | .010 |
| 4.0 | .080 | .065 | .053 | .045 | .039 | .034 | .031 | .027 | .025 | .023 | .021 | .015 | .010 | .008 |
| 4.2 | .074 | .059 | .048 | .040 | .034 | .030 | .026 | .024 | .021 | .019 | .018 | .013 | .008 | .006 |
| 4.4 | .068 | .053 | .043 | .036 | .030 | .026 | .023 | .020 | .018 | .016 | .015 | .010 | .006 | .005 |
| 4.6 | .063 | .049 | .039 | .032 | .027 | .023 | .020 | .018 | .016 | .014 | .013 | .009 | .005 | .004 |
| 4.8 | .058 | .044 | .035 | .029 | .024 | .020 | .017 | .015 | .013 | .012 | .011 | .007 | .004 | .003 |

**Table VIIIa** Tail-Probabilities of the $F$-Distribution (*continued*)
4 d.f. in numerator

| F | Denominator degrees of freedom | | | | | | | | | | | | | |
|---|---|---|---|---|---|---|---|---|---|---|---|---|---|---|
| | 5 | 6 | 7 | 8 | 9 | 10 | 11 | 12 | 13 | 14 | 15 | 20 | 30 | 40 |
| 5.0 | .054 | .041 | .032 | .026 | .021 | .018 | .015 | .013 | .012 | .010 | .009 | .006 | .003 | .002 |
| 5.2 | .050 | .037 | .029 | .023 | .019 | .016 | .013 | .012 | .010 | .009 | .008 | .005 | .003 | .002 |
| 5.4 | .046 | .034 | .026 | .021 | .017 | .014 | .013 | .010 | .009 | .008 | .007 | .004 | .002 | .001 |
| 5.6 | .043 | .032 | .024 | .019 | .015 | .012 | .010 | .009 | .008 | .007 | .006 | .003 | .002 | .001 |
| 5.8 | .040 | .029 | .022 | .017 | .014 | .011 | .009 | .008 | .107 | .006 | .005 | .003 | .001 | .001 |
| 6.0 | .038 | .027 | .020 | .016 | .012 | .010 | .008 | .007 | .006 | .005 | .004 | .002 | .001 | .001 |
| 6.2 | .036 | .025 | .019 | .014 | .011 | .009 | .007 | .006 | .005 | .004 | .004 | .002 | .001 | .001 |
| 6.4 | .033 | .023 | .017 | .013 | .010 | .008 | .007 | .005 | .004 | .004 | .003 | .002 | .001 | .000 |
| 6.6 | .031 | .022 | .016 | .012 | .009 | .007 | .006 | .005 | .004 | .003 | .003 | .001 | .001 | .000 |
| 6.8 | .030 | .020 | .015 | .011 | .008 | .007 | .005 | .004 | .004 | .003 | .002 | .001 | .001 | .000 |
| 7.0 | .028 | .019 | .014 | .010 | .008 | .006 | .005 | .004 | .003 | .003 | .002 | .001 | .000 | .000 |
| 7.2 | .026 | .018 | .013 | .009 | .007 | .005 | .004 | .003 | .003 | .002 | .002 | .001 | .000 | .000 |
| 7.4 | .025 | .017 | .012 | .009 | .006 | .005 | .004 | .003 | .002 | .002 | .002 | .001 | .000 | .000 |
| 7.6 | .024 | .016 | .011 | .008 | .006 | .004 | .003 | .003 | .002 | .002 | .001 | .001 | .000 | .000 |
| 7.8 | .022 | .015 | .010 | .006 | .005 | .004 | .003 | .002 | .002 | .002 | .001 | .001 | .000 | .000 |
| 8.0 | .021 | .014 | .009 | .007 | .005 | .004 | .003 | .002 | .002 | .001 | .001 | .001 | .000 | .000 |
| 8.2 | .020 | .013 | .009 | .006 | .005 | .003 | .003 | .002 | .002 | .001 | .001 | .000 | .000 | .000 |
| 8.4 | .019 | .012 | .008 | .006 | .004 | .003 | .002 | .002 | .001 | .001 | .001 | .000 | .000 | .000 |
| 8.6 | .018 | .012 | .008 | .005 | .004 | .003 | .002 | .002 | .001 | .001 | .001 | .000 | .000 | .000 |
| 8.8 | .017 | .011 | .007 | .006 | .004 | .003 | .002 | .001 | .001 | .001 | .001 | .000 | .000 | .000 |
| 9.0 | .017 | .010 | .007 | .005 | .003 | .002 | .002 | .001 | .001 | .001 | .001 | .000 | .000 | .000 |
| 9.5 | .015 | .009 | .006 | .004 | .003 | .002 | .001 | .001 | .001 | .001 | .000 | .000 | .000 | .000 |
| 10.0 | .013 | .008 | .005 | .003 | .002 | .002 | .001 | .001 | .001 | .000 | .000 | .000 | .000 | .000 |
| 10.5 | .012 | .007 | .004 | .003 | .002 | .001 | .001 | .001 | .001 | .000 | .000 | .000 | .000 | .000 |
| 11.0 | .011 | .006 | .004 | .002 | .002 | .001 | .001 | .001 | .000 | .000 | .000 | .000 | .000 | .000 |
| 11.5 | .010 | .006 | .003 | .002 | .001 | .001 | .001 | .000 | .000 | .000 | .000 | .000 | .000 | .000 |
| 12.0 | .009 | .005 | .003 | .002 | .001 | .001 | .001 | .000 | .000 | .000 | .000 | .000 | .000 | .000 |
| 13.0 | .007 | .004 | .002 | .001 | .001 | .001 | .000 | .000 | .000 | .000 | .000 | .000 | .000 | .000 |
| 15.0 | .005 | .003 | .002 | .001 | .001 | .000 | .000 | .000 | .000 | .000 | .000 | .000 | .000 | .000 |
| 20.0 | .003 | .001 | .001 | .000 | .000 | .000 | .000 | .000 | .000 | .000 | .000 | .000 | .000 | .000 |

5 d.f. in numerator

| F | 5 | 6 | 7 | 8 | 9 | 10 | 11 | 12 | 13 | 14 | 15 | 20 | 30 | 40 |
|---|---|---|---|---|---|---|---|---|---|---|---|---|---|---|
| 2.0 | .233 | .212 | .196 | .183 | .173 | .164 | .157 | .151 | .146 | .141 | .137 | .123 | .107 | .100 |
| 2.2 | .204 | .182 | .166 | .154 | .144 | .135 | .128 | .122 | .117 | .113 | .109 | .095 | .081 | .073 |
| 2.4 | .179 | .158 | .142 | .130 | .120 | .112 | .105 | .099 | .095 | .090 | .087 | .074 | .060 | .004 |
| 2.6 | .159 | .138 | .123 | .110 | .101 | .093 | .087 | .081 | .077 | .073 | .069 | .057 | .045 | .040 |
| 2.8 | .141 | .121 | .106 | .094 | .085 | .078 | .072 | .067 | .063 | .059 | .056 | .045 | .034 | .029 |

(*continues*)

**Table VIIIa** (*continued*)

5 d.f. in numerator

| F | 5 | 6 | 7 | 8 | 9 | Denominator degrees of freedom 10 | 11 | 12 | 13 | 14 | 15 | 20 | 30 | 40 |
|---|---|---|---|---|---|---|---|---|---|---|---|---|---|---|
| 3.0 | .127 | .107 | .092 | .081 | .072 | .066 | .060 | .055 | .051 | .048 | .045 | .035 | .026 | .022 |
| 3.2 | .114 | .095 | .081 | .070 | .062 | .055 | .050 | .046 | .042 | .039 | .037 | .028 | .020 | .016 |
| 3.4 | .103 | .084 | .071 | .061 | .053 | .047 | .042 | .038 | .034 | .032 | .030 | .022 | .015 | .012 |
| 3.6 | .093 | .075 | .062 | .053 | .046 | .040 | .036 | .032 | .029 | .027 | .024 | .017 | .011 | .009 |
| 3.8 | .085 | .067 | .055 | .046 | .040 | .034 | .030 | .027 | .024 | .022 | .020 | .014 | .009 | .007 |
| 4.0 | .077 | .061 | .049 | .041 | .035 | .030 | .026 | .023 | .020 | .018 | .017 | .011 | .007 | .005 |
| 4.2 | .071 | .055 | .044 | .036 | .030 | .026 | .022 | .019 | .017 | .015 | .014 | .009 | .005 | .004 |
| 4.4 | .065 | .050 | .039 | .032 | .026 | .022 | .019 | .017 | .015 | .013 | .011 | .007 | .004 | .003 |
| 4.6 | .060 | .045 | .035 | .028 | .023 | .019 | .016 | .014 | .012 | .011 | .010 | .006 | .003 | .002 |
| 4.8 | .055 | .041 | .032 | .025 | .020 | .017 | .014 | .012 | .010 | .009 | .008 | .005 | .002 | .002 |
| 5.0 | .051 | .038 | .029 | .023 | .018 | .015 | .012 | .010 | .009 | .008 | .007 | .004 | .002 | .001 |
| 5.2 | .047 | .034 | .026 | .020 | .016 | .013 | .011 | .009 | .008 | .007 | .006 | .003 | .001 | .001 |
| 5.4 | .044 | .032 | .024 | .018 | .014 | .012 | .009 | .008 | .007 | .006 | .005 | .003 | .001 | .001 |
| 5.6 | .041 | .029 | .022 | .016 | .013 | .010 | .008 | .007 | .006 | .005 | .004 | .002 | .001 | .001 |
| 5.8 | .038 | .027 | .020 | .015 | .011 | .009 | .007 | .006 | .005 | .004 | .004 | .002 | .001 | .001 |
| 6.0 | .036 | .025 | .018 | .013 | .010 | .008 | .006 | .005 | .004 | .004 | .003 | .001 | .000 | .000 |
| 6.2 | .033 | .023 | .017 | .012 | .009 | .007 | .006 | .005 | .004 | .003 | .003 | .001 | .000 | .000 |
| 6.4 | .031 | .021 | .015 | .011 | .008 | .006 | .005 | .004 | .003 | .003 | .002 | .001 | .000 | .000 |
| 6.6 | .029 | .020 | .014 | .010 | .008 | .006 | .005 | .004 | .003 | .002 | .002 | .001 | .000 | .000 |
| 6.8 | .028 | .019 | .013 | .009 | .007 | .005 | .004 | .003 | .003 | .002 | .002 | .001 | .000 | .000 |
| 7.0 | .026 | .017 | .012 | .008 | .006 | .005 | .004 | .003 | .002 | .002 | .001 | .001 | .000 | .000 |
| 7.2 | .025 | .016 | .011 | .008 | .006 | .004 | .003 | .002 | .002 | .002 | .001 | .001 | .000 | .000 |
| 7.4 | .023 | .015 | .010 | .007 | .005 | .004 | .003 | .002 | .002 | .001 | .001 | .000 | .000 | .000 |
| 7.6 | .022 | .014 | .009 | .007 | .005 | .003 | .003 | .002 | .002 | .001 | .001 | .000 | .000 | .000 |
| 7.8 | .021 | .013 | .009 | .006 | .004 | .003 | .002 | .002 | .001 | .001 | .001 | .000 | .000 | .000 |
| 8.0 | .020 | .012 | .008 | .006 | .004 | .003 | .002 | .002 | .001 | .001 | .001 | .000 | .000 | .000 |
| 8.2 | .019 | .012 | .008 | .005 | .004 | .003 | .002 | .001 | .001 | .001 | .001 | .000 | .000 | .000 |
| 8.4 | .018 | .011 | .007 | .005 | .003 | .002 | .002 | .001 | .001 | .001 | .001 | .000 | .000 | .000 |
| 8.6 | .017 | .010 | .007 | .004 | .003 | .002 | .002 | .001 | .001 | .001 | .001 | .000 | .000 | .000 |
| 8.8 | .016 | .010 | .006 | .004 | .003 | .002 | .001 | .001 | .001 | .001 | .001 | .000 | .000 | .000 |
| 9.0 | .015 | .009 | .006 | .004 | .003 | .002 | .001 | .001 | .001 | .001 | .000 | .000 | .000 | .000 |
| 9.5 | .014 | .008 | .005 | .003 | .002 | .001 | .001 | .001 | .001 | .000 | .000 | .000 | .000 | .000 |
| 10.0 | .012 | .007 | .004 | .003 | .002 | .001 | .001 | .001 | .000 | .000 | .000 | .000 | .000 | .000 |
| 10.5 | .011 | .006 | .004 | .002 | .001 | .001 | .001 | .000 | .000 | .000 | .000 | .000 | .000 | .000 |
| 11.0 | .010 | .006 | .003 | .002 | .001 | .001 | .001 | .000 | .000 | .000 | .000 | .000 | .000 | .000 |
| 11.5 | .009 | .005 | .003 | .002 | .001 | .001 | .000 | .000 | .000 | .000 | .000 | .000 | .000 | .000 |
| 12.0 | .008 | .004 | .003 | .001 | .001 | .000 | .000 | .000 | .000 | .000 | .000 | .000 | .000 | .000 |
| 13.0 | .007 | .004 | .002 | .001 | .001 | .000 | .000 | .000 | .000 | .000 | .000 | .000 | .000 | .000 |
| 15.0 | .005 | .002 | .001 | .001 | .000 | .000 | .000 | .000 | .000 | .000 | .000 | .000 | .000 | .000 |
| 20.0 | .004 | .002 | .001 | .001 | .000 | .000 | .000 | .000 | .000 | .000 | .000 | .000 | .000 | .000 |

**Table VIIIb** Ninety-Fifth Percentiles of the $F$-Distribution

| | | | | | | Numerator degrees of freedom | | | | | | | | |
|---|---|---|---|---|---|---|---|---|---|---|---|---|---|---|
| | 1 | 2 | 3 | 4 | 5 | 6 | 8 | 10 | 12 | 15 | 20 | 24 | 30 |
| 1 | 161 | 200 | 216 | 225 | 230 | 234 | 239 | 242 | 244 | 246 | 248 | 249 | 250 |
| 2 | 18.5 | 19.0 | 19.2 | 19.2 | 19.3 | 19.3 | 19.4 | 19.4 | 19.4 | 19.4 | 19.4 | 19.5 | 19.5 |
| 3 | 10.1 | 9.55 | 9.28 | 9.12 | 9.01 | 8.94 | 8.85 | 8.79 | 8.74 | 8.70 | 8.66 | 8.64 | 8.62 |
| 4 | 7.71 | 6.94 | 6.59 | 6.39 | 6.26 | 6.16 | 6.04 | 5.96 | 5.91 | 5.86 | 5.80 | 5.77 | 5.75 |
| 5 | 6.61 | 5.79 | 5.41 | 5.19 | 5.05 | 4.95 | 4.82 | 4.74 | 4.68 | 4.62 | 4.56 | 4.53 | 4.50 |
| 6 | 5.99 | 5.14 | 4.76 | 4.53 | 4.39 | 4.28 | 4.15 | 4.06 | 4.00 | 3.94 | 3.87 | 3.84 | 3.81 |
| 7 | 5.59 | 4.74 | 4.35 | 4.12 | 3.97 | 3.87 | 3.73 | 3.64 | 3.57 | 3.51 | 3.44 | 3.41 | 3.38 |
| 8 | 5.32 | 4.46 | 4.07 | 3.84 | 3.69 | 3.58 | 3.44 | 3.35 | 3.28 | 3.22 | 3.15 | 3.12 | 3.08 |
| 9 | 5.12 | 4.26 | 3.86 | 3.63 | 3.48 | 3.37 | 3.23 | 3.14 | 3.07 | 3.01 | 2.94 | 2.90 | 2.86 |
| 10 | 4.96 | 4.10 | 3.71 | 3.48 | 3.33 | 3.22 | 3.07 | 2.98 | 2.91 | 2.85 | 2.77 | 2.74 | 2.70 |
| 11 | 4.84 | 3.98 | 3.59 | 3.36 | 3.20 | 3.09 | 2.95 | 2.85 | 2.79 | 2.72 | 2.65 | 2.61 | 2.57 |
| 12 | 4.75 | 3.89 | 3.49 | 3.26 | 3.11 | 3.00 | 2.85 | 2.75 | 2.69 | 2.62 | 2.54 | 2.51 | 2.47 |
| 13 | 4.67 | 3.81 | 3.41 | 3.18 | 3.03 | 2.92 | 2.77 | 2.67 | 2.60 | 2.53 | 2.46 | 2.42 | 2.38 |
| 14 | 4.60 | 3.74 | 3.34 | 3.11 | 2.96 | 2.85 | 2.70 | 2.60 | 2.53 | 2.46 | 2.39 | 2.35 | 2.31 |
| 15 | 4.54 | 3.68 | 3.29 | 3.06 | 2.90 | 2.79 | 2.64 | 2.54 | 2.48 | 2.40 | 2.33 | 2.29 | 2.25 |
| 16 | 4.49 | 3.63 | 3.24 | 3.01 | 2.85 | 2.74 | 2.59 | 2.49 | 2.42 | 2.35 | 2.28 | 2.24 | 2.19 |
| 17 | 4.45 | 3.59 | 3.20 | 2.96 | 2.81 | 2.70 | 2.55 | 2.45 | 2.38 | 2.31 | 2.23 | 2.19 | 2.15 |
| 18 | 4.41 | 3.55 | 3.16 | 2.93 | 2.77 | 2.66 | 2.51 | 2.41 | 2.34 | 2.27 | 2.19 | 2.15 | 2.11 |
| 19 | 4.38 | 3.52 | 3.13 | 2.90 | 2.74 | 2.63 | 2.48 | 2.38 | 2.31 | 2.23 | 2.16 | 2.11 | 2.07 |
| 20 | 4.35 | 3.49 | 3.10 | 2.87 | 2.71 | 2.60 | 2.45 | 2.35 | 2.28 | 2.20 | 2.12 | 2.08 | 2.04 |
| 21 | 4.32 | 3.47 | 3.07 | 2.84 | 2.68 | 2.57 | 2.42 | 2.32 | 2.25 | 2.18 | 2.10 | 2.05 | 2.01 |
| 22 | 4.30 | 3.44 | 3.05 | 2.82 | 2.66 | 2.55 | 2.40 | 2.30 | 2.23 | 2.15 | 2.07 | 2.03 | 1.98 |
| 23 | 4.28 | 3.42 | 3.03 | 2.80 | 2.64 | 2.53 | 2.37 | 2.27 | 2.20 | 2.13 | 2.05 | 2.01 | 1.96 |
| 24 | 4.26 | 3.40 | 3.01 | 2.78 | 2.62 | 2.51 | 2.36 | 2.25 | 2.18 | 2.11 | 2.03 | 1.98 | 1.94 |
| 25 | 4.24 | 3.39 | 2.99 | 2.76 | 2.60 | 2.49 | 2.34 | 2.24 | 2.16 | 2.09 | 2.01 | 1.96 | 1.92 |
| 30 | 4.17 | 3.32 | 2.92 | 2.69 | 2.53 | 2.42 | 2.27 | 2.16 | 2.09 | 2.01 | 1.93 | 1.89 | 1.84 |
| 40 | 4.08 | 3.23 | 2.84 | 2.61 | 2.45 | 2.34 | 2.18 | 2.08 | 2.00 | 1.92 | 1.84 | 1.79 | 1.74 |
| 60 | 4.00 | 3.15 | 2.76 | 2.53 | 2.37 | 2.25 | 2.10 | 1.99 | 1.92 | 1.84 | 1.75 | 1.70 | 1.65 |

*Denominator degrees of freedom*

Note: The fifth percentiles are obtainable as follows:

$$F_{.05}(r, s) = \frac{1}{F_{.95}(s, r)}.$$

$$\left[ \text{For example, } F_{.05}(3, 6) = \frac{1}{F_{.95}(6, 3)} = \frac{1}{8.94}. \right]$$

**Table VIIIb** Ninety-Ninth Percentiles of the $F$-Distribution

| | | | | | | *Numerator degrees of freedom* | | | | | | | | |
|---|---|---|---|---|---|---|---|---|---|---|---|---|---|---|
| | 1 | 2 | 3 | 4 | 5 | 6 | 8 | 10 | 12 | 15 | 20 | 24 | 30 |
| 1 | 4050 | 5000 | 5400 | 5620 | 5760 | 5860 | 5980 | 6060 | 6110 | 6160 | 6210 | 6235 | 6260 |
| 2 | 98.5 | 99.0 | 99.2 | 99.2 | 99.3 | 99.3 | 99.4 | 99.4 | 99.4 | 99.4 | 99.4 | 99.5 | 99.5 |
| 3 | 34.1 | 30.8 | 29.5 | 28.7 | 28.2 | 27.9 | 27.5 | 27.3 | 27.1 | 26.9 | 26.7 | 26.6 | 26.5 |
| 4 | 21.2 | 18.0 | 16.7 | 16.0 | 15.5 | 15.2 | 14.8 | 14.5 | 14.4 | 14.2 | 14.0 | 13.9 | 13.8 |
| 5 | 16.3 | 13.3 | 12.1 | 11.4 | 11.0 | 10.7 | 10.3 | 10.1 | 9.89 | 9.72 | 9.55 | 9.47 | 9.38 |
| 6 | 13.7 | 10.9 | 9.78 | 9.15 | 8.75 | 8.47 | 8.10 | 7.87 | 7.72 | 7.56 | 7.40 | 7.31 | 7.23 |
| 7 | 12.2 | 9.55 | 8.45 | 7.85 | 7.46 | 7.19 | 6.84 | 6.62 | 6.47 | 6.31 | 6.16 | 6.07 | 5.99 |
| 8 | 11.3 | 8.65 | 7.59 | 7.01 | 6.63 | 6.37 | 6.03 | 5.81 | 5.67 | 5.52 | 5.36 | 5.28 | 5.20 |
| 9 | 10.6 | 8.02 | 6.99 | 6.42 | 6.06 | 5.80 | 5.47 | 5.26 | 5.11 | 4.96 | 4.81 | 4.73 | 4.65 |
| 10 | 10.0 | 7.56 | 6.55 | 5.99 | 5.64 | 5.39 | 5.06 | 4.85 | 4.71 | 4.56 | 4.41 | 4.33 | 4.25 |
| 11 | 9.65 | 7.21 | 6.22 | 5.67 | 5.32 | 5.07 | 4.74 | 4.54 | 4.40 | 4.25 | 4.10 | 4.02 | 3.94 |
| 12 | 9.33 | 6.93 | 5.95 | 5.41 | 5.06 | 4.82 | 4.50 | 4.30 | 4.16 | 4.01 | 3.86 | 3.78 | 3.70 |
| 13 | 9.07 | 6.70 | 5.74 | 5.21 | 4.86 | 4.62 | 4.30 | 4.10 | 3.96 | 3.82 | 3.66 | 3.59 | 3.51 |
| 14 | 8.86 | 6.51 | 5.56 | 5.04 | 4.69 | 4.46 | 4.14 | 3.94 | 3.80 | 3.66 | 3.51 | 3.43 | 3.35 |
| 15 | 8.68 | 6.36 | 5.42 | 4.89 | 4.56 | 4.32 | 4.00 | 3.80 | 3.67 | 3.52 | 3.37 | 3.29 | 3.21 |
| 16 | 8.53 | 6.23 | 5.29 | 4.77 | 4.44 | 4.20 | 3.89 | 3.69 | 3.55 | 3.41 | 3.26 | 3.18 | 3.10 |
| 17 | 8.40 | 6.11 | 5.18 | 4.67 | 4.34 | 4.10 | 3.79 | 3.59 | 3.46 | 3.31 | 3.16 | 3.08 | 3.00 |
| 18 | 8.29 | 6.01 | 5.09 | 4.58 | 4.25 | 4.01 | 3.71 | 3.51 | 3.37 | 3.23 | 3.08 | 3.00 | 2.92 |
| 19 | 8.18 | 5.93 | 5.01 | 4.50 | 4.17 | 3.94 | 3.63 | 3.43 | 3.30 | 3.15 | 3.00 | 2.92 | 2.84 |
| 20 | 8.10 | 5.85 | 4.94 | 4.43 | 4.10 | 3.87 | 3.56 | 3.37 | 3.23 | 3.09 | 2.94 | 2.86 | 2.78 |
| 21 | 8.02 | 5.78 | 4.87 | 4.37 | 4.04 | 3.81 | 3.51 | 3.31 | 3.17 | 3.03 | 2.88 | 2.80 | 2.72 |
| 22 | 7.95 | 5.72 | 4.82 | 4.31 | 3.99 | 3.76 | 3.45 | 3.26 | 3.12 | 2.98 | 2.83 | 2.75 | 2.67 |
| 23 | 7.88 | 5.66 | 4.76 | 4.26 | 3.94 | 3.71 | 3.41 | 3.21 | 3.07 | 2.93 | 2.87 | 2.70 | 2.62 |
| 24 | 7.82 | 5.61 | 4.72 | 4.22 | 3.90 | 3.67 | 3.36 | 3.17 | 3.03 | 2.89 | 2.74 | 2.66 | 2.58 |
| 25 | 7.77 | 5.57 | 4.68 | 4.18 | 3.86 | 3.63 | 3.32 | 3.13 | 2.99 | 2.85 | 2.70 | 2.62 | 2.54 |
| 30 | 7.56 | 5.39 | 4.51 | 4.02 | 3.70 | 3.47 | 3.17 | 2.98 | 2.84 | 2.70 | 2.55 | 2.47 | 2.39 |
| 40 | 7.31 | 5.18 | 4.31 | 3.83 | 3.51 | 3.29 | 2.99 | 2.80 | 2.66 | 2.52 | 2.37 | 2.29 | 2.20 |
| 60 | 7.08 | 4.98 | 4.13 | 3.65 | 3.34 | 3.12 | 2.82 | 2.63 | 2.50 | 2.35 | 2.20 | 2.12 | 2.03 |

*Denominator degrees of freedom*

**Table IXa** Tail-Probabilities for the One-Sample Kolmogorov-Smirnov Statistic
$$nD_n = n \sup_x |F_n(x) - F_0(x)|$$

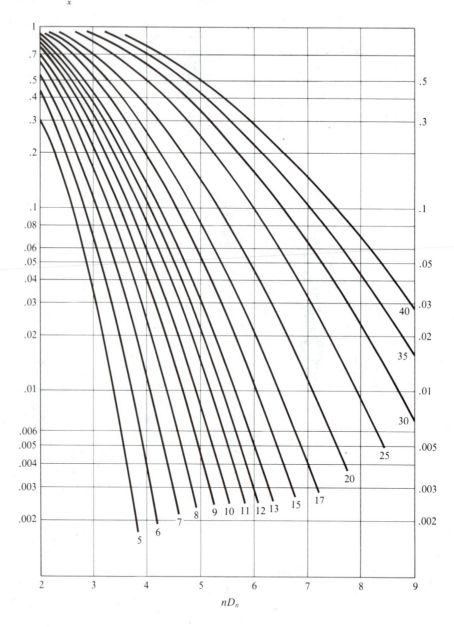

**Table IXb** Critical Values for the One-Sample
Kolmogorov-Smirnov Statistic

| Sample size (n) | Significance level | | | | |
|---|---|---|---|---|---|
| | .20 | .15 | .10 | .05 | .01 |
| 1 | .900 | .925 | .950 | .975 | .995 |
| 2 | .684 | .726 | .776 | .842 | .929 |
| 3 | .565 | .597 | .642 | .708 | .829 |
| 4 | .494 | .525 | .564 | .624 | .734 |
| 5 | .446 | .474 | .510 | .563 | .669 |
| 6 | .410 | .436 | .470 | .521 | .618 |
| 7 | .381 | .405 | .438 | .486 | .577 |
| 8 | .358 | .381 | .411 | .457 | .543 |
| 9 | .339 | .360 | .388 | .432 | .514 |
| 10 | .322 | .342 | .368 | .409 | .486 |
| 11 | .307 | .326 | .352 | .391 | .468 |
| 12 | .295 | .313 | .338 | .375 | .450 |
| 13 | .284 | .302 | .325 | .361 | .433 |
| 14 | .274 | .292 | .314 | .349 | .418 |
| 15 | .266 | .283 | .304 | .338 | .404 |
| 16 | .258 | .274 | .295 | .328 | .391 |
| 17 | .250 | .266 | .286 | .318 | .380 |
| 18 | .244 | .259 | .278 | .309 | .370 |
| 19 | .237 | .252 | .272 | .301 | .361 |
| 20 | .231 | .246 | .264 | .294 | .352 |
| 25 | .21 | .22 | .24 | .264 | .32 |
| 30 | .19 | .20 | .22 | .242 | .29 |
| 35 | .18 | .19 | .21 | .23 | .27 |
| 40 | | | | .21 | .25 |
| 50 | | | | .19 | .23 |
| 60 | | | | .17 | .21 |
| 70 | | | | .16 | .19 |
| 80 | | | | .15 | .18 |
| 90 | | | | .14 | |
| 100 | | | | .14 | |
| Asymptotic formula: | $\dfrac{1.07}{\sqrt{n}}$ | $\dfrac{1.14}{\sqrt{n}}$ | $\dfrac{1.22}{\sqrt{n}}$ | $\dfrac{1.36}{\sqrt{n}}$ | $\dfrac{1.63}{\sqrt{n}}$ |

Reject the hypothetical distribution if $D_n = \max|F_n(x) - F(x)|$ exceeds the tabulated value.

(For $\alpha = .01$ and .05, asymptotic formulas give values that are too high—by 1.5% for $n = 80$.)

This table is taken from F. J. Massey, Jr., "The Kolmogorov–Smirnov test for goodness of fit," *J. Am. Stat. Assn. 46* (1951), 68–78 except that certain corrections and additional entries are from Z. W. Birnbaum, "Numerical tabulation of the distribution of Kolmogorov's statistic for finite sample size," *J. Am. Stat. Assn.* 47 (1952), 425–441 with the kind permission of the authors and the *J. Am. Stat. Assn.*

**Table Xa** Tail-Probabilities for the Two-Sample Kolmogorov-Smirnov Statistic

*Case 1: P-values when sample sizes are equal.*
Independent, random samples from $F$ and $G$, each of size $n$.

Statistic: $X = n \cdot \sup|F_n - G_n|$.

| $x$ $\backslash$ $n$ | 4 | 5 | 6 | 7 | 8 | 9 | 10 | 11 | 12 | 15 | 20 | $x$ |
|---|---|---|---|---|---|---|---|---|---|---|---|---|
| 2 | .771 | .873 | .931 | .963 | .980 | .989 | .994 | .997 | .998 | 1.00 | 1.00 | 2 |
| 3 | .229 | .357 | .474 | .575 | .660 | .730 | .787 | .833 | .869 | .938 | .983 | 3 |
| 4 | .029 | .079 | .143 | .212 | .283 | .352 | .418 | .479 | .536 | .678 | .832 | 4 |
| 5 |  | .008 | .026 | .053 | .087 | .126 | .168 | .211 | .256 | .386 | .571 | 5 |
| 6 |  |  | .002 | .008 | .019 | .034 | .052 | .075 | .100 | .184 | .336 | 6 |
| 7 |  |  |  | .001 | .002 | .006 | .012 | .021 | .031 | .075 | .175 | 7 |
| 8 |  |  |  |  | .000 | .001 | .002 | .004 | .008 | .026 | .081 | 8 |
| 9 |  |  |  |  |  | .000 | .000 | .001 | .001 | .008 | .034 | 9 |
| 10 |  |  |  |  |  |  | .000 | .000 | .000 | .002 | .012 | 10 |
| 11 |  |  |  |  |  |  |  | .000 | .000 | .000 | .004 | 11 |
| 12 |  |  |  |  |  |  |  |  | .000 | .000 | .001 | 12 |
| 13 |  |  |  |  |  |  |  |  |  | .000 | .000 | 13 |

*Case 2: P-values for large samples.*
Independent random samples: size $m$ from $F$, size $n$ from $G$.

Statistic: $Y = \sqrt{\dfrac{mn}{m+n}} \cdot \sup|F_m - G_n|$.

| $y$ | $P$ | $y$ | $P$ | $y$ | $P$ | $y$ | $P$ | $y$ | $P$ |
|---|---|---|---|---|---|---|---|---|---|
| 1.00 | .270 | 1.20 | .112 | 1.40 | .040 | 1.60 | .012 | 1.80 | .003 |
| 1.01 | .259 | 1.21 | .107 | 1.41 | .038 | 1.61 | .011 | 1.81 | .003 |
| 1.02 | .249 | 1.22 | .102 | 1.42 | .035 | 1.62 | .011 | 1.82 | .003 |
| 1.03 | .239 | 1.23 | .097 | 1.43 | .033 | 1.63 | .010 | 1.83 | .002 |
| 1.04 | .230 | 1.24 | .092 | 1.44 | .032 | 1.64 | .009 | 1.84 | .002 |
| 1.05 | .220 | 1.25 | .088 | 1.45 | .030 | 1.65 | .009 | 1.85 | .002 |
| 1.06 | .211 | 1.26 | .084 | 1.46 | .028 | 1.66 | .008 | 1.86 | .002 |
| 1.07 | .202 | 1.27 | .079 | 1.47 | .027 | 1.67 | .008 | 1.87 | .002 |
| 1.08 | .194 | 1.28 | .075 | 1.48 | .025 | 1.68 | .007 | 1.88 | .002 |
| 1.09 | .186 | 1.29 | .072 | 1.49 | .024 | 1.69 | .007 | 1.89 | .002 |
| 1.10 | .178 | 1.30 | .068 | 1.50 | .022 | 1.70 | .006 | 1.90 | .001 |
| 1.11 | .170 | 1.31 | .065 | 1.51 | .021 | 1.71 | .006 | 1.91 | .001 |
| 1.12 | .163 | 1.32 | .061 | 1.52 | .020 | 1.72 | .005 | 1.92 | .001 |
| 1.13 | .155 | 1.33 | .058 | 1.53 | .019 | 1.73 | .005 | 1.93 | .001 |
| 1.14 | .149 | 1.34 | .055 | 1.54 | .017 | 1.74 | .005 | 1.94 | .001 |
| 1.15 | .142 | 1.35 | .052 | 1.55 | .016 | 1.75 | .004 | 1.95 | .001 |
| 1.16 | .136 | 1.36 | .049 | 1.56 | .015 | 1.76 | .004 | 1.96 | .001 |
| 1.17 | .129 | 1.37 | .047 | 1.57 | .014 | 1.77 | .004 | 1.97 | .001 |
| 1.18 | .123 | 1.38 | .044 | 1.58 | .014 | 1.78 | .004 | 1.98 | .001 |
| 1.19 | .118 | 1.39 | .038 | 1.59 | .013 | 1.79 | .003 | 1.99 | .001 |

**Table Xb** Critical Values for the Two-Sample Kolmogorov-Smirnov Statistic

| Sample size $n_2$ | Sample size $n_1$: 1 | 2 | 3 | 4 | 5 | 6 | 7 | 8 | 9 | 10 | 12 | 15 |
|---|---|---|---|---|---|---|---|---|---|---|---|---|
| 1 | *<br>* | *<br>* | *<br>* | *<br>* | *<br>* | *<br>* | *<br>* | *<br>* | *<br>* | *<br>* | | |
| 2 | | *<br>* | *<br>* | *<br>* | *<br>* | *<br>* | *<br>* | 7/8<br>* | 16/18<br>* | 9/10<br>* | | |
| 3 | | | *<br>* | *<br>* | 12/15<br>* | 5/6<br>* | 18/21<br>* | 18/24<br>* | 7/9<br>8/9 | | 9/12<br>11/12 | |
| 4 | | | | 3/4<br>* | 16/20<br>* | 9/12<br>10/12 | 21/28<br>24/28 | 6/8<br>7/8 | 27/36<br>32/36 | 14/20<br>16/20 | 8/12<br>10/12 | |
| 5 | | | | | 4/5<br>4/5 | 20/30<br>25/30 | 25/35<br>30/35 | 27/40<br>32/40 | 31/45<br>36/45 | 7/10<br>8/10 | | 10/15<br>11/15 |
| 6 | | | | | | 4/6<br>5/6 | 29/42<br>35/42 | 16/24<br>18/24 | 12/18<br>14/18 | 19/30<br>22/30 | 7/12<br>9/12 | |
| 7 | | | | | | | 5/7<br>5/7 | 35/56<br>42/56 | 40/63<br>47/63 | 43/70<br>53/70 | | |
| 8 | | | | | | | | 5/8<br>6/8 | 45/72<br>54/72 | 23/40<br>28/40 | 14/24<br>16/24 | |
| 9 | | | | | | | | | 5/9<br>6/9 | 52/90<br>62/90 | 20/36<br>24/36 | |
| 10 | | | | | | | | | | 6/10<br>7/10 | | 15/30<br>19/30 |
| 12 | | | | | | | | | | | 6/12<br>7/12 | 30/60<br>35/60 |
| 15 | | | | | | | | | | | | 7/15<br>8/15 |

Notes: 1. Reject $H_0$ if $D = \max|F_{n_2}(x) - F_{n_1}(x)|$ exceeds the tabulated value. The upper value gives a level at most .05 and the lower at most .01.

2. Where * appears, do not reject $H_0$ at the given level.

3. For large values of $n_1$ and $n_2$, the following approximate formulas may be used:

$$\alpha = .05\colon\ 1.36\sqrt{\frac{n_1 + n_2}{n_1 n_2}}.$$

$$\alpha = .01\colon\ 1.63\sqrt{\frac{n_1 + n_2}{n_1 n_2}}.$$

**Table XI** Critical Values for the Lilliefors
Distribution (for testing normality)

| $n$ | Right tail-probability | | | | | |
|---|---|---|---|---|---|---|
| | .20 | .15 | .10 | .05 | .01 | .001 |
| 5 | .289 | .303 | .319 | .343 | .397 | .439 |
| 6 | .269 | .281 | .297 | .323 | .371 | .424 |
| 7 | .252 | .264 | .280 | .304 | .351 | .402 |
| 8 | .239 | .250 | .265 | .288 | .333 | .384 |
| 9 | .227 | .238 | .252 | .274 | .317 | .365 |
| 10 | .217 | .228 | .241 | .262 | .304 | .352 |
| 11 | .208 | .218 | .231 | .251 | .291 | .338 |
| 12 | .200 | .210 | .222 | .242 | .281 | .325 |
| 13 | .193 | .202 | .215 | .234 | .271 | .314 |
| 14 | .187 | .196 | .208 | .226 | .262 | .305 |
| 15 | .181 | .190 | .201 | .219 | .254 | .296 |
| 16 | .176 | .184 | .195 | .213 | .247 | .287 |
| 17 | .171 | .179 | .190 | .207 | .240 | .279 |
| 18 | .167 | .175 | .185 | .202 | .234 | .273 |
| 19 | .163 | .170 | .181 | .197 | .228 | .266 |
| 20 | .159 | .166 | .176 | .192 | .223 | .260 |
| 25 | .143 | .150 | .159 | .173 | .201 | .236 |
| 30 | .131 | .138 | .146 | .159 | .185 | .217 |
| 40 | .115 | .120 | .128 | .139 | .162 | .189 |
| 100 | .074 | .077 | .082 | .089 | .104 | .122 |
| 400 | .037 | .039 | .041 | .045 | .052 | .061 |

Assumes that in calculating the K–S statistic, the population
mean and variance are replaced by the sample mean and
variance, respectively.

**Table XII** Expected Values of Order Statistics from a Standard Normal Population

| $n$ \ $i$ | $n-9$ | $n-8$ | $n-7$ | $n-6$ | $n-5$ | $n-4$ | $n-3$ | $n-2$ | $n-1$ | $n$ | $i$ \ $n$ |
|---|---|---|---|---|---|---|---|---|---|---|---|
| 2 | | | | | | | | | | .564 | 2 |
| 3 | | | | | | | | | | .846 | 3 |
| 4 | | | | | | | | | .297 | 1.029 | 4 |
| 5 | | | | | | | | | .495 | 1.163 | 5 |
| 6 | | | | | | | | .202 | .642 | 1.267 | 6 |
| 7 | | | | | | | | .353 | .757 | 1.352 | 7 |
| 8 | | | | | | | .153 | .473 | .852 | 1.424 | 8 |
| 9 | | | | | | | .275 | .572 | .932 | 1.485 | 9 |
| 10 | | | | | | .123 | .376 | .656 | 1.001 | 1.539 | 10 |
| 11 | | | | | | .225 | .462 | .729 | 1.062 | 1.586 | 11 |
| 12 | | | | | .103 | .312 | .537 | .793 | 1.116 | 1.629 | 12 |
| 13 | | | | | .190 | .388 | .603 | .850 | 1.164 | 1.668 | 13 |
| 14 | | | | .088 | .267 | .456 | .662 | .901 | 1.208 | 1.703 | 14 |
| 15 | | | | .165 | .335 | .516 | .715 | .948 | 1.248 | 1.736 | 15 |
| 16 | | | .077 | .234 | .396 | .570 | .763 | .990 | 1.285 | 1.766 | 16 |
| 17 | | | .146 | .295 | .451 | .619 | .807 | 1.030 | 1.319 | 1.794 | 17 |
| 18 | | .069 | .208 | .351 | .502 | .665 | .848 | 1.066 | 1.350 | 1.820 | 18 |
| 19 | | .131 | .264 | .402 | .548 | .707 | .886 | 1.099 | 1.380 | 1.844 | 19 |
| 20 | .062 | .187 | .315 | .448 | .590 | .745 | .921 | 1.131 | 1.408 | 1.867 | 20 |

Notes: 1. Assume random sample of size $n$.
2. Entries are expected locations of $X_{(i)}$, the $i$th smallest observation.
3. Entries for the lower half are the negatives of the entries for the corresponding upper half. For example, for $n = 5$, expected locations are $-1.163, -.495, 0, .495, 1.163$.

**Table XIII** Distribution of the Standardized Range $W = R/\sigma$ (assuming a normal population)

|  | Sample size | | | | | | | | | | | |
|---|---|---|---|---|---|---|---|---|---|---|---|---|
|  | 2 | 3 | 4 | 5 | 6 | 7 | 8 | 9 | 10 | 12 | 15 |
| $E(W)$ | 1.128 | 1.693 | 2.059 | 2.326 | 2.534 | 2.704 | 2.847 | 2.970 | 3.078 | 3.258 | 3.472 |
| $\sigma_W$ | .853 | .888 | .880 | .864 | .848 | .833 | .820 | .808 | .797 | .778 | .755 |
| $W_{.005}$ | .01 | .13 | .34 | .55 | .75 | .92 | 1.08 | 1.21 | 1.33 | 1.55 | 1.80 |
| $W_{.01}$ | .02 | .19 | .43 | .66 | .87 | 1.05 | 1.20 | 1.34 | 1.47 | 1.68 | 1.93 |
| $W_{.025}$ | .04 | .30 | .59 | .85 | 1.06 | 1.25 | 1.41 | 1.55 | 1.67 | 1.88 | 2.14 |
| $W_{.05}$ | .09 | .43 | .76 | 1.03 | 1.25 | 1.44 | 1.60 | 1.74 | 1.86 | 2.07 | 2.32 |
| $W_{.1}$ | .18 | .62 | .98 | 1.26 | 1.49 | 1.68 | 1.83 | 1.97 | 2.09 | 2.30 | 2.54 |
| $W_{.2}$ | .36 | .90 | 1.29 | 1.57 | 1.80 | 1.99 | 2.14 | 2.28 | 2.39 | 2.59 | 2.83 |
| $W_{.3}$ | .55 | 1.14 | 1.53 | 1.82 | 2.04 | 2.22 | 2.38 | 2.51 | 2.62 | 2.82 | 3.04 |
| $W_{.4}$ | .74 | 1.36 | 1.76 | 2.04 | 2.26 | 2.44 | 2.59 | 2.71 | 2.83 | 3.01 | 3.23 |
| $W_{.5}$ | .95 | 1.59 | 1.98 | 2.26 | 2.47 | 2.65 | 2.79 | 2.92 | 3.02 | 3.21 | 3.42 |
| $W_{.6}$ | 1.20 | 1.83 | 2.21 | 2.48 | 2.69 | 2.86 | 3.00 | 3.12 | 3.23 | 3.41 | 3.62 |
| $W_{.7}$ | 1.47 | 2.09 | 2.47 | 2.73 | 2.94 | 3.10 | 3.24 | 3.35 | 3.46 | 3.63 | 3.83 |
| $W_{.8}$ | 1.81 | 2.42 | 2.78 | 3.04 | 3.23 | 3.39 | 3.52 | 3.63 | 3.73 | 3.90 | 4.09 |
| $W_{.9}$ | 2.33 | 2.90 | 3.24 | 3.48 | 3.66 | 3.81 | 3.93 | 4.04 | 4.13 | 4.29 | 4.47 |
| $W_{.95}$ | 2.77 | 3.31 | 3.63 | 3.86 | 4.03 | 4.17 | 4.29 | 4.39 | 4.47 | 4.62 | 4.80 |
| $W_{.975}$ | 3.17 | 3.68 | 3.98 | 4.20 | 4.36 | 4.49 | 4.61 | 4.70 | 4.79 | 4.92 | 5.09 |
| $W_{.99}$ | 3.64 | 4.12 | 4.40 | 4.60 | 4.76 | 4.88 | 4.99 | 5.08 | 5.16 | 5.29 | 5.45 |
| $W_{.995}$ | 3.97 | 4.42 | 4.69 | 4.89 | 5.03 | 5.15 | 5.26 | 5.34 | 5.42 | 5.54 | 5.70 |

**Table XIV** Selected Percentage
Points of the Shapiro-Wilk $W$

|     | Significance level | | | |
| --- | --- | --- | --- | --- |
| $n$ | .01 | .05 | .10 | .50 |
| 5 | .675 | .777 | .817 | .922 |
| 10 | .776 | .842 | .869 | .940 |
| 15 | .815 | .878 | .903 | .954 |
| 20 | .858 | .902 | .921 | .962 |
| 35 | .919 | .943 | .952 | .976 |
| 50 | .935 | .953 | .963 | .981 |
| 75 | .956 | .969 | .973 | .986 |
| 99 | .967 | .976 | .980 | .989 |

**Table XV** Random Digits

| | | | | | | | | | |
|---|---|---|---|---|---|---|---|---|---|
| 42916 | 50199 | 26435 | 97117 | 77100 | 62919 | 74498 | 14252 | 11052 | 70038 |
| 49019 | 02101 | 14580 | 14421 | 58592 | 30885 | 60248 | 29783 | 39125 | 97534 |
| 04421 | 62261 | 52644 | 36493 | 53146 | 31906 | 00208 | 98915 | 27613 | 58180 |
| 74606 | 07765 | 21788 | 03093 | 69158 | 44498 | 51540 | 61267 | 70550 | 90599 |
| 76288 | 24031 | 13826 | 61989 | 54283 | 95614 | 20378 | 35853 | 86644 | 68259 |
| | | | | | | | | | |
| 42866 | 46273 | 43621 | 93636 | 23582 | 59351 | 29828 | 53006 | 06004 | 00427 |
| 99017 | 74447 | 14581 | 32223 | 89571 | 38437 | 43037 | 17654 | 32705 | 02726 |
| 67245 | 80759 | 07378 | 06307 | 51311 | 52458 | 57898 | 15213 | 72105 | 18792 |
| 29317 | 02377 | 60654 | 51918 | 97109 | 38972 | 71750 | 81431 | 69776 | 00892 |
| 56457 | 56692 | 88071 | 93055 | 31559 | 77054 | 33921 | 24189 | 47537 | 18470 |
| | | | | | | | | | |
| 66908 | 96815 | 00106 | 47915 | 72072 | 34460 | 04085 | 74036 | 99640 | 88672 |
| 83966 | 92418 | 68500 | 70046 | 30009 | 99166 | 49224 | 68804 | 34733 | 69265 |
| 53196 | 82252 | 58476 | 40657 | 09612 | 15380 | 70717 | 33052 | 93954 | 14642 |
| 63291 | 73919 | 67613 | 81329 | 27561 | 97499 | 79346 | 28385 | 20829 | 73829 |
| 62205 | 91166 | 04127 | 19669 | 17699 | 31072 | 16918 | 81168 | 72908 | 00561 |
| | | | | | | | | | |
| 83502 | 34546 | 70327 | 79999 | 26659 | 68085 | 43541 | 69983 | 09041 | 05677 |
| 20293 | 65765 | 45954 | 12799 | 49028 | 44691 | 19957 | 40928 | 81503 | 07030 |
| 72932 | 94622 | 89404 | 69024 | 73518 | 29828 | 35482 | 83798 | 92363 | 13918 |
| 69803 | 06247 | 23872 | 32055 | 36776 | 77634 | 01444 | 88377 | 50827 | 83716 |
| 01155 | 81380 | 11691 | 18090 | 13236 | 34313 | 13390 | 31223 | 64796 | 40116 |
| | | | | | | | | | |
| 44290 | 82296 | 81987 | 09423 | 44272 | 24414 | 43248 | 50536 | 52161 | 18884 |
| 16980 | 43552 | 32970 | 87214 | 99340 | 79058 | 70912 | 03514 | 87351 | 05102 |
| 73249 | 52463 | 51467 | 18602 | 28336 | 41484 | 49543 | 74121 | 04575 | 78007 |
| 06050 | 29975 | 60715 | 02040 | 12974 | 02831 | 52032 | 69726 | 67679 | 13772 |
| 34216 | 50564 | 74588 | 70102 | 62585 | 25511 | 38134 | 13802 | 98334 | 76947 |
| | | | | | | | | | |
| 04607 | 52269 | 21767 | 98347 | 69224 | 44987 | 31255 | 00344 | 60841 | 53970 |
| 92738 | 66714 | 58465 | 83216 | 95109 | 31032 | 99817 | 18844 | 31514 | 44004 |
| 76152 | 98002 | 84257 | 47518 | 53932 | 46337 | 96349 | 17004 | 81135 | 26247 |
| 24405 | 52117 | 41434 | 82281 | 02756 | 40000 | 26893 | 71507 | 55783 | 78195 |
| 99046 | 61444 | 59911 | 58255 | 45299 | 60971 | 72833 | 61883 | 52645 | 60945 |
| | | | | | | | | | |
| 26312 | 73154 | 21070 | 90104 | 42013 | 27302 | 55283 | 13166 | 14051 | 81929 |
| 36315 | 59502 | 91215 | 86654 | 44578 | 04159 | 63389 | 43516 | 48971 | 40922 |
| 52467 | 19775 | 71391 | 63601 | 84377 | 63350 | 59557 | 74397 | 06289 | 74426 |
| 66790 | 72193 | 63999 | 20307 | 47423 | 55164 | 93870 | 43783 | 06851 | 90065 |
| 16427 | 71681 | 64661 | 59249 | 74118 | 46257 | 69308 | 31035 | 64498 | 19592 |
| | | | | | | | | | |
| 63988 | 01319 | 15012 | 95770 | 82029 | 99778 | 81793 | 73836 | 11528 | 81863 |
| 67468 | 22553 | 71756 | 30281 | 28244 | 58696 | 72161 | 46240 | 63452 | 56485 |
| 60477 | 14463 | 49722 | 95808 | 73193 | 37865 | 84147 | 46004 | 43753 | 92444 |
| 95384 | 28822 | 12047 | 59393 | 14588 | 22723 | 64262 | 93653 | 00284 | 05594 |
| 51396 | 45671 | 08283 | 96848 | 27039 | 20852 | 38008 | 65531 | 65322 | 51775 |

**Table XV** (*continued*)

| | | | | | | | | | |
|---|---|---|---|---|---|---|---|---|---|
| 70321 | 26394 | 01403 | 77390 | 52111 | 27816 | 33570 | 28064 | 41906 | 81867 |
| 98710 | 50639 | 43559 | 34442 | 25514 | 32178 | 83688 | 31018 | 11232 | 70459 |
| 61664 | 16238 | 04228 | 33224 | 18550 | 02255 | 34597 | 64773 | 97872 | 28450 |
| 12906 | 19628 | 77265 | 38578 | 00958 | 67476 | 92199 | 70519 | 32591 | 80452 |
| 07633 | 02489 | 78236 | 70986 | 74294 | 29591 | 31175 | 20817 | 64727 | 70957 |
| 35933 | 31203 | 16796 | 66581 | 55006 | 90733 | 07198 | 65126 | 54346 | 42214 |
| 57652 | 46065 | 59420 | 33920 | 44589 | 70899 | 41795 | 86683 | 27317 | 74817 |
| 86860 | 69306 | 49382 | 48964 | 92022 | 98252 | 47414 | 05190 | 66648 | 35104 |
| 54447 | 02332 | 11406 | 27021 | 60064 | 70307 | 42155 | 15810 | 08324 | 36194 |
| 69865 | 39302 | 09057 | 46982 | 14177 | 94534 | 90536 | 44442 | 43337 | 16371 |
| 40500 | 21406 | 00571 | 87320 | 81683 | 42788 | 86367 | 44686 | 22159 | 67015 |
| 35892 | 49668 | 83991 | 72088 | 30210 | 74009 | 86370 | 97956 | 02132 | 93512 |
| 54819 | 26094 | 51409 | 21485 | 94764 | 85806 | 13393 | 48543 | 07042 | 76538 |
| 64224 | 47909 | 09994 | 23750 | 17351 | 52141 | 30486 | 60380 | 86546 | 66606 |
| 36913 | 58173 | 45709 | 83679 | 82617 | 23381 | 09603 | 61107 | 00566 | 06572 |
| 64745 | 10614 | 86371 | 43244 | 97154 | 10397 | 50975 | 68006 | 20045 | 16942 |
| 25536 | 74031 | 31807 | 70133 | 78790 | 40341 | 68730 | 39635 | 39013 | 66841 |
| 44043 | 96215 | 21270 | 59427 | 25034 | 40645 | 84741 | 52083 | 54503 | 36861 |
| 27659 | 95463 | 53847 | 40921 | 70116 | 61536 | 56756 | 08967 | 31079 | 20097 |
| 76014 | 99818 | 16606 | 19713 | 66904 | 27106 | 24874 | 96701 | 73287 | 76772 |
| 06073 | 57343 | 51428 | 91171 | 28299 | 17520 | 64903 | 04177 | 36071 | 94952 |
| 59008 | 28543 | 11576 | 74547 | 13260 | 20688 | 41261 | 02780 | 06633 | 37536 |
| 08844 | 95774 | 49323 | 30448 | 14154 | 83379 | 71259 | 23302 | 68402 | 43750 |
| 88505 | 15575 | 44927 | 06584 | 29867 | 21541 | 65763 | 12154 | 86616 | 79877 |
| 73259 | 68626 | 98962 | 68548 | 86576 | 48046 | 51755 | 64995 | 03661 | 64585 |
| 81550 | 46798 | 49319 | 50206 | 22024 | 05175 | 12923 | 23427 | 55915 | 91723 |
| 55831 | 83784 | 81034 | 86779 | 34622 | 84570 | 18960 | 48798 | 42970 | 95789 |
| 39465 | 82353 | 68905 | 44234 | 18244 | 54345 | 05592 | 89361 | 14644 | 67924 |
| 66415 | 89349 | 88530 | 72096 | 44459 | 05258 | 48317 | 48866 | 56886 | 90458 |
| 75889 | 04514 | 37227 | 11302 | 04667 | 02129 | 80414 | 86289 | 15887 | 87380 |
| 50749 | 83220 | 50529 | 20619 | 11606 | 36531 | 23409 | 78122 | 19566 | 76564 |
| 33045 | 66703 | 30017 | 35347 | 35038 | 12952 | 13971 | 03922 | 98702 | 11786 |
| 38388 | 69556 | 76728 | 60535 | 59961 | 23634 | 42211 | 98387 | 34880 | 27755 |
| 93182 | 99040 | 96390 | 65989 | 38375 | 03652 | 59657 | 57431 | 24666 | 11061 |
| 64713 | 85185 | 72849 | 58611 | 31220 | 26657 | 77056 | 24553 | 24993 | 05210 |
| 89024 | 32054 | 46997 | 92652 | 28363 | 98992 | 22593 | 97710 | 47766 | 37646 |
| 93573 | 95502 | 33790 | 92973 | 27766 | 62671 | 89698 | 10877 | 73893 | 41004 |
| 96035 | 18795 | 48080 | 59666 | 30241 | 35233 | 87353 | 43647 | 13404 | 41982 |
| 19264 | 29229 | 61369 | 08309 | 39383 | 42305 | 25944 | 13577 | 51545 | 68990 |
| 69801 | 37145 | 79189 | 55897 | 57793 | 66816 | 21930 | 56771 | 79296 | 73793 |

(*continues*)

**Table XV** Random Digits (*continued*)

| | | | | | | | | | |
|---|---|---|---|---|---|---|---|---|---|
| 21632 | 42301 | 23693 | 72641 | 56310 | 85576 | 03004 | 25669 | 69221 | 32996 |
| 23040 | 65782 | 23712 | 13414 | 10758 | 15590 | 97298 | 74246 | 51511 | 46900 |
| 36795 | 38292 | 03852 | 06384 | 84421 | 03446 | 91670 | 45312 | 27609 | 87034 |
| 06683 | 83891 | 88991 | 16533 | 09197 | 31427 | 60384 | 48525 | 90978 | 46107 |
| 21693 | 12956 | 21804 | 46558 | 37682 | 81207 | 85840 | 53238 | 35026 | 04835 |
| | | | | | | | | | |
| 53264 | 41376 | 17783 | 64756 | 39278 | 25403 | 33042 | 20954 | 31193 | 24247 |
| 45911 | 92453 | 25370 | 86602 | 48574 | 57865 | 26436 | 16122 | 76614 | 17028 |
| 21262 | 59718 | 77821 | 14036 | 31033 | 90563 | 45410 | 15158 | 90209 | 84089 |
| 38053 | 60780 | 54166 | 14255 | 33120 | 27171 | 71798 | 91214 | 80040 | 56699 |
| 12475 | 40193 | 59415 | 04769 | 75920 | 01036 | 02692 | 75862 | 16612 | 73670 |
| | | | | | | | | | |
| 61182 | 03305 | 90334 | 00187 | 91659 | 28063 | 75684 | 50017 | 82643 | 09282 |
| 77376 | 85469 | 08164 | 05584 | 36623 | 82597 | 83859 | 03435 | 98460 | 70095 |
| 80257 | 04381 | 06501 | 08924 | 35514 | 14297 | 54373 | 71369 | 05172 | 15955 |
| 82441 | 04636 | 48215 | 06821 | 03385 | 17663 | 40107 | 55679 | 30366 | 42390 |
| 95895 | 16083 | 58499 | 17176 | 55993 | 51034 | 49296 | 04010 | 78974 | 35930 |
| | | | | | | | | | |
| 02019 | 96226 | 27167 | 68245 | 53109 | 59037 | 37843 | 79243 | 10262 | 58797 |
| 61490 | 82590 | 52411 | 54783 | 29447 | 94551 | 30026 | 97959 | 93939 | 73217 |
| 82573 | 62154 | 78291 | 33728 | 39102 | 11484 | 86210 | 43794 | 73553 | 87435 |
| 01110 | 77108 | 56521 | 78610 | 08254 | 01842 | 43068 | 70415 | 79195 | 26136 |
| 49786 | 47279 | 38471 | 20379 | 54704 | 86614 | 91138 | 51595 | 50818 | 80186 |
| | | | | | | | | | |
| 95809 | 54837 | 55978 | 10534 | 46194 | 00273 | 03659 | 57186 | 73342 | 95949 |
| 76742 | 70505 | 64773 | 48334 | 00869 | 80439 | 69374 | 35279 | 99952 | 85860 |
| 03880 | 30798 | 40515 | 66819 | 40691 | 72678 | 17590 | 76085 | 62741 | 93844 |
| 81045 | 58617 | 37788 | 64693 | 50968 | 31853 | 95733 | 08068 | 21988 | 60613 |
| 54802 | 96997 | 52909 | 14310 | 08726 | 09630 | 49081 | 66952 | 05603 | 08950 |
| | | | | | | | | | |
| 63119 | 68055 | 76641 | 87635 | 64835 | 06121 | 86006 | 76257 | 51695 | 44571 |
| 85397 | 39692 | 66765 | 50318 | 33763 | 45429 | 30943 | 20128 | 14439 | 68279 |
| 06618 | 14101 | 87706 | 77153 | 54866 | 00025 | 34092 | 92939 | 12528 | 24763 |
| 19525 | 93122 | 11658 | 06188 | 43735 | 43104 | 18115 | 28815 | 21863 | 22218 |
| 49494 | 16854 | 95248 | 00045 | 40357 | 73893 | 50732 | 11319 | 38804 | 15121 |
| | | | | | | | | | |
| 51694 | 84242 | 33341 | 77153 | 71970 | 66070 | 80879 | 10293 | 70875 | 61168 |
| 01838 | 72046 | 40042 | 59287 | 78115 | 74332 | 36858 | 17687 | 14357 | 58846 |
| 32697 | 10332 | 92643 | 06454 | 39300 | 20099 | 61461 | 15730 | 29333 | 95548 |
| 75128 | 04855 | 96418 | 26636 | 53328 | 69758 | 17597 | 56658 | 81043 | 40374 |
| 15675 | 20566 | 45945 | 10123 | 21679 | 38139 | 95843 | 76372 | 78669 | 14598 |
| | | | | | | | | | |
| 78142 | 72095 | 47327 | 43718 | 48286 | 88374 | 65046 | 27199 | 50484 | 03834 |
| 72869 | 28546 | 22578 | 94059 | 56817 | 88443 | 65557 | 75239 | 38101 | 17180 |
| 13896 | 62838 | 09470 | 31133 | 65941 | 84219 | 23017 | 34539 | 77391 | 52502 |
| 95808 | 40466 | 39870 | 79974 | 71187 | 02420 | 23124 | 15714 | 91874 | 86307 |
| 52320 | 60822 | 38657 | 81962 | 32388 | 50425 | 53231 | 62797 | 95490 | 87063 |

**Table XV** (*continued*)

| | | | | | | | | | |
|---|---|---|---|---|---|---|---|---|---|
| 03564 | 46703 | 30528 | 28041 | 86108 | 99297 | 31593 | 21021 | 12451 | 90445 |
| 89432 | 52921 | 28068 | 71091 | 12944 | 06524 | 42605 | 02606 | 69417 | 81733 |
| 88573 | 55150 | 01443 | 97336 | 79910 | 49014 | 02237 | 85000 | 32344 | 45649 |
| 32920 | 01391 | 01105 | 15435 | 10918 | 92181 | 03839 | 92364 | 84229 | 83989 |
| 52704 | 39386 | 81791 | 35616 | 97616 | 64947 | 60456 | 16196 | 79527 | 43770 |
| | | | | | | | | | |
| 02696 | 86377 | 34209 | 38850 | 43712 | 58088 | 58490 | 42162 | 16423 | 79089 |
| 83961 | 43893 | 81108 | 79331 | 27601 | 30995 | 25447 | 05835 | 26029 | 01069 |
| 21914 | 31443 | 85624 | 29878 | 97401 | 66466 | 88421 | 76385 | 65526 | 93134 |
| 60215 | 43656 | 42638 | 13774 | 87380 | 32166 | 44914 | 57637 | 95151 | 08573 |
| 17644 | 23867 | 35765 | 75634 | 22484 | 86921 | 29597 | 94523 | 39661 | 15403 |
| | | | | | | | | | |
| 69499 | 38275 | 93129 | 99455 | 06429 | 10947 | 62748 | 09375 | 53925 | 65096 |
| 96761 | 14313 | 79554 | 48204 | 56142 | 39889 | 98293 | 07233 | 25422 | 43510 |
| 42115 | 20104 | 10771 | 93968 | 76480 | 82630 | 09458 | 50774 | 00461 | 17435 |
| 96628 | 79221 | 70360 | 47978 | 68880 | 91249 | 42500 | 92943 | 89942 | 94929 |
| 98056 | 88721 | 38743 | 37395 | 23774 | 29013 | 39877 | 56221 | 08293 | 74795 |
| | | | | | | | | | |
| 62011 | 34646 | 99276 | 37811 | 30494 | 51236 | 30385 | 22514 | 77077 | 68381 |
| 97562 | 82265 | 27078 | 02950 | 45701 | 53691 | 27376 | 17196 | 53122 | 40779 |
| 81485 | 72983 | 95838 | 93212 | 68260 | 21176 | 33964 | 32478 | 98334 | 81713 |
| 53465 | 74671 | 88519 | 84254 | 65937 | 56020 | 45728 | 52449 | 17785 | 75868 |
| 14640 | 29533 | 35425 | 50917 | 85742 | 38691 | 31928 | 68477 | 70081 | 31907 |
| | | | | | | | | | |
| 58410 | 98236 | 76474 | 16076 | 17250 | 91650 | 83632 | 54718 | 16705 | 22827 |
| 26780 | 38018 | 96714 | 32836 | 11929 | 56912 | 71592 | 29622 | 71248 | 49260 |
| 74439 | 29381 | 62148 | 13205 | 68606 | 03817 | 35829 | 21987 | 40162 | 35558 |
| 51015 | 49183 | 15384 | 28173 | 53705 | 96163 | 72306 | 10015 | 63078 | 95319 |
| 28516 | 56674 | 30562 | 96465 | 17886 | 00360 | 11265 | 05653 | 51383 | 85153 |
| | | | | | | | | | |
| 79501 | 69898 | 55076 | 54853 | 66742 | 70410 | 44434 | 15140 | 88331 | 75362 |
| 35468 | 52850 | 17797 | 78112 | 42126 | 33055 | 99776 | 40129 | 70370 | 27342 |
| 29763 | 07791 | 20976 | 69285 | 32965 | 95201 | 96582 | 71055 | 16511 | 13122 |
| 76537 | 32386 | 84442 | 97095 | 31922 | 39406 | 56418 | 76857 | 51158 | 43193 |
| 74519 | 89378 | 58353 | 83848 | 02802 | 06046 | 74264 | 76358 | 08642 | 31973 |
| | | | | | | | | | |
| 94988 | 12022 | 77021 | 60277 | 39048 | 03087 | 18920 | 98682 | 26756 | 05107 |
| 72363 | 40974 | 09594 | 10276 | 09631 | 43203 | 13227 | 90021 | 35899 | 21515 |
| 74967 | 66480 | 83894 | 82989 | 24784 | 42757 | 24447 | 44970 | 60048 | 26514 |
| 26236 | 32399 | 81419 | 47377 | 93952 | 89101 | 79748 | 03446 | 23212 | 10489 |
| 05632 | 68465 | 67842 | 85597 | 02094 | 42059 | 86912 | 23145 | 43060 | 94694 |
| | | | | | | | | | |
| 67352 | 41392 | 17545 | 30949 | 87565 | 83820 | 19827 | 34043 | 00575 | 23260 |
| 92727 | 35027 | 03117 | 80848 | 74559 | 96797 | 08118 | 72948 | 91838 | 00281 |
| 18223 | 91136 | 39695 | 39943 | 77413 | 48937 | 32672 | 94704 | 99738 | 50907 |
| 80723 | 91394 | 02992 | 11530 | 67845 | 05881 | 76173 | 18594 | 91937 | 74215 |
| 75007 | 85671 | 88211 | 55080 | 15581 | 02685 | 07889 | 26594 | 47083 | 76723 |

(*continues*)

**Table XV** Random Digits (*continued*)

| | | | | | | | | | |
|---|---|---|---|---|---|---|---|---|---|
| 60050 | 80463 | 30926 | 74970 | 38951 | 14928 | 81875 | 61424 | 62060 | 96004 |
| 69715 | 28522 | 73974 | 99491 | 50647 | 20252 | 44455 | 66593 | 23255 | 94807 |
| 92096 | 43555 | 48882 | 60717 | 07963 | 39375 | 21441 | 66090 | 80430 | 68380 |
| 35482 | 63353 | 08086 | 66635 | 71009 | 95777 | 70335 | 65808 | 69105 | 42800 |
| 24879 | 78061 | 38949 | 21123 | 28430 | 72627 | 04565 | 14741 | 85781 | 20795 |
| 68985 | 60486 | 58133 | 07709 | 25899 | 68531 | 10370 | 46536 | 46506 | 86675 |
| 96601 | 96785 | 20850 | 70389 | 74637 | 34020 | 61780 | 33461 | 32496 | 51247 |
| 66706 | 67664 | 93292 | 05934 | 71050 | 68192 | 51898 | 18872 | 01371 | 95558 |
| 39273 | 41912 | 40198 | 36441 | 89472 | 38835 | 41709 | 85397 | 57429 | 33822 |
| 42539 | 21771 | 58672 | 71421 | 92528 | 67229 | 57837 | 89729 | 97494 | 40943 |
| 78209 | 77315 | 08393 | 95809 | 15832 | 31381 | 83170 | 32933 | 90911 | 49431 |
| 31279 | 02627 | 93411 | 96192 | 88570 | 88861 | 79463 | 64823 | 93331 | 97785 |
| 76915 | 11168 | 58452 | 30237 | 43211 | 88094 | 40120 | 44502 | 89799 | 09877 |
| 31714 | 15972 | 03620 | 07957 | 84828 | 01328 | 66806 | 45975 | 01001 | 72953 |
| 23131 | 71925 | 82240 | 57451 | 86216 | 08900 | 42868 | 39434 | 74956 | 93714 |
| 95186 | 13811 | 57341 | 15008 | 70542 | 51583 | 01563 | 50348 | 01403 | 32881 |
| 69801 | 54360 | 53265 | 70858 | 65549 | 02535 | 43657 | 75928 | 00109 | 59990 |
| 22293 | 14758 | 37440 | 49589 | 96421 | 74696 | 46442 | 27647 | 63950 | 60872 |
| 91834 | 04476 | 12776 | 34486 | 64027 | 15943 | 12307 | 36791 | 13600 | 32570 |
| 18615 | 53505 | 27034 | 34479 | 81642 | 20618 | 19992 | 12006 | 37023 | 48116 |
| 97224 | 65695 | 75788 | 47328 | 86654 | 52342 | 31619 | 97675 | 40129 | 57606 |
| 26686 | 27899 | 76013 | 33828 | 00554 | 35016 | 56081 | 74251 | 51761 | 20546 |
| 71559 | 57727 | 59162 | 88726 | 75150 | 37162 | 90733 | 79311 | 48085 | 67398 |
| 46858 | 08197 | 42531 | 34583 | 46155 | 77480 | 13712 | 15607 | 61265 | 81522 |
| 68979 | 78824 | 23553 | 16776 | 72208 | 12765 | 33682 | 68422 | 53269 | 99473 |
| 82578 | 98058 | 70623 | 49929 | 23105 | 14041 | 96639 | 89657 | 53337 | 31527 |
| 26830 | 21649 | 96350 | 76934 | 95421 | 23398 | 24309 | 97283 | 55317 | 06715 |
| 66780 | 42434 | 36053 | 11518 | 10114 | 29978 | 35558 | 73063 | 47919 | 41275 |
| 26850 | 83856 | 07550 | 60184 | 71509 | 33950 | 19410 | 57673 | 82122 | 98268 |
| 83552 | 46929 | 39407 | 19738 | 08959 | 95800 | 66562 | 70100 | 16777 | 17829 |
| 69664 | 34188 | 44958 | 12585 | 14608 | 48201 | 21973 | 64338 | 56105 | 01633 |
| 91846 | 59409 | 44903 | 39135 | 09758 | 79483 | 93450 | 39965 | 19354 | 55190 |
| 94766 | 30727 | 36436 | 87104 | 45035 | 27194 | 40273 | 71694 | 57000 | 40477 |
| 06482 | 16026 | 77039 | 23656 | 00584 | 01539 | 43904 | 13910 | 79229 | 91014 |
| 38098 | 51630 | 27394 | 80495 | 28942 | 12426 | 84937 | 10365 | 44686 | 73746 |
| 42624 | 08957 | 40485 | 75135 | 35223 | 18951 | 86245 | 86562 | 52403 | 37080 |
| 75639 | 11741 | 34530 | 96298 | 61180 | 04574 | 41471 | 76100 | 49195 | 68552 |
| 58681 | 84924 | 85898 | 94144 | 40948 | 88720 | 92349 | 75081 | 72752 | 03225 |
| 49384 | 67921 | 07641 | 03287 | 85245 | 06555 | 59403 | 71346 | 25280 | 03515 |
| 02176 | 23783 | 96594 | 05593 | 94006 | 41335 | 81326 | 28049 | 25784 | 22043 |

Reprinted with permission from *A Million Random Digits with 100,000 Normal Deviates*, Rand Corporation, Santa Monica, Calif.

# Answers to Selected Problems

## Chapter 1

**1. (a)** $\{6, N6, NN6, NNN6, ...\}$  **(b)** $\{M1, M2, M3, M4, F1, F2, F3, F4\}$

**(c)** $\{$Democrat, Republican, Other, None$\}$

**(d)** An interval of real numbers from 50 to 400 should suffice.

**(e)** The set of nonnegative real numbers

**2. (a)** $E = \{Ht, Th, Tt\}$, $F = \{Hh, Ht, Th\}$, $G = \{Hh, Th\}$.

**(b)** $EG^c = \{Ht, Tt\}$, $EFG = \{Th\}$, $E^c \cup F^c = \{Hh, Tt\}$, $(EF)^c = \{Hh, Tt\}$,

$E \cup FG = \Omega = \{Hh, Ht, Tt, Th\}$.  **(c)** 16

**5. (a)** $E$  **(b)** $\emptyset$  **(c)** $\Omega$  **(d)** $E$  **(e)** $E$  **(f)** $E$  **(g)** $\Omega$  **(h)** $\emptyset$

**8. (a)** $3/4$  **(b)** $5/13$  **(c)** $5/26$  **(d)** $1/2$  **(e)** $21/26$  **(f)** $2/13$

**10. (a)** $2/5$  **(b)** $3/5$  **12.** $3/10$  **14.** $12$

**16. (a)** $(15)_4 = 32,760$.  **(b)** $\binom{15}{4} = 1365$.

**18. (a)** $4^{20} \doteq 1.0995 \times 10^{12}$.  **(b)** $4 \times 3^{19} \doteq 4.649 \times 10^9$.

**20. (a)** $2^{10}$  **(b)** $252$  **(c)** $1/512$  **22.** $(12)_8 = 19,958,400$.

**24. (a)** $5,040$  **(b)** $720$  **(c)** $288$  **(d)** $144$  **(e)** $3600$  **25.** $n(n-1)$

**26. (a)** $20,358,520$  **(b)** Approx. $1.9 \times 10^{14}$  **28. (a)** $120$  **(b)** $112$  **(c)** $64$

**29.** $1,023$  **31.** $15$  **34.** $1/56$  **35. (a)** $\frac{1}{1,352,000}$ for each  **(b)** $\frac{20}{1,352,000}$

**36. (a)** $13/45$  **(b)** $11/15$  **(c)** $1$  **(d)** $4/10$  **38. (a)** $5/14$  **(b)** $2/7$

**40.** Approximately $6.44 \times 10^{-7}$  **42.** $.504$

**44. (a)** $56$  **(b)** $1/28$  **(c)** $1/14$  **(d)** $3/7$  **(e)** $3/7$

**45. (a)** $.0211 \left(\frac{54,912}{2,598,960}\right)$  **(b)** $.00144 \left(\frac{3,744}{2,598,960}\right)$

**47. (a)** $y^5 + 5xy^4 + 10x^2y^3 + 10x^3y^2 + 5x^4y + x^5$

**(b)** $z^3 + 3z^2y + 3z^2x + 6xyz + 3zy^2 + 3zx^2 + 3x^2y + 3xy^2 + y^3 + x^3$

**49.** $1, 6, 15, 20, 15, 6, 1$  **50. (a)** $56$  **(b)** $105$  **(c)** $2520$

**51. (a)** $1/6$  **(b)** $\left(\frac{5}{6}\right)^{k-1}\left(\frac{1}{6}\right)$, $k = 1, 2, ...$  **52. (a)** $.38$  **(b)** $.91$  **(c)** $.68$  **(d)** $.32$

**R-1.** **(a)** $\emptyset$ **(b)** $E$ **R-3.** 6! **R-4.** 34650 **R-5.** .91 **R-6.** 3/5
**R-7.** $495p^4q^8$ **R-8.** **(a)** 32760 **(b)** 3003 **(c)** 33/91 **(d)** 1/5 **(e)** 4/15
**R-9.** **(a)** 3/8 **(b)** 1/56 **(c)** 55/56 **(d)** 1/4 **(e)** 5/168 **(f)** 5/28
**R-10.** .2 **R-11.** **(a)** $\frac{1}{54979155}$ **(b)** .00014187 (approx. 1:7048) **(c)** .5266

# Chapter 2

**1.**

| $x$ | 2 | 3 | 4 | 5 | 6 | 7 | 8 | 9 | 10 | 11 | 12 |
|---|---|---|---|---|---|---|---|---|---|---|---|
| $f(x)$ | $\frac{1}{36}$ | $\frac{2}{36}$ | $\frac{3}{36}$ | $\frac{4}{36}$ | $\frac{5}{36}$ | $\frac{6}{36}$ | $\frac{5}{36}$ | $\frac{4}{36}$ | $\frac{3}{36}$ | $\frac{2}{36}$ | $\frac{1}{36}$ |

**2.**

| $y$ | $-5$ | $-4$ | $-3$ | $-2$ | $-1$ | 0 | 1 | 2 | 3 | 4 | 5 |
|---|---|---|---|---|---|---|---|---|---|---|---|
| $f(y)$ | $\frac{1}{36}$ | $\frac{2}{36}$ | $\frac{3}{36}$ | $\frac{4}{36}$ | $\frac{5}{36}$ | $\frac{6}{36}$ | $\frac{5}{36}$ | $\frac{4}{36}$ | $\frac{3}{36}$ | $\frac{2}{36}$ | $\frac{1}{36}$ |

**4.** $\omega = 12, 13, 14, 15, 23, 24, 25, 34, 35, 45$

**(a)**

| $s$ | 3 | 4 | 5 | 6 | 7 | 8 | 9 |
|---|---|---|---|---|---|---|---|
| $f(s)$ | $\frac{1}{10}$ | $\frac{1}{10}$ | $\frac{2}{10}$ | $\frac{2}{10}$ | $\frac{2}{10}$ | $\frac{1}{10}$ | $\frac{1}{10}$ |

**(b)**

| $d$ | 1 | 2 | 3 | 4 |
|---|---|---|---|---|
| $f(d)$ | $\frac{4}{10}$ | $\frac{3}{10}$ | $\frac{2}{10}$ | $\frac{1}{10}$ |

**(c)**

| $u$ | 2 | 3 | 4 | 5 |
|---|---|---|---|---|
| $f(u)$ | $\frac{1}{10}$ | $\frac{2}{10}$ | $\frac{3}{10}$ | $\frac{4}{10}$ |

**5.** **(a)** .5161 **(b)** .2577 **(c)** .5652 **(d)** .1538
**7.** **(a)** 4/20 **(b)** 4/20 **(c)** 2/20 **(d)** 6/20
**9.** **(a)** $f_Y(1) = .6, f_Y(2) = .4, f_X(1) = .2, f_X(2) = .3, f_X(3) = .5$
**(b)** .8 **(c)** $f(2) = .2, f(4) = .7, f(5) = .1.$
**11.** **(a)** 1/13 **(b)** 1/4 **(c)** 1/2 **(d)** 1 **(e)** 1/13 **(f)** 1/2
**13.** **(a)** $\frac{600}{2109} = .2845.$ **(b)** .24 **15.** **(a)** $\frac{11}{100}$ **(b)** $\frac{357}{494}$ **(c)** 1/4
**17.** **(a)** 1/5 **(b)** 1/9 **18.** 1/25 **20.** 3/5
**23.** **(b)** **(c)** $f(0) = 2/12, f(1) = 5/12, f(2) = 5/12.$

|  |  | $x$ | | | |
|---|---|---|---|---|---|
|  |  | 1 | 2 | 3 | 4 |
| | 0 | $\frac{1}{8}$ | $\frac{1}{24}$ | 0 | 0 |
| $y$ | 1 | $\frac{1}{8}$ | $\frac{1}{6}$ | $\frac{1}{8}$ | 0 |
| | 2 | 0 | $\frac{1}{24}$ | $\frac{1}{8}$ | $\frac{1}{4}$ |

**(d)** 3/5

**25.** 2/47 **28.** .2266
**30.** **(a)** .276 **(b)** .446 **31.** **(a)** 1/36 **(b)** 1/6 **(c)** 1/12 **32.** .2015
**33.** $P$(at least one hit in 20 games) $= .0111.$ **35.** No: If $W = 13, Z = 10.$
**38.** **(a)** 25/36 **(b)** 2/3 **40.** 91/216 **44.** .24
**49.** **(a)** 8/13 **(b)** 2/3 **(c)** 10/39
**R-1.** **(a)** $X$ or $Y$: $f(0) = .2, f(1) = f(2) = .4$ **(b)** .4
**(c)** $(1, 2, 3)$ with prob.'s $(1/4, 1/2, 1/4)$ resp. **(d)** yes **(e)** no
**(f)** $f(2) = f(6) = .1, f(4) = f(5) = .4.$
**R-2.** **(a)** 0, .9 **(b)** .2, .7 **R-3.** **(a)** 6/25 **(b)** 2/5 **(c)** 2/5 **(d)** 7/13

**R-4.**  10/13     **R-5.**  **(a)**  yes     **(b)**  no, no

**R-6.**  It would result in $P(AB) < P(ABC)$, which is impossible.

**R-7.**  **(a)**  .7     **(b)**  .6     **R-8.**  $f(j, k) = 2^{-(j+k)}$, $j, k = 1, 2, \ldots$

**R-9.**  $P[E(F \cup G)] = P(E)[P(F) + P(G) - P(FG)]$, etc.

**R-10.**  **(a)**  1/7     **(b)**  1/3     **R-11.**  **(a)**  4/12     **(b)**  4/10

**R-12.**  3/4     **R-13.**  $p^3(1 - p)^2$

**R-15.**  **(a)**  $f(0) = .6, f(1) = .3, f(2) = .1.$     **(b)**  $f(0) = 1/2, f(1) = 1/3, f(2) = 1/6.$
**(c)**  $f(0) = .3, f(1) = .4, f(2) = .3.$

---

# Chapter 3

**1.**  **(a)**  2     **(b)**  0     **(c)**  5.6     **2.**  6, 2, 4     **3.**  **(a)**  1/2     **(b)**  13/22

**5.**  **(b)**  3/2, 3/8     **7.**  **(a)**  $\binom{k-1}{2}/\binom{8}{3}$, $k = 3, 4, \ldots, 8$     **(b)**  13/14     **(c)**  27/4

**9.**  **(a)**  1/4     **(b)**  1/4     **(c)**  1     **12.**  4     **14.**  .866     **15.**  1.31     **17.**  .584

**20.**  4/3, 5/9     **22.**  3/4     **23.**  **(a)**  .08     **(b)**  .6     **(c)**  1.01     **(d)**  29/46

**24.**  **(a)**  1     **(b)**  4     **(c)**  0     **(d)**  1     **(e)**  2     **(f)**  25     **28.**  **(a)**  1/48     **(b)**  1

**29.**  .209     **31.**  $1 - p + pt$     **32.**  $(1 - p + pt)^3$, $3p^2(1 - p)$

**34.**  **(a)**  $t/(2 - t)$     **(b)**  $[t/(2 - t)]^2$, 1/4

**R-1.**  **(a)**  2     **(b)**  $.6 + .1t + .3t^2$     **(c)**  .6     **(d)**  3.24     **R-2.**  4

**R-3.**  **(a)**  3/16, 2/9     **(b)**  2/3     **(c)**  0     **(d)**  17/144

**R-4.**  **(a)**  5     **(b)**  5     **(c)**  $-1$     **(d)**  $-\dfrac{1}{3\sqrt{5}}$

**R-5.**  **(a)**  $f(0) = .3, f(1) = .2, f(2) = .5.$     **(b)**  .76     **R-6.**  5/4     **R-7.**  0

**R-8.**  **(a)**  .1     **(b)**  $-1/3$     **(c)**  2/3     **(d)**  1     **(e)**  .6     **R-9.**  **(b)**  4/3

**R-10.**  **(b)**  .15     **(c)**  5/9

**R-12.**  **(a)**

| $y$ | 0 | 1 | 2 | 3 | 4 | 5 | 6 |
|-----|-----|-----|-----|-----|-----|-----|-----|
| $f(y)$ | .04 | .16 | .28 | .28 | .17 | .06 | .01 |

---

# Chapter 4

**1.**  **(a)**  .2461     **(b)**  .1719     **(c)**  5, $\sqrt{10/4}$

**2.**  **(a)**  .3826     **(b)**  .8131     **(c)**  .5695     **(d)**  .8

**4.**  **(a)**  5, 25     **(b)**  9.68     **(c)**  .0099     **(d)**  .415

**6.**  10, 40, 60, 40, 10, resp.     **8.**  $n \geq 7$

**10.**  **(a)**  1.25     **(b)**  .740     **(c)**  .0677     **(d)**  .0726     **12.**  **(a)**  .7636     **(b)**  1     **(c)**  .1697

**14.**  **(a)**  one in 1,947,792     **(b)**  .000092     **(c)**  .00335     **(d)**  .3048     **(e)**  $5.30

**15.**  **(a)**  .0229     **(b)**  10     **(c)**  30     **17.**  **(a)**  6     **(b)**  3.5

**19.**  **(a)**  17.26     **(b)**  35/128     **(c)**  24     **21.**  25/3     **23.**  .1151     **25.**  .0521

**27.**  .0736     **29.**  **(a)**  .223     **(b)**  .3353     **(c)**  .934

**31. (a)** $\dfrac{\binom{950}{20}+\binom{50}{1}\binom{950}{19}}{\binom{100}{20}} = .736043$    **(b)** .73584    **(c)** .735759    **(d)** $\Phi\left(\frac{.5}{.95}\right) \doteq .70$

**33. (a)** .593    **(b)** .135    **(c)** .003    **(d)** .385

**35. (a)** 1.8    **(b)** .0273    **(c)** .025    **(d)** .884    **37.** 44.33

**40. (a)** .0098    **(b)** .01344    **(c)** .238    **(d)** .401    **(e)** .393

**42. (a)** .0335    **(b)** .0209    **(c)** .114

**44. (a)** $\mathrm{Bin}(n,\,p_1)$    **(b)** $\mathrm{Bin}(n,\,p_1+p_2)$    **(c)** $\mathrm{Bin}(n-k,\,p)$, where $p = \frac{p_1}{p_1+p_3}$    **(d)** $\mathrm{Bin}\left(k,\,\frac{p_1}{p_1+p_2}\right)$

**47.** $\left\{\dfrac{pt}{1-(1-p)t}\right\}^r$

**R-1.** .159    **R-2. (a)** 150    **(b)** 4, 3.84    **(c)** $\mathrm{Ber}(100,.04)$    **(d)** .785    **(e)** .322

**R-3. (a)** .385    **(b)** .301

**R-4. (a)** $\left\{\dfrac{pt}{1-(1-p)t}\right\}^4$    **(b)** $4/p, 4(1-p)/p^2$    **(c)** $20p^4(1-p)^3$

**R-5. (a)** 10, 10    **(b)** .647

**R-6.** .0513    **R-7. (a)** .064    **(b)** 3.5    **(c)** $\mathrm{Bin}(10,.7)$

**R-8. (a)** 10 correct (25%)    **(b)** .292

**R-9. (a)** .715    **(b)** $\frac{1}{2}e^{-1}$    **(c)** 1/2    **(d)** .046    **R-10. (a)** .3087    **(b)** .283

**R-11. (a)** $e^{-1}, e^{-1}, 1-e^{-1}$    **(b)** .695, .253, .305

# Chapter 5

**1. (a)** 1/8    **(b)** 1/2    **(c)** 1/2    **(d)** $x_0 = 1/2$    **3. (a)** 0, 1    **(b)** $1/\sqrt{2}$

**5. (a)** $x^2, 0 < x < 1$    **(b)** $y^2/144, 0 < y < 12$

**6. (a)** 1/2    **(b)** $(y+1)/2, -1 < y < 1.$    **(c)** $z^2, 0 < z < 1$

**8. (a)** 1/6    **(b)** 1/3    **(c)** 2/3    **(d)** 1/2    **(e)** 5/6    **(f)** 0

**10.** 2/5    **11.** $f(x) = 1/2, 0 < x < 2$ [i.e., $X \sim \mathcal{U}(0, 2)$].

**13. (a)** $f(x) = 1/2$ for $0 < x < 1$, and 1/4 for $1 < x < 3$.    **(b)** 1/2

**15. (a)** 1    **(b)** 1/4

**(c)** On $(-1, 0)$, $F(x) = (1+x)^2/2$; on $(0, 1)$, $F(x) = 1 - (1-x)^2/2$.

**17.** On $(0, 1)$, $f(y) = 2/3$; on $(1, 2)$, $f(y) = 2(2-y)/3$ (and 0 elsewhere).    **19.** .03183

**21. (a)** $f(x) = 1/2, -1 < x < 1.$    **(b)** $f_Y(y) = 1, 0 < y < 1.$    **(c)** $\frac{1}{2\sqrt{z}}, 0 < z < 1.$

**23. (a)** $\sqrt{2}$    **(b)** 4/3    **(c)** .9    **(d)** .8    **25. (a)** 3/4    **(b)** 0    **(c)** 1/3    **(d)** 1/6

**27. (a)** $2^{-1/3}$    **(b)** 3/20    **(c)** .45    **30. (a)** 1/2    **(b)** 1    **31.** $2\sqrt{2/\pi}$

**33.** $1/\sqrt{5}$    **35.** 3/8    **37.** 3/80    **38. (a)** 1.25    **(b)** 22.16

**39. (a)** $F(x) = 5x^4 - 4x^5, 0 < x < 1$    **(b)** 2/3    **(c)** 2/63    **(d)** 1/12    **41.** 64/45

**43. (a)** $f_X(u) = f_Y(u) = 2(1-u), 0 < u < 1.$    **(b)** 1/12    **(c)** 2/3    **(d)** 3/4    **(e)** 1/2    **(f)** 1/4

**44.** Each is $\mathcal{U}(0, 1)$.

**47. (a)** $f_X(u) = f_Y(u) = \frac{1}{2} + u, 0 < u < 1.$    **(b)** 7/6    **(c)** 1/3    **(d)** $z^3/3$

**49. (a)** 3/4    **(b)** $f_X(u) = f_Y(u) = 1/2, -1 < u < 1.$    **(c)** Yes

**51.** 0    **53. (a)** $F_Z(z) = z^2, f_Z(z) = 2z, 0 < z < 1.$    **(c)** $-1/2$

**54.**  $-\sqrt{(1-\rho_{X,Y})/2}$

**57.**  **(a)** $f_X(u) = f_Y(u) = 1 - |u|,\ |u| < 1.$  **(b)** no, yes  **(c)** $1/3$  **(d)** $(1+z)/2,\ -1 < z < 1.$

**59.**  **(a)** $3/2$  **(b)** $1/4$  **(c)** $1/6$  **(d)** $-1/6$

**61.**  **(a)** $2^n x_1 x_2 \cdots x_n$ for $0 < x_i < 1$ (all $i$)

**62.**  **(a)** $\frac{x+y}{1/2+y},\ 0 < x < 1.$  **(b)** $\frac{2+3y}{3+6y}$  **64.** $f(y\mid x) = \frac{1}{1-x},\ 0 < y < 1 - x.$

**66.**  **(a)** $ye^{-y(x+1)},\ x > 0,\ y > 0$  **(b)** $f_X(x) = \frac{1}{(1+x)^2},\ x > 0.$  **(c)** $f_{Y|x}(y) = (1+x)^2 y e^{-(1+x)y}.$

**68.**  **(a)** $3y^2, 0 < y < 1$  **(b)** $8x, 0 < x < 1/2$

**69.**  **(a)** 0 for $k$ odd, $k!$ for $k$ even  **(b)** 2  **71.** $\psi(t) = \frac{e^{\theta t} - 1}{\theta t}$

**73.**  $\psi(t) = (1-t)^{-1}, t < 1;\ E(X^k) = k!$

**R-1.**  **(a)** $1/2$  **(b)** 0  **(c)** $1/4$

**R-2.**  **(a)** $y^2/3$ for $0 < y < 1$, and $\frac{1}{3}(2y-1)$ for $1 < y < 2$

**(b)** $\sqrt{3}/2$  **(c)** $5/4$  **(d)** $11/9$  **(e)** $37/162$

**R-3.**  **(a)** $1/2$  **(b)** $1/3, 0 < v < 3$  **(c)** $F_U(u) = u, 0 < u < 1$

**R-4.**  $f(y) = 2y, 0 < y < 1$  **R-5.** $3e^{-3y}, y > 0$

**R-6.**  **(a)** 0  **(b)** .529  **(c)** 0  **(d)** 5  **R-7.** **(a)** $1/\sqrt{3}$  **(b)** 2

**R-8.**  **(a)** $1/2$  **(b)** $\frac{2}{\pi}\sqrt{1-x^2}, -1 < x < 1.$

**R-9.**  **(a)** $uve^{-(u+v)}, u > 0, v > 0$  **(b)** 0  **(c)** $(1-t)^{-4}, t < 1$

**R-10.**  **(a)** $2x, 0 < x < 1$  **(b)** $x/2$  **(c)** $1/2$  **(d)** $1/4$

**R-11.**  **(a)** 0  **(b)** $1/2$  **(c)** $1, -\frac{1}{2} < y < \frac{1}{2}$  **(d)** 0  **(e)** $5/6$

**R-12.**  **(b)** $1 - \exp\{-x^2/2\}, 0 < x.$  **(c)** 2  **R-13.** .0222

**R-14.**  **(a)** 6  **(b)** $\mathcal{U}(0, 1 - x - y)$  **(c)** (same for all) $3(1-u)^2,\ 0 < u < 1$

**R-15.**  **(a)** 1  **(b)** 9  **R-16.** **(a)** 0  **(b)** .1  **(c)** .122  **(d)** .588

# Chapter 6

**1.**  **(a)** 4  **(b)** $-1.5$  **2.** **(a)** .1587  **(b)** .6826

**4.**  **(a)** .181  **(b)** $19.1, 25.3$  **(c)** $1/2$  **(d)** .55

**6.**  **(a)** $-6, 64$  **(b)** $\exp[-6t + 32t^2]$  **8.** 14

**10.**  **(a)** $\mathcal{N}(11, 7)$  **(b)** $\mathcal{N}(-1, 5)$  **(c)** $\mathcal{N}(1, 10)$  **12.** .8944

**14.**  **(a)** .135  **(b)** .393  **(c)** .323  **16.** **(a)** $5/3$  **(b)** .015  **(c)** $5/6$

**18.**  $\lambda^n e^{-\lambda\Sigma x_i}, x_i > 0$  **19.** **(a)** $1871.25$  **(b)** $354.375$  **(c)** $37.6$

**21.**  $\mathrm{Gam}(\alpha_1 + \alpha_2, \lambda)$  **24.** **(a)** .04  **(b)** .900

**26.**  **(a)** $(1 - t/\lambda)^{-n}$  **(b)** $(1 - 2t)^{-n}; \mathrm{chi}^2(2n)$  **(c)** .963

**28.**  $\mathrm{chi}^2(3)$  **30.** **(a)** $(1+t)^{-1}$  **(b)** (doesn't exist)  **(c)** $\frac{1+t_0}{1+t+t_0}$  **(d)** $\frac{1}{1+t}$

**32.**  $f(t) = 2e^{-2t}, t > 0;\ ET = 1/2.$  **34.** $R_1(R_2 + R_3 - R_2 R_3)$

**R-1.**  **(a)** $\sqrt{\pi}$  **(b)** $239.2$  **R-2.** **(a)** $1/\sqrt{8\pi}$  **(b)** $\mathcal{N}(0, 1)$  **(c)** $e^4$

**R-3.**  **(a)** $\mathcal{N}(5n, 4n)$  **(b)** $(8\pi)^{-n/2}\exp\{-\frac{1}{8}\sum(x_i - 5)^2\}$  **(c)** 0

**R-4.**  **(a)** 30 min.  **(b)** $1 - e^{-1}$  **(c)** .446

**R-5.**  **(a)** .0104  **(b)** .261  **(c)** $\mathcal{N}(0, 1)$

**R-6.** **(a)** $\mu = 2, \sigma^2 = 4$.    **(b)** .1587    **R-7.** Gam$(3, 1/4)$

**R-8.** **(a)** 29    **(b)** $\mathcal{N}(14, 16)$    **(c)** chi$^2(1)$    **R-9.** **(a)** .433    **(b)** $1 - e^{-2}$

**R-10.** **(a)** $\exp(-x_1 - x_2 - x_3)$ for all $x_i > 0$.    **(b)** $\frac{1}{2} y^2 e^{-y}$, $y > 0$.

# Chapter 7

**1.** **(b)**                                                   **2.** **(a)** 13/19    **(b)** 5/9    **(c)** 30/45

|      |     | Class |   |    |   |   |
|------|-----|-------|---|----|---|---|
|      |     | 1     | 2 | 3  | 4 | 5 |
| Mar. | Y   | 1     | 4 | 6  | 4 | 0 |
|      | N   | 2     | 8 | 13 | 5 | 2 |

**4.**

|      |     | Class |    |    |    |   |
|------|-----|-------|----|----|----|---|
|      |     | 1     | 2  | 3  | 4  | 5 |
| Sex  | M   | 0     | 7  | 22 | 12 | 7 |
|      | F   | 3     | 12 | 19 | 9  | 2 |

**6.**

| psi | Freq. |
|-----|-------|
| 364 | 2     |
| 377 | 2     |
| 390 | 12    |
| 403 | 17    |
| 416 | 8     |
| 429 | 6     |
| 442 | 2     |
| 455 | 1     |

(both are "correct")

**9.** **(a)**

| 8  | 846     |
|----|---------|
| 9  | 432     |
| 10 | 4878768 |
| 11 | 24466   |
| 12 | 084083  |
| 13 | 13322   |
| 14 | 4       |

**13.** **(b)** 7.35, 14.55    **14.** 114 oz    **16.** 152.5, 68    **17.** 79

**18.** **(a)** 3.1, 5.75, 7.35, 9.8, 26    **(b)** 84, 106, 114, 128, 144

**22.** **(a)** 405.08    **(b)** 404.28    **23.** 144.8, 22.095    **24.** 1.155

**25.** **(a)** 8.42, 7.425    **(b)** 4.473    **(c)** 4.05    **27.** 18.82    **28.** 184, 570

**31.** 10, 5    **33.** .798    **35.** .784    **37.** **(a)** 5/6    **(b)** 0    **42.** .996, .950

**R-1.** 16    **R-2.** $\pm 3$

**R-4.** **(a)**

| Stem | Leaves (units) |
|------|----------------|
| 1    | 0              |
| 1    | 67589          |
| 2    | 214311         |
| 2    | 9567           |
| 3    | 0103           |

**(b)** 22.5, 23, 9.5    **(c)** 22.9, 6.07

**R-5.** $\overline{X} = 15.32, \overline{Y} = 63.6, S_X = 8.08075, S_Y = 21.002, r = -.9294$

**R-6.** **(b)** $r = .716$    **R-7.** **(a)** $\overline{D} = .3$    **(b)** $r = .9966$
**R-8.** **(a)** $S_U = 3,\ S_V = 10$    **(b)** .8    **R-9.** **(a)** 13.55    **(b)** 2.8

**R10.** **(a)**

| | | |
|---:|:---:|:---|
| | 3 | 3 |
| 5 | 4 | |
| 9644 | 5 | 3666 |
| 999886666410 | 6 | 11156666779 |
| 866543221 | 7 | 11113446888889 |
| 0 | 8 | 56668 |
| 9 | 9 | 22 |
| | 10 | 9 |

**(b)**   Five-number summaries, for box-plot construction:
Never: 3.3, 6.6, 7.1, 7.8, 10.9. Smoked: 4.5, 6.25, 6.85, 7.35, 9.9.

# Chapter 8

**1.** **(a)** $f(x_1, ..., x_4) = 1/16,\ -1 < x_i < 1$    **(b)** $f(x_1, ..., x_4) = 16e^{-2\Sigma x_i},\ x_i > 0$
**3.** $\lambda^3 e^{-13\lambda},\ \lambda > 0$    **5.** For $0 < p < 1$: **(a)** $p^n(1-p)^{\Sigma x_i - n}$    **(b)** $p^3(1-p)^{27}$
**7.** **(a), (b)** $M(M-1)(12-M)(11-M)$    **(c)** $M = 6$    **9.** **(a)** $\sum X_i^2$    **(b), (c)** $\sum X_i$
**11.** $(\sum X_i, \sum Y_i)$    **13.** $(X_{(1)}, X_{(n)})$    **15.** $X_{(n)}$    **17.** $\sum X_i$
**18.** **(a)** $f_{\overline{X}}(u) = 1/15$ for $u = 24, 25, 27, 30, 36, 39, 41$, $f_{\overline{X}}(u) = 2/15,\ u = 23, 26, 28, 37$
**(b)** $E(\overline{X}) = EX = 30$, $\text{var}\overline{X} = 36.27$

**19.** **(c)** $f(s) = \begin{cases} .1,\ s = 6, 7, 11, 12 \\ .2,\ s = 8, 9, 10 \end{cases}$

**22.** **(a)** $\mathcal{N}(n\mu, n\sigma^2)$    **(b)** $\text{Poi}(nm)$    **(c)** $\text{Gam}(n, \lambda)$    **23.** $\frac{nk}{nk+1}$
**24.** .5426    **27.** **(a)** $f_{\tilde{X}}(u) = 30u^2(1-u)^2, 0 < u < 1$    **(b)** $1/2, 1/28$
**31.** $\frac{n(n-1)}{(b-a)^n}\ (v-u)^{n-2},\ a < u < v < b$
**34.** **(a)** $E(\overline{X}) = \mu\ (= 10?),\ \sigma_{\overline{X}} = 6/\sqrt{20}$    **(b)** .136    **36.** .0439
**38.** **(a)** .003    **(b)** .058, with cont. corr.   [**(a)** .005    **(b)** .0422 without cont. corr.]
**40.** .233    **42.** 190    **46.** .035    **47.** $\text{Gam}(x+1, 4)$
**49.** **(a)** $\frac{1}{B(7, 5)}\ p^6(1-p)^4$    **(b)** $\frac{1}{B(8, 9)}\ p^7(1-p)^8$    **(c)** $\frac{1}{B(12, 6)}\ p^{11}(1-p)^5$
**52.** **(a)** $1/4$    **54.** $\frac{rs}{(r+s)^2(r+s+1)}$    **55.** $\mathcal{N}(514.7, 96.97)$
**R-1.** $81/32$
**R-2.** **(a)** $\theta^n(\prod X_i)^{\theta-1}$    **(b)** $\prod X_i$    **(c)** $F_{X_{(n)}}(u) = u^{n\theta},\ 0 < u < 1$.    **(d)** $\frac{n\theta}{n\theta+1}$
**R-3.** **(a)** $\theta_4$    **(b)** $2/3, 0, 1/3, 0$ for $i = 0, 1, 2, 3$ resp.— the same for all $\theta$.
**R-4.** **(a)** .26    **(b)** $6/13$    **R-5.** **(a)** $\theta, \theta/\sqrt{2n}$    **(b)** $\mathcal{N}(\theta, \frac{\theta^2}{2n})$
**R-6.** **(a)** $\lambda^{20} e^{-12\lambda},\ \lambda > 0$    **(b)** $\frac{13^{21}}{20!}\ \lambda^{20} e^{-13\lambda},\ \lambda > 0$.    **(c)** $\text{Gam}(20, \lambda)$
**R-7.** .0082    **R-8.** .0062
**R-9.** **(a)** $\sum X_i^2$    **(b)** $2n\theta$    **(c)** $T/\theta \sim \text{Gam}(n, \frac{1}{2})$ [which is $\text{chi}^2(2n)$]

**R-10.** **(a)** .0456    **(b)** .1177

**R-11.** **(a)** $\sigma^{-n}\exp\left\{-\frac{1}{2\sigma^2}\sum(Y_i - \beta x_i)^2\right\}$    **(b)** $(\sum x_i Y_i,\ \sum Y_i^2)$

**R-12.** **(a)** $\frac{k}{n+1}$    **(b)** $\frac{1}{n+1}$

**R-13.** **(a)** $\lambda^n e^{-\lambda \sum X_i},\ \lambda > 0$    **(b)** $\frac{(\sum X_i + \beta)^{n+\alpha}}{\Gamma(n+\alpha)} \lambda^{n+\alpha-1} e^{-(\sum X_i + \beta)\lambda},\ \lambda > 0.$    **(c)** $\frac{(n+\alpha)(\sum X_i + \beta)^{n+\alpha}}{(\sum X_i + \beta + y)^{n+\alpha+1}}$

**(d)** $\frac{\sum X_i + \beta}{n + \alpha - 1}$

**R-14.** **(b)** $\frac{1}{n+1}$.    **R-15.** **(a)** $1/\pi$    **(b)** $1/2$    **(c)** $3/4$

---

# Chapter 9

---

**1.** .42 (.0127)    **3.** 603.2 (33.60)    **4.** **(a)** $-\frac{\theta}{n+1}$    **(b)** $\frac{2\theta^2}{(n+1)(n+2)}$

**6.** .0114    **8.** **(a)** 25    **(b)** 100    **9.** **(a)** 2500    **(b)** 625    **13.** $2\Phi\left(-\frac{\epsilon}{\sigma/\sqrt{n}}\right)$

**17.** $18.31 \pm .29$    **18.** **(a)** $.41 \pm .0127$    **(b)** $.41 \pm .0249$    **20.** $404.3 \pm 4.40$

**23.** $89.14 \pm 12.53,\ 89.14 \pm 17.22$    **25.** **(a)** .9544    **(b)** .934

**27.** **(b)** $\frac{1}{1+k^2/n}\left\{\hat{p} + \frac{k^2}{2n} \pm k\sqrt{\frac{\hat{p}(1-\hat{p})}{n} + \frac{k^2}{4n^2}}\right\}$, where $k = 1.96$.

**29.** .38 (or .35, with continuity correction)    **30.** **(a)** .048    **(b)** $.113 \pm .124$

**32.** $30 \pm 2.1$    **34.** .0616    **36.** **(a)** $2.78 < \sigma < 7.87$    **(b)** $2.34 < \sigma < 7.10$

**37.** $\frac{\sum X_i^2}{26.1} < \sigma^2 < \frac{\sum X_i^2}{5.63}$    **40.** **(a), (b)** $\frac{n}{\sum X_i^2}$    **41.** $\frac{1}{n}\sum X_i^2$    **43.** $\overline{X}$

**44.** $(X_{(1)}, X_{(n)})$    **46.** $(\sum Y_i / \sum X_i)^{1/2}$ [Yes, but it can be shown not sufficient.]

**48.** $\hat{\beta} = \frac{\sum c_i Y_i - \frac{1}{n}\sum c_i \sum Y_i}{\sum c_i^2 - \frac{1}{n}(\sum c_i)^2}$, $\hat{\alpha} = \overline{Y} - \hat{\beta}\overline{c}$, $\hat{\sigma}^2 = \frac{1}{n}\sum(Y_i - \hat{\alpha} - \hat{\beta}c_i)^2$.

**49.** **(a)** 8/17    **(b)** 2/3    **52.** **(a)** $\alpha/\lambda$    **(b)** 2/3    **54.** 7/11    **57.** **(a)** $\frac{n}{pq}$

**61.** $\frac{\lambda}{n-1}$, $\frac{\lambda^2}{(n-1)^2(n-1)}$

**R-1.** 225    **R-2.** **(a)** $1,112$    **(b)** $.30 \pm .0269$    **R-3.** 1.5    **R-4.** $1/\theta^2$

**R-5.** **(a)** $T$ is unbiased and sufficient, so efficient.    **(b)** $\sum X_i^2 / \theta \sim \text{chi}^2(n)$, free of $\theta$

**R-6.** **(a)** $1/\lambda$    **(b)** $n/\lambda$    **(c)** eff. $= 1$    **R-7.** **(a)** 10/11    **(b)** .88

**R-8.** **(a)** $\frac{1}{2n\lambda^2}$    **(b)** $\hat{\theta} = \overline{X}/2 = 1/\hat{\lambda}$    **(c)** $1/\overline{X}$    **(d)** 22/13

**R-9.** **(a)** $\frac{1}{2n}\sum X_i^2$    **(b)** $\frac{2}{\pi}\overline{X}^2$    **(c)** $X^2/\theta \sim \text{chi}^2(2)$    **(d)** $\theta^2/n$

**R-11.** $.04 < \theta < 4.86$: **(a)** .963    **(b)** .957

---

# Chapter 10

---

**1.** .123 (.15 without continuity correction)    **2.** **(a)** .0197    **(b)** 1.18    **(c)** .303

**3.** $P = .09$ (2-sided)    **5.** **(a)** $5^n/9^Y$    **(b)** Large values.

**7.** **(a)** $\text{chi}^2(n-1)$    **(b)** .044    **8.** Some evidence that $\mu < 500$ $(Z = 1.77, P = .038)$.

**10.** $Z = 2.97, P = .0015$ (strong evidence against $H_0$)

13.  $Z = 3, P = .0013$ (1-sided—strong evidence against $H_0$).
14.  (a)  .0228      (b)  .0228     16.  $Z \doteq -6$  (strong evidence against $H_0$).
17.  $T = -1.27$ (6 d.f.)  $P = .242$ (2-sided).
19.  $T = 1.97$ (7 d.f.)  $P = .043$ (1-sided).
21.  $T = 1.64$ (7 d.f.)  $P = .146$ (2-sided)     23.  $R_- = 6$, 2-sided $P = .11$
25.  (a)  1-sided $P = .29$      (b)  $P \doteq .07$  ($R_- = 23.5$)      (c)  $T \doteq 1.88$ (12 d.f.), $P \doteq .04$
27.  (a)  256      (b)  7/256
29.  (a)  Outlier suggests heavy tail.      (b) & (c)   2-sided $P = 1/256$.
30.  (a)  $\mathcal{N}(513, 128)$      (b)  .125      (c)  same as (b)      **32.**   .01123
R-1.  (a)  [with $H_0$: $\mu = 0$] $P = .354$      (b)  .269
R-2.  (a)  .0228      (b)  .045 [this is .0228 doubled: 1-sided $P$ for $\chi^2$ is 2-sided for $Z$].
R-4.  (a)  $T = 5.63$ (8 d.f.), $P \doteq .000$.      (b)  $P = 1/2^9$
R-5.  $t = .77$ (7 d.f.), $P > .2$.      R-8.   $8S^2/9 = 15.73$ [chi$^2$(8)], $P \doteq .046$.
R-9.  .002, .0095      **R-10.**  1/2      **R-11.**  (a)  .693      (b)  .520

# Chapter 11

1.  .0035      2.  (a)  $R_1$      (b)  $R_2$      3.  (a)  $\alpha_1 = .3174, \alpha_2 = .1096$.      (b)  .157, .344
5.  (a)  $R = [\overline{X} > 10.164 \text{ or } \overline{X} < 9.836]$.      (b)  .10

7.  (a)

| $K$ | $-1$ | 0 | 1 | 2 | 3 | 4 | 5 |
|---|---|---|---|---|---|---|---|
| $\alpha$ | 1 | .9688 | .8125 | .5000 | .1875 | .0313 | 0 |
| $\beta$ | 0 | .0000 | .0005 | .0086 | .0815 | .4095 | 1 |

8.  $p^7(120q^3 + 45pq^2 + 10p^2q + p^3)$, where $q = 1 - p$.
9.  $R_1$: $\Phi(2\mu - 21) + \Phi(19 - 2\mu)$,  $R_2$: $\Phi(2\mu - 21.6) + \Phi(18.4 - 2\mu)$
11.  (a)  .15, $\frac{1}{2} + \frac{1}{\pi}\text{Arctan}(\theta - 2)$      (b)  .37, $\frac{1}{\pi}[\text{Arctan}(\theta - 1.5) - \text{Arctan}(\theta + 1.5)]$
13.  $\Phi(10\mu - 101.645) + \Phi(98.36 - 10\mu)$      14.  $1 - (1 - p)^{10}$
16.  (a)  .073      (b)  $\Phi(48.5 - 1.25\mu)$      (c)  $n = 28, K = 40.76$
18.  (a)  $\overline{X} > K$      (b)  $\alpha = 1 - \Phi(2K), \beta = \Phi(2K - 2)$.      19.  $\overline{X} > K$
21.  $\overline{X} > K$      23.  $|\overline{X} - \mu_0| > K$      25.  $ue^{-u} < K$, where $u = \lambda_0\overline{X}$
26.  $\emptyset, \{z_3\}, \{z_3, z_1\}, \{z_3, z_1, z_4\}, \Omega$
28.  (b)  $\Lambda = 1$ if $\overline{X} > 0$, and $\Lambda = e^{-n\overline{X}^2/2}$ if $\overline{X} > 0$.
31.  (a)  Accept $H_0$ $(1 < 1.5)$.      (b)  Reject $H_0$ $(9 < 1018)$.
R-1.  (a)  $\sum X_i^2 > K$.      (b)  $K \doteq 7.78$.      (c)  $\beta \doteq .25$.      (d)  $n = 23, K = 41.6$.
R-2.  (a)  $\sum X_i > K$.      (b)  .133, .220      R-3.   $\Phi(2\mu - 12)$
R-4.  (a)  .035      (b)  $\pi(\sigma^2) = 1 - F_{\chi^2(9)}(36/\sigma^2)$.
R-5.  (a)  Posterior odds are 4:30:9 for $z_1$, 6:5:0 for $z_2$, 2:5:12 for $z_3$, 8:10:9 for $z_4$.
(b)  Reject $H_0$ if $Z = z_1, z_3$, accept $H_0$ if $Z = z_2$ or $z_4$.
R-6.  $\Lambda = (e\widehat{\theta}e^{-\widehat{\theta}})^n$; $R$ consists of two regions: $\widehat{\theta} > B$ and $\widehat{\theta} < A$, where $A$ and $B$ are the points where $ue^{-u} = K$.
R-7.  $\Lambda = (\widehat{\sigma}_0^2/\widehat{\sigma}^2)^{-n/2}$.

**R-8.** **(a)** $p_1^{15} p_2^{15} p_3^{30}$; $1/3^{60}$, $2^{90}/3^{60}$     **(b)** $-2 \log \Lambda = 7.067$ [chi$^2(2)$], $P \doteq .03$.

# Chapter 12

**1.** $Z = 2.15$, $P = .016$ (some evidence of a difference).

**3.** $Z = 1.96$, (1-sided) $P = .025$ (marginal).

**4.** $Z = 1.5$, (1-sided) $P = .067$ (little or no evidence of a difference).

**6.** $Z = .595$ (no evidence against $H_0$)     **8.** $Z = 5.2$ (strong evidence against $H_0$)

**10.** $Z = 3.44$, (1-sided) $P \doteq .0002$ (strong evidence against $H_0$).

**12.** $T = 3.9$ (39 d.f.) [or $Z \doteq 4$] $P = .000$.

**14.** $T = 1.1$ (10 d.f.), (1-sided) $P \doteq .15$.

**16.** $R = 77$, $Z = -2.78$, (1-sided) $P \doteq .0027$.

**18.** Exact: .051, normal approximation: .0501     **20.** $\hat{\theta} = \frac{n_1 + n_2 - 2}{n_1 + n_2} S_p^2$

**22.** $T = 4.4$ (3 d.f.), 1-sided $P \doteq .01$.     **24.** $R_- = 4$, 1-sided $P = .007$.

**26.** **(a)** $P = .0156$.     **(b)** $R_- = 0$, $P = .0156$.     **(c)** $T \doteq 5.8$, $P = .0011$.

**27.** $F = 1.13 (28, 19)$ d.f., $P > .05$     **30.** **(a)** $\frac{1}{m-2}$     **(b)** $\frac{k}{m-2}$     **(c)** $\frac{m}{m-2}$

**R-1.** $Z = -2.89$, $P \doteq .0019$.     **R-2.** **(a)** $R_1 = 17$, $P = .016$.     **(b)** $\frac{4}{\binom{10}{5}} = .01587$.

**R-3.** **(a)** $\bar{D} \sim \mathcal{N}\left(\delta, \frac{1}{n_1} + \frac{1}{n_2}\right)$     **(b)** .9772

**R-4.** $Z = .895$, $P = .188$ (no reason to doubt $H_0$).

**R-5.** $T = .71$ (10 d.f.), $P > .1$; $R_1 = 37$, $P > .216$ (computer: $P = .285$).

**R-6.** $T = 1.32$ (26 d.f.), 1-sided $P = .10$.

**R-7.** **(a)** $T = 2.85$ (9 d.f.), $P = .0095$.     **(b)** $R_- = 6$. $P = .014$.     **(c)** $P = .172$.

**R-8.** $Z = 11.8$, $P \doteq .000$.     **R-9.** $Z = 2.88$, $P = .002$.     **R-10.** **(a)** $1/2$     **(b)** $5/6$

**R-11.** $T = 1.25$ (13 d.f.) $\doteq .117$ [or $R_- = 36$, $P = .163$], inconclusive.

**R-12.** $Z = 1.42$, (one-sided) $P \doteq .078$.

# Chapter 13

**1.** $\chi^2 = 14.2$ (3 d.f.), $P < .005$.

**3.** **(a)** $\chi^2 = 15.7$ (2 d.f.), $P < .002$.     **(b)** $\chi^2 = .24$ (2 d.f.); data are not inconsistent with $H_0$.

**5.** With last two cells grouped: $\chi^2 \doteq 2$ (1 d.f.), $P > .138$.

**7.** **(a)** $\chi^2 = 4.55$ (3 d.f.), $P = .22$.     **(b)** $\chi^2 = 2.25$ (5 d.f.), $P \doteq .8$.

**9.** $D_{10} = .19$, $P \doteq .7$ (no reason to doubt normality).

**11.** **(a)** $D_n \doteq .103$, $P \doteq .20$ (from asymptotic formula in Table IXb).

**(b)** $D_n \doteq .02$ (data not inconsistent with $H_0$).

**14.** $D = .20$, $P > .5$.

**16.** BOD: $W = .340$ (strong evidence against $H_0$), COD: $W = .88$, $.01 < P < .05$.

**17.** **(a)** $\chi^2 = 7.9$ (1 d.f.), 2-sided $P \doteq .005$.     **(b)** $Z \doteq -2.8$, 1-sided $P \doteq .003$.

**18.** $\chi^2 = 45$ (2 d.f.), $P < .001$.     **21.**   $\chi^2 = 16.2$ (3 d.f.), $P < .001$.

**23.** **(a)**  $\chi^2 = 15.7$ (1 d.f.), $P < .005$.     **(b)**   $Z^2 = 15.64$ [equivalent to (a)].

**25.** $\chi^2 = 8.44$ (3 d.f.), $P = .038$.     **26.**   $-2 \log \Lambda \doteq 15$ (3 d.f.), $P < .005$.

**28.** $-2 \log \Lambda \doteq 16.0$ (1 d.f.), $P < .001$.

**30.** **(a)** $\hat{p} = \frac{2n_{11} + n_{12} + n_{21}}{2n}$.   **(b)**   $-2 \log \Lambda = 19.70$ (1 d.f.).   **(c)**   $-2 \log \Lambda = 22.68$ ( $= 19.70 + 2.98$).

**R-1.** $\chi^2 = 40/9$ (3 d.f.), $P \doteq .22$.     **R-2.**   $D_n = .40$, $P \doteq .055$.

**R-3.** $Z \doteq 3.8$, $P = .0001$.     **R-4.** **(a)**   $\chi^2 = 5.48$ (2 d.f.), $P \doteq .07$     **(b)**   $-\log \Lambda = 5.49$ (2 d.f.).

**R-5.** **(a)** $\theta^{100}(1 - \theta)^{100}$, $0 < \theta < 1$.     **(b)**   $1/2$     **(c)**   $\chi^2 = 8.16$, $P = .017$.

**R-6.** $W = .79$, $.01 < P < .05$  (closer to .01).

**R-7.** $Z \doteq 2.485$, 1-sided $P \doteq .0065$  ($\chi^2 = 6.17$, $P \doteq .013$).

**R-8.** $W \doteq .93$, $P \doteq .10$.     **R-9.**   $Z = 1.374$, $P = .085$  ($\chi^2 \doteq 1.89$, $P > .14$ from Table Vb).

# Chapter 14

**1.** $F = 2.61$.

**3.** **(a)**

| Source | d.f. | SS | MS | F | P |
|---|---|---|---|---|---|
| Variety | 4 | .502 | .126 | 1.12 | .40 |
| Error | 10 | 1.12 | .112 | – | – |
| Total | 14 | 1.62 | – | – | |

**5.** SSE $= 2719.75$, $F = 3.33$ (1, 26) d.f.;  $P \doteq .08$.

**6.** $F = 6.51$  (5, 18) d.f., $.001 < P < .002$.     **8.**   Both $= \tau_i$.

**14.** **(a)**  $F = 15.9$ (3, 32) d.f., $P < .001$.     **(b)**  S  <u>T  C</u>  J     ($\alpha = .05$).

**(c)** Same conclusion as in (b).

**16.** **(a)** and **(b)**:   <u>C   A   E</u>   D   B

**17.**

| Source | d.f. | SS | MS | F | P |
|---|---|---|---|---|---|
| A | 2 | 294 | 147 | 49 | .0015 |
| B | 2 | 6 | 3 | 1 | .44 |
| Error | 4 | 12 | 3 | – | – |
| Total | 8 | 312 | – | – | |

**19.**

| Source | d.f. | SS | MS | F | P |
|---|---|---|---|---|---|
| Detergent | 2 | 186 | 93 | 4.77 | .087 |
| Temperature | 2 | 42 | 2.1 | 1.08 | .42 |
| Error | 4 | 78 | 19.5 | – | – |
| Total | 8 | 306 | – | – | |

**21.** $F = 3.61$  (7, 16) d.f., $.01 < P < .05$.

**R-1.** $F = 4.54$ (2, 79) d.f., $P < .016$.     **R-2.**   If $n_1 = n_2 \cdots = n_k$.

**R-4.** Yes: too severe round-off before subtractions.

**R-7.** **(a)**  $F = 17.48$ ( $= t^2$).     **(b)**   $t = 4.18$ (22 d.f.), $P \doteq .000$.

# Chapter 15

1.  **(a)** $y = \frac{1}{14}(4 + 9x)$.    **(b)**  $23/14$      **2.**  $y = 736 + 127.6\,x$, $r = .996$.
4.  $y = .383 - .104(\log x)$.      **6.**  $y = 3.95 - 5.55\,x + 1.75\,x^2$.
10.  **(a)**  s.e.$(\widehat{\alpha}) = S_\epsilon \sqrt{\frac{1}{n} + \frac{\bar{x}^2}{\mathrm{SS}_{xx}}}$ .      **(b)**  $\widehat{\alpha} \pm t_{.975}(n-2) \times$ s.e.$(\widehat{\alpha})$.
15.  $t = -2.08$ (7 d.f.), $P = .076$.      **19.**  **(a)**  $.499$      **(b)**  $(-.092, 1.804)$
22.  **(a)**  $y = \widehat{\beta} x$, where $\widehat{\beta} = \sum x_i y_i / \sum x_i^2$.      **(b)**  $\widehat{\beta} \sim \mathcal{N}(\beta,\, \sigma^2 / \sum x_i^2)$.
**(d)**  Reject $H_0$ when $\frac{(\widehat{\beta} - \beta_0)^2}{\widehat{\sigma}^2 / \sum x_i^2} > C$.
24.  **(a)**  $-.45 + .0055\,x$      **(b)**  $3.84$      **(c)**  $.168$      **(d)**  $Z_Y$ is smaller $(2 < 2.67)$.
26.  **(a) and (b):**  $\frac{2}{3}(1 - x)$.
29.  **(a)**  $.885$      **(b)**  $\log D = -15.619 + 1.9545(\log V)$.      **(c)**  $-2.1177$      **(d)**  $1.133$
35.  $\mathcal{N}(2, 2)$, $\mathcal{N}(-2, 4)$      **37.**  $\frac{1}{2\pi\sqrt{5}} \exp\left\{ -\frac{9}{10} \left( x^2 - \frac{4x(y-4)}{9} + \frac{(y-4)^2}{9} \right) \right\}$
39.  **(a)**  $E(Y \mid x) = \frac{1}{2}(x + 2)$, $E(X \mid y) = y - 1$.
**(b)**  $\mu_X = 0$, $\mu_Y = 1$, var $X = 2$, var $Y = 1$, $\sigma_{X,Y} = 1$, $\rho = 1/\sqrt{2}$.
40.  Bivariate normal: $\mu_X = \mu_Y = 0$, var $X = 5$, var $Y = 10$, $\sigma_{X,Y} = 1$.
41.  **(a)**  means: $-8, 6$; variances: $8, 18$; $\rho = -1/3$.      **(b)**  $-\frac{2}{9}(y + 30)$
43.  Greater:  predicted $Y$ is $\bar{Y} - 1.4 S_Y > \bar{Y} - 2 S_Y$.
47.  **(a)**  $C = -4158 + 64.46 D + .219 S$.      **(b)**  $F = 6.78$ $(2, 6)$ d.f., $P < .029$.
**(c)**  $R^2 = .6932$.      **(d)**  $T = .2194/.1304 = 1.68$ (6 d.f.), 2-sided $P = .1434$.
R-1.  **(b)**  $.4757$      **(c)**  $-26.32 < \beta < -21.72$.      **(d)**  $3.03\%$
R-2.  **(a)**  $y = 1.636 - .9022x$.      **(b)**  Residuals: $.376, -.522, .106, .039$      **(c)**  $.91$
R-3.  **(b)**  $r = .992$.      **(c)**  $y = -3.3665 + 7.006\,x$.      **(d)**  $35.2$
R-4.  **(a)**  $64.4$, s.e. $= 8.16$.      **R-9.**  $r \doteq .99$; $C = .0416 + .0076\,T$.
R-10.  $r = .190$, $t \doteq .55$ (8 d.f.— not significant).

# Index